McGraw-Hill

Dictionary of
Chemistry

Second
Edition

McGraw-Hill

New York   Chicago   San Francisco   Lisbon   London   Madrid
Mexico City   Milan   New Delhi   San Juan   Seoul   Singapore
Sydney   Toronto

The McGraw·Hill Companies

6 7 8 DOH 14 13 12

ISBN 0-07-141046-5

 This book is printed on recycled, acid-free paper containing a minimum of 50% recycled, de-inked fiber.

This book was set in Helvetica Bold and Novarese Book by the Clarinda Company, Clarinda, Iowa. It was printed and bound by RR Donnelley, The Lakeside Press.

Library of Congress Cataloging-in-Publication Data

McGraw-Hill dictionary of chemistry — 2nd. ed.
     p.   cm.
  "All text in this dictionary was published previously in the McGraw-Hill dictionary of scientific and technical terms, sixth edition, c2003..."
— T.p. verso.
  ISBN 0-07-141046-5 (alk, paper)
  1. Chemistry—Dictionaries. I. Title: Dictionary of chemistry. II. Title: Chemisty. III. McGraw-Hill dictionary of scientific and technical terms. 6th ed.

QD5.M36   2003
540′.3—dc21   2002033183

On the cover: Blown-glass light sculpture filled with an inert gas mixture and lighted with high-frequency static electricity. (*Courtesy of Mundy Hepburn*)

Contents

Preface v
Staff vi
How to Use the Dictionary vii
Fields and Their Scope ix
Pronunciation Key x
A–Z Terms 1–414
Appendix 415–431
 Equivalents of commonly used units for the U.S.
  Customary System and the metric system 417
 Conversion factors for the U.S. Customary System,
  metric system, and International System 418
 Defining fixed points of the International Temperature
  Scale of 1990 (ITS-90) 422
 Primary thermometry methods 423
 Periodic table 424
 Electrochemical series of the elements 425
 Average electronegativities from the thermochemical data 426
 Standard atomic weights 427
 Principal organic functional groups 429
 Compounds containing functional groups 430
 Physical properties of some organic solvents 431

Preface

The McGraw-Hill Dictionary of Chemistry provides a compendium of 8000 terms that are central to chemistry and related fields of science and technology. The coverage in this Second Edition is focused on the the areas of analytical chemisty, general chemistry, inorganic chemistry, organic chemistry, physical chemistry, and spectroscopy, with new terms added and others revised as necessary.

Chemistry deals with the composition, properties, and structure of matter. Its various branches analyze composition and properties, and study the changes that occur in matter, the underlying processes, the energetics of these processes, and the rates at which they occur. Thus, the terms contained in this Dictionary may be used in virtually all areas of science, for example, biochemistry, geochemistry, and cosmochemistry, and in many areas of technology.

All of the definitions are drawn from the McGraw-Hill Dictionary of Scientific and Technical Terms, Sixth Edition (2003). Each definition is classified according to the field with which it is primarily associated; if it is used in more than one area, it is identified by the general label [CHEM]. The pronunciation of each term is provided, along with synonyms, acronyms, and abbreviations where appropriate. A guide to the use of the Dictionary appears on pages vii-viii, explaining the alphabetical organization of terms, the format of the book, cross referencing, chemical formulas, and how synonyms, variant spellings, abbreviations, and similar information are handled. The Pronunciation Key is provided on page x. The Appendix provides conversion tables for commonly used scientific units as well as other listings of chemical data.

It is the editors' hope that the Second Edition of the McGraw-Hill Dictionary of Chemistry will serve the needs of scientists, engineers, students, teachers, librarians, and writers for high-quality information, and that it will contribute to scientific literacy and communication.

Mark D. Licker
Publisher

Staff

Mark D. Licker, Publisher—Science

Elizabeth Geller, Managing Editor
Jonathan Weil, Senior Staff Editor
David Blumel, Staff Editor
Alyssa Rappaport, Staff Editor
Charles Wagner, Digital Content Manager
Renee Taylor, Editorial Assistant

Roger Kasunic, Vice President—Editing, Design, and Production

Joe Faulk, Editing Manager
Frank Kotowski, Jr., Senior Editing Supervisor

Ron Lane, Art Director

Thomas G. Kowalczyk, Production Manager
Pamela A. Pelton, Senior Production Supervisor

Henry F. Beechhold, Pronunciation Editor
Professor Emeritus of English
Former Chairman, Linguistics Program
The College of New Jersey
Trenton, New Jersey

How to Use the Dictionary

ALPHABETIZATION. The terms in the McGraw-Hill *Dictionary of Chemistry*, Second Edition, are alphabetized on a letter-by-letter basis; word spacing, hyphen, comma, solidus, and apostrophe in a term are ignored in the sequencing. Also ignored in the sequencing of terms (usually chemical compounds) are italic elements, numbers, small capitals, and Greek letters. For example, the following terms appear within alphabet letter "A":

amino alcohol	***para*-aminophenol**
1-aminoanthraquinone	***n*-amylamine**
γ-aminobutyric acid	**4-AP**

FORMAT. The basic format for a defining entry provides the term in boldface, the field is small capitals, and the single definition in lightface:

    term [FIELD]   Definition

A term may be followed by multiple definitions, each introduced by a boldface number:

    term [FIELD]   **1.** Definition. **2.** Definition. **3.** Definition.

A term may have definitions in two or more fields:

    term [PHYS CHEM]   Definition.   [SPECT]   Definition.

A simple cross-reference entry appears as:

    term   *See* another term.

A cross reference may also appear in combination with definitions:

    term [PHYS CHEM]   Definition.   [SPECT]   *See* another term.

CROSS REFERENCING. A cross-reference entry directs the user to the defining entry. For example, the user looking up "arachic acid" finds:

    arachic acid   *See* eicosanoic acid.

The user then turns to the "E" terms for the definition. Cross references are also made from variant spellings, acronyms, abbreviations, and symbols.

    AES   *See* Auger electron spectroscopy.
    aluminium   *See* aluminum.
    at. wt   *See* atomic weight.
    Au   *See* gold.

CHEMICAL FORMULAS. Chemistry definitions may include either an empirical formula (say, for abietic acid, $C_{20}H_{30}O_2$) or a line formula (for acrylonitrile, CH_2CHCN), whichever is appropriate.

ALSO KNOWN AS . . . , etc. A definition may conclude with a mention of a synonym of the term, a variant spelling, an abbreviation for the term, or other such information, introduced by "Also known as . . . ," "Also spelled . . . ," "Abbreviated . . . ," "Symbolized . . . ," "Derived from . . . ." When a term has more than one definition, the positioning of any of these phrases conveys the extent of applicability. For example:

      term [PHYS CHEM] **1.** Definition. Also known as synonym. **2.** Definition. Symbolized T.

In the above arrangement, "Also known as . . ." applies only to the first definition; "Symbolized . . ." applies only to the second definition.

      term [PHYS CHEM] **1.** Definition. **2.** Definition. [SPECT] Definition. Also known as synonym.

In the above arrangement, "Also known as . . ." applies only to the second field.

      term [PHYS CHEM] Also known as synonym. **1.** Definition. **2.** Definition. [SPECT] Definition.

In the above arrangement, "Also known as . . ." applies only to both definitions in the first field.

      term Also known as synonym. [PHYS CHEM] **1.** Definition. **2.** Definition. [SPECT] Definition.

In the above arrangement, "Also known as . . ." applies to all definitions in both fields.

Fields and Their Scope

[ANALY CHEM] **analytical chemistry**—The science of the characterization and measurement of chemicals; qualitative analysis is concerned with the description of chemical composition in terms of elements, compounds, or structural units, whereas quantitative analysis is concerned with the measurement of amount.

[CHEM] **chemistry**—The scientific study of the properties, composition, and structure of matter, the changes in structure and composition of matter, and accompanying energy changes.

[INORG CHEM] **inorganic chemistry**—The branch of chemistry that deals with reactions and properties of all chemical elements and their compounds, excluding hydrocarbons but usually including carbides and other simple carbon compounds (such as CO_2, CO, and HCN).

[ORG CHEM] **organic chemistry**—The study of the structure, preparation, properties, and reactions of carbon compounds.

[PHYS CHEM] **physical chemistry**—The branch of chemistry that deals with the interpretation of chemical phenomena and properties in terms of the underlying physical processes, and with the development of techniques for their investigation.

[SPECT] **spectroscopy**—The branch of physics concerned with the production, measurement, and interpretation of electromagnetic spectra arising from either emission or absorption of radiant energy by various substances.

Pronunciation Key

Vowels

a	as in bat, that
ā	as in bait, crate
ä	as in bother, father
e	as in bet, net
ē	as in beet, treat
i	as in bit, skit
ī	as in bite, light
ō	as in boat, note
ò	as in bought, taut
u̇	as in book, pull
ü	as in boot, pool
ə	as in but, sofa
au̇	as in crowd, power
òi	as in boil, spoil
yə	as in formula, spectacular
yü	as in fuel, mule

Semivowels/Semiconsonants

w	as in wind, twin
y	as in yet, onion

Stress (Accent)

'	precedes syllable with primary stress
ˌ	precedes syllable with secondary stress
¦	precedes syllable with variable or indeterminate primary/secondary stress

Consonants

b	as in bib, dribble
ch	as in charge, stretch
d	as in dog, bad
f	as in fix, safe
g	as in good, signal
h	as in hand, behind
j	as in joint, digit
k	as in cast, brick
k̲	as in Bach (used rarely)
l	as in loud, bell
m	as in mild, summer
n	as in new, dent
n̲	indicates nasalization of preceding vowel
ŋ	as in ring, single
p	as in pier, slip
r	as in red, scar
s	as in sign, post
sh	as in sugar, shoe
t	as in timid, cat
th	as in thin, breath
t̲h̲	as in then, breathe
v	as in veil, weave
z	as in zoo, cruise
zh	as in beige, treasure

Syllabication

· Indicates syllable boundary when following syllable is unstressed

A

abalyn [ORG CHEM] A liquid rosin that is a methyl ester of abietic acid; prepared by treating rosin with methyl alcohol; used as a plasticizer. { 'ab·ə,lin }

Abegg's rule [CHEM] An empirical rule, holding for a large number of elements, that the sum of the maximum positive and negative valencies of an element equals eight. { 'ä·begz 'rül }

Abel tester [PHYS CHEM] A laboratory instrument used in testing the flash point of kerosine and other volatile oils having flash points below 120°F (49°C); the oil is contained in a closed cup which is heated by a fixed flame below and a movable flame above. { ä'bəl 'tes·tər }

abietic acid [ORG CHEM] $C_{20}H_{30}O_2$ A tricyclic, crystalline acid obtained from rosin; used in making esters for plasticizers. { ,a·bē'et·ik 'as·əd }

ab initio computation [PHYS CHEM] Computation of the geometry of a molecule solely from a knowledge of its composition and molecular structure as derived from the solution of the Schrödinger equation for the given molecule. { ,ab ə¦nish·ē·ō ,käm·pyə'tā·shən }

Abney mounting [SPECT] A modification of the Rowland mounting in which only the slit is moved to observe different parts of the spectrum. { 'ab·nē ,maúnt·iŋ }

ABS See acrylonitrile butadiene styrene resin.

absolute alcohol [ORG CHEM] Ethyl alcohol that contains no more than 1% water. Also known as anhydrous alcohol. { 'ab·sə,lüt 'al·kə·hól }

absolute boiling point [CHEM] The boiling point of a substance expressed in the unit of an absolute temperature scale. { 'ab·sə,lüt 'bóil·iŋ ,póint }

absolute configuration [ORG CHEM] The three-dimensional arrangement of substituents around a chiral center in a molecule. Also known as absolute stereochemistry. { 'ab·sə,lüt kən,fig·yə'rā·shən }

absolute density See absolute gravity. { 'ab·sə,lüt 'dens·ə·dē }

absolute detection limit [ANALY CHEM] The smallest amount of an element or compound that is detectable in or on a given sample; expressed in terms of mass units or numbers of atoms or molecules. { 'ab·sə,lüt di'tek·shən ,lim·ət }

absolute gravity [CHEM] Density or specific gravity of a fluid reduced to standard conditions; for example, with gases, to 760 mmHg pressure and 0°C temperature. Also known as absolute density. { 'ab·sə,lüt 'grav·ə·dē }

absolute method [ANALY CHEM] A method of chemical analysis that bases characterization completely on standards defined in terms of physical properties. { 'ab·sə,lüt 'meth·əd }

absolute reaction rate [PHYS CHEM] The rate of a chemical reaction as calculated by means of the (statistical-mechanics) theory of absolute reaction rates. { 'ab·sə,lüt rē'ak·shən ,rāt }

absolute stereochemistry See absolute configuration. { 'ab·sə,lüt ,ster·ē·ō'kem·ə·strē }

absorb [CHEM] To take up a substance in bulk. { əb'sórb }

absorbance [PHYS CHEM] The common logarithm of the reciprocal of the transmittance of a pure solvent. Also known as absorbancy; extinction. { əb'sór·bəns }

absorbancy See absorbance. { əb'sór·bən·sē }

absorbency [CHEM] Penetration of one substance into another. { əb'sór·bən·sē }

absorbency index See absorptivity. { əb'sór·bən·sē 'in·deks }

absorptiometer [ANALY CHEM] **1.** An instrument equipped with a filter system or other simple dispersing system to measure the absorption of nearly monochromatic radiation in the visible range by a gas or a liquid, and so determine the concentration of the absorbing constituents in the gas or liquid. **2.** A device for regulating the thickness of a liquid in spectrophotometry. { əb,sȯrp·tē'ä·məd·ər }

absorptiometric analysis [ANALY CHEM] Chemical analysis of a gas or a liquid by measurement of the peak electromagnetic absorption wavelengths that are unique to a specific material or element. { əb,sȯrp·tē·ə'met·rik ə'nal·ə·sis }

absorption [CHEM] The taking up of matter in bulk by other matter, as in dissolving of a gas by a liquid. { əb'sȯrp·shən }

absorption constant See absorptivity. { əb'sȯrp·shən ,käns·tənt }

absorption edge [SPECT] The wavelength corresponding to a discontinuity in the variation of the absorption coefficient of a substance with the wavelength of the radiation. Also known as absorption limit. { əb'sȯrp·shən ,ej }

absorption limit See absorption edge. { əb'sȯrp·shən 'lim·ət }

absorption line [SPECT] A minute range of wavelength or frequency in the electromagnetic spectrum within which radiant energy is absorbed by the medium through which it is passing. { əb'sȯrp·shən ,līn }

absorption peak [SPECT] A wavelength of maximum electromagnetic absorption by a chemical sample; used to identify specific elements, radicals, or compounds. { əb'sȯrp·shən ,pēk }

absorption spectrophotometer [SPECT] An instrument used to measure the relative intensity of absorption spectral lines and bands. Also known as difference spectrophotometer. { əb'sȯrp·shən ,spek·trə·fə'täm·ə·dər }

absorption spectroscopy [SPECT] An instrumental technique for determining the concentration and structure of a substance by measuring the intensity of electromagnetic radiation it absorbs at various wavelengths. { əb'sȯrp·shən ,spek'träs·kə·pē }

absorption spectrum [SPECT] A plot of how much radiation a sample absorbs over a range of wavelengths; the spectrum can be a plot of either absorbance or transmittance versus wavelength, frequency, or wavenumber. { əb'sȯrp·shən ,spek·trəm }

absorption tube [CHEM] A tube filled with a solid absorbent and used to absorb gases and vapors. { əb'sȯrp·shən ,tüb }

absorptive power See absorptivity. { əb'sȯrp·tiv ,paū·ər }

absorptivity [ANALY CHEM] The constant a in the Beer's law relation A = abc, where A is the absorbance, b the path length, and c the concentration of solution. Also known as absorptive power. Formerly known as absorbency index; absorption constant; extinction coefficient. { əb,sȯrp'tiv·əd·ē }

abstraction reaction [CHEM] A bimolecular chemical reaction in which an atom that is either neutral or charged is removed from a molecular entity. { ab'strak·shən rē,ak·shən }

Ac See actinium.

acaroid resin [ORG CHEM] A gum resin from aloelike trees of the genus Xanthorrhoea in Australia and Tasmania; used in varnishes and inks. Also known as gum accroides; yacca gum. { 'a·kə,rȯid 'rez·ən }

accelerator mass spectrometer [SPECT] A combination of a mass spectrometer and an accelerator that can be used to measure the natural abundances of very rare radioactive isotopes. { ak¦sel·ə,rād·ər ¦mas spek'träm·əd·ər }

accelofilter [CHEM] A filtration device that uses a vacuum or pressure to draw or force the liquid through the filter to increase the rate of filtration. { ak'sel·ō,fil·tər }

acceptor [CHEM] **1.** A chemical whose reaction rate with another chemical increases because the other substance undergoes another reaction. **2.** A species that accepts electrons, protons, electron pairs, or molecules such as dyes. { ak'sep·tər }

accessory element See trace element. { ak'ses·ə·rē 'el·ə·mənt }

acenaphthene [ORG CHEM] $C_{12}H_{10}$ An unsaturated hydrocarbon whose colorless crystals melt at 92°C; insoluble in water; used as a dye intermediate and as an agent for inducing polyploidy. { ,as·ə'naf·thēn }

acenaphthequinone [ORG CHEM] $C_{10}H_6(CO)_2$ A three-ring hydrocarbon in the form of yellow needles melting at 261–263°C; insoluble in water and soluble in alcohol; used in dye synthesis. { ,as·ə¦naf·thə·kwa¦nōn }

acene [ORG CHEM] Any condensed polycyclic compound with fused rings in a linear arrangement; for example, anthracene. { ə'sēn }

acenocoumarin See acenocoumarol. { ə,sēn·ə'kü·mə·rən }

acenocoumarol [ORG CHEM] $C_{19}H_{15}NO_6$ A tasteless, odorless, white, crystalline powder with a melting point of 197°C; slightly soluble in water and organic solvents; used as an anticoagulant. Also known as acenocoumarin. { ə,sēn·ə'kü·mə·rol }

acephate [ORG CHEM] $C_4H_{10}NO_3PS$ A white solid with a melting point of 72–80°C; very soluble in water; used as an insecticide for a wide range of aphids and foliage pests. { 'as·ə·fāt }

acephatemet [ORG CHEM] $CH_3OCH_3SPONH_2$ A white, crystalline solid with a melting point of 39–41°C; limited solubility in water; used as an insecticide to control cutworms and borers on vegetables. { as·ə'fāt·mət }

acetal [ORG CHEM] **1.** $CH_3CH(OC_2H_5)_2$ A colorless, flammable liquid used as a solvent and in manufacture of perfumes. Also known as 1,1-diethoxyethane. **2.** Any one of a class of compounds formed by the addition of alcohols to aldehydes. { 'as·ə,tal }

acetaldehyde [ORG CHEM] C_2H_4O A colorless, flammable liquid used chiefly to manufacture acetic acid. { ,as·əd'al·də,hīd }

acetaldehyde cyanohydrin See lactonitrile. { ,as·əd'al·də,hīd ,sī·ə·nō'hīd·rən }

acetal resins [ORG CHEM] Linear, synthetic resins produced by the polymerization of formaldehyde (acetal homopolymers) or of formaldehyde with trioxane (acetal copolymers); hard, tough plastics used as substitutes for metals. Also known as polyacetals. { 'as·ə,təl 'rez·ənz }

acetamide [ORG CHEM] CH_3CONH_2 The crystalline, colorless amide of acetic acid, used in organic synthesis and as a solvent. { ə'sed·ə,mīd }

acetamidine hydrochloride [ORG CHEM] $C_2H_6N_2·HCl$ Deliquescent crystals that are long prisms with a melting point reported as either 174°C or 164–166°C; soluble in water and alcohol; used in the synthesis of imidazoles, pyrimidines, and triazines. { ə·sed'am·ə,dēn hī·drə'klò,rīd }

acetaminophen [ORG CHEM] $C_8H_9O_2N$ Large monoclinic prisms with a melting point of 169–170°C; soluble in organic solvents such as methanol and ethanol; used in the manufacture of azo dyes and photographic chemicals, and as an analgesic and antipyretic. { ə,sēd·ə'mēn·ə·fən }

acetanilide [ORG CHEM] An odorless compound in the form of white, shining, crystalline leaflets or a white, crystalline powder with a melting point of 114–116°C; soluble in hot water, alcohol, ether, chloroform, acetone, glycerol, and benzene; used as a rubber accelerator, in the manufacture of dyestuffs and intermediates, as a precursor in penicillin manufacture, and as a painkiller. { ,a·səd'an·ə,līd }

acetate [ORG CHEM] One of two species derived from acetic acid, CH_3COOH; one type is the acetate ion, CH_3COO^-; the second type is a compound whose structure contains the acetate ion, such as ethyl acetate. { 'as·ə,tāt }

acetate dye [CHEM] **1.** Any of a group of water-insoluble azo or anthroquinone dyes used for dyeing acetate fibers. **2.** Any of a group of water-insoluble amino azo dyes that are treated with formaldehyde and bisulfate to make them water-soluble. { 'as·ə,tāt ,dī }

acetate of lime [ORG CHEM] Calcium acetate made from pyroligneous acid and a water suspension of calcium hydroxide. { 'as·ə,tāt əv 'līm }

acetenyl See ethinyl. { ə'sed·ə,nil }

acetic acid [ORG CHEM] CH_3COOH **1.** A clear, colorless liquid or crystalline mass with a pungent odor, miscible with water or alcohol; crystallizes in deliquescent needles; a component of vinegar. Also known as ethanoic acid. **2.** A mixture of the normal and acetic salts; used as a mordant in the dyeing of wool. { ə'sēd·ik 'as·əd }

acetic anhydride [ORG CHEM] $(CH_3CO)_2O$ A liquid with a pungent odor that combines with water to form acetic acid; used as an acetylating agent. { ə'sēd·ik an'hīd,rīd }

acetic ester See ethyl acetate. { ə'sēd·ik 'es·tər }

3

acetic ether See ethyl acetate. { ə'sēd·ik 'ē·thər }
acetidin See ethyl acetate. { ə'sed·ə·din }
acetin [ORG CHEM] C₃H₅(OH)₂OOCCH₃ A thick, colorless, hygroscopic liquid with a boiling point of 158°C, made by heating glycerol and strong acetic acid; soluble in water and alcohol; used in tanning, as a dye solvent and food additive, and in explosives. Also spelled acetine. { 'as·ə·tin }
acetine See acetin. { 'as·ə,tēn }
acetoacetate [ORG CHEM] A salt which contains the CH₃COCH₂COO radical; derived from acetoacetic acid. { ¦as·ə,tō·'as·ə,tāt }
acetoacetic acid [ORG CHEM] CH₃COCH₂COOH A colorless liquid miscible with water; derived from β-hydroxybutyric acid in the body. { ¦as·ə,tō,ə'sēd·ik 'as·əd }
acetoacetic ester See ethyl acetoacetate. { ¦as·ə,tō,ə'sēd·ik 'es·tər }
acetoamidoacetic acid See aceturic acid. { ¦as·ə,tō¦am·ə,dō·ə'sēd·ik 'as·əd }
acetoin [ORG CHEM] CH₃COCHOHCH₃ A slightly yellow liquid, melting point 15°C, used as an aroma carrier in the preparation of flavors and essences; produced by fermentation or from diacetyl by partial reduction with zinc and acid. { ə'sed·ə·wən }
acetol [ORG CHEM] CH₃COCH₂OH A colorless liquid soluble in water; a reducing agent. { 'as·ə·tōl }
acetolysis [ORG CHEM] Decomposition of an organic molecule through the action of acetic acid or acetic anhydride. { ,as·ə'täl·ə·səs }
acetone [ORG CHEM] CH₃COCH₃ A colorless, volatile, extremely flammable liquid, miscible with water; used as a solvent and reagent. Also known as 2-propanone. { 'as·ə,tōn }
acetone cyanohydrin [ORG CHEM] (CH₃)₂COHCN A colorless liquid obtained from condensation of acetone with hydrocyanic acid; used as an insecticide or as an organic chemical intermediate. { 'as·ə,tōn sī,ə·nō'hīd·rən }
acetone glucose See acetone sugar. { 'as·ə,tōn 'glü·kōs }
acetone number [CHEM] A ratio used to estimate the degree of polymerization of materials such as drying oils; it is the weight in grams of acetone added to 100 grams of a drying oil to cause an insoluble phase to form. { 'as·ə·tōn 'nəm·bər }
acetone pyrolysis [ORG CHEM] Thermal decomposition of acetone into ketene. { 'as·ə·tōn pī'räl·ə·səs }
acetone-sodium bisulfite [ORG CHEM] (CH₃)₂C(OH)SO₃Na Crystals that have a slight sulfur dioxide odor and slightly fatty feel; freely soluble in water, decomposed by acids; used in photography and in textile dyeing and printing. { 'as·ə·tōn 'sōd·ē·əm ,bī'səl,fāt }
acetone sugar [ORG CHEM] Any reducing sugar that contains acetone; examples are 1,2-monoacetone-D-glucofuranose and 1,2,6-diacetone-D-glucofuranose. Also known as acetone glucose. { 'as·ə·tōn 'shùg·ər }
acetonitrile [ORG CHEM] CH₃CN A colorless liquid soluble in water; used in organic synthesis. { ,as·ə·tō'nī,tril }
acetonylacetone [ORG CHEM] CH₃COCH₂CH₂COCH₃ A colorless liquid with a boiling point of 192.2°C; soluble in water; used as a solvent and as an intermediate for pharmaceuticals and photographic chemicals. { ,as·ə,tän·əl'as·ə,tōn }
acetophenone [ORG CHEM] C₆H₅COCH₃ Colorless crystals with a melting point of 19.6°C and a specific gravity of 1.028; used as a chemical intermediate. { ,as·ə,tä·fə'nōn }
acetostearin [ORG CHEM] A general term for monoglycerides of stearic acid acetylated with acetic anhydride; used as a protective food coating and as plasticizers for waxes and synthetic resins to improve low-temperature characteristics. { ə,sē·dō'stēr·ən }
acetoxime [ORG CHEM] (CH₃)₂CNOH Colorless crystals with a chlorallike odor and a melting point of 61°C; soluble in alcohol, ethers, and water; used in organic synthesis and as a solvent for cellulose ethers. { ,as·ə'täk,sēm }
aceturic acid [ORG CHEM] CH₃CONHCHCH₂COOH Long, needlelike crystals with a melting point of 206–208°C; soluble in water and alcohol; forms stable salts with organic bases; used in medicine. { ¦as·ə¦tür·ik 'as·əd }
acetyl [ORG CHEM] CH₃CO− A two-carbon organic radical containing a methyl group and a carbonyl group. { ə'sēd·əl }

4

α-acetylacetanilide *See* acetoacetic acid. { 'al·fə əˌsed·əlˌa·səd'an·əˌlīd }

acetylacetone [ORG CHEM] $CH_3COCH_2OCCH_3$ A colorless liquid with a pleasant odor and a boiling point of 140.5°C; soluble in water; used as a solvent, lubricant additive, paint drier, and pesticide. { əˌsed·əl'as·əˌtōn }

acetylating agent [ORG CHEM] A reagent, such as acetic anhydride, capable of bonding an acetyl group onto an organic molecule. { ə'sed·əlˌāt·iŋ ˌā·jənt }

acetylation [ORG CHEM] The process of bonding an acetyl group onto an organic molecule. { əˌsed·əl'ā·shən }

acetyl benzoyl peroxide [ORG CHEM] $C_6H_5CO·O_2·OCCH_3$ White crystals with a melting point of 36.6°C; moderately soluble in ether, chloroform, carbon tetrachloride, and water; used as a germicide and disinfectant. { ə'sed·əl 'ben·zȯil pə'räkˌsīd }

acetyl bromide [ORG CHEM] CH_3COBr A colorless, fuming liquid with a boiling point of 81°C; soluble in ether, chloroform, and benzene; used in organic synthesis and dye manufacture. { ə'sed·əl 'brōˌmīd }

α-acetylbutyrolactone [ORG CHEM] $C_6H_8O_3$ A liquid with an esterlike odor; soluble in water; used in the synthesis of 3,4-disubstituted pyridines. { ˌal·fə əˌsed·əlˌbyüd·ə·rō'lak·tōn }

acetyl chloride [ORG CHEM] CH_3COCl A colorless, fuming liquid with a boiling point of 51–52°C; soluble in ether, acetone, and acetic acid; used in organic synthesis, and in the manufacture of dyestuffs and pharmaceuticals. { ə'sed·əl 'klȯˌrīd }

acetylene [ORG CHEM] C_2H_2 A colorless, highly flammable gas that is explosive when compressed; the simplest compound containing a triple bond; used in organic synthesis and as a welding fuel. Also known as ethyne. { ə'sed·əlˌēn }

acetylene black [ORG CHEM] A form of carbon with high electrical conductivity; made by decomposing acetylene by heat. { ə'sed·əlˌēn 'blak }

acetylene series [ORG CHEM] A series of unsaturated aliphatic hydrocarbons, each containing at least one triple bond and having the general formula C_nH_{2n-2}. { ə'sed·əlˌēn 'sir·ēz }

acetylene tetrabromide [ORG CHEM] $CHBr_2CHBr_2$ A yellowish liquid with a boiling point of 239–242°C; soluble in alcohol and ether; used for separating minerals and as a solvent. { ə'sed·əlˌēn ˌte·trə'brōˌmīd }

acetylenic [ORG CHEM] Pertaining to acetylene or being like acetylene, such as having a triple bond. { əˌsed·ə'len·ik }

acetylenyl *See* ethinyl. { əˌsed·ə'len·əl }

N-acetylethanolamine [ORG CHEM] $CH_3CONHC_2H_4OH$ A brown, viscous liquid with a boiling range of 150–152°C; soluble in alcohol, ether, and water; used as a plasticizer, humectant, high-boiling solvent, and textile conditioner. { ˌen əˌsed·əlˌeth·ə'näl·əˌmēn }

acetylide [ORG CHEM] A compound formed from acetylene with the H atoms replaced by metals, as in cuprous acetylide (Cu_2C_2). { ə'sed·əlˌīd }

acetyl iodide [ORG CHEM] CH_3COI A colorless, transparent, fuming liquid with a boiling point of 105–108°C; soluble in ether and benzene; used in organic synthesis. { ə'sed·əl 'ī·əˌdīd }

acetylisoeugenol [ORG CHEM] $C_6H_3(CHCHCH_3)(OCH_3)(OCOCH_3)$ White crystals with a clovelike odor and a congealing point of 77°C; used in perfumery and flavoring. { əˌsed·əlˌī·sō'yü·jəˌnȯl }

acetyl number [ANALY CHEM] A measure of free hydroxyl groups in fats or oils determined by the amount of potassium hydroxide used to neutralize the acetic acid formed by saponification of acetylated fat or oil. { ə'sed·əl ˌnəm·bər }

acetyl peroxide [ORG CHEM] $(CH_3CO)_2O_2$ Colorless crystals with a melting point of 30°C; soluble in alcohol and ether; used as an initiator and catalyst for resins. { ə'sed·əl pə'räkˌsīd }

acetyl propionyl [ORG CHEM] $CH_3COCOCH_2CH_3$ A yellow liquid with a boiling point of 106–110°C; used in butterscotch- and chocolate-type flavors. { ə'sed·əl 'prō·əˌnil }

acetylsalicylic acid [ORG CHEM] $CH_3COOC_6H_4COOH$ A white, crystalline, weakly acidic

5

N-acetylsulfanilyl chloride

substance, with melting point 137°C; slightly soluble in water; used medicinally as an antipyretic. Also known by trade name aspirin. { ə¦sed·əl¦sal·ə¦sil·ik 'as·əd }

N-acetylsulfanilyl chloride [ORG CHEM] C₈H₈ClNO₃S Thick, light tan prisms ranging to brown powder or fine crystals with a melting point of 149°C; soluble in benzene, chloroform, and ether; used as an intermediate in the preparation of sulfanilamide and its derivatives. Abbreviated ASC. { ¦en ə¦sed·əl·səl'fan·ə·lil 'klȯr,īd }

acetylurea [ORG CHEM] CH₃CONHCONH₂ Crystals that are colorless and are slightly soluble in water. { ə,sēd·əl,yù'rē·ə }

acetyl valeryl [ORG CHEM] CH₃COCOC₄H₉ A yellow liquid used for cheese, butter, and other flavors. Also known as heptadione-2,3. { ə'sēd·əl ,val·ə,ril }

achiral molecules [ORG CHEM] Molecules which are superposable to their mirror images. { ¦ā,kī·rəl 'mäl·ə,kyülz }

acid [CHEM] **1.** Any of a class of chemical compounds whose aqueous solutions turn blue litmus paper red, react with and dissolve certain metals to form salts, and react with bases to form salts. **2.** A compound capable of transferring a hydrogen ion in solution. **3.** A substance that ionizes in solution to yield the positive ion of the solvent. **4.** A molecule or ion that combines with another molecule or ion by forming a covalent bond with two electrons from the other species. { 'as·əd }

π-acid [ORG CHEM] An acid that readily forms stable complexes with aromatic systems. { 'pī 'as·əd }

acid acceptor [ORG CHEM] A stabilizer compound added to plastic and resin polymers to combine with trace amounts of acids formed by decomposition of the polymers. { 'as·əd ək'sep·tər }

acid alcohol [ORG CHEM] A compound containing both a carboxyl group (−COOH) and an alcohol group (−CH₂OH, =CHOH, or ≡COH). { 'as·əd 'al·kə·hȯl }

acid amide [ORG CHEM] A compound derived from an acid in which the hydroxyl group (−OH) of the carboxyl group (−COOH) has been replaced by an amino group (−NH₂) or a substituted amino group (−NHR or −NHR₂). { 'as·əd 'a,mīd }

acid anhydride [CHEM] An acid with one or more molecules of water removed; for example, SO₃ is the acid anhydride of H₂SO₄, sulfuric acid. { 'as·əd ,an'hīd,rīd }

acid azide [ORG CHEM] **1.** A compound in which the hydroxy group of a carboxylic acid is replaced by the azido group (−NH₃). **2.** An acyl or aroyl derivative of hydrazoic acid. Also known as acyl azide. { 'as·əd 'ā,zīd }

acid-base catalysis [CHEM] The increase in speed of certain chemical reactions due to the presence of acids and bases. { 'as·əd 'bās kə'tal·ə·sis }

acid-base equilibrium [CHEM] The condition when acidic and basic ions in a solution exactly neutralize each other; that is, the pH is 7. { 'as·əd 'bās ,ik·wə'lib·rē·əm }

acid-base indicator [ANALY CHEM] A substance that reveals, through characteristic color changes, the degree of acidity or basicity of solutions. { 'as·əd 'bās 'in·də,kād·ər }

acid-base pair [CHEM] A concept in the Brönsted theory of acids and bases; the pair consists of the source of the proton (acid) and the base generated by the transfer of the proton. { 'as·əd 'bās 'pär }

acid-base titration [ANALY CHEM] A titration in which an acid of known concentration is added to a solution of base of unknown concentration, or the converse. { 'as·əd 'bās tī'trā·shən }

acid calcium phosphate See calcium phosphate. { 'as·əd 'kal·sē·əm 'fäs,fāt }

acid cell [PHYS CHEM] An electrolytic cell whose electrolyte is an acid. { 'as·əd ,sel }

acid chloride [ORG CHEM] A compound containing the radical −COCl; an example is benzoyl chloride. { 'as·əd 'klȯr,īd }

acid disproportionation [CHEM] The self-oxidation of a sample of an oxidized element to the next higher oxidation state and then a corresponding reduction to lower oxidation states. { 'as·əd ,dis·prə,pȯr·shə'nā·shən }

acid dye [ORG CHEM] Any of a group of sodium salts of sulfonic and carboxylic acids used to dye natural and synthetic fibers, leather, and paper. { 'as·əd ,dī }

acid electrolyte [INORG CHEM] A compound, such as sulfuric acid, that dissociates into

6

ions when dissolved, forming an acidic solution that conducts an electric current.
{ 'as·əd ə'lek·trə,līt }

acid halide [ORG CHEM] A compound of the type RCOX, where R is an alkyl or aryl radical and X is a halogen. { 'as·əd 'ha,līd }

acid heat test [ANALY CHEM] The determination of degree of unsaturation of organic compounds by reacting with sulfuric acid and measuring the heat of reaction. { 'as· əd 'hēt ,test }

acidic [CHEM] **1.** Pertaining to an acid or to its properties. **2.** Forming an acid during a chemical process. { ə'sid·ik }

acidic dye [ORG CHEM] An organic anion that binds to and stains positively charged macromolecules. { ə¦sid·ik 'dī }

acidic group [ORG CHEM] The radical COOH, present in organic acids. { ə'sid·ik 'grüp }

acidic oxide [INORG CHEM] An oxygen compound of a nonmetal, for example, SO_2 or P_2O_5, which yields an oxyacid with water. { ə'sid·ik 'äk,sīd }

acidic titrant [ANALY CHEM] An acid solution of known concentration used to determine the basicity of another solution by titration. { ə'sid·ik 'tī·trənt }

acidification [CHEM] Addition of an acid to a solution until the pH falls below 7. { ə,sid·ə·fə'kā·shən }

acidimeter [ANALY CHEM] An apparatus or a standard solution used to determine the amount of acid in a sample. { ,as·ə'dim·ə·tər }

acidimetry [ANALY CHEM] The titration of an acid with a standard solution of base. { ,as·ə'dim·ə·trē }

acidity [CHEM] The state of being acid. { ə'sid·ə·tē }

acidity function [CHEM] A quantitative scale for measuring the acidity of a solvent system; usually established over a range of compositions. { ə'sid·əd·ē ,fəŋk·shən }

acid number See acid value. { 'as·əd ¦nəm·bər }

acidolysis [ORG CHEM] A chemical reaction involving the decomposition of a molecule, with the addition of the elements of an acid to the molecule; the reaction is comparable to hydrolysis or alcoholysis, in which water or alcohol, respectively, is used in place of the acid. Also known as acyl exchange. { ,as·ə'däl·ə·səs }

acid phosphate [INORG CHEM] A mono- or dihydric phosphate; for example, M_2HPO_4 or MH_2PO_4, where M represents a metal atom. { 'as·əd 'fäs,fāt }

acid potassium phthalate See potassium biphthalate. { 'as·əd pə'tas·ē·əm 'tha,lāt }

acid potassium sulfate See potassium bisulfate. { 'as·əd pə'tas·ē·əm 'səl,fāt }

acid reaction [CHEM] A chemical reaction produced by an acid. { 'as·əd rē'ak·shən }

acid salt [CHEM] A compound derived from an acid and base in which only a part of the hydrogen is replaced by a basic radical; for example, the acid sulfate $NaHSO_4$. { 'as·əd ,sölt }

acid sodium tartrate See sodium bitartrate. { 'as·əd sōd·ē·əm 'tär,trāt }

acid solution [CHEM] An aqueous solution containing more hydrogen ions than hydroxyl ions. { 'as·əd sə'lü·shən }

acid tartrate See bitartrate. { 'as·əd 'tär,trāt }

acid value Also known as acid number. [CHEM] The acidity of a solution expressed in terms of normality. [ORG CHEM] A number indicating the amount of nonesterified fatty acid present in a sample of fat or fatty oil as determined by alkaline titration. { 'as·əd 'val·yü }

aconitic acid [ORG CHEM] $C_6H_6O_6$ A white, crystalline organic acid found in sugarcane and sugarbeet; obtained during manufacture of sugar. { ,ak·ə'nid·ik 'as·əd }

Acree's reaction [ANALY CHEM] A test for protein in which a violet ring appears when concentrated sulfuric acid is introduced below a mixture of the unknown solution and a formaldehyde solution containing a trace of ferric chloride. { 'ak·rēz rē'ak·shən }

acridine [ORG CHEM] $(C_6H_4)_2$NCH A typical member of a group of organic heterocyclic compounds containing benzene rings fused to the 2,3 and 5,6 positions of pyridine; derivatives include dyes and medicines. { 'ak·rə,dēn }

acridine dye [ORG CHEM] Any of a class of basic dyes containing the acridine nucleus that bind to deoxyribonucleic acid. { 'ak·rə,dēn ¦dī }

acridine orange [ORG CHEM] A dye with an affinity for nucleic acids; the complexes of

acriflavine

nucleic acid and dye fluoresce orange with RNA and green with DNA when observed in the fluorescence microscope. { 'ak·rə,dēn 'är·inj }

acriflavine [ORG CHEM] $C_{14}H_{14}N_3Cl$ A yellow acridine dye obtained from proflavine by methylation in the form of red crystals; used as an antiseptic in solution. { ,ak· rə'flā,vēn }

acrolein [ORG CHEM] $CH_2{=}CHCHO$ A colorless to yellow liquid with a pungent odor and a boiling point of 52.7°C; soluble in water, alcohol, and ether; used in organic synthesis, pharmaceuticals manufacture, and as an herbicide and tear gas. { ə'krōl· ē·ən }

acrolein cyanohydrin [ORG CHEM] $CH_2{:}CHCH(OH)CN$ A liquid soluble in water and boiling at 165°C; copolymerizes with ethylene and acrylonitrile; used to modify synthetic resins. { ə'krōl·ē·ən ,sī·ə·nō'hī·drən }

acrolein dimer [ORG CHEM] $C_6H_8O_2$ A flammable, water-soluble liquid used as an intermediate for resins, dyestuffs, and pharmaceuticals. { ə'krōl·ē·ən 'dī·mər }

acrolein test [ANALY CHEM] A test for the presence of glycerin or fats; a sample is heated with potassium bisulfate, and acrolein is released if the test is positive. { ə'krōl·ē·ən ,test }

acrylamide [ORG CHEM] $CH_2CHCONH_2$ Colorless, odorless crystals with a melting point of 84.5°C; soluble in water, alcohol, and acetone; used in organic synthesis, polymerization, sewage treatment, ore processing, and permanent press fabrics. { ə'kril· ə,mīd }

acrylamide copolymer [ORG CHEM] A thermosetting resin formed of acrylamide with other resins, such as the acrylic resins. { ə'kril·ə,mīd kō'päl·ə·mər }

acrylate [ORG CHEM] **1.** A salt or ester of acrylic acid. **2.** See acrylate resin. { 'ak· rə,lāt }

acrylate resin [ORG CHEM] Acrylic acid or ester polymer with a $-CH_2-CH(COOR)-$ structure; used in paints, sizings and finishes for paper and textiles, adhesives, and plastics. Also known as acrylate. { 'ak·rə,lāt 'rez·ən }

acrylic acid [ORG CHEM] $CH_2CHCOOH$ An easily polymerized, colorless, corrosive liquid used as a monomer for acrylate resins. { ə'kril·ik 'as·əd }

acrylic ester [ORG CHEM] An ester of acrylic acid. { ə'kril·ik 'es·tər }

acrylic resin [ORG CHEM] A thermoplastic synthetic organic polymer made by the polymerization of acrylic derivatives such as acrylic acid, methacrylic acid, ethyl acrylate, and methyl acrylate; used for adhesives, protective coatings, and finishes. { ə'kril· ik 'rez·ən }

acrylic rubber [ORG CHEM] Synthetic rubber containing acrylonitrile; for example, nitrile rubber. { ə'kril·ik 'rəb·ər }

acrylonitrile [ORG CHEM] CH_2CHCN A colorless liquid compound used in the manufacture of acrylic rubber and fibers. Also known as vinylcyanide. { ,ak·rə,lō'nī·trəl }

acrylonitrile butadiene styrene resin [ORG CHEM] A polymer made by blending acrylonitrile-styrene copolymer with a butadiene-acrylonitrile rubber or by interpolymerizing polybutadiene with styrene and acrylonitrile; combines the advantages of hardness and strength of the vinyl resin component with the toughness and impact resistance of the rubbery component. Abbreviated ABS. { ,ak·rə,lō'nī·trəl ,byüd· ə'dī,ēn 'stī·rēn 'rez·ən }

acrylonitrile copolymer [ORG CHEM] Oil-resistant synthetic rubber made by polymerization of acrylonitrile with compounds such as butadiene or acrylic acid. { ,ak· rə,lō'nī·trəl kō'päl·ə·mər }

actinide series [CHEM] The group of elements of atomic number 89 through 103. Also known as actinoid elements. { 'ak·tə,nīd 'sir,ēz }

actinism [CHEM] The production of chemical changes in a substance upon which electromagnetic radiation is incident. { 'ak·tə'niz·əm }

actinium [CHEM] A radioactive element, symbol Ac, atomic number 89; its longest-lived isotope is ^{227}Ac with a half-life of 21.7 years; the element is trivalent; chief use is, in equilibrium with its decay products, as a source of alpha rays. { ak'tin·ē·əm }

actinochemistry [CHEM] A branch of chemistry concerned with chemical reactions produced by light or other radiation. { ,ak·tə·nō'kem·ə·strē }

8

actinoid elements *See* actinide series. { 'ak·tə,nȯid 'el·ə·məns }

activated complex [PHYS CHEM] An energetically excited state which is intermediate between reactants and products in a chemical reaction. Also known as transition state. { 'ak·tə,vād·əd 'käm·pleks }

activation [CHEM] Treatment of a substance by heat, radiation, or activating reagent to produce a more complete or rapid chemical or physical change. { ,ak·tə'vā·shən }

activation energy [PHYS CHEM] The energy, in excess over the ground state, which must be added to an atomic or molecular system to allow a particular process to take place. { ,ak·tə'vā·shən 'en·ər·jē }

activator [CHEM] **1.** A substance that increases the effectiveness of a rubber vulcanization accelerator; for example, zinc oxide or litharge. **2.** A trace quantity of a substance that imparts luminescence to crystals; for example, silver or copper in zinc sulfide or cadmium sulfide pigments. { 'ak·tə,vād·ər }

active center [CHEM] **1.** Any one of the points on the surface of a catalyst at which the chemical reaction is initiated or takes place. **2.** *See* active site. { 'ak·tiv 'sen·tər }

active site [CHEM] The effective site at which a given heterogeneous catalytic reaction can take place. Also known as active center. { 'ak·tiv 'sīt }

active solid [CHEM] A porous solid possessing adsorptive properties and used for chromatographic separations. { 'ak·tiv 'säl·əd }

activity [PHYS CHEM] A thermodynamic function that correlates changes in the chemical potential with changes in experimentally measurable quantities, such as concentrations or partial pressures, through relations formally equivalent to those for ideal systems. { ,ak'tiv·əd·ē }

activity coefficient [PHYS CHEM] A characteristic of a quantity expressing the deviation of a solution from ideal thermodynamic behavior; often used in connection with electrolytes. { ,ak'tiv·əd·ē ,kō·ə'fish·ənt }

activity series [CHEM] A series of elements that have similar properties—for example, metals—arranged in descending order of chemical activity. { ak'tiv·əd·ē ,sir·ēz }

actol *See* silver lactate. { 'ak,tȯl }

acyclic compound [ORG CHEM] A chemical compound with an open-chain molecular structure rather than a ring-shaped structure; for example, the alkane series. { ā'sik·lik 'käm,paund }

acyl [ORG CHEM] A radical formed from an organic acid by removal of a hydroxyl group; the general formula is RCO, where R may be aliphatic, alicyclic, or aromatic. { 'a·səl }

acylation [ORG CHEM] Any process whereby the acyl group is incorporated into a molecule by substitution. { ,as·ə'lā·shən }

acyl azide *See* acid azide. { 'a·səl 'ā,zīd }

acylcarbene [ORG CHEM] A carbene radical in which at least one of the groups attached to the divalent carbon is an acyl group; for example, acetylcarbene. { ,a·səl'kär,bēn }

acyl exchange *See* acidolysis. { 'a·səl iks'chānj }

acyl halide [ORG CHEM] One of a large group of organic substances containing the halocarbonyl group; for example, acyl fluoride. { 'a·səl 'hal,īd }

acylnitrene [ORG CHEM] A nitrene in which the nitrogen is covalently bonded to an acyl group. { ,a·səl'nī,trēn }

acyloin [ORG CHEM] An organic compound that may be synthesized by condensation of aldehydes; an example is benzoin, $C_6H_5COCHOHC_6H_5$. { ə'sil·ə·wən }

acyloin condensation [ORG CHEM] The reaction of an aliphatic ester with metallic sodium to form intermediates converted by hydrolysis into aliphatic α-hydroxy-ketones called acyloins. { ə'sil·ə·wən ,kän,den'sā·shən }

adamantane [ORG CHEM] A $C_{10}H_{16}$ alicyclic hydrocarbon whose structure has the same arrangement of carbon atoms as does the basic unit of the diamond lattice. { ,ad·ə'man,tān }

adamsite [ORG CHEM] $C_6H_4 \cdot NH \cdot C_6H_4 \cdot AsCl$ A yellow crystalline arsenical; used in leather tanning and in warfare and riot control to produce skin and eye irritation, chest distress, and nausea; U.S. Army code is DM. Also known as diphenylaminechloroarsine; phenarsazine chloride. { 'a·dəm,zīt }

9

adatom

adatom [PHYS CHEM] An atom adsorbed on a surface so that it will migrate over the surface. { 'ad,ad·əm }

addition agent [PHYS CHEM] A substance added to a plating solution to change characteristics of the deposited substances. { ə'di·shən ,ā·jənt }

addition polymer [ORG CHEM] A polymer formed by the chain addition of unsaturated monomer molecules, such as olefins, with one another without the formation of a by-product, as water; examples are polyethylene, polypropylene, and polystyrene. Also known as addition resin. { ə'di·shən 'päl·ə·mər }

addition polymerization [ORG CHEM] A reaction initiated by an anion, cation, or radical in which a large number of monomer units are added rapidly (a chain reaction) until terminated by some mechanism, forming a high-molecular-weight polymer in a very short time; an example is the free-radical polymerization of propylene to polypropylene. { ə¦dish·ən pə,lim·ə·rə'zā·shən }

addition reaction [ORG CHEM] A type of reaction of unsaturated hydrocarbons with hydrogen, halogens, halogen acids, and other reagents, so that no change in valency is observed and the organic compound forms a more complex one. { ə'di·shən rē'ak·shən }

addition resin See addition polymer. { ə'di·shən 'rez·ən }

adduct [CHEM] **1.** A chemical compound that forms from chemical addition of two species; for example, reaction of butadiene with styrene forms an adduct, 4-phenyl-1-cyclohexene. **2.** The complex compound formed by association of an inclusion complex. { 'a,dəkt }

adiabatic approximation See Born-Oppenheimer approximation. { ¦ad·ē·ə¦bad·ik ə,präk·sə'mā·shən }

adiabatic calorimeter [PHYS CHEM] An instrument used to study chemical reactions which have a minimum loss of heat. { ¦ad·ē·ə¦bad·ik ,kal·ə'rim·əd·ər }

adiabatic flame temperature [PHYS CHEM] The highest possible temperature of combustion obtained under the conditions that the burning occurs in an adiabatic vessel, that it is complete, and that dissociation does not occur. { ¦ad·ē·ə¦bad·ik ¦flām 'tem·prə·chər }

adipate [ORG CHEM] Salt produced by reaction of adipic acid with a basic compound. { 'ad·ə,pāt }

adipic acid [ORG CHEM] $HOOC(CH_2)_4COOH$ A colorless crystalline dicarboxylic acid, sparingly soluble in water; used in nylon manufacture. { ə'dip·ik 'as·əd }

adiponitrile [ORG CHEM] $NC(CH_2)_4CN$ The high-boiling liquid dinitrile of adipic acid; used to make nylon intermediates. { ,ad·ə·pō'nī·trəl }

adjective dye [CHEM] Any dye that needs a mordant. { ə'jek·tiv ,dī }

adsorbate [CHEM] A solid, liquid, or gas which is adsorbed as molecules, atoms, or ions by such substances as charcoal, silica, metals, water, and mercury. { ad'sȯr,bāt }

adsorbent [CHEM] A solid or liquid that adsorbs other substances; for example, charcoal, silica, metals, water, and mercury. { ad'sȯr·bənt }

adsorption [CHEM] The surface retention of solid, liquid, or gas molecules, atoms, or ions by a solid or liquid, as opposed to absorbtion, the penetration of substances into the bulk of the solid or liquid. { ad'sȯrp·shən }

adsorption catalysis [PHYS CHEM] A catalytic reaction in which the catalyst is an adsorbent. { ad'sorp·shən kə'tal·ə·səs }

adsorption chromatography [ANALY CHEM] Separation of a chemical mixture (gas or liquid) by passing it over an adsorbent bed which adsorbs different compounds at different rates. { ad'sȯrp·shən ,krō·mə'täg·rə·fē }

adsorption complex [CHEM] An entity consisting of an adsorbate and that portion of the adsorbent to which it is bound. { ad'sȯrp·shən ,käm,pleks }

adsorption indicator [ANALY CHEM] An indicator used in solutions to detect slight excess of a substance or ion; precipitate becomes colored when the indicator is adsorbed. An example is fluorescein. { ad'sȯrp·shən ,in·də,kād·ər }

adsorption isobar [PHYS CHEM] A graph showing how adsorption varies with some parameter, such as temperature, while holding pressure constant. { ad'sȯrp·shən 'ī·sō,bär }

10

adsorption isotherm |PHYS CHEM| The relationship between the gas pressure p and the amount w, in grams, of a gas or vapor taken up per gram of solid at a constant temperature. { ad'sȯrp·shən 'ī·sō,thərm }

adsorption potential |PHYS CHEM| A change in the chemical potential that occurs as an ion moves from a gas or solution phase to the surface of an adsorbent. { ad'sȯrp· shən pə,ten·chəl }

aeration cell |PHYS CHEM| An electrolytic cell whose electromotive force is due to electrodes of the same material located in different concentrations of dissolved air. Also known as oxygen cell. { e'rā·shən ,sel }

aerogel |CHEM| A porous solid formed from a gel by replacing the liquid with a gas with little change in volume so that the solid is highly porous. { 'e·rō,jel }

aerosol |CHEM| A suspension of small particles in a gas; the particles may be solid or liquid or a mixture of both; aerosols are formed by the conversion of gases to particles, the disintegration of liquids or solids, or the suspension of powdered material. { 'e·rə,sȯl }

AES See Auger electron spectroscopy.

affinity |CHEM| The extent to which a substance or functional group can enter into a chemical reaction with a given agent. Also known as chemical affinity. { ə'fin·əd·ē }

affinity chromatography |ANALY CHEM| A chromatographic technique that utilizes the ability of biological molecules to bend to certain ligands specifically and reversibly; used in protein biochemistry. { ə'fin·əd·ē ,krō·mə'täg·rə·fē }

Ag See silver.

agaric acid |ORG CHEM| $C_{19}H_{36}(OH)(COOH)_3$ An acid with melting point 141°C; soluble in water, insoluble in benzene; used as an irritant. Also known as agaricin. { ə'gar· ik 'as·əd }

agaricin See agaric acid. { ə'gar·ə·sən }

agavose |ORG CHEM| $C_{12}H_{22}O_{11}$ A sugar found in the juice of the agave tree; used in medicine as a diuretic and laxative. { 'ag·ə,vōs }

aggregate |CHEM| A group of atoms or molecules that are held together in any way, for example, a micelle. { 'ag·rə·gət }

aggregation |CHEM| A process that results in the formation of aggregates. { ,ag· rə'gā·shən }

aging |CHEM| All irreversible structural changes that occur in a precipitate after it has formed. { 'āj·iŋ }

agostic |ORG CHEM| A three-center, two-electron bonding interaction in which a hydrogen atom is bonded to both a carbon atom and a metal atom, such as the interaction of a CH bond and an unsaturated transition-metal compound. { ə'gäs·tik }

air |CHEM| A predominantly mechanical mixture of a variety of individual gases forming the earth's enveloping atmosphere. { er }

air deficiency |CHEM| Insufficient air in an air-fuel mixture causing either incomplete fuel oxidation or lack of ignition. { 'er di,fish·ən·sē }

air-fuel ratio |CHEM| The ratio of air to fuel by weight or volume which is significant for proper oxidative combustion of the fuel. { 'er 'fyül ,rā·shō }

air line |SPECT| Lines in a spectrum due to the excitation of air molecules by spark discharges, and not ordinarily present in arc discharges. { 'er ,līn }

air-sensitive crystal |CHEM| A crystal that decomposes when exposed to air. { 'er ¦sen·səd·iv 'krist·əl }

air-slaked |CHEM| Having the property of a substance, such as lime, that has been at least partially converted to a carbonate by exposure to air. { 'er ,slākt }

ajmaline |ORG CHEM| $C_{20}H_{26}N_2O_2$ An amber, crystalline alkaloid obtained from *Rauwolfia* plants, especially R. *serpentina*. { 'aj·mə,lēn }

Al See aluminum.

alanyl |ORG CHEM| The radical CH_3CHNH_2CO-; occurs in, for example, alanyl alanine, a dipeptide. { 'al·ə,nil }

alchemy |CHEM| A speculative chemical system having as its central aims the transmutation of base metals to gold and the discovery of the philosopher's stone. { 'al· kə·mē } alcogel

alcogel [CHEM] A gel formed by an alcosol. { 'al·kə¦jel }
alcohol [ORG CHEM] Any member of a class of organic compounds in which a hydrogen atom of a hydrocarbon has been replaced by a hydroxy (−OH) group. { 'al·kə¦hȯl }
alcoholate [ORG CHEM] A compound formed by the reaction of an alcohol with an alkali metal. Also known as alkoxide. { ¦al·kə'hȯ¦lāt }
alcoholysis [ORG CHEM] The breaking of a carbon-to-carbon bond by addition of an alcohol. { ¦al·kə'hȯl·ə·səs }
alcosol [CHEM] Mixture of an alcohol and a colloid. { 'al·kə¦sȯl }
aldehyde [ORG CHEM] One of a class of organic compounds containing the CHO radical. { 'al·də¦hīd }
aldehyde ammonia [ORG CHEM] $CH_3CHOHNH_2$ A white, crystalline solid with a melting point of 97°C; soluble in water and alcohol; used in organic synthesis and as a vulcanization accelerator. { 'al·də¦hīd ə'mō·nyə }
aldehyde polymer [ORG CHEM] Any of the plastics based on aldehydes, such as formaldehyde, acetaldehyde, butyraldehyde, or acrylic aldehyde (acrolein). { 'al·də¦hīd 'päl·ə·mər }
aldicarb [ORG CHEM] $C_7H_{14}N_2O_2S$ A colorless, crystalline compound with a melting point of 100°C; used as an insecticide, miticide, and nematicide to treat soil for cotton, sugarbeets, potatoes, peanuts, and ornamentals. { 'al·də¦kärb }
aldohexose [ORG CHEM] A hexose, such as glucose or mannose, containing the aldehyde group. { ¦al·dō'hek¦sōs }
aldol [ORG CHEM] $CH_3CH(OH)CH_2CHO$ A colorless, thick liquid with a boiling point of 83°C; used in manufacturing rubber age resistors, accelerators, and vulcanizers. { 'al¦dȯl }
aldol condensation [ORG CHEM] Formation of a β-hydroxycarbonyl compound by the condensation of an aldehyde or a ketone in the presence of an acid or base catalyst. Also known as aldol reaction. { 'al¦dȯl ¦kän·dən'sā·shən }
aldol reaction See aldol condensation. { 'al¦dȯl rē'ak·shən }
aldose [ORG CHEM] A class of monosaccharide sugars; the molecule contains an aldehyde group. { 'al¦dōs }
Aldrin [ORG CHEM] $C_{12}H_8Cl_6$ Trade name for a water-insoluble, white, crystalline compound, consisting mainly of chlorinated dimethanonaphthalene; used as a pesticide. { 'al·drən }
alfin catalyst [ORG CHEM] A catalyst derived from reaction of an alkali alcoholate with an olefin halide; used to convert olefins (for example, ethylene, propylene, or butylenes) into polyolefin polymers. { 'al·fin 'kad·ə¦list }
algin See sodium alginate. { 'al·jən }
alginic acid [ORG CHEM] $(C_6H_8O_6)_n$ An insoluble colloidal acid obtained from brown marine algae; it is hard when dry and absorbent when moist. Also known as algin. { al'jin·ik 'as·əd }
alginic acid sodium salt See sodium alginate. { al'jin·ik 'as·əd 'sōd·ē·əm 'sȯlt }
alicyclic [ORG CHEM] **1.** Having the properties of both aliphatic and cyclic substances. **2.** Referring to a class of saturated hydrocarbon compounds whose structures contain one ring. Also known as cycloaliphatic; cycloalkane.Any one of the compounds of the alicyclic class. Also known as cyclane. { ¦al·ə¦sī·klik }
aliphatic [ORG CHEM] Of or pertaining to any organic compound of hydrogen and carbon characterized by a straight chain of the carbon atoms; three subgroups of such compounds are alkanes, alkenes, and alkynes. { ¦al·ə¦fad·ik }
aliphatic acid [ORG CHEM] Any organic acid derived from aliphatic hydrocarbons. { ¦al·ə¦fad·ik 'as·əd }
aliphatic acid ester [ORG CHEM] Any organic ester derived from aliphatic acids. { ¦al·ə¦fad·ik 'as·əd 'es·tər }
aliphatic polycyclic hydrocarbon [ORG CHEM] A hydrocarbon compound in which at least two of the aliphatic structures are cyclic or closed. { ¦al·ə¦fad·ik ¦pä·lə'sī·klik ¦hī·drə'kär·bən }

12

aliphatic polyene compound [ORG CHEM] Any unsaturated aliphatic or alicyclic compound with more than four carbons in the chain and with at least two double bonds; for example, hexadiene. { ¦al·ə¦fad·ik 'päl·ē‚ēn ‚käm‚paúnd }

aliphatic series [ORG CHEM] A series of open-chained carbon-hydrogen compounds; the two major classes are the series with saturated bonds and with the unsaturated. { ¦al·ə¦fad·ik 'sir·ēz }

aliquant [CHEM] A part of a sample that has been divided into a set of equal parts plus a smaller remainder part. { 'al·ə‚kwänt }

aliquot [CHEM] A part of a sample that has been divided into exactly equal parts with no remainder. { 'al·ə‚kwät }

alizarin [ORG CHEM] $C_{14}H_6O_2(OH)_2$ An orange crystalline compound, insoluble in cold water; made synthetically from anthraquinone; used in the manufacture of dyes and red pigments. { ə'liz·ə·rən }

alizarin dye [ORG CHEM] Sodium salts of sulfonic acids derived from alizarin. { ə'liz· ə·rən 'dī }

alizarin red [ORG CHEM] Any of several red dyes derived from anthraquinone. { ə'liz· ə·rən 'red }

alkadiene See diene. { ‚al·kə'dī‚ēn }

alkalescence [CHEM] The property of a substance that is alkaline, that is, having a pH greater than 7. { ‚al·kə'les·əns }

alkali [CHEM] Any compound having highly basic qualities. { 'al·kə‚lī }

alkali-aggregate reaction [CHEM] The chemical reaction of an aggregate with the alkali in a cement, resulting in a weakening of the concrete. { 'al·kə‚lī 'ag·rə·gət rē'ak· shən }

alkali alcoholate [ORG CHEM] A compound formed from an alcohol and an alkali metal base; the alkali metal replaces the hydrogen in the hydroxyl group. { 'al·kə‚lī ‚al· kə'hò‚lät }

alkali blue [ORG CHEM] The sodium salt of triphenylrosanilinesulfonic acid; used as an indicator. { 'al·kə‚lī 'blü }

alkalide [INORG CHEM] A member of a class of crystalline salts with an alkali metal atom. { 'al·kə‚līd }

alkali metal [CHEM] Any of the elements of group I in the periodic table: lithium, sodium, potassium, rubidium, cesium, and francium. { 'al·kə‚lī ‚med·əl }

alkalimeter [ANALY CHEM] **1.** An apparatus for measuring the quantity of alkali in a solid or liquid. **2.** An apparatus for measuring the quantity of carbon dioxide formed in a reaction. { ‚al·kə'lim·əd·ər }

alkalimetry [ANALY CHEM] Quantitative measurement of the concentration of bases or the quantity of one free base in a solution; techniques include titration and other analytical methods. { ‚al·kə'lim·ə·trē }

alkaline [CHEM] **1.** Having properties of an alkali. **2.** Having a pH greater than 7. { 'al·kə‚līn }

alkaline earth [INORG CHEM] An oxide of an element of group 2 in the periodic table, such as barium, calcium, and strontium. Also known as alkaline-earth oxide. { ¦al· kə‚līn 'ərth }

alkaline-earth metals [CHEM] The heaviest members of group 2 in the periodic table; usually calcium, strontium, magnesium, and barium. { ¦al·kə‚līn 'ərth 'med·əlz }

alkaline-earth oxide See alkaline earth. { ¦al·kə‚līn 'ərth 'äk‚sīd }

alkalinity [CHEM] The property of having excess hydroxide ions in solution. { ‚al· kə'lin·ə·dē }

alkaloid [ORG CHEM] One of a group of nitrogenous bases of plant origin, such as nicotine, cocaine, and morphine. { 'al·kə‚lòid }

alkalometry [ANALY CHEM] The measurement of the quantity of alkaloids present in a substance. { ‚al·kə'läm·ə·trē }

alkamine [ORG CHEM] A compound that has both the alcohol and amino groups. Also known as amino alcohol. { 'al·kə‚mēn }

alkane [ORG CHEM] A member of a series of saturated aliphatic hydrocarbons having

the empirical formula C_nH_{2n+2}. Also known as paraffin; paraffinic hydrocarbon.
{ 'al,kān }

alkannin |ORG CHEM| $C_{16}H_{16}O_5$ A red powder, the coloring ingredient of alkanet; soluble in alcohol, benzene, ether, and oils; used as a coloring agent for fats and oils, wines, and wax. { al'ka·nən }

alkanolamine |ORG CHEM| One of a group of viscous, water-soluble amino alcohols of the aliphatic series. { ,al·kə'näl·ə,mēn }

alkene |ORG CHEM| One of a class of unsaturated aliphatic hydrocarbons containing one or more carbon-to-carbon double bonds. { 'al,kēn }

alkoxide See alcoholate. { al'käk,sīd }

alkoxy |ORG CHEM| An alkyl radical attached to a molecule by oxygen, such as the ethoxy radical. { al'käk·sē }

alkyd resin |ORG CHEM| A class of adhesive resins made from unsaturated acids and glycerol. { 'al·kəd 'rez·ən }

alkyl |ORG CHEM| An organic group that results from removal of a hydrogen atom from an acyclic, saturated hydrocarbon; may be represented in a chemical formula by R–. { 'al,kil }

alkylamine |ORG CHEM| A compound consisting of an alkyl group attached to the nitrogen of an amine; an example is ethylamine, $C_2H_5NH_2$. { ¦al·kəl·ə¦mēn }

alkylaryl sulfonates |ORG CHEM| General name for alkylbenzene sulfonates. { ¦al·kəl·ə¦rəl 'səl·fə,nāts }

alkylate |ORG CHEM| A product of the alkylation process in petroleum refining. { 'al·kə,lāt }

alkylation |ORG CHEM| A chemical process in which an alkyl radical is introduced into an organic compound by substitution or addition. { ,al·kə'lā·shən }

alkylbenzene sulfonates |ORG CHEM| Widely used nonbiodegradable detergents, commonly dodecylbenzene or tridecylbenzene sulfonates. { ¦al·kəl¦ben,zēn 'səl·fə,nāts }

alkylene |ORG CHEM| An organic radical formed from an unsaturated aliphatic hydrocarbon; for example, the ethylene radical C_2H_3–. { 'al·kə,lēn }

alkyl halide |ORG CHEM| A compound consisting of an alkyl group and a halogen; an example is ethylbromide. { 'al·kəl 'hāl,īd }

alkyloxonium ion |ORG CHEM| $(ROH_2)^+$ An oxonium ion containing one alkyl group. { ¦al·kil,äk¦sō·nē·əm 'ī,än }

alkyne |ORG CHEM| One of a group of organic compounds containing a carbon-to-carbon triple bond. { 'al,kīn }

allelochemistry |CHEM| The science of compounds synthesized by one organism that stimulate or inhibit other organisms. { ə¦lē·lō¦kem·ə·strē }

allene |ORG CHEM| C_3H_4 An unsaturated aliphatic hydrocarbon with two double bonds. Also known as propadiene. { ə'lēn }

allethrin |ORG CHEM| An insecticide, a synthetic pyrethroid, more effective than pyrethrin. { 'al·ə·thrən }

allidochlor |ORG CHEM| $C_8H_{12}NOCl$ An amber liquid having slight solubility in water; used as a preemergence herbicide for vegetable crops, soybeans, sorghum, and ornamentals. { ə'lid·ə,klòr }

allo- |CHEM| Prefix applied to the stabler form of two isomers. { 'a·lō }

allotriomorphism See allotropy. { ə¦lä·trē·ə¦mor,fiz·əm }

allotrope |CHEM| A form of an element showing allotropy. { 'a·lə,trōp }

allotropism See allotropy. { ,a·lə'trä,piz·əm }

allotropy |CHEM| The assumption by an element of two or more different forms or structures which are most frequently stable in different temperature ranges, such as different crystalline forms of carbon as charcoal, graphite, or diamond. Also known as allotriomorphism; allotropism. { ə'lä·trə·pē }

allulose |ORG CHEM| $CH_2OHCO(CHOH)_3CH_2OH$ A constituent of cane sugar molasses; it is nonfermentable. { 'al·yə,lōs }

allyl- |ORG CHEM| A prefix used in names of compounds whose structure contains an allyl cation. { 'al·əl }

allylacetone |ORG CHEM| $CH_2CHCH_2CH_2COCH_3$ A colorless liquid, soluble in water and organic solvents; used in pharmaceutical synthesis, perfumes, fungicides, and insecticides. { ,al·əl'as·ə,tōn }

allyl alcohol [ORG CHEM] CH_2CHCH_2OH Colorless, pungent liquid, boiling at 96°C; soluble in water; made from allyl chloride by hydrolysis. { 'al·əl 'al·kə,hȯl }

allylamine [ORG CHEM] $CH_2CHCH_2NH_2$ A yellow oil that is miscible with water; boils at 58°C; prepared from mustard oil. { ¦al·əl·ə¦mēn }

allyl bromide [ORG CHEM] C_3H_5Br A colorless to light yellow, irritating toxic liquid with a boiling point of 71.3°C; soluble in organic solvents; used in organic synthesis and for the manufacture of synthetic perfumes. { 'al·əl 'brō,mīd }

allyl cation [ORG CHEM] A carbonium cation with a structure usually represented as $CH_2=CH-CH_2^+$; attachment site is the saturated carbon atom. { 'al·əl 'kat,ī·ən }

allyl chloride [ORG CHEM] CH_2CHCH_2Cl A volatile, pungent, toxic, flammable, colorless liquid, boiling at 46°C; insoluble in water; made by chlorination of propylene at high temperatures. { 'al·əl 'klȯr,īd }

allyl cyanide [ORG CHEM] C_4H_5N A liquid with an onionlike odor and a boiling point of 119°C; slightly soluble in water; used as a cross-linking agent in polymerization. { 'al·əl 'sī·ə,nīd }

allylene [ORG CHEM] $CH_3C:CH$ An acetylenic, three-carbon hydrocarbon; a colorless gas boiling at −24°C; soluble in ether. Also known as propyne. { 'al·ə,lēn }

allylic hydrogen [ORG CHEM] In an organic molecule, a hydrogen attached to a carbon atom that is adjacent to a double bond. { ə'lil·ik 'hī·drə·jən }

allylic substitution [ORG CHEM] A reaction occurring at position 1 of an allylic system (with the double bond between positions 2 and 3) in which the incoming group is attached to the same atom (position 1) as the leaving group, or the incoming group is attached at position 3, with the double bond moving from positions 2 and 3 to positions 1 and 2. { ə¦lil·ik ,səb·stə'tü·shən }

allyl isothiocyanate [ORG CHEM] $CH_2CH:CH_2NCS$ A pungent, colorless to pale-yellow liquid; soluble in alcohol, slightly soluble in water; irritating odor; boiling point 152°C; used as a fumigant and as a poison gas. Also known as mustard oil. { 'al·əl ¦ī·sō,thī·ō'sī·ə,nāt }

allyl mercaptan [ORG CHEM] CH_2CHCH_2SH A colorless liquid with a boiling point of 67–68°C; soluble in ether and alcohol; used as intermediate in pharmaceutical manufacture. { 'al·əl mər'kap,tan }

allyl plastic See allyl resin. { 'al·əl 'plas·tik }

allyl resin [ORG CHEM] Any of a class of thermosetting synthetic resins derived from esters of allyl alcohol or allyl chloride; used in making cast and laminated products. Also known as allyl plastic. { 'al·əl 'rez·ən }

allyl sulfide [ORG CHEM] $(CH_2CHCH_2)_2S$ A colorless liquid with a garliclike odor and a boiling point of 139°C; used in synthetic oil of garlic. { 'al·əl 'səl,fīd }

allylthiourea [ORG CHEM] $C_3H_5NHCSNH_2$ A white, crystalline solid that melts at 78°C; soluble in water; used as a corrosion inhibitor. { ,al·əl,thī·ō,yu'rē·ə }

allyltrichlorosilane [ORG CHEM] $CH_2CHCH_2SiCl_3$ A pungent, colorless liquid with a boiling point of 117.5°C; used as an intermediate for silicones. { ,al·əl,trī,klȯr'äs·ə,lān }

allylurea [ORG CHEM] $C_4H_8N_2O$ Crystals with a melting point of 85°C; freely soluble in water and alcohol; used to manufacture allylthiourea and other corrosion inhibitors. { ,al·əl·yu'rē·ə }

allyxycarb [ORG CHEM] $C_{16}H_{22}N_2O_2$ A yellow, crystalline compound used as an insecticide for fruit orchards, vegetable crops, rice, and citrus. { ə'liks·ə,karb }

alpha cellulose [ORG CHEM] A highly refined, insoluble cellulose from which sugars, pectin, and other soluble materials have been removed. Also known as chemical cellulose. { 'al·fə 'sel·yə,lōs }

alpha olefin [ORG CHEM] An olefin where the unsaturation (double bond) is at the alpha position, that is, between the two end carbons of the carbon chain. { 'al·fə 'ō·lə,fən }

alpha position [ORG CHEM] In chemical nomenclature, the position of a substituting group of atoms in the main group of a molecule; for example, in a straight-chain

15

alternant hydrocarbon

compound such as α-hydroxypropionic acid ($CH_3CHOHCOOH$), the hydroxyl radical is in the alpha position. { 'al·fə pə,zish·ən }

alternant hydrocarbon [ORG CHEM] A member of a class of conjugated molecules whose carbon atoms can be divided into two sets so that members of one set are formally bonded only to members of the other set. { 'ȯl·tər·nənt ‚hī·drə'kär·bən }

alternating copolymer [ORG CHEM] A polymer formed of two different monomer molecules that alternate in sequence in the polymer chain. { 'ȯl·tər,nād·iŋ ‚kō'päl·ə·mər }

alternation of multiplicities law [CHEM] The law that the periodic table arranges the elements in such a sequence that their number of orbital electrons, and hence their multiplicities, alternates between even and odd numbers. { ‚ȯl·tər'nā·shən əv ‚məl·tə'plis·əd·ēz ‚lȯ }

alum [INORG CHEM] **1.** Any of a group of double sulfates of trivalent metals such as aluminum, chromium, or iron and a univalent metal such as potassium or sodium. **2.** *See* aluminum sulfate; ammonium aluminum sulfate; potassium aluminum sulfate. { 'al·əm }

alumina [INORG CHEM] Al_2O_3 The native form of aluminum oxide occurring as corundum or in hydrated forms, as a powder or crystalline substance. { ə'lüm·ə·nə }

aluminate [INORG CHEM] A negative ion usually assigned the formula AlO_2^- and derived from aluminum hydroxide. { ə'lüm·ə,nāt }

alumina trihydrate [INORG CHEM] $Al_2O_3·3H_2O$, or $Al(OH)_3$ A white powder; insoluble in water, soluble in hydrochloric or sulfuric acid or sodium hydroxide; used in the manufacture of ceramic glasses and in paper coating. Also known as aluminum hydrate; aluminum hydroxide; hydrated alumina; hydrated aluminum oxide. { ə'lüm·ə·nə ‚trī'hī,drāt }

aluminium *See* aluminum. { ‚al·yü'min·ē·əm }

aluminon [ORG CHEM] $C_{22}H_{23}N_3O_9$ A yellowish-brown, glassy powder that is freely soluble in water; used for the detection and colorimetric estimation of aluminum in foods, water, and tissues, and as a pharyngeal aerosol spray. { ə'lüm·ə,nän }

aluminosilicate [INORG CHEM] $3Al_2O_3·2SiO_2$ A colorless, crystalline combination of silicate and aluminate in the form of rhombic crystals. { ə'lüm·ə,nō'sil·ə,kāt }

aluminum [CHEM] A chemical element, symbol Al, atomic number 13, and atomic weight 26.9815. Also spelled aluminium. { ə'lüm·ə·nəm }

aluminum acetate [ORG CHEM] $Al(CH_3COO)_3$ A white, amorphous powder that is soluble in water; used in aqueous solution as an antiseptic. { ə'lüm·ə·nəm 'as·ə,tāt }

aluminum ammonium sulfate *See* ammonium aluminum sulfate. { ə'lüm·ə·nəm ə'mon·ē·əm 'səl,fāt }

aluminum borohydride [ORG CHEM] $Al(BH_4)_3$ A volatile liquid with a boiling point of 44.5°C; used in organic synthesis and as a jet fuel additive. { ə'lüm·ə·nəm bȯr·ō'hī,drīd }

aluminum chloride [INORG CHEM] $AlCl_3$ or Al_2Cl_6 A deliquescent compound in the form of white to colorless hexagonal crystals; fumes in air and reacts explosively with water; used as a catalyst. { ə'lüm·ə·nəm 'klȯr,īd }

aluminum fluoride [INORG CHEM] $AlF_3·3^1/_2H_2O$ A white, crystalline powder, insoluble in cold water. { ə'lüm·ə·nəm 'flür,īd }

aluminum fluosilicate [INORG CHEM] $Al_2(SiF_6)_3$ A white powder that is soluble in hot water; used for artificial gems, enamels, and glass. Also known as aluminum silicofluoride. { ə'lüm·ə·nəm ‚flü·ə'sil·i·kət }

aluminum halide [INORG CHEM] A compound of aluminum with a halogen element, such as aluminum chloride. { ə'lüm·ə·nəm 'ha,līd }

aluminum hydrate *See* alumina trihydrate. { ə'lüm·ə·nəm 'hī·drāt }

aluminum hydroxide *See* alumina trihydrate. { ə'lüm·ə·nəm hī'dräk,sīd }

aluminum monostearate [ORG CHEM] $Al(OH)_2[OOC(CH_2)_{16}CH_3]$ A white to yellowish-white powder with a melting point of 155°C; used in the manufacture of medicine, paint, and ink, in waterproofing, and as a plastics stabilizer. { ə'lüm·ə·nəm ‚män·ō'stir,āt }

aluminum nitrate [INORG CHEM] $Al(NO_3)_3·9H_2O$ White, deliquescent crystals with a melting point of 73°C; soluble in alcohol and acetone; used as a mordant for textiles, in leather tanning, and as a catalyst in petroleum refining. { ə'lüm·ə·nəm 'nī‚trāt }

aluminum oleate [ORG CHEM] A soaplike compound of aluminum and oleic acid, used in lubricating oils and greases to improve their viscosity. { ə'lüm·ə·nəm 'ō·lē‚āt }

aluminum orthophosphate [INORG CHEM] AlPO₄ White crystals, melting above 1500°C; insoluble in water, soluble in acids and bases; useful in ceramics, paints, pulp, and paper. Also known as aluminum phosphate. { ə'lüm·ə·nəm ‚òr·thō'fäs‚fāt }

aluminum oxide [INORG CHEM] Al₂O₃ A compound in the form of a white powder or colorless hexagonal crystals; melts at 2020°C; insoluble in water; used in aluminum production, paper, spark plugs, absorbing gases, light bulbs, artificial gems, and manufacture of abrasives, refractories, ceramics, and electrical insulators. { ə'lüm·ə·nəm 'äk‚sīd }

aluminum palmitate [ORG CHEM] Al(C₁₆H₃₁O₂)·H₂O An aluminum soap used in waterproofing fabrics, paper, and leather and as a drier in paints. { ə'lüm·ə·nəm 'päm·ə‚tāt }

aluminum phosphate See aluminum orthophosphate. { ə'lüm·ə·nəm 'fäs‚fāt }

aluminum potassium sulfate See potassium aluminum sulfate. { ə'lüm·ə·nəm pə'tas·ē·əm 'səl‚fāt }

aluminum silicate [INORG CHEM] Al₂(SiO₃)₃ A white solid that is insoluble in water; used as a refractory in glassmaking. { ə'lüm·ə·nəm 'sil·ə‚kāt }

aluminum silicofluoride See aluminum fluosilicate. { ə'lüm·ə·nəm ‚sil·ə·kō¦flùr‚īd }

aluminum soap [ORG CHEM] Any of various salts of higher carboxylic acids and aluminum that are insoluble in water and soluble in oils; used in lubricating greases, paints, varnishes, and waterproofing substances. { ə'lüm·ə·nəm 'sōp }

aluminum sodium sulfate [INORG CHEM] AlNa(SO₄)₂· 12H₂O Colorless crystals with an astringent taste and a melting point of 61°C; soluble in water; used as a mordant and for waterproofing textiles, as a food additive, and for matches, tanning, ceramics, engraving, and water purification. Abbreviated SAS. Also known as porous alum; soda alum; sodium aluminum sulfate. { ə'lüm·ə·nəm 'sōd·ē·əm 'səl‚fāt }

aluminum stearate [ORG CHEM] Al(C₁₇H₃₅COO)₃ An aluminum soap in the form of a white powder that is insoluble in water and soluble in oils; used for waterproofing fabrics and concrete and as a drier in paints and varnishes. { ə'lüm·ə·nəm 'stir‚āt }

aluminum sulfate [INORG CHEM] Al₂(SO₄)₃·18H₂O A colorless salt in the form of monoclinic crystals that decompose in heat and are soluble in water; used in papermaking, water purification, and tanning, and as a mordant in dyeing. Also known as alum. { ə'lüm·ə·nəm 'səl‚fāt }

aluminum triacetate [ORG CHEM] Al(C₂H₃O₂)₃ A white solid that is very slightly soluble in cold water. { ə'lüm·ə·nəm ‚trī'as·ə‚tāt }

Am See americium.

ambident [ORG CHEM] A chemical species or molecule that possesses two alternative reactive sites, either of which can bond in a reaction; examples include cyanate ions, thiosulfate ions, oxime anions, and enolate ions. Also known as ambidentate. { 'am·bə·dənt }

ambidentate See ambident. { ‚am·bə'den‚tāt }

americium [CHEM] A chemical element, symbol Am, atomic number 95; the mass number of the isotope with the longest half-life is 243. { ‚am·ə'ris·ē·əm }

americyl ion [INORG CHEM] A dioxo monocation of americium, with the formula (AmO₂)⁻. { ə'mer·ə·səl 'ī‚än }

Ames test [ANALY CHEM] A bioassay that uses a set of histidine auxotrophic mutants of *Salmonella typhimurium* for detecting mutagenic and possibly carcinogenic compounds. { 'āmz ‚test }

amicron [PHYS CHEM] A particle having a size of 10⁻⁷ centimeter or less, which is a size in a system of classification of particle sizes in colloid chemistry. { ā'mī‚krän }

amidation [ORG CHEM] The process of forming an amide; for example, in the laboratory benzyl reacts with methyl amine to form N-methylbenzamide. { ‚am·ə‚dā·shən }

amide [ORG CHEM] One of a class of organic compounds containing the CONH₂ radical. { 'am‚īd }

amide hydrolysis

amide hydrolysis [ORG CHEM] The cleavage of an amide into its constitutive acid and amine fragments by a net addition of water. { 'am,īd hī'dräl·ə·səs }

amidine [ORG CHEM] A compound which contains the radical $CNHNH_2$. { 'am·ə,dēn }

amido [ORG CHEM] Indicating the NH_2 radical when it is present in a molecule with the CO radical. { ə'mē,dō }

amidol [ORG CHEM] $C_6H_3(NH_2)_2OH \cdot HCl$ A grayish-white crystalline salt; soluble in water, slightly soluble in alcohol; used as a developer in photography and as an analytical reagent. { 'am·i,dȯl }

amidourea hydrochloride See semicarbazide hydrochloride. { ¦am·ə·dō·yü'rē·ə 'hī·drə'klȯr,īd }

amination [ORG CHEM] **1.** The preparation of amines. **2.** A process in which the amino group (=NH_2) is introduced into organic molecules. { ,am·ə'nā·shən }

amine [ORG CHEM] One of a class of organic compounds which can be considered to be derived from ammonia by replacement of one or more hydrogens by functional groups. { ə'mēn }

amino-, amin- [CHEM] Having the property of a compound in which the group NH_2 is attached to a radical other than an acid radical. { ə'mē,nō }

amino alcohol See alkamine. { ə'mē,nō 'al·kə,hȯl }

1-aminoanthraquinone [ORG CHEM] $C_{14}H_9NO_2$ Ruby-red crystals with a melting point of 250°C; freely soluble in alcohol, benzene, chloroform, ether, glacial acetic acid, and hydrochloric acid; used in the manufacture of dyes and pharmaceuticals. { ¦wən ə¦mē·nō,an·thrə·kwē'nōn }

2-amino-1-butanol [ORG CHEM] $CH_3CH_2CH(NH_2)$ CH_2OH A liquid miscible with water, soluble in alcohols; used in the synthesis of surface-active agents, vulcanizing accelerators, and pharmaceuticals, and as an emulsifying agent for such products as cosmetic creams and lotions. { ¦tü ə¦mē·nō ¦wən ¦byüt·ən,ȯl }

γ-aminobutyric acid [ORG CHEM] $H_2NCH_2CH_2CH_2COOH$ Crystals which are either leaflets or needles, with a melting point of 202°C; thought to be a central nervous system postsynaptic inhibitory transmitter. Abbreviated GABA. { ¦gam·ə ə¦mē·nō,byü¦tir·ik 'as·əd }

ε-aminocaproic acid [ORG CHEM] $C_6H_{13}NO_2$ Crystals with a melting point of 204–206°C; freely soluble in water; used as an antifibrinolytic agent and a spacer for affinity chromatography. { ¦ep·sə,lən ə¦mē·nō·kə¦prō·ik 'as·əd }

aminocarb [ORG CHEM] $C_{11}H_{16}N_2O_2$ A tan, crystalline compound with a melting point of 93–94°C; slightly soluble in water; used as an insecticide for control of forest insects and pests of cotton, tomatoes, tobacco, and fruit crops. { ə'mē·nō,kärb }

aminocide See succinic acid 2,2-dimethylhydrazide. { ə'mē·nō,sīd }

aminodiborane [INORG CHEM] Any compound derived from diborane (B_2H_6) in which one H of the bridge has been replaced by NH_2. { ə¦mē·nō,dī¦bȯr,ān }

3-amino-2,5-dichlorobenzoic acid [ORG CHEM] $C_7H_5O_2NCl_2$ A white solid with a melting point of 200–201°C; solubility in water is 700 parts per million at 20°C; used as a preemergence herbicide for soybeans, corn, and sweet potatoes. { ¦thrē ə¦mē·nō ¦tü ¦fīv dī,klȯr·ə,ben¦zō·ik 'as·əd }

aminoethane See ethyl amine. { ə'mē·nō,eth·ən }

amino group [ORG CHEM] A functional group (−NH_2) formed by the loss of a hydrogen atom from ammonia. { ə'mē·nō ,grüp }

2-amino-2-methyl-1,3-propanediol [ORG CHEM] $HOCH_2C(CH_3)(NH_2)CH_2OH$ Crystals with a melting point of 109–111°C; soluble in water and alcohol; used in the synthesis of surface-active agents, pharmaceuticals, and vulcanizers, and as an emulsifying agent for cosmetics, leather dressings, polishes, and cleaning compounds. { ¦tü ə'mē·nō ¦tü ¦meth·əl ¦wən ¦thrē ¦prō,pān'dī,ȯl }

3-amino-2-naphthoic acid [ORG CHEM] $H_2NC_{10}H_6COOH$ Yellow crystals in the shape of scales with a melting point of 214°C; soluble in alcohol and ether; used in the determination of copper, nickel, and cobalt. { ¦thrē ə'mē·nō ¦tü naf'thō·ik 'as·əd }

1-amino-2-naphthol-4-sulfonic acid [ORG CHEM] $H_2NC_{10}H_5(OH)SO_3H$ White or gray, needlelike crystals; soluble in hot sodium bisulfite solutions; used in the manufacture of azo dyes. { ¦wən ə'mē·nō ¦tü 'naf,thȯl ¦fȯr səl'fän·ik 'as·əd }

18

2-amino-5-naphthol-7-sulfonic acid [ORG CHEM] C$_{10}$H$_5$NH$_2$OHSO$_3$H Gray or white needles that are soluble in hot water; used as a dye intermediate. { ¦tü ə'mē·nō ¦fīv 'naf,thȯl ¦sev·ən səl'fän·ik 'as·əd }

amino nitrogen [CHEM] Nitrogen combined with hydrogen in the amino group. Also known as ammonia nitrogen. { ə'mē·nō 'nī·trə·jən }

aminophenol [ORG CHEM] A type of compound containing the NH$_2$ and OH groups joined to the benzene ring; examples are para-aminophenol and ortho-hydroxylaniline. { ə,me·nō'fē,nȯl }

para-aminophenol [ORG CHEM] p-HOC$_6$H$_4$NO$_2$ A phenol in which an amino (−NH$_2$) group is located on the benzene ring of carbon atoms para (p) to the hydroxyl (−OH) group; used as a photographic developer and as an intermediate in dye manufacture. { ¦par·ə ə,mē·nō'fē·nȯl }

3-aminophthalic hydrazide See luminol. { ¦thrē ə¦mē·nō¦thal·ik 'hī·drə·zīd }

2-aminopropane See isopropylamine. { ¦tü ə,mē·nō'prō·pān }

3-aminopyridine See β-aminopyridine. { ¦thrē ə,mē·nō'pī·rə,dēn }

4-aminopyridine [ORG CHEM] C$_5$H$_6$N$_2$ White crystals with a melting point of 158.9°C; soluble in water; used as a repellent for birds. Abbreviated 4-AP. { ¦fȯr ə,mē·nō'pī·rə,dēn }

β-aminopyridine [ORG CHEM] C$_5$H$_6$N$_2$ Crystals with a melting point of 64°C; soluble in water, alcohol, and benzene; used in drug and dye manufacture. Also known as 3-aminopyridine. { ¦bād·ə ə,mē·nō'pī·rə,dēn }

amino resin [ORG CHEM] A type of resin prepared by condensation polymerization, with an aldehyde, of a compound containing an amino group. { ə'mē·nō 'rez·ən }

2-aminothiazole [ORG CHEM] C$_3$H$_4$N$_2$S Pale-yellow crystals that melt at 92°C; soluble in cold water, slightly soluble in ethyl alcohol; used as an intermediate in the synthesis of sulfathiazole. { ¦tü ə,mē·nō'thī·ə,zōl }

aminotriazole [ORG CHEM] C$_2$H$_4$N$_4$ Crystals with a melting point of 159°C; soluble in water, methanol, chloroform, and ethanol; used as an herbicide, cotton plant defoliant, and growth regulator for annual grasses and broadleaf and aquatic weeds. Abbreviated ATA. { ə¦mē·nō'trī·ə,zōl }

ammine [INORG CHEM] One of a group of complex compounds formed by coordination of ammonia molecules with metal ions. { 'a,mēn }

ammonation [INORG CHEM] A reaction in which ammonia is added to other molecules or ions by covalent bond formation utilizing the unshared pair of electrons on the nitrogen atom, or through ion-dipole electrostatic interactions. { ,a·mə'nā·shən }

ammonia [INORG CHEM] NH$_3$ A colorless gaseous alkaline compound that is very soluble in water, has a characteristic pungent odor, is lighter than air, and is formed as a result of the decomposition of most nitrogenous organic material; used as a fertilizer and as a chemical intermediate. { ə'mōn·yə }

ammonia alum See ammonium aluminum sulfate. { ə'mōn·yə 'al·əm }

ammoniac See ammoniacal. { ə'mōn·ē,ak }

ammoniacal [INORG CHEM] Pertaining to ammonia or its properties. Also known as ammoniac. { ¦a·mə¦nī·ə·kəl }

ammonia dynamite [CHEM] Dynamite with part of the nitroglycerin replaced by ammonium nitrate. { ə'mōn·yə 'dī·nə,mīt }

ammoniated mercuric chloride See ammoniated mercury. { ə¦mōn·ē·ād·əd mər¦kyür·ik ¦klȯr,īd }

ammoniated mercury [INORG CHEM] HgNH$_2$Cl A white powder that darkens on light exposure; insoluble in water and alcohol, soluble in ammonium carbonate solutions and in warm acids; used in pharmaceuticals and as a local anti-infective in medicine. { ə'mōn·ē·ād·əd 'mər·kyə·rē }

ammoniated superphosphate [INORG CHEM] A fertilizer containing 5 parts of ammonia to 100 parts of superphosphate. { ə'mōn·ē·ād·əd ,sü·pər'fäs,fāt }

ammoniation [CHEM] Treating or combining with ammonia. { ə,mōn·ē'ā·shən }

ammonia water [CHEM] A water solution of ammonia; a clear colorless liquid that is basic because of dissociation of NH$_4$OH to produce hydroxide ions; used as a reagent, solvent, and neutralizing agent. { ə'mōn·yə ,wȯd·ər }

ammonification [CHEM] Addition of ammonia or ammonia compounds, especially to the soil. { ə,män·ə·fə'kā·shən }

ammonium acetate [ORG CHEM] **1.** CH_3COONH_4 A normal salt formed by the neutralization of acetic acid with ammonium hydroxide; a white, crystalline, deliquescent material used in solution for the standardization of electrodes for hydrogen ions. **2.** $CH_3COONH_4 \cdot CH_3COOH$ An acid salt resulting from the distillation of the neutral salt or from its solution in hot acetic acid; crystallizes in deliquescent needles. **3.** A mixture of the normal and acetic salts; used as a mordant in the dyeing of wool. { ə'mōn·yəm 'as·ə,tāt }

ammonium alginate [ORG CHEM] $(C_6H_7O_6 \cdot NH_4)_n$ A high-molecular-weight, hydrophilic colloid; used as a thickening agent/stabilizer in ice cream, cheese, canned fruits, and other food products. { ə'mōn·yəm 'al·jə,nāt }

ammonium alum See ammonium aluminum sulfate. { ə'mōn·yəm 'al·əm }

ammonium aluminum sulfate [INORG CHEM] $NH_4Al(SO_4)_2 \cdot 12H_2O$ Colorless, odorless crystals that are soluble in water; used in manufacturing medicines and baking powder, dyeing, papermaking, and tanning. Also known as alum; aluminum ammonium sulfate; ammonia alum; ammonium alum. { ə'mōn·yəm ə¦lü·mə·nəm 'səl,fāt }

ammonium benzoate [ORG CHEM] $NH_4C_7H_5O_2$ A salt of benzoic acid prepared as a coarse, white powder; used as a preservative in certain adhesives and rubber latex. { ə'mōn·yəm 'ben·zə,wāt }

ammonium bicarbonate [INORG CHEM] NH_4HCO_3 White, crystalline, water-soluble salt; used in baking powders and in fire-extinguishing mixtures. Also known as ammonium hydrogen carbonate. { ə'mōn·yəm bī'kär·bə,nāt }

ammonium bichromate See ammonium dichromate. { ə'mōn·yəm bī'krō,māt }

ammonium bifluoride [INORG CHEM] $NH_4F \cdot HF$ A salt that crystallizes in the orthorhombic system and is soluble in water; prepared in the form of white flakes from ammonia treated with hydrogen fluoride; used in solution as a fungicide and wood preservative. Also known as ammonium acid fluoride; ammonium hydrogen fluoride. { ə'mōn·yəm bī'flür,īd }

ammonium bitartrate [ORG CHEM] $NH_4HC_4H_4O_6$ Colorless crystals that are soluble in water; used to make baking powder and to detect calcium. Also known as monoammonium tartrate. { ə'mōn·yəm bī'tär,trāt }

ammonium borate [INORG CHEM] NH_4BO_3 A white, crystalline, water-soluble salt which decomposes at 198°C; used as a fire retardant on fabrics. { ə'mōn·yəm 'bòr,āt }

ammonium bromide [INORG CHEM] NH_4Br An ammonium halide that crystallizes in the cubic system; made by the reaction of ammonia with hydrobromic acid or bromine; used in photography and for pharmaceutical preparations (sedatives). { ə'mōn·yəm 'brō,mīd }

ammonium carbamate [INORG CHEM] $NH_4NH_2CO_2$ A salt that forms colorless, rhombic crystals, which are very soluble in cold water; an important, unstable intermediate in the manufacture of urea; found in commercial ammonium carbonate. { ə'mōn·yəm 'kär·bə,māt }

ammonium carbonate [INORG CHEM] **1.** $(NH_4)_2CO_3$ The normal ammonium salt of carbonic acid, prepared by passing gaseous carbon dioxide into an aqueous solution of ammonia and allowing the vapors (ammonia, carbon dioxide, water) to crystallize. **2.** $NH_4HCO_3 \cdot NH_2COONH_4$ A white, crystalline double salt of ammonium bicarbonate and ammonium carbamate produced commercially; the principal ingredient of smelling salts. { ə'mōn·yəm 'kär·bə,nāt }

ammonium chloride [INORG CHEM] NH_4Cl A white crystalline salt that occurs naturally as a sublimation product of volcanic action or is manufactured; used as an electrolyte in dry cells, as a flux for soldering, tinning, and galvanizing, and as an expectorant. { ə'mōn·yəm 'klòr,īd }

ammonium chromate [INORG CHEM] $(NH_4)_2CrO_4$ A salt that forms yellow, monoclinic crystals; made from ammonium hydroxide and ammonium dichromate; used in photography as a sensitizer for gelatin coatings. { ə'mōn·yəm 'krō,māt }

ammonium citrate [ORG CHEM] $(NH_4)_2HC_6H_5O_7$ White, granular material; used as a reagent. { ə'mōn·yəm 'sī,trāt }

ammonium dichromate [INORG CHEM] $(NH_4)_2Cr_2O_7$ A salt that forms orange, monoclinic crystals; made from ammonium sulfate and sodium dichromate; soluble in water and alcohol; ignites readily; used in photography, lithography, pyrotechnics, and dyeing. Also known as ammonium bichromate. { ə'mōn·yəm dī'krō,māt }

ammonium fluoride [INORG CHEM] NH_4F A white, unstable, crystalline salt with a strong odor of ammonia; soluble in cold water; used in analytical chemistry, glass etching, and wood preservation, and as a textile mordant. { ə'mōn·yəm 'flúr,īd }

ammonium fluosilicate [INORG CHEM] $(NH_4)_2SiF_6$ A toxic, white, crystalline powder; soluble in alcohol and water; used for mothproofing, glass etching, and electroplating. Also known as ammonium silicofluoride. { ə'mōn·yəm ,flü·ə'sil·ə,kāt }

ammonium formate [ORG CHEM] HCO_2NH_4 Deliquescent crystals or granules with a melting point of 116°C; soluble in water and alcohol; used in analytical chemistry to precipitate base metals from salts of the noble metals. { ə'mōn·yəm 'fȯr·,māt }

ammonium gluconate [ORG CHEM] $NH_4C_6H_{11}O_7$ A white, crystalline powder made from gluconic acid and ammonia; soluble in water; used as an emulsifier for cheese and salad dressing and as a catalyst in textile printing. { ə'mōn·yəm 'glü·kə,nāt }

ammonium halide [INORG CHEM] A compound with the ammonium ion bonded to an ion formed from one of the halogen elements. { ə'mōn·yəm 'hal,īd }

ammonium hydrogen carbonate See ammonium bicarbonate. { ə'mōn·yəm 'hi·drə·jən 'kär·bə,nāt }

ammonium hydrogen fluoride See ammonium bifluoride. { ə'mōn·yəm 'hi·drə·jən 'flúr,īd }

ammonium hydroxide [INORG CHEM] NH_4OH A hydrate of ammonia, crystalline below −79°C; it is a weak base known only in solution as ammonia water. Also known as aqua ammonia. { ə'mōn·yəm ,hī'dräk,sīd }

ammonium iodide [INORG CHEM] NH_4I A salt prepared from ammonia and hydrogen iodide or iodine; it forms colorless, regular crystals which sublime when heated; used in photography and for pharmaceutical preparations. { ə'mōn·yəm 'ī·ə,dīd }

ammonium lactate [ORG CHEM] $NH_4C_3H_5O_3$ A yellow, syrupy liquid used in finishing leather. { ə'mōn·yəm 'lak,tāt }

ammonium lineolate [ORG CHEM] $C_{17}H_{31}COONH_4$ A soft, pasty material used as an emulsifying agent in various industrial applications. { ə'mōn·yəm lə'nē·ə,lāt }

ammonium metatungstate [ORG CHEM] $(NH_4)_6H_2W_{12}O_{40}$ A white powder, soluble in water, used for electroplating. { ə'mōn·yəm ,med·ə'təŋ,stāt }

ammonium molybdate [INORG CHEM] $(NH_4)_2MoO_4$ White, crystalline salt used as an analytic reagent, as a precipitant of phosphoric acid, and in pigments. { ə'mōn·yəm mə'lib,dāt }

ammonium nickel sulfate See nickel ammonium sulfate. { ə'mōn·yəm |nik·əl 'səl,fāt }

ammonium nitrate [INORG CHEM] NH_4NO_3 A colorless crystalline salt; very insensitive and stable high explosive; also used as a fertilizer. { ə'mōn·yəm 'nī,trāt }

ammonium oxalate [ORG CHEM] $(NH_4)_2C_2O_4·H_2O$ A salt in the form of colorless, rhombic crystals. { ə'mōn·yəm 'äk·sə,lāt }

ammonium perchlorate [INORG CHEM] NH_4ClO_4 A salt that forms colorless or white rhombic and regular crystals, which are soluble in water; it decomposes at 150°C, and the reaction is explosive at higher temperatures. { ə'mōn·yəm pər'klȯr,āt }

ammonium persulfate [INORG CHEM] $(NH_4)_2S_2O_8$ White crystals which decompose on melting; soluble in water; used as an oxidizing agent and bleaching agent, and in etching, electroplating, food preservation, and aniline dyes. { ə'mōn·yəm pər'səl,fāt }

ammonium phosphate [INORG CHEM] $(NH_4)_2HPO_4$ A salt of ammonia and phosphoric acid that forms white monoclinic crystals, which are soluble in water; used as a fertilizer and fire retardant. { ə'mōn·yəm 'fäs,fāt }

ammonium picrate [ORG CHEM] $NH_4C_6H_2O(NO_2)_3$ Compound with stable yellow and metastable red forms of orthorhombic crystals; used as a military explosive for armor-piercing shells. { ə'mōn·yəm 'pik,rāt }

ammonium salt [INORG CHEM] A product of a reaction between ammonia and various

21

acids; examples are ammonium chloride and ammonium nitrate. { ə'mōn·yəm 'sȯlt }

ammonium silicofluoride *See* ammonium fluosilicate. { ə'mōn·yəm ¦sil·ə·kō'flür‚īd }

ammonium soap [ORG CHEM] A product from reaction of a fatty acid with ammonium hydroxide; used in toiletry preparations such as soaps and in emulsions. { ə'mōn·yəm 'sōp }

ammonium stearate [ORG CHEM] $C_{17}H_{35}COONH_4$ A tan, waxlike substance with a melting point of 73–75°C; used in cosmetics and for waterproofing cements, paper, textiles, and other materials. { ə'mōn·yəm 'stir‚āt }

ammonium sulfamate [INORG CHEM] $NH_4OSO_2NH_2$ White crystals with a melting point of 130°C; soluble in water; used for flameproofing textiles, in electroplating, and as an herbicide to control woody plant species. { ə'mōn·yəm 'səl·fə‚māt }

ammonium sulfate [INORG CHEM] $(NH_4)_2SO_4$ Colorless, rhombic crystals which melt at 140°C and are soluble in water. { ə'mōn·yəm 'səl‚fāt }

ammonium sulfide [INORG CHEM] $(NH_4)_2S$ Yellow crystals, stable only when dry and below 0°C; decomposes on melting; soluble in water and alcohol; used in photographic developers and for coloring brasses and bronzes. { ə'mōn·yəm 'səl‚fīd }

ammonium tartrate [ORG CHEM] $C_4H_{12}N_2O_6$ Colorless, monoclinic crystals; used in textiles and in medicine. { ə'mōn·yəm 'tär‚trāt }

ammonium thiocyanate [ORG CHEM] NH_4SCN Colorless, deliquescent crystals with a melting point of 149.6°C; soluble in water, acetone, alcohol, and ammonia; used in analytical chemistry, freezing solutions, fabric dyeing, electroplating, photography, and steel pickling. { ə'mōn·yəm ‚thī·ō'sī·ə‚nāt }

ammonium vanadate [INORG CHEM] NH_4VO_3 A white to yellow, water-soluble, crystalline powder; used in inks and as a paint drier and textile mordant. { ə'mōn·yəm 'van·ə‚dāt }

ammonolysis [CHEM] **1.** A dissociation reaction of the ammonia molecule producing H^+ and NH_2^- species. **2.** Breaking of a bond by addition of ammonia. { ‚a·mə'näl·ə·səs }

amount of substance [CHEM] A measure of the number of elementary entities present in a substance or system; usually measured in moles. { ə'maünt əv 'səb·stəns }

amperometric titration [PHYS CHEM] A titration that involves measuring an electric current or changes in current during the course of the titration. { ¦am·pə·rə¦me·trik tī'trā·shən }

amperometry [PHYS CHEM] Chemical analysis by techniques which involve measuring electric currents. { ‚am·pə'rä·me·trē }

amphipathic molecule [ORG CHEM] A molecule having both hydrophilic and hydrophobic groups; examples are wetting agents and membrane lipids such as phosphoglycerides. { ‚am·fə'path·ik 'mäl·ə‚kyül }

amphiphile [CHEM] A molecule which has a polar head attached to a long hydrophobic tail. { 'am·fə‚fīl }

amphiprotic *See* amphoteric. { ¦am·fə¦präd·ik }

ampholyte [CHEM] An amphoteric electrolyte. { 'am·fə‚līt }

ampholytic detergent [CHEM] A detergent that is cationic in acidic solutions and anionic in basic solutions. { ¦am·fə¦lid·ik di'tər·jənt }

amphoteric [CHEM] Having both acidic and basic characteristics. Also known as amphiprotic. { ¦am·fə¦ter·ik }

amphoterism [CHEM] The property of being able to react either as an acid or a base. { am'fäd·ə‚riz·əm }

amyl [ORG CHEM] Any of the eight isomeric arrangements of the radical C_5H_{11} or a mixture of them. Also known as pentyl. { 'am·əl }

amyl acetate [ORG CHEM] $CH_3COO(CH_2)_2CH(CH_3)_2$ A colorless liquid, boiling at 142°C; soluble in alcohol and ether, slightly soluble in water; used in flavors and perfumes. Also known as banana oil; isoamyl acetate. { 'am·əl 'as·ə‚tāt }

amyl alcohol [ORG CHEM] **1.** A colorless liquid that is a mixture of isomeric alcohols. **2.** An optically active liquid composed of isopentyl alcohol and active amyl alcohol. { 'am·əl 'al·kə‚hȯl }

n-amylamine |ORG CHEM| $C_5H_{11}NH_2$ A colorless liquid with a boiling point of 104.4°C; soluble in water, alcohol, and ether; used in dyestuffs, insecticides, synthetic detergents, corrosion inhibitors, and pharmaceuticals, and as a gasoline additive. { ¦en ə'mil·ə‚mēn }

amyl benzoate *See* isoamyl benzoate. { 'am·əl 'ben·zə‚wāt }

amylene |ORG CHEM| C_5H_{10} A highly flammable liquid with a low boiling point, 37.5–38.5°C; often a component of petroleum. Also known as 2-methyl-2-butene. { 'am·ə‚lēn }

amyl ether |ORG CHEM| **1.** Either of two isomeric compounds, n-amyl ether or isoamyl ether; both may be represented by the formula $(C_5H_{11})_2O$. **2.** A mixture mainly of isoamyl ether and n-amyl ether formed in preparation of amyl alcohols from amyl chloride; very slightly soluble in water; used mainly as a solvent. { 'am·əl 'ē·thər }

amyl mercaptan |ORG CHEM| $C_5H_{11}SH$ A colorless to light yellow liquid with a boiling range of 104–130°C; soluble in alcohol; used in odorant for detecting gas line leaks. { 'am·əl mər'kap‚tan }

amyl nitrate |ORG CHEM| $C_5H_{11}ONO_2$ An ester of amyl alcohol added to diesel fuel to raise the cetane number. { 'am·əl 'nī‚trāt }

amyl nitrite |ORG CHEM| $(CH_3)_2CH(CH_2)_2NO_2$ A yellow liquid; soluble in alcohol, very slightly soluble in water; fruity odor; it is flammable and the vapor is explosive; used in medicine and perfumes. Also known as isoamyl nitrite. { 'am·əl 'nī‚trīt }

amyl propionate |ORG CHEM| $CH_3CH_2COOC_5H_{11}$ A colorless liquid with an applelike odor and a distillation range of 135–175°C; used in perfumes, lacquers, and flavors. { 'am·əl 'prō·pē·ə‚nāt }

amyl salicylate |ORG CHEM| $C_6H_4OHCOOC_5H_{11}$ A clear liquid that occasionally has a yellow tinge; boils at 280°C; soluble in alcohol, insoluble in water; used in soap and perfumes. Also known as isoamyl salicylate. { 'am·əl sə'lis·ə‚lāt }

amyl xanthate |ORG CHEM| A salt formed by replacing the hydrogen attached to the sulfur in amylxanthic acid by a metal; used as collector agent in the flotation of certain minerals. { 'am·əl 'zan‚thāt }

anabasine |ORG CHEM| A colorless, liquid alkaloid extracted from the plants *Anabasis aphylla* and *Nicotiana glauca*; boiling point is 105°C; soluble in alcohol and ether; used as an insecticide. { ə'na·bə‚sēn }

anagyrine |ORG CHEM| $C_{15}H_{20}N_2O$ A toxic alkaloid found in several species of *Lupinus* in the western United States; acute poisoning produces nervousness, depression, loss of muscular control, convulsions, and coma. { ‚an·ə'jī‚rēn }

analog |CHEM| A compound whose structure is similar to that of another compound but whose composition differs by one element. { 'an·əl‚äg }

analysis |ANALY CHEM| The determination of the composition of a substance. { ə'nal·ə·səs }

analysis line |SPECT| The spectral line used in determining the concentration of an element in spectrographic analysis. { ə'nal·ə·səs ‚līn }

analyte |ANALY CHEM| **1.** The sample being analyzed. **2.** The specific component that is being measured in a chemical analysis. { 'an·ə‚līt }

analytical blank *See* blank. { ‚an·əl'id·ə·kəl 'blaŋk }

analytical chemistry |CHEM| The branch of chemistry dealing with techniques which yield any type of information about chemical systems. { ‚an·əl'id·ə·kəl 'kem·ə·strē }

analytical distillation |ANALY CHEM| Precise resolution of a volatile liquid mixture into its components; the mixture is vaporized by heat or vacuum, and the vaporized components are recondensed into liquids at their respective boiling points. { ‚an·əl'id·ə·kəl ‚dis·tə'lā·shən }

analytical extraction |ANALY CHEM| Precise transfer of one or more components of a mixture (liquid to liquid, gas to liquid, solid to liquid) by contacting the mixture with a solvent in which the component of interest is preferentially soluble. { ‚an·əl'id·ə·kəl ik'strak·shən }

anaphoresis |PHYS CHEM| Upon application of an electric field, the movement of positively charged colloidal particles or macromolecules suspended in a liquid toward the anode. { ¦an·ə·fə'rē·səs }

23

anatabine |ORG CHEM| C₁₀H₁₂N₂ An alkaloid found in tobacco. { ə'nad·ə,bēn }
anchimeric assistance |ORG CHEM| The participation by a neighboring group in the rate-determining step of a reaction; most often encountered in reactions of carbocation intermediates. Also known as neighboring-group participation. { ¦aŋ·kə¦mer· ik ə'sis·təns }
anchored catalyst See immobilized catalyst. { 'aŋ·kərd 'kad·ə,list }
anethole |ORG CHEM| C₁₀H₁₂O White crystals that melt at 22.5°C; very slightly soluble in water; affected by light; odor resembles oil of anise; used in perfumes and flavors, and as a sensitizer in color-bleaching processes in color photography. { 'an·ə,thòl }
angle-resolved photoelectron spectroscopy |SPECT| A type of photoelectron spectroscopy which measures the kinetic energies of photoelectrons emitted from a solid surface and the angles at which they are emitted relative to the surface. Abbreviated ARPES. { 'aŋ·gəl ri'zälvd ¦fōd·ō·ə'lek,trän ,spek'träs·kə·pē }
anharmonic oscillator spectrum |SPECT| A molecular spectrum which is significantly affected by anharmonicity of the forces between atoms in the molecule. { ,an· här¦män·ik ¦äs·ə,läd·ər ¦spek·trəm }
anhydride |CHEM| A compound formed from an acid by removal of water. { an'hī,drīd }
anhydrous |CHEM| Being without water, especially water of hydration. { an'hī·drəs }
anhydrous alcohol See absolute alcohol. { an'hī·drəs 'al·kə,hòl }
anhydrous ammonia |INORG CHEM| Liquid ammonia, a colorless liquid boiling at −33.3°C. { an'hī·drəs ə'mōn·yə }
anhydrous ferric chloride See ferric chloride. { an'hī·drəs ,fer·ik 'klòr,īd }
anhydrous hydrogen chloride |INORG CHEM| HCl Hazardous, toxic, colorless gas used in polymerization, isomerization, alkylation, nitration, and chlorination reactions; becomes hydrochloric acid in aqueous solutions. { an'hī·drəs ,hī·drə·jən 'klòr,īd }
anhydrous phosphoric acid See phosphoric anhydride. { an'hī·dres fä'sfòrik 'as·əd }
anhydrous plumbic acid See lead dioxide. { an'hī·dres 'pləmb·ik 'as·əd }
anhydrous sodium carbonate See soda ash. { an'hi·dres ,sōd·ē·əm 'kärb·ə,nāt }
anhydrous sodium sulfate |INORG CHEM| Na₂SO₄ Water-soluble, white crystals with bitter, salty taste; melts at 888°C; used in the manufacture of glass, paper, pharmaceuticals, and textiles, and as an analytical reagent. { an'hī·drəs ,sōd·ē·əm 'səl,fāt }
anilazine |ORG CHEM| C₉H₅Cl₃N₄ A tan solid with a melting point of 159–160°C; used for fungal diseases of lawns, turf, and vegetable crops. { ə'nil·ə,zēn }
anilide |ORG CHEM| A compound that has the C₆H₅NH₂−group; an example is benzanilide, C₆H₅NHCOC₆H₅. { 'an·əl,īd }
aniline |ORG CHEM| C₆H₅NH₂ An aromatic amine compound that is a pale brown liquid at room temperature; used in the dye, pharmaceutical, and rubber industries. { 'an· əl·ən }
aniline black |ORG CHEM| A black dye produced on certain textiles, such as cotton, by oxidizing aniline or aniline hydrochloride. { 'an·əl·ən 'blak }
aniline N,N-dimethyl See N,N-dimethylaniline. { 'an·əl·ən ¦en ¦en di'meth·əl }
aniline dye |ORG CHEM| A dye derived from aniline. { 'an·əl·ən 'dī }
aniline-formaldehyde resin |CHEM| A thermoplastic resin made by polymerizing aniline and formaldehyde. { 'an·əl·ən ,fòr'mal·də,hīd ,rez·ən }
aniline hydrochloride |ORG CHEM| C₆H₅NH₂·HCl White crystals, although sometimes the commercial variety has a greenish tinge; melting point 198°C; soluble in water and ethanol; used in dye manufacture, dyeing, and printing. { 'an·əl·ən ,hī·drə'klòr,īd }
animal black |CHEM| Finely divided carbon made by calcination of animal bones or ivory; used for pigments, decolorizers, and purifying agents; varieties include bone black and ivory black. { 'an·ə·məl ,blak }
animal charcoal |CHEM| Charcoal obtained by the destructive distillation of animal matter at high temperatures; used to adsorb organic coloring matter. { 'an·ə·məl 'chär,kōl }
anion |CHEM| An ion that is negatively charged. { 'a,nī·ən }
anion exchange |CHEM| A type of ion exchange in which the immobilized functional groups on the solid resin are positive. { 'a,nī·ən iks'chānj }

anionic polymerization [ORG CHEM] A type of polymerization in which Lewis bases, such as alkali metals and metallic alkyls, act as catalysts. { ˈaˌnīˈän·ik pəˌlim·əˈrəˈzā·shən }

anionotropy [CHEM] The breaking off of an ion such as hydroxyl or bromide from a molecule so that a positive ion remains in a state of dynamic equilibrium. { ˈaˌnī·əˈnä·trə·pē }

anisaldehyde [ORG CHEM] $C_6H_4(OCH_3)CHO$ A compound with melting point 2.5°C, boiling point 249.5°C; insoluble in water, soluble in alcohol and ether; used in perfumery and flavoring, and as an intermediate in production of antihistamines. { ˈaˌnəsˈalˈdəˌhīd }

anisic acid [ORG CHEM] $CH_3OC_6H_4COOH$ White crystals or powder with a melting point of 184°C; soluble in alcohol and ether; used in medicine and as an insect repellent and ovicide. { əˈnis·ik ˈas·əd }

anisic alcohol [ORG CHEM] $C_8H_{10}O_2$ A colorless liquid that boils in the range 255–265°C; it is obtained by reduction of anisic aldehyde; used in perfumery, and as an intermediate in the manufacture of pharmaceuticals. { əˈnis·ik ˈal·kəˌhȯl }

anisole [ORG CHEM] $C_6H_5OCH_3$ A colorless liquid that is soluble in ether and alcohol, insoluble in water; boiling point is 155°C; vapors are highly toxic; used as a solvent and in perfumery. { ˈan·əˌsōl }

annular atoms [ORG CHEM] The atoms in a cyclic compound that are members of the ring. { ˈan·yə·lər ˈad·əmz }

annulene [ORG CHEM] One of a group of monocyclic conjugated hydrocarbons which have the general formula $[-CH=CH-]_n$. { ˈan·yəˌlēn }

anode [PHYS CHEM] The positive terminal of an electrolytic cell. { ˈaˌnōd }

anode-corrosion efficiency [PHYS CHEM] The ratio of actual weight loss of an anode due to corrosion to the theoretical loss as calculated by Faraday's law. { ˈaˌnōd kəˈrō·zhən iˌfish·ən·sē }

anode effect [PHYS CHEM] A condition produced by polarization of the anode in the electrolysis of fused salts and characterized by a sudden increase in voltage and a corresponding decrease in amperage. { ˈaˌnōd iˌfekt }

anode film [CHEM] The portion of solution in immediate contact with the anode. { ˈaˌnōd ˌfilm }

anodic polarization [PHYS CHEM] The change in potential of an anode caused by current flow. { əˈnäd·ik pō·lə·rəˈzā·shən }

anolyte [CHEM] The part of the electrolyte at or near the anode that is changed in composition by the reactions at the anode. { ˈan·əˌlīt }

anomalous Zeeman effect [SPECT] A type of splitting of spectral lines of a light source in a magnetic field which occurs for any line arising from a combination of terms of multiplicity greater than one; due to a nonclassical magnetic behavior of the electron spin. { əˈnäm·ə·ləs ˈzāˌmän iˌfekt }

anomer [ORG CHEM] One of a pair of isomers of cyclic carbohydrates; resulting from creation of a new point of symmetry when a rearrangement of the atoms occurs at the aldehyde or ketone position. { ˈan·ə·mər }

antacid [CHEM] Any substance that counteracts or neutralizes acidity. { ˈantˈas·əd }

antarafacial [ORG CHEM] The stereochemistry when, simultaneously, two sigma bonds are formed or broken on the opposite faces of the component pi systems, such as in a cycloaddition reaction. { ˌan·tə·rəˈfā·shəl }

anthracene [ORG CHEM] $C_{14}H_{10}$ A crystalline tricyclic aromatic hydrocarbon, colorless when pure, melting at 218°C and boiling at 342°C; obtained in the distillation of coal tar; used as an important source of dyestuffs, and in coating applications. { ˈan·thrəˌsēn }

anthracene violet See gallein. { ˈan·thrəˌsēn ˈvī·lət }

anthraciferous coal [ORG CHEM] Anthracite-hard coal containing or yielding anthracene. { ˈan·thrəˌsif·ə·rəs ˈkōl }

anthranilic acid [ORG CHEM] o-$NH_2C_6H_4COOH$ A white or pale yellow, crystalline acid melting at 146°C; used as an intermediate in the manufacture of dyes, pharmaceuticals, and perfumes. { ˈanˈthrəˌlinˈik ˈas·əd }

anthrapurpurin

anthrapurpurin [ORG CHEM] $C_6H_3OH(CO)_2C_6H_2(OH)_2$ Orange-yellow, crystalline needles with a melting point of 369°C; soluble in alcohol and alkalies; used in dyeing. { ‚an·thrə'pər·pə‚rin }

anthraquinone [ORG CHEM] $C_6H_4(CO)_2C_6H_4$ Yellow crystalline diketone that is insoluble in water; used in the manufacture of dyes. { ‚an·thrə·kwi'nōn }

anthrone [ORG CHEM] $C_{14}H_{10}O$ Colorless needles with a melting point of 156°C; soluble in alcohol, benzene, and hot sodium hydroxide; used as a reagent for carbohydrates. { 'an‚thrōn }

anti [ORG CHEM] In stereochemistry, on the opposite side of a reference plane; for example, the stereochemical outcome of an addition reaction where the new bonds are on the opposite side of the original pi bond is called anti addition. { an‚tē }

antiaromatic [CHEM] A cyclic compound with delocalized electrons that does not obey Hückel's rule, and is much less stable than similar nonaromatic compounds. { ‚an·tē‚ar·ə'mad·ik }

anticatalyst [CHEM] A material that slows down the action of a catalyst; an example is lead, which inhibits the action of platinum. { ¦an·tē'kad·əl·ist }

antifoaming agent [ORG CHEM] A substance, such as silicones, organic phosphates, and alcohols, that inhibits the formation of bubbles in a liquid during its agitation by reducing its surface tension. { ¦an·tē¦fōm·iŋ ‚ā·jənt }

antifreeze [CHEM] A substance added to a liquid to lower its freezing point; the principal automotive antifreeze component is ethylene glycol. { 'an·tē‚frēz }

antimonate [CHEM] The radical $[Sb(OH)_6]^-$ in salts derived from antimony pentoxide, Sb_4O_{10}, and bases. { 'an·tə·mə‚nāt }

antimonic [CHEM] Derived from or pertaining to pentavalent antimony. { ¦an·tə¦män·ik }

antimonide [INORG CHEM] A binary compound of antimony with a more positive compound, for example, H_5Sb. Also known as stibide. { 'an·tə·mə‚nīd }

antimonous [CHEM] Pertaining to antinomy, especially trivalent antimony. { 'an·tə·mə·nəs }

antimony [CHEM] A chemical element, symbol Sb, atomic number 51, atomic weight 121.75. { 'an·tə‚mō·nē }

antimonyl [CHEM] The inorganic radical SbO^-. { 'an·tə·mə‚nil }

antimonyl potassium tartrate See tartar emetic. { 'an·tə·mə‚nil pə'tas·ē·əm 'tär‚trāt }

antimony(III) oxide [INORG CHEM] Sb_2O_3 Colorless, rhombic crystals, melting at 656°C; insoluble in water; powerful reducing agent. { 'an·tə‚mō·nē ‚thrē 'äk‚sīd }

antimony pentachloride [INORG CHEM] $SbCl_5$ A reddish-yellow, oily liquid; hygroscopic, it solidifies after moisture is absorbed and decomposes in excess water; soluble in hydrochloric acid and chloroform; used in analytical testing for cesium and alkaloids, for dyeing, and as an intermediary in synthesis. Also known as antimony perchloride. { 'an·tə‚mō·nē ‚pent·ə'klòr‚īd }

antimony pentafluoride [INORG CHEM] SbF_5 A corrosive, hygroscopic, moderately viscous fluid; reacts violently with water; forms a clear solution with glacial acetic acid; used in the fluorination of organic compounds. { 'an·tə‚mō·nē ‚pent·ə'flùr‚īd }

antimony pentasulfide [INORG CHEM] Sb_2S_5 An orange-yellow powder; soluble in alkali, soluble in concentrated hydrochloric acid, with hydrogen sulfide as a by-product, and insoluble in water; used as a red pigment. Also known as antimony persulfide; antimony red; golden antimony sulfide. { 'an·tə‚mō·nē ‚pent·ə'səl‚fīd }

antimony perchloride See antimony pentachloride. { 'an·tə‚mō·nē ‚per'klòr‚īd }

antimony persulfide See antimony pentasulfide. { 'an·tə‚mō·nē ‚per'səl‚fīd }

antimony red See antimony pentasulfide. { 'an·tə‚mō·nē 'red }

antimony sodiate See sodium antimonate. { 'an·tə‚mō·nē 'sō·dē·āt }

antimony sulfate [INORG CHEM] $Sb_2(SO_4)_3$ Antimony(III) sulfate, a white, deliquescent powder; soluble in acids. { 'an·tə‚mō·nē 'səl‚fāt }

antimony trichloride [INORG CHEM] $SbCl_3$ Hygroscopic, colorless, crystalline mass; fumes slightly in air, is soluble in alcohol and acetone, and forms antimony oxychloride in water; used as a mordant, as a chlorinating agent, and in fireproofing textiles. { 'an·tə‚mō·nē ‚trī'klòr‚īd }

26

antimony trisulfide [INORG CHEM] Sb_2S_3 Black and orange-red rhombic crystals; soluble in concentrated hydrochloric acid and sulfide solutions, insoluble in water; melting point 546°C; used as a pigment, and in matches and pyrotechnics. { 'an·tə,mō·nē ,trī'səl,fīd }

antimony yellow *See* lead antimonite. { 'an·tə,mō·nē 'ye·lō }

antioxidant [PHYS CHEM] A substance that, when present at a lower concentration than that of the oxidizable substrate, significantly inhibits or delays oxidative processes, while being itself oxidized. In primary antioxidants, antioxidative activity is implemented by the donation of an electron or hydrogen atom to a radical derivative, and in secondary antioxidants by the removal of an oxidative catalyst and the consequent prevention of the initiation of oxidation. Antioxidants are used in polymers to prevent degradation, and in foods, beverages, and cosmetic products to inhibit deterioration and spoilage. { ,an·tē'äk·sə·dənt }

anti-Stokes lines [SPECT] Lines of radiated frequencies which are higher than the frequency of the exciting incident light. { ,an·tē'stōks ,līnz }

ANTU *See* 1-(1-naphthyl)-2-thiourea.

4-AP *See* 4-aminopyridine.

apo- [CHEM] A prefix that denotes formation from or relationship to another chemical compound. { 'ap·ō *or* 'ap·ə }

apoatropine [ORG CHEM] $C_{17}H_{21}NO_2$ An alkaloid melting at 61°C with decomposition of the compound; highly toxic; obtained by dehydrating atropine. { ,ap·ō'a·trə,pēn }

apodization [SPECT] A mathematical transformation carried out on data received from an interferometer to alter the instrument's response function before the Fourier transformation is calculated to obtain the spectrum. { ,a·pə·də'zā·shən }

apparent concentration [ANALY CHEM] The value of analyte concentration obtained when the interference is not considered. { ə'par·ənt ,kän·sən'trā·shən }

aprotic solvent [CHEM] A solvent that does not yield or accept a proton. { ā'präd·ik 'säl·vənt }

aqua [CHEM] Latin for water. { 'äk·wə }

aqua ammonia *See* ammonium hydroxide. { 'äk·wə ə'mōn·ē·ə }

aquafortis *See* nitric acid. { 'äk·wə'fȯrd·əs }

aquametry [ANALY CHEM] Analytical processes to measure the water present in materials; methods include Karl Fischer titration, reactions with acid chlorides and anhydrides, oven drying, distillation, and chromatography. { ə'kwäm·ə·trē }

aqua regia [INORG CHEM] A fuming, highly corrosive, volatile liquid with a suffocating odor made by mixing 1 part concentrated nitric acid and 3 parts concentrated hydrochloric acid; reacts with all metals, including silver and gold. { äk·wə 'rē·jə }

aquasol *See* hydrosol. { 'äk·wə,sȯl }

aquation [CHEM] Formation of a complex that contains water by replacement of other coordinated groups in the complex. { ə'kwā·shən }

aqueous electron *See* hydrated electron. { 'äk·wē·əs i'lek,trän }

aqueous solution [CHEM] A solution with the solvent as water. { 'äk·wē·əs sə'lü·shən }

aquo ion [CHEM] Any ion containing one or more water molecules. { 'a·kwō 'ī,än }

Ar *See* argon.

arabine *See* harman. { 'ar·ə,bēn }

arabite *See* arabitol. { 'ar·ə,bīt }

arabitol [ORG CHEM] $CH_2OH(CHOH)_3CH_2OH$ An alcohol that is derived from arabinose; a sweet, colorless crystalline material present in D and L forms; soluble in water; melts at 103°C. Also known as arabite. { ə'rab·ə,tȯl }

arachic acid *See* eicosanoic acid. { ə'rak·ik 'as·əd }

arachidic acid *See* eicosanoic acid. { ,a·rə'kid·ik 'as·əd }

aralkyl [ORG CHEM] A radical in which an aryl group is substituted for an alkyl H atom. Derived from arylated alkyl. { a'ral,kil }

arbutin [ORG CHEM] $C_{12}H_{16}O_7$ A bitter glycoside from the bearberry and certain other plants; sometimes used as a urinary antiseptic. { är'byüt·ən }

arc spectrum [SPECT] The spectrum of a neutral atom, as opposed to that of a molecule

arecoline

or an ion; it is usually produced by vaporizing the substance in an electric arc; designated by the roman numeral I following the symbol for the element, for example, Hel. { 'ärk ,spek·trəm }

arecoline [ORG CHEM] $C_8H_{13}O_2N$ An alkaloid from the betel nut; an oily, colorless liquid with a boiling point of 209°C; soluble in water, ethanol, and ether; combustible; used as a medicine. { ə'rek·ə,lēn }

arene *See* aromatic hydrocarbon. { 'a,rēn }

argentic [CHEM] Relating to or containing silver. { är'jen·tik }

argentic oxide *See* silver suboxide. { är'jen·tik 'äk,sīd }

argentocyanides [INORG CHEM] Complexes formed, for example, in the cyanidation of silver ores and in electroplating, when silver cyanide reacts with solutions of soluble metal cyanides. Also known as dicyanoargentates. { är,jen·tō'sī·ə,nīdz }

argentometry [ANALY CHEM] A volumetric analysis that employs precipitation of insoluble silver salts; the salts may be chromates or chlorides. { ,är·jən'täm·ə·trē }

argentum [CHEM] Latin for silver. { är'jen·təm }

argon [CHEM] A chemical element, symbol Ar, atomic number 18, atomic weight 39.998. { 'ar,gän }

aristolochic acid [ORG CHEM] $C_{17}H_{11}NO_7$ Crystals in the form of shiny brown leaflets that decompose at 281–286°C; soluble in alcohol, chloroform, acetone, ether, acetic acid, and aniline; used as an aromatic bitter. Also known as aristolochine. { ə¦ris·tə¦läk·ik 'as·əd }

aristolochine *See* aristolochic acid. { a,ris'täl·ə,kēn }

armchair nanotube [PHYS CHEM] A carbon nanotube formed from a graphite sheet that is rolled up so that the edge is in the shape of armchairs. { ,ärm,chār 'nan·ō,tüb }

Armstrong's acid *See* naphthalene-1,5-disulfonic acid. { 'ärm,strȯŋz 'as·əd }

Arndt-Eistert synthesis [ORG CHEM] A method of increasing the length of an aliphatic acid by one carbon by reacting diazomethane with acid chloride. { ¦ärnt ¦ī·stərt ,sin·thə·səs }

aromatic [ORG CHEM] **1.** Pertaining to or characterized by the presence of at least one benzene ring. **2.** Describing those compounds having physical and chemical properties resembling those of benzene. { ¦ar·ə¦mad·ik }

aromatic alcohol [ORG CHEM] Any of the compounds containing the hydroxyl group in a side chain to a benzene ring, such as benzyl alcohol. { ¦ar·ə¦mad·ik 'al·kə,hȯl }

aromatic aldehyde [ORG CHEM] An aromatic compound containing the CHO radical, such as benzaldehyde. { ¦ar·ə¦mad·ik 'al·də,hīd }

aromatic amine [ORG CHEM] An organic compound that contains one or more amino groups joined to an aromatic structure. { ¦ar·ə¦mad·ik 'am,ēn }

aromatic hydrocarbon [ORG CHEM] A member of the class of hydrocarbons, of which benzene is the first member, consisting of assemblages of cyclic conjugated carbon atoms and characterized by large resonance energies. Also known as arene. { ¦ar·ə¦mad·ik ¦hī·drə'kär·bən }

aromatic ketone [ORG CHEM] An aromatic compound containing the −CO radical, such as acetophenone. { ¦ar·ə¦mad·ik 'kē,tōn }

aromatic nucleus [ORG CHEM] The six-carbon ring characteristic of benzene and related series, or condensed six-carbon rings of naphthalene, anthracene, and so forth. { ¦ar·ə¦mad·ik 'nü·klē·əs }

aroyl [ORG CHEM] The radical RCO, where R is an aromatic (benzoyl, napthoyl) group. { 'ar·ə·wəl }

aroylation [ORG CHEM] A reaction in which the aroyl group is incorporated into a molecule by substitution. { ,ar·ə·wə'lā·shən }

ARPES *See* angle-resolved photoelectron spectroscopy.

Arrhenius equation [PHYS CHEM] The relationship that the specific reaction rate constant k equals the frequency factor constant s times exp $(-\Delta H_{act}/RT)$, where ΔH_{act} is the heat of activation, R the gas constant, and T the absolute temperature. { ar'rā·nē·əs i'kwä·zhən }

arsenate [INORG CHEM] **1.** AsO_4^{3-} A negative ion derived from orthoarsenic acid, $H_3AsO_4·\frac{1}{2}H_2O$. **2.** A salt or ester of arsenic acid. { 'ärs·ən,āt }

28

arsenic [CHEM] A chemical element, symbol As, atomic number 33, atomic weight 74.9216. { 'ärs·ən·ik }

arsenic acid [INORG CHEM] $H_3AsO_4\cdot{}^1/_2H_2O$ White, poisonous crystals, soluble in water and alcohol; used in manufacturing insecticides, glass, and arsenates and as a defoliant. Also known as orthoarsenic acid. { är'sen·ik 'as·əd }

arsenical [CHEM] **1.** Pertaining to arsenic. **2.** A compound that contains arsenic. { ar'sen·ə·kəl }

arsenic disulfide [INORG CHEM] As_2S_2 Red, orange, or black monoclinic crystals, insoluble in water; used in fireworks; occurs naturally as realgar. { 'ärs·ən·ik dī'səl‚fīd }

arsenic oxide [INORG CHEM] **1.** An oxide of arsenic. **2.** See arsenic pentoxide; arsenic trioxide. { 'ärs·ən·ik 'äk‚sīd }

arsenic pentasulfide [INORG CHEM] As_2S_5 Yellow crystals that are insoluble in water and readily decompose to the trisulfide and sulfur; used as a pigment. { 'ärs·ən·ik ‚pent·ə'səl‚fīd }

arsenic pentoxide [INORG CHEM] As_2O_5 A white, deliquescent compound that decomposes by heat and is soluble in water. Also known as arsenic oxide. { 'ärs·ən·ik ‚pent'äk‚sīd }

arsenic trichloride [INORG CHEM] $AsCl_3$ An oily, colorless liquid that dissolves in water; used in ceramics, organic chemical syntheses, and in the preparation of pharmaceuticals. { 'ärs·ən·ik ‚tri'klȯr‚īd }

arsenic trioxide [INORG CHEM] As_2O_3 A toxic compound, slightly soluble in water; octahedral crystals change to the monoclinic form by heating at 200°C; occurs naturally as arsenolite and claudetite; used in small quantities in some medicinal preparations. Also known as arsenic oxide; arsenious acid. { 'ärs·ən·ik ‚tri'äk‚sīd }

arsenic trisulfide [INORG CHEM] As_2S_3 An acidic compound in the form of yellow or red monoclinic crystals with a melting point at 300°C; occurs as the mineral orpiment; used as a pigment. { 'ärs·ən·ik ‚tri'səl‚fīd }

arsenide [CHEM] A binary compound of negative, trivalent arsenic; for example, H_3As or GaAs. { 'ärs·ən‚īd }

arsenin [ORG CHEM] A heterocyclic organic compound composed of a six-membered ring system in which the carbon atoms are unsaturated and the unique heteroatom is arsenic, with no nitrogen atoms present. { är'sen·ən }

arsenious acid See arsenic trioxide. { är'sēn·ē·əs 'as·əd }

arsenite [INORG CHEM] **1.** $AsO_3{}^{3-}$ A negative ion derived from aqueous solutions of As_4O_6. **2.** A salt or ester of arsenious acid. { 'är·sə‚nīt }

arsenobenzene [ORG CHEM] $C_6H_5As:AsC_6H_5$ White needles that melt at 212°C; insoluble in cold water, soluble in benzene; derivatives have some use in medicine. { ‚ärs·ən·ō'ben‚zēn }

arseno compound [ORG CHEM] A compound containing an As-As bond with the general formula $(RAs)_n$, where R represents a functional group; structures are cyclic or long-chain polymers. { ‚ärs·ən·ō 'käm‚paůnd }

arsenous oxide See arsenic trioxide. { 'är·sə·nəs 'äk·sīd }

arsine [INORG CHEM] H_3As A colorless, highly poisonous gas with an unpleasant odor. { är'sēn }

arsinic acid [INORG CHEM] An acid of general formula R_2AsO_2H; derived from trivalent arsenic; an example is cacodylic acid, or dimethylarsinic acid, $(CH_3)_2AsO_2H$. { är'sin·ik 'as·əd }

arsonic acid [INORG CHEM] An acid derived from orthoarsenic acid, $OAs(OH)_3$; the type formula is generally considered to be $RAsO(OH)_2$; an example is para-aminobenzenearsonic acid, $NH_2C_6H_4AsO(OH)_2$. { är'sän·ik 'as·əd }

arsonium [INORG CHEM] $-AsH_4$ A radical which may be considered analogous to the ammonium radical in that a compound such as AsH_4OH may form. { är'sōn·ē·əm }

artificial camphor See terpene hydrochloride. { ‚ärd·ə‚fish·əl 'kam·fər }

artificial gold See stannic sulfide. { ‚ärd·ə‚fish·əl 'gōld }

artificial malachite See copper carbonate. { ‚ärd·ə‚fish·əl 'mal·ə‚kīt }

artificial neroli oil See methyl anthranilate. { ‚ärd·ə‚fish·əl nə'rōl·ē ‚ȯil }

artificial scheelite See calcium tungstate. { ‚ärd·ə‚fish·əl 'shā‚līt }

aryl [ORG CHEM] An organic group derived from an aromatic hydrocarbon by removal of one hydrogen. { 'ar·əl }

aryl acid [ORG CHEM] An organic acid that has an aryl group. { 'ar·əl 'as·əd }

arylamine [ORG CHEM] An organic compound formed from an aromatic hydrocarbon that has at least one amine group joined to it, such as aniline. { 'ar·əl·ə,mēn }

arylated alkyl *See* aralkyl. { ar·ə'lād·əd 'al·kəl }

aryl compound [ORG CHEM] Molecules with the six-carbon aromatic ring structure characteristic of benzene or compounds derived from aromatics. { 'ar·əl ˌkäm,paůnd }

aryl diazo compound [ORG CHEM] A diazo compound bonded to the ring structure characteristic of benzene or any other aromatic derivative. { 'ar·əl dī'az·ō ˌkäm,paůnd }

arylene [ORG CHEM] A radical that is bivalent and formed by removal of hydrogen from two carbon sites on an aromatic nucleus. { 'ar·ə,lēn }

aryl halide [ORG CHEM] An aromatic derivative in which a ring hydrogen has been replaced by a halide atom. { 'ar·əl 'hal,īd }

arylide [ORG CHEM] A compound formed from a metal and an aryl group, for example, PbR_4, where R is the aryl group. { 'ar·ə,līd }

aryloxy compound [ORG CHEM] One of a group of compounds useful as organic weed killers, such as 2,4-dichlorophenoxyacetic acid (2,4-D). { ˌar·əlˈäk·sē ˌkäm,paůnd }

aryne [ORG CHEM] An aromatic species in which two adjacent atoms of a ring lack substituents, with two orbitals each missing an electron. Also known as benzyne. { 'a,rīn }

As *See* arsenic.

asarone [ORG CHEM] $C_{12}H_{16}O_3$ A crystalline substance with melting point 67°C; insoluble in water, soluble in alcohol; found in plants of the genus *Asarum*; used as a constituent in essential oils such as calumus oil. { 'as·ə,rōn }

ascaridole [ORG CHEM] $C_{10}H_{16}O_2$ A terpene peroxide, explosive when heated; used as an initiator in polymerization. { ə'skar·ə,dōl }

ascending chromatography [ANALY CHEM] A technique for the analysis of mixtures of two or more compounds in which the mobile phase (sample and carrier) rises through the fixed phase. { ə'send·iŋ ˌkrō·mə'täg·rə·fē }

ash [CHEM] The incombustible matter remaining after a substance has been incinerated. { ash }

ashing [ANALY CHEM] An analytical process in which the chemical material being analyzed is oven-heated to leave only noncombustible ash. { 'ash·iŋ }

aspartame [ORG CHEM] $C_{14}H_{18}N_2O_5$ A dipeptide ester about 160 times sweeter than sucrose in aqueous solution; used as a low-calorie sweetener. { a'spär,tām }

aspirin *See* acetylsalicylic acid. { 'as·prən }

assay [ANALY CHEM] Qualitative or quantitative determination of the components of a material, as an ore or a drug. { 'a,sā }

association [CHEM] Combination or correlation of substances or functions. { ə,sō·sē'ā·shən }

A stage [ORG CHEM] An early stage in a thermosetting resin reaction characterized by linear structure, solubility, and fusibility of the material. { 'ā ˌstāj }

astatine [CHEM] A radioactive chemical element, symbol At, atomic number 85, the heaviest of the halogen elements. { 'as·tə,tēn }

asterism [SPECT] A star-shaped pattern sometimes seen in x-ray spectrophotographs. { 'as·tə,riz·əm }

astigmatic mounting [SPECT] A mounting designed to minimize the astigmatism of a concave diffraction grating. { ˌa·stig,mad·ik 'móunt·iŋ }

astronomical spectrograph [SPECT] An instrument used to photograph spectra of stars. { ˌas·trə'näm·ə·kəl 'spek·trə,graf }

astronomical spectroscopy [SPECT] The use of spectrographs in conjunction with telescopes to obtain observational data on the velocities and physical conditions of astronomical objects. { ˌas·trə'näm·ə·kəl ˌspek'träs·kə·pē }

asymmetric carbon atom [ORG CHEM] A carbon atom with four different atoms or groups of atoms bonded to it. Also known as chiral carbon atom; stereogenic center.   { ¦ā·sə¦me·trik ¦kär·bən 'ad·əm }

asymmetric synthesis   [ORG CHEM] Chemical synthesis of a pure enantiomer, or of an enantiomorphic mixture in which one enantiomer predominates, without the use of resolution.   { ¦ā·sə¦me·trik 'sin·thə·səs }

asymmetry   [PHYS CHEM] The geometrical design of a molecule, atom, or ion that cannot be divided into like portions by one or more hypothetical planes. Also known as molecular asymmetry.   { ¦ā'sim·ə·trə }

asymmetry effect   [PHYS CHEM] The asymmetrical distribution of the ion cloud around an ion that results from the finite relaxation time for the ion cloud when a voltage is applied; leads to a reduction in ion mobility.   { ā'sim·ə·trē i,fekt }

At See astatine.

ATA See aminotriazole.

atactic   [ORG CHEM] Of the configuration for a polymer, having the opposite steric configurations for the carbon atoms of the polymer chain occur in equal frequency and more or less at random.   { ā'tak·tik }

atom   [CHEM] The individual structure which constitutes the basic unit of any chemical element.   { 'ad·əm }

atom cluster   [PHYS CHEM] An assembly of between three and a few thousand atoms or molecules that are weakly bound together and have properties intermediate between those of the isolated atom or molecule and the bulk or solid-state material.   { 'ad·əm ,kləs·tər }

atomic absorption spectroscopy   [SPECT] An instrumental technique for detecting concentrations of atoms to parts per million by measuring the amount of light absorbed by atoms or ions vaporized in a flame or an electrical furnace.   { ə¦tä·mik əb¦sörp·shən ,spek'träs·kə·pē }

atomic connectivity   [PHYS CHEM] The specific pattern of chemical bonds between atoms in a molecule.   { ə¦täm·ik kə,nek'tiv·əd·ē }

atomic emission spectroscopy   [SPECT] A form of atomic spectroscopy in which one observes the emission of light at discrete wavelengths by atoms which have been electronically excited by collisions with other atoms and molecules in a hot gas.   { ə¦täm·ik ə¦mish·ən spek'träs·kə·pē }

atomic fluorescence spectroscopy   [SPECT] A form of atomic spectroscopy in which the sample atoms are first excited by absorbing radiation from an external source containing the element to be detected, and the intensity of radiation emitted at characteristic wavelengths during transitions of these atoms back to the ground state is observed.   { ə¦täm·ik flü¦res·əns spek'träs·kə·pē }

atomic heat capacity   [PHYS CHEM] The heat capacity of a gram-atomic weight of an element.   { ə'täm·ik 'hēt kə'pas·əd·ē }

atomic hydrogen   [CHEM] Gaseous hydrogen whose molecules are dissociated into atoms.   { ə'täm·ik 'hī·drə·jən }

atomicity   [CHEM] The number of atoms in a molecule of a compound.   { ,ad·ə'mis·əd·ē }

atomic percent   [CHEM] The number of atoms of an element in 100 atoms representative of a substance.   { ə'täm·ik pər'sent }

atomic photoelectric effect   See photoionization.   { ə'täm·ik ,fōd·ō·i'lek·trik i'fekt }

atomic polarization   [PHYS CHEM] Polarization of a material arising from the change in dipole moment accompanying the stretching of chemical bonds between unlike atoms in molecules.   { ə'täm·ik ,pōl·ə·rə'zā·shən }

atomic radius   [PHYS CHEM] Also known as covalent radius. **1.** Half the distance between the nuclei of two like atoms that are covalently bonded. **2.** The experimentally determined radius of an atom in a covalently bonded compound.   { ə'täm·ik 'rād·ē·əs }

atomic spectroscopy   [SPECT] The branch of physics concerned with the production, measurement, and interpretation of spectra arising from either emission or absorption of electromagnetic radiation by atoms.   { ə'täm·ik ,spek'träs·kə·pē }

31

atomic spectrum

atomic spectrum [SPECT] The spectrum of radiations due to transitions between energy levels in an atom, either absorption or emission. { ə'täm·ik 'spek·trəm }

atomic theory [CHEM] The assumption that matter is composed of particles called atoms and that these are the limit to which matter can be subdivided. { ə'täm·ik 'thē·ə·rē }

atomic volume [PHYS CHEM] The volume occupied by 1 gram-atom of an element in the solid state. { ə'täm·ik 'väl·yəm }

atomic weight [CHEM] The relative mass of an atom based on a scale in which a specific carbon atom (carbon-12) is assigned a mass value of 12. Abbreviated at. wt. Also known as relative atomic mass. { ə'täm·ik 'wāt }

atomization [ANALY CHEM] In flame spectrometry, conversion of a volatilized sample into free atoms. [CHEM] A process in which the chemical bonds in a molecule are broken to yield separated (free) atoms. { ˌad·ə·mə'zā·shən }

atoms-in-molecules method [PHYS CHEM] The description of the electronic structure of a molecule as a perturbation of the isolated states of its constituent atoms. { ¦ad·əmz in 'mäl·ə¦kyülz ˌmeth·əd }

atom trap trace analysis [ANALY CHEM] An atom-counting method in which individual atoms of a chosen isotope are captured and detected with a laser trap. { ¦ad·əm ¦trap 'träs ə¦nal·ə·səs }

ATR See attenuated total reflectance.

atrazine [CHEM] C₈H₁₄ClN₅ A white crystalline compound widely used as a photosynthesis-inhibiting herbicide for weeds. { 'a·trə,zēn }

atropisomer [ORG CHEM] One of two conformations of a molecule whose interconversion is slow enough to allow separation and isolation under predetermined conditions. { ¦a·trō¦pīz·ə·mər }

attachment [ORG CHEM] The conversion of a molecular entity into another molecular structure solely by formation of a single two-center bond with another molecular entity and no other changes in bonding. { ə'tach·mənt }

attenuated total reflectance [SPECT] A method of spectrophotometric analysis based on the reflection of energy at the interface of two media which have different refractive indices and are in optical contact with each other. Abbreviated ATR. Also known as frustrated internal reflectance; internal reflectance spectroscopy. { ə'ten·yə,wād·əd 'tōd·əl ri'flek·təns }

at. wt See atomic weight.

Au See gold.

Aufbau principle [CHEM] A description of the building up of the elements in which the structure of each in sequence is obtained by simultaneously adding one positive charge (proton) to the nucleus of the atom and one negative charge (electron) to an atomic orbital. { 'aůf,baů ¦prin·sə·pəl }

Auger electron spectroscopy [SPECT] The energy analysis of Auger electrons produced when an excited atom relaxes by a radiationless process after ionization by a high-energy electron, ion, or x-ray beam. Abbreviated AES. { ō'zhā i'lek,trän spek'träs·kə·pē }

auramine hydrochloride [ORG CHEM] C₁₇H₂₂ClN₃·H₂O A compound melting at 267°C; very soluble in water, soluble in ethanol; used as a dye and an antiseptic. Also known as yellow pyoktanin. { 'òr·ə,mēn ,hī·drə'klòr,īd }

aurantia [ORG CHEM] C₁₂H₈N₈O₁₂ An orange aniline dye, used in stains in biology and in some photographic filters. { ò'ranch·ə }

aurantiin See naringin. { ò'ran·tē·ən }

auric oxide See gold oxide. { 'òr·ik 'äk,sīd }

aurin [ORG CHEM] C₁₉H₁₄O₃ A derivative of triphenylmethane; solid with red-brown color with green luster; melting point about 220°C; insoluble in water; used as a dye intermediate. { 'òr·ən }

auroral line [SPECT] A prominent green line in the spectrum of the aurora at a wavelength of 5577 angstroms, resulting from a certain forbidden transition of oxygen. { ə'ròr·əl ,līn }

32

autoacceleration [ORG CHEM] The increase in polymerization rate and molecular weight of certain vinyl monomers during bulk polymerization. { ¦ȯd·ō·ik¸sel·ə'rā·shən }

autocatalysis [CHEM] A catalytic reaction started by the products of a reaction that was itself catalytic. { ¦ȯd·ō·kə'tal·ə·səs }

autogenous ignition temperature See ignition temperature. { ȯ'täj·ə·nəs ig'nish·ən ¸tem·prə·chər }

autoignition temperature [CHEM] The temperature at which a material (solid, liquid, or gas) will self-ignite and sustain combustion in air without an external spark or flame. { ¦ȯd·ō·ig¦nish·ən 'tem·prə·chər }

automatic titrator [ANALY CHEM] **1.** Titration with quantitative reaction and measured flow of reactant. **2.** Electrically generated reactant with potentiometric, ampherometric, or colorimetric end-point or null-point determination. { ¦ȯd·ə¦mad·ik 'tī¸trād·ər }

autooxidation See autoxidation.

autopoisoning See self-poisoning. { ¸ȯd·ō'pȯiz·ən·iŋ }

autoprotolysis [CHEM] Transfer of a proton from one molecule to another of the same substance. { ¸ȯd·ō·prə'täl·ə·səs }

autoprotolysis constant [CHEM] A constant denoting the equilibrium condition for the autoprotolysis reaction. { ¸ȯd·ō·prə'täl·ə·səs 'kän·stənt }

autoracemization [ORG CHEM] A racemization process that occurs spontaneously. { ¦ȯd·ō¸rā·sə·mə'zā·shən }

autoxidation [CHEM] Also known as autooxidation. **1.** The slow, flameless combustion of materials by reaction with oxygen. **2.** An oxidation reaction that is self-catalyzed and spontaneous. **3.** An oxidation reaction begun only by an inductor. { ȯ¦täk·sə'dā·shən }

auxiliary electrode [PHYS CHEM] An electrode in an electrochemical cell used for transfer of electric current to the test electrode. { ȯg'zil·yə·rē i'lek¸trōd }

auxochrome [CHEM] Any substituent group such as $-NH_2$ and $-OH$ which, by affecting the spectral regions of strong absorption in chromophores, enhance the ability of the chromogen to act as a dye. { 'ȯk·sə¸krōm }

available chlorine [CHEM] The quantity of chlorine released by a bleaching powder when treated with acid. { ə'vāl·ə·bəl 'klȯr¸ēn }

average bond dissociation energy [PHYS CHEM] The average value of the bond dissociation energies associated with the homolytic cleavage of several bonds of a set of equivalent bonds of a molecule. Also known as bond energy. { ¦av·rij ¦bänd di·sō·sē'ā·shən ¸en·ər·jē }

average molecular weight [ORG CHEM] The calculated number to average the molecular weights of the varying-length polymer chains present in a polymer mixture. { 'av·rij mə'lek·yə·lər 'wāt }

azacrown ether [ORG CHEM] A crown ether that has nitrogen donor atoms as well as oxygen donor atoms to coordinate to the metal iron. { 'az·ə¸kraún 'ē·thər }

9-azafluorene See carbazole. { ¦nīn ə'za·¸flȯr·ēn }

azelaic acid [ORG CHEM] $HOOC(CH_2)_7CCOH$ Colorless leaflets; melting point 106.5°C; a dicarboxylic acid useful in lacquers, alkyd salts, organic synthesis, and formation of polyamides. { ¦az·ə¦lā·ik 'as·əd }

azelate [ORG CHEM] A salt of azelaic acid, for example, sodium azelate. { 'az·əl¸āt }

azeotrope See azeotropic mixture. { ā'zē·ə¸trōp }

azeotropic mixture [CHEM] A solution of two or more liquids, the composition of which does not change upon distillation. Also known as azeotrope. { ¦a¸zē·ə¦träp·ik 'miks·chər }

azide [ORG CHEM] One of several types of compounds containing the $-N_3$ group and derived from hydrazoic acid, HN_3. { 'ā¸zīd }

azimino See diazoamine. { ā'zim·ē·nō }

azine [ORG CHEM] A compound of six atoms in a ring; at least one of the atoms is nitrogen, and the ring structure resembles benzene; an example is pyridine. { 'ā¸zēn }

azine dyes

azine dyes [ORG CHEM] Benzene-type dyes derived from phenazine; members of the group, such as nigrosines and safranines, are quite varied in application. { 'ā,zēn ¦dīz }

aziridine See ethyleneimine. { ə'zir·ə,dēn }

azlactone [ORG CHEM] A compound that is an anhydride of α-acylamino acid; the basic ring structure is the 5-oxazolone type. { az'lak,tōn }

azo- [ORG CHEM] A prefix indicating the group —N=N—. { 'a·zō }

azobenzene [ORG CHEM] $C_6H_5N_2C_6H_5$ A compound existing in cis and trans geometric isomers; the cis form melts at 71°C; the trans form comprises orange-red leaflets, melting at 68.5°C; used in manufacture of dyes and accelerators for rubbers. { ¦a·zō'ben,zēn }

2,2'-azobisisobutyronitrile [ORG CHEM] $C_8H_{12}N_4$ Crystals that decompose at 107°C; soluble in methanol and in ethanol; used as an initiator of free radical reactions and as a blowing agent for plastics and elastomers. { ¦tü ¦tü,prīm ,a·zō·bī,ī·sō,byüd·ə·rə'nī·trəl }

azo compound [ORG CHEM] A compound having two organic groups separated by an azo group (—N=N—). { 'ā·zō ,käm,paúnd }

azo dyes [ORG CHEM] Widely used commercial dyestuffs derived from amino compounds, with the —N— chromophore group; can be made as acid, basic, direct, or mordant dyes. { 'a·zō ,dīz }

azoic dye [ORG CHEM] A water-insoluble azo dye that is formed by coupling of the components on a fiber. Also known as ice color; ingrain color. { a'zō·ik 'dī }

azole [ORG CHEM] One of a class of organic compounds with a five-membered N-heterocycle containing two double bonds; an example is 1,2,4-triazole. { 'ā,zōl }

azotometer See nitrometer. { ,az·ə'täm·əd·ər }

azoxybenzene [ORG CHEM] $C_6H_5NO=N—C_6H_5$ A compound existing in cis and trans forms; the cis form melts at 87°C; the trans form comprises yellow crystals, melting at 36°C, insoluble in water, soluble in ethanol. { ə¦zäk·sē'ben,zēn }

azoxy compound [ORG CHEM] A compound having an oxygen atom bonded to one of the nitrogen atoms of an azo compound. { ā'zäk·sē ,käm,paúnd }

azulene [ORG CHEM] $C_{16}H_{26}O$ The blue coloring matter of wormwood and other essential oils; an oily, blue liquid, boiling at 170°C; insoluble in water; used in cosmetics. { 'azh·ə,lēn }

B

B *See* boron.

Ba *See* barium.

Babo's law [PHYS CHEM] A law stating that the relative lowering of a solvent's vapor pressure by a solute is the same at all temperatures. { 'bä,bōz ,lȯ }

backflash [CHEM] Rapid combustion of a material occurring in an area that the reaction was not intended for. { 'bak,flash }

back titration [CHEM] A titration to return to the end point which was passed. { 'bak tī'trā·shən }

Badger's rule [PHYS CHEM] An empirical relationship between the stretching force constant for a molecular bond and the bond length. { 'baj·ərz ,rül }

baeckeol [ORG CHEM] $C_{13}H_{18}O_4$ A phenolic ketone that is crystalline and pale yellow; found in oils from plants of species of the myrtle family. { 'bāk·ē,ōl }

Baeyer strain theory [ORG CHEM] The theory that the relative stability of penta- and hexamethylene ring compounds is caused by a propitious bond angle between carbons and a lack of bond strain. { 'bā·ər 'strān ,thē·ə·rē }

baking soda *See* sodium bicarbonate. { 'bāk·iŋ ,sōd·ə }

balance [CHEM] To bring a chemical equation into balance so that reaction substances and reaction products obey the laws of conservation of mass and charge. { 'bal·əns }

Balmer continuum [SPECT] A continuous range of wavelengths (or wave numbers or frequencies) in the spectrum of hydrogen at wavelengths less than the Balmer limit, resulting from transitions between states with principal quantum number $n = 2$ and states in which the single electron is freed from the atom. { ¦bäl·mər kən'tin·yə·wəm }

Balmer discontinuity *See* Balmer jump. { 'bȯl·mər dis,känt·ən'ü·əd·ē }

Balmer formula [SPECT] An equation for the wavelengths of the spectral lines of hydrogen, $1/\lambda = R[(1/m^2) - (1/n^2)]$, where λ is the wavelength, R is the Rydberg constant, and m and n are positive integers (with n larger than m) that give the principal quantum numbers of the states between which occur the transition giving rise to the line. { 'bȯl·mər ,fȯr·myə·lə }

Balmer jump [SPECT] The sudden decrease in the intensity of the continuous spectrum of hydrogen at the Balmer limit. Also known as Balmer discontinuity. { 'bȯl·mər ,jəmp }

Balmer limit [SPECT] The limiting wavelength toward which the lines of the Balmer series crowd and beyond which they merge into a continuum, at approximately 365 nanometers. { 'bȯl·mər ,lim·ət }

Balmer lines [SPECT] Lines in the hydrogen spectrum, produced by transitions between $n = 2$ and $n > 2$ levels either in emission or in absorption; here n is the principal quantum number. { 'bȯl·mər ,līnz }

Balmer series [SPECT] The set of Balmer lines. { 'bȯl·mər ¦sir·ēz }

Bamberger's formula [ORG CHEM] A structural formula for naphthalene that shows the valencies of the benzene rings pointing toward the centers. { 'bäm,bər·gərz 'fȯr·myə·lə }

banana oil [ORG CHEM] **1.** A solution of nitrocellulose in amyl acetate having a bananalike odor. **2.** *See* amyl acetate. { bə'nan·ə ,ȯil }

band [ANALY CHEM] The position and spread of a solute within a series of tubes in a

band head

liquid-liquid extraction procedure. Also known as zone. |SPECT| *See* band spectrum. { band }

band head |SPECT| A location on the spectrogram of a molecule at which the lines of a band pile up. { 'band ,hed }

band spectrum |SPECT| A spectrum consisting of groups or bands of closely spaced lines in emission or absorption, characteristic of molecular gases and chemical compounds. Also known as band. { 'band ,spek·trəm }

barban [ORG CHEM] $C_{11}H_9O_2NCl_2$ A white, crystalline compound with a melting point of 75–76°C; used as a postemergence herbicide of wild oats in barley, flax, lentil, mustard, and peas. { 'bär,ban }

barbital |ORG CHEM| $C_8H_{12}N_2O_3$ A compound crystallizing in needlelike form from water; has a faintly bitter taste; melting point 188–192°C; used to make sodium barbital, a long-duration hypnotic and sedative. { 'bär·bə,tol }

barbituric acid [ORG CHEM] $C_4H_4O_3N_2$ 2,4,6-Trioxypyrimidine, the parent compound of the barbiturates; colorless crystals melting at 245°C, slightly soluble in water. { |bär·bə|túr·ik 'as·əd }

Barfoed's test [ANALY CHEM] A test for monosaccharides conducted in an acid solution; cupric acetate is reduced to cuprous oxide, a red precipitate. { 'bär·fúts ,test }

barium [CHEM] A chemical element, symbol Ba, with atomic number 56 and atomic weight of 137.34. { 'bar·ē·əm }

barium acetate [INORG CHEM] $Ba(C_2H_3O_2)_2·H_2O$ A barium salt made by treating barium sulfide or barium carbonate with acetic acids; it forms colorless, triclinic crystals that decompose upon heating; used as a reagent for sulfates and chromates. { 'bar·ē·əm 'as·ə,tāt }

barium azide [INORG CHEM] $Ba(N_3)_2$ A crystalline compound soluble in water; used in high explosives. { 'bar·ē·əm 'ā,zīd }

barium binoxide *See* barium peroxide. { 'bar·ē·əm bī'näk,sīd }

barium bromate [INORG CHEM] $Ba(BrO_3)_2·H_2O$ A poisonous compound that forms colorless, monoclinic crystals, decomposing at 260°C; used for preparing other bromates. { 'bar·ē·əm 'brō,māt }

barium bromide [INORG CHEM] $BaBr_2·2H_2O$ Colorless crystals soluble in water and alcohol; used in photographic compounds. { 'bar·ē·əm 'brō,mīd }

barium carbonate [INORG CHEM] $BaCO_3$ A white powder with a melting point of 174°C; soluble in acids (except sulfuric acid); used in rodenticides, ceramic flux, optical glass, and television picture tubes. { 'bar·ē·əm 'kär·bə·nət }

barium chlorate [INORG CHEM] $Ba(ClO_3)_2·H_2O$ A salt prepared by the reaction of barium chloride and sodium chlorate; it forms colorless, monoclinic crystals, soluble in water; used in pyrotechnics. { 'bar·ē·əm 'klor,āt }

barium chloride [INORG CHEM] $BaCl_2$ A toxic salt obtained as colorless, water-soluble cubic crystals, melting at 963°C; used as a rat poison, in metal surface treatment, and as a laboratory reagent. { 'bar·ē·əm 'klor,īd }

barium chromate [INORG CHEM] $BaCrO_4$ A toxic salt that forms yellow, rhombic crystals, insoluble in water; used as a pigment in overglazes. { 'bar·ē·əm 'krō,māt }

barium citrate [ORG CHEM] $Ba_3(C_6H_5O_7)_2·2H_2O$ A grayish-white, toxic, crystalline powder; used as a stabilizer for latex paints. { 'bar·ē·əm 'sī,trāt }

barium cyanide [ORG CHEM] $Ba(CN)_2$ A white, crystalline powder; soluble in water and alcohol; used in metallurgy and electroplating. { 'bar·ē·əm 'sī·ə,nīd }

barium dioxide *See* barium peroxide. { 'bar·ē·əm dī'äk,sīd }

barium fluoride [INORG CHEM] BaF_2 Colorless, cubic crystals, slightly soluble in water; used in enamels. { 'bar·ē·əm flúr,īd }

barium fluosilicate [INORG CHEM] $BaSiF_6H$ A white, crystalline powder; insoluble in water; used in ceramics and insecticides. Also known as barium silicofluoride. { 'bar·ē·əm ,flü·ə'sil·ə,kāt }

barium hydroxide [INORG CHEM] $Ba(OH)_2·8H_2O$ Colorless, monoclinic crystals, melting at 78°C; soluble in water, insoluble in acetone; used for fat saponification and fusing of silicates. { 'bar·ē·əm hī'dräk,sīd }

barium hyposulfite *See* barium thiosulfate. { 'bar·ē·əm ,hī·pō'səl,fīt }

36

barium manganate [INORG CHEM] BaMnO$_4$ A toxic, emerald-green powder which is used as a paint pigment. Also known as Cassel green; manganese green. { 'bar·ē·əm 'maŋ·gə,nāt }

barium mercury iodide *See* mercuric barium iodide. { 'bar·ē·əm 'mər·kyə·rē 'ī·ə,dīd }

barium molybdate [INORG CHEM] BaMoO$_4$ A toxic, white powder with a melting point of approximately 1600°C; used in electronic and optical equipment and as a paint pigment. { 'bar·ē·əm mə'lib,dāt }

barium monosulfide [INORG CHEM] BaS A colorless, cubic crystal that is soluble in water; used in pigments. { 'bar·ē·əm ,män·ō'səl,fīd }

barium monoxide *See* barium oxide. { 'bar·ē·əm mə'näk ,sīd }

barium nitrate [INORG CHEM] Ba(NO$_3$)$_2$ A toxic salt occurring as colorless, cubic crystals, melting at 592°C, and soluble in water; used as a reagent, in explosives, and in pyrotechnics. Also known as nitrobarite. { 'bar·ē·əm 'nī,trāt }

barium oxide [INORG CHEM] BaO A white to yellow powder that melts at 1923°C; it forms the hydroxide with water; may be used as a dehydrating agent. Also known as barium monoxide; barium protoxide. { 'bar·ē·əm 'äk,sīd }

barium perchlorate [INORG CHEM] Ba(ClO$_4$)$_2$·4H$_2$O Tetrahydrate variety which forms colorless hexagons; used in pyrotechnics. { 'bar·ē·əm pər'klȯr,āt }

barium permanganate [INORG CHEM] Ba(MnO$_4$)$_2$ Brownish-violet, toxic crystals; soluble in water; used as a disinfectant. { 'bar·ē·əm pər'maŋ·gə,nāt }

barium peroxide [INORG CHEM] BaO$_2$ A compound formed as white toxic powder, insoluble in water; used as a bleach and in the glass industry. Also known as barium binoxide; barium dioxide; barium superoxide. { 'bar·ē·əm pər'äk,sīd }

barium protoxide *See* barium oxide. { 'bar·ē·əm prō'täk,sīd }

barium silicide [INORG CHEM] BaSi$_2$ A compound that has the appearance of metal-gray lumps; melts at white heat; used in metallurgy to deoxidize steel. { 'bar·ē·əm 'sil·ə,sīd }

barium stearate [ORG CHEM] Ba(C$_{18}$H$_{35}$O$_2$)$_2$ A white, crystalline solid; melting point 160°C; used as a lubricant in manufacturing plastics and rubbers, in greases, and in plastics as a stabilizer against deterioration caused by heat and light. { 'bar·ē·əm 'stir,āt }

barium sulfate [INORG CHEM] BaSO$_4$ A salt occurring in the form of white, rhombic crystals, insoluble in water; used as a white pigment, as an opaque contrast medium for roentgenographic processes, and as an antidiarrheal. { 'bar·ē·əm 'səl,fāt }

barium sulfite [INORG CHEM] BaSO$_3$ A toxic, white powder; soluble in dilute hydrochloric acid; used in paper manufacturing. { 'bar·ē·əm 'səl,fīt }

barium superoxide *See* barium peroxide. { 'bar·ē·əm ,sü·pər'äk,sīd }

barium tetrasulfide [INORG CHEM] BaS$_4$·H$_2$O Red or yellow, rhombic crystals, soluble in water. { 'bar·ē·əm ,te·trə'səl,fīd }

barium thiocyanate [INORG CHEM] Ba(SCN)·2H$_2$O White crystals that deliquesce; used in dyeing and in photography. { 'bar·ē·əm ,thī·ō'sī·ə,nāt }

barium thiosulfate [INORG CHEM] BaS$_2$O$_3$·H$_2$O A white powder that decomposes upon heating; used to make explosives and in matches. Also known as barium hyposulfite. { 'bar·ē·əm ,thī·ō'səl,fāt }

barium titanate [INORG CHEM] BaTiO$_3$ A grayish powder that is insoluble in water but soluble in concentrated sulfuric acid; used as a ferroelectric ceramic. { 'bar·ē·əm 'tī·tə,nāt }

barium tungstate [INORG CHEM] BaWO$_4$ A toxic, white powder used as a pigment and in x-ray photography. Also known as barium white; barium wolframate; tungstate white; wolfram white. { 'bar·ē·əm 'təŋ,stāt }

barium white *See* barium tungstate. { 'bar·ē·əm 'wīt }

barium wolframate *See* barium tungstate. { 'bar·ē·əm 'wùl·frə,māt }

Barlow's rule [PHYS CHEM] The rule that the volume occupied by the atoms in a given molecule is proportional to the valences of the atoms, using the lowest valency values. { 'bär,lōz ,rül }

Bart reaction [ORG CHEM] Formation of an aryl arsonic acid by treating the aryl diazo

baryta water

compound with trivalent arsenic compounds, such as sodium arsenite. { 'bärt rē'ak·shən }

baryta water [CHEM] A solution of barium hydroxide. { bə'rīd·ə 'wȯd·ər }

base [CHEM] Any chemical species, ionic or molecular, capable of accepting or receiving a proton (hydrogen ion) from another substance; the other substance acts as an acid in giving of the proton. Also known as Brønsted base. { bās }

base-line technique [ANALY CHEM] A method for measurement of absorption peaks for quantitative analysis of chemical compounds in which a base line is drawn tangent to the spectrum background; the distance from the base line to the absorption peak is the absorbence due to the sample under study. { 'bās ‚līn tek'nēk }

base peak [SPECT] The tallest peak in a mass spectrum; it is assigned a relative intensity value of 100, and lesser peaks are reported as a percentage of it. { 'bās ‚pēk }

basic [CHEM] Of a chemical species that has the properties of a base. { 'bā·sik }

basic copper carbonate See copper carbonate. { 'bā·sik 'käp·ər 'kär·bə‚nāt }

basic group [CHEM] A chemical group (for example, OH⁻) which, when freed by ionization in solution, produces a pH greater than 7. { 'bā·sik ‚grüp }

basic oxide [INORG CHEM] A metallic oxide that is a base, or that forms a hydroxide when combined with water, such as sodium oxide to sodium hydroxide. { 'bā·sik 'äk‚sīd }

basic salt [INORG CHEM] A compound that is a base and a salt because it contains elements of both, for example, copper carbonate hydroxide, Cu₂(OH)₂CO₃. { 'bā·sik 'sȯlt }

basic titrant [CHEM] A standard solution of a base used for titration. { 'bā·sik 'tī·trənt }

basis metal See base metal. { 'bā·səs 'med·əl }

bathochromic shift [PHYS CHEM] The shift of the fluorescence of a compound toward the red part of the spectrum due to the presence of a bathochrome radical in the molecule. { ‚bath·ō‚krō'mad·ik 'shift }

battery depolarizer See depolarizer. { 'bad·ə·rē ‚dē'pōl·ə‚rīz·ər }

battery electrolyte [PHYS CHEM] A liquid, paste, or other conducting medium in a battery, in which the flow of electric current takes place by migration of ions. { 'bad·ə·rē i'lek·trə‚līt }

battery manganese See manganese dioxide. { 'bad·ə·rē ‚maŋ·gə‚nēs }

Baumé hydrometer scale [PHYS CHEM] A calibration scale for liquids that is reducible to specific gravity by the following formulas: for liquids heavier than water, specific gravity = 145 ÷ (145 − n) [at 60°F]; for liquids lighter than water, specific gravity = 140 ÷ (130 + n) [at 60°F]; n is the reading on the Baumé scale, in degrees Baumé; Baumé is abbreviated Bé. { bō'mā hī'dräm·əd·ər ‚skāl }

BBC See bromobenzylcyanide.

Be See beryllium.

Bé See Baumé hydrometer scale.

bead test [ANALY CHEM] In mineral identification, a test in which borax is fused to a transparent bead, by heating in a blowpipe flame, in a small loop formed by platinum wire; when suitable minerals are melted in this bead, characteristic glassy colors are produced in an oxidizing or reducing flame and serve to identify elements. { 'bēd ‚test }

beam attenuator [SPECT] An attachment to the spectrophotometer that reduces reference to beam energy to accommodate undersized chemical samples. { 'bēm ə'ten·yə‚wād·ər }

beam-condensing unit [SPECT] An attachment to the spectrophotometer that condenses and remagnifies the beam to provide reduced radiation at the sample. { 'bēm kən'den·siŋ ‚yü·nət }

bebeerine [ORG CHEM] C₃₆H₃₈N₂O₆ An alkaloid derived from the bark of the tropical tree Nectandra rodiaei; the dextro form is soluble in acetone, the levo form is soluble in benzene and is an antipyretic; the dextro form is also known as chondrodendrin; the levo, as curine. { bə'bi‚rēn }

38

Béchamp reduction [ORG CHEM] Reduction of nitro groups to amino groups by the use of ferrous salts or iron and dilute acid. { bā'shän ri'dək·shən }

Beckmann rearrangement [ORG CHEM] An intramolecular change of a ketoxime into its isomeric amide when treated with phosphorus pentachloride. { 'bek·män rē· ə'ränj·mənt }

bed [CHEM] The ion-exchange resin contained in the column in an ion-exchange system. { bed }

Beer-Lambert-Bouguer law *See* Bouguer-Lambert-Beer law. { ¦bā·ər ¦läm·bərt bü'ger ‚lȯ }

Beer's law [PHYS CHEM] The law which states that the absorption of light by a solution changes exponentially with the concentration, all else remaining the same. { 'bā· ərz ‚lȯ }

behenic acid *See* docosanoic acid. { bə'hen·ik 'as·əd }

behenyl alcohol [ORG CHEM] $CH_3(CH_2)_{20}CH_2OH$ A saturated fatty alcohol; colorless, waxy solid with a melting point of 71°C; soluble in ethanol and chloroform; used for synthetic fibers and lubricants. Also known as 1-docosanol. { bə'hen·əl 'al·kə‚hȯl }

bempa [ORG CHEM] $C_6H_{18}N_3PO$ A white solid soluble in water; used as chemosterilant for insects. Also known as hexamethylphosphorictriamide. { 'bem·pə }

Benedict equation of state [PHYS CHEM] An empirical equation relating pressures, temperatures, and volumes for gases and gas mixtures; superseded by the Benedict-Webb-Rubin equation of state. { 'ben·ə‚dikt i'kwā·zhən əv 'stāt }

Benedict's solution [ANALY CHEM] A solution of potassium and sodium tartrates, copper sulfate, and sodium carbonate; used to detect reducing sugars. { 'ben·ə‚diks sə'lü·shən }

benequinox [ORG CHEM] $C_{13}H_{11}N_3O_2$ A yellow-brown powder that decomposes at 195°C; used as a fungicide for grain seeds and seedlings. { ben'ē·kwə‚näks }

benomyl [ORG CHEM] $C_{14}H_{18}N_4O_3$ Methyl-l-butylcarbamoyl-2-benzimidazole carbamate; a fungicide used to control plant disease. { 'ben·ə‚mil }

bensulide [ORG CHEM] $C_{14}H_{24}O_4NPS_3$ An S-(O,O-diisopropyl phosphorodithioate) ester of N-(2-mercaptoethyl)-benzenesulfonamide; an amber liquid slightly soluble in water; melting point is 34.4°C; used as a preemergent herbicide for annual grasses and for broadleaf weeds in lawns and vegetable and cotton crops. { 'ben·sə‚līd }

benthiocarb [ORG CHEM] $C_{12}H_{16}NOCl$ An amber liquid with a boiling point of 126–129°C; slightly soluble in water; used as an herbicide to control aquatic weeds in rice crops. { ben'thī·ō‚kärb }

benzadox [ORG CHEM] $C_6H_5CONHOCH_2COOH$ White crystals with a melting point of 140°C; soluble in water; used as an herbicide to control kochia in sugarbeets. { 'ben·zə‚däks }

benzal chloride [ORG CHEM] $C_6H_5CHCl_2$ A colorless liquid that is refractive and fumes in air; boiling point 207°C; used to make benzaldehyde and cinnamic acid. { ‚benz· əl 'klȯr‚īd }

benzaldehyde [ORG CHEM] C_6H_5CHO A colorless, liquid aldehyde, boiling at 170°C and possessing the odor of bitter almonds; used as a flavoring agent and an intermediate in chemical syntheses. { benz'al·də‚hīd }

benzaldoxime [ORG CHEM] C_6H_5CHNOH An oxime of benzaldehyde; the antiisomeric form melts at 130°C, the syn form at 34°C; both forms are soluble in ethyl alcohol and ether; used in synthesis of other organic compounds. { ‚benz·əl'däk‚sēm }

benzalkonium [ORG CHEM] $C_6H_5CH_2N(CH_3)_2R^+$ An organic radical in which R may range from C_8H_{17} to $C_{18}H_{37}$; found in surfactants, as the chloride salt. { ‚benz·əl'kōn·ē·əm }

benzalkonium chloride [ORG CHEM] $C_6H_5CH_2(CH_3)_2NRCl$ A yellow-white powder soluble in water; used as a fungicide and bactericide; the R is a mixture of alkyls from C_8H_{17} to $C_{18}H_{37}$. { ‚benz·əl'kōn·ē·əm 'klȯr‚īd }

benzamide [ORG CHEM] $C_6H_5CONH_2$ A compound with melting point 132.5° to 133.5°C; slightly soluble in water, soluble in ethyl alcohol and carbon tetrachloride; used in chemical synthesis. { ben'za‚mīd }

benzanilide [ORG CHEM] $C_6H_5CONHC_6H_5$ Leaflet crystals with a melting point of 163°C; soluble in alcohol; used to manufacture dyes and perfumes. { benz'an·ə‚līd }

benzanthracene

benzanthracene |ORG CHEM| $C_{18}H_{14}$ A weakly carcinogenic material that is isomeric with naphthacene; melting point 162°C; insoluble in water, soluble in benzene. { benz'an·thrə,sēn }

benzanthrone |ORG CHEM| $C_{17}H_{10}O$ A compound with melting point 170°C; insoluble in water; used in dye manufacture. { benz'an,thrōn }

benzene |ORG CHEM| C_6H_6 A colorless, liquid, flammable, aromatic hydrocarbon that boils at 80.1°C and freezes at 5.4–5.5°C; used to manufacture styrene and phenol. Also known as benzol. { 'ben,zēn }

benzenediazonium chloride |ORG CHEM| $C_6H_5N(N)Cl$ An ionic salt soluble in water; used as a dye intermediate. { ,ben,zēn,dī·ə'zōn·ē·əm 'klȯr,īd }

benzenephosphorus dichloride |ORG CHEM| $C_6H_5PCl_2$ An irritating, colorless liquid with a boiling point of 224.6°C; soluble in inert organic solvents; used in organic synthesis and oil additives. { 'ben,zēn'fäs·fə·rəs ,dī'klȯr,īd }

benzene ring |ORG CHEM| The six-carbon ring structure found in benzene, C_6H_6, and in organic compounds formed from benzene by replacement of one or more hydrogen atoms by other chemical atoms or radicals. { 'ben,zēn ,riŋ }

benzene series |ORG CHEM| A series of carbon-hydrogen compounds based on the benzene ring, with the general formula C_nH_{2n-6}, where n is 6 or more; examples are benzene, C_6H_6, toluene, C_7H_8, and xylene, C_8H_{10}. { 'ben,zēn ,sir·ēz }

benzenesulfonate |ORG CHEM| Any salt or ester of benzenesulfonic acid. { ¦ben ,zēn'səl·fə,nāt }

benzenesulfonic acid |ORG CHEM| $C_6H_5SO_3H$ An organosulfur compound, strongly acidic, water soluble, nonvolatile, and hygroscopic; used in the manufacture of detergents and phenols. { ¦ben,zēn,səl'fän·ik 'as·əd }

1,2,4-benzenetricarboxylic acid |ORG CHEM| $C_6H_3(COOH)_3$ Crystals with a melting point of 218–220°C; crystallizes from acetic acid or from dilute alcohol; used as an intermediate in the preparation of adhesives, plasticizers, dyes, inks, and resins. { ¦wən ¦tü ¦fȯr ¦ben,zēn·trī¦kär·bäk¦sil·ik 'as·əd }

1,2,4-benzenetriol |ORG CHEM| $C_6H_3(OH)_3$ Monoprismatic leaflets with a melting point of 141°C; freely soluble in water, ether, alcohol, and ethylacetate; used in gas analysis. { ¦wən ¦tü ¦fȯr ¦ben,zēn'trī,ȯl }

benzenoid |ORG CHEM| Any substance which has the electronic character of benzene. { 'ben·zə,nȯid }

benzhydrol |ORG CHEM| $(C_6H_5)_2CHOH$ Colorless needles; melting point 69°C; slightly soluble in water, very soluble in ethanol and ether; used in preparation of other organic compounds including antihistamines. { benz'hī,drȯl }

benzidine |ORG CHEM| $NH_2C_6H_4C_6H_4NH_2$ An aromatic amine with a melting point of 128°C; used as an intermediate in syntheses of direct dyes for cotton. { 'ben·zə,dēn }

benzil |ORG CHEM| $C_6H_5COCOC_6H_5$ A yellow powder; melting point 95°C; insoluble in water, soluble in ethanol, ether, and benzene; used in organic synthesis. { 'ben,zil }

benzilic acid |ORG CHEM| $(C_6H_5)_2C(OH)CO_2H$ A white, crystalline acid, synthesized by heating benzil with alcohol and potassium hydroxide; used in organic synthesis. { ben'zil·ik 'as·əd }

benzimidazole |ORG CHEM| $C_7H_6N_2$ Colorless crystals; melting point 170°C; slightly soluble in water, soluble in ethanol; used in organic synthesis. { ,ben·zə'mid·ə,zōl }

benzoate |ORG CHEM| A salt or ester of benzoic acid, formed by replacing the acidic hydrogen of the carboxyl group with a metal or organic radical. { 'ben·zə,wāt }

benzocaine See ethyl-*para*-aminobenzoate. { 'ben·zə,kān }

benzodihydropyrone |ORG CHEM| $C_9H_8O_2$ A white to light yellow, oily liquid having a sweet odor; soluble in alcohol, chloroform, and ether; used in perfumery. { 'ben·zō·dī,hī·dra'pī,rōn }

benzoic acid |ORG CHEM| C_6H_5COOH An aromatic carboxylic acid that melts at 122.4°C, boils at 250°C, and is slightly soluble in water and relatively soluble in alcohol and ether; derivatives are valuable in industry, commerce, and medicine. { ben'zō·ik 'as·əd }

benzoic anhydride |ORG CHEM| $(C_6H_5CO)_2O$ An acid anhydride that melts at 42°C, boils at 360°C, and crystallizes in colorless prisms; used in synthesis of a variety of organic chemicals, including some dyes. { ben'zō·ik an'hī,drīd }

benzoin [ORG CHEM] $C_{14}H_{12}O_2I$ An optically active compound; white or yellowish crystals, melting point 137°C; soluble in acetone, slightly soluble in water; used in organic synthesis. { 'ben·zə·wən }

α-benzoin oxime [ORG CHEM] $C_6H_5CH(OH)C(NOH)C_6H_5$ Prisms crystallized from benzene; melting point is 151–152°C; soluble in alcohol and in aqueous ammonium hydroxide solution; used in the detection and determination of copper, molybdenum, and tungsten. { 'al·fə 'ben·zə·wən 'äk,sēm }

benzol See benzene. { 'ben,zōl }

benzomate [ORG CHEM] $C_{18}H_{18}O_5N$ A white solid that melts at 71.5–73°C; used as a wettable powder as a miticide. { 'ben·zə,māt }

benzonitrile [ORG CHEM] C_6H_5CN A colorless liquid with an almond odor; made by heating benzoic acid with lead thiocyanate and used in the synthesis of organic chemicals. Also known as phenyl cyanide. { |ben·zō|nī·trəl }

benzophenone [ORG CHEM] $C_6H_5COC_6H_6$ A diphenyl ketone, boiling point 305.9°C, occurring in four polymorphic forms (α, β, γ, and δ) each with different melting point; used as a constituent of synthetic perfumes and as a chemical intermediate. Also known as diphenyl ketone; phenyl ketone. { |ben·zō·fə'nōn }

benzopyrene [ORG CHEM] $C_{20}H_{12}$ A five-ring aromatic hydrocarbon found in coal tar, in cigarette smoke, and as a product of incomplete combustion; yellow crystals with a melting point of 179°C; soluble in benzene, toluene, and xylene. { |ben·zō|pī,rēn }

1,2-benzopyrone See coumarin. { |wən |tü |ben·zō|pī,rōn }

5,6-benzoquinoline [ORG CHEM] $C_{13}H_9N$ Crystals which are soluble in dilute acids, alcohol, ether, or benzene; melting point is 93°C; used as a reagent for the determination of cadmium. { |fīv |siks |ben·zō'kwin·əl,ēn }

benzoquinone See quinone. { |ben·zō,kwə'nōn }

benzoresorcinol [ORG CHEM] $C_{13}H_{10}O_3$ A compound crystallizing as needles from hot-water solution; used in paints and plastics as an ultraviolet light absorber. Also known as resbenzophenone. { |ben·zō·ri'sȯr·sə,nȯl }

benzosulfimide See saccharin. { |ben·zō'səl·fə,mīd }

benzothiazole [ORG CHEM] C_6H_4SCHN A thiazole fused to a benzene ring; can be made by ring closure from o-amino thiophenols and acid chlorides; derivatives are important industrial products. { |ben·zō'thī·ə,zȯl }

4-benzothienyl-N-methylcarbamate [ORG CHEM] $C_{10}H_9NO_2S$ A white powder compound with a melting point of 128°C; used as an insecticide for crop insects. { |fȯr |ben·zō'thī·ə,nil |en|meth·əl'kär·bə,māt }

benzothiofuran See thianaphthene. { |ben,zō|thī,ō'fyü,ran }

1,2,3-benzotriazole [ORG CHEM] $C_6H_5N_3$ A compound with melting point 98.5°C; soluble in ethanol, insoluble in water; derivatives are ultraviolet absorbers; used as a chemical intermediate. { |wən |tü |thrē |ben·zō'trī·ə,zōl }

benzotrichloride [ORG CHEM] $C_6H_5CCl_3$ A colorless to yellow liquid that fumes upon exposure to air; has penetrating odor; insoluble in water, soluble in ethanol and ether; used to make dyes. { |ben·zō,trī'klȯr,īd }

benzotrifluoride [ORG CHEM] Colorless liquid, boiling point 102.1°C; used for dyes and pharmaceuticals, as solvent and vulcanizing agent, in insecticides. { |ben·zō,trī'flúr,īd }

benzoyl [ORG CHEM] The radical $C_6H_5ICO^-$ found, for example, in benzoyl chloride. { 'ben·zə·wəl }

benzoylation [ORG CHEM] Introduction of the aryl radical (C_6H_5CO) into a molecule. { ,ben·zō·ə'lā·shən }

benzoyl chloride [ORG CHEM] C_6H_5COCl Colorless liquid whose vapor induces tears; soluble in ether, decomposes in water; used as an intermediate in chemical synthesis. { 'ben·zə·wəl 'klȯr,īd }

benzoyl chloride 2,4,6-trichlorophenylhydrazone [ORG CHEM] $C_6H_5CCIN_2HC_6H_2Cl_3$ A white to yellow solid with a melting point of 96.5–98°C, insoluble in water; used

41

benzoyl peroxide

as an anthelminthic for citrus.   { 'ben·zə·wəl 'klȯr,īd ¦tü ¦fȯr ¦siks ,trī·klȯr·ə,fen· əl'hī·drə,zōn }

benzoyl peroxide   [ORG CHEM] $(C_6H_5CO)_2O_2$ A white, crystalline solid; melting point 103–105°C; explodes when heated above 105°C; slightly soluble in water, soluble in organic solvents; used as a bleaching and drying agent and a polymerization catalyst.   { 'ben·zə·wəl pə'räk,sīd }

benzoylpropethyl   [ORG CHEM] $C_{18}H_{17}Cl_2NO_3$ An off-white, crystalline compound with a melting point of 72°C; used as a preemergence herbicide for control of wild oats.   { ¦ben·zə·wəl¦prō·pə·thəl }

3,4-benzpyrene   [ORG CHEM] $C_{20}H_{12}$ A polycyclic hydrocarbon; a chemical carcinogen that will cause skin cancer in many species when applied in low dosage.   { ¦thrē ¦fȯr ,benz'pī,rēn }

benzthiazuron   [ORG CHEM] $C_9H_9N_3SO$ A white powder that decomposes at 287°C; slightly soluble in water; used as a preemergent herbicide for sugarbeets and fodder beet crops.   { ,benz,thī'az·yə,rän }

benzyl   [ORG CHEM] The radical $C_6H_5CH_2{}^-$ found, for example, in benzyl alcohol, $C_6H_5CH_2OH$.   { 'ben·zəl }

benzyl acetate   [ORG CHEM] $C_6H_5CH_2OOCCH_3$ A colorless liquid with a flowery odor; used in perfumes and flavorings and as a solvent for plastics and resins, inks, and polishes.   Also known as phenylmethyl acetate.   { 'ben·zəl 'as·ə,tāt }

benzylacetone   [ORG CHEM] $C_6H_5(CH_2)_2COCH_3$ A liquid with a melting point of 233–234°C; used as an attractant to trap melon flies.   { ¦ben·zəl'as·ə,tōn }

benzyl alcohol   [ORG CHEM] $C_6H_5CH_2OH$ An alcohol that melts at 15.3°C, boils at 205.8°C, and is soluble in water and readily soluble in alcohol and ether; valued for the esters it forms with acetic, benzoic, and sebacic acids and used in the soap, perfume, and flavor industries.   Also known as phenylmethanol.   { 'ben·zəl 'al·kə,hȯl }

benzylamine   [ORG CHEM] $C_6H_5CH_2NH_2$ A liquid that is soluble in water, ethanol, and ether; boils at 185°C (770 mmHg) and at 84°C (24 mmHg); it is toxic; used as a chemical intermediate in dye production.   Also known as aminotoluene.   { ¦ben·zəl'am,ēn }

benzyl benzoate   [ORG CHEM] $C_6H_5COOCH_2C_6H_5$ An oily, colorless liquid ester; used as an antispasmodic drug and as a scabicide.   { 'ben·zəl 'ben·zə,wāt }

benzyl bromide   [ORG CHEM] $C_6H_5CH_2Br$ A toxic, irritating, corrosive clear liquid with a boiling point of 198–199°C; acts as a lacrimator; soluble in alcohol, benzene, and ether; used to make foaming and frothing agents.   { 'ben·zəl 'brō,mīd }

benzyl chloride   [ORG CHEM] $C_6H_5CH_2Cl$ A colorless liquid with a pungent odor produced by the chlorination of toluene.   { 'ben·zəl 'klȯr,īd }

benzyl chloroformate   [ORG CHEM] $C_8H_7ClO_2$ An oily liquid with an acrid odor which causes eyes to tear; boiling point is 103°C (20 mmHg pressure); used to block the amino group in peptide synthesis.   { 'ben·zəl ,klȯr·ə'fȯr,māt }

benzyl cinnamate   [ORG CHEM] $C_8H_7COOCH_2C_6H_5$ White crystals; melting point 39°C; insoluble in water, soluble in ethanol; used in perfumery.   { 'ben·zəl 'sin·ə,māt }

benzyl cyanide   [ORG CHEM] $C_6H_5CH_2CN$ A toxic, colorless liquid; insoluble in water, soluble in alcohol and ethanol; boils at 234°C; used in organic synthesis.   { 'ben·zəl 'sī·ə,nīd }

benzyl ether   [ORG CHEM] $(C_6H_5CH_2)_2O$ A liquid unstable at room temperature; boiling point 295–298°C; used in perfumes and as a plasticizer for nitrocellulose.   Also known as dibenzyl ether.   { 'ben·zəl 'ē·thər }

benzyl ethyl ether   [ORG CHEM] $C_6H_5CH_2OC_2H_5$ A colorless, oily, combustible liquid with a boiling point of 185°C; used in organic synthesis and as a flavoring.   { 'ben·zəl 'eth·əl 'ē·thər }

benzyl fluoride   [ORG CHEM] $C_6H_5CH_2F$ A toxic, irritating, colorless liquid with a boiling point of 139.8°C at 753 millimeters of mercury; used in organic synthesis.   { 'ben·zəl 'flu̇r,īd }

benzyl formate   [ORG CHEM] $C_6H_5CH_2OOCH$ A colorless liquid with a fruity-spicy odor and a boiling point of 203°C; used in perfumes and as a flavoring.   { 'ben·zəl 'fȯr,māt }

benzylideneacetone   [ORG CHEM] $C_6H_5CH=CHCOCH_3$ A crystalline compound soluble

in alcohol, benzene, chloroform, and ether; melting point is 41–45°C; used in perfume manufacture and in organic synthesis.   { ben¦zil·ə,dēn'as·ə,tōn }
benzyl isoeugenol   [ORG CHEM] $CH_3CHCHC_6H_3(OCH_3)OCH_2C_6H_5$ A white, crystalline compound with a floral odor; soluble in alcohol and ether; used in perfumery.   { 'ben· zəl ¦ī·sō'yü·jə,nȯl }
benzyl mercaptan   [ORG CHEM] $C_6H_5CH_2SH$ A colorless liquid with a boiling point of 195°C; soluble in alcohol and carbon disulfide; used as an odorant and for flavoring.   { 'ben·zəl mər'kap·tan }
benzyl penicillinic acid   [ORG CHEM] $C_{16}H_{18}N_2O_4S$ An amorphous white powder extracted with ether or chloroform from an acidified aqueous solution of benzyl penicillin.   { 'ben·zəl ¦pen·ə·sə¦lin·ik 'as·əd }
benzyl propionate   [ORG CHEM] $C_2H_5COOCH_2C_6H_5$ A combustible liquid with a sweet odor and a boiling point of 220°C; used in perfumes and for flavoring.   { 'ben·zəl 'prō·pē·ə,nāt }
benzyl salicylate   [ORG CHEM] $C_{14}H_{12}O_3$ A thick liquid with a slight, pleasant odor; used as a fixer in perfumery and in sunburn preparations.   { 'ben·zəl sə'lis·ə,lāt }
benzyne   [ORG CHEM] C_6H_4 A chemical species whose structure consists of an aromatic ring in which four carbon atoms are bonded to hydrogen atoms and two adjacent carbon atoms lack substitutents; a member of a class of compounds known as arynes.   { 'ben,zīn }
berbamine   [ORG CHEM] $C_{37}H_{40}N_2O_6$ An alkaloid; melting point 170°C; slightly soluble in water, soluble in alcohol and ether.   { 'bər·bə,mēn }
berberine   [ORG CHEM] $C_{20}H_{19}NO_5$ A toxic compound; melting point 145°C; the anhydrous form is insoluble in water, soluble in alcohol and ether.   { 'bər·bə,rēn }
Berg's diver method   See diver method.   { 'bərgz 'dīv·ər ,meth·əd }
berkelium   [CHEM] A radioactive element, symbol Bk, atomic number 97, the eighth member of the actinide series; properties resemble those of the rare-earth cerium.   { 'bər·klē·əm }
Berthelot equation   [PHYS CHEM] A form of the equation of state which relates the temperature, pressure, and volume of a gas with the gas constant.   { 'ber·tə·lō i'kwā·zhən }
Berthelot-Thomsen principle   [PHYS CHEM] The principle that of all chemical reactions possible, the one developing the greatest amount of heat will take place, with certain obvious exceptions such as changes of state.   { 'ber·tə·lō ¦täm·sən ,prin·sə·pəl }
berthollide   [CHEM] A compound whose solid phase exhibits a range of composition.   { 'bər·thə,līd }
beryllate   [INORG CHEM] **1.** $BeO_2{}^{2-}$ An ion containing beryllium and oxygen. **2.** A salt produced by the reaction of a strong alkali such as sodium hydroxide with beryllium oxide.   { 'ber·ə,lāt }
beryllia   See beryllium oxide.   { bə'ril·ē·ə }
beryllide   [INORG CHEM] A chemical combination of beryllium with a metal, such as zirconium or tantalum.   { bə'ril·ə,dē }
beryllium   [CHEM] A chemical element, symbol Be, atomic number 4, atomic weight 9.0122.   { bə'ril·ē·əm }
beryllium fluoride   [INORG CHEM] BeF_2 A hygroscopic, amorphous solid with a melting point of 800°C; soluble in water; used in beryllium metallurgy.   { bə'ril·ē·əm ¦flúr,īd }
beryllium nitrate   [INORG CHEM] $Be(NO_3)_2 \cdot 3H_2O$ A compound that forms colorless, deliquescent crystals that are soluble in water; used to introduce beryllium oxide into materials used in incandescent mantles.   { bə'ril·ē·əm 'nī,trāt }
beryllium nitride   [INORG CHEM] Be_3N_2 Refractory, white crystals with a melting point of 2200±40°C; used in the manufacture of radioactive carbon-14 and in experimental rocket fuels.   { bə'ril·ē·əm ¦nī,trīd }
beryllium oxide   [INORG CHEM] BeO An amorphous white powder, insoluble in water; used to make beryllium salts and as a refractory.   Also known as beryllia.   { bə'ril· ē·əm 'äk,sīd }
betaine   [ORG CHEM] $C_5H_{11}O_2N$ An alkaloid; very soluble in water, soluble in ethyl alcohol

and methanol; the hydrochloride is used as a source of hydrogen chloride and in medicine. Also known as lycine; oxyneurine. { 'bēd·ə,ēn }

beta-ray spectrometer [SPECT] An instrument used to determine the energy distribution of beta particles and secondary electrons. Also known as beta spectrometer. { 'bād·ə ,rā spek'träm·əd·ər }

beta spectrometer *See* beta-ray spectrometer. { 'bād·ə spek'träm·əd·ər }

BET equation *See* Brunauer-Emmett-Teller equation. { ¦bē¦ē¦tē i'kwā·zhən }

betula oil *See* methyl salicylate. { 'bech·ə·lə ,ȯil }

betulinic acid [ORG CHEM] $C_{30}H_{48}O_3$ A dibasic acid, slightly soluble in water, ethyl alcohol, and acetone. { ¦bech·ə¦lin·ik 'as·əd }

Bh *See* bohrium.

BHA *See* butylated hydroxyanisole.

BHC *See* 1,2,3,4,5,6-hexachlorocyclohexane.

BHT *See* butylated hydroxytoluene.

Bi *See* bismuth.

biacetyl *See* diacetyl. { ¦bī·ə'sēd·əl }

biamperometry [ANALY CHEM] Amperometric titration that uses two polarizing or indicating electrodes to detect the end point of a redox reaction between the substance being titrated and the titrant. { ¦bī,am·pə'räm·ə·trē }

bias [ANALY CHEM] A systematic error occurring in a chemical measurement that is inherent in the method itself or caused by some artifact in the system, such as a temperature effect. { 'bī·əs }

bibenzyl [ORG CHEM] $C_{14}H_{14}$ A hydrocarbon consisting of two benzene rings attached to ethane. Also known as dibenzyl. { ¦bī'ben·zil }

bicarbonate [INORG CHEM] A salt obtained by the neutralization of one hydrogen in carbonic acid. { bī'kär·bə,nət }

bicarbonate of soda *See* sodium bicarbonate. { bī¦kär·bə·nət əv 'sō·də }

bichloride of mercury *See* mercuric chloride. { bī'klȯr,īd əv 'mər·kyə·rē }

bichromate *See* dichromate. { ,bī'krō,māt }

bicuculine [ORG CHEM] $C_{20}H_{17}NO_6$ A convulsant alkaloid found in plants of the family Fumariaceae. { bī'kü·kyə,lēn }

bicyclic compound [ORG CHEM] A compound having two rings which share a pair of bridgehead carbon atoms. { bī'sik·lik 'käm,paůnd }

bidentate ligand [INORG CHEM] A chelating agent having two groups capable of attachment to a metal ion. { bī'den,tāt 'lig·ənd }

Biebrich red *See* scarlet red. { 'bē,brik 'red }

bifenox [ORG CHEM] $C_{14}H_9Cl_2NO_5$ A tan, crystalline compound with a melting point of 84–86°C; insoluble in water; used as a preemergence herbicide for weed control in soybeans, corn, and sorghum, and as a pre- and postemergence herbicide in rice and small greens. { bī'fen,äks }

bifluoride [INORG CHEM] An acid fluoride whose formula has the form MHF_2; an example is sodium bifluoride, $NaHF_2$. { bī'flůr,īd }

bifunctional catalyst [CHEM] A catalytic substance that possesses two catalytic sites and thus is capable of catalyzing two different types of reactions. Also known as dual-function catalyst. { ¦bī¦fəŋk·shən·əl 'kad·ə,list }

bifunctional chelating agent [ORG CHEM] A reagent with a molecular structure that contains a strong metal-chelating group and a chemically reactive functional group. { ¦bī¦fəŋk·shən·əl 'kē,lād·iŋ ,ā·jənt }

bilateral slit [SPECT] A slit for spectrometers and spectrographs that is bounded by two metal strips which can be moved symmetrically, allowing the distance between them to be adjusted with great precision. { ¦bī,lad·ə·rəl 'slit }

bilayer [CHEM] A layer two molecules thick, such as that formed on the surface of the aqueous phase by phospholipids in aqueous solution. { 'bī,lā·ər }

bimolecular [CHEM] Referring to two molecules. { ¦bī·mə'lek·yə·lər }

bimolecular reaction [CHEM] A chemical transformation or change involving two molecules. { ¦bī·mə'lek·yə·lər rē'ak·shən }

binapacryl [ORG CHEM] $C_{15}H_{18}O_6N_2$ A light tan solid with a melting point of 68–69°C; insoluble in water; used for powdery mildew and for mites on fruits. { bə'nap·ə,kril }

binary compound [CHEM] A compound that has two elements; it may contain two or more atoms; examples are KCl and $AlCl_3$. { 'bīn·ə·rē 'käm,paund }

bioassay [ANALY CHEM] A method for quantitatively determining the concentration of a substance by its effect on the growth of a suitable animal, plant, or microorganism under controlled conditions. { ¦bī·ō'as,ā }

bioautography [ANALY CHEM] A bioassay based upon the ability of some compounds (for example, vitamin B_{12}) to enhance the growth of some organisms or compounds and to repress the growth of others; used to assay certain antibiotics. { ¦bī·ō,ò'täg·rə·fē }

biochemistry [CHEM] The study of chemical substances occurring in living organisms and the reactions and methods for identifying these substances. { ¦bī·ō'kem·ə·strē }

biologic artifact [ORG CHEM] An organic compound with a chemical structure that demonstrates the compound's derivation from living matter. { ¦bī·ə¦läj·ik 'ard·ə,fakt }

biomimetic catalyst [ORG CHEM] A synthetic compound that can simulate the mode of action of a natural enzyme by catalyzing a reaction at ambient conditions. { ¦bī·ō·mə'med·ik 'kad·ə,list }

biosensor [ANALY CHEM] An analytical device that converts the concentration of an analyte in an appropriate sample into an electrical signal by means of a biologically derived sensing element intimately connected to, or integrated into, a transducer. { ¦bī·ō¦sen·sər }

biphenyl [ORG CHEM] $C_{12}H_{10}$ A white or slightly yellow crystalline hydrocarbon, melting point 70.0°C, boiling point 255.9°C, and density 1.9896, which gives plates or monoclinic prismatic crystals; used as a heat-transfer medium and as a raw material for chlorinated diphenyls. Also known as diphenyl; phenylbenzene. { bī'fen·əl }

***para*-biphenylamine** [ORG CHEM] $C_{12}H_{11}N$ Leaflets with a melting point of 53°C; readily soluble in hot water, alcohol, and chloroform; used in the detection of sulfates and also as a carcinogen in cancer research. { ¦par·ə ,bī·fə'nil·ə,mēn }

2,2'-bipyridine *See* 2,2'-dipyridyl. { ¦tü ¦tü·prīm ¦bī'pir·ə,dēn }

biradical [CHEM] A chemical species having two independent odd-electron sites. { bī'rad·ə·kəl }

Birge-Sponer extrapolation [SPECT] A method of calculating the dissociation limit of a diatomic molecule when the convergence limit cannot be observed directly, based on the assumption that vibrational energy levels converge to a limit for a finite value of the vibrational quantum number. { ¦bir·gə 'spōn·ər ik,strap·ə'lā·shən }

bis- [CHEM] A prefix indicating doubled or twice. { bis }

2,2-bis(*para*-chlorophenyl)-1,1-dichloroethane [ORG CHEM] $C_{14}H_{10}Cl_4$ A colorless, crystalline compound with a melting point of 109–111°C; insoluble in water; used as an insecticide on fruits and vegetables. Also known as DDD; TDE. { ¦tü ¦tü ,bis 'par·ə ,klòr·ə'fen·əl ¦wən ¦wən di,klòrō'e,thān }

bismuth [CHEM] A metallic element, symbol Bi, of atomic number 83 and atomic weight 208.980. { 'biz·məth }

bismuthate [INORG CHEM] A compound of bismuth in which the bismuth has a valence of +5; an example is sodium bismuthate, $NaBiO_3$. { 'biz·mə,thāt }

bismuth carbonate *See* bismuth subcarbonate. { 'bis·məth 'kär·bə,nāt }

bismuth chloride [INORG CHEM] $BiCl_3$ A deliquescent material that melts at 230–232°C and decomposes in water to form the oxychloride; used to make bismuth salts. Also known as bismuth trichloride. { 'biz·məth 'klòr,īd }

bismuth chromate [INORG CHEM] $Bi_2O_3 \cdot Cr_2O_3$ An orange-red powder, soluble in alkalies and acids; used as a pigment. { 'biz·məth 'krō,māt }

bismuth citrate [ORG CHEM] $BiC_6H_5O_7$ A salt of citric acid that forms white crystals, insoluble in water; used as an astringent. { 'biz·məth 'sī,trāt }

bismuth hydroxide [INORG CHEM] $Bi(OH)_3$ A water-insoluble, white powder; precipitated by hydroxyl ion from bismuth salt solutions. { 'biz məth hī'dräk,sīd }

bismuth iodide [INORG CHEM] BiI_3 A bismuth halide that sublimes in grayish-black

bismuth nitrate

hexagonal crystals melting at 408°C, insoluble in water; used in analytical chemistry.
{ 'biz·məth 'ī·ə,dīd }

bismuth nitrate [INORG CHEM] $Bi(NO_3)_3 \cdot 5H_2O$ White, triclinic crystals that decompose in water; used as an astringent and antiseptic. { 'biz·məth 'nī,trāt }

bismuth oleate [ORG CHEM] $Bi(C_{17}H_{33}COO)_3$ A salt of oleic acid obtained as yellow granules; used in medicines to treat skin diseases. { 'biz·məth 'ō·lē,āt }

bismuth oxide *See* bismuth trioxide. { 'biz·məth 'äk,sīd }

bismuth oxycarbonate *See* bismuth subcarbonate. { 'biz·məth ,äk·sē'kär·bə,nāt }

bismuth oxychloride [INORG CHEM] BiOCl A white powder; insoluble in water, soluble in acid; a toxic material if ingested; used in pigments and cosmetics. { 'biz·məth ,äk·sē'klȯr,īd }

bismuth phenate [ORG CHEM] $C_6H_5O \cdot Bi(OH)_2$ An odorless, tasteless, gray-white powder; used in medicine. { 'biz·məth 'fen,āt }

bismuth potassium tartrate *See* potassium bismuth tartrate. { 'biz·məth pe'tas·ē·əm 'tär,trāt }

bismuth pyrogallate [ORG CHEM] $Bi(OH)C_6H_3(OH)O_2$ An odorless, tasteless, yellowish-green, amorphous powder; used in medicine as intestinal antiseptic and dusting powder. { 'biz·məth ¦pī·rō'gal,āt }

bismuth subcarbonate [INORG CHEM] $(BiO)_2CO_3$ or $Bi_2O_3 \cdot CO_2 \cdot {}^1/_2H_2O$ A white powder; dissolves in hydrochloric or nitric acid, insoluble in alcohol and water; used as opacifier in x-ray diagnosis, in ceramic glass, and in enamel fluxes. { 'biz·məth səb'kär·bə,nāt }

bismuth subgallate [ORG CHEM] $C_6H_2(OH)_3COOBi(OH)_2$ A yellow powder; dissolves in dilute alkali solutions, but is insoluble in water, ether, and alcohol; used in medicine. { 'biz·məth ,səb'gal,āt }

bismuth subnitrate [INORG CHEM] $4BiNO_3(OH)_2 \cdot BiO(OH)$ A white, hygroscopic powder; used in bismuth salts, perfumes, cosmetics, ceramic enamels, pharmaceuticals, and analytical chemistry. { 'biz·məth ,səb'ni,trāt }

bismuth subsalicylate [INORG CHEM] $Bi(C_7H_5O)_3Bi_2O_3$ A white powder that is insoluble in ethanol and water; used in medicine and as a fungicide for tobacco crops. { 'biz·məth ,səb·sə'lis·ə,lāt }

bismuth telluride [INORG CHEM] Bi_2Te_3 Gray, hexagonal platelets with a melting point of 573°C; used for semiconductors, thermoelectric cooling, and power generation applications. { 'biz·məth 'tel·yə,rīd }

bismuth trichloride *See* bismuth chloride.

bismuth trioxide [INORG CHEM] Bi_2O_3 A yellow powder; melting point 820°C; insoluble in water, dissolves in acid; used to make enamels and to color ceramics. Also known as bismuth oxide; bismuth yellow. { 'biz·məth trī'äk,sīd }

bismuth yellow *See* bismuth trioxide. { 'biz·məth 'yel·ō }

bisphenol A [ORG CHEM] $(CH_3)_2C(C_6H_5OH)_2$ Brown crystals that are insoluble in water; used in the production of phenolic and epoxy resins. { bī'sfēn·ȯl 'ā }

bistable system [CHEM] A chemical system with two relatively stable states which permits an oscillation between domination by one of these states to domination by the other. { ¦bī¦stā·bəl 'sis·təm }

bisulfate [INORG CHEM] A compound that has the HSO_4^- radical; derived from sulfuric acid. { bī'səl,fāt }

bitartrate [ORG CHEM] A salt with the radical $HC_4H_4O_6^-$. Also known as acid tartrate. { bī'tär,trāt }

bithionol [ORG CHEM] A halogenated form of bisphenol used as an ingredient in germicidal soaps and as a medicine in the treatment of clonorchiases. { bī'thī·ə,nȯl }

biuret [ORG CHEM] $NH_2CONHCONH_2$ Colorless needles that are soluble in hot water and decompose at 190°C; a condensation product of urea. { ,bī·ya'ret }

bivalent [CHEM] Possessing a valence of two. { bī'vā·lənt }

bixin [ORG CHEM] $C_{25}H_{30}O_4$ A carotenoid acid occurring in the seeds of *Bixa orellano*; used as a fat and food coloring agent. { 'bik·sən }

Bk *See* berkelium.

black [CHEM] Fine particles of impure carbon that are made by the incomplete burning

46

of carbon compounds, such as natural gas, naphthas, acetylene, bones, ivory, and vegetables. { blak }

black cyanide *See* calcium cyanide. { ¦blak 'sī·ə,nīd }

black iron oxide *See* ferrous oxide. { 'blak 'ī·ərn 'äk,sīd }

Blagden's law [PHYS CHEM] The law that the lowering of a solution's freezing point is proportional to the amount of dissolved substance. { 'blag·dən ,lȯ }

blanc fixe [INORG CHEM] BaSO₄ A commercial name for barium sulfate, with some use in pure form in the paint, paper, and pigment industries as a pigment extender. { ,bläŋk 'fēks }

Blanc rule [ORG CHEM] The rule that glutaric and succinic acids yield cyclic anhydrides on pyrolysis, while adipic and pimelic acids yield cyclic ketones; there are certain exceptions. { 'bläŋk ,rül }

blank [ANALY CHEM] In a chemical analysis, the measured value that is obtained in the absence of a specified component of a sample and that reflects contamination from sources external to the component; it is deducted from the value obtained when the test is performed with the specified component present. Also known as analytical blank. { blaŋk }

blasticidin-S [ORG CHEM] A compound with a melting point of 235–236°C; soluble in water; used as a fungicide for rice crops. { ,blas'tis·ə·dən 'es }

bleaching agent [CHEM] An oxidizing or reducing chemical such as sodium hypochlorite, sulfur dioxide, sodium acid sulfite, or hydrogen peroxide. { 'blēch·iŋ ,ā·jənt }

bleed [CHEM] Diffusion of coloring matter from a substance. { blēd }

blind sample [ANALY CHEM] In chemical analysis, a selected sample whose composition is unknown except to the person submitting it; used to test the validity of the measurement process. { ¦blīnd 'samp·əl }

block copolymer [ORG CHEM] A copolymer in which the like monomer units occur in relatively long alternate sequences on a chain. Also known as block polymer. { 'bläk kō'päl·ə·mər }

blocking [CHEM] Undesired adhesion of granular particles; often occurs with damp powders or plastic pellets in storage bins or during movement through conduits. { 'bläk·iŋ }

blocking group [ORG CHEM] In peptide synthesis, a group that is reacted with a free amino or carboxyl group on an amino acid to prevent its taking part in subsequent formation of peptide bonds. { 'bläk·iŋ ,grüp }

block polymer *See* block copolymer. { 'bläk 'päl·ə·mər }

blowpipe reaction analysis [ANALY CHEM] A method of analysis in which a blowpipe is used to heat and decompose a compound or mineral; a characteristic color appears in the flame or a colored crust appears on charcoal. { 'blō,pīp rē'ak·shən ə'nal·ə·səs }

blue tetrazolium [ORG CHEM] C₄₀H₃₂Cl₂N₈O₂ Lemon yellow crystals that decompose at 242–245°C; soluble in chloroform, ethanol, and methanol; used in seed germination research, as a stain for molds and bacteria, and in histochemical studies. { ¦blü te·trə'zōl·ē·əm }

BNOA *See* β-naphthoxyacetic acid.

boat [CHEM] A platinum or ceramic vessel for holding a substance for analysis by combustion. { bōt }

boat conformation [ORG CHEM] A boat-shaped conformation in space which can be assumed by cyclohexane or similar compounds; a relatively unstable form. { 'bōt ,kän·fər'mā·shən }

Boettger's test [ANALY CHEM] A test for the presence of saccharides, utilizing the reduction of bismuth subnitrate to metallic bismuth, a precipitate. { 'bet,gərz ,test }

bohrium [CHEM] A synthetic chemical element, symbolized Bh, atomic number 107; the fifteenth transuranium element. { 'bȯr·ē·əm }

boiler compound [CHEM] Any chemical used to treat boiler water to prevent corrosion, the fouling of heat-absorbing surfaces, foaming, and the contamination of steam. { 'bȯil·ər ,kam,paúnd }

boiler scale [CHEM] Deposits from silica and other contaminants in boiler water that

boiling

form on the internal surfaces of heat-absorbing components, increase metal temperatures, and result in eventual failure of the pressure parts because of overheating.
Also known as scale. { 'bȯil·ər ‚skāl }

boiling [PHYS CHEM] The transition of a substance from the liquid to the gaseous
phase, taking place at a single temperature in pure substances and over a range of
temperatures in mixtures. { 'bȯil·iŋ }

boiling point [PHYS CHEM] Abbreviated bp. **1.** The temperature at which the transition
from the liquid to the gaseous phase occurs in a pure substance at fixed pressure.
2. See bubble point. { 'bȯil·iŋ ‚pȯint }

boiling-point elevation [CHEM] The raising of the normal boiling point of a pure liquid
compound by the presence of a dissolved substance, the elevation being in direct
relation to the dissolved substance's molecular weight. { 'bȯil·iŋ ‚pȯint el·ə'vā·
shən }

boiling range [CHEM] The temperature range of a laboratory distillation of an oil from
start until evaporation is complete. { 'bȯil·iŋ ‚rānj }

boletic acid See fumaric acid. { bə'led·ik 'as·əd }

bollseye See sodium cacodylate. { 'bōlz‚ī }

bond [CHEM] The strong attractive force that holds together atoms in molecules and
crystalline salts. Also known as chemical bond. { bänd }

bond angle [PHYS CHEM] The angle between bonds sharing a common atom. Also
known as valence angle. { 'bänd ‚aŋ·gəl }

bond dissociation energy [PHYS CHEM] The change in enthalpy that occurs with the
homolytic cleavage of a chemical bond under conditions of standard state. { ‚bänd
di‚sō·sē¦ā·shən 'en·ər·jē }

bond distance [PHYS CHEM] The distance separating the two nuclei of two atoms
bonded to each other in a molecule. Also known as bond length. { 'bänd ‚dis·təns }

bonded-phase chromatography [ANALY CHEM] A type of high-pressure liquid chromatography which employs a stable, chemically bonded stationary phase. { ¦bän·dəd
‚fāz ‚krō·mə'täg·rə·fē }

bond energy [PHYS CHEM] **1.** The average value of specific bond dissociation energies
that have been measured from different molecules of a given type. **2.** See average
bond dissociation energy. { 'bänd ‚en·ər·jē }

bond hybridization [CHEM] The linear combination of two or more simple atomic
orbitals. { ¦bänd ‚hī·brəd·ə'zā·shən }

bonding [CHEM] The joining together of atoms to form molecules or crystalline salts.
{ 'bän·diŋ }

bonding electron [PHYS CHEM] An electron whose orbit spans the entire molecule and
so assists in holding it together. { 'bän·diŋ i'lek‚trän }

bonding orbital [PHYS CHEM] A molecular orbital formed by a bonding electron whose
energy decreases as the nuclei are brought closer together, resulting in a net attraction
and chemical bonding. { 'bän·diŋ 'ȯr·bəd·əl }

bond length See bond distance. { 'bänd ‚leŋkth }

bond-line formula [ORG CHEM] A representation of a molecule in which bonds are
represented by lines, carbon atoms are represented by line ends and intersections,
and atoms other than hydrogen and carbon are represented by their elemental
symbols, as is hydrogen when it is bonded to an atom other than hydrogen or
carbon. Also known as carbon-skeleton formula; line-segment formula. { 'bänd
‚līn ‚fȯr·myə·lə }

bond migration [CHEM] The movement of a bond to a different position within the
same molecular entity. { 'bänd mī‚grā·shən }

bond moment [PHYS CHEM] The degree of polarity of a chemical bond as calculated
from the value of the force of the response of the bond when the bond is subjected
to an electric field. { 'bänd ‚mō·mənt }

bond strength [CHEM] The strength with which a chemical bond holds two atoms
together; conventionally measured in terms of the amount of energy, in kilocalories
per mole, required to break the bond. { 'bänd ‚streŋkth }

bone ash [CHEM] A white ash consisting primarily of tribasic calcium phosphate obtained by burning bones in air; used in cleaning jewelry and in some pottery. { 'bōn ,ash }

boracic acid See boric acid. { bə'ras·ik 'as·əd }

borane |INORG CHEM| **1.** A class of binary compounds of boron and hydrogen; boranes are used as fuels. Also known as boron hydride. **2.** A substance which may be considered a derivative of a boron-hydrogen compound, such as BCl_3 and $B_{10}H_{12}I_2$. { 'bȯ,rän }

borate |CHEM| **1.** A generic term referring to salts or esters of boric acid. **2.** Related to boric oxide, B_2O_3, or commonly to only the salts of orthoboric acid, H_3BO_3. { 'bȯ,rāt }

boration See hydroboration. { bȯ'rā·shən }

borazole |INORG CHEM| $B_3N_3H_6$ A colorless liquid boiling at 53°C; with water it hydrolyzes to form boron hydrides; the borazole molecule is the inorganic analog of the benzene molecule. { 'bȯr·ə,zōl }

borazon |INORG CHEM| A form of boron nitride with a zinc blende structure produced by subjecting the ordinary form to high pressure and temperature. { 'bȯr·ə,zän }

boric acid |INORG CHEM| H_3BO_3 An acid derived from boric oxide in the form of white, triclinic crystals, melting at 185°C, soluble in water. Also known as boracic acid; orthoboric acid. { ¦bȯr·ik 'as·əd }

boric acid ester |ORG CHEM| Any compound readily hydrolyzed to yield boric acid and the respective alcohol; for example, trimethyl borate hydrolyzes to boric acid and methyl alcohol. { ¦bȯr·ik 'as·əd 'es·tər }

boric oxide |INORG CHEM| B_2O_3 A trioxide of boron obtained as rhombic crystals melting at 460°C; used as an intermediate in the production of boron halides and metallic borides and as a thermal neutron absorber in nuclear engineering. Also known as boron oxide. { ¦bȯr·ik 'äk,sīd }

boride |INORG CHEM| A binary compound of boron and a metal formed by heating a mixture of the two elements. { 'bȯr,īd }

borneol |ORG CHEM| $C_{10}H_{17}OH$ White lumps with camphor odor; insoluble in water, soluble in alcohol; melting point 203°C; used in perfumes, medicine, and chemical synthesis. { 'bȯr·nē,ȯl }

Born equation |PHYS CHEM| An equation for determining the free energy of solvation of an ion in terms of the Avogadro number, the ionic valency, the ion's electronic charge, the dielectric constant of the electrolytic, and the ionic radius. { 'bȯrn i'kwā·zhən }

Born-Oppenheimer approximation |PHYS CHEM| The approximation, used in the Born-Oppenheimer method, that the electronic wave functions and energy levels at any instant depend only on the positions of the nuclei at that instant and not on the motions of the nuclei. Also known as adiabatic approximation. { ¦bȯrn 'äp·ən,hī·mər ə,präk·sə,mā·shən }

Born-Oppenheimer method |PHYS CHEM| A method for calculating the force constants between atoms by assuming that the electron motion is so fast compared with the nuclear motions that the electrons follow the motions of the nuclei adiabatically. { ¦bȯrn 'äp·ən,hīm·ər ,meth·əd }

bornyl acetate |ORG CHEM| $C_{10}H_{17}OOCCH_3$ A colorless liquid that forms crystals at 10°C; has characteristic piny-camphoraceous odor; used in perfumes and for flavoring. { 'bȯrn·əl 'as·ə,tāt }

bornyl isovalerate |ORG CHEM| $C_{10}H_{17}OOC_5H_9$ An aromatic fluid with a boiling point of 255–260°C; soluble in alcohol and ether; used in medicine and as a flavoring. { 'bȯrn·əl ¦ī·sō'val·ə,rāt }

boron |CHEM| A chemical element, symbol B, atomic number 5, atomic weight 10.811; it has three valence electrons and is nonmetallic. { 'bȯ,rän }

boron carbide |ORG CHEM| Any compound of boron and carbon, especially B_4C (used as an abrasive, alloying agent, and neutron absorber). { 'bȯ,rän 'kär,bīd }

boron fiber |CHEM| Fiber produced by vapor-deposition methods; used in various composite materials to impart a balance of strength and stiffness. Also known as boron filament. { ¦bȯ,ran ,fī·bər }

boron filament See boron fiber. { 'bȯ,rän ,fil·ə·mənt }

boron fluoride

boron fluoride [INORG CHEM] BF_3 A colorless pungent gas in a dry atmosphere; used in industry as an acidic catalyst for polymerizations, esterifications, and alkylations. Also known as boron trifluoride. { 'bȯ,rän 'flu̇r,īd }

boron hydride *See* borane. { 'bȯ,rän 'hī,drīd }

boron nitride [INORG CHEM] BN A binary compound of boron and nitrogen, especially a white, fluffy powder with high chemical and thermal stability and high electrical resistance. { 'bȯ,rän 'nī,trīd }

boron nitride fiber [INORG CHEM] Inorganic, high-strength fiber, made of boron nitride, that is resistant to chemicals and electricity but susceptible to oxidation above 1600°F (870°C); used in composite structures for yarns, fibers, and woven products. { 'bȯ,rän 'nī,trīd 'fī·bər }

boron oxide *See* boric oxide. { 'bȯ,rän 'äk,sīd }

boron polymer [ORG CHEM] Macromolecules formed by polymerization of compounds containing, for example, boron-nitrogen, boron-phosphorus, or boron-arsenic bonds. { 'bȯ,rän 'päl·ə·mər }

boron trichloride [INORG CHEM] BCl_3 A colorless liquid used as a catalyst and in refining of aluminum, magnesium, zinc, and copper. { 'bȯ,rän trī'klȯr,īd }

boron triethoxide *See* ethyl borate. { 'bȯ,rän ,trī·ə'thäk,sīd } ·

boron triethyl *See* triethylborane. { 'bȯ,rän trī'eth·əl }

boron trifluoride *See* boron fluoride. { 'bȯ,rän trī'flu̇r,īd }

boron trifluoride etherate [ORG CHEM] $C_4H_{10}BF_3O$ A fuming liquid hydrolyzed by air immediately; boiling point is 125.7°C; used as a catalyst in reactions involving condensation, dehydration, polymerization, alkylation, and acetylation. { 'bȯ,rän trī'flu̇r,īd 'ē·thə,rāt }

bottom steam [CHEM] Steam piped into the bottom of the still during oil distillation. { 'bäd·əm ,stēm }

boturon [ORG CHEM] $C_{12}H_{13}N_2OCl$ A white solid with a melting point of 145–146°C; used as pre- and postemergence herbicide in cereals, orchards, and vineyards. Also known as butyron. { 'bäch·ə,rän }

Bouguer-Lambert-Beer law [ANALY CHEM] The intensity of a beam of monochromatic radiation in an absorbing medium decreases exponentially with penetration distance. Also known as Beer-Lambert-Bouguer law; Lambert-Beer law. { bü'ger ¦läm·bert ¦ber ,lȯ }

Bouguer-Lambert law [ANALY CHEM] The law that the change in intensity of light transmitted through an absorbing substance is related exponentially to the thickness of the absorbing medium and a constant which depends on the sample and the wavelength of the light. Also known as Lambert's law. { bü'ger ¦läm·bərt ,lȯ }

boundary line [PHYS CHEM] On a phase diagram, the line along which any two phase areas adjoin in a binary system, or the line along which any two liquidus surfaces intersect in a ternary system. { 'bau̇n·drē ,līn }

boundary value component *See* perfectly mobile component. { 'bau̇n·drē ,val·yü kəm,pō·nənt }

boundary wavelength *See* quantum limit. { ¦bau̇n·drē 'wāv,leŋkth }

bound water [CHEM] Water that is a portion of a system such as tissues or soil and does not form ice crystals until the material's temperature is lowered to about −20°C. { ¦bau̇nd 'wȯd·ər }

Bouvealt-Blanc method [ORG CHEM] A laboratory method for preparing alcohols by reduction of esters utilizing sodium dissolved in alcohol. { ¦bü,vō ¦blän ,meth·əd }

bp *See* boiling point.

Br *See* bromine.

BRA *See* β-resorcylic acid.

Brackett series [SPECT] A series of lines in the infrared spectrum of atomic hydrogen whose wave numbers are given by $R_H|(1/16)| − |(1/n^2)|$, where R_H is the Rydberg constant for hydrogen and n is any integer greater than 4. { 'brak·ət ,sir·ēz }

braking effects [PHYS CHEM] The electrophoretic effect and the asymmetry effect, which together control the speed with which ions drift in a strong electrolyte. { 'brāk·iŋ i,feks }

branch *See* side chain. { branch }

branched chain *See* side chain. { 'brancht 'chān }

branched polymer [ORG CHEM] A polymer chain having branch points that connect three or more chain segments; examples include graft copolymers, star polymers, comb polymers, and dendritic polymers. { ¦brancht 'päl·ə·mər }

bridge [ORG CHEM] A connection between two different parts of a molecule consisting of a valence bond, an atom, or an unbranched chain of atoms. { brij }

bridged intermediate *See* bridged ion. { 'brijd in·tər'mēd·ē·ət }

bridged ion [ORG CHEM] A reactive intermediate in which an atom from one of the reactants is bonded partially to each of two carbon atoms of a reactant containing a double carbon-to-carbon bond. Also known as bridged intermediate; cyclic ion. { 'brijd 'ī·ən }

bridging ligand [ORG CHEM] A ligand in which an atom or molecular species which is able to exist independently is simultaneously bonded to two or more metal atoms. { 'brij·iŋ ¸līg·ənd }

bright-line spectrum [SPECT] An emission spectrum made up of bright lines on a dark background. { 'brīt ¸līn 'spek·trəm }

broadening of spectral lines [SPECT] A widening of spectral lines by collision or pressure broadening, or possibly by Doppler effect. { 'bród·ən·iŋ əv ¦spek·trəl 'līn }

Broenner's acid *See* Brönner's acid. { 'bren·ərz 'as·əd }

bromacetone [ORG CHEM] $CH_2BrCOCH_3$ A colorless liquid which is a powerful irritant and lacrimator; used as tear gas and to make other chemicals. { ¸brōm'as·ə¸tōn }

bromacil [ORG CHEM] 5-Bromo-3-sec-butyl-6-methyluracil, a soil sterilant; general at high dosage and selective at low. { 'brom·ə¸sil }

bromadiolone [ORG CHEM] $C_{30}H_{23}BrO_4$ A rodenticide. { ¸brō·mə'dī·ə¸lōn }

bromate [CHEM] **1.** BrO_3^- A negative ion derived from bromic acid, $HBrO_3$ **2.** A salt of bromic acid. **3.** $C_9H_9ClO_3$ A light brown solid with a melting point of 118–119°C; used as a herbicide to control weeds in crops such as flax, cereals, and legumes. { 'brō¸māt }

bromcresol green *See* bromocresol green. { ¸brōm'krē¸sól 'grēn }

bromcresol purple *See* bromocresol purple. { brōm'krē¸sól 'pər·pəl }

bromethalin [ORG CHEM] $C_{14}H_7Br_3F_3N_3O_4$ A rodenticide. { ¸brō·mə'thal·ən }

bromeosin *See* eosin. { ¸brōm'ē·ə·sən }

bromic acid [INORG CHEM] $HBrO_3$ A liquid, colorless to slightly yellow; boils with decomposition at 100°C; used in dyes and as a chemical intermediate. { 'brō·mik 'as·əd }

bromide [CHEM] A compound derived from hydrobromic acid, HBr, with the bromine atom in the 1-oxidation state. { 'brō¸mīd }

brominating agent [CHEM] A compound capable of introducing bromine into a molecule; examples are phosphorus tribromide, bromine chloride, and aluminum tribromide. { 'brō·mə¸nād·iŋ ¸ā·jənt }

bromination [CHEM] The process of introducing bromine into a molecule. { ¸brō·mə'nā·shən }

bromine [CHEM] A chemical element, symbol Br, atomic number 35, atomic weight 79.904; used to make dibromide ethylene and in organic synthesis and plastics. { 'brō¸mēn }

bromine number [ANALY CHEM] The amount of bromine absorbed by a fatty oil; indicates the purity of the oil and degree of unsaturation. { 'brō¸mēn ¸nəm·bər }

bromine trifluoride [CHEM] BrF_3 A liquid with a boiling point of 135°C. { 'brō¸mēn ¸trī'flúr¸īd }

bromine water [CHEM] An aqueous saturated solution of bromine used as a reagent wherever a dilute solution of bromine is needed. { 'brō¸mēn ¸wód·ər }

bromo- [CHEM] A prefix that indicates the presence of bromine in a molecule. { 'brō·mō }

N-bromoacetamide [ORG CHEM] $CH_3CONHBr$ Needlelike crystals with a melting point of 102–105°C; soluble in warm water and cold ether; used as a brominating agent and in the oxidation of primary and secondary alcohols. { ¦en ¦brō·mō·ə'sed·ə¸mīd }

para-bromoacetanilide [ORG CHEM] C_8H_8BrNO Crystals with a melting point of 168°C; soluble in benzene, chloroform, and ethyl acetate; insoluble in cold water; used as an analgesic and antipyretic. { ¦par·ə ¦brō·mō₁a·səd'an·əl₁īd }

bromoacetone [ORG CHEM] $BrCH_2COCH_3$ A colorless liquid used as a lacrimatory agent. { ¦brō·mō'as·ə₁tōn }

bromo acid *See* eosin. { 'brō·mō 'as·əd }

bromoalkane [ORG CHEM] An aliphatic hydrocarbon with bromine bonded to it. { ¦brō·mō'al₁kān }

para-bromoaniline [ORG CHEM] $BrC_6H_4NH_2$ Rhombic crystals with a melting point of 66–66.5°C; soluble in alcohol and in ether; used in the preparation of azo dyes and dihydroquinazolines. { ¦par·ə ¦brō·mō'an·ə·lēn }

para-bromoanisole [ORG CHEM] C_7H_7BrO Crystals which melt at 9–10°C; used in disinfectants. { ¦par·ə ¦brō·mō'an·ə₁sōl }

bromobenzene [ORG CHEM] C_6H_5Br A heavy, colorless liquid with a pleasant odor; used as a solvent, in motor fuels and top-cylinder compounds, and to make other chemicals. { ¦brō·mō'ben₁zēn }

para-bromobenzyl bromide [ORG CHEM] $BrC_6H_4CH_2Br$ Crystals with an aromatic odor and a melting point of 61°C; soluble in cold and hot alcohol, water, and ether; used to identify aromatic carboxylic acids. { ¦par·ə ₁brō·mo'benz·əl 'brō₁mīd }

bromobenzylcyanide [ORG CHEM] $C_6H_5CHBrCN$ A light yellow oily compound used as a tear gas for training and for riot control. Abbreviated BBC. { ¦brō·mō¦benz·əl'sī·ə₁nīd }

bromochloromethane [ORG CHEM] $BrCH_2Cl$ A clear, colorless liquid with a boiling point of 67°C; volatile, soluble in organic solvents, with a chloroform-like odor; used in fire extinguishers. { ¦brō·mō¦klȯr·ō'me₁thān }

bromochloroprene [ORG CHEM] $CHCl=CHCH_2Br$ A compound used as a nematicide and soil fumigant. { ₁brō·mō'klȯr·ə₁prēn }

bromocresol green [ORG CHEM] Tetrabromo-*m*-cresol sulfonphthalein, a gray powder soluble in water or alcohol; used as an indicator between pH 4.5 (yellow) and 5.5 (blue). Also known as bromcresol green. { ¦brō·mō'krē₁sȯl 'grēn }

bromocresol purple [ORG CHEM] Dibromo-*o*-cresol sulfonphthalein, a yellow powder soluble in water; used as an indicator between pH 5.2 (yellow) and 6.8 (purple). Also known as bromcresol purple. { ¦brō·mō'krē₁sȯl ₁pər·pəl }

bromocriptine [ORG CHEM] $C_{32}H_{40}BrN_5O_5$ A polypeptide alkaloid that is a derivative of the ergotoxin group of ergot alkaloids and is a dopamine receptor agonist. { ₁brō·mō'krip₁tēn }

bromocyclen [ORG CHEM] $C_8H_5BrCl_6$ A compound used as an insecticide for wheat crops. { ₁brō·mō'sī·klən }

bromofenoxim [ORG CHEM] $C_{13}H_7N_3O_6Br_2$ A cream-colored powder with melting point 196–197°C; slightly soluble in water; used as herbicide to control weeds in cereal crops. { ¦brō·mō·fə'näk·səm }

bromoform [ORG CHEM] $CHBr_3$ A colorless liquid, slightly soluble in water; used in the separation of minerals. { ₁brō·mə'fȯrm }

1-bromonaphthalene [ORG CHEM] $C_{10}H_7Br$ An oily liquid that is slightly soluble in water and miscible with chloroform, benzene, ether, and alcohol; used in the determination of index of refraction of crystals and for refractometric fat determination. { ¦wən ¦brō·mō'naf·thə₁lēn }

bromonium ion [ORG CHEM] A halonium ion in which the halogen is bromine; occurs as a bridged structure. { brə'mōn·ē·əm 'ī·ən }

1-bromooctane [ORG CHEM] $CH_3(CH_2)_6CH_2Br$ Colorless liquid that is miscible with ether and alcohol; boiling point is 198–200°C; used in organic synthesis. { ¦wən ¦brō·mō'äk₁tān }

para-bromophenacyl bromide [ORG CHEM] $C_8H_6Br_2O$ Crystals with a melting point of 109–110°C; soluble in warm alcohol; used in the identification of carboxylic acids and as a protecting reagent for acids and phenols. { ¦par·ə ¦brō·mō·fə'nas·əl 'brō₁mīd }

para-bromophenylhydrazine [ORG CHEM] $C_6H_7BrN_2$ Needlelike crystals with a melting point of 108–109°C; soluble in benzene, ether, chloroform, and alcohol; used in the preparation of indoleacetic acid derivatives and in the study of transosazonation of sugar phenylosazones. { ¦par·ə ¦brō·mō·fen·əl'hī·drə,zēn }

bromophos [ORG CHEM] $C_8H_8SPBrCl_2O_3$ A yellow, crystalline compound with a melting point of 54°C; used as an insecticide and miticide for livestock, household insects, flies, and lice. { 'brō·mə,fäs }

bromopicrin [ORG CHEM] CBr_3NO_2 Prismatic crystals with a melting point of 103°C; soluble in alcohol, benzene, and ether; used for military poison gas. Also known as nitrobromoform. { ,brō·mō'pik·rən }

N-bromosuccinimide [ORG CHEM] $C_4H_4BrNO_2$ Orthorhombic bisphenoidal crystals with a melting point of 173–175°C; used in the bromination of olefins. { ¦en ¦brō·mō,sək 'sin·ə,mīd }

bromotrifluoroethylene [ORG CHEM] $BrFC:CF_2$ A colorless gas with a freezing point of −168°C and a boiling point of −58°C; soluble in chloroform; used as a refrigerant, in hardening of metals, and as a low-toxicity fire extinguisher. Abbreviated BFE. { ¦brō·mō·trī¦flúr·ō'eth·ə,lēn }

bromotrifluoromethane [ORG CHEM] $CBrF_3$ Fluorine compound that has a molecular weight of 148.93, melting point −180°C, boiling point −59°C; used as a fire-extinguishing agent. { ¦brō·mō·trī¦flúr·ō'me,thān }

α-bromo-meta-xylene [ORG CHEM] $CH_3C_6H_4CH_2Br$ A liquid that is a powerful lacrimator; soluble in alcohol and ether; used in organic synthesis and chemical warfare. { ¦al·fə ¦brō·mō ¦med·ə 'zī,lēn }

bromoxynil [ORG CHEM] $C_7H_3OBr_2N$ A colorless solid with a melting point of 194–195°C; slightly soluble in water; used as a herbicide in wheat, barley, oats, rye, and seeded turf. { ,brō'mäk·sə·nil }

bromoxynil octanoate [ORG CHEM] $C_{15}H_{17}Br_2NO_2$ A pale brown liquid, insoluble in water; melting point is 45–46°C; used to control broadleaf weeds. { ,brō'mäk·sə· nil ,äk'tan·ə,wāt }

bromthymol blue [ORG CHEM] An acid-base indicator in the pH range 6.0 to 7.6; color change is yellow to blue. { ,brōm'thī,mól 'blü }

Brönner's acid [ORG CHEM] $C_{10}H_6(NH_2)SO_3H$ A colorless, water-soluble naphthylamine sulfonic acid that forms needle crystals; used in dyes. Also spelled Broenner's acid. { 'bren·ərz 'as·əd }

Brønsted acid [CHEM] A chemical species which can act as a source of protons. Also known as proton acid; protonic acid. { 'brən·steth or 'bren,sted 'as·əd }

Brønsted base See base. { 'brən·steth ,bās }

Brønsted-Lowry theory [CHEM] A theory that all acid-base reactions consist simply of the transfer of a proton from one base to another. Also known as Brønsted theory. { ¦brən·steth ¦laú·rē ,thē·ə·rē }

Brønsted theory See Brønsted-Lowry theory. { 'brən·steth ,thē·ə·rē }

brown lead oxide See lead dioxide. { ¦braún ¦led 'äk,sīd }

brown-ring test [ANALY CHEM] A common qualitative test for the nitrate ion; a brown ring forms at the juncture of a dilute ferrous sulfate solution layered on top of concentrated sulfuric acid if the upper layer contains nitrate ion. { 'braún,riŋ ,test }

broxyquinoline [ORG CHEM] $C_9H_5Br_2NO$ Crystals with a melting point of 196°C; soluble in acetic acid, chloroform, benzene, and alcohol; used as a reagent for copper, iron, and other metals. { ,bräk·si'kwin·ə,lēn }

brucine [ORG CHEM] $C_{23}H_{26}N_2O_4$ A poisonous alkaloid from the seeds of plant species such as Nux vomica; used in alcohol as a denaturant. { 'brü,sīn }

Brunauer-Emmett-Teller equation [PHYS CHEM] An extension of the Langmuir isotherm equation in the study of sorption; used for surface area determinations by computing the monolayer area. Abbreviated BET equation. { ¦brü,naúr ¦em·ət ¦tel·ər i'kwā· zhən }

B stage [ORG CHEM] An intermediate stage in a thermosetting resin reaction in which the plastic softens but does not fuse when heated, and swells but does not dissolve in contact with certain liquids. { 'bē ,stāj }

bubble point [PHYS CHEM] In a solution of two or more components, the temperature

53

Bucherer reaction

at which the first bubbles of gas appear. Also known as boiling point. { 'bäb·əl ,pȯint }

Bucherer reaction [ORG CHEM] A method of preparation of polynuclear primary aromatic amines; for example, α-naphthylamine is obtained by heating β-naphthol in an autoclave with a solution of ammonia and ammonium sulfite. { 'bük·ər·ər rē'ak·shən }

buckminsterfullerene [CHEM] C_{60} The most abundant and most stable of the fullerenes, containing 60 carbon atoms in a highly spherical arrangement; named in honor of R. Buckminster Fuller, a practitioner of geodesic dome architecture. Also known as buckyball. { ¦bək,min·stər'fül·ə,rēn }

buckyball See buckminsterfullerene. { 'bək·ē,bȯl }

buffer [CHEM] A solution selected or prepared to minimize changes in hydrogen ion concentration which would otherwise occur as a result of a chemical reaction. Also known as buffer solution. { 'bəf·ər }

buffer capacity [CHEM] The relative ability of a buffer solution to resist pH change upon addition of an acid or a base. { 'bəf·ər kə'pas·əd·ē }

buffer solution See buffer. { 'bəf·ər sə'lü·shən }

bufotenine [ORG CHEM] $C_{12}H_{16}N_2O$ An active pressor agent found in the skin of the common toad; a toxic alkaloid with epinephrinelike biological activity. { ,byü·fə'te,nēn }

bulk sample See gross sample. { ¦bəlk ¦sam·pəl }

bulk sampling [ANALY CHEM] The taking of samples in arbitrary, irregular units rather than discrete units of uniform size for chemical analysis. { ¦bəlk ¦sam·pliŋ }

bullvalene [ORG CHEM] **1.** A compound, molecular formula $C_{10}H_{10}$, that does not have a permanent structure, but has more than 1,200,000 equivalent structures. **2.** A fluxional compound. { 'bül·və,lēn }

bumping [CHEM] Uneven boiling of a liquid caused by irregular rapid escape of large bubbles of highly volatile components as the liquid mixture is heated. { 'bəm·piŋ }

Bunsen-Kirchhoff law [SPECT] The law that every element has a characteristic emission spectrum of bright lines and absorption spectrum of dark lines. { ¦bən·sən 'kir ,kȯf ,lȯ }

buret [CHEM] A graduated glass tube used to deliver variable volumes of liquid; usually equipped with a stopcock to control the liquid flow. { byü'ret }

burning velocity [CHEM] The normal velocity of the region of combustion reaction (reaction zone) relative to nonturbulent unburned gas, in the combustion of a flammable mixture. { 'bər·niŋ və'läs·əd·ē }

burnt lime See calcium oxide. { ¦bərnt 'līm }

Burstein effect [SPECT] The shift of the absorption edge in the spectrum of a semiconductor to higher energies at high carrier densities in the semiconductor. { 'bər,stīn i,fekt }

1,3-butadiene [ORG CHEM] C_4H_6 A colorless gas, boiling point −4.41°C, a major product of the petrochemical industry; used in the manufacture of synthetic rubber, latex paints, and nylon. { ¦wən ¦thrē ,byüd·ə'dī·ēn }

butadiene dimer [ORG CHEM] C_8H_{12} The third ingredient in ethylene-propylene-terpolymer (EPT) synthetic rubbers; isomers include 3-methyl-1,4,6-heptatriene, vinylcyclohexene, and cyclooctadiene. { ,byüd·ə'dī·ēn 'dī·mər }

butadiene rubber See polybutadiene. { ,byüd·ə'dī·ēn 'rəb·ər }

butane [ORG CHEM] C_4H_{10} An alkane of which there are two isomers, n and isobutane; occurs in natural gas and is produced by cracking petroleum. { 'byü,tān }

2,3-butanediol [ORG CHEM] $CH_3CHOHCHOHCH_3$ A major fermentation product of several species of bacteria. { ¦tü ¦thrē ,byüd·ə·nēd·ē,ȯl }

butanol [ORG CHEM] Any one of four isomeric alcohols having the formula C_4H_9OH; colorless, toxic liquids soluble in most organic liquids. Also known as butyl alcohol. { 'byüt·ən,ȯl }

butazolidine See phenylbutazone. { ,byüd·ə'zäl·ə,dēn }

butene-1 [ORG CHEM] $CH_3CH_2CHCH_2$ A colorless, highly flammable gas; insoluble in

54

water, soluble in organic solvents; used to produce polybutenes, butadiene aldehydes, and other organic derivatives. { 'byü,tēn 'wən }

butene-2 [ORG CHEM] $CH_3CHCHCH_3$ A colorless, highly flammable gas, used to make butadiene and in the synthesis of four- and five-carbon organic molecules; the cis form, boiling point 3.7°C, is insoluble in water, soluble in organic solvents, and is also known as high-boiling butene-2; the trans form, boiling point 0.88°C, is insoluble in water, soluble in most organic solvents, and is also known as low-boiling butene-2. { 'byü,tēn 'tü }

butopyronoxyl [ORG CHEM] $C_{12}H_{18}O_4$ A yellow to amber liquid with a boiling point of 256–260°C; miscible with ether, glacial acetic acid, alcohol, and chloroform; used as an insect repellent for skin and clothing. { ,byüd·ə,pī·rə'näk·səl }

2-butoxyethanol [ORG CHEM] $HOCH_2CH_2OC_4H_9$ A liquid with a boiling point of 171–172°C; soluble in most organic solvents and water; used in dry cleaning as a solvent for nitrocellulose, albumin, resins, oil, and grease. { ¦tü ,byü,täk·sē'eth·ə,nól }

butyl [ORG CHEM] Any of the four variations of the hydrocarbon radical C_4H_9: $CH_3CH_2CH_2CH_2-$, $(CH_3)_2CHCH_2-$, $CH_3CH_2CHCH_3-$, and $(CH_3)_3C-$. { 'byüd·əl }

butyl acetate [ORG CHEM] $CH_3COOC_4H_9$ A colorless liquid slightly soluble in water; used as a solvent. { 'byüd·əl 'as·ə,tāt }

butyl acetoacetate [ORG CHEM] $C_8H_{14}O_3$ A colorless liquid with a boiling point of 213.9°C; soluble in alcohol and ether; used for synthesis of dyestuffs and pharmaceuticals. { ¦byüd·əl ¦as·ə·tō'as·ə,tāt }

butyl acrylate [ORG CHEM] $CH_2CHCOOC_4H_9$ A colorless liquid that is nearly insoluble in water and polymerizes readily upon heating; used as an intermediate for organic synthesis, polymers, and copolymers. { ¦byüd·əl 'ak·rə,lāt }

butyl alcohol See butanol. { ¦byüd·əl 'al·kə,hól }

n-butylamine [ORG CHEM] $C_4H_9NH_2$ A colorless, flammable liquid; miscible with water and ethanol; used as an intermediate in organic synthesis and to make insecticides, emulsifying agents, and pharmaceuticals. { ¦en ¦byüd·əl·ə¦mēn }

sec-butylamine [ORG CHEM] $CH_3CHNH_2C_2H_5$ A flammable, colorless liquid; boils in the range 63–68°C; may be used as an intermediate in organic synthesis. { ¦sek ¦byüd·əl·ə¦mēn }

tert-butylamine [ORG CHEM] $(CH_3)_3CNH_2$ A flammable liquid; boiling range 63–68°C; may be used in organic synthesis as an intermediate. { ¦tərt ¦byüd·əl·ə¦mēn }

butylate [ORG CHEM] $C_{11}H_{23}NOS$ A colorless liquid used as an herbicide for preplant control of weeds in corn. { 'byüd·əl,āt }

butylated hydroxyanisole [ORG CHEM] $(CH_3)_3CC_6H_3OH(OCH_3)$ An antioxidant consisting chiefly of a mixture of 2- and 3-tert-butyl-4-hydroxyanisole and used to control rancidity of lard and animal fats in foods. Abbreviated BHA. { 'byüd·əl,ād·əd hī,dräk·sē'an·ə,sól }

butylated hydroxytoluene [ORG CHEM] $[(CH_3)_3C]_2C_6H_2(CH_3)OH$ Crystals with a melting point of 72°C; soluble in toluene, methanol, and ethanol; used as an antioxidant in foods, in petroleum products, and for synthetic rubbers. Abbreviated BHT. { 'byüd·əl,ād·əd hī,dräk·sē'täl·yə,wēn }

butylbenzene [ORG CHEM] $C_6H_5C_4H_9$ A colorless liquid used as a raw material for organic synthesis, especially for insecticides; forms are normal (1-phenylbutane), secondary (2-phenylbutane), and tertiary (2-methyl-2-phenylpropane). { ¦byüd·əl,ben,zēn }

N-sec-butyl-4-tert-butyl-2,6-dinitroaniline [ORG CHEM] $C_{14}H_{21}N_3O_4$ Orange crystals with a melting point of 60–61°C; solubility in water is 1.0 part per million at 24°C; used as a preemergence herbicide. { ¦en ¦sek 'byüd·əl ¦fór ¦tərt 'byüd·əl ¦tü ¦siks ,dī,nī·trō'an·ə,lēn }

butyl carbinol [ORG CHEM] $(CH_3)_3CCH_2OH$ Colorless crystals that melt at 52°C; slightly soluble in water. { 'byüd·əl 'kär·bə,nól }

butyl chloride [ORG CHEM] C_4H_9Cl A colorless liquid used as an alkylating agent in organic synthesis, as a solvent, and as an anthelminthic; forms are normal (1-chlorobutane), secondary, and iso or tertiary. { 'byüd·əl 'klór,īd }

tert-butyl chloroacetate [ORG CHEM] $ClCH_2COOC(CH_3)_3$ A liquid with a boiling point

butyl citrate

of 155°C; hydrolyzes to *tert*-butyl alcohol and chloroacetic acid; used in glycidic ester condensation. { ¦tәrt ¦byüd·әl ‚klȯr·ȯ'as·ә‚tāt }

butyl citrate [ORG CHEM] $C_3H_5O(COOC_4H_9)_3$ A colorless, odorless, nonvolatile liquid, almost insoluble in water; used as a plasticizer, solvent for cellulose nitrate, and antifoam agent. { 'byüd·әl 'sī‚trāt }

butyl diglycol carbonate [ORG CHEM] $(C_4H_9OCO_2CH_2CH_2)_2O$ A colorless, combustible liquid with a boiling range of 164–166°C; used as a plasticizer and solvent and in pharmaceuticals and lubricants manufacture. { 'byüd·әl dī¦glī‚kȯl 'kär·bә·nāt }

butylene [ORG CHEM] Any of three isomeric alkene hydrocarbons with the formula C_4H_8; all are flammable and easily liquefied gases. { 'byüd·ә‚lēn }

1,3-butylene glycol [ORG CHEM] $HOCH_2CH_2CH(OH)CH_3$ A viscous, colorless, hygroscopic liquid; soluble in water and alcohol; used as a solvent, food additive, and flavoring, and for plasticizers and polyurethanes. { ¦wәn ¦thrē 'byüd·ә‚lēn 'glī‚kȯl }

1,4-butylene glycol [ORG CHEM] $HOCH_2CH_2CH_2CH_2OH$ A colorless, combustible, oily liquid with a boiling point of 230°C; soluble in alcohol; used as a solvent and humectant, and in plastics and pharmaceuticals manufacture. { ¦wәn ¦fȯr 'byüd·ә‚lēn 'glī‚kȯl }

1,2-butylene oxide [ORG CHEM] $H_2COCHCH_2CH_3$ A colorless, water-soluble liquid with a boiling point of 63°C; used as an intermediate for various polymers. { ¦wәn ¦tü 'byüd·ә‚lēn 'äk‚sīd }

butyl ether [ORG CHEM] $C_8H_{18}O$ A colorless liquid, boiling at 142°C, and almost insoluble in water; used as an extracting agent, as a medium for Grignard and other reactions, and for purifying other solvents. { ¦byüd·әl 'ē·thәr }

butyl formate [ORG CHEM] $HCOOC_4H_9$ An ester of formic acid and butyl alcohol. { 'byüd·әl 'fȯr‚māt }

***tert*-butylhydroperoxide** [ORG CHEM] $(CH_3)_3COOH$ A liquid soluble in organic solvents; used as a catalyst in polymerization reactions, to introduce the peroxy group into organic molecules. { ¦tәrt ¦byüd·әl‚hī·drō·pә'räk‚sīd }

butyl lactate [ORG CHEM] $CH_3CHOHCOOC_4H_9$ A stable liquid, water-white and nontoxic, miscible with many solvents; used as a solvent for resins and gums, in lacquers and varnishes, and as a chemical intermediate. { 'byüd·әl 'lak‚tāt }

butyl mercaptan [ORG CHEM] C_4H_9SH A colorless, odorous liquid, a component of skunk secretion; used commercially as a gas-odorizing agent. { 'byüd·әl mәr'kap·tan }

butyl oleate [ORG CHEM] $C_{22}H_{42}O_2$ A butyl ester of oleic acid; used as a plasticizer. { 'byüd·әl 'ō·lē‚āt }

***para-tert*-butylphenol** [ORG CHEM] $(CH_3)_3CC_6H_4OH$ Needlelike crystals with a melting point of 98°C; soluble in alcohol and ether; used as an intermediate in production of varnish and lacquer resins, an additive in motor oil, and an ingredient in deemulsifiers in oil fields. { ¦par·ә ¦tәrt ¦byüd·әl'fē‚nȯl }

butyl propionate [ORG CHEM] $C_2H_5COOC_4H_9$ A colorless aromatic liquid; used in fruit essences. { 'byüd·әl 'prō·pē·ә‚nāt }

butyl stearate [ORG CHEM] $C_{17}H_{35}COOC_4H_9$ A liquid that solidifies at approximately 19°C; mixes with vegetable oils and is soluble in alcohol and ethers but insoluble in water; used as a lubricant, in polishes, as a plasticizer, and as a dye solvent. { 'byüd·әl 'stir‚āt }

butynedial [ORG CHEM] $HOCH_2C:CCH_2OH$ White crystals with a melting point of 58°C; soluble in water, aqueous acids, alcohol, and acetone; used as a corrosion inhibitor, defoliant, electroplate brightener, and polymerization accelerator. { ‚byüd·ә'nēd·ē·әl }

butyraldehyde [ORG CHEM] $CH_3(CH_2)_2CHO$ A colorless liquid boiling at 75.7°C; soluble in ether and alcohol, insoluble in water; derived from the oxo process. { ¦byüd·әr'al·dә‚hīd }

butyrate [ORG CHEM] An ester or salt of butyric acid containing the $C_4H_7O_2$ radical. { 'byüd·ә‚rāt }

butyric acid [ORG CHEM] $CH_3CH_2CH_2COOH$ A colorless, combustible liquid with boiling point 163.5°C (757 mmHg); soluble in water, alcohol, and ether; used in synthesis of flavors, in pharmaceuticals, and in emulsifying agents. { byü'tir·ik 'as·әd }

56

butyric anhydride [ORG CHEM] $C_8H_{14}O_3$ A colorless liquid that decomposes in water to form butyric acid; exists in two isomeric forms.   { byü'tir·ik an'hī,drīd }

butyrolactone [ORG CHEM] $C_4H_6O_2$ A liquid, the anhydride of butyric acid; used as a solvent in the manufacture of plastics.   { ¦byüd·ə·rō'lak,tōn }

butyronitrile [ORG CHEM] $CH_3(CH_2)_2CN$ A toxic, colorless liquid with a boiling point of 116–117.7°C; soluble in alcohol and ether; used in industrial, chemical, and pharmaceutical products, and in poultry medicines.   { 'byüd·ə,rän'ī,tril }

C

C *See* carbon.

Ca *See* calcium.

Cabannes' factor [ANALY CHEM] An equational factor to correct for the depolarization effect of the horizontal components of scattered light during the determination of molecular weight by optical methods. { kə'bänz ,fak·tər }

cacodyl [ORG CHEM] $(CH_3)_2As^-$ A radical found in, for example, cacodylic acid, $(CH_3)_2A$-sOOH. { 'kak·ə,dil }

cacodylate [ORG CHEM] Any salt of cacodylic acid. { ,kak·ə'di,lāt }

cacodylic acid [ORG CHEM] $(CH_3)_2AsOOH$ Colorless crystals that melt at 200°C; soluble in alcohol and water; used as a herbicide. { ¦kak·ə¦dil·ik 'as·əd }

cacotheline [ORG CHEM] $C_{20}H_{22}N_2O_5(NO_2)_2$ An azoic compound used as a metal indicator in chelometric titrations. { kə'käth·ə,lēn }

cadalene [ORG CHEM] $C_{15}H_{18}$ A colorless liquid which boils at 291–292°C (720 mmHg; 95,990 pascals) and which is a substituted naphthalene. { 'kad·əl,ēn }

cadinene [ORG CHEM] $C_{15}H_{24}$ A colorless liquid that boils at 274.5°C, and is a terpene derived from cubeb oil, cade oil, juniper berry oil, and other essential oils. { 'kad·ən,ēn }

cadmium [CHEM] A chemical element, symbol Cd, atomic number 48, atomic weight 112.40. { 'kad·mē·əm }

cadmium acetate [ORG CHEM] $Cd(OOCCH_3)_2·3H_2O$ A compound that forms colorless monoclinic crystals, soluble in water and in alcohol; used for chemical testing for sulfides, selenides, and tellurides and for producing iridescent effects on porcelain. { 'kad·mē·əm 'as·ə,tāt }

cadmium bromate [INORG CHEM] $Cd(BrO_3)_2$ Colorless powder, soluble in water; used as an analytical reagent. { 'kad·mē·əm 'brō,māt }

cadmium bromide [INORG CHEM] $CdBr_2$ A compound produced as a yellow crystalline powder, soluble in water and alcohol; used in photography, process engraving, and lithography. { 'kad·mē·əm 'brō,mīd }

cadmium carbonate [INORG CHEM] $CdCO_3$ A white crystalline powder, insoluble in water, soluble in acids and potassium cyanide; used as a starting compound for other cadmium salts. { 'kad·mē·əm 'kär·bə,nāt }

cadmium chlorate [INORG CHEM] $CdClO_3$ White crystals, soluble in water; a highly toxic material. { 'kad·mē·əm 'klȯr,āt }

cadmium chloride [INORG CHEM] $CdCl_2$ A cadmium halide in the form of colorless crystals, soluble in water, methanol, and ethanol; used in photography, in dyeing and calico printing, and as a solution to precipitate sulfides. { 'kad·mē·əm 'klȯr,īd }

cadmium fluoride [INORG CHEM] CdF_2 A crystalline compound with a melting point of 1110°C; soluble in water and acids; used for electronic and optical applications and as a starting material for laser crystals. { 'kad·mē·əm 'flu̇r,īd }

cadmium hydroxide [INORG CHEM] $Cd(OH)_2$ A white powder, soluble in dilute acids; used to prepare negative electrodes for cadmium-nickel storage batteries. { 'kad·mē·əm hī'dräk,sīd }

cadmium iodide [INORG CHEM] CdI_2 A cadmium halide that forms lustrous, white, hexagonal scales, consisting of two water-soluble allotropes, used in photography, in process engraving, and formerly as an antiseptic. { 'kad·mē·əm 'ī·ə,dīd }

cadmium nitrate

cadmium nitrate [CHEM] $Cd(NO_3)_2 \cdot 4H_2O$ White, hygroscopic crystals, soluble in water, alcohol, and liquid ammonia; used to give a reddish-yellow luster to glass and porcelain ware. { 'kad·mē·əm 'nī,trāt }

cadmium oxide [INORG CHEM] CdO In the cubic form, a brown, amorphous powder, insoluble in water, soluble in acids and ammonia salts; used for cadmium plating baths and in the manufacture of paint pigments. { 'kad·mē·əm 'äk,sīd }

cadmium potassium iodide See potassium tetraiodocadmate. { 'kad·mē·əm pə'tas·ē· əm 'ī·ə,dīd }

cadmium sulfate [INORG CHEM] $CdSO_4$ A compound that forms colorless, efflorescent crystals, soluble in water; used as an antiseptic and astringent, in the treatment of syphilis, gonorrhea, and rheumatism, and as a detector of hydrogen sulfide and fumaric acid. { 'kad·mē·əm 'səl,fāt }

cadmium sulfide [INORG CHEM] CdS A compound with two forms: orange, insoluble in water, used as a pigment, and also known as orange cadmium; light yellow, hexagonal crystals, insoluble in water, and also known as cadmium yellow. { 'kad· mē·əm 'səl,fīd }

cadmium telluride [INORG CHEM] CdTe Brownish-black, cubic crystals with a melting point of 1090°C; soluble, with decomposition, in nitric acid; used for semiconductors. { 'kad·mē·əm 'tel·yə,rīd }

cadmium tungstate [INORG CHEM] $CdWO_4$ White or yellow crystals or powder; soluble in ammonium hydroxide and alkali cyanides; used in fluorescent paint, x-ray screens, and scintillation counters. { 'kad·mē·əm 'təŋ,stāt }

caffeic acid [ORG CHEM] $C_9H_8O_4$ A yellow crystalline acid that melts at 223–225°C with decomposition; soluble in water and alcohol. { ka'fē·ik 'as·əd }

caffeine [ORG CHEM] $C_8H_{10}O_2N_4 \cdot H_2O$ An alkaloid found in a large number of plants, such as tea, coffee, cola, and mate. { 'kaf,ēn }

cage [PHYS CHEM] An aggregate of molecules in the condensed phase that surrounds fragments formed by thermal or photochemical dissociation or pairs of molecules in a solution that have collided without reacting. { kāj }

cage effect [PHYS CHEM] A phenomenon involving the dissociation of molecules unable to move apart rapidly because of the presence of other molecules, with the result that the dissociation products may recombine. { 'kāj i,fekt }

cage hydrocarbon [ORG CHEM] A compound composed of only carbon and hydrogen atoms that contains three or more rings arranged topologically so as to enclose a volume of space; in general, the space within a cage hydrocarbon is too small to accommodate even a proton. { ¦kāj ,hī·drə'kär·bən }

Cailletet and Mathias law [PHYS CHEM] The law that describes the relationship between the mean density of a liquid and its saturated vapor at that temperature as being a linear function of the temperature. { kī·ə'tā an mə'thī·əs ,lò }

cajeputol See eucalyptol. { 'kaj·ə·pə,tòl }

calabarine See physostigmine. { kə'lab·ə,rēn }

calcined gypsum See plaster of paris. { 'kal,sīnd 'jip·səm }

calcined soda See soda ash. { 'kal,sīnd 'sō·də }

calcium [CHEM] A chemical element, symbol Ca, atomic number 20, atomic weight 40.08; used in metallurgy as an alloying agent for aluminum-bearing metal, as an aid in removing bismuth from lead, and as a deoxidizer in steel manufacture, and also used as a cathode coating in some types of photo tubes. { 'kal·sē·əm }

calcium acetate [ORG CHEM] $Ca(C_2H_3O_2)_2$ A compound that crystallizes as colorless needles that are soluble in water; formerly used as an important source of acetone and acetic acid; now used as a mordant and as a stabilizer of plastics. { 'kal·sē· əm 'as·ə,tāt }

calcium acrylate [ORG CHEM] $(CH_2CHCOO)_2Ca$ Free-flowing, water-soluble white powder used for soil stabilization, oil-well sealing, and ion exchange and as a binder for clay products and foundry molds. { 'kal·sē·əm 'ak·rə,lāt }

calcium arsenate [INORG CHEM] $Ca_3(AsO_4)_2$ An arsenic compound used as an insecticide to control cotton pests. { 'kal·sē·əm 'ärs·ən,āt }

calcium arsenite [INORG CHEM] $Ca_3(AsO_3)_2$ White granules that are soluble in water; used as an insecticide. { 'kal·sē·əm 'ärs·ən‚īt }

calcium bisulfite [INORG CHEM] $Ca(HSO_3)_2$ A white powder, used as an antiseptic and in the sulfite pulping process. { 'kal·sē·əm bī'səl‚fīt }

calcium bromide [INORG CHEM] $CaBr_2$ A deliquescent salt in the form of colorless hexagonal crystals that are soluble in water and absolute alcohol. { 'kal·sē·əm 'brō‚mīd }

calcium carbide [INORG CHEM] CaC_2 An alkaline earth carbide obtained in the pure form as transparent crystals that decompose in water; used to make acetylene gas. { 'kal·sē·əm 'kär‚bīd }

calcium carbonate [INORG CHEM] $CaCO_3$ White rhombohedrons or a white powder; occurs naturally as calcite; used in paint manufacture, as a dentifrice, as an anticaking medium for table salt, and in manufacture of rubber tires. { 'kal·sē·əm 'kär·bə‚nāt }

calcium chlorate [INORG CHEM] $Ca(ClO_3)_2·2H_2O$ White monoclinic crystals, decomposed by heating. { 'kal·sē·əm 'klȯr‚āt }

calcium chloride [INORG CHEM] $CaCl_2$ A colorless, deliquescent powder that is soluble in water and ethanol; used as an antifreeze and as an antidust agent. { 'kal·sē·əm 'klȯr‚īd }

calcium chromate [INORG CHEM] $CaCrO_4·2H_2O$ Yellow, monoclinic crystals that are slightly soluble in water; used to make other pigments. { 'kal·sē·əm 'krō‚māt }

calcium cyanamide [INORG CHEM] $CaCN_2$ In pure form, colorless rhombohedral crystals, the commercial form being a gray material containing 55–70% $CaCN_2$; used as a fertilizer, weed killer, and defoliant. { 'kal·sē·əm sī'an·ə‚mīd }

calcium cyanide [INORG CHEM] $Ca(CN)_2$ In pure form, a white powder that gives off hydrogen cyanide in air at normal humidity; prepared commercially in impure black or gray flakes; used as an insecticide and rodenticide. Also known as black cyanide. { 'kal·sē·əm 'sī·ə‚nīd }

calcium cyclamate [ORG CHEM] $C_{12}H_{24}O_6N_2S_2Ca_2H_2O$ White crystals with a very sweet taste, soluble in water; has been used as a low-calorie sweetening agent. { 'kal·sē·əm 'sī·klə‚māt }

calcium dihydrogen phosphate See calcium phosphate. { 'kal·sē·əm dī'hī·drə·jən 'fäs‚fāt }

calcium fluoride [INORG CHEM] CaF_2 Colorless, cubic crystals that are slightly soluble in water and soluble in ammonium salt solutions; used in etching glass and preparing hydrofluoric acid. { 'kal·sē·əm 'flùr‚īd }

calcium gluconate [ORG CHEM] $Ca(C_6H_{11}O_7)_2·H_2O$ White powder that loses water at 120°C; soluble in hot water but less soluble in cold water, insoluble in acetic acid and alcohol; used in medicine, as a foaming agent, and as a buffer in foods. { 'kal·sē·əm 'glü·kə‚nāt }

calcium hardness [CHEM] Presence of calcium ions in water, from dissolved carbonates and bicarbonates; treated in boiler water by introducing sodium phosphate. { 'kal·sē·əm ‚härd·nəs }

calcium hydride [INORG CHEM] CaH_2 In pure form, white crystals that are insoluble in water; used in the production of chromium, titanium, and zirconium in the Hydromet process. { 'kal·sē·əm 'hī‚drīd }

calcium hydrogen phosphate See calcium phosphate. { ¦kal·sē·əm ¦hī·drə·jən 'fäs‚fāt }

calcium hydroxide [INORG CHEM] $Ca(OH)_2$ White crystals, slightly soluble in water; used in cement, mortar, and manufacture of calcium salts. Also known as hydrated lime. { 'kal·sē·əm hī'dräk‚sīd }

calcium hypochlorite [INORG CHEM] $Ca(OCl)_2·4H_2O$ A white powder, used as a bleaching agent and disinfectant for swimming pools. { 'kal·sē·əm hī·pō'klȯr‚īt }

calcium iodide [INORG CHEM] CaI_2 A yellow, hygroscopic powder that is very soluble in water; used in photography. { 'kal·sē·əm 'ī·ə‚dīd }

calcium iodobehenate [ORG CHEM] $Ca(OOCC_{21}H_{42}I)_2$ A yellowish powder that is soluble in warm chloroform; used in feed additives. { 'kal·sē·əm ¦ī·ə·dō¦be·ə‚nāt }

calcium lactate [ORG CHEM] $Ca(C_3H_5O_3)_2·5H_2O$ A salt of lactic acid in the form of white

calcium naphthenate

crystals that are soluble in water; used in calcium therapy and as a blood coagulant. { 'kal·sē·əm 'lak,tāt }

calcium naphthenate [ORG CHEM] Calcium derivative of cycloparaffin hydrocarbon (generally cyclopentane or cyclohexane base) that is a light, sticky, water-insoluble mass; used as a hardening agent in plastic compounds, in waterproofing, adhesives, wood fillers, and varnishes. { 'kal·se·əm 'naf·thə,nāt }

calcium nitrate [INORG CHEM] Ca(NO₃)₂·4H₂O Colorless, monoclinic crystals that are soluble in water; the anhydrous salt is very deliquescent; used as a fertilizer and in explosives. Also known as nitrocalcite. { 'kal·se·əm 'nī,trāt }

calcium orthoarsenate [ORG CHEM] Ca₃(AsO₄)₂ A white powder, insoluble in water; used as a preemergence insecticide and herbicide for turf. { 'kal·se·əm ¦ȯr·thō'ärs·ən,āt }

calcium oxalate [INORG CHEM] CaC₂O₄·H₂O A salt of oxalic acid in the form of white crystals that are insoluble in water. { 'kal·se·əm 'äk·sə,lāt }

calcium oxide [INORG CHEM] CaO A caustic white solid sparingly soluble in water; the commercial form is prepared by roasting calcium carbonate limestone in kilns until all the carbon dioxide is driven off; used as a refractory, in pulp and paper manufacture, and as a flux in manufacture of steel. Also known as burnt lime; calx; caustic lime. { 'kal·se·əm 'äk,sīd }

calcium pantothenate [ORG CHEM] (C₉H₁₆NO₅)₂Ca White slightly hygroscopic powder; soluble in water, insoluble in chloroform and ether; melts at 170–172°C; found in either the dextro or levo form or as a racemate; used in nutrition and in animal feed. { 'kal·se·əm pan·tə'the,nāt }

calcium peroxide [INORG CHEM] CaO₂ A cream-colored powder that decomposes in water; used as an antiseptic and a detergent. { 'kal·se·əm pə'räk,sīd }

calcium phosphate [INORG CHEM] **1.** Any phosphate of calcium. **2.** Any of the following three calcium orthophosphates, all of which are white or colorless in pure form: Ca(H₂PO₄)₂ is used as a fertilizer, as a plastics stabilizer, and in baking powder, and is also known as acid calcium phosphate, calcium dihydrogen phosphate, monobasic calcium phosphate, monocalcium phosphate; CaHPO₄ is used in pharmaceuticals, animal feeds, and toothpastes, and is also known as calcium hydrogen phosphate, dibasic calcium phosphate, dicalcium orthophosphate, dicalcium phosphate; Ca₃(PO₄)₂ is used as a fertilizer, and is also known as tribasic calcium phosphate, tricalcium phosphate. { 'kal·se·əm 'fäs,fāt }

calcium plumbate [INORG CHEM] Ca(PbO₃)₂ Orange crystals that are insoluble in cold water but decompose in hot water; used as an oxidizer in the manufacture of glass and matches. { 'kal·se·əm 'pləm,bāt }

calcium plumbite [INORG CHEM] CaPbO₂ Colorless crystals that are slightly soluble in water. { 'kal·se·əm 'pləm,bīt }

calcium pyrophosphate [INORG CHEM] Ca₂P₂O₇ White, abrasive powder, used in dentifrice polishes, in metal polishes, and as a food supplement. { 'kal·se·əm ,pī·rō'fäs,fāt }

calcium resinate [ORG CHEM] Yellowish white, amorphous powder that is soluble in acid, insoluble in water; made by boiling rosin with calcium hydroxide and filtering, or by fusion of melted rosin with hydrated lime; used for waterproofing, leather tanning, and the manufacture of paint driers and enamels. Also known as limed rosin. { 'kal·se·əm 'rez·ən,āt }

calcium reversal lines [SPECT] Narrow calcium emission lines that appear as bright lines in the center of broad calcium absorption bands in the spectra of certain stars. { 'kal·se·əm ri'vər·səl ,līnz }

calcium silicate [INORG CHEM] Any of three silicates of calcium: tricalcium silicate, Ca₃SiO₅; dicalcium silicate, Ca₂SiO₄; calcium metasilicate, CaSiO₃. { 'kal·se·əm 'sil·ə,kāt }

calcium stearate [ORG CHEM] Ca(C₁₈H₃₅O₂)₂ A metallic soap produced as a white powder that is insoluble in water but slightly soluble in petroleum, benzene, and toluene. { 'kal·se·əm 'stir,āt }

calcium sulfate [INORG CHEM] **1.** CaSO₄ A white crystalline salt, insoluble in water; used in Keene's cement, in pigments, as a paper filler, and as a drying agent.

2. Either of two hydrated forms of the salt: the dihydrate, $CaSO_4 \cdot 2H_2O$, and the hemihydrate, $CaSO_4 \cdot 1/2H_2O$.   { 'kal·se·əm 'səl,fāt }

calcium sulfide   [INORG CHEM] CaS In pure form, white cubic crystals, slightly soluble in water; used as a base for luminescent materials.   Also known as hepar calcies; sulfurated lime.   { 'kal·se·əm 'səl,fīd }

calcium sulfite   [INORG CHEM] $CaSO_3 \cdot 2H_2O$ A white powder that is soluble in dilute sulfurous acid; may be dehydrated at 150°C to the anhydrous salt; used in the sulfite process for the manufacture of wood pulp.   { 'kal·se·əm 'səl,fīt }

calcium tungstate   [INORG CHEM] $CaWO_4$ White, tetragonal crystals, slightly soluble in water; used in manufacture of luminous paints.   Also known as artificial scheelite; calcium wolframate.   { 'kal·se·əm 'təŋ,stāt }

calcium wolframate   See calcium tungstate.   { 'kal·se·əm 'wùl·frə,māt }

calculation-based molecular modeling   [PHYS CHEM] The use of computers, together with theoretical chemistry and mathematical expressions, to describe the structure of molecules and predict the most favorable conformation of a molecule or to calculate the energy of interaction between two molecules.   { ,kal·kyə'lā·shən ¦bäst mə'lek·yə·lər 'mäd·əl·iŋ }

calibrant   [ANALY CHEM] In chemical analysis, a substance used to calibrate the response of a measurement system to the analyte.   { 'kal·ə·brənt }

calibration reference   [ANALY CHEM] Any of the standards of various types that indicate whether an analytical instrument or procedure is working within prescribed limits; examples are test solutions used with pH meters, and solutions with known concentrations (standard solutions) used with spectrophotometers.   { 'kal·ə,brā·shən ,ref·rəns }

californium   [CHEM] A chemical element, symbol Cf, atomic number 98; all isotopes are radioactive.   { ,kal·ə'fòr·nē·əm }

calixarene   [ORG CHEM] A cyclic structure containing the group $(-Ar-CH_2-)_n$, where Ar represents an aryl group.   { kə'lik·sə,rēn }

calmagite   [ORG CHEM] $C_{17}H_{14}N_2O_5S$ A compound crystallizing from acetone as red crystals that are soluble in water; used as an indicator in the titration of calcium or magnesium with EDTA.   { 'kal·mə,jīt }

calomel electrode   [PHYS CHEM] A reference electrode of known potential consisting of mercury, mercury chloride (calomel), and potassium chloride solution; used to measure pH and electromotive force.   Also known as calomel half-cell; calomel reference electrode.   { 'kal·ə·məl i'lek,trōd }

calomel half-cell   See calomel electrode.   { 'kal·ə·məl 'haf ,sel }

calomel reference electrode   See calomel electrode.   { 'kal·ə·məl 'ref·rəns i'lek,trōd }

calorimetric titration   See thermometric titration.   { kə¦lòr·ə¦me·trik tī'trā·shən }

calx   See calcium oxide.   { kalks }

camphane   [ORG CHEM] $C_{10}H_{18}$ An alicyclic hydrocarbon; white crystals, soluble in alcohol, with a melting point of 158–159°C.   { 'kam,fān }

camphene   [ORG CHEM] $C_{10}H_{16}$ A bicyclic terpene used as raw material in the synthesis of insecticides such as toxaphene and camphor.   { 'kam,fēn }

camphor   [ORG CHEM] $C_{10}H_{16}O$ A bicyclic saturated terpene ketone that exists in optically active dextro and levo forms and as a racemate; the dextro form is obtained from the wood and bark of the camphor tree, the levo form is found in some essential oils, and the inactive form is obtained from an Asiatic chrysanthemum or made synthetically from certain terpenes.   { 'kam·fər }

d-camphorsulfonic acid   [ORG CHEM] $C_{10}H_{16}O_4S$ A compound crystallizing as prisms from ethyl acetate or glacial acetic acid; slightly soluble in glacial acetic acid and in ethyl acetate; used in the resolution of optically active isomers.   Also known as Reychler's acid.   { ¦dē ¦kam·fər,səl'fän·ik 'as·əd }

cane sugar   [ORG CHEM] Sucrose derived from sugarcane.   { 'kān ,shùg·ər }

cannabidiol   [ORG CHEM] $C_{21}H_{28}(OH)_2$ A constituent of cannabis which, upon isomerization to a tetrahydrocannabinol, has some of the physiologic activity of marijuana.   { ¦kan·ə·bə'dī,òl }

cannabinoid   [ORG CHEM] Any one of the various chemical constituents of cannabis

63

cannabinol

(marijuana), that is, the isomeric tetrahydrocannabinols, cannabinol, and cannabidiol. { kə'nab·ə,nȯid }

cannabinol [ORG CHEM] $C_{21}H_{26}O_2$ A physiologically inactive phenol formed by spontaneous dehydrogenation of tetrahydrocannabinol from cannabis. { 'kan·ə·bə,nȯl }

cannabiscetin *See* myricetin. { ¦kan·ə'bis·ə,tēn }

Cannizzaro reaction [ORG CHEM] The reaction in which aldehydes that do not have a hydrogen attached to the carbon adjacent to the carbonyl group, upon encountering strong alkali, readily form an alcohol and an acid salt. { kän·it'sär·ō rē'ak·shən }

canonical form [ORG CHEM] **1.** A resonance structure for a cyclic compound in which the bonds do not intersect. **2.** *See* contributing structure. { kə'nän·ə·kəl ,fȯrm }

canonical structure *See* contributing structure. { kə'nän·ə·kəl 'strak·chər }

cantharides camphor *See* cantharidin. { kan'thar·ə,dēz 'kam·fər }

cantharidin [ORG CHEM] $C_{10}H_{12}O_4$ Colorless crystals that melt at 218°C; slightly soluble in acetone, chloroform, alcohol, and water; used in veterinary medicine. Also known as cantharides camphor. { kan'thar·ə·dən }

capacity [ANALY CHEM] In chromatography, a measurement used in ion-exchange systems to express the adsorption ability of the ion-exchange materials. { kə'pas·əd·ē }

capillary column [ANALY CHEM] One of the long, narrow (100 meters by 0.2–0.5 millimeter or 330 feet by 0.008–0.02 inch) columns used for capillary gas chromatography. Also known as open tubular column. { 'kap·ə,ler·ē ,käl·əm }

capillary condensation [PHYS CHEM] Condensation of an adsorbed vapor within the pores of the adsorbate. { 'kap·ə,ler·ē ,kän,den'sā·shən }

capillary electrochromatography [ANALY CHEM] A separation technique in which analytes are transported through a small-diameter packed column by electroosmosis (electrically induced flow of the mobile phase) by applying a high potential (5–30 kilovolts) across the column. { ¦kap·ə·ler·ē i,lek·trō,krō·mə'täg·rə·fē }

capillary electrophoresis [ANALY CHEM] A technique for separating substances from a fluid substrate; the sample is placed in a capillary tube which is then subjected to a high-voltage current that separates its chemical constituents. { ¦kap·ə·ler·ē i,lek·trō·fə'rē·səs }

capillary gas chromatography [ANALY CHEM] A highly efficient type of gas chromatography in which the gaseous sample passes through capillary tubes with internal diameters between 0.2 and 0.5 millimeter and lengths up to 100 meters, and adsorption takes place on a medium that is spread on the inner walls of these tubes. { 'kap·ə,ler·ē ¦gas krō·mə'täg·rə·fē }

capillary gel electrophoresis [ANALY CHEM] A form of capillary electrophoresis in which a polyacrylamide gel (or other polymeric material) is placed inside the capillary and separation is based on size and charge; often used to separate oligonucleotides and proteins. { ¦kap·ə·ler·ē ¦jel i,lek·trō·fə'rē·səs }

capillary zone electrophoresis [ANALY CHEM] A type of capillary electrophoresis in which the capillary is filled with a homogenous buffer, and compounds are separated on the basis of their relative charge and size. { ¦kap·ə·ler·ē ¦zōn i,lek·trō,fə'rē·səs }

caprate [ORG CHEM] Any of the salts of capric acid, containing the group $C_9H_{19}COO-$. { 'ka,prāt }

capric acid [ORG CHEM] $CH_3(CH_2)_8COOH$ A fatty acid found in oils and animal fats. { 'ka,prik 'as·əd }

capric anhydride [ORG CHEM] $(CH_3(CH_2)_8CO)_2O$ White crystals that are insoluble in water; used as a chemical intermediate. { 'ka,prik an'hī,drīd }

caproamide [ORG CHEM] $CH_3(CH_2)_4CONH_2$ An amide, melting point 100–101°C; used as a chemical intermediate. { ¦ka·prō'am,īd }

caproic acid [ORG CHEM] $CH_3(CH_2)_4COOH$ A colorless liquid fatty acid found in oils and animal fats; used in synthesizing pharmaceuticals and flavors. { kə'prō·ik 'as·əd }

caproic anhydride [ORG CHEM] $[CH_3(CH_2)_4COO]_2$ White crystals that are insoluble in water, melting point −40.6°C, boiling point 241–243°C. { kə'prō·ik an'hī,drīd }

caprolactam [ORG CHEM] $(CH_2)_5NH·CO$ White flakes, melting point 68–69°C, made from cyclohexanone; used to make synthetic fiber, particularly nylon-6. { ¦ka·prō ¦lak·təm }

64

carbodiimide

ε-caprolactone [ORG CHEM] $CH_2(CH_2)_4NHCO$ White crystals, used to make synthetic fibers, plastics, films, coatings, and plasticizers; its vapors or fine crystals are respiratory irritants. { ¦ā·də ¦ka·prō¦lak,tōn }

caprylamide [ORG CHEM] $CH_3(CH_2)_6CONH_2$ An amine, melting point 105–110°C; decomposes above 200°C; used as a chemical intermediate. { kə'pril·ə,mīd }

capryl compounds [ORG CHEM] A misnomer for octyl compounds; that is, the term octyl halide is preferred for caprylic halides, and octanoic acid for caprylic acid. { 'ka,prəl ,käm,paúnz }

1-caprylene See 1-octene. { ¦wən 'kap·rə,lēn }

caprylic acid [ORG CHEM] $C_8H_{16}O_2$ A liquid fatty acid occurring in butter, coconut oil, and other fats and oils. { kə'pril·ik 'as·əd }

caprylic anhydride [ORG CHEM] $[CH_3(CH_2)_6CO]_2O$ A white solid that melts at −1°C; used as a chemical intermediate. { kə'pril·ik an'hī,drīd }

capsaicin [ORG CHEM] $C_{18}H_{27}O_3N$ A toxic material extracted from capsicum. { kap'sā·ə·sən }

captan [ORG CHEM] $C_9H_8O_2NSCl_3$ A buff to white solid with a melting point of 175°C; used as a fungicide for diseases of fruits, vegetables, and flowers. { 'kap,tan }

carbamate [ORG CHEM] An ester of carbamic acid. { 'kär·bə,māt }

carbamide See urea. { 'kär·bə,mīd }

carbamoyl [ORG CHEM] The radical NH_2CO, formed from carbamic acid. { kär'bam·ə,wil }

carbanilide [ORG CHEM] $(NHC_6H_5)CO(NHC_6H_5)$ Colorless crystals that are very slightly soluble in water, and dissolve in ether and alcohol; used in organic synthesis. { kär·bə'nil,īd }

carbanion [CHEM] One of the charged fragments which arise on heterolytic cleavage of a covalent bond involving carbon; the fragment carries an unshared pair of electrons and bears a negative charge. { ¦kärb'an,ī·ən }

carbaryl [ORG CHEM] $C_{12}H_{11}NO_2$ A colorless, crystalline compound with a melting point of 142°C; used as an insecticide for crops, forests, lawns, poultry, and pets. { 'kär·bə,ril }

carbazide See carbodihydrazide. { 'kär·bə,zīd }

carbazole [ORG CHEM] One of a group of organic heterocyclic compounds containing a dibenzopyrrole system. Also known as 9-azafluorene. { 'kär·bə,zōl }

carbene [ORG CHEM] A compound of carbon which exhibits two valences to a carbon atom; the two valence electrons are distributed in the same valence; an example is CH_2. { 'kär,bēn }

carbenium ion [ORG CHEM] A cation in which the charged atom is carbon; for example, R_2C^+, where R is an organic group. { kär'bē·nē·əm ,ī·ən }

carbenoid species [ORG CHEM] A species that is not a free carbene but has the characteristics of a carbene when participating in a chemical reaction. { 'kär·bə,nóid ,spē·shēz }

carbide [INORG CHEM] A binary compound of carbon with an element more electropositive than carbon; carbon-hydrogen compounds are excluded. { 'kär,bīd }

carbinol [ORG CHEM] **1.** A primary alcohol with general formula RCH_2OH. **2.** The radical CH_2OH of primary alcohols. **3.** An alcohol derived from methanol. { 'kär·bə,nól }

carbinyl See methyl. { 'kär·bə,nil }

carbocation [ORG CHEM] A positively charged ion whose charge resides, at least in part, on a carbon atom or group of carbon atoms. { ¦kär·bō'kat,ī·ən }

carbocyclic compound [ORG CHEM] A compound with a homocyclic ring in which all the ring atoms are carbon, for example, benzene. { ¦kär·bō¦si·klik 'kam,paúnd }

carbodihydrazide [ORG CHEM] $CO(NHNH_2)_2$ Colorless crystals that melt at 154°C; very soluble in alcohol and water; used in photographic chemicals. { ¦kär·bō,dī'hī·drə,zīd }

carbodiimide [ORG CHEM] **1.** HN=C=NH An unstable tautomer of cyanamide. **2.** Any compound with the general formula RN=C=NR which is a formal derivative of carbodiimide. { ¦kär·bō'dī·ə,mīd }

65

carbofuran

carbofuran [ORG CHEM] $C_{12}H_{15}NO_3$ A white solid with a melting point of 150–152°C; soluble in water; used as an insecticide, miticide, and nematicide in many crops. { ‚kär·bō'fyúr‚än }

carbohydrate gum [ORG CHEM] A polysaccharide which produces a gel of a viscous solution when it is dispersed in water at low concentrations; examples are agar, guar gum, xanthan gum, gum arabic, and sodium carboxymethyl cellulose. { ‚kär·bō'hī‚drāt 'gəm }

carbolic acid See phenol. { kär'bäl·ik 'as·əd }

carbon [CHEM] A nonmetallic chemical element, symbol C, atomic number 6, atomic weight 12.01115; occurs freely as diamond, graphite, and coal. { 'kär·bən }

carbonate [CHEM] **1.** An ester or salt of carbonic acid. **2.** A compound containing the carbonate ($CO_3{}^{2-}$) ion. **3.** Containing carbonates. { 'kär·bə·nət }

carbonation [CHEM] Conversion to a carbonate. { ‚kär·bə'nā·shən }

carbon black [CHEM] **1.** An amorphous form of carbon produced commercially by thermal or oxidative decomposition of hydrocarbons and used principally in rubber goods, pigments, and printer's ink. **2.** See gas black. { ¦kär·bən ¦blak }

carbon dioxide [INORG CHEM] CO_2 A colorless, odorless, tasteless gas about 1.5 times as dense as air. { ¦kär·bən dī'äk‚sīd }

carbon dioxide absorption tube [ANALY CHEM] An absorbent-packed tube used to capture the carbon dioxide formed during the microdetermination of carbon-hydrogen by the Pragl combustion procedure. { ¦kär·bən dī'äk‚sīd əb'sörp·shən ‚tüb }

carbon disulfide [ORG CHEM] CS_2 A sulfide, used as a solvent for oils, fats, and rubbers and in paint removers. { ¦kär·bən dī'səl‚fīd }

carbon film [ANALY CHEM] Carbon deposited by evaporation onto a specimen to protect and prepare it for electron microscopy. { ¦kär·bən 'film }

carbon-hydrogen analyzer [ANALY CHEM] A device used in the quantitative analysis of the carbon and hydrogen content of organic compounds. { ¦kär·bən ¦hī·drə·jən 'an·ə‚līz·ər }

carbonic acid [INORG CHEM] H_2CO_3 The acid formed by combination of carbon dioxide and water. { kär'bän·ik 'as·əd }

carbonium ion [ORG CHEM] A carbocation which has a positively charged carbon with a coordination number greater than 3. { kär'bōn·ē·əm ‚ī·ən }

carbonization [CHEM] The conversion of a carbon-containing substance to carbon or a carbon residue as the destructive distillation of coal by heat in the absence of air, yielding a solid residue with a higher percentage of carbon than the original coal; carried on for the production of coke and of fuel gas. { ‚kär·bə·nə'zā·shən }

carbon molecular sieve [CHEM] A molecular sieve that utilizes a special type of activated carbon for the adsorbent. { ¦kär·bən mə¦lek·yə·lər 'siv }

carbon monoxide [INORG CHEM] CO A colorless, odorless gas resulting from the incomplete oxidation of carbon; found, for example, in mines and automobile exhaust; poisonous to animals. { ¦kär·bən mə'näk‚sīd }

carbon nanotubes [CHEM] Cylindrical molecules (sealed at both ends with a convex arrangement of atoms) composed of carbon with a diameter of around 1 nanometer and lengths up to a few micrometers. Single-walled nanotubes may be conducting or semiconducting, depending on the diameter and chirality of the tube. Multiwall nanotubes containing coaxial shells of the elemental single-wall nanotubes are also possible. { ‚kär·bən 'nan·ō‚tübz }

carbon number [ANALY CHEM] The number of carbon atoms in a material under analysis; plotted against chromatographic retention volume for compound identification. { 'kär·bən ‚nəm·bər }

carbon replication [ANALY CHEM] A faithful carbon-film, mold of a specimen surface (for example, powders, bones, or crystals) which is thin enough to be studied by electron microscopy. { 'kär·bən ‚rep·lə'kā·shən }

carbon-skeleton formula See bond-line formula. { ¦kr·bən ¦skel·ə·tən ‚för·myə·lə }

carbon suboxide [INORG CHEM] C_3O_2 A colorless lacrimatory gas having an unpleasant odor with a boiling point of −6.8°C. { 'kär·bən səb'äk‚sīd }

66

carbon tetrachloride [ORG CHEM] CCl_4 Colorless dense liquid, specific gravity 1.595, slightly soluble in water; used as a dry-cleaning agent. { 'kär·bən te·trə'klȯr‚īd }

carbon tetrafluoride [ORG CHEM] CF_4 A colorless gas with a boiling point of $-126°C$; used as a refrigerant. Also known as tetrafluoromethane. { 'kär·bən te·trə'flu̇r‚īd }

carbonyl [ORG CHEM] A functional group found in organic compounds in which a carbon atom is doubly bonded to an oxygen atom ($-CO-$). Also known as carbonyl group. { 'kär·bə‚nil }

carbonylation [CHEM] Introduction of a carbonyl radical into a molecule. { kär‚bän·əl'ā·shən }

carbonyl bromide [ORG CHEM] $COBr_2$ A poisonous liquid boiling at 187.83°C; may be used by the military as a toxic suffocant. { 'kär·bə‚nil 'brō‚mīd }

carbonyl compound [ORG CHEM] A compound containing the carbonyl group (CO). { 'kär·bə‚nil ‚käm‚pau̇nd }

N,N'-carbonyldiimidazole** [ORG CHEM] $C_7H_6N_4O$ Crystals with a melting point of 115.5–116°C; hydrolyzed by water very quickly; used in the synthesis of peptides. { ¦en ¦en‚prīm 'kar·bə‚nil‚dī·i'mid·ə‚zōl }

carbonyl fluoride [ORG CHEM] COF_2 A colorless gas that is soluble in water; used in organic synthesis. { 'kär·bə‚nil 'flu̇r‚īd }

carbonyl group See carbonyl. { 'kär·bə‚nil ‚grüp }

carbophenothion [ORG CHEM] $C_{11}H_{16}ClO_2PS_3$ An amber liquid used to control pests on fruits, nuts, vegetables, and fiber crops. { ¦kär·bō¦fēn·ō'thī‚än }

carborane [ORG CHEM] **1.** Any of a class of compounds containing boron, carbon, and hydrogen. **2.** $B_{10}C_2H_{12}$ A specific member of the class. { 'kär·bə‚rān }

carboxin [ORG CHEM] $C_{12}H_{13}NO_2S$ An off-white solid with a melting point of 91.5–92.5°C; used to treat seeds of barley, oats, wheat, corn, and cotton for fungus diseases. Also known as DCMO. { kär'bäk·sən }

carboxy group [ORG CHEM] $-COOH$ The functional group of carboxylic acid. Also known as carboxyl group. { kär'bäk·sē ‚grüp }

carboxylate anion [ORG CHEM] An anion with the general formula $(RCO_2)^-$, which is formed when the hydrogen attached to the carboxyl group of a carboxylic acid is removed. { kär'bäk·sə‚lāt 'an‚ī·ən }

carboxylation [ORG CHEM] Addition of a carboxyl group into a molecule. { kär‚bäk·sə'lā·shən }

carboxyl group See carboxy group. { kär'bäk·səl ‚grüp }

carboxylic [CHEM] Having chemical properties resembling those of carboxylic acid. { ¦kär‚bäk¦sil·ik }

carboxylic acid [ORG CHEM] Any of a family of organic acids characterized by the presence of one or more carboxyl groups. { ¦kär‚bäk¦sil·ik 'as·əd }

carboxymethylcellulose [ORG CHEM] An acid ether derivative of cellulose used as a sodium salt; a white, odorless, bulky solid used as a stabilizer and emulsifier; negatively charged resin used in ion-exchange chromatography as a cation exchanger. Also known as cellulose gum. { kär‚bäk·sē¦meth·əl 'sel·yə‚lōs }

carbyne [CHEM] Elemental carbon in a triply bonded form. { 'kär‚bīn }

carcerand [ORG CHEM] A macrocyclic compound capable of including organic guest molecules. { 'kär·sə·rənd }

δ-3-carene [ORG CHEM] $C_{10}H_{16}$ A clear, colorless, combustible terpene liquid, stable to about 250°C; used as a solvent and in chemical synthesis. { ¦del·tə ¦thrē 'ka‚rēn }

Carius method [ANALY CHEM] A procedure used to analyze organic compounds for sulfur, halogens, and phosphorus that involves heating the sample with fuming nitric acid in a sealed tube. { 'kär·ē·əs ‚meth·əd }

carminic acid [ORG CHEM] $C_{22}H_{20}O_{13}$ A glucosidal hydroxyanthrapurin that is derived from cochineal; a red crystalline dye used as a stain for biological materials. Also known as cochinilin. { kär'min·ik 'as·əd }

carnaubic acid [ORG CHEM] $C_{24}H_{48}O_2$ An acid found in carnauba wax and beef kidney. { kär'nȯ·bik 'as·əd }

Carnot's reagent [CHEM] A solution of sodium bismuth thiosulfate in alcohol used for determining potassium. { kar'noz rē'ā·jənt }

67

Caro's acid [INORG CHEM] H_2SO_5 A white solid melting at about 45°C, formed during the acid hydrolysis of peroxydisulfates.   { 'kä·roz 'as·əd }
carrageenan [ORG CHEM] A polysaccharide derived from the red seaweed (Rhodophyceae) and used chiefly as an emulsifying, gelling, and stabilizing agent and as a viscosity builder in foods, cosmetics, and pharmaceuticals.   Also spelled carrageenin.   { ‚kar·ə'gē·nən }
carrageenin See carrageenan.   { ‚kar·ə'gē·nən }
carrier [CHEM] A substance that, when associated with a trace of another substance, will carry the trace with it through a chemical or physical process.   { 'kar·ē·ər }
carrier gas [ANALY CHEM] In gas chromatography, a gas used as an eluant for extracting the sample from the column as the gas passes through.   Also known as eluant gas.   { 'kar·ē·ər ‚gas }
carvacrol [ORG CHEM] $(CH_3)_2CHC_6H_3(CH_3)OH$ A colorless liquid, boiling at 237°C; used in perfumes, flavorings, and fungicides.   { 'kär·və‚król }
carvol See carvone.   { 'kär‚vòl }
carvone [ORG CHEM] $C_{10}H_{14}O$ A liquid ketone that boils at 231°C; soluble in water and alcohol; it is optically active and occurs naturally in both dextro and levo forms; used in flavorings and perfumery.   Also known as carvol.   { 'kär‚vōn }
caryophyllene [ORG CHEM] $C_{15}H_{24}$ A liquid sesquiterpene that is found in some essential oils, particularly clove oil.   { ‚kar·ē·ō'fī‚lēn }
caryophyllin [ORG CHEM] $C_{30}H_{48}O_3$ A ketone, soluble in alcohol, extracted from oil of cloves.   { ‚kar·ē·ō'fil·ən }
cascade molecule See dendrimer.   { ka¦skād 'mäl·ə‚kyül }
casein [ORG CHEM] The protein of milk; a white solid soluble in acids.   { 'ka‚sēn }
casein-formaldehyde [ORG CHEM] A modified natural polymer.   { 'ka‚sēn fór'mal·də‚hīd }
Cassel green See barium manganate.   { 'kas·əl ‚grēn }
castor oil acid See ricinoleic acid.   { 'kas·tər ¦òil 'as·əd }
cata-condensed polycyclic [ORG CHEM] An aromatic compound in which no more than two rings have a single carbon atom in common.   { ‚kad·ə·kən'denst ‚päl·i'sī·klik }
catalysis [CHEM] A phenomenon in which a relatively small amount of substance augments the rate of a chemical reaction without itself being consumed.   { kə'tal·ə·səs }
catalyst [CHEM] Substance that alters the velocity of a chemical reaction and may be recovered essentially unaltered in form and amount at the end of the reaction.   { 'kad·əl·əst }
catalyst carrier [CHEM] A neutral material used to support a catalyst, such as activated carbon, diatomaceous earth, or activated alumina.   { 'kad·əl·əst ‚kar·ē·ər }
catalyst selectivity [CHEM] **1.** The relative activity of a catalyst in reference to a particular compound in a mixture.   **2.** The relative rate of a single reactant in competing reactions.   { 'kad·əl·əst sə‚lek'tiv·əd·ē }
cataphoresis See electrophoresis.   { ‚kad·ə·fə'rē·səs }
catechol [ORG CHEM] One of a group of three isomeric dihydroxy benzenes in which the two hydroxyl groups are ortho to each other.   Also known as catechin; pyrocatechol; pyrocatechuic acid.   { 'kad·ə‚kól }
catenane [ORG CHEM] A supramolecular species consisting of mechanically interlocked macrocyclic rings.   { 'kat·ən‚ān }
catenation [CHEM] Formation of a chain structure by the bonding of atoms of the same element, for example, carbon in the hydrocarbons.   { ‚kat·ən'ā·shən }
cathode [PHYS CHEM] The electrode at which reduction takes place in an electrochemical cell, that is, a cell through which electrons are being forced.   { 'kath‚ōd }
cathodic polarization [PHYS CHEM] Portion of electric cell polarization occurring at the cathode.   { kə'thäd·ik ‚pō·lə·rə'zā·shən }
catholyte [CHEM] Electrolyte adjacent to the cathode in an electrolytic cell.   { 'kath·ə‚līt }
cation [CHEM] A positively charged atom or group of atoms, or a radical which moves to the negative pole (cathode) during electrolysis.   { 'kat‚ī·ən }

cation analysis [ANALY CHEM] Qualitative analysis for cations in aqueous solution. { 'kat,ī·ən ə'nal·ə·səs }

cation exchange [CHEM] A chemical reaction in which hydrated cations of a solid are exchanged, equivalent for equivalent, for cations of like charge in solution. { 'kat,ī· ən iks'chānj }

cation exchange resin [ORG CHEM] A highly polymerized synthetic organic compound consisting of a large, nondiffusible anion and a simple, diffusible cation, which later can be exchanged for a cation in the medium in which the resin is placed. { 'kat,ī· ən iks'chānj 'rez·ən }

cationic detergent [CHEM] A member of a group of detergents that have molecules containing a quaternary ammonium salt cation with a group of 12 to 24 carbon atoms attached to the nitrogen atom in the cation; an example is alkyltrimethyl ammonium bromide. { ,kad·ē'än·ik di'tər·jənt }

cationic hetero atom [CHEM] A positively charged atom, other than carbon, in an otherwise carbon atomic chain or ring. { ,kad·ē'än·ik 'hed·ə·rō 'ad·əm }

cationic polymerization [ORG CHEM] A type of polymerization in which Lewis acids act as catalysts. { ,kad·ē'än·ik pə,lim·ə·rə'zā·shən }

cationic reagent [CHEM] A surface-active agent with active positive ions used for ore beneficiation (flotation via flocculation); an example of a cationic reagent is cetyl trimethyl ammonium bromide. { ,kad·ē'än·ik rē'ā·jənt }

cationtrophy [CHEM] The breaking off of an ion, such as a hydrogen ion or metal ion, from a molecule so that a negative ion remains in equilibrium. { ,kad·ē'än·trə·fē }

caustic [CHEM] **1.** Burning or corrosive. **2.** A hydroxide of a light metal. { 'kȯ·stik }

caustic alcohol See sodium ethylate. { 'kȯ·stik 'al·kə,hȯl }

causticity [CHEM] The property of being caustic. { kȯ'stis·əd·ē }

caustic lime See calcium oxide. { 'kȯ·stik'līm }

caustic potash See potassium hydroxide. { 'kȯ·stik 'päd,ash }

caustic soda See sodium hydroxide. { 'kȯ·stik 'sōd·ə }

caustic wash [CHEM] **1.** Treating a product with a solution of caustic soda to remove impurities. **2.** The solution itself. { 'kȯ·stik 'wäsh }

cavitation [CHEM] Emulsification produced by disruption of a liquid into a liquid-gas two-phase system, when the hydrodynamic pressure of the liquid is reduced to the vapor pressure. { ,kav·ə'tā·shən }

cavity ringdown laser absorption spectroscopy [SPECT] A direct absorption technique used for measuring short-lived species and for trace-gas analysis in which the rate of decay of light, injected with a pulsed laser and trapped in a cavity formed by two highly reflective mirrors, is measured, allowing the calculation of the amount of light absorbed by the sample. Abbreviated CRLAS. { ¦kav·əd·ē ¦riŋ,daún ¦lā·zər əb¦sȯrp· shən ,spek'träs·kə·pē }

Cd See cadmium.

Ce See cerium.

ceiling temperature [ORG CHEM] For addition (chain) polymerization, the temperature at which the propagation and depropagation rates are equal, that is, the net rate of polymer formation is zero. Above the ceiling temperture, depolymerization, an unzipping reaction to reform monomer, occurs. { 'sēl·iŋ ,tem·prə·chər }

cell [PHYS CHEM] A cup, jar, or vessel containing electrolyte solutions and metal electrodes to produce an electric current (conductiometric or potentiometric) or for electrolysis (electrolytic). { sel }

cell constant [PHYS CHEM] The ratio of distance between conductance-titration electrodes to the area of the electrodes, measured from the determined resistance of a solution of known specific conductance. { 'sel ,kän·stənt }

cellobiose [ORG CHEM] $C_{12}H_{22}O_{11}$ A disaccharide which does not occur freely in nature or as a glucoside; a unit of cellulose and lichenin; crystallizes as minute water-soluble crystals from alcohol. Also known as cellose. { ,sel·ō'bī,ōs }

cellose See cellobiose. { 'se,lōs }

cellosolve [ORG CHEM] $C_2H_5OCH_2CH_2OH$ An important industrial chemical used in

69

varnish removers, in cleaning solutions, and as a solvent for paints, varnishes, and plastics. Also known as 2-ethoxyethanol. { 'sel·ə‚sälv }

α-cellulose *See* alpha cellulose. { ¦al·fə 'sel·yə‚lōs }

cellulose acetate [ORG CHEM] An acetic acid ester of cellulose; a tough, flexible, slow-burning, and long-lasting thermoplastic material used as the base for magnetic tape and movie film, in acetate rayon, as a plastic film in food packaging, in lacquers, and for molded receiver cabinets. { 'sel·yə‚lōs 'as·ə‚tāt }

cellulose acetate butyrate [ORG CHEM] An ester of cellulose formed by the action of a mixture of acetic acid and butyric acid and their anhydrides on purified cellulose; has high impact resistance, clarity, and weatherability; used in making plastic film, lacquer, lenses, and outdoor signs. { 'sel·yə‚lōs 'as·ə‚tāt 'byüd·ə‚rāt }

cellulose diacetate [ORG CHEM] The ester formed by esterification of two hydroxyl groups of a cellulose molecule with acetic acid. { 'sel·yə‚lōs dī'as·ə‚tāt }

cellulose ester [ORG CHEM] Cellulose in which the free hydroxyl groups have been replaced wholly or in part by acidic groups. { 'sel·yə‚lōs 'es·tər }

cellulose ether [ORG CHEM] The product of the partial or complete etherification of the hydroxyl groups in a cellulose molecule. { 'sel·yə‚lōs 'ē·thər }

cellulose fiber [ORG CHEM] Any fiber based on esters or ethers of cellulose. { 'sel·yə‚lōs 'fī·bər }

cellulose gum *See* carboxymethyl cellulose. { 'sel·yə‚lōs 'gəm }

cellulose methyl ether *See* methylcellulose. { 'sel·yə‚lōs 'meth·əl ‚ē·thər }

cellulose nitrate [ORG CHEM] Any of several esters of nitric acid, produced by treating cotton or some other form of cellulose with a mixture of nitric and sulfuric acids; used as explosive and propellant. Also known as nitrocellulose; nitrocotton. { 'sel·yə‚lōs 'nī‚trāt }

cellulose propionate [ORG CHEM] An ester of cellulose and propionic acid. { 'sel·yə‚lōs 'prō·pē·ə‚nāt }

cellulose triacetate [ORG CHEM] A cellulose resin formed by the complete esterification of the cellulose by acetic acid; used as a base in protective coatings. { 'sel·yə‚lōs trī'as·ə‚tāt }

cellulose xanthate [ORG CHEM] A compound formed by reaction of soda cellulose (prepared by treating cellulose with strong sodium hydroxide solution) with carbon disulfide. { 'sel·yə‚lōs 'zan‚thāt }

cellulosic [ORG CHEM] Any of the derivatives of cellulose, such as cellulose acetate. { ‚sel·yə'lō·sik }

cellulosic resin [ORG CHEM] Any resin based on cellulose compounds such as esters and ethers. { 'sel·yə‚lōs 'rez·ən }

cementation [CHEM] The setting of a plastic material. { ‚sē‚men'tā·shən }

centrifugation potentials [PHYS CHEM] Electric potential differences between points at different distances from the axis of rotation of a colloidal solution that is being rapidly rotated in a centrifuge. { sen‚trif·ə'gā·shən pə‚ten·chəlz }

centrifuge tube [ANALY CHEM] Calibrated, tube-shaped glass container used with laboratory centrifuges for volumetric analysis of separable (solid-liquid or immiscible liquid) samples. { 'sen·trə‚fyüj ‚tüb }

CEPHA *See* ethephon. { 'sef·ə }

cephaeline [ORG CHEM] $C_{14}H_{19}O_2N$ An alkaloid, slightly soluble in water, extracted from the root of ipecac; used as an emetic. { sə'fā·ə‚lēn }

ceramide [ORG CHEM] Any of a group of amides formed by linking a fatty acid to sphingosine. { 'ser·ə‚mīd }

cerate [ORG CHEM] A metallic salt or soap made from lard. { 'sir‚āt }

ceria *See* ceric oxide. { 'ser·ē·ə }

ceric oxide [INORG CHEM] CeO_2 A pale-yellow to white powder; soluble in sulfuric acid, insoluble in dilute acid and water; used in ceramics and as a polish for optical glass. Also known as ceria; cerium dioxide; cerium oxide. { 'sir·ik 'äk‚sīd }

ceric sulfate [INORG CHEM] $Ce(SO_4)_2 \cdot 4H_2O$ Yellow needles forming a basic salt with excess water; used in waterproofing, mildew-proofing, and in dyeing and printing textiles. { 'sir·ik 'səl‚fāt }

cerinic acid *See* cerotic acid. { sə'rēn·ik 'as·əd }
cerium [CHEM] A chemical element, symbol Ce, atomic number 58, atomic weight
140.12; a rare-earth metal, used as a getter in the metal industry, as an opacifier
and polisher in the glass industry, in Welsbach gas mantles, in cored carbon arcs,
and as a liquid-liquid extraction agent to remove fission products from spent uranium
fuel. { 'sir·ē·əm }
cerium dioxide *See* ceric oxide. { 'sir·ē·əm dī'äk,sīd }
cerium fluoride [INORG CHEM] CeF₃ White hexagonal crystals, melting point 1460°C;
used in arc carbons to increase the brilliance of carbon-arc lamps. { 'sir·ē·əm
'flúr,īd }
cerium oxide *See* ceric oxide. { 'sir·ē·əm 'äk,sīd }
cerium stearate [ORG CHEM] Ce(C₁₈H₃₅O₂)₂ White, waxy, inert powder, melting point
100–110°C; used in waterproofing compounds. { 'sir·ē·əm 'stir,āt }
cerotic acid [ORG CHEM] CH₃(CH₂)₂₄COOH A fatty acid derived from carnauba wax or
beeswax; melts at 87.7°C. Also known as cerinic acid; hexacosanoic acid. { sə'räd·
ik 'as·əd }
certified reference material [ANALY CHEM] A reference material, one or more of whose
property values are certified by a technically valid procedure, for which a certificate
or other documentation has been issued by an appropriate certifying agency. { 'sərd·
ə,fīd 'ref·rəns ¦mə'tir·ē·əl }
ceryl alcohol [ORG CHEM] C₂₆H₅₃OH An alcohol derived from Chinese wax, melting at
79°C and insoluble in water. { 'sir·əl 'al·kə,hòl }
cesium [CHEM] A chemical element, symbol Cs, atomic number 55, atomic weight
132.905. { 'sē·zē·əm }
cesium bromide [INORG CHEM] CsBr A colorless, crystalline powder with a melting
point of 636°C; soluble in water; used in medicine, for infrared spectroscopy, and
in scintillation counters. { 'sē·zē·əm 'brō,mīd }
cesium carbonate [INORG CHEM] Cs₂CO₃ A white, hygroscopic, crystalline powder; soluble in water; used in specialty glasses. { 'sē·zē·əm 'kär·bə,nāt }
cesium chloride [INORG CHEM] CsCl Colorless cuboid crystals, melting point 646°C;
used in filaments of radio tubes to increase sensitivity, in photoelectric cells, and
for photosensitive deposit on cathodes. { 'sē·zē·əm 'klòr,īd }
cesium fluoride [INORG CHEM] CsF Toxic, irritating, deliquescent crystals with a melting
point of 682°C; soluble in water and methanol; used in medicine, mineral water, and
brewing. { 'sē·zē·əm 'flúr,īd }
cesium hydroxide [INORG CHEM] CsOH Colorless or yellow, fused crystalline mass with
a melting point of 272.3°C; soluble in water; used as electrolyte in alkaline storage
batteries at subzero temperatures. { 'sē·zē·əm ¸hī'dräk,sīd }
cesium iodide [INORG CHEM] CsI A colorless, deliquescent, crystalline powder with a
melting point of 621°C; soluble in water and alcohol; crystals used for infrared
spectroscopy. { 'sē·zē·əm 'ī·ə,dīd }
cesium perchlorate [INORG CHEM] CsClO₄ A crystalline solid with a melting point of
250°C; soluble in water; used in optics and for specialty glasses. { 'sē·zē·əm
pər'klòr,āt }
cesium sulfate [INORG CHEM] Cs₂SO₄ Colorless crystals with a melting point of 1010°C;
soluble in water; used for brewing and in mineral waters. { 'sē·zē·əm 'səl,fāt }
cetane *See* n-hexadecane. { 'sē,tān }
cetane-number improver [CHEM] A chemical which has the effect of increasing a diesel
fuel's cetane number; examples are nitrates, nitroalkanes, nitrocarbonates, and perox-
ides. { 'sē,tān ¸nəm·bər im'prüv·ər }
cetin [ORG CHEM] C₁₅H₃₁COOC₁₆H₃₃ A white, crystalline, waxy substance with a melting
point of 50°C; soluble in alcohol and ether; used as a base for ointments and
emulsions and in the manufacture of soaps and candles. { 'sēt·ən }
cetrimonium bromide [ORG CHEM] CH₃(CH₂)₁₅N(CH₃)₃Br Crystals with a melting point
of 237–243°C; soluble in alcohol, water, and sparingly in acetone; used as a cationic
detergent, antiseptic, and precipitant for nucleic acids and mucopolysaccharides.
{ ¸se·trə'mōn·ē·əm 'brō,mīd }

71

cetyl

cetyl |ORG CHEM| The radical represented as $C_{16}H_{33}-$. { 'sēd·əl }
cetyl alcohol |ORG CHEM| $C_{15}H_{33}OH$ A colorless wax, insoluble in water although a solution in kerosine forms an insoluble film on water. { 'sēd·əl 'al·kə,hȯl }
cetyl vinyl ether |ORG CHEM| $C_{16}H_{33}OCO:CH_2$ A colorless liquid with a boiling point of 142°C; may be copolymerized with unsaturated monomers to make internally plasticized resins. { 'sēd·əl 'vīn·əl 'ē·thər }
Cf See californium.
CFC See chlorofluorocarbon.
chain |CHEM| A structure in which similar atoms are linked by bonds. { chān }
chain balance |ANALY CHEM| An analytical balance with one end of a fine gold chain suspended from the beam and the other fastened to a device which moves over a graduated vernier scale. { 'chān ,bal·əns }
chain isomerism |ORG CHEM| A type of molecular isomerism seen in carbon compounds; as the number of carbon atoms in the molecule increases, the linkage between the atoms may be a straight chain or branched chains producing isomers that differ from each other by possessing different carbon skeletons. { 'chān ,ī'säm·ə,riz·əm }
chain reaction |CHEM| A chemical reaction in which many molecules undergo chemical reaction after one molecule becomes activated. { ¦chān rē'ak·shən }
chain scission |ORG CHEM| The cleavage of polymer chains, as in natural rubber as a result of heating. { 'chān ,sizh·ən }
chain transfer |ORG CHEM| The abstraction of an atom from another molecule (initiator, monomer, polymer, or solvent) by the radical end of a growing (addition) polymer, which simultaneously terminates the polymer chain and creates a new radical capable of chain polymerization; also occurs in cationic polymerization. { 'chān ,tranz·fər }
chair conformation See chair form. { 'cher ,kän·fər,mā·shən }
chair form |PHYS CHEM| A particular nonplanar conformation of a cyclic molecule with more than five atoms in the ring; for example, in the chair form of cyclohexane, the hydrogens are staggered and directed perpendicularly to the mean plane of the carbons (axial conformation, a) or equatorially to the center of the mean plane (equatorial conformation, e). Also known as chair conformation. { 'cher ,fȯrm }
chalcogen |INORG CHEM| Any of the elements that form group 16 of the periodic table; included are oxygen, sulfur, selenium, tellurium, and polonium. { 'kal·kə·jən }
chalcogenide |INORG CHEM| A binary compound containing a chalcogen and a more electropositive element or radical. { 'kal·kə·jə,nīd }
chalking |CHEM| 1. Treating with chalk. 2. Forming a powder which is easily rubbed off. { 'chȯk·iŋ }
chamber acid |INORG CHEM| Sulfuric acid made by the obsolete chamber process. { 'chām·bər 'as·əd }
chance cause |ANALY CHEM| A cause for variability in a measurement process that occurs randomly and unpredictably and for unknown reasons. { ¦chans 'kȯz }
channel black See gas black. { 'chan·əl ,blak }
channeling |ANALY CHEM| In chromatography, furrows or breaks in an ion-exchange bed which permit a solution to run through without having contact with active groups elsewhere in the bed. { 'chan·əl·iŋ }
characteristic loss spectroscopy |SPECT| A branch of electron spectroscopy in which a solid surface is bombarded with monochromatic electrons, and backscattered particles which have lost an amount of energy equal to the core-level binding energy are detected. Abbreviated CLS. { ,kar·ik·tə'ris·tik 'lȯs ,spek'träs·kə·pē }
charge-delocalized ion |ORG CHEM| A charged species in which the charge is distributed over more than one atom. { 'chärj dē'lōk·əl,īzd 'ī·ən }
charged species |CHEM| A chemical entity in which the overall total of electrons is unequal to the overall total of protons. { 'chärjd 'spē·shēz }
charge-localized ion |ORG CHEM| A charged species in which the charge is centered on a single atom. { 'chärj ,lō·kə,līzd 'ī·ən }
charge population |CHEM| The net electric charge on a specified atom in a molecule

72

that, while it cannot be observed physically, can be determined by a prescribed definition. { 'chärj ,päp·yə,lā·shən }

charge transfer [PHYS CHEM] The process in which an ion takes an electron from a neutral atom, with a resultant transfer of charge. { 'chärj ,tranz·fər }

charge-transfer complexes [CHEM] Compounds in which electrons move between molecules. { 'chärj ,tranz·fər 'käm·plek·səs }

chavicol [ORG CHEM] $C_3H_5C_6H_4OH$ A colorless phenol that is liquid at room temperature; boils at 230°C; soluble in alcohol and water; found in many essential oils. { 'chav·ə,kȯl }

check sample See control sample. { 'chek ,sam·pəl }

check standard [ANALY CHEM] In physical calibration, an artifact that is measured at specified intervals. { 'chek ,stan·dərd }

chelate [ORG CHEM] A molecular structure in which a heterocyclic ring can be formed by the unshared electrons of neighboring atoms. { 'kē,lāt }

chelating agent [ORG CHEM] An organic compound in which atoms form more than one coordinate bond with metals in solution. { 'ke,lād·iŋ ,ā·jənt }

chelating resin [ORG CHEM] Any of the ion-exchange resins with unusually high selectivity for specific cations; for example, phenol-formaldehyde resin with 8-quinolinol replacing part of the phenol, particularly selective for copper, nickel, cobalt, and iron(III). { 'ke,lād·iŋ 'rez·ən }

chelation [ORG CHEM] A chemical process involving formation of a heterocyclic ring compound which contains at least one metal cation or hydrogen ion in the ring. { kē'lā·shən }

chelerythrine [ORG CHEM] $C_{21}H_{17}O_4H$ A poisonous, crystalline alkaloid, slightly soluble in alcohol; it is derived from the seeds of the herb celandine (*Chelidonium majus*) and has narcotic properties. { ,kel·ə'rī,thrēn }

cheletropic reaction [PHYS CHEM] A chemical reaction involving the elimination of a molecule in which two sigma bonds terminating at a single atom are made or broken. { ,kel·ə'trä·pik rē'ak·shən }

chelidonic acid [ORG CHEM] $C_7H_4O_6$ A pyran isolated from the perennial herb celandine (*Chelidonium majus*). { ¦kel·ə¦dän·ik 'as·əd }

chelometry [ANALY CHEM] Analytical technique involving the formation of 1:1 soluble chelates when a metal ion is titrated with aminopolycarboxylate and polyamine reagents; a form of complexiometric titration. { ke'läm·ə·trē }

chemical [CHEM] **1.** Related to the science of chemistry. **2.** A substance characterized by definite molecular composition. { 'kem·i·kəl }

chemical affinity See affinity. { 'kem·i·kəl ə'fin·əd·ē }

chemical bond See bond. { 'kem·i·kəl ¦bänd }

chemical cellulose See alpha cellulose. { 'kem·i·kəl 'sel·yə,lōs }

chemical compound See compound. { 'kem·i·kəl 'käm,paúnd }

chemical dating [ANALY CHEM] The determination of the relative or absolute age of minerals and of ancient objects and materials by measurement of their chemical compositions. { 'kem·i·kəl 'dād·iŋ }

chemical deposition [CHEM] Precipitation of a metal from a solution of a salt by introducing another metal. { 'kem·i·kəl ,dep·ə'zish·ən }

chemical dynamics [PHYS CHEM] A branch of physical chemistry that seeks to explain time-dependent phenomena, such as energy transfer and chemical reactions, in terms of the detailed motion of the nuclei and electrons that constitute the system. { 'kem·ə·kəl dī'nam·iks }

chemical element See element. { 'kem·i·kəl 'el·ə·mənt }

chemical energy [PHYS CHEM] Energy of a chemical compound which, by the law of conservation of energy, must undergo a change equal and opposite to the change of heat energy in a reaction; the rearrangement of the atoms in reacting compounds to produce new compounds causes a change in chemical energy. { 'kem·i·kəl 'en·ər·jē }

chemical equilibrium [CHEM] A condition in which a chemical reaction is occurring at equal rates in its forward and reverse directions, so that the concentrations of

73

chemical exchange process

the reacting substances do not change with time. Also known as equilibrium. { 'kem·i·kəl ˌē·kwə'lib·rē·əm }

chemical exchange process [CHEM] A method of separating isotopes of the lighter elements by the repetition of a process of chemical change which involves exchange of the isotopes. { 'kem·i·kəl iks'chānj ˌpräs·əs }

chemical flux [CHEM] In a chemical reaction, the amount of a given substance per unit volume transformed per unit time. Also known as chemiflux. { 'kem·ə·kəl 'fləks }

chemical formula [CHEM] A notation utilizing chemical symbols and numbers to indicate the chemical composition of a pure substance; examples are CH_4 for methane and HCl for hydrogen chloride. { 'kem·i·kəl 'fȯr·myə·lə }

chemical indicator [ANALY CHEM] **1.** A substance whose physical appearance is altered at or near the end point of a chemical titration. **2.** A substance whose color varies as the concentration of hydrogen ions in the solution to which it is added varies. Also known as indicator. { 'kem·i·kəl 'in·dəˌkād·ər }

chemical inhibitor [CHEM] A substance capable of stopping or retarding a chemical reaction. { 'kem·i·kəl in'hib·əd·ər }

chemical kinetics [PHYS CHEM] That branch of physical chemistry concerned with the mechanisms and rates of chemical reactions. Also known as reaction kinetics. { 'kem·i·kəl kə'ned·iks }

chemically pure [CHEM] Without impurities detectable by analysis. Abbreviated cp. { 'kem·ik·lē 'pyu̇r }

chemical microscopy [ANALY CHEM] Application of the microscope to the solution of chemical problems. { 'kem·i·kəl mī'kräs·kə·pē }

chemical polarity [PHYS CHEM] Tendency of a molecule, or compound, to be attracted or repelled by electrical charges because of an asymmetrical arrangement of atoms around the nucleus. { 'kem·i·kəl pə'lar·əd·ē }

chemical potential [PHYS CHEM] In a thermodynamic system of several constituents, the rate of change of the Gibbs function of the system with respect to the change in the number of moles of a particular constituent. { 'kem·i·kəl pə'ten·chəl }

chemical purity See purity. { 'kem·ə·kə 'pyu̇r·ə·dē }

chemical reaction [CHEM] A change in which a substance (or substances) is changed into one or more new substances; there is only a minute change, Δm, in the mass of the system, given by $\Delta E = \Delta mc^2$, where ΔE is the energy emitted or absorbed and c is the speed of light. { 'kem·i·kəl rē'ak·shən }

chemical reactivity [CHEM] The tendency of two or more chemicals to react to form one or more products differing from the reactants. { 'kem·i·kəl rē,ak'tiv·əd·ē }

chemical relaxation [CHEM] The readjustment of a chemical system to a new equilibrium after the equilibrium of a chemical reaction is disturbed by a sudden change, particularly in an external parameter such as pressure or temperature. { 'kem·ə·kəl ˌrē,lak'sā·shən }

chemical shift [PHYS CHEM] Shift in a nuclear magnetic-resonance spectrum resulting from diamagnetic shielding of the nuclei by the surrounding electrons. { 'kem·i·kəl 'shift }

chemical species See species. { 'kem·i·kəl 'spē,shēz }

chemical symbol [CHEM] A notation for one of the chemical elements, consisting of letters; for example Ne, O, C, and Na represent neon, oxygen, carbon, and sodium. { 'kem·i·kəl 'sim·bəl }

chemical synthesis [CHEM] The formation of one chemical compound from another. { 'kem·i·kəl 'sin·thə·səs }

chemical thermodynamics [PHYS CHEM] The application of thermodynamic principles to problems of chemical interest. { 'kem·i·kəl ˌthər·mō·də'nam·iks }

chemiclearance [CHEM] The use of chemical analysis to establish the safe use of a substance. { 'kem·iˌklir·əns }

chemiflux See chemical flux. { 'kem·ə,fləks }

chemi-ionization [CHEM] Ionization that occurs as a result of the collison of a particle with a neutral species, usually excited, such as a metastable atom. { ˌkem·ē,ī·ə·nə'zā·shən }

chemiluminescence [PHYS CHEM] Emission of light as a result of a chemical reaction without an apparent change in temperature. { ‚kem·i‚lüm·ə'nes·əns }

chemionics [CHEM] The chemistry of molecular components and devices that operate on photons, electrons, and ions. { ‚kem·ē'än·iks }

chemiosmosis [CHEM] A chemical reaction occurring through an intervening semipermeable membrane. Also known as chemosmosis. { ‚kem·ē‚äs‚mō·səs }

chemisorption [PHYS CHEM] A chemical adsorption process in which weak chemical bonds are formed between gas or liquid molecules and a solid surface. { 'kem·i‚sorp·shən }

chemist [CHEM] A scientist specializing in chemistry. { 'kem·əst }

chemometrics [ANALY CHEM] The use of statistics and mathematics for experimental design and analysis of chemical data. { ‚kē·mō'me·triks }

chemoselectivity [ORG CHEM] The preferential reaction of a chemical reagent with one functional group in the presence of other similar functional groups; for example, a chemoselective reducing agent might reduce an aldehyde but not a ketone. { ‚kē·mō‚si·lek'tiv·əd·ē }

chemosmosis *See* chemiosmosis. { ‚kem‚äs'mō·səs }

chinaldine *See* quinaldine. { ki'näl‚dēn }

Chinese vermilion *See* mercuric sulfide. { chīn'nēz vər'mil·yən }

Chinese white [CHEM] A term used in the paint industry for zinc oxide and kaolin used as a white pigment. Also known as zinc white. { chīn'nēz 'wīt }

chinic acid *See* quinic acid. { 'kin·ik 'as·əd }

chinoidine *See* quinoidine. { ki'noÏ‚dēn }

chinone *See* quinone. { kin'ōn }

chiral carbon atom *See* asymmetric carbon atom. { ‚kī·rəl ‚kär·bən 'ad·əm }

chiral center [ORG CHEM] An atom in a molecule that is attached to four different groups. { 'kī·rəl 'sen·tər }

chirality [CHEM] The handedness of an asymmetric molecule. { kī'ral·əd·ē }

chiral molecules [CHEM] Molecules which are not superposable with their mirror images. { 'kī·rəl 'mäl·ə‚kyülz }

chiral nanotube [PHYS CHEM] A carbon nanotube formed from a graphite sheet that is rolled up so that the succession of hexagons of carbon atoms on a particular cylinder makes an angle with the axis of the nanotube. { ‚kī·rəl 'nan·ō‚tüb }

chloflurecol methyl ester [ORG CHEM] $C_{15}H_{11}ClO_3$ A white, crystalline compound with a melting point of 152°C; slight solubility in water; used as a growth regulator for grass and weeds. { klō'flür·ə‚kòl 'meth·əl 'es·tər }

chloral [ORG CHEM] CCl_3CHO A colorless, oily liquid soluble in water; used industrially to prepare DDT; a hypnotic. Also known as trichloroacetic aldehyde; trichloroethanal. { 'klòr·əl }

chloralase [ORG CHEM] $C_8H_{11}Cl_3O_6$ Colorless, water-soluble crystals, melting at 185°C; made by heating chloral with dextrose; used as a hypnotic. { 'klòr·ə‚lās }

chloral hydrate [ORG CHEM] $CCl_3CH(OH)_2$ Colorless, deliquescent needles with slightly bitter caustic taste, soluble in water; a hypnotic. Also known as crystalline chloral; hydrated chloral. { 'klòr·əl 'hī‚drāt }

chloralkane [ORG CHEM] Chlorinated aliphatic hydrocarbon of the methane series (C_nH_{2n+2}). { klòr'al‚kān }

chloralosane *See* chloralose. { ‚klòr·ə'lō‚sän }

chloralose [ORG CHEM] $C_8H_{11}O_6Cl_3$ A crystalline compound with a melting point of 178°C; used as a repellent for birds. Also known as glucochloralose. { 'klòr·ə‚lōs }

α-chloralose [ORG CHEM] $C_8H_{11}O_6Cl_3$ Needlelike crystals with a melting point of 87°C; soluble in glacial acetic acid and ether; used on seed grains as a bird repellent and as a hypnotic for animals. Also known as chloralosane; glucochloral. { 'al·fə ‚klòr·ə‚lōs }

chloramine T [ORG CHEM] $CH_3C_6H_4SO_2NClNa·3H_2O$ A white, crystalline powder that decomposes slowly in air, freeing chlorine; used as an antiseptic, a germicide, and an oxidizing agent and chlorinating agent. { 'klòr·ə‚mēn 'tē }

chloranil

chloranil [ORG CHEM] C₆Cl₄O₂ Yellow leaflets melting at 290°C; soluble in organic solvents; made from phenol by treatment with potassium chloride and hydrochloric acid; used as an agricultural fungicide and as an oxidizing agent in the manufacture of dyes. { klȯr'an·əl }

chloranilic acid [ORG CHEM] C₆H₂Cl₂O₄ A relatively strong dibasic acid whose crystals are red and melt between 283 and 284°C; used in spectrophotometry. { klȯr·ə'nil·ik 'as·əd }

chlorate [INORG CHEM] ClO₃⁻ **1.** A negative ion derived from chloric acid. **2.** A salt of chloric acid. { 'klȯr,āt }

chlorbenside [ORG CHEM] C₁₃H₁₀SCl₂ White crystals with a melting point of 72°C; used as a miticide for spider mites on fruit trees and ornamentals. { klȯr'ben,sīd }

chlorbromuron [ORG CHEM] C₉H₁₀ONBrCl A white solid with a melting point of 94–96°C; used as a pre- and postemergence herbicide for annual grass and for broadleaf weeds on crops, soybeans, and Irish potatoes. { ,klȯr·brə'myü·rən }

chlordan *See* chlordane. { 'klȯr,dan }

chlordane [ORG CHEM] C₁₀H₆Cl₈ A volatile liquid insecticide; a chlorinated hexahydromethanoindene. Also spelled chlordan. { 'klȯr,dān }

chlordimeform [ORG CHEM] C₁₀H₁₃ClN₂ A tan-colored solid, melting point 35°C; used as a miticide and insecticide for fruits, vegetables, and cotton. { ,klȯr'dī·mə,fȯrm }

chlorendic acid [ORG CHEM] C₉H₄Cl₆O₄ White, fine crystals used in fire-resistant polyester resins and as an intermediate for dyes, fungicides, and insecticides. { klȯr'en·dik 'as·əd }

chlorendic anhydride [ORG CHEM] C₉H₂Cl₆O₃ White, fine crystals used in fire-resistant polyester resins, in hardening epoxy resins, and as a chemical intermediate. { klȯr'en·dik an'hī,drīd }

chlorfenethol [ORG CHEM] C₁₄H₁₂Cl₂O A colorless, crystalline compound with a melting point of 69.5–70°C; insoluble in water; used for control of mites in ornamentals and shrub trees. { ,klȯr'fen·ə,thȯl }

chlorfenpropmethyl [ORG CHEM] C₁₀H₁₀OCl₂ A colorless to brown liquid used as a postemergence herbicide of wild oats, cereals, fodder beets, sugarbeets, and peas. { ¦klȯr·fən,präp'meth·əl }

chlorfensulfide [ORG CHEM] C₁₂H₆Cl₄N₂S A yellow, crystalline compound with a melting point of 123.5–124°C; used as a miticide for citrus. { ,klȯr·fən'səl,fīd }

chlorfenvinphos [ORG CHEM] C₁₂H₁₄Cl₃O₄P An amber liquid with a boiling point of 168–170°C; used as an insecticide for ticks, flies, lice, and mites on cattle. { ,klȯr·fən'vin,fäs }

chlorhydrin *See* chlorohydrin. { klȯr'hī·drən }

chloric acid [INORG CHEM] HClO₃ A compound that exists only in solution and as chlorate salts; breaks down at 40°C. { 'klȯr·ik 'as·əd }

chloride [CHEM] **1.** A compound which is derived from hydrochloric acid and contains the chlorine atom in the −1 oxidation state. **2.** In general, any binary compound containing chloride. { 'klȯr,īd }

chloride benzilate *See* lachesne. { 'klȯr,īd 'ben·zə,lāt }

chloridization *See* chlorination. { ,klȯr·ə·də'zā·shən }

chlorimide *See* dichloramine. { 'klȯr·ə,mīd }

chlorinated paraffin [ORG CHEM] One of a group of chlorine derivatives of paraffin compounds. { 'klȯr·ə,nād·əd 'par·ə·fən }

chlorination [CHEM] **1.** Introduction of chlorine into a compound. Also known as chloridization. **2.** Water sterilization by chlorine gas. { ,klȯr·ə'nā·shən }

chlorine [CHEM] A chemical element, symbol Cl, atomic number 17, atomic weight 35.453; used in manufacture of solvents, insecticides, and many non-chlorine-containing compounds, and to bleach paper and pulp. { 'klȯr,ēn }

chlorine dioxide [INORG CHEM] ClO₂ A green gas used to bleach cellulose and to treat water. { 'klȯr,ēn dī'äk,sīd }

chlorine water [CHEM] A clear, yellowish liquid used as a deodorizer, antiseptic, and disinfectant. { 'klȯr,ēn ,wȯd·ər }

chlorite [INORG CHEM] A salt of chlorous acid. { 'klȯr,īt }

76

chloritization |CHEM| The introduction of, production of, replacement by, or conversion into chlorite. { ,klȯr·əd·ə'za·shən }

chlormephos [ORG CHEM] C₅H₁₂O₂S₂ClP A liquid used as an insecticide for soil. { 'klȯr·mə,fäs }

chloro- [ORG CHEM] A prefix describing an organic compound which contains chlorine atoms substituted for hydrogen. { 'klȯr·ō }

chloroacetic acid |ORG CHEM| ClCH₂COOH White or colorless, deliquescent crystals that are soluble in water, ether, chloroform, benzene, and alcohol; used as an herbicide and in the manufacture of dyes and other organic molecules. { ¦klȯr·ə¦sēd·ik 'as·əd }

chloroacetic anhydride [ORG CHEM] C₄H₄Cl₂O₃ Crystals with a melting point of 46°C; soluble in chloroform and ether; used in the preparation of cellulose chloracetates and in the N-acetylation of amino acids in alkaline solution. { ¦klȯr·ə¦sēd·ik an'hī,drīd }

chloroacetone [ORG CHEM] CH₃COCH₂Cl Pungent, colorless liquid used as military tear gas and in organic synthesis. { klȯr'as·ə,tōn }

chloroacetonitrile [ORG CHEM] ClCH₂CN A colorless liquid with a pungent odor; soluble in hydrocarbons and alcohols; used as a fumigant. { ¦klȯr·ō,as·ə·'tän·ə·trəl }

chloroacetophenone [ORG CHEM] C₆H₅COCH₂Cl Rhombic crystals melting at 59°C; an intermediate in organic synthesis. { ¦klȯr·ō,as·ə'tä·fə,nōn }

chloroacrolein [ORG CHEM] H₂C:ClCHO A colorless liquid with a boiling point of 29–31°C; used as a tear gas. { ,klȯr·ō·ə'krō·lē·ən }

chlorobenzaldehyde [ORG CHEM] C₆H₄CHOCl A colorless to yellowish liquid (ortho form) or powder (para form) with a boiling range of 209–215°C; soluble in alcohol, ether, and acetone; used in dye manufacture. { ,klȯr·ō,ben'zal·də,hīd }

chlorobenzene [ORG CHEM] C₆H₅Cl A colorless, mobile, volatile liquid with an almondlike odor; used to produce phenol, DDT, and aniline. { ,klȯr·ō'ben,zēn }

chlorobenzilate [ORG CHEM] C₁₆H₁₄Cl₂O₃ A yellow-brown, viscous liquid with a melting point of 35–37°C; used as a miticide in agriculture and horticulture. { ,klȯr·ō'ben·zə,lāt }

para-chlorobenzoic acid [ORG CHEM] ClC₆H₄COOH A white powder with a melting point of 238°C; soluble in methanol, absolute alcohol, and ether; used in the manufacture of dyes, fungicides, and pharamaceuticals. { ¦par·ə ¦klȯr·ō,ben'zō·ik 'as·əd }

chlorobenzoyl chloride [ORG CHEM] ClC₆H₄COCl A colorless liquid with a boiling range of 227–239°C; soluble in alcohol, acetone, and water; used in dye and pharmaceuticals manufacture. { ,klȯr·ō'ben,zȯil 'klȯr,īd }

chlorobenzyl chloride [ORG CHEM] ClC₆H₄CH₂Cl A colorless liquid with a boiling range of 216–222°C; soluble in acetone, alcohol, and ether; used in the manufacture of organic chemicals. { ,klor·ō'ben·zil 'klȯr,īd }

chlorobutadiene See chloroprene. { ,klor·ō,byüd·ə'dī,ēn }

chlorobutanol [ORG CHEM] Cl₃CC(CH₃)₂OH Colorless to white crystals with a melting point of 78°C; soluble in alcohol, glycerol, ether, and chloroform; used as a plasticizer and a preservative for biological solutions. { ,klȯr·ō'byüt·ən,ȯl }

chlorocarbon [ORG CHEM] A compound of chlorine and carbon only, such as carbon tetrachloride, CCl₄. { ¦klȯr·ō'kär·bən }

chlorochromic anhydride See chromyl chloride. { ¦klȯr·ō¦krō·mik an'hī,drīd }

1,1,1-chlorodifluoroethane [ORG CHEM] CH₃CClF₂ A colorless gas with a boiling point of −130.8°C; used as a refrigerant, solvent, and aerosol propellant. { ¦wən ¦wən ¦wən ,klȯr·ō,dī,flür·ō'eth,ān }

chlorodifluoromethane [ORG CHEM] CHClF₂ A colorless gas with a boiling point of −40.8°C and freezing point of −160°C; used as an aerosol propellant and refrigerant. { klȯr·ō·dī·flür·ō'meth·ān }

1-chloro-2,4-dinitrobenzene [ORG CHEM] C₆H₃ClN₂O₄ Yellow crystals with a melting point of 52–54°C; soluble in hot alcohol, ether, and benzene; used as a reagent in the determination of pyridine compounds such as nicotinic acid, and nicotinamide. { ¦wən ,klȯr·ō ¦tü ¦fȯr dī,nī·trō'ben,zēn }

chloroethane See ethyl chloride. { ,klȯr·ō'eth,ān }

chloroethene See vinyl chloride. { ,klor·ō'eth,ēn }

chloroethyl alcohol

chloroethyl alcohol *See* ethylene chlorohydrin.   { ¦klȯr·ō'eth·əl 'al·kə,hȯl }
chlorofluorocarbon   [ORG CHEM] A compound consisting of chlorine, fluorine, and carbon; has the potential to destroy ozone in the stratosphere.   Abbreviated CFC. Also known as fluorochlorocarbon.   { ¦klȯr·ə¦flu̇r·ə,kär·bən }
chlorofluoromethane   [ORG CHEM] A compound consisting of chlorine, fluorine, and carbon, has the potential to destroy ozone in the stratosphere.   Also known as fluorochlorocarbon (FCC).   Abbreviated CFC.   { ¦klȯr·ə,flu̇r·ə'meth,ān }
chloroform   [ORG CHEM] CHCl₃ A colorless, sweet-smelling, nonflammable liquid; used at one time as an anesthetic.   Also known as trichloromethane.   { 'klȯr·ə,fȯrm }
chlorohydrin   [ORG CHEM] Any of the compounds derived from a group of glycols or polyhydroxy alcohols by chlorine substitution for part of the hydroxyl groups.   Also spelled chlorhydrin.   { ¦klȯr·ə'hī·drən }
chlorohydrocarbon   [ORG CHEM] A carbon- and hydrogen-containing compound with chlorine substituted for some hydrogen in the molecule.   { ¦klȯr·ō'hī·drə,kär·bən }
chlorohydroquinone   [ORG CHEM] ClC₆H₃(OH)₂ White to light tan crystals with a melting point of 100°C; soluble in water and alcohol; used as a photographic developer and bactericide and for dyestuffs.   { ¦klȯr·ō,hī·drə·kwin'ōn }
5-chloro-8-hydroxyquinoline   [ORG CHEM] C₉H₆ClNO Crystals with a melting point of 130°C; used as a fungicide and bactericide.   { ¦fīv ¦klȯr·ō ¦āt hī,dräk·sē'kwin·ə,lēn }
chloromethane   [ORG CHEM] CH₃Cl A colorless, noncorrosive, liquefiable gas which condenses to a colorless liquid; used as a refrigerant, and as a catalyst carrier in manufacture of butyl rubber.   Also known as methyl chloride.   { ¦klȯr·ō'meth,ān }
1-chloronaphthalene   [ORG CHEM] C₁₀H₇Cl An oily liquid used as an immersion medium in the microscopic determination of refractive index of crystals and as a solvent for oils, fats, and DDT.   { ¦wən ,klȯr·ō'naf·thə,lēn }
chloronium ion   [ORG CHEM] A halonium ion in which the halogen is chlorine; sometimes occurs as a bridged form.   { klə'rōn·ē·əm 'ī·ən }
chlorophenol red   [ORG CHEM] C₁₉H₁₂Cl₂O₅S A dye that is used as an acid-base indicator; yellow in acid solution, red in basic solution.   Also spelled chlorphenol red.   { ¦klȯr·ə'fen·ȯl 'red }
chloropicrin   [INORG CHEM] CCl₃NO₂ A colorless liquid with a sweet odor whose vapor is very irritating to the lungs and causes vomiting, coughing, and crying; used as a soil fumigant.   Also known as nitrochloroform; trichloronitromethane.   { ,klȯr·ō'pik·rən }
chloroplatinate   [INORG CHEM] **1.** A double salt of platinic chloride and another chloride. **2.** A salt of chloroplatinic acid.   Also known as platinochloride.   { ,klȯr·ō'plat·ən,āt }
chloroplatinic acid   [INORG CHEM] H₂PtCl₆ An acid obtained as red-brown deliquescent crystals; used in chemical analysis.   Also known as platinic chloride.   { ¦klȯr·ə·plə'tin·ik 'as·əd }
chloroprene   [ORG CHEM] C₄H₅Cl A colorless liquid which polymerizes to chloroprene resin.   Also known as chlorobutadiene.   { 'klȯr·ə,prēn }
chloroprene resin   [ORG CHEM] A polymer of chloroprene used to form materials resembling natural rubber.   { 'klȯr·ə,prēn 'rez·ən }
chloropropane   [ORG CHEM] Propane molecules with chlorine substituted in various amounts for the hydrogen atoms.   { ¦klȯr·ō'prō,pān }
3-chloro-1,2-propanediol   [ORG CHEM] ClCH₂CH(OH)CH₂OH A sweetish-tasting liquid that has a tendency to turn a straw color; soluble in ether, alcohol, and water; used to manufacture dye intermediates and to lower the freezing point of dynamite.   { ¦thrē ,klȯr·ō ¦wən ¦tü ¦prō,pān'dī,ȯl }
chloropropene   [CHEM] Propene molecules with chlorine substituted for some hydrogen atoms.   { ¦klȯr·ə'prō,pēn }
β-chloropropionitrile   [ORG CHEM] ClCH₂CH₂CN A liquid with an acrid odor; miscible with various organic solvents such as ethanol, ether, and acetone; used in polymer synthesis and in the synthesis of pharmaceuticals.   { ¦bād·ə ,klȯr·ō¦prō·pē·ō'nī,trəl }
N-chlorosuccinimide   [ORG CHEM] C₄H₄ClNO₂ Orthorhombic crystals with the smell of chlorine; melting point is 150–151°C; soluble in water, benzene, and alcohol; used as a chlorinating agent.   { ¦en ,klȯr·ō·sək'sin·ə,mīd }

chlorosulfonic acid [INORG CHEM] ClSO$_2$OH A fuming liquid that decomposes in water to sulfuric acid and hydrochloric acid; used in pharmaceuticals, pesticides, and dyes, and as a chemical intermediate. { ¦klȯr·ō·səl'fän·ik 'as·əd }

chlorothalonil [ORG CHEM] C$_8$Cl$_4$N$_2$ Colorless crystals with a melting point of 250–251°C; used as a fungicide for crops, turf, and ornamental flowers. { ˌklȯr·ə'thal·ə·nəl }

chlorothymol [ORG CHEM] CH$_3$C$_6$H$_2$(OH)(C$_3$H$_7$)Cl White crystals melting at 59–61°C; soluble in benzene alcohol, insoluble in water; used as a bactericide. { ˌklȯr·ə'thī,mȯl }

***ortho*-chlorotoluene** [ORG CHEM] CH$_3$C$_6$H$_4$Cl A liquid with a boiling point of 158.97°C; soluble in alcohol, chloroform, benzene, and ether; used in organic synthesis, as a solvent, and as an intermediate in dyestuff manufacture. { ¦ȯr·thō ˌklȯr·ō'täl·yə,wēn }

chlorotrifluoroethylene polymer [ORG CHEM] A colorless, noninflammable, heat-resistant resin, soluble in most organic solvents, and with a high impact strength; can be made into transparent filling and thin sheets; used for chemical piping, fittings, and insulation for wire and cables, and in electronic components. Also known as fluorothene; polytrifluorochloroethylene resin. { ¦klȯr·ō·trī¦flu̇r·ō'eth·əl,en 'päl·ə·mər }

chlorotrifluoromethane [ORG CHEM] CClF$_3$ A colorless gas having a boiling point of −81.4°C and a freezing point of −181°C; used as a dielectric and aerospace clinical, refrigerant, and aerosol propellant, and for metals hardening and pharmaceuticals manufacture. { ¦klȯr·ō·trī¦flu̇r·ō'me,thān }

chloroxine [ORG CHEM] C$_9$H$_5$Cl$_2$NO Crystals with a melting point of 179–180°C; soluble in benzene and in sodium and potassium hydroxides; used as an analytical reagent. { klə'räk·sən }

4-chloro-3,5-xylenol [ORG CHEM] ClC$_6$H$_2$(CH$_3$)$_2$OH Crystals with a melting point of 115.5°C; soluble in water, 95% alcohol, benzene, terpenes, ether, and alkali hydroxides; used as an antiseptic and germicide and to stop mildew; used in humans as a topical and urinary antiseptic and as a topical antiseptic in animals. { ¦fȯr ¦klȯr·ō ¦thrē ¦fīv 'zī·lə,nȯl }

chlorphenol red *See* chlorophenol red. { klȯr'fē,nȯl 'red }

chlorthiamid [ORG CHEM] C$_7$H$_5$Cl$_2$NS An off-white, crystalline compound with a melting point of 151–152°C; used as a herbicide for selective weed control in industrial sites. { klȯr'thī·ə,mid }

cholesteric material [PHYS CHEM] A liquid crystal material in which the elongated molecules are parallel to each other within the plane of a layer, but the direction of orientation is twisted slightly from layer to layer to form a helix through the layers. { kə'les·tə·rik mə'tir·ē·əl }

cholesteric phase [PHYS CHEM] A form of the nematic phase of a liquid crystal in which the molecules are spiral. { kə'les·tə·rik ˌfāz }

choline succinate dichloride dihydrate *See* succinylcholine chloride. { 'kō,lēn 'sək·sə,nāt di'klȯr,īd di'hī,drāt }

chondrodendrin *See* bebeerine. { ¦kän·drō¦den·drən }

Christiansen effect [ANALY CHEM] Monochromatic transparency effect when finely powdered substances, such as glass or quartz, are immersed in a liquid having the same refractive index. { 'kris·chən·sən i'fekt }

chromate [INORG CHEM] **1.** CrO$_4^{2-}$ **2.** An ion derived from the unstable acid H$_2$CrO$_4$. **3.** A salt or ester of chromic acid. { 'krō,māt }

chromatogram [ANALY CHEM] The pattern formed by zones of separated pigments and of colorless substance in chromatographic procedures. { krō'mad·ə,gram }

chromatograph [ANALY CHEM] To employ chromatography to separate substances. { krō'mad·ə,graf }

chromatographic adsorption [ANALY CHEM] Preferential adsorption of chemical compounds (gases or liquids) in an ascending molecular-weight sequence onto a solid adsorbent material, such as activated carbon, alumina, or silica gel; used for analysis and separation of chemical mixtures. { krō¦mad·ə¦graf·ik ad'sȯrp·shən }

chromatographic bed

chromatographic bed [ANALY CHEM] Any of the different configurations in which the stationary phase is contained. { krō|mad·ə|graf·ik 'bed }

chromatography [ANALY CHEM] A method of separating and analyzing mixtures of chemical substances by chromatographic adsorption. { ‚krō·mə'täg·rə·fē }

chrome alum [INORG CHEM] KCr(SO$_4$)$_2$·12H$_2$O An alum obtained as purple crystals and used as a mordant, in tanning, and in photography in the fixing bath. Also known as potassium chromium sulfate. { 'krōm 'al·əm }

chrome dye [CHEM] One of a class of acid dyes used on wool with a chromium compound as mordant. { |krōm |dī }

chrome green See chromic oxide. { |krōm ‚grēn }

chrome red [CHEM] **1.** A pigment containing basic lead chromate. **2.** Any of several mordant acid dyes. { |krōm 'red }

chrome yellow [CHEM] **1.** A yellow pigment composed of normal lead chromate, PbCrO$_4$, or other lead compounds. **2.** Any of several mordant acid dyes. { |krōm 'yel·ō }

chromic acid [INORG CHEM] H$_2$CrO$_4$ The hydrate of CrO$_3$; exists only as salts or in solution. { |krō·mik 'as·əd }

chromic chloride [INORG CHEM] CrCl$_3$ Crystals that are pinkish violet shimmering plates, almost insoluble in water, but easily soluble in presence of minute traces of chromous chloride; used in calico printing, as a mordant for cotton and silk. { |krō·mik 'klȯr‚īd }

chromic fluoride [INORG CHEM] CrF$_3$·4H$_2$O Crystals that are green, soluble in water; used in dyeing cottons. { |krō·mik 'flȯr‚īd }

chromic hydroxide [INORG CHEM] Cr(OH)$_3$·2H$_2$O Gray-green, gelatinous precipitate formed when a base is added to a chromic salt; the precipitate dries to a bluish, amorphous powder; prepared as an intermediate in the manufacture of other soluble chromium salts. { |krō·mik hī'dräk‚sīd }

chromic nitrate [INORG CHEM] Cr(NO$_3$)$_3$·9H$_2$O Purple, rhombic crystals that are soluble in water; used as a mordant in textile dyeing. { |krō·mik 'nī‚trāt }

chromic oxide [INORG CHEM] Cr$_2$O$_3$ A dark green, amorphous powder, forming hexagonal crystals on heating that are insoluble in water or acids; used as a pigment to color glass and ceramic ware and as a catalyst. Also known as chrome green. { |krō·mik 'äk‚sīd }

chromium [CHEM] A metallic chemical element, symbol Cr, atomic number 24, atomic weight 51.996. { 'krō·mē·əm }

chromium carbide [INORG CHEM] Cr$_3$C$_2$ Orthorhombic crystals with a melting point of 1890°C; resistant to oxidation, acids, and alkalies; used for hot-extrusion dies, in spray-coating materials, and as a component for pumps and valves. { 'krō·mē·əm 'kär‚bīd }

chromium chloride [INORG CHEM] A group of compounds of chromium and chloride; chromium may be in the +2, +3, or +6 oxidation state. { 'krō·mē·əm 'klȯr‚īd }

chromium dioxide [INORG CHEM] Cr$_2$O$_2$ Black, acicular crystals; a semiconducting material with strong magnetic properties used in recording tapes. { 'krō·mē·əm dī'äk‚sīd }

chromium oxide [INORG CHEM] A compound of chromium and oxygen; chromium may be in the +2, +3, or +6 oxidation state. { 'krō·mē·əm 'äk‚sīd }

chromium oxychloride See chromyl chloride. { 'krō·mē·əm äk·sē'klȯr‚īd }

chromium stearate [ORG CHEM] Cr(C$_{18}$H$_{35}$O$_2$)$_3$ A dark-green powder, melting at 95–100°C; used in greases, ceramics, and plastics. { 'krō·mē·əm 'stir‚āt }

chromometer See colorimeter. { krə'mäm·əd·ər }

chromophore [CHEM] An arrangement of atoms that gives rise to color in many organic substances. { 'krō·mə‚fȯr }

chromotropic acid [ORG CHEM] C$_{10}$H$_8$O$_8$S$_2$ White, needlelike crystals that are soluble in water; used as an analytical reagent and azo dye intermediate. { |krō·mə|träp·ik 'as·əd }

chromyl chloride [INORG CHEM] CrO$_2$Cl$_2$ A dark-red, toxic, fuming liquid that boils at 116°C; reacts with water to form chromic acid; used to make dyes and chromium

complexes. Also known as chlorochromic anhydride; chromium oxychloride.
{ 'krō·məl 'klȯr,īd }

chronoamperometry [ANALY CHEM] Electroanalysis by measuring at a working electrode the rate of change of current versus time during a titration; the potential is controlled. { ¦krän·ō,am·pə'räm·ə·trē }

chronocoulometry [ANALY CHEM] The study of electrode surface properties, such as surface area. { ¦krä·nō,kü'läm·ə·trē }

chronopotentiometry [ANALY CHEM] Electroanalysis based on the measurement at a working electrode of the rate of change in potential versus time; the current is controlled. { ¦krän·ō·pə,ten·chē'äm·ə·trē }

chrysazin See 1,8-dihydroxyanthraquinone. { 'krī·sə·sən }

chrysene [ORG CHEM] $C_{18}H_{12}$ An organic, polynuclear hydrocarbon which when pure gives a bluish fluorescence; a component of short afterglow or luminescent paint. { 'krī ,sēn }

chrysoidine [ORG CHEM] $C_6H_5NNC_6H_3(NH_2)_2 \cdot HCl$ Large, black crystals or a red-brown powder that melts at 117°C; soluble in water and alcohol; used as an orange dye for silk and cotton. { kri'sō·ə,dēn }

chrysophanic acid [ORG CHEM] $C_{15}H_{10}O_4$ Yellow leaves that melt at 196°C; soluble in ether, chloroform, and hot alcohol; extracted from senna leaves and rhubarb root; used in medicine as a mild laxative. { ¦kris·ō'fan·ik 'as·əd }

Chugaev reaction [ORG CHEM] The thermal decomposition of methyl esters of xanthates to yield olefins without rearrangement. { chü'gä·əv rē,ak·shən }

cigarette burning [CHEM] In rocket propellants, black powder, gasless delay elements, and pyrotechnic candles, the type of burning induced in a solid grain by permitting burning on one end only, so that the burning progresses in the direction of the longitudinal axis. { 'sig·ə,ret ¦bərn·iŋ }

cincholepidine See lepidine. { ,siŋ·kə'lep·ə,dēn }

cinchonamine [ORG CHEM] $C_{19}H_{24}N_2O$ A yellow, crystalline, water-insoluble alkaloid that melts at 184°C; derived from the bark of *Remijia purdieana*, a member of the madder family of shrubs. { siŋ'kän·ə,mēn }

cinchonine [ORG CHEM] $C_{19}H_{22}N_2O$ A colorless, crystalline alkaloid that melts at about 245°C; extracted from cinchona bark, it is used as a substitute for quinine and as a spot reagent for bismuth. { 'siŋ·kə,nēn }

cineol See eucalyptol. { 'sin·ē,ȯl }

cinnamate [ORG CHEM] A salt of cinnamic acid, containing the functional group $C_9H_7O_2-$. { 'sin·ə,māt }

cinnamic acid [ORG CHEM] $C_6H_5CHCHCOOH$ Colorless, monoclinic acid; forms scales, slightly soluble in water; found in natural balsams. { sə'nam·ik 'as·əd }

cinnamic alcohol [ORG CHEM] $C_6H_5CH:CHCH_2OH$ White needles that congeal upon heating and are soluble in alcohol; used in perfumery. { sə'nam·ik 'al·kə,hȯl }

cinnamic aldehyde [ORG CHEM] $C_6H_5CH:CHCHO$ A yellow oil with a cinnamon odor, sweet taste, and a boiling point of 248°C; used in flavors and perfumes. { sə'nam·ik 'al·də,hīd }

cinnamoyl chloride [ORG CHEM] $C_6H_5CHCHCOCl$ Yellow crystals that melt at 35°C, and decompose in water; used as a chemical intermediate. { 'sin·ə,mȯil 'klȯr,īd }

circular chromatography See radial chromatography. { 'sər·kyə·lər ,krō·mə'täg·rə·fē }

circular paper chromatography [ANALY CHEM] A paper chromatographic technique in which migration from a spot in the sheet takes place in 360° so that zones separate as a series of concentric rings. { 'sər·kyə·lər 'pā·pər ,krō·mə'täg·rə·fē }

cis [ORG CHEM] A descriptive term indicating a form of isomerism in which atoms are located on the same side of an asymmetric molecule. { sis }

cis-trans isomerism [ORG CHEM] A type of geometrical isomerism found in alkenic systems in which it is possible for each of the doubly bonded carbons to carry two different atoms or groups; two similar atoms or groups may be on the same side (cis) or on opposite sides (trans) of a plane bisecting the alkenic carbons and perpendicular to the plane of the alkenic system. { 'si¦stranz ī'säm·ə,riz·əm }

citraconic acid

citraconic acid [ORG CHEM] $C_5H_6O_4$ A dicarboxylic acid; hygroscopic crystals that melt at 91°C; derived from citric acid by heating.   { ¦si·trə¦kän·ik 'as·əd }

citral [ORG CHEM] $C_{10}H_{16}O$ A pale-yellow liquid that in commerce is a mixture of two isomeric forms, alpha and beta; insoluble in water, soluble in glycerin or benzyl benzoate; used in perfumery and as an intermediate to form other compounds. Also known as geranial; geranialdehyde.   { 'si,tral }

citronellal hydrate See hydroxycitronellal.   { ¸si·trə'nel·əl 'hī,drāt }

citronellol [ORG CHEM] $C_{10}H_{19}OH$ A liquid derived from citronella oil; soluble in alcohol; used in perfumery.   { ¸si·trə'nel,ȯl }

Cl See chlorine.

Claisen condensation [ORG CHEM] **1.** Condensation, in the presence of sodium ethoxide, of esters or of esters and ketones to form β-dicarbonyl compounds. **2.** Condensation of arylaldehydes and acylphenones with esters or ketones in the presence of sodium ethoxide to yield unsaturated esters. Also known as Claisen reaction.   { 'klās·ən känd·ən'sā·shən }

Claisen flask [CHEM] A glass flask with a U-shaped neck, used for distillation.   { 'klās· ən ¸flask }

Claisen reaction See Claisen condensation.   { 'klās·ən ri'äk·shən }

Claisen rearrangement [ORG CHEM] A thermally induced sigmatrophic shift in which an allyl phenyl ether is rearranged to yield an *ortho*-allylphenol.   { 'klā·sən ¸rē· ə'rānj·mənt }

Claisen-Schmidt condensation [ORG CHEM] A reaction employed for preparation of unsaturated aldehydes and ketones by condensation of aromatic aldehydes with aliphatic aldehydes or ketones in the presence of sodium hydroxide.   { ¦klās·ən ¦shmit känd·ən'sā·shən }

Clark degree See English degree.   { 'klärk də,grē }

clathrate [CHEM] An inclusion compound in which the guest species is enclosed on all sides by the species forming the crystal lattice. Also known as cage compound; inclusion compound.   { 'klath,rāt }

clathrochelate [INORG CHEM] A type of coordination compound containing a metal ion both coordinately saturated and encapsulated by a single ligand.   { ¦klath·rō'kē,lāt }

Cleveland open-cup tester [ANALY CHEM] A laboratory apparatus used to determine flash point and fire point of petroleum products.   { 'klev·lənd ¸ō·pən 'kəp ¸test·ər }

CLS See characteristic loss spectroscopy.

Cm See curium.

Co See cobalt.

coacervate [CHEM] An aggregate of colloidal droplets bound together by the force of electrostatic attraction.   { kō'as·ər,vāt }

coacervation [CHEM] The separation, by addition of a third component, of an aqueous solution of a macromolecule colloid (polymer) into two liquid phases, one of which is colloid-rich (the coacervate) and the other an aqueous solution of the coacervating agent (the equilibrium liquid).   { kō,as·ər'vā·shən }

coagulant [CHEM] An agent that causes coagulation.   { kō'ag·yə·lənt }

coagulation [CHEM] A separation or precipitation from a dispersed state of suspensoid particles resulting from their growth; may result from prolonged heating, addition of an electrolyte, or from a condensation reaction between solute and solvent; an example is the setting of a gel.   { kō,ag·yə'lā·shən }

coalescent [CHEM] Chemical additive used in immiscible liquid-liquid mixtures to cause small droplets of the suspended liquid to unite, preparatory to removal from the carrier liquid.   { ¸kō·ə'les·ənt }

coal-tar dye [ORG CHEM] Dye made from a coal-tar hydrocarbon or a derivative such as benzene, toluene, xylene, naphthalene, or aniline.   { 'kōl ¸tär ¸dī }

cobalt [CHEM] A metallic element, symbol Co, atomic number 27, atomic weight 58.93; used chiefly in alloys.   { 'kō,bȯlt }

cobalt blue [CHEM] A green-blue pigment formed of alumina and cobalt oxide. Also known as cobalt ultramarine; king's blue.   { ¦kō,bȯlt ¦blü }

cobalt bromide See cobaltous bromide.   { 'kō,bȯlt 'brō,mīd }

cobalt chloride See cobaltous chloride. { 'kō,bȯlt 'klȯr,īd }

cobaltic fluoride See cobalt trifluoride. { kə'bȯl·tik 'flůr,īd }

cobalt nitrate See cobaltous nitrate. { 'kō,bȯlt 'nī,trāt }

cobaltous acetate [ORG CHEM] $Co(C_2H_3O_2)_2 \cdot 4H_2O$ Reddish-violet, deliquescent crystals; soluble in water, alcohol, and acids; used in paint and varnish driers, for anodizing, and as a feed additive mineral supplement. Also known as cobalt acetate. { kō'bȯl·təs 'as·ə,tāt }

cobaltous bromide [INORG CHEM] $CoBr_2 \cdot 6H_2O$ Red-violet crystals with a melting point of 47–48°C; soluble in water, alcohol, and ether; used in hygrometers. Also known as cobalt bromide. { kō'bȯl·təs 'brō,mīd }

cobaltous chloride [INORG CHEM] $CoCl_2$ or $CoCl_2 \cdot 6H_2O$ A compound whose anhydrous form consists of blue crystals and sublimes when heated, and whose hydrated form consists of red crystals and melts at 86.8°C; both forms are used as an absorbent for ammonia in dyes and as a catalyst. Also known as cobalt chloride. { kō'bȯl·təs 'klȯr,īd }

cobaltous fluorosilicate [INORG CHEM] $CoSiF_6 \cdot H_2O$ A water-soluble, orange-red powder, used in toothpastes. { kō'bȯl·təs ¦flůr·ō'sil·ə,kāt }

cobaltous nitrate [INORG CHEM] $Co(NO_3)_2 \cdot 6H_2O$ A red crystalline compound with a melting point of 56°C; soluble in organic solvents; used in sympathetic inks, as an additive to soils and animal feeds, and for vitamin preparations and hair dyes. Also known as cobalt nitrate. { kō'bȯl·təs 'nī,trāt }

cobalt oxide [INORG CHEM] CoO A grayish brown powder that decomposes at 1935°C, insoluble in water; used as a colorant in ceramics and in manufacture of glass. { 'kō,bȯlt 'äk,sīd }

cobalt potassium nitrite [INORG CHEM] $K_3Co(NO_2)_6$ A yellow powder which decomposes at the melting point of 200°C; used in medicine and as a yellow pigment. Also known as cobalt yellow; Fischer's salt; potassium cobaltinitrite. { 'kō,bȯlt pə'tas·ē·əm 'nī,trīt }

cobalt sulfate [INORG CHEM] Any compound of either divalent or trivalent cobalt and the sulfate group; anhydrous cobaltous sulfate, $CoSO_4$, contains divalent cobalt, has a melting point of 96.8°C, is soluble in methanol, and is utilized to prepare pigments and cobalt salts; cobaltic sulfate, $Co_2(SO_4)_3 \cdot 18H_2O$, contains trivalent cobalt, is soluble in sulfuric acid, and functions as an oxidizing agent. { 'kō,bȯlt 'səl,fāt }

cobalt trifluoride [INORG CHEM] CoF_3 A brownish powder that reacts with water to form a precipitate of cobaltic hydroxide; used as a fluorinating agent. Also known as cobaltic fluoride. { 'kō,bȯlt trī'flůr,īd }

cobalt ultramarine See cobalt blue. { 'kō,bȯlt ,əl·trə·mə'rēn }

cobalt yellow See cobalt potassium nitrite. { 'kō,bȯlt 'yel·ō }

cochineal [CHEM] A red dye made of the dried bodies of the female cochineal insect (*Coccus cacti*), found in Central America and Mexico; used as a biological stain and indicator. { 'käch·ə,nēl }

cochineal solution [ANALY CHEM] An indicator in acid-base titration. { 'käch·ə,nēl sə'lü·shən }

cochinilin See carminic acid. { kō'chin·ə·lən }

cocodyl oxide [ORG CHEM] $(CH_3)_2AsOAs(CH_3)_2$ A liquid that has an obnoxious odor; slightly soluble in water, soluble in alcohol and ether; boils at 150°C. Also known as alkarsine; bisdimethyl arsenic oxide; dicacodyl oxide. { 'kō·kə·dəl 'äk,sīd }

codimer [ORG CHEM] **1.** A copolymer formed from the polymerization of two dissimilar olefin molecules. **2.** The product of polymerization of isobutylene with one of the two normal butylenes. { ¦kō'dī·mər }

coelute [ANALY CHEM] In chromatography, two or more chemical compounds that do not separate. { 'kō·ə,lüt }

cognac oil See ethyl enanthate. { 'kōn,yak ,ȯil }

coherent precipitate [PHYS CHEM] A precipitate that is a continuation of the lattice structure of the solvent and has no phase or grain boundary. { kō'hir·ənt prə'sip·ə,tāt }

83

coion |ANALY CHEM| Any of the small ions entering a solid ion exchanger and having the same charge as that of the fixed ions.   { kō'ī͵än }

colchicine |ORG CHEM| $C_{22}H_{25}O_6N$ An alkaloid extracted from the stem of the autumn crocus; used experimentally to inhibit spindle formation and delay centromere division, and medicinally in the treatment of gout.   { 'käl·chə͵sēn }

colcothar |INORG CHEM| Red ferric oxide made by heating ferrous sulfate in the air; used as a pigment and as an abrasive in polishing glass.   { 'käl·kə͵thär }

collection trap |ANALY CHEM| Cooled device to collect gas-chromatographic eluent, holding it for subsequent compound-identification analysis.   { kə'lek·shən ͵trap }

2,4,6-collidine |ORG CHEM| $(CH_3)_3C_5H_2N$ A liquid boiling at 170.4°C; slightly soluble in water, soluble in alcohol; used as a chemical intermediate.   { ͵tü ͵fòr ͵siks 'käl·ə͵dēn }

colligative properties |PHYS CHEM| Properties dependent on the number of molecules but not their nature.   { kə'lig·ə·div ͵präp·ərd·ēz }

collision broadening See collision line-broadening.   { kə'lizh·ən ͵bròd·ən·iŋ }

collision diameter |PHYS CHEM| The distance between the centers of two molecules taking part in a collision at the time of their closest approach.   { kə'lizh·ən dī͵am·əd·ər }

collision line-broadening |SPECT| Spreading of a spectral line due to interruption of the radiation process when the radiator collides with another particle.   Also known as collision broadening.   { kə'lizh·ən 'līn ͵bròd·ən·iŋ }

collision theory |PHYS CHEM| Theory of chemical reaction proposing that the rate of product formation is equal to the number of reactant-molecule collisions multiplied by a factor that corrects for low-energy-level collisions.   { kə'lizh·ən ͵thē·ə·rē }

collodion |ORG CHEM| Cellulose nitrate deposited from a solution of 60% ether and 40% alcohol, used for making fibers and film and in membranes for dialysis.   { kə'lōd·ē·ən }

collodion cotton See pyroxylin.   { kə'lōd·ē·ən ͵kät·ən }

collodion replication |ANALY CHEM| Production of a faithful collodion-film mold of a specimen surface (for example, powders, bones, microorganisms, crystals) which is sufficiently thin to be studied by electron microscopy.   { kə'lōd·ē·ən rep·lə'kā·shən }

colloid |CHEM| The phase of a colloidal system made up of particles having dimensions of 10–10,000 angstroms (1–1000 nanometers) and which is dispersed in a different phase.   { 'käl͵òid }

colloidal crystal |CHEM| A periodic array of suspended colloidal particles that can arise spontaneously in a monodisperse colloidal system under appropriate conditions.   { kə'lòid·əl 'krist·əl }

colloidal dispersion See colloidal system.   { kə'lòid·əl dis'pər·zhən }

colloidal electrolyte |PHYS CHEM| An electrolyte that yields at least one type of ion in the colloidal size range.   { kə'lòid·əl i'lek·trə͵līt }

colloidal suspension See colloidal system.   { kə'lòid·əl səs'pen·shən }

colloidal system |CHEM| An intimate mixture of two substances, one of which, called the dispersed phase (or colloid), is uniformly distributed in a finely divided state through the second substance, called the dispersion medium (or dispersing medium); the dispersion medium or dispersed phase may be a gas, liquid, or solid.   Also known as colloidal dispersion; colloidal suspension.   { kə'lòid·əl 'sis·təm }

colloid chemistry |PHYS CHEM| The scientific study of matter whose size is approximately 10 to 10,000 angstroms (1 to 1000 nanometers), and which exists as a suspension in a continuous medium, especially a liquid, solid, or gaseous substance.   { ͵käl͵òid 'kem·ə·strē }

color comparator |ANALY CHEM| A photoelectric instrument that compares an unknown color with that of a standard color sample for matching purposes.   Also known as photoelectric color comparator.   { 'kəl·ər kəm'par·əd·ər }

colorimeter |ANALY CHEM| A device for measuring concentration of a known constituent in solution by comparison with colors of a few solutions of known concentration of that constituent.   Also known as chromometer.   { ͵kəl·ə'rim·əd·ər }

color stability |CHEM| Resistance of materials to change in color that can be caused by light or aging, as of petroleum or whiskey.   { 'kəl·ər stə'bil·əd·ē }

color standard [ANALY CHEM] Liquid solution of known chemical composition and concentration, hence of known and standardized color, used for optical analysis of samples of unknown strength. { 'kəl·ər ˌstan·dərd }

color test [ANALY CHEM] The quantitative analysis of a substance by comparing the intensity of the color produced in a sample by a reagent with a standard color produced similarly in a solution of known strength. { 'kəl·ər ˌtest }

color throw [ANALY CHEM] In an ion-exchange process, discoloration of the liquid passing through the bed. { 'kəl·ər ˌthrō }

columbium See niobium. { kə'ləm·bē·əm }

column [ANALY CHEM] In chromatography, a tube holding the stationary phase through which the mobile phase is passed. { 'käl·əm }

column bleed [ANALY CHEM] The loss of carrier liquid during gas chromatography due to evaporation into the gas under analysis. { 'käl·əm ˌblēd }

column chromatography [ANALY CHEM] Chromatographic technique of two general types: packed columns usually contain either a granular adsorbent or a granular support material coated with a thin layer of high-boiling solvent (partitioning liquid); open-tubular columns contain a thin film of partitioning liquid on the column walls and have an opening so that gas can pass through the center of the column. { 'käl· əm ˌkrō·mə'täg·rə·fē }

column development chromatography [ANALY CHEM] Columnar apparatus for separating or concentrating one or more components from a physical mixture by use of adsorbent packing; as the specimen percolates along the length of the adsorbent, its various components are preferentially held at different rates, effecting a separation. { 'käl·əm də'vel·əp·mənt ˌkrō·mə'täg·rə·fē }

combination principle See Ritz's combination principle. { ˌkäm·bə'nā·shən ˌprin·sə· pəl }

combination reaction [CHEM] A chemical reaction in which two reactions combine to form a single product. { ˌkäm·bə'nā·shən rē̩ak·shən }

combination vibration [SPECT] A vibration of a polyatomic molecule involving the simultaneous excitation of two or more normal vibrations. { ˌkäm·bə'nā·shən vī'brā·shən }

combinatorial chemistry [ORG CHEM] A method for reacting a small number of chemicals to produce simultaneously a very large number of compounds, called libraries, which are screened to identify useful products such as drug candidates. { kəm̩bīn· ə̦tȯr·ē·əl 'kem·ə·strē }

combined carbon [CHEM] Carbon that is chemically combined within a compound, as contrasted with free or uncombined elemental carbon. { kəm'bīnd 'kär·bən }

combined cyanide [ORG CHEM] The cyanide portion of a complex ion composed of cyanide and a metal. { kəm'bīnd 'sī·ə̦nīd }

combining-volumes principle [CHEM] The principle that when gases take part in chemical reactions the volumes of the reacting gases and those of the products (if gaseous) are in the ratio of small whole numbers, provided that all measurements are made at the same temperature and pressure. Also known as Gay-Lussac's law of volumes. { kəm̦bīn·iŋ ̦väl·yəmz ˌprin·sə·pəl }

combining weight [CHEM] The weight of an element that chemically combines with 8 grams of oxygen or its equivalent. { kəm'bīn·iŋ ˌwāt }

comb polymer [ORG CHEM] A macromolecule in which the main chain has one long branch per repeat unit. { ̦kōm 'päl·ə·mər }

combustion [CHEM] The burning of gas, liquid, or solid, in which the fuel is oxidized, evolving heat and often light. { kəm'bəs·chən }

combustion efficiency [CHEM] The ratio of heat actually developed in a combustion process to the heat that would be released if the combustion were perfect. { kəm'bəs·chən i'fish·ən·sē }

combustion furnace [ANALY CHEM] A heating device used in the analysis of organic compounds for elements. { kəm'bəs·chən ˌfər·nəs }

combustion rate [CHEM] The rate of burning of any substance. { kəm'bəs·chən ˌrāt }

85

combustion train

combustion train |ANALY CHEM| The arrangement of apparatus for elementary organic analysis. { kəm'bəs·chən ˌtrān }

combustion tube |ANALY CHEM| A glass, silica, or porcelain tube, resistant to high temperatures, that is a component of a combustion train. { kəm'bəs·chən ˌtüb }

combustion wave |CHEM| **1.** A zone of burning propagated through a combustible medium. **2.** The zoned, reacting, gaseous material formed when an explosive mixture is ignited. { kəm'bəs·chən ˌwāv }

common cause |ANALY CHEM| A cause of variability in a measurement process that is inherent in and common to the process itself. { 'käm·ən 'kȯz }

common-ion effect |CHEM| The lowering of the degree of ionization of a compound when another ionizable compound is added to a solution; the compound added has a common ion with the other compound. { ¦käm·ən ¦ī,än i'fekt }

common salt See sodium chloride. { ¦käm·ən 'sȯlt }

comonomer |CHEM| One of the compounds used to produce a specific polymeric product. { ¦kō'män·ə·mər }

comparator |ANALY CHEM| An instrument used to determine the concentration of a solution by comparing the intensity of color with a series of standard colors. { kəm'par·əd·ər }

comparator-densitometer |ANALY CHEM| Device that projects a labeled spectrum onto a screen adjacent to an enlarged image of the spectrum to be analyzed, allowing visual comparison. { ¦kəm'par·əd·ər den·sə'täm·əd·ər }

comparison spectrum |SPECT| A line spectrum whose wavelengths are accurately known, and which is matched with another spectrum to determine the wavelengths of the latter. { kəm'par·ə·sən ˌspek·trəm }

compatibilizer |ORG CHEM| Any polymeric interfacial agent that facilitates formation of uniform blends of normally immiscible polymers with desirable end properties. { kəm'pad·ə·bə,līz·ər }

competing equilibria condition |CHEM| The competition for a reactant in a complex chemical system in which several reactions are taking place at the same time. { kəm'pēd·iŋ ˌē·kwə'lib·rē·ə kən,dish·ən }

complete combustion |CHEM| Combustion in which the entire quantity of oxidizable constituents of a fuel is reacted. { kəm'plēt kəm'bəs·chən }

complexation See complexing. { ˌkäm,plek'sā·shən }

complexation analysis |ANALY CHEM| The determination of the ligand/metal ratio in a coordination complex. { ˌkäm·plek'sā·shən ə,nal·ə·səs }

complexation indicator See metal ion indicator. { ˌkäm·plek'sā·shən ˌin·də,kād·ər }

complexation reaction |CHEM| A chemical reaction that takes place between a metal ion and a molecular or ionic entity known as a ligand that contains at least one atom with an unshared pair of electrons. { ˌkäm·plek'sā·shən rē,ak·shən }

complex chemical reaction |CHEM| A chemical system in which a number of chemical reactions take place simultaneously, including reversible reactions, consecutive reactions, and concurrent or side reactions. { 'käm,pleks ¦kem·i·kəl rē'ak·shən }

complex compound |CHEM| Any of a group of chemical compounds in which a part of the molecular bonding is of the coordinate type. Also known as coordination complex. { 'käm,pleks 'käm,paúnd }

compleximetric titration See complexometric titration. { kəm,plek·sə¦me·trik tī'trā·shən }

complexing |CHEM| Formation of a complex compound. Also known as complexation. { 'käm,plek·siŋ }

complexing agent |CHEM| A substance capable of forming a complex compound with another material in solution. { 'käm,plek·siŋ ˌā·jənt }

complex ion |CHEM| A complex, electrically charged group of atoms or radical, for example, Cu(NH$_3$)$_2^{+2}$. { 'käm,pleks 'ī,än }

complexometric titration |ANALY CHEM| A technique of volumetric analysis in which the formation of a colored complex is used to indicate the end point of a titration. Also known as chelatometry. Also spelled compleximetric titration. { kəm¦plek·sə¦me·trik ˌtī'trā·shən }

86

complex salt [INORG CHEM] A class of salts in which there are no detectable quantities of each of the metal ions existing in solution; an example is $K_3Fe(CN)_6$, which in solution has K^+ but no Fe^{3+} because Fe is strongly bound in the complex ion, $Fe(CN)_6^{3-}$. { 'käm,pleks 'sȯlt }

component [CHEM] **1.** A part of a mixture. **2.** The smallest number of chemical substances which are able to form all the constituents of a system in whatever proportion they may be present. { kəm'pō·nənt }

component-substances law [CHEM] The law that each substance, singly or in mixture, composing a material exhibits specific properties that are independent of the other substances in that material. { kəm'pō·nənt 'sub·stən·səs ‚lȯ }

composite sample [ANALY CHEM] A sample comprising two or more increments selected to represent the material being analyzed. { kəm'päz·ət 'sam·pəl }

composition [CHEM] The elements or compounds making up a material or produced from it by analysis. { ‚käm·pə'zish·ən }

compound [CHEM] A substance whose molecules consist of unlike atoms and whose constituents cannot be separated by physical means. Also known as chemical compound. { 'käm,paúnd }

Compton rule [PHYS CHEM] An empirical law stating that the heat of fusion of an element times its atomic weight divided by its melting point in degrees Kelvin equals approximately 2. { 'käm·tən ‚rül }

computational chemistry [CHEM] The use of calculations to predict molecular structure, properties, and reactions. { ‚käm·pyə'tā·shən·əl 'kem·ə·strē }

concave grating [SPECT] A reflection grating which both collimates and focuses the light falling upon it, made by spacing straight grooves equally along the chord of a concave spherical or paraboloid mirror surface. Also known as Rowland grating. { 'kän,kāv 'grād·iŋ }

concentrate [CHEM] To increase the amount of a dissolved substance by evaporation. { 'kän·sən,trāt }

concentration [CHEM] In solutions, the mass, volume, or number of moles of solute present in proportion to the amount of solvent or total solution. { ‚kän·sən'trā·shən }

concentration cell [PHYS CHEM] **1.** Electrochemical cell for potentiometric measurement of ionic concentrations where the electrode potential electromotive force produced is determined as the difference in emf between a known cell (concentration) and the unknown cell. **2.** An electrolytic cell in which the electromotive force is due to a difference in electrolyte concentrations at the anode and the cathode. { ‚kän·sən'trā·shən ‚sel }

concentration gradient [CHEM] The graded difference in the concentration of a solute throughout the solvent phase. { ‚kän·sən'trā·shən ‚grād·ē·ənt }

concentration polarization [PHYS CHEM] That part of the polarization of an electrolytic cell resulting from changes in the electrolyte concentration due to the passage of current through the solution. { ‚kän·sən'trā·shən ‚pō·lə·rə'zā·shən }

concentration potential [CHEM] Tendency for a univalent electrolyte to concentrate in a specific region of a solution. { ‚kän·sən'trā·shən pə'ten·shəl }

concentration scale [CHEM] Any of several numerical systems defining the quantitative relation of the components of a mixture; for solutions, concentration is expressed as the mass, volume, or number of moles of solute present in proportion to the amount of solvent or total solution. { ‚kän·sən'trā·shən ‚skāl }

concerted reaction [ORG CHEM] A reaction in which there is a simultaneous occurrence of bond making and bond breaking. { kən'sərd·əd rē'ak·shən }

concomitant [ANALY CHEM] Any species in a material undergoing chemical analysis other than the analyte or the solvent in which the sample is dissolved. { kən'käm·ə·tənt }

condensable vapors [CHEM] Gases or vapors which when subjected to appropriately altered conditions of temperature or pressure become liquids. { kən'den·sə·bəl 'va·pərz }

condensation [CHEM] Transformation from a gas to a liquid. { ‚kän·dən'sā·shən }

87

condensation polymer

condensation polymer |ORG CHEM| A high-molecular-weight compound formed by condensation polymerization. { ˌkän·dən'sā·shən 'päl·ə·mər }

condensation polymerization |ORG CHEM| The stepwise reaction between functional groups of reactants in which a high-molecular-weight polymer is formed only after a large number of steps, for example, the reaction of dicarboxylic acids with diamines to form a polyamide. { ˌkän·dən'sā·shən pəˌlim·ə·rə'zā·shən }

condensation reaction |CHEM| One of a class of chemical reactions involving a combination between molecules or between parts of the same molecule. { ˌkän·dən'sā·shən rē'ak·shən }

condensation resin |ORG CHEM| A resin formed by polycondensation. { ˌkän·dən'sā·shən 'rez·ən }

condensation temperature |ANALY CHEM| In boiling-point determination, the temperature established on the bulb of a thermometer on which a thin moving film of liquid coexists with vapor from which the liquid has condensed, the vapor phase being replenished at the moment of measurement from a boiling-liquid phase. { ˌkän·dən'sā·shən 'tem·prə·chər }

condensed phase |PHYS CHEM| Either the solid or liquid phase of a material. { kən'denst ˌfāz }

condensed structural formula |CHEM| A structural representation of a compound that includes all of the atoms present in a molecule or other chemical entity but represents only certain bonds as lines in order to emphasize a structural characteristic. { kənˌdenst ˌstrək·chər·əl 'fȯr·myə·lə }

condensed system |PHYS CHEM| A chemical system in which the vapor pressure is negligible or in which the pressure maintained on the system is greater than the vapor pressure of any portion. { kən'denst 'sis·təm }

conductance coefficient |PHYS CHEM| The ratio of the equivalent conductance of an electrolyte, at a given concentration of solute, to the limiting equivalent conductance of the electrolyte as the concentration of the electrolyte approaches 0. { kənˌdak·təns ˌkō·ə'fish·ənt }

conductimetry |CHEM| The scientific study of conductance measurements of solutions; to avoid electrolytic complications, conductance measurements are usually taken with alternating current. { kän·dək'tim·ə·trē }

conductometric titration |ANALY CHEM| A titration in which electrical conductance of a solution is measured during the course of the titration. { kənˌdək·tə'me,trik tī'trā·shən }

configuration |CHEM| The three-dimensional spatial arrangement of atoms in a stable or isolable molecule. { kənˌfig·yə'rā·shən }

configuration interaction |PHYS CHEM| Interaction between two different possible arrangements of the electrons in an atom (or molecule); the resulting electron distribution, energy levels, and transitions differ from what would occur in the absence of the interaction. { kənˌfig·yə'rā·shən in·tər'ak·shən }

conformation |ORG CHEM| In a molecule, a specific orientation of the atoms that varies from other possible orientations by rotation or rotations about single bonds; generally in mobile equilibrium with other conformations of the same structure. Also known as conformational isomer; conformer. { kän·fər'mā·shən }

conformational analysis |PHYS CHEM| The determination of the arrangement in space of the constituent atoms of a molecule that may rotate about a single bond. { kän·fər'mā·shən·əl ə'nal·ə·səs }

conformational isomer See conformation. { kän·fər'mā·shən·əl 'ī·sə·mər }

conformer See conformation. { kən'fȯr·mər }

congener |CHEM| A chemical substance that is related to another substance, such as a derivative of a compound or an element belonging to the same family as another element in the periodic table. { 'kän·jə·nər }

conglomerate See racemate. { kən'gläm·ə·rət }

congo red |ORG CHEM| $C_{32}H_{22}N_6Na_2O_6S_2$ An azo dye, sodium diphenyldiazo-bis-α-naphthylamine sulfonate, used as a biological stain and as an acid-base indicator; it is red in alkaline solution and blue in acid solution. { 'käŋ·gō 'red }

88

conjugate acid-base pair [CHEM] An acid and a base related by the ability of the acid to generate the base by loss of a proton. { 'kän·jə·gət ¦as·əd ¦bās 'per }

conjugated diene [ORG CHEM] An acyclic hydrocarbon with a molecular structure containing two carbon-carbon double bonds separated by a single bond. { 'kän·jə₁gäd· əd 'dī₁ēn }

conjugated polyene [ORG CHEM] An acyclic hydrocarbon with a molecular structure containing alternating carbon-carbon double and single bonds. { 'kän·jə₁gäd·əd 'päl·ē₁ēn }

conservation of orbital symmetry See Woodward-Hoffmann rule. { ₁kän·sər'vā·shən əv 'ȯr·bəd·əl 'sim·ə·trē }

consolute [CHEM] Of or pertaining to liquids that are perfectly miscible in all proportions under certain conditions. { 'kan·sə₁lüt }

constant-current electrolysis [CHEM] Electrolysis in which a constant current flows through the cell; used in electrodeposition analysis. { ¦kän·stənt 'kər·ənt i₁lek'träl·ə·səs }

constant-current titration See potentiometric titration. { ¦kän·stənt 'kər·ənt tī'trā·shən }

constant-deviation spectrometer [SPECT] A spectrometer in which the collimator and telescope are held fixed and the observed wavelength is varied by rotating the prism or diffraction grating. { ¦kän·stənt ₁dē·vē¦ā·shən spek'träm·əd·ər }

constant-potential electrolysis [CHEM] Electrolysis in which a constant voltage is applied to the cell; used in electrodeposition analysis. { ¦kän·stənt pə'ten·chəl i₁lek'träl·ə·səs }

constant series See displacement series. { 'kän·stənt 'sir₁ēz }

constitutional isomers [ORG CHEM] Isomers which differ in the manner in which their atoms are linked. Also known as structural isomers. { ₁kän·stə'tü·shən·əl 'T·sə·mərz }

constitutional unit [CHEM] An atom or group of atoms that is part of a chain in a polymer or oligomer. { ₁kän·stə'tü·shən·əl 'yü·nət }

constitutive property [CHEM] Any physical or chemical property that depends on the constitution or structure of the molecule. { 'kän·stə₁tüd·iv 'präp·ərd·ē }

contact acid [INORG CHEM] Sulfuric acid produced by the contact process. { 'kän₁takt 'as·əd }

contact ion pair [ORG CHEM] An ion pair composed of individual ions which keep their stereochemical configuration; no solvent molecules separate the cation and anion. Also known as intimate ion pair. { ¦kän₁takt 'T·ən ₁per }

contemporary carbon [CHEM] The isotopic carbon content of living matter, based on the assumption of a natural proportion of carbon-14. { kən'tem·pə₁rer·ē 'kär·bən }

continuous phase [CHEM] The liquid in a disperse system in which solids are suspended or droplets of another liquid are dispersed. Also known as dispersion medium, external phase. { kən¦tin·yə·wəs 'fāz }

continuous spectrum [SPECT] A radiation spectrum which is continuously distributed over a frequency region without being broken up into lines or bands. { kən¦tin·yə·wəs 'spek·trəm }

continuous titrator [ANALY CHEM] A titrator so equipped that a reservoir refills the buret. { kən¦tin·yə·wəs 'tī₁träd·ər }

contributing structure [ORG CHEM] A structural formula that is one of a set of formulas, each contributing to the total wave function of a molecule. Also known as canonical form; canonical structure. { kən¦trib·yəd·ing 'strək·chər }

control sample [ANALY CHEM] A material of known composition that is analyzed along with test samples in order to evaluate the accuracy of an analytical procedure. Also known as check sample. { kən'trōl ₁sam·pəl }

convergence limit [SPECT] **1.** The short-wavelength limit of a set of spectral lines that obey a Rydberg series formula; equivalently, the long-wavelength limit of the continuous spectrum corresponding to ionization from or recombination to a given state. **2.** The wavelength at which the difference between successive vibrational bands in a molecular spectrum decreases to 0. { kən'vər·jəns ₁lim·ət }

89

convergence pressure [PHYS CHEM] The pressure at which the different constant-temperature K (liquid-vapor equilibrium) factors for each member of a two-component system converge to unity. { kən'vər·jəns ˌpresh·ər }

conversion [CHEM] Change of a compound from one isomeric form to another. { kən'vər·zhən }

cool flame [CHEM] A faint, luminous phenomenon observed when, for example, a mixture of ether vapor and oxygen is slowly heated; it proceeds by diffusion of reactive molecules which initiate chemical processes as they go. { ¦kül ¦flām }

coordinate bond See coordinate valence. { kō'órd·ən·ət 'bänd }

coordinated complex See coordination compound. { kō'órd·ən‚ād·əd 'käm‚pleks }

coordinate valence [CHEM] A chemical bond between two atoms in which a shared pair of electrons forms the bond and the pair has been supplied by one of the two atoms. Also known as coordinate bond; dative bond. { kō'órd·ən·ət 'vā·ləns }

coordination chemistry [CHEM] The chemistry of metal ions in their interactions with other molecules or ions. { kō‚órd·ən'ā·shən 'kem·ə·strē }

coordination complex See complex compound. { kō‚órd·ən'ā·shən 'käm‚pleks }

coordination compound [CHEM] A compound with a central atom or ion and a group of ions or molecules surrounding it. Also known as coordinated complex; Werner complex. { kō‚órd·ən'ā·shən ‚käm‚paúnd }

coordination polygon [CHEM] The symmetrical polygonal chemical structure of simple polyatomic aggregates having coordination numbers of 4 or less. { kō‚órd·ən'ā·shən ‚päl·i‚gän }

coordination polyhedron [CHEM] The symmetrical polyhedral chemical structure of relatively simple polyatomic aggregates having coordination numbers of 4 to 8. { kō‚órd·ən'ā·shən ‚päl·i·hē·drən }

coordination polymer [ORG CHEM] Organic addition polymer that is neither free-radical nor simply ionic; prepared by catalysts that combine an organometallic (for example, triethyl aluminum) and a transition metal compound (for example, $TiCl_4$). { kō‚órd·ən'ā·shən ‚päl·ə·mər }

copolymer [ORG CHEM] A mixed polymer, the product of polymerization of two or more substances at the same time. { kō'päl·i·mər }

copolymerization [CHEM] A polymerization reaction that forms a copolymer. { ‚kō·pə‚lim·ə·rə'zā·shən }

copper [CHEM] A chemical element, symbol Cu, atomic number 29, atomic weight 63.546. { 'käp·ər }

copper acetate See cupric acetate. { 'käp·ər 'as·ə‚tāt }

copper arsenate [INORG CHEM] $Cu_3(AsO_4)_2\cdot4H_2O$ or $Cu_5H_2(AsO_4)_4\cdot2H_2O$ Bluish powder, soluble in ammonium hydroxide and dilute acids, insoluble in water and alcohol; used as a fungicide and insecticide. { 'käp·ər 'ärs·ən‚āt }

copper arsenite [INORG CHEM] $CuHAsO_3$ A toxic, light green powder which is soluble in acids and decomposes at the melting point; used as a pigment and insecticide. Also known as copper orthoarsenite; cupric arsenite; Scheele's green. { 'käp·ər 'ärs·ən‚īt }

copperas See ferrous sulfate. { 'käp·ə·rəs }

copper blue See mountain blue. { 'käp·ər ¦blü }

copper bromide See cupric bromide; cuprous bromide. { 'käp·ər 'brō‚mīd }

copper carbonate [INORG CHEM] $Cu_2(OH)_2CO_3$ A toxic, green powder; decomposes at 200°C and is soluble in acids; used in pigments and pyrotechnics and as a fungicide and feed additive. Also known as artificial malachite; cupric carbonate; mineral green. { 'käp·ər 'kär·bə‚nāt }

copper chloride See cupric chloride; cuprous chloride. { 'käp·ər 'klór‚īd }

copper chromate See cupric chromate. { 'käp·ər 'krō‚māt }

copper cyanide See cupric cyanide. { 'käp·ər sī·ə‚nīd }

copper-8-quinolinolate [ORG CHEM] $C_{18}H_{14}N_2O_2Cu$ A khaki-colored, water-insoluble solid used as a fungicide in fruit-handling equipment. { 'käp·ər ¦āt ¦kwīn·ə¦lin·ə‚lāt }

copper fluoride See cupric fluoride; cuprous fluoride. { 'käp·ər 'flúr‚īd }

copper gluconate [ORG CHEM] [CH₂OH(CHOH)₄COO]₂Cu A light blue, crystalline powder; soluble in water; used in medicine and as a dietary supplement. Also known as cupric gluconate. { 'käp·ər 'glü·kə,nāt }

copper hydroxide See cupric hydroxide. { 'käp·ər hī'dräk,sīd }

copper nitrite See cupric nitrate. { 'käp·ər 'nī,trīt }

copper number [ANALY CHEM] The number of milligrams of copper obtained by the reduction of Benedict's or Fehling's solution by 1 gram of carbohydrate. { 'käp·ər ,nəm·bər }

copper oleate [ORG CHEM] Cu[OOC(CH₂)₇CH=CH(CH₂)₇CH₃]₂ A green-blue liquid, used as a fungicide for fruits and vegetables. { 'käp·ər 'ō·lē,āt }

copperon See cupferron. { 'käp·ə,rän }

copper orthoarsenite See copper arsenite. { 'käp·ər ,ȯr·thō'ärs·ən,īt }

copper oxide See cupric oxide; cuprous oxide. { 'käp·ər 'äk,sīd }

copper resinate [ORG CHEM] Poisonous green powder, soluble in oils and ether, insoluble in water; made by heating rosin oil with copper sulfate, followed by filtering and drying of the resultant solids; used as a metal-paint preservative and insecticide. { 'käp·ər 'rez·ən,āt }

copper sulfate See cupric sulfate. { 'käp·ər 'səl,fāt }

copper sulfide [INORG CHEM] CuS Black, monoclinic or hexagonal crystals that break down at 220°C; used in paints on ship bottoms to prevent fouling. { 'käp·ər 'səl,fīd }

coprecipitation [CHEM] Simultaneous precipitation of more than one substance. { ¦kō·prə,sip·ə'tā·shən }

coriandrol See linalool. { ,kȯr·ē'an,drȯl }

Coriolis operator [SPECT] An operator which gives a large contribution to the energy of an axially symmetric molecule arising from the interaction between vibration and rotation when two vibrations have equal or nearly equal frequencies. { kȯr·ē'ō·ləs ,äp·ə,rād·ər }

Coriolis resonance interactions [SPECT] Perturbation of two vibrations of a polyatomic molecule, having nearly equal frequencies, on each other, due to the energy contribution of the Coriolis operator. { kȯr·ē'ō·ləs 'rez·ən·əns ,in·tər,ak·shənz }

corresponding states [PHYS CHEM] The condition when two or more substances are at the same reduced pressures, the same reduced temperatures, and the same reduced volumes. { ,kär·ə'spänd·iŋ 'stāts }

corrosion inhibitor [PHYS CHEM] A compound or material deposited as a film on a metal surface that either provides physical protection against corrosive attack or reduces the open-circuit potential difference between local anodes and cathodes and increases the polarization of the former. { kə'rōzh·ən in,hib·əd·ər }

corrosive sublimate See mercuric chloride. { kə'rō·siv 'səb·lə,māt }

cotectic [PHYS CHEM] Referring to conditions of pressure, temperature, and composition under which two or more solid phases crystallize at the same time, with no resorption, from a single liquid over a finite range of decreasing temperature. { kō'tek·tik }

cotectic crystallization [PHYS CHEM] Simultaneous crystallization of two or more solid phases from a single liquid over a finite range of falling temperature without resorption. { kō'tek·tik ,krist·əl·ə'zā·shən }

cotinine [ORG CHEM] The major metabolic product of nicotine which is excreted in the urine; used as a marker for environmental tobacco smoke. { 'kōt·ən,ēn }

Cotton effect [ANALY CHEM] The characteristic wavelength dependence of the optical rotatory dispersion curve or the circular dichroism curve or both in the vicinity of an absorption band. { 'kät·ən i'fekt }

coudé spectrograph [SPECT] A stationary spectrograph that is attached to the tube of a coudé telescope. { kü'dā 'spek·trə,graf }

coudé spectroscopy [SPECT] The production and investigation of astronomical spectra using a coudé spectrograph. { kü'dā spek'träs·kə·pē }

Coulomb energy [PHYS CHEM] The energy associated with the electrostatic interaction between two or more electron distributions in terms of which the actual electron distribution of a covalent bond is described. { 'kü,läm ,en·ər·jē }

91

coulometer

coulometer [PHYS CHEM] An electrolytic cell for the precise measurement of electrical quantities or current intensity by quantitative determination of chemical substances produced or consumed. Also known as voltameter. { kü'läm·əd·ər }

coulometric analysis [ANALY CHEM] A technique in which the amount of a substance is determined quantitatively by measuring the total amount of electricity required to deplete a solution of the substance. { ‚kü·lə'me·trik ə'nal·ə·səs }

coulometric titration [ANALY CHEM] The slow electrolytic generation of a soluble species which is capable of reacting quantitatively with the substance sought; some independent property must be observed to establish the equivalence point in the reaction. { ‚kü·lə'me·trik tī'trā·shən }

coulometry [ANALY CHEM] A determination of the amount of an electrolyte released during electrolysis by measuring the number of coulombs used. { kə'läm·ə·trē }

coulostatic analysis [PHYS CHEM] An electrochemical technique involving the application of a very short, large pulse of current to the electrode; the pulse charges the capacitive electrode-solution interface to a new potential, then the circuit is opened, and the return of the working electrode potential to its initial value is monitored; the current necessary to discharge the electrode interface comes from the electrolysis of electroactive species in solution; the change in electrode potential versus time results in a plot, the shape of which is proportional to concentration. { ‚kü·lə'stad·ik ə'nal·ə·səs }

coumachlor [ORG CHEM] $C_{19}H_{15}ClO_4$ A white, crystalline compound with a melting point of 169–171°C; insoluble in water; used as a rodenticide. { 'kü·mə‚klȯr }

coumarin [ORG CHEM] $C_9H_6O_2$ The anhydride of o-coumaric acid; a toxic, white, crystalline lactone found in many plants and made synthetically; used in making perfume and soap. Also known as 1,2-benzopyrone. { 'kü·mə·rən }

coumarone [ORG CHEM] C_8H_6O A colorless liquid, boiling point 169°C. { 'kü·mə‚rōn }

coumarone-indene resin [ORG CHEM] A synthetic resin prepared by polymerization of coumarone and indene. { 'kü·mə‚rōn 'in‚dēn ‚rez·ən }

coumatetralyl [ORG CHEM] $C_{19}H_{16}O_3$ A yellow-white, crystalline compound with a melting point of 172–176°C; slightly soluble in water; used as a rodenticide. { 'kü·mə¦te·trə‚lil }

count [CHEM] An ionizing event. { kaünt }

countercurrent cascade [ANALY CHEM] An extraction process involving the introduction of a sample, all at once, into a continuously flowing countercurrent system where both phases are moving in opposite directions and are continuously at equilibrium. { 'kaünt·ər‚kər·ənt kas'kād }

counterion [PHYS CHEM] In a solution, an ion with a charge opposite to that of another ion included in the ionic makeup of the solution. { 'kaünt·ər‚ī‚än }

couple [CHEM] Joining of two molecules. { 'kəp·əl }

coupled reaction [CHEM] A reaction which involves two oxidants with a single reductant, where one reaction taken alone would be thermodynamically unfavorable. { 'kəp·əld rē'ak·shən }

coupling agent [CHEM] A substance that can react with both reinforcement and matrix components of a composite material to form a binding link at their interface. { 'kəp·liŋ ‚ā·jənt }

covalence [CHEM] The number of covalent bonds which an atom can form. { kō'vā·ləns }

covalent bond [CHEM] A bond in which each atom of a bound pair contributes one electron to form a pair of electrons. Also known as electron pair bond. { kō'vā·lənt 'bänd }

covalent hydride [INORG CHEM] A compound formed from a nonmetal and hydrogen, for example, H_2S and NH_3. { kō'vā·lənt 'hī‚drīd }

covalent radius See atomic radius. { kō'vā·lənt 'rād·ē·əs }

Cox chart [CHEM] A straight-line graph of the logarithm of vapor pressure against a special nonuniform temperature scale; vapor pressure-temperature lines for many substances intersect at a common point on the Cox chart. { 'käks ‚chärt }

cp See chemically pure.

Cr *See* chromium.
crack [CHEM] To break a compound into simpler molecules. { krak }
cream of tartar *See* potassium bitartrate. { 'krēm əv 'tärd·ər }
creosol [ORG CHEM] $CH_3O(CH_3)C_6H_3OH$ A combination of isomers, derived from coal tar or petroleum; a yellowish liquid with a phenolic odor; used as a disinfectant, in the manufacture of resins, and in flotation of ore. Also known as hydroxymethylbenzene; methyl phenol. { 'krē·ə,sōl }
cresol [ORG CHEM] $CH_3C_6H_4OH$ One of three poisonous, colorless isomeric methyl phenols: o-cresol, m-cresol, p-cresol; used in the production of phenolic resins, tricresyl phosphate, disinfectants, and solvents. { 'krē,sól }
cresol red [ORG CHEM] $C_{21}H_{18}O_5S$ A compound derived from o-cresol and used as an acid-base indicator; color change is yellow to red at pH 0.4 to 1.8, or 7.0 to 8.8, depending on preparation. { 'krē,sól 'red }
cricondenbar [PHYS CHEM] Maximum pressure at which two phases (for example, liquid and vapor) can coexist. { krə'kän·dən,bär }
cricondentherm [PHYS CHEM] Maximum temperature at which two phases (for example, liquid and vapor) can coexist. { krə'kän·dən,thərm }
critical absorption wavelength [SPECT] The wavelength, characteristic of a given electron energy level in an atom of a specified element, at which an absorption discontinuity occurs. { 'krid·ə·kəl əb'sórp·shən 'wāv,leŋkth }
critical condensation temperature [PHYS CHEM] The temperature at which the sublimand of a sublimed solid recondenses; used to analyze solid mixtures, analogous to liquid distillation. Also known as true condensing point. { 'krid·ə·kəl ,kän·dən'sā·shən ,tem·prə·chər }
critical constant [PHYS CHEM] A characteristic temperature, pressure, and specific volume of a gas above which it cannot be liquefied. { 'krid·ə·kəl 'kän·stənt }
critical current density [PHYS CHEM] The amount of current per unit area of electrode at which an abrupt change occurs in a variable of an electrolytic process. { 'krid·ə·kəl 'kər·ənt ,den·səd·ē }
critical density [CHEM] The density of a substance exhibited at its critical temperature and critical pressure. { 'krid·ə·kəl 'den·səd·ē }
critical line *See* critical locus. { 'krid·ə·kəl 'līn }
critical locus [PHYS CHEM] The line connecting the critical points of a series of liquid-gas phase-boundary loops for multicomponent mixtures plotted on a pressure versus temperature graph. Also known as critical line. { 'krid·ə·kəl 'lō·kəs }
critical micelle concentration [PHYS CHEM] The concentration of a micelle (oriented molecular arrangement of an electrically charged colloidal particle or ion) at which the rate of increase of electrical conductance with increase in concentration levels off or proceeds at a much slower rate. { 'krid·ə·kəl mi'sel ,kän·sən'trā·shən }
critical phenomena [PHYS CHEM] Physical properties of liquids and gases at the critical point (conditions at which two phases are just about to become one); for example, critical pressure is that needed to condense a gas at the critical temperature, and above the critical temperature the gas cannot be liquefied at any pressure. { 'krid·ə·kəl fə'näm·ə·nə }
critical point [PHYS CHEM] **1.** The temperature and pressure at which two phases of a substance in equilibrium with each other become identical, forming one phase. **2.** The temperature and pressure at which two ordinarily partially miscible liquids are consolute. { 'krid·ə·kəl 'póint }
critical properties [PHYS CHEM] Physical and thermodynamic properties of materials at conditions of critical temperature, pressure, and volume, that is, at the critical point. { 'krid·ə·kəl 'präp·ərd·ēz }
critical solution temperature [PHYS CHEM] The temperature at which a mixture of two liquids, immiscible at ordinary temperatures, ceases to separate into two phases. { 'krid·ə·kəl sə'lü·shən ,tem·prə·chər }
critical state [PHYS CHEM] Unique condition of pressure, temperature, and composition wherein all properties of coexisting vapor and liquid become identical. { 'krid·ə·kəl 'stāt }

93

critical temperature [PHYS CHEM] The temperature of the liquid-vapor critical point, that is, the temperature above which the substance has no liquid-vapor transition. Symbolized T_c. { 'krid·ə·kəl 'tem·prə·chər }

CRLAS See cavity ringdown laser absorption spectroscopy.

crosscurrent extraction [ANALY CHEM] Procedure of batchwise liquid-liquid extraction in a separatory funnel; solvent is added to the sample in the funnel, which is then shaken, and the extract phase is allowed to coalesce, then is drawn off. { 'krȯs,kər·ənt ik'strak·shən }

crosslink [ORG CHEM] The covalent bonds between adjacent polymer chains that lock the chains in place. { 'krȯs ,liŋk }

crosslinking [ORG CHEM] The setting up of chemical links between the molecular chains of polymers. { 'krȯs ,liŋk·iŋ }

crotonaldehyde [ORG CHEM] C_3H_5CHO A colorless liquid boiling at 104°C, soluble in water; vapors are lacrimatory; used as an intermediate in manufacture of n-butyl alcohol and quinaldine. Also known as propylene aldehyde. { ¦krōt·ən'al·də,hīd }

crotonic acid [ORG CHEM] C_3H_5COOH An unsaturated acid, with colorless, monoclinic crystals, soluble in water; used in the preparation of synthetic resins, plasticizers, and pharmaceuticals. { krō'tän·ik 'as·əd }

crown ether [ORG CHEM] A macrocyclic polyether whose structure exhibits a conformation with a so-called hole capable of trapping cations by coordination with a lone pair of electrons on the oxygen atoms. { 'kraún ,ē·thər }

crufomate [ORG CHEM] $C_{12}H_{19}ClNO_3P$ A white, crystalline compound, with a melting point of 61.8°C, which is insoluble in water; used both internally and externally for cattle parasites. { 'krü·fə,māt }

cryohydrate [CHEM] A salt that contains water of crystallization at low temperatures. Also known as cryosel. { ¦krī·ō'hī,drāt }

cryohydric point [PHYS CHEM] The eutectic point of an aqueous salt solution. { ,krī·ō'hī,drik ,pȯint }

cryoscopic constant [ANALY CHEM] Equation constant expressed in degrees per mole of pure solvent; used to calculate the freezing-point-depression effects of a solute. { ¦krī·ə¦skäp·ik 'kän·stənt }

cryoscopy [ANALY CHEM] A phase-equilibrium technique to determine molecular weight and other properties of a solute by dissolving it in a liquid solvent and then ascertaining the solvent's freezing point. { krī'äs·kə,pē }

cryosel See cryohydrate. { 'krī·ə,sel }

cryptand [ORG CHEM] A macropolycyclic polyazo-polyether, where the three-coordinate nitrogen atoms provide the vertices of a three-dimensional structure. { 'krip,tand }

cryptate [ORG CHEM] The adduct formed between a cryptand and a guest (cation, anion, or neutral species) molecular entity. { 'krip,tāt }

crystal aerugo See cupric acetate. { 'krist·əl ē'rü·gō }

crystal field theory [PHYS CHEM] The theory which assumes that the ligands of a coordination compound are the sources of negative charge which perturb the energy levels of the central metal ion and thus subject the metal ion to an electric field analogous to that within an ionic crystalline lattice. { ¦krist·əl 'fēld ,thē·ə·rē }

crystal grating [SPECT] A diffraction grating for gamma rays or x-rays which uses the equally spaced lattice planes of a crystal. { 'krist·əl ,grād·iŋ }

crystalline chloral See chloral hydrate. { 'kris·tə·lən 'klȯr·əl }

crystalline polymer [CHEM] A polymer whose sections of adjacent chains are packed in a regular array. { 'kris·tə·lən 'päl·i·mər }

crystallinity [ORG CHEM] The degree to which polymer molecules are oriented into repeating patterns. { ,kris·tə'lin·əd·ē }

crystal monochromator [SPECT] A spectrometer in which a collimated beam of slow neutrons from a reactor is incident on a single crystal of copper, lead, or other element mounted on a divided circle. { ¦krist·əl ,män·ə¦krō,mād·ər }

crystals of Venus See cupric acetate. { 'krist·əlz əv 'vē·nəs }

crystal violet See methyl violet. { ¦krist·əl 'vī·lət }

crystogen See cystamine. { 'kris·tə·jən }

Cs See cesium.

C stage [ORG CHEM] The final stage in a thermosetting resin reaction in which the material is relatively insoluble and infusible; the resin in a fully cured thermoset molding is in this stage. Also known as resite. { 'sē ˌstāj }

Cu See copper.

cubane [ORG CHEM] C_8H_8 (polycyclic octane) A cage hydrocarbon with carbons and their bonds at the corners and along the edges, forming a cube. { 'kyü·bān }

cumene [ORG CHEM] $C_6H_5CH(CH_3)_2$ A colorless, oily benzenoid hydrocarbon cooling at 152.4°C; used as an additive for high-octane motor fuel. { 'kyüˌmēn }

cumene hydroperoxide [ORG CHEM] $C_6H_5C(CH_3)_2OOH$ An isopropyl hydroperoxide of cumene; an oily liquid, used to make phenol and acetone. { 'kyüˌmēn ˌhī·drō·pə'räkˌsīd }

cumidine [ORG CHEM] $C_9H_{13}N$ A colorless, water-insoluble liquid, boiling at 225°C. { 'kyü·məˌdēn }

cumulated double bonds [CHEM] Two double bonds on the same carbon atom, as in ⟩C=C=C⟨. { 'kyü·myəˌlād·əd ˌdəb·əl 'bändz }

cumulative double bonds [ORG CHEM] Double bonds joining at least three contiguous carbon atoms in a single structure, for example, $H_2C=C=CH_2$ (allene). Also known as twinned double bonds. { 'kyü·myə·ləd·iv 'dəb·əl 'bänz }

cumulene [ORG CHEM] A compound with a molecular structure which contains two or more double bonds in succession. { 'kyü·myəˌlēn }

cupferron [ORG CHEM] $NH_4ONONC_6H_5$ A colorless salt that forms crystals with a melting point of 164°C; its acid solution is a precipitating reagent. Also known as copperon. { 'kəp·fəˌrän }

cupreine [ORG CHEM] $C_{19}H_{22}O_2N_2·H_2O$ Colorless, anhydrous crystals with a melting point of 198°C; soluble in chloroform and ether; used in medicine. Also known as hydroxycinchonine. { 'kyü·prē·ēn }

cupric [CHEM] The divalent ion of copper. { 'kyü·prik }

cupric acetate [ORG CHEM] $Cu(C_2H_3O_2)_2·H_2O$ Blue-green crystals, soluble in water; used as a raw material to make paris green. Also known as copper acetate; crystal aerugo; crystals of Venus; verdigris. { 'kyü·prik 'as·əˌtāt }

cupric arsenite See copper arsenite. { 'kyü·prik ärs·ənˌīt }

cupric bromide [INORG CHEM] $CuBr_2$ Black prismatic crystals; used in photography as an intensifier and in organic synthesis as a brominating agent. Also known as copper bromide. { 'kyü·prik 'brōˌmīd }

cupric carbonate See copper carbonate. { 'kyü·prik 'kär·bəˌnāt }

cupric chloride [INORG CHEM] **1.** Also known as copper chloride. **2.** $CuCl_2$ Yellowish-brown, deliquescent powder soluble in water, alcohol, and ammonium chloride. **3.** $CuCl_2·H_2O$ A dihydrate of cupric chloride forming green crystals soluble in water; used as a mordant in dyeing and printing textile fabrics and in the refining of copper, gold, and silver. { 'kyü·prik 'klôrˌīd }

cupric chromate [INORG CHEM] $CuCrO_4$ A yellow liquid, used as a mordant. Also known as copper chromate. { 'kyü·prik 'krōˌmāt }

cupric cyanide [INORG CHEM] $Cu(CN)_2$ A green powder, insoluble in water; used in electroplating copper on iron. Also known as copper cyanide. { 'kyü·prik 'sī·əˌnīd }

cupric fluoride [INORG CHEM] CuF_2 White crystalline powder used in ceramics and in the preparation of brazing and soldering fluxes. Also known as copper fluoride. { 'kyü·prik 'flùrˌīd }

cupric gluconate See copper gluconate. { 'kyü·prik 'glü·kəˌnāt }

cupric hydroxide [INORG CHEM] $Cu(OH)_2$ Blue macro- or microscopic crystals; used as a mordant and pigment, in manufacture of many copper salts, and for staining paper. Also known as copper hydroxide. { 'kyü·prik hī'dräkˌsīd }

cupric nitrate [INORG CHEM] $Cu(NO_3)_2·3H_2O$ Green powder or blue crystals soluble in water; used in electroplating copper on iron. Also known as copper nitrate. { 'kyü·prik 'nīˌtrāt }

cupric oxide [INORG CHEM] CuO Black, monoclinic crystals, insoluble in water; used in making fibers and ceramics, and in organic and gas analyses. Also known as copper oxide.   { 'kyü·prik 'äk,sīd }

cupric sulfate   [INORG CHEM] CuSO₄ A water-soluble salt used in copper-plating baths; crystallizes as hydrous copper sulfate, which is blue. Also known as copper sulfate. { 'kyü·prik 'səl,fāt }

cuprous bromide   [INORG CHEM] Cu₂Br₂ White or gray crystals slightly soluble in cold water. Also known as copper bromide.   { 'kyü·prəs 'brō,mīd }

cuprous chloride   [INORG CHEM] CuCl or Cu₂Cl₂ Green, tetrahedral crystals, insoluble in water. Also known as copper chloride; resin of copper.   { 'kyü·prəs 'klȯr,īd }

cuprous fluoride   [INORG CHEM] Cu₂F₂ Red, crystalline powder, melting point 908°C. Also known as copper fluoride.   { 'kyü·prəs 'flür,īd }

cuprous oxide   [INORG CHEM] Cu₂O An oxide of copper found in nature as cuprite and formed on copper by heat; used chiefly as a pigment and as a fungicide. Also known as copper oxide.   { 'kyü·prəs 'äk,sīd }

curare   [ORG CHEM] Poisonous extract from the plant *Strychnos toxifera* containing a mixture of alkaloids that produce paralysis of the voluntary muscles by acting on synaptic junctions; used as an adjunct to anesthesia in surgery.   { kyü'rä·rē }

cure   [CHEM] To change the properties of a resin material by chemical polycondensation or addition reactions.   { kyür }

curine *See* bebeerine.   { 'kyü,rēn }

curing temperature   [CHEM] That temperature at which a resin or adhesive is subjected to curing.   { 'kyür·iŋ ,tem·prə·chər }

curing time   [CHEM] The period of time in which a part is subjected to heat or pressure to cure the resin.   { 'kyür·iŋ ,tīm }

curium   [CHEM] An element, symbol Cm, atomic number 96; the isotope of mass 244 is the principal source of this artificially produced element.   { 'kyür·ē·əm }

current efficiency   [PHYS CHEM] The ratio of the amount of electricity, in coulombs, theoretically required to yield a given quantity of material in an electrochemical process, to the amount actually consumed.   { 'kər·ənt i,fish·ən·sē }

Curtius reaction   [ORG CHEM] A laboratory method for degrading a carboxylic acid to a primary amine by converting the acid to an acyl azide to give products which can be hydrolyzed to amines.   { 'kərd·ē·əs rē,ak·shən }

cyanalcohol *See* cyanohydrin.   { ,sī·ən'al·kə,hȯl }

cyanamide   [INORG CHEM] NHCNH An acidic compound that forms colorless needles, melting at 46°C, soluble in water. Also known as urea anhydride.   { sī'an·ə,mid }

cyanate   [INORG CHEM] A salt or ester of cyanic acid containing the radical CNO. { 'sī·ə,nāt }

cyanazine   [ORG CHEM] C₉H₁₃N₆Cl A white solid with a melting point of 166.5–167°C; used as a pre- and postemergence herbicide for corn, sorghum, soybeans, alfalfa, cotton, and wheat.   { sī'an·ə,zēn }

cyanic acid   [ORG CHEM] HCNO A colorless, poisonous liquid, which polymerizes to cyamelide and fulminic acid.   { sī'an·ik 'as·əd }

cyanidation   [CHEM] Joining of cyanide to an atom or molecule.   { ,sī·ə·nə'dā·shən }

cyanide   [INORG CHEM] Any of a group of compounds containing the CN group and derived from hydrogen cyanide, HCN.   { 'sī·ə,nīd }

cyanine dye   [ORG CHEM] C₂₉H₃₅N₂I Green metallic crystals, soluble in water; unstable to light, the dye is used in the photography industry as a chemical sensitizer for film. Also known as iodocyanin; quinoline blue.   { 'sī·ə·nən ,dī }

cyano-   [CHEM] Combining form indicating the radical CN.   { 'sī·ə·nō }

cyanoacetamide   [ORG CHEM] C₃H₄N₂O Needlelike crystals with a melting point of 119.5°C; soluble in water; used in organic synthesis. Also known as malonamide nitrile.   { ¦sī·ə·nō·ə'sed·ə·mīd }

cyanoacetic acid   [ORG CHEM] NCCH₂COOH Hygroscopic crystals with a melting point of 66°C; decomposes at 160°C; soluble in ether, water, and alcohol; used in the synthesis of intermediates and in the commercial preparation of barbital. Also known as malonic mononitrile.   { ¦sī·ə·nō·ə'sēd·ik 'as·əd }

cyanocarbon |ORG CHEM| A derivative of hydrocarbon in which all of the hydrogen atoms are replaced by the CN group. { ¦sī·ə·nō'kär·bən }

cyano complex [CHEM] A coordination compound containing the CN group. { sī'an· ō 'käm,pleks }

cyanoethylation |ORG CHEM| A chemical reaction involving the addition of acrylonitrile to compounds with a reactive hydrogen. { ¦sī·ə·nō,e·thə'lā·shən }

2-cyanoethanol See ethylene cyanohydrin. { ¦tü ¦sī·ə·nō'eth·ə,nól }

cyanogen [CHEM] A univalent radical, CN. |INORG CHEM| C_2N_2 A colorless, highly toxic gas with a pungent odor; a starting material for the production of complex thiocyanates used as insecticides. Also known as dicyanogen. { sī'an·ə·jən }

cyanogen bromide |INORG CHEM| CNBr White crystals melting at 52°C, vaporizing at 61.3°C, and having toxic fumes that affect nerve centers; used in the synthesis of organic compounds and as a fumigant. { sī'an·ə·jən 'brō,mīd }

cyanogen chloride |INORG CHEM| ClCN A poisonous, colorless gas or liquid, soluble in water; used in organic synthesis. { sī'an·ə·jən 'klór,īd }

cyanogen fluoride |INORG CHEM| CNF A toxic, colorless gas, used as a tear gas. { sī'an· ə·jən 'flúr,īd }

cyanogen iodide See iodine cyanide. { sī'an·ə·jən 'ī·ə,dīd }

cyanohydrin |ORG CHEM| A compound containing the radicals CN and OH. Also known as cyanalcohol. { 'sī·ə·nō'hī·drən }

cyanophosphos |ORG CHEM| $C_{15}H_{14}NO_2PS$ A white, crystalline solid with a melting point of 83°C; used as an insecticide to control larval pests on rice and vegetables. { ,sī·ə·nō'fäs,fós }

cyanoplatinate See platinocyanide. { ¦sī·ə·nō'plat·ən,āt }

cyanuric acid |ORG CHEM| $HOC(NCOH)_2N\cdot2H_2O$ Colorless, monoclinic crystals, slightly soluble in water; formed by polymerization of cyanic acid. Also known as pyrolithic acid. { ¦sī·ə¦nůr·ik 'as·əd }

cyclamate |ORG CHEM| The calcium or sodium salt of cyclohexylsulfamate, an artificial sweetener. { 'sī·klə,māt }

cyclane See alicyclic. { 'sī,klān }

cyclethrin |ORG CHEM| $C_{21}H_{28}O_3$ A viscous, brown liquid, soluble in organic solvents; used as an insecticide. { sī'klē·thrən }

cyclic amide |ORG CHEM| An amide arranged in a ring of carbon atoms. { 'sīk·lik 'a,mīd }

cyclic anhydride |ORG CHEM| A ring compound formed by the removal of water from a compound; an example is phthalic anhydride. { 'sīk·lik an'hī,drīd }

cyclic chronopotentiometry |ANALY CHEM| An analytic electrochemical method in which instantaneous current reversal is imposed at the working electrode, and its potential is monitored with time. { 'sīk·lik ¦krän·ō·pə,ten·chē'äm·ə·trē }

cyclic coil See random coil. { 'sīk·lik 'kóil }

cyclic compound |ORG CHEM| A compound that contains a ring of atoms. { 'sīk· lik 'käm,paúnd }

cyclic ion See bridged ion. { 'sīk·lik 'ī·ən }

cyclic voltammetry |PHYS CHEM| An electrochemical technique for studying variable potential at an electrode involving application of a triangular potential sweep, allowing one to sweep back through the potential region just covered. { 'sīk·lik vōl'täm·ə·trē }

cyclitol |ORG CHEM| A cycloalkane that contains one hydroxyl group on each of three or more of the atoms constituting the ring. { 'sī·klə,tól }

cyclization |ORG CHEM| Changing an open-chain hydrocarbon to a closed ring. { ,sī· klə'zā·shən }

cycloaddition |ORG CHEM| A reaction in which unsaturated molecules combine to form a cyclic compound. { ¦sī·klō·ə'dish·ən }

cycloaliphatic See alicyclic. { ¦sī·klō·al·ə'fad·ik }

cycloalkane See alicyclic. { ¦sī·klō'al,kān }

cycloalkene |ORG CHEM| An unsaturated, monocyclic hydrocarbon having the formula C_nH_{2n-2}. Also known as cycloolefin. { ¦sī·klō'al,kēn }

cycloalkylaryl compound

cycloalkylaryl compound [ORG CHEM] A compound with a multiringed molecular structure containing both aromatic and saturated rings. { ¦sī·klō¦al·kə'lar·əl 'käm,paúnd }

cycloalkyne [ORG CHEM] A cyclic compound containing one or more triple bonds between carbon atoms. { ¦sī·klō'al,kīn }

cycloate [ORG CHEM] $C_{11}H_{21}NOS$ A yellow liquid with limited solubility in water; boiling point is 145–146°C; used as an herbicide to control weeds in sugarbeets, spinach, and table beets. { 'sī·klə,wāt }

cyclobutadiene [ORG CHEM] C_4H_4 A cyclic compound containing two alternate double bonds; used in organic synthesis. Also known as butene. { ¦sī·klō,byüd·ə'dī,ēn }

cyclobutane [ORG CHEM] C_4H_8 An alicyclic hydrocarbon, boiling point 11°C; synthesized as a condensable gas; used in organic synthesis. Also known as tetramethylene. { ¦sī·klō'byü,tān }

cyclobutene [ORG CHEM] C_4H_6 An asymmetrical cyclic hydrocarbon occurring in several isomeric forms. Also known as cyclobutylene. { ¦sī·klō'byü,tēn }

cyclodiolefin [ORG CHEM] A cycloalkene with two double bonds; sometimes included with alkenes, cycloalkenes, and hydrocarbons containing more than one ethylene bond as olefins in a generic sense. { ¦sī·klō·dī'ō·lə·fən }

cyclododecatriene [ORG CHEM] $C_{12}H_{18}$ One of two cyclic hydrocarbons with three double bonds; the two forms are stereoisomeric; used to make nylon-6 and nylon-12. { ¦sī·klō,dō·dek·ə'trī,ēn }

cycloelimination [ORG CHEM] The reverse of cycloaddition. Also known as cycloreversion. { ,sī·klō·i,lim·ə'nā·shən }

1,3-cyclohexadiene [ORG CHEM] C_6H_8 A partly saturated benzene compound with two double bonds; used in organic synthesis. { ¦wən ¦thrē ¦sī·klō,hek·sə'dī,ēn }

cyclohexane [ORG CHEM] C_6H_{12} A colorless liquid that is a cyclic hydrocarbon synthesized by hydrogenation of benzene; used in organic synthesis. Also known as hexamethylene. { ¦sī·klō'hek,sān }

cyclohexanol [ORG CHEM] $C_6H_{11}OH$ An oily, colorless, hygroscopic liquid with a camphorlike odor and a boiling point of 160.9°C; used in soapmaking, insecticides, dry cleaning, plasticizers, and germicides. Also known as hexahydrophenol. { ¦sī·klō'hek·sə,nól }

cyclohexanone [ORG CHEM] $C_6H_{10}O$ An oily liquid with an odor suggesting peppermint and acetone; soluble in alcohol, ether, and other organic solvents; used as an industrial solvent, in the production of adipic acid, and in the preparation of cyclohexanone resins. { ¦sī·klō'hek·sə,nōn }

cyclohexene [ORG CHEM] C_6H_{10} A compound that occurs in coal tar; a liquid that is used as an alkylation component; used in the manufacture of hexahydrobenzoic acid, adipic acid, and maleic acid. { ¦sī·klō'hek,sēn }

cyclohexylamine [ORG CHEM] $C_6H_{11}NH_2$ A liquid with a strong, fishy, amine odor; miscible with water and common organic solvents; used in organic synthesis and in the manufacture of plasticizers, rubber chemicals, corrosion inhibitors, dye-stuffs, dry-cleaning soaps, and emulsifying agents. { ¦sī·klō·hek'sil·ə,mēn }

cycloidal mass spectrometer [SPECT] Small mass spectrometer of limited mass range fitted with a special-type analyzer that generates a cycloidal-path beam of the sample mass. { sī'klóid·əl 'mas spek'träm·əd·ər }

cyclonite [ORG CHEM] $(CH_2)_3N_3(NO_2)_3$ A white, crystalline explosive, consisting of hexahydrotrinitrotriazine, and having high sensitivity and brisance; mixed with other explosives or substances. Abbreviated RDX. Also known as cyclotrimethylenetrinitramine. { 'sī·klə,nīt }

1,5-cyclooctadiene [ORG CHEM] C_8H_{12} A cyclic hydrocarbon with two double bonds; prepared from butadiene and used to make cyclooctene and cyclooctane, which are intermediates for the production of plastics, fibers, and so on. { ¦wən ¦fīv ¦sī·klō,äk·tə'dī,ēn }

cyclooctane [ORG CHEM] $(CH_2)_8$ A cyclic alkane melting at 9.5°C; used as an intermediate in production of plastics, fibers, adhesives, and coatings. Also known as octomethylene. { ¦sī·klō'äk,tān }

98

cyclooctatetraene [ORG CHEM] C_8H_8 A cyclic olefin with alternate double bonds; highly reactive; rearranges to styrene. { ¦sī·klō¦äk·tə'te·trə‚ēn }

cycloolefin *See* cycloalkene. { ¦sī·klō'ō·lə·fən }

cycloparaffin *See* alicyclic. { ¦sī·klō'par·ə·fən }

1,3-cyclopentadiene [ORG CHEM] C_5H_6 A colorless liquid boiling at 41.5°C; used to make resins. { ¦wən ¦thrē ¦sī·klō‚pen·tə'dī‚ēn }

cyclopentadienyl anion [ORG CHEM] $C_5H_5^-$ A radical formed from cyclopentadiene. { ¦sī·klō‚pen·tə‚dī'e·nil 'an‚ī·ən }

cyclopentane [ORG CHEM] C_5H_{10} A cyclic hydrocarbon that is a colorless liquid; present in crude petroleum, it is converted during refining to aromatics which improve antiknock and combustion properties of gasoline. { ¦sī·klō'pen‚tān }

cyclopentanoid [ORG CHEM] A compound whose key structural unit consists of five carbon atoms arranged in a ring. { ¦sī·klō'pen·tə‚nȯid }

cyclopentanol [ORG CHEM] C_5H_9OH A colorless liquid boiling at 139°C; used as a solvent for perfumes and pharmaceuticals. Also known as cyclopentyl alcohol. { ¦sī·klō'pen·tə‚nȯl }

cyclopentanone [ORG CHEM] C_5H_8O A saturated monoketone; a colorless liquid boiling at 130°C; used as an intermediate in pharmaceutical preparation. { ¦sī·klō'pen·tə‚nōn }

cyclopentene [ORG CHEM] $(CH_2)_3CHCH$ A colorless liquid boiling at 45°C; used as a chemical intermediate in petroleum chemistry. { ¦sī·klō'pen‚tēn }

cyclopentenylundecylic acid *See* hydnocarpic acid. { ¦sī·klō'pen·tə‚nil‚ən·də'sil·ik 'as·əd }

cyclopentyl alcohol *See* cyclopentanol. { ¦sī·klō'pent·əl 'al·kə‚hȯl }

cyclophane [ORG CHEM] A molecule composed of an aromatic ring (most frequently a benzene ring) and an aliphatic unit which forms a bridge between two (or more) positions of the aromatic ring. { 'sī·klə‚fān }

cyclopropane [ORG CHEM] C_3H_6 A colorless gas, insoluble in water; used as an anesthetic. { ¦sī·klō'prō‚pān }

cyclopropanoid [ORG CHEM] A compound whose key structural unit consists of three carbon atoms arranged in a ring. { ¦sī·klō'prō·pə‚nȯid }

cycloreversion *See* cycloelimination. { ‚sī·klə·ri'ver·zhən }

cyclotrimethylenetrinitramine *See* cyclonite. { ¦sī·klō‚trī¦meth·əl‚ēn‚trī'nī·trə‚mēn }

cyhexatin [ORG CHEM] $C_{18}H_{34}OSn$ A whitish solid, insoluble in water; used as a miticide to control plant-feeding mites. { sī'hek·sə·tən }

cymene [ORG CHEM] Any of the isomeric hydrocarbons metacymene, paracymene, and orthocymene; paracymene is a liquid that is colorless, has a pleasant odor, and is made from oil of cumin or oil of wild thyme. { 'sī‚mēn }

cystamine [ORG CHEM] $(CH_2)_6N_4$ A white, crystalline powder, melting at 280°C; used to make synthetic resins. Also known as aminiform; crystogen; cystamine methenamine; hexamethylene tetramine; urotropin. { 'sis·tə‚mēn }

cystamine methenamine *See* cystamine. { 'sis·tə‚mēn mə'then·ə‚mēn }

Czerny-Turner spectrograph [SPECT] A spectrograph used chiefly in laboratory work, which has a plane reflection grating and spherical reflectors for the collimator and camera. { ¦cher·nē¦tərn·ər 'spek·trə‚graf }

99

D

2,4-D *See* 2,4-dichlorophenoxyacetic acid.

dalapon [ORG CHEM] Generic name for 2,2-dichloropropionic acid; a liquid with a boiling point of 185–190°C at 760 mmHg; soluble in water, alcohol, and ether; used as a herbicide. { 'dal·ə,pän }

Dalton's atomic theory [CHEM] Theory forming the basis of accepted modern atomic theory, according to which matter is made of particles called atoms, reactions must take place between atoms or groups of atoms, and atoms of the same element are all alike but differ from atoms of another element. { 'dȯl·tənz ə,täm·ik 'thē·ə·rē }

Daniell cell [PHYS CHEM] A primary cell with a constant electromotive force of 1.1 volts, having a copper electrode in a copper sulfate solution and a zinc electrode in dilute sulfuric acid or zinc sulfate, the solutions separated by a porous partition or by gravity. { 'dan·yəl ,sel }

dansyl chloride [ORG CHEM] $(CH_3)_2NC_{10}H_6SO_2Cl$ A reagent for fluorescent labeling of amines, amino acids, proteins, and phenols. { 'dans·əl 'klȯr,īd }

DAP *See* diallyl phthalate.

dark-line spectrum [SPECT] The absorption spectrum that results when white light passes through a substance, consisting of dark lines against a bright background. { 'därk ,līn 'spek·trəm }

Darzen's procedure [ORG CHEM] Preparation of alkyl halides by refluxing a molecule of an alcohol with a molecule of thionyl chloride in the presence of a molecule of pyridine. { 'där·zənz prə,sē·jər }

Darzen's reaction [ORG CHEM] Condensation of aldehydes and ketones with α-haloesters to produce glycidic esters. { 'där·zənz rē,ak·shən }

dative bond *See* coordinate valence. { ¦dād·iv 'bänd }

dazomet [ORG CHEM] $C_5H_{10}N_2S_2$ A white, crystalline compound that decomposes at 100°C; used as a herbicide and nematicide for soil fungi and nematodes, weeds, and soil insects. Also known as tetrahydro-3,5-dimethyl-2H-1,3,5-thiadiazine-6-thione. { 'dā· zə·mət }

Db *See* dubnium.

DBCP *See* dibromochloropropane.

d-block element [CHEM] A transition element occupying the first, second, and third long periods of the periodic table. { 'dē ,bläk ,el·ə·mənt }

DCB *See* 1,4-dichlorobutane.

DCC *See* dicyclohexylcarbodiimide.

DCCI *See* dicyclohexylcarbodiimide.

DCMO *See* carboxin.

DCNA *See* 2,6-dichloro-4-nitroaniline.

DCPA *See* dimethyl-2,3,5,6-tetrachloroterephthalate.

DDD *See* 2,2-bis(*para*-chlorophenyl)-1,1-dichloroethane.

DDT [ORG CHEM] Common name for an insecticide; melting point 108.5°C, insoluble in water, very soluble in ethanol and acetone, colorless, and odorless; especially useful against agricultural pests, flies, lice, and mosquitoes. Also known as dichlorodiphenyltrichloroethane.

DDTA *See* derivative differential thermal analysis.

DDVP *See* dichlorvos.

DEA

DEA *See* diethanolamine.

deacetylation [ORG CHEM] The removal of an acetyl group from a molecule. { ,dē·ə,sēd·əl'ā·shən }

deacidification [CHEM] **1.** Removal of acid. **2.** A process for reducing acidity. { ,dē·ə,sid·ə·fə'kā·shən }

deactivation [CHEM] **1.** Rendering inactive, as of a catalyst. **2.** Loss of radioactivity. { dē,ak·tə'vā·shən }

deacylation [ORG CHEM] Removal of an acyl group from a compound. { dē,as·ə'lā·shən }

DEAE-cellulose *See* diethylaminoethyl cellulose. { 'dē,ē'ā,ē 'sel·yə,lōs }

dealkalization [CHEM] **1.** Removal of alkali. **2.** Reduction of alkalinity, as in the process of neutralization. { dē,al·kə·lə'zā·shən }

dealkylate [CHEM] To remove alkyl groups from a compound. { dē'al·kə,lāt }

dealuminization [CHEM] Removal of aluminum. { ,dē·ə,lü·mə·nə'zā·shən }

deamidation [ORG CHEM] Removal of the amido group from a molecule. { dē,am·ə'dā·shən }

deamination [ORG CHEM] Removal of an amino group from a molecule. { dē,am·ə'nā·shən }

deashing [CHEM] A form of deionization in which inorganic salts are removed from solution by the adsorption of both the anions and cations by ion-exchange resins. { dē'ash·iŋ }

debenzylation [ORG CHEM] Removal from a molecule of the benzyl group. { dē,ben·zə'lā·shən }

de Brun-van Eckstein rearrangement [ORG CHEM] The isomerization of an aldose or ketose when mixed with aqueous calcium hydroxide to form a mixture of various monosaccharides and unfermented ketoses; used to prepare certain ketoses. { də¦brün van'ek,shtīn ,rē·ə,rānj·mənt }

Debye-Falkenhagen effect [PHYS CHEM] The increase in the conductance of an electrolytic solution when the applied voltage has a very high frequency. { də¦bī 'fäl·kən,häg·ən i,fekt }

Debye force *See* induction force. { də'bī ,förs }

Debye-Hückel theory [PHYS CHEM] A theory of the behavior of strong electrolytes, according to which each ion is surrounded by an ionic atmosphere of charges of the opposite sign whose behavior retards the movement of ions when a current is passed through the medium. { də¦bī 'hik·əl ,thē·ə·rē }

Debye relaxation time [PHYS CHEM] According to the Debye-Hückel theory, the time required for the ionic atmosphere of a charge to reach equilibrium in a current-carrying electrolyte, during which time the motion of the charge is retarded. { də'bī ,rē,lak'sā·shən ,tīm }

decaborane (14) [INORG CHEM] $B_{10}H_{14}$ A binary compound of boron and hydrogen that is relatively stable at room temperature; melting point 99.5°C, boiling point 213°C. { ¦dek·ə¦bòr,ān 'fōr¦tēn }

decahydrate [CHEM] A compound that has 10 water molecules. { ,dek·ə'hī,drāt }

decahydronaphthalene [ORG CHEM] $C_{10}H_{18}$ A liquid hydrocarbon, used in some paints and lacquers as a solvent. { ,dek·ə,hī·drō'naf·thə,lēn }

decalcification [CHEM] Loss or removal of calcium or calcium compounds from a calcified material such as bone or soil. { dē,kal·sə·fə'kā·shən }

decane [ORG CHEM] $C_{10}H_{22}$ Any of several saturated aliphatic hydrocarbons, especially $CH_3(CH_2)_8CH_3$. { 'dē,kān }

decanol *See* decyl alcohol. { 'dek·ə,nòl }

decarbonize [CHEM] To remove carbon by chemical means. { dē'kär·bə,nīz }

decarboxylate [ORG CHEM] To remove the carboxyl radical, especially from amino acids and protein. { ,dē·kär'bäk·sə,lāt }

decavanadate [INORG CHEM] A deep-orange polyvanadate ($V_{10}O_{28}{}^{6-}$), composed of 10 fused VO_6 octahedra. { ,dē·kə 'van·ə,dāt }

dechlorination [CHEM] Removal of chlorine from a substance. { dē,klòr·ə'nā·shən }

decinormal [CHEM] Pertaining to a chemical solution that is one-tenth normality in reference to a 1 normal solution. { ¦des·ə'nȯr·məl }

decolorizing carbon [CHEM] Porous or finely divided carbon (activated or bone) with large surface area; used to adsorb colored impurities from liquids, such as lube oils. { dē¦kəl·ə¦rīz·iŋ 'kär·bən }

decomposition [CHEM] The more or less permanent structural breakdown of a molecule into simpler molecules or atoms. { dē¦käm·pə'zish·ən }

decomposition potential [PHYS CHEM] The electrode potential at which the electrolysis current begins to increase appreciably. Also known as decomposition voltage. { dē¦käm·pə'zish·ən pə¦ten·chəl }

decomposition voltage See decomposition potential. { dē¦käm·pə'zish·ən ¦vōl·tij }

decyl [ORG CHEM] An isomeric grouping of univalent radicals, all with formulas $C_{10}H_{21}$, and derived from the decanes by removing one hydrogen. { 'des·əl }

decyl acetate [ORG CHEM] $CH_3(CH_2)_9OOCCH_3$ Perfumery liquid with a floral orangerose aroma. { 'des·əl 'as·ə¦tāt }

decyl alcohol [ORG CHEM] $C_{10}H_{21}OH$ A colorless oil, boiling at 231°C; used in plasticizers, synthetic lubricants, and detergents. Also known as decanol. { 'des·əl 'al·kə¦hȯl }

decyl aldehyde [ORG CHEM] $CH_3(CH_2)_8CHO$ A liquid aldehyde, found in essential oils; used in flavorings and perfumes. { 'des·əl 'al·də¦hīd }

decylene [ORG CHEM] Any of a group of isomeric hydrocarbons with formula $C_{10}H_{20}$; the group is part of the ethylene series. { 'des·ə¦lēn }

decyltrichlorosilane [ORG CHEM] n-$C_{10}H_{21}SiCl_3$ An organochlorosilane that boils at 183°C at 84 mmHg; used in coupling agents or primers to obtain improved bonding between organic polymers and mineral surfaces. { ¦des·əl·trī¦klȯr·ō'sī¦lān }

DEET See diethyltoluamide.

definite-composition law [CHEM] The law that a given chemical compound always contains the same elements in the same fixed proportions by weight. Also known as definite-proportions law. { ¦def·ə·nət ¦käm·pə'zish·ən ¦lȯ }

definite-proportions law See definite-composition law. { ¦def·ə·nət prə'pȯr·shənz ¦lȯ }

deflagrating spoon [CHEM] A long-handled spoon used in chemistry to demonstrate deflagration. { 'def·lə¦grād·iŋ ¦spün }

deflagration [CHEM] A chemical reaction accompanied by vigorous evolution of heat, flame, sparks, or spattering of burning particles. { ¦def·lə'grā·shən }

deflocculant [CHEM] An agent that causes deflocculation; examples are sodium carbonate and other basic materials used to deflocculate clay slips. { dē'fläk·yə·lənt }

defluorination [CHEM] Removal of fluorine. { dē¦flür·ə'nā·shən }

degasser See getter. { dē'gas·ər }

degradation [ORG CHEM] Conversion of an organic compound to one containing a smaller number of carbon atoms. { ¦deg·rə'dā·shən }

degree [CHEM] Any one of several units for measuring hardness of water, such as the English or Clark degree, the French degree, and the German degree. { di'grē }

degree of crystallinity [ORG CHEM] In a fairly large sample of a polymer, the fraction that consists of regions showing long-range three-dimensional order. { di'grē əv ¦kris·tə'lin·əd·ē }

degree of freedom [PHYS CHEM] Any one of the variables, including pressure, temperature, composition, and specific volume, which must be specified to define the state of a system. { di'grē əv 'frē·dəm }

degree of polymerization [ORG CHEM] The number of structural units in the average polymer molecule in a particular sample. Abbreviated D.P. { di'grē əv pə¦lim·ə·rə'zā·shən }

dehalogenate [ORG CHEM] To remove halogen atoms from a molecule. { dē'hal·ə·jə¦nāt }

dehydration [CHEM] Removal of water from any substance. [ORG CHEM] An elimination reaction in which a molecule loses both a hydroxyl group (OH) and a hydrogen atom (H) that was bonded to an adjacent carbon. { ¦dē·hī'drā·shən }

dehydrator

dehydrator [CHEM] A substance that removes water from a material; an example is sulfuric acid. { dē'hī,drād·ər }

dehydroacetic acid [ORG CHEM] $C_8H_8O_4$ Crystals that melt at 108.5°C and are insoluble in water, soluble in acetone; used as a fungicide and bactericide. Abbreviated DHA. { dē¦hī·drō·ə¦sēd·ik 'as·əd }

dehydroascorbic acid [ORG CHEM] $C_6H_6O_6$ A relatively inactive acid resulting from elimination of two hydrogen atoms from ascorbic acid when the latter is oxidized by air or other agents; has potential ascorbic acid activity. { dē¦hī·drō·ə¦skòr·bik 'as·əd }

dehydrocholic acid [ORG CHEM] $C_{24}H_{34}O_5$ A white powder melting at 231–240°C, very slightly soluble in water; used as a pharmaceutical intermediate and in medicine. { dē¦hī·drə¦käl·ik 'as·əd }

dehydroepiandrosterone [ORG CHEM] $C_{19}H_{28}O_2$ Dimorphous crystals with a melting point of 140–141°C, or leaflet crystals with a melting point of 152–153°C; soluble in alcohol, benzene, and ether; used as an androgen. { dē¦hī·drō·e,pē·ən'dräs·tə ,rōn }

dehydrogenation [CHEM] Removal of hydrogen from a compound. { dē¦hī·drə·jə'nā·shən }

dehydrohalogenation [CHEM] Removal of hydrogen and a halogen from a compound. { dē¦hī·dro,hal·ə·jə'nā·shən }

deionization [CHEM] An ion-exchange process in which all charged species or ionizable organic and inorganic salts are removed from solution. { dē,ī·ən·ə'zā·shən }

de la Tour method [ANALY CHEM] Measurement of critical temperature, involving sealing the sample in a tube and heating it; the temperature at which the meniscus disappears is the critical temperature. { del·ə'tür ,meth·əd }

Delepine reaction [ORG CHEM] Slow ammonolysis of alkyl halides in acid to primary amines in the presence of hexamethylenetetramine. { 'del·ə,pīn rē,ak·shən }

deliquescence [PHYS CHEM] The absorption of atmospheric water vapor by a crystalline solid until the crystal eventually dissolves into a saturated solution. { del·ə'kwes·əns }

delocalized bond [CHEM] A type of molecular bonding in which the electron density of delocalized electrons is regarded as being spread over several atoms or over the whole molecule. Also known as nonlocalized bond. { dē'lō·kə,līzd 'bänd }

delphidenolon See myricetin. { ¦del·fə¦den·ə,län }

demal [CHEM] A unit of concentration, equal to the concentration of a solution in which 1 gram-equivalent of solute is dissolved in 1 cubic decimeter of solvent. { 'dem·əl }

demasking [CHEM] A process by which a masked substance is made capable of undergoing its usual reactions; can be brought about by a displacement reaction involving addition of, for example, another cation that reacts more strongly with the masking ligand and liberates the masked ion. { dē'mask·iŋ }

demethylation [ORG CHEM] Removal of the methyl group from a compound. { de,meth·ə'lā·shən }

demeton-S-methyl [ORG CHEM] $C_6H_{15}O_3PS_2$ An oily liquid with a 0.3% solubility in water; used as an insecticide and miticide to control aphids. { 'dem·ə,tän ¦es 'meth·əl }

demeton-S-methyl sulfoxide [ORG CHEM] $C_6H_{15}O_4PS_2$ A clear, amber liquid; limited solubility in water; used as an insecticide and miticide for pests of vegetable, fruit, and field crops, ornamental flowers, shrubs, and trees. { 'dem·ə,tän ¦es 'meth·əl səl'fäk,sīd }

Demjanov rearrangement [ORG CHEM] A structural rearrangement that accompanies treatment of certain primary aliphatic amines with nitrous acid; the amine will undergo a ring contraction or expansion. { dem'yä·nòf rē·ə'ränj·mənt }

denaturant [CHEM] An inert, bad-tasting, or poisonous chemical substance added to a product such as ethyl alcohol to make it unfit for human consumption. { dē'nā·chə·rənt }

denature [CHEM] **1.** To change a protein by heating it or treating it with alkali or acid so that the original properties such as solubility are changed as a result of the protein's molecular structure being changed in some way. **2.** To add a denaturant, such as methyl alcohol, to grain alcohol to make the grain alcohol poisonous and unfit for human consumption. { dē'nā·chər }

denatured alcohol [CHEM] Ethyl alcohol containing a poisonous substance, such as methyl alcohol or benzene, which makes it unfit for human consumption. { dē'nā·chərd 'al·kə‚hȯl }

dendrimer [ORG CHEM] **1.** A polymer with a well-defined core, an interior, and peripheral surface components constructed in a concentric ordered fashion around the core. **2.** A polymer having a regular branched structure. Also known as cascade molecule; dendritic polymer; dendron; starburst polymer. { 'den·drə·mər }

dendritic macromolecule [ORG CHEM] A macromolecule whose structure is characterized by a high degree of branching that originates from a single focal point (core). { den'drid·ik ‚mak·rō'mäl·ə‚kyül }

dendritic polymer See dendrimer. { den¦drid·ik 'päl·ə·mər }

dendrochemistry [CHEM] The analysis of the chemical composition of tree rings for naturally occurring or human-manufactured chemicals, especially the mineral elements, to understand the impact of pollution in the air, or surface-water or groundwater supply in ecosystems, or to detect environmental changes over time. { ‚den·drō'kem·i·strē }

dendron See dendrimer. { 'den‚drän }

denitration [CHEM] Removal of nitrates or nitrogen. Also known as denitrification. { dē‚nī'trā·shən }

denitrification See denitration. { dē‚nī·trə·fə'kā·shən }

density gradient centrifugation [ANALY CHEM] Separation of particles according to density by employing a gradient of varying densities; at equilibrium each particle settles in the gradient at a point equal to its density. { 'den·səd·ē ¦grād·ē·ənt sen‚trif·ə'gā·shən }

deoxidant See deoxidizer. { dē'äk·sə·dənt }

deoxidation [CHEM] **1.** The condition of a molecule's being deoxidized. **2.** The process of deoxidizing. { dē‚äk·sə'dā·shən }

deoxidize [CHEM] **1.** To remove oxygen by any of several processes. **2.** To reduce from the state of an oxide. { dē'äk·sə‚dīz }

deoxidizer [CHEM] Any substance which reduces the amount of oxygen in a substance, especially a metal, or reduces oxide compounds. Also known as deoxidant. { dē'äk·sə‚dīz·ər }

deoxygenation [CHEM] Removal of oxygen from a substance, such as blood or polluted water. { dē‚äk·sə·jə'nā·shən }

2,4-DEP See tris|2-(2,4-dichlorophenoxy)ethyl|phosphite.

DEPC See diethyl pyrocarbonate.

depolarizer [PHYS CHEM] A substance added to the electrolyte of a primary cell to prevent excessive buildup of hydrogen bubbles by combining chemically with the hydrogen gas as it forms. Also known as battery depolarizer. { dē'pō·lə‚rīz·ər }

depolymerization [ORG CHEM] Decomposition of macromolecular compounds into relatively simple compounds. { ‚dē·pə‚lim·ə·rə'zā·shən }

deposition potential [PHYS CHEM] The smallest potential which can produce electrolytic deposition when applied to an electrolytic cell. { ‚dep·ə'zish·ən pə'ten·chəl }

deproteinize [ORG CHEM] To remove protein from a substance. { dē'prō‚tē‚nīz }

depside [ORG CHEM] One of a class of esters that form from the joining of two or more molecules of phenolic carboxylic acid. { 'dep‚sīd }

depsidone [ORG CHEM] One of a class of compounds that consists of esters such as depsides, but are also cyclic ethers. { 'dep·sə‚dōn }

derichment [ANALY CHEM] In gravimetric analysis by coprecipitation of salts, a system with λ less than unity, when λ is the logarithmic distribution coefficient expressed by the ratio of the logarithms of the ratios of the initial and final solution concentrations of the two salts. { dē'rich·mənt }

derivative [CHEM] A substance that is made from another substance. { də'riv·əd·iv }

105

derivative differential thermal analysis

derivative differential thermal analysis [ANALY CHEM] A method for precise determination in thermograms of slight temperature changes by taking the first derivative of the differential thermal analysis curve (thermogram) which plots time versus differential temperature as measured by a differential thermocouple. Also known as DDTA. { də'riv·əd·iv dif·ə'ren·chəl 'thər·məl ə'nal·ə·səs }

derivative polarography [ANALY CHEM] Polarography technique in which the rate of change of current with respect to applied potential is measured as a function of the applied potential (di/dE versus E, where i is current and E is applied potential). { də'riv·əd·iv ,pō·lə'räg·rə·fē }

derivative thermometric titration [ANALY CHEM] The use of a special resistance-capacitance network to record first and second derivatives of a thermometric titration curve (temperature versus weight change upon heating) to produce a sharp end-point peak. { də'riv·əd·iv thər·mə'me·trik tī'trā·shən }

descending chromatography [ANALY CHEM] A type of paper chromatography in which the sample-carrying solvent mixture is fed to the top of the developing chamber, being separated as it works downward. { di'sen·diŋ krō·mə'täg·rə·fē }

desiccant *See* drying agent. { 'des·i·kənt }

designated volume [ANALY CHEM] The volume of an item of volumetric glassware as calibrated at a given temperature, frequently 20°C (68°F). { ¦dez·ig,nād·əd 'väl·yəm }

desmetryn [ORG CHEM] $C_9H_{17}N_5S$ A white, crystalline compound with a melting point of 84–86°C; used as a postemergence herbicide for broadleaf and grassy weeds. { dez'me·trən }

desorption [PHYS CHEM] The process of removing a sorbed substance by the reverse of adsorption or absorption. { dē'sȯrp·shən }

destructive distillation [ORG CHEM] Decomposition of organic compounds by heat without the presence of air. { di'strək·tiv dis·tə'lā·shən }

desulfonation [ORG CHEM] Removal of the sulfonate group from an organic molecule. { dē,səl·fə'nā·shən }

desyl [ORG CHEM] The functional group $C_6H_5CO-CH(C_6H_5)-$; may be formed from desoxybenzoin. { 'des·əl }

DET *See* diethyltoluamide.

detection limit [ANALY CHEM] In chemical analysis, the minimum amount of a particular component that can be determined by a single measurement with a stated confidence level. { di'tek·shən ,lim·ət }

detergent alkylate *See* dodecylbenzene. { di'tər·jənt 'al·kə,lāt }

determination [ANALY CHEM] The finding of the value of a chemical or physical property of a compound, such as reaction-rate determination or specific-gravity determination. { də,tər·mə'nā·shən }

detonation [CHEM] An exothermic chemical reaction that propagates with such rapidity that the rate of advance of the reaction zone into the unreacted material exceeds the velocity of sound in the unreacted material; that is, the advancing reaction zone is preceded by a shock wave. { ,det·ən'ā·shən }

deuteration [CHEM] The addition of deuterium to a chemical compound. { ,düd·ər'ā·shən }

deuteride [CHEM] A hydride in which the hydrogen is deuterium. { 'düd·ə,rīd }

deuterium [CHEM] The isotope of the element hydrogen with one neutron and one proton in the nucleus; atomic weight 2.0144. Designated D, d, H^2, or ^2H. { dü'tir·ē·əm }

deuterium oxide *See* heavy water. { dü'tir·ē·əm 'äk,sīd }

developed dye [CHEM] A direct azo dye that can be further diazotized by a developer after application to the fiber; it couples with the fiber to form colorfast shades. Also known as diazo dye. { də'vel·əpt 'dī }

developer [CHEM] An organic compound which interacts on a textile fiber to develop a dye. { də'vel·əp·ər }

development [ANALY CHEM] In the separation of mixtures by paper chromatography or thin-layer chromatography, the production of colored derivatives of the solutes

106

by spraying the stationary phase with selective reagents in order to establish the location of individual substances. { də'vel·əp·mənt }

devitrification [CHEM] The process by which the glassy texture of a material is converted into a crystalline texture. { dē,vi·trə·fə'kā·shən }

devrinol [ORG CHEM] $C_{17}H_{21}O_2N$ A brown solid with a melting point of 68.5–70.5°C; slight solubility in water; used as a herbicide for crops. Also known as 2-(α-naphthoxy)-N,N-diethylpropionamide. { 'dev·rə,nól }

Dewar structure [ORG CHEM] A structural formula for benzene that contains a bond between opposite atoms. { 'dü·ər ,strək·chər }

dew point [CHEM] The temperature and pressure at which a gas begins to condense to a liquid. { 'dü ,póint }

dextrinization [ORG CHEM] Any process that involves dextrinizing. { ,dek·strə·nə'zā·shən }

dextrinize [ORG CHEM] To convert a starch into dextrins. { 'dek·strə,nīz }

dextro See dextrorotatory enantiomer. { 'dek,strō }

dextropimaric acid [ORG CHEM] $C_{19}H_{29}COOH$ A compound found in particular in oleoresins of pine trees. { ¦dek·strō·pə'mar·ik 'as·əd }

dextrorotatory enantiomer [ORG CHEM] An optically active substance that rotates the plane of plane-polarized light clockwise. Symbolized *d*. Also known as dextro. { ,dek·strō'rōd·ə,tòr·ē ə¦nan·tē¦ō·mər }

dezincification [CHEM] Removal of zinc. { dē,ziŋk·ə·fə'kā·shən }

DHA See dehydroacetic acid; dihydroxyacetone.

Di See didymium.

diacetate [ORG CHEM] An ester or salt that contains two acetate groups. { dī'as·ə,tāt }

diacetic acid See acetoacetic acid. { dī·ə'sēd·ik 'as·əd }

diacetic ether See ethyl acetoacetate. { dī·ə'sēd,ik 'ē·thər }

diacetin [ORG CHEM] $C_3H_5(OH)(CH_3COO)_2$ A colorless, hygroscopic liquid that is soluble in water, alcohol, ether, and benzene; boiling point 259°C; used as a plasticizer and softening agent and as a solvent. Also known as glyceryl diacetate. { dī'as·əd·ən }

diacetone alcohol [ORG CHEM] $CH_3COCH_2C(CH_3)_2OH$ A colorless liquid used as a solvent for nitrocellulose and resins. { dī'as·ə,tōn 'al·kə,hòl }

diacetyl [ORG CHEM] **1.** $CH_3COCOCH_3$ A yellowish-green liquid with a boiling point of 88°C; has a strong odor that resembles quinone; occurs naturally in bay oil and butter and is produced from methyl ethyl ketone or by a special fermentation of glucose; used as an aroma carrier in food manufacturing. Also known as biacetyl. **2.** A prefix indicating two acetyl groups. { dī'as·əd·əl }

diacetylurea [ORG CHEM] $C_5H_8O_3N_2$ An acyl derivative of urea containing two acetyl groups. { dī¦as·əd·əl·yü'rē·ə }

diacid [CHEM] An acid that has two acidic hydrogen atoms; an example is oxalic acid. { dī'as·əd }

dialdehyde [ORG CHEM] A molecule that has two aldehyde groups, such as dialdehyde starch. { dī'al·də,hīd }

dialifor [ORG CHEM] $C_{14}H_{17}ClNO_4S_2P$ A white, crystalline compound with a melting point of 67–69°C; insoluble in water; used to control pests in citrus fruits, grapes, and pecans. { dī'al·ə,fòr }

dialkyl [ORG CHEM] A molecule that has two alkyl groups. { dī'al·kəl }

dialkyl amine [ORG CHEM] An amine that has two alkyl groups bonded to the amino nitrogen. { dī'al·kəl 'a,mēn }

diallyl phthalate [ORG CHEM] $C_6H_4(COOCH_2CH:CH_2)_2$ A colorless, oily liquid with a boiling range of 158–165°C; used as a plasticizer and for polymerization. Abbreviated DAP. { dī'al·əl 'tha,lāt }

dialuric acid [ORG CHEM] $C_4H_4N_2O$ An acid that is derived by oxidation of uric acid or by the reduction of alloxan; may be used in organic synthesis. { ¦dī·ə¦lúr·ik 'as·əd }

dialysis [PHYS CHEM] A process of selective diffusion through a membrane; usually used to separate low-molecular-weight solutes which diffuse through the membrane from the colloidal and high-molecular-weight solutes which do not. { dī'al·ə·səs }

107

dialyzate

dialyzate [CHEM] The material that does not diffuse through the membrane during dialysis; alternatively, it may be considered the material that has diffused. { 'dī'al·ə,zāt }

diamide [ORG CHEM] A molecule that has two amide ($-CONH_2$) groups. { 'dī·ə,mīd }

diamidine [ORG CHEM] A molecule that has two amidine ($-C=NHNH_2$) groups. { dī'am·ə,dēn }

diamine [ORG CHEM] Any compound containing two amino groups. { 'dī·ə,mēn }

diamino [ORG CHEM] A term used in chemical nomenclature to indicate the presence in a molecule of two amino ($-NH_2$) groups. { dī'am·ə,nō }

3,5-diaminobenzoic acid [ORG CHEM] $C_7H_8N_2O_2$ Monohydrate crystals with a melting point of 228°C; soluble in organic solvents such as alcohol and benzene; used in the detection and determination of nitrites. { |thrē |fīv dī|am·ə,nō,ben'zō·ik 'as·əd }

2,7-diaminofluorene [ORG CHEM] $C_{13}H_{12}N_2$ A compound crystallizing as needlelike crystals from water; the melting point is 165°C; soluble in alcohol; used to detect bromide, chloride, nitrate, persulfate, cadmium, zinc, copper, and cobalt. Also known as 2,7-fluorenediamine. { |tü |sev·ən dī|am·ə,nō'flür,ēn } .

diamyl phenol [ORG CHEM] $(C_5H_{11})_2C_6H_3OH$ A straw-colored liquid with a boiling range of 280–295°C; used in synthetic resins, lubricating oil additives, plasticizers, detergents, and fungicides. { dī'am·əl 'fē,nòl }

diamyl sulfide [ORG CHEM] $(C_5H_{11})_2S$ A combustible, yellow liquid with a distillation range of 170–180°C; used as a flotation agent and an odorant. { dī'am·əl 'səl,fīd }

diarsine [ORG CHEM] An arsenic compound containing an As-As bond with the general formula $(R_2As)_2$, where R represents a functional group such as CH_3. { dī'är,sēn }

diarylamine [ORG CHEM] A molecule that contains an amine group and two aryl groups joined to the amino nitrogen. { |dī·ə'ril·ə,mēn }

diastereoisomer [ORG CHEM] One of a pair of optical isomers which are not mirror images of each other. Also known as diastereomer. { |dī·ə,ster·ē·ō'ī·sə· mər }

diastereomer See diastereoisomer. { |dī·ə|ster·ē'ō·mər }

diastereotopic ligand [ORG CHEM] A ligand whose replacement or addition gives rise to diastereomers. { |dī·ə,ster·ē·ə'täp·ik 'līg·ənd }

diatomic [CHEM] Consisting of two atoms. { |dī·ə'täm·ik }

diazine [ORG CHEM] **1.** A hydrocarbon consisting of an unsaturated hexatomic ring of two nitrogen atoms and four carbons. **2.** Suffix indicating a ring compound with two nitrogen atoms. { 'dī·ə,zēn }

diazinon [ORG CHEM] $C_{12}H_{21}N_2O_3PS$ A light amber to dark brown liquid with a boiling point of 83–84°C; used as an insecticide for soil and household pests, and as an insecticide and nematicide for fruits and vegetables. { dī'a·zə,nòn }

diazoalkane [ORG CHEM] A compound with the general formula $R_2C=N_2$ in which two hydrogen atoms of an alkane molecule have been replaced by a diazo group. { dī|a·zō'al,kān }

diazoamine [ORG CHEM] The grouping $-N=NNH-$. Also known as azimino. { dī|a·zō'a,mēn }

diazoaminobenzene [ORG CHEM] $C_6H_5NNNHC_6H_5$ Golden yellow scales with a melting point of 96°C; soluble in alcohol, ether, and benzene; used for dyes and insecticides. { dī|a·zō|am·ə·nō'ben,zēn }

diazoate [ORG CHEM] A salt with molecular formula of the type $C_6H_5N=NOOM$, where M is a nonvalent metal. { dī'a·zə,wāt }

diazo compound [ORG CHEM] An organic compound containing the radical $-N=N-$. { 'dī'a·zō 'käm,paùnd }

diazo dye See developed dye. { dī'a·zō ,dī }

diazo group [ORG CHEM] A functional group with the formula $=N_2$. { dī'a·zō ,grüp }

diazoic acid [ORG CHEM] $C_6H_5N=NOOH$ An isomeric form of phenylnitramine. { 'dī·ə,zō·ik 'as·əd }

diazole [ORG CHEM] A cyclic hydrocarbon with five atoms in the ring, two of which are nitrogen atoms and three are carbon. { 'dī·ə,zōl }

diazomethane [ORG CHEM] CH_2N_2 A poisonous gas used in organic synthesis to methylate compounds. { dī|a·zō'me,thān }

diazonium [ORG CHEM] The grouping =N≡N. { ˌdī·ə'zō·nē·əm }
diazonium salts [ORG CHEM] Compounds of the type R·X·N:N, where R represents an alkyl or aryl group and X represents an anion such as a halide. { ˌdī·ə'zō·nē·əm 'sȯls }
diazo oxide [ORG CHEM] An organic molecule or a grouping of organic molecules that have a diazo group and an oxygen atom joined to ortho positions of an aromatic nucleus. Also known as diazophenol. { dī'a·zō 'äk‚sīd }
diazophenol See diazo oxide. { dī¦a·zō'fe‚nȯl }
diazo process See diazotization. { dī'a·zō ‚präs·əs }
diazosulfonate [ORG CHEM] A salt formed from diazosulfonic acid. { dī¦a·zō 'səl·fə‚nāt }
diazosulfonic acid [ORG CHEM] C₆H₅N=NSO₃H Any of a group of aromatic acids containing the diazo group bonded to the sulfonic acid group. { dī¦a·zō·səl'fän·ik 'as·əd }
diazotization [ORG CHEM] Reaction between a primary aromatic amine and nitrous acid to give a diazo compound. Also known as diazo process. { dī‚az·ət·ə'zā·shən }
dibasic [CHEM] **1.** Compounds containing two hydrogens that may be replaced by a monovalent metal or radical. **2.** An alcohol that has two hydroxyl groups, for example, ethylene glycol. { dī'bās·ik }
dibasic acid [CHEM] An acid having two hydrogen atoms capable of replacement by two basic atoms or radicals. { dī'bās·ik 'as·əd }
dibasic calcium phosphate See calcium phosphate. { dī'bās·ik 'kal·sē·əm 'fäs‚fāt }
dibasic magnesium citrate [ORG CHEM] MgHC₆H₅O₇·5H₂O A white or yellowish powder soluble in water; used as a dietary supplement or in medicine. { dī'bās·ik mag'nē·zē·əm 'sī‚trāt }
dibenzyl See bibenzyl. { dī'ben·zil }
dibenzyl disulfide [ORG CHEM] C₆H₅CH₂SSCH₂C₆H₅ A compound crystallizing in leaflets with a melting point of 71–72°C; soluble in hot methanol, benzene, ether, and hot ethanol; used as an antioxidant in compounding of rubber and as an additive to silicone oils. { dī'ben·zil dī'səl‚fīd }
dibenzyl ether See benzyl ether. { dī'ben·zil 'ē·thər }
diborane [INORG CHEM] B₂H₆ A colorless, volatile compound that is soluble in ether; boiling point −92.5°C, melting point −165.5°C; can be used to produce pentaborane and decaborane, proposed for use as rocket fuels; also used to synthesize organic boron compounds. { dī'bȯr‚ān }
dibromide [CHEM] Indicating the presence of two bromine atoms in a molecule. { dī'brō‚mīd }
dibromo- [CHEM] A prefix indicating two bromine atoms. { dī'brō·mō }
dibromochloropropane [ORG CHEM] C₃H₅Br₂Cl A light yellow liquid with a boiling point of 195°C; used as a nematicide for crops. Abbreviated DBCP. { dī¦brō·mō‚klȯr·ə'prō‚pān }
dibromodifluoromethane [ORG CHEM] CF₂Br₂ A colorless, heavy liquid with a boiling point of 24.5°C; soluble in methanol and ether; used in the synthesis of dyes and pharmaceuticals and as a fire-extinguishing agent. { dī¦brō·mō·dī¦flür· ō'me‚thān }
dibromomethane See methylene bromide. { dī¦brō·mō'me‚thān }
2,6-dibromoquinone-4-chlorimide [ORG CHEM] C₆H₂Br₂ClNO Yellow prisms, soluble in water; used as a reagent for phenol and phosphatases. { ¦tü ¦siks dī¦brō·mō·kwə'nōn ¦fȯr 'klȯr·ə‚mīd }
3,5-dibromosalicylaldehyde [ORG CHEM] Br₂C₆H₂(OH)CHO Pale yellow crystals with a melting point of 86°C; readily soluble in ether, chloroform, benzene, alcohol, and glacial acetic acid; used as an antibacterial agent. { ¦thrē ¦fīv dī¦brō·mō¦sal·ə·səl'al·də‚hīd }
dibucaine [ORG CHEM] C₂₀H₂₉O₂N₃ A local anesthetic used both as the base and the hydrochloride salt. { 'dī·byə‚kān }
dibutyl [ORG CHEM] Indicating the presence of two butyl groupings bonded through a third atom or group in a molecule. { dī'byüd·əl }
dibutyl amine [ORG CHEM] C₈H₁₉N A colorless, clear liquid with amine aroma; either di-*n*-butylamine, (C₄H₉)₂NH, boiling at 160°C, insoluble in water, soluble in hydrocarbon

dibutyl maleate

solvents, or di-*sec*-butylamine, $(CH_3CHCH_2CH_3)_2NH$, boiling at 133°C, flammable; used in the manufacture of dyes. { dī'byüd·əl 'a‚mēn }

dibutyl maleate [ORG CHEM] $C_4H_9OOCCHCHCOOC_4H_9$ Oily liquid used for copolymers and plasticizers and as a chemical intermediate. { dī'byüd·əl 'mal·ē‚āt }

dibutyl oxalate [ORG CHEM] $(COOC_4H_9)_2$ High-boiling, water-white liquid with mild odor, used as a solvent and in organic synthesis. { dī'byüd·əl 'äk·sə‚lāt }

dibutyl phthalate [ORG CHEM] $C_{16}H_{22}O_4$ A colorless liquid, used as a plasticizer and insect repellent. { dī'byüd·əl 'tha‚lāt }

dibutyl succinate [ORG CHEM] $C_{12}H_{22}O_4$ A colorless liquid, insoluble in water; used as a repellent for cattle flies, cockroaches, and ants around barns. { dī'byüd·əl 'sək·sə‚nāt }

dibutyl tartrate [ORG CHEM] $(COOC_4H_9)_2(CHOH)_2$ Liquid used as a solvent and plasticizer for cellulosics and as a lubricant. { dī'byüd·əl 'tär‚trāt }

dicalcium [CHEM] A molecule containing two atoms of calcium. { dī'kal·sē·əm }

dicalcium orthophosphate See calcium phosphate. { dī'kal·sē·əm ‚ór·thō'fäs‚fāt }

dicalcium phosphate See calcium phosphate. { dī'kal·sē·əm 'fäs‚fāt }

dicarbocyanine [ORG CHEM] **1.** A member of a group of dyes termed the cyanine dyes; the structure consists of two heterocyclic rings joined to the five-carbon chain: $=CH-CH=CH-CH=CH-$. **2.** A particular dicarbocyanine dye containing two quinoline heterocyclic rings. { dī‚kär·bə‚sī·ə‚nēn }

dicarboxylic acid [ORG CHEM] A compound with two carboxyl groups. { dī‚kär·bäk‚sil·ik 'as·əd }

dication [CHEM] A doubly charged cation with the general formula X^{2+}. { dī'kat‚ī·ən }

dichlobenil [ORG CHEM] $C_7H_3Cl_2N$ A colorless, crystalline compound with a melting point of 139–145°C; used as a herbicide to control weeds in orchards and nurseries. { dī'klō·bə·nəl }

dichlofenthion [ORG CHEM] $C_{10}H_{13}Cl_2O_3PS$ A white, liquid compound, insoluble in water; used as an insecticide and nematicide for ornamentals, flowers, and lawns. { dī‚klō·fən'thī‚än }

dichlofluanid [ORG CHEM] $C_9H_{11}Cl_2FN_2O_2S$ A white powder with a melting point of 105–105.6°C; insoluble in water; used as a fungicide for fruits, garden crops, and ornamental flowers. { ‚dī·klō·flü'an·əd }

dichlone [ORG CHEM] $C_{10}H_4O_2Cl_2$ A yellow, crystalline compound, used as a fungicide for foliage and as an algicide. { 'dī‚klōn }

dichloramine [INORG CHEM] **1.** NH_2Cl_2 An unstable molecule considered to be formed from ammonia by action of chlorine. Also known as chlorimide. **2.** Any chloramine with two chlorine atoms joined to the nitrogen atom. { dī'klór·ə‚mēn }

dichloride [CHEM] Any inorganic salt or organic compound that has two chloride atoms in its molecule. { dī'klór‚īd }

dichloroacetic acid [ORG CHEM] $CHCl_2COOH$ A strong liquid acid, formed by chlorinating acetic acid; used in organic synthesis. { dī‚klór·ō·ə‚sēd·ik 'as·əd }

dichlorobenzene [ORG CHEM] $C_6H_4Cl_2$ Any of a group of substitution products of benzene and two atoms of chlorine; the three forms are *meta*-dichlorobenzene, colorless liquid boiling at 172°C, soluble in alcohol and ether, insoluble in water, or *ortho*-, colorless liquid boiling at 179°C, used as a solvent and chemical intermediate, or *para*-, volatile white crystals, insoluble in water, soluble in organic solvents, used as a germicide, insecticide, and chemical intermediate. { dī‚klór·ō'ben‚zēn }

1,4-dichlorobutane [ORG CHEM] $Cl(CH_2)_4Cl$ A colorless, flammable liquid with a pleasant odor, boiling point 155°C; soluble in organic solvents; used in organic synthesis, including adiponitrile. Abbreviated DCB. { ‚wən ‚fór dī‚klór·ō'byü‚tān }

dichlorodiethylsulfide See mustard gas. { dī‚klór·ō·dī‚eth·əl'səl‚fīd }

dichlorodifluoromethane [ORG CHEM] CCl_2F_2 A nontoxic, nonflammable, colorless gas made from carbon tetrachloride; boiling point −30°C; used as a refrigerant and as a propellant in aerosols. { dī‚klór·ō·dī‚flür·ō'me‚th aman }

***p,p'*-dichlorodiphenylmethyl carbinol** [ORG CHEM] $(ClC_6H_4)_2C(CH_3)OH$ Crystals with a melting point of 69–69.5°C; soluble in organic solvents; used as an insecticide. Abbreviated DMC. { ‚pē ‚pē‚prīm dī‚klór·ō·dī‚fen·əl‚meth·əl 'kär·bə‚nól }

110

dichlorodiphenyltrichloroethane See DDT. { dī¦klȯr·ō·dī¦fen·əl·trī¦klȯr·ō'e‚thān }

sym-dichloroethylene [ORG CHEM] CHClCHCl Colorless, toxic liquid with pleasant aroma, boiling at 59°C; decomposes in light, air, and moisture; soluble in organic solvents, insoluble in water; exists in cis and trans forms; used as solvent, in medicine, and for chemical synthesis. { ¦sim dī¦klȯr·ō'eth·ə‚lēn }

dichloroethyl ether [ORG CHEM] ClCH₂CH₂OCH₂CH₂Cl A colorless liquid insoluble in water, soluble in organic solvents; used as a solvent in paints, varnishes, lacquers, and as a soil fumigant. { dī‚klȯr·ō'eth·əl'ē·thər }

dichlorofluoromethane [ORG CHEM] CHCl₂F A colorless, heavy gas with a boiling point of 8.9°C and a freezing point of −135°C; soluble in alcohol and ether; used in fire extinguishers and as a solvent, refrigerant, and aerosol propellant. Also known as fluorocarbon 21; fluorodichloromethane. { dī¦klȯr·ō¦flür·ō'me‚thān }

α-dichlorohydrin [ORG CHEM] CH₂ClCHOHCH₂Cl Unstable liquid, the commercial product consisting of a mixture of two isomers; used as a solvent and a chemical intermediate. Abbreviated GDCH. { ¦al·fə dī‚klȯr·ō'hī·drən }

2,6-dichloro-4-nitroaniline [ORG CHEM] C₆H₄Cl₂N₂O₂ A yellow, crystalline compound that melts at 192–194°C; used as a fungicide for fruits, vegetables, and ornamental flowers. Abbreviated DCNA. { ¦tü ¦siks dī'klȯr·ō ¦fȯr ‚nī·trō'an·ə‚lēn }

dichloropentane [ORG CHEM] C₅H₁₀Cl₂ Mixed dichloro derivatives of normal pentane and isopentane; clear, light-yellow liquid used as solvent, paint and varnish remover, insecticide, and soil fumigant. { dī¦klȯr·ō'pen‚tān }

dichlorophen [ORG CHEM] C₁₃H₁₀Cl₂O₂ A white, crystalline compound with a melting point of 177–178°C; used as an agricultural fungicide, germicide in soaps, and antihelminthic drug in humans. { dī'klȯr·ə·fən }

2,4-dichlorophenoxyacetic acid [ORG CHEM] Cl₂C₆H₃OCH₂COOH Yellow crystals, melting at 142°C; used as a herbicide and pesticide. Abbreviated 2,4-D. { ¦tü ¦fȯr dī¦klȯr·ō·fə¦näk·sē·ə'sēd·ik 'as·əd }

1,3-dichloro-2-propanol [ORG CHEM] ClCH₂CHOH CH₂Cl A liquid soluble in water and miscible with alcohol and ether; used as a solvent for nitrocellulose and hard resins, as a binder for watercolors, in the production of photographic lacquer, and in the determination of vitamin A. { ¦wən ¦thrē dī¦klȯr·ō ¦tü 'prō·pə‚nȯl }

dichlorotoluene [ORG CHEM] C₇H₆Cl₂ A colorless liquid, soluble in organic solvents, insoluble in water; isomers are 2,4-CH₃C₆H₃Cl₂, boiling at 200–202°C, and 3,4-(CH₃C₆H₃Cl₂), boiling at 209°C; used as solvent and chemical intermediate. { dī¦klȯr·ō'täl·yə‚wēn }

dichlorprop [ORG CHEM] C₉H₈Cl₂O₃ A colorless, crystalline solid with a melting point of 117–118°C; used as a herbicide and fumigant for brush control on rangeland and rights-of-way. Abbreviated 2,4-DP. { dī'klȯr‚präp }

dichlorvos [ORG CHEM] C₄H₇O₄Cl₂P An amber liquid, used as an insecticide and miticide on public health pests, stored products, and flies on cattle. Abbreviated DDVP. { dī'klȯr‚väs }

dichromate [INORG CHEM] A salt of dichromic acid, usually orange or red. Also known as bichromate. { dī'krō‚māt }

dichromatic dye [CHEM] Dye or indicator in which different colors are seen, depending upon the thickness of the solution. { dī·krə'mād·ik 'dī }

dichromic [CHEM] Pertaining to a molecule with two atoms of chromium. { dī'krō·mik }

dichromic acid [INORG CHEM] H₂Cr₂O₇ An acid known only in solution, especially in the form of dichromates. { dī'krō·mik 'as·əd }

dicovalent carbon See divalent carbon. { ‚dī·kō'vā·lənt 'kär·bən }

dicrotophos [ORG CHEM] C₈H₁₆O₂P The dimethyl phosphate of 3-hydroxy-N,N-dimethyl-cis-crotonamide; a brown liquid with a boiling point of 400°C; miscible with water; used as an insecticide and miticide for cotton, soybeans, seeds, and ornamental flowers. { dī'kräd·ə‚fäs }

dicyandiamide [ORG CHEM] NH₂C(NH)(NHCN) White crystals with a melting range of 207–209°C; soluble in water and alcohol; used in fertilizers, explosives, oil well

111

dicyanide

drilling muds, pharmaceuticals, and dyestuffs. Also known as cyanoguanidine. { dī,sī·ən'dī·ə·məd }

dicyanide [CHEM] A salt that has two cyanide groups. { dī'sī·ə,nīd }

dicyanoargentates I See argentocyanides. { dī¦sī·ə·nō,är·jən'tād·ēz 'wən }

dicyclohexylamine [ORG CHEM] $(C_6H_{11})_2NH$ A clear, colorless liquid with a boiling point of 256°C; used for insecticides, corrosion inhibitors, antioxidants, and detergents, and as a plasticizer and catalyst. { dī¦sī·klō,hek'sil·ə,mēn }

dicyclohexylcarbodiimide [ORG CHEM] $C_{13}H_{22}N_2$ Crystals with a melting point of 35–36°C; used in peptide synthesis. Abbreviated DCC; DCCI. { dī¦sī·klō¦hek·səl¦kär·bō'dī·ə·məd }

DIDA See diisodecyl adipate.

didodecyl ether See dilauryl ether. { dī'dō·də·səl 'ē·thər }

DIDP See diisodecyl phthalate.

didymium [CHEM] A mixture of the rare-earth elements praseodymium and neodymium. Abbreviated Di. { dī'dim·ē·əm }

dieldrin [ORG CHEM] $C_{12}H_8Cl_6O$ A white, crystalline contact insecticide obtained by oxidation of aldrin; used in mothproofing carpets and other furnishings. { 'dēl·drən }

dielectric vapor detector [ANALY CHEM] Apparatus to measure the change in the dielectric constant of gases or gas mixtures; used as a detector in gas chromatographs to sense changes in carrier gas. { 'dī·ə'lek·trik 'vā·pər di,tek·tər }

dielectrophoresis [PHYS CHEM] The ability of an uncharged material to move when subjected to an electric field. { ¦dī·ə,lek·trō·fə'rē·səs }

Diels-Alder reaction [ORG CHEM] The 1,4 addition of a conjugated diolefin to a compound, known as a dienophile, containing a double or triple bond; the dienophile may be activated by conjugation with a second double bond or with an electron acceptor. { ¦dēlz ¦äl·dər rē,ak·shən }

diene [ORG CHEM] One of a class of organic compounds containing two ethylenic linkages (carbon-to-carbon double bonds) in the molecules. Also known as alkadiene; diolefin. { 'dī,ēn }

diene resin [ORG CHEM] Material containing the diene group of double bonds that may polymerize. { 'dī,ēn 'rez·ən }

diene value [ORG CHEM] A number that represents the amount of conjugated bonds in a fatty acid or fat. { 'dī,ēn ,val·yü }

dienophile [ORG CHEM] The alkene component of a reaction between an alkene and a diene. { dī'en·ə,fīl }

diester [ORG CHEM] A compound containing two ester groupings. { ¦dī'es·tər }

diethanolamine [ORG CHEM] $(HOCH_2CH_2)_2NH$ Colorless, water-soluble, deliquescent crystals, or liquid boiling at 217°C; soluble in alcohol and acetone, insoluble in ether and benzene; used in detergents, as an absorbent of acid gases, and as a chemical intermediate. Also known as DEA. { di·ə·thə'näl·ə,mēn }

diether [ORG CHEM] A molecule that has two oxygen atoms with ether bonds. { dī'ē·thər }

1,1-diethoxyethane See acetal. { ¦wən ¦wən ,dī·ə¦thäk·sē'e,thān }

diethyl [ORG CHEM] Pertaining to a molecule with two ethyl groups. { dī'eth·əl }

diethyl adipate [ORG CHEM] $C_2H_5OCO(CH_2)_4OCOC_2H_5$ Water-insoluble, colorless liquid, boiling at 245°C; used as a plasticizer. { dī'eth·əl 'ad·ə,pāt }

diethylamine [ORG CHEM] $(C_2H_5)_2NH$ Water-soluble, colorless liquid with ammonia aroma, boiling at 56°C; used in rubber chemicals and pharmaceuticals and as a solvent and flotation agent. { ,dī,eth·əl'a,mēn }

diethylaminoethyl cellulose [ORG CHEM] A positively charged resin used in ion-exchange chromatography; an anion exchanger. Also known as DEAE-cellulose. { ,dī,eth·əl¦am·ə·nō'eth·əl 'sel·ya,lōs }

5,5-diethylbarbituric acid See barbital. { ¦fīv ¦fīv ,dī,eth·əl'bär·bə'tür·ik 'as·əd }

diethylbenzene [ORG CHEM] $C_6H_4(C_2H_5)_2$ Colorless liquid, boiling at 180–185°C; soluble in organic solvents, insoluble in water; usually a mixture of three isomers, which are 1,2- (or *ortho*-diethylbenzene), boiling at 183°C, and 1,3- (or *meta*-), boiling at 181°C, and 1,4- (or *para*-), boiling at 184°C; used as a solvent. { ,dī,eth·əl'ben,zēn }

112

diethylcarbamazine [ORG CHEM] $C_{16}H_{29}O_8N_3$ White, water-soluble, hygroscopic crystals, melting at 136°C; used as an anthelminthic. { ,dī,eth·əl,kär'bam·ə,zēn }

diethyl carbinol [ORG CHEM] $(CH_3CH_2)_2CHOH$ Colorless, alcohol-soluble liquid, boiling at 116°C; slightly soluble in water; used in pharmaceuticals and as a solvent and flotation agent. Also known as sec-*n*-amyl alcohol. { ,dī'eth·əl 'kär·bə,nól }

diethyl carbonate [ORG CHEM] $(C_2H_5)_2CO_3$ Stable, colorless liquid with mild aroma, boiling at 126°C; soluble with most organic solvents; used as a solvent and for chemical synthesis. Also known as ethyl carbonate. { ,dī'eth·əl 'kär·bə,nāt }

diethylene glycol [ORG CHEM] $CH_2OHCH_2OCH_2CH_2OH$ Clear, hygroscopic, water-soluble liquid, boiling at 245°C; soluble in many organic solvents; used as a softener, conditioner, lubricant, and solvent, and in antifreezes and cosmetics. { dī'eth·ə,lēn 'glī,kól }

diethylene glycol monoethyl ether [ORG CHEM] $C_6H_{14}O_3$ A hygroscopic liquid used as a solvent for cellulose esters and in lacquers, varnishes, and enamels. { dī'eth·ə,lēn 'glī,kól ,män·ō'eth·əl 'ē·thər }

diethylenetriamine [ORG CHEM] $(NH_2C_2H_4)_2NH$ A yellow, hygroscopic liquid with a boiling point of 206.7°C; soluble in water and hydrocarbons; used as a solvent, saponification agent, and fuel component. { dī'eth·ə,lēn'trī·ə,mēn }

diethyl ether [ORG CHEM] $C_4H_{10}O$ A colorless liquid, slightly soluble in water; used as a reagent and solvent. Also known as ethyl ether; ethyl oxide; ethylic ether. { dī'eth·əl'ē·thər }

diethyl maleate [ORG CHEM] $(HCCOOC_2H_5)_2$ Clear, colorless liquid, boiling at 225°C; slightly soluble in water, soluble in most organic solvents; used as a chemical intermediate. { ,dī'eth·əl 'mal·ē,āt }

diethyl phosphite [ORG CHEM] $(C_2H_5O)_2HPO$ A colorless liquid with a boiling point of 138°C; soluble in water and common organic solvents; used as a paint solvent, antioxidant, and reducing agent. { dī'eth·əl 'fäs,fīt }

diethyl phthalate [ORG CHEM] $C_6H_4(CO_2C_2H_5)_2$ Clear, colorless, odorless liquid with bitter taste, boiling at 298°C; soluble in alcohols, ketones, esters, and aromatic hydrocarbons, partly soluble in aliphatic solvents; used as a cellulosic solvent, wetting agent, alcohol denaturant, mosquito repellent, and in perfumes. { dī'eth·əl 'tha,lāt }

diethyl pyrocarbonate [ORG CHEM] $C_6H_{10}O_5$ A viscous liquid, soluble in alcohols, esters, and ketones; used as a gentle esterifying agent, as a preservative for fruit juices, soft drinks, and wines, and as an inhibitor for ribonuclease. Abbreviated DEPC. { ,dī'eth·əl ,pī·rō'kär·bə,nāt }

diethyl succinate [ORG CHEM] $(CH_2COOC_2H_5)_2$ Water-white liquid with pleasant aroma, boiling at 216°C; soluble in alcohol and ether, slightly soluble in water; used as a chemical intermediate and plasticizer. { ,dī'eth·əl 'sək·sə,nāt }

diethyl sulfate [ORG CHEM] $(C_2H_5)_2SO_4$ A colorless oil with a peppermint odor, and boiling at 208°C; used as an intermediate in organic synthesis. Also known as ethyl sulfate. { ,dī'eth·əl 'səl,fāt }

diethyl sulfide See ethyl sulfide. { ,dī'eth·əl 'səl,fīd }

diethyltoluamide [ORG CHEM] $C_{12}H_{17}ON$ A liquid whose color ranges from off-white to light yellow; used as an insect repellent for people and clothing. Also known as DEET; DET; N,N-diethyl-*meta*-toluamide. { ,dī'eth·əl,täl·yü'a,mīd }

difference spectrophotometer See absorption spectrophotometer. { 'dif·rəns ,spek·trə·fə'täm·əd·ər }

differential aeration cell [PHYS CHEM] An electrolytic cell whose electromotive force derives from a difference in concentration of atmospheric oxygen at one electrode with reference to another electrode of the same material. Also known as oxygen concentration cell. { ,dif·ə'ren·chəl e'rā·shən ,sel }

differential ebuliometer [ANALY CHEM] Apparatus for precise and simultaneous measurement of both the boiling temperature of a liquid and the condensation temperature of the vapors of the boiling liquid. { ,dif·ə'ren·chəl ə,bü·lē'äm·əd· ər }

differential heat of dilution See heat of dilution. { dif·ə'ren·chəl 'hēt əv də'lü·shən }

differential polarography [ANALY CHEM] Technique of polarographic analysis which measures the difference in current flowing between two identical dropping-mercury

differential reaction rate

electrodes at the same potential but in different solutions. { ˌdif·ə'ren·chəl ˌpō·lə'räg·rə·fē }

differential reaction rate [PHYS CHEM] The order of a chemical reaction expressed as a differential equation with respect to time; for example, $dx/dt = k(a - x)$ for first order, $dx/dt = k(a - x)(b - x)$ for second order, and so on, where k is the specific rate constant, a is the concentration of reactant A, b is the concentration of reactant B, and dx/dt is the rate of change in concentration for time t. { ˌdif·əˈren·chəl rēˈak·shən ˌrāt }

differential scanning calorimetry [ANALY CHEM] A method in which a sample and a reference are individually heated (by separately controlled resistance heaters, at a predetermined rate), and enthalpic (heat-generating or -absorbing) processes are detected as differences in electrical energy supplied to either the sample or the reference material to maintain this heating rate. This difference in electrical energy, in milliwatts per second, of the heat flow into or out of the sample is due to the occurrence of a physical or chemical process. { ˌdif·əˈren·chəl ˈskan·iŋ ˌkal·əˈrim·ə·trē }

differential spectrophotometry [SPECT] Spectrophotometric analysis of a sample when a solution of the major component of the sample is placed in the reference cell; the recorded spectrum represents the difference between the sample and the reference cell. { ˌdif·əˈren·chəl ˌspek·trō·fəˈtäm·ə·trē }

differential thermometric titration [ANALY CHEM] Thermometric titration in which titrant is added simultaneously to the reaction mixture and to a blank in identically equipped cells. { ˌdif·əˈren·chəl ˈthər·məˌmeˈtrik tīˈträ·shən }

diffraction grating [SPECT] An optical device consisting of an assembly of narrow slits or grooves which produce a large number of beams that can interfere to produce spectra. Also known as grating. { diˈfrak·shən ˌgrād·iŋ }

diffraction spectrum [SPECT] Parallel light and dark or colored bands of light produced by diffraction. { diˈfrak·shən ˌspek·trəm }

diffuse series [SPECT] A series occurring in the spectra of many atoms having one, two, or three electrons in the outer shell, in which the total orbital angular momentum quantum number changes from 2 to 1. { dəˈfyüs ˈsir·ēz }

diffuse spectrum [SPECT] Any spectrum having lines which are very broad even when there is no possibility of line broadening by collisions. { dəˈfyüs ˈspek·trəm }

diffusion current [ANALY CHEM] In polarography with a dropping-mercury electrode, the flow that is controled by the rate of diffusion of the active solution species across the concentration gradient produced by the removal of ions or molecule at the electrode surface. { dəˈfyü·zhən ˌkər·ənt }

diffusion flame [CHEM] A long gas flame that radiates uniformly over its length and precipitates free carbon uniformly. { dəˈfyü·zhən ˌflām }

diffusion potential [PHYS CHEM] A potential difference across the boundary between electrolytic solutions with different compositions. Also known as liquid junction potential. { dəˈfyü·zhən pəˌten·chəl }

diffusivity analysis [ANALY CHEM] Analysis of difficult-to-separate materials in solution by diffusion effects, using, for example, dialysis, electrodialysis, interferometry, amperometric titration, polarography, or voltammetry. { dif·yüˈziv·əd·ē əˈnal·ə·səs }

difunctional molecule [ORG CHEM] An organic structure possessing two sites that are highly reactive. { dīˈfəŋk·shən·əl ˈmäl·əˌkyül }

digallic acid See tannic acid. { dīˈgal·ik ˈas·əd }

digitoxigenin [ORG CHEM] $C_{23}H_{34}O_4$ The steroid aglycone formed by removal of three molecules of the sugar digitoxose from digitoxin. { ˌdij·əˌtäk·səˈjen·ən }

digitoxin [ORG CHEM] $C_{41}H_{64}O_{13}$ A poisonous steroid glycoside found as the most active principle of digitalis, from the foxglove leaf. { ˌdij·əˈtäk·sən }

diglycerol [ORG CHEM] A compound that is a diester of glycerol. { dīˈglis·əˌrōl }

diglycine See iminodiacetic acid. { dīˈglīˌsēn }

diglycolic acid [ORG CHEM] $O(CH_2COOH)_2$ A white powder that forms a monohydrate; used in the manufacture of plasticizers and in organic synthesis, and to break emulsions. { dī·glīˈkäl·ik ˈas·əd }

diglycol laurate [ORG CHEM] $C_{11}H_{23}COOC_2H_4OC_2H_4OH$ A light, straw-colored, oily liquid; soluble in methanol, ethanol, toluene, and mineral oil; used in emulsions and as an antifoaming agent. { dī'glī,kȯl 'lȯr,āt }

diglycol stearate [ORG CHEM] $(C_{17}H_{35}COOC_2H_4)_2O$ A white, waxy solid with a melting point of 54–55°C; used as an emulsifying agent, suspending medium for powders in the manufacture of polishes, and thickening agent, and in pharmaceuticals. { dī'glī,kȯl 'stir,āt }

digoxin [ORG CHEM] $C_{41}H_{64}O_{14}$ A crystalline steroid obtained from a foxglove leaf (*Digitalis lanata*); similar to digitalis in pharmacological effects. { dī'gäk·sən }

dihalide [CHEM] A molecule containing two atoms of halogen combined with a radical or element. { dī'ha,līd }

dihexy *See* dodecane. { dī'hek·sē }

dihydrate [CHEM] A compound with two molecules of water of hydration. { dī'hī,drāt }

dihydrazone [ORG CHEM] A molecule containing two hydrazone radicals. { dī'hī·drə,zōn }

dihydro- [CHEM] A prefix indicating combination with two atoms of hydrogen. { dī¦hī·drō }

dihydrochloride [CHEM] A compound containing two molecules of hydrochloric acid. { dī¦hī·drə'klȯr,īd }

dihydroxy [CHEM] A molecule containing two hydroxyl groups. { ¦dī,hī¦dräk·sē }

dihydroxyacetone [ORG CHEM] $(HOCH_2)_2CO$ A colorless, crystalline solid with a melting point of 80°C; soluble in water and alcohol; used in medicine, fungicides, plasticizers, and cosmetics. Abbreviated DHA. { ¦dī,hī¦dräk·sē'as·ə,tōn }

2,4′-dihydroxyacetophenone [ORG CHEM] $(HO)_2C_6H_3COCH_3$ Needlelike or leafletlike crystals with a melting point of 145–147°C; soluble in pyridine, warm alcohol, and glacial acetic acid; used as a reagent for the determination of iron. { ¦tü ¦fȯr,prīm ¦dī,hī¦dräk·sē¦as·ə,tä·fə'nōn }

dihydroxy alcohol *See* glycol. { ¦dī,hī¦dräk·sē 'al·kə,hȯl }

1,8-dihydroxyanthraquinone [ORG CHEM] $C_{14}H_8O_4$ Orange, needlelike crystals that dissolve in glacial acetic acid; used as an intermediate in the commercial preparation of indanthrene and alizarin dyestuffs. Also known as chrysazin. { ¦wən ¦āt ¦dī,hī¦dräk·sē,an·thrə·kwə'nōn }

2,2′-dihydroxy-4,4′-dimethoxybenzophenone [ORG CHEM] $[CH_3OC_6H_3(OH)]_2CO$ Crystals with a melting point of 139–140°C; used in paint and plastics as a light absorber. { ¦tü ¦tü,prīm ¦dī,hī¦dräk·sē ¦fȯr ¦fȯr,prīm ,dī·mə¦thäk·sē¦ben·zō·fə'nōn }

dihydroxymaleic acid [ORG CHEM] $C_4H_4O_6$ Crystals soluble in alcohol; used in the detection of titanium and fluorides. { ¦dī,hī¦dräk·sē,mə'lā·ik 'as·əd }

diiodomethane *See* methylene iodide. { dī¦ī·ə,dō'me,thān }

3,5-diiodosalicylic acid [ORG CHEM] $C_7H_4I_2O_3$ Crystals with a sweetish, bitter taste and a melting point of 235–236°C; soluble in most organic solvents; used as a source of iodine in foods and a growth promoter in poultry, hog, and cattle feeds. { ¦thrē ¦fīv dī¦ī·ə,dō,sal·ə'sil·ik 'as·əd }

diisobutylene [ORG CHEM] C_8H_{16} Any one of a number of isomers, but most often 2,4,4-trimethylpentene-1 and 2,4,4-trimethylpentene-2; used in alkylation and as a chemical intermediate. { dī¦ī,sō'byüd·əl,ēn }

diisobutyl ketone [ORG CHEM] $(CH_3)_2CHCH_2COCH_2CH(CH_3)_2$ Stable liquid, boiling at 168°C; soluble in most organic liquids; toxic and flammable; used as a solvent, in lacquers and coatings, and as a chemical intermediate. { dī¦ī,sō'byüd·əl 'kē,tōn }

diisocyanate [ORG CHEM] A compound that contains two NCO (isocyanate) groups; used to produce polyurethane foams, resins, and rubber. { dī¦ī,sō'sī·ə,nāt }

diisodecyl adipate [ORG CHEM] $(C_{10}H_{21}OOC)_2(CH_2)_4$ A light-colored, oily liquid with a boiling range of 239–246°C; used as a primary plasticizer for polymers. Abbreviated DIDA. { dī¦ī,sō'de,sel 'ad·ə,pāt }

diisodecyl phthalate [ORG CHEM] $C_6H_4(COOC_{10}H_{21})_2$ A clear liquid with a boiling point of 250–257°C; used as a plasticizer. Abbreviated DIDP. { dī¦ī,sō'de,sel 'tha,lāt }

diisopropanolamine [ORG CHEM] $(CH_3CHOHCH_2)_2NH$ A white, crystalline solid with a

diisopropyl

boiling point of 248.7°C; used as an emulsifying agent for polishes, insecticides, and water paints. Abbreviated DIPA. { dīˌīˌsō,prō·pəˈnäl·əˌmēn }

diisopropyl [ORG CHEM] **1.** A molecule containing two isopropyl groups. **2.** See 2,3-dimethylbutane. { dīˌīˌsōˈprō·pəl }

diisopropyl ether See isopropyl ether. { dīˌīˌsōˈprō·pəl 'ēˌthər }

diketene [ORG CHEM] CH₃COCHCO A colorless, readily polymerized liquid with pungent aroma; insoluble in water, soluble in organic solvents; used as a chemical intermediate. { dī'kēˌtēn }

diketone [ORG CHEM] A molecule containing two ketone carbonyl groups. { dī'kēˌtōn }

diketopiperazine [ORG CHEM] **1.** C₄H₆N₂O₂ A compound formed by dehydration of two molecules of glycine. **2.** Any of the cyclic molecules formed from α-amino acids other than glycine or by partial hydrolysis of protein. { dīˌkē·dō·pi'per·əˌzēn }

dilactone [ORG CHEM] A molecule that contains two lactone groups. { dī'lākˌtōn }

dilatancy [CHEM] The property of a viscous suspension which sets solid under the influence of pressure. { dī'lāt·ən·sē }

dilatant [CHEM] A material with the ability to increase in volume when its shape is changed. { dī'lātˌənt }

dilauryl ether [ORG CHEM] (C₁₂H₂₅)₂NH A liquid with a boiling point of 190–195°C; used for electrical insulators, water repellents, and antistatic agents. Also known as didodecyl ether. { dī'lòr·əl 'ē·thər }

dilauryl thiodipropionate [ORG CHEM] (C₁₂H₂₅OOCCH₂CH₂)₂S White flakes with a melting point of 40°C; soluble in most organic solvents; used as an antioxidant, plasticizer, and preservative, and in food wraps and edible fats and oils. { dī'lòr·əl ˌthī·ō,dī'prō·pē·əˌnāt }

dilinoleic acid [ORG CHEM] C₃₄H₆₂(COOH)₂ A light yellow, viscous liquid used as an emulsifying agent and shellac substitute. { dīˌlin·əˌlā·ik 'as·əd }

diluent [CHEM] An inert substance added to some other substance or solution so that the volume of the latter substance is increased and its concentration per unit volume is decreased. { 'dil·yə·wənt }

dilute [CHEM] To make less concentrated. { dī'lüt }

dilution [CHEM] Increasing the proportion of solvent to solute in any solution and thereby decreasing the concentration of the solute per unit volume. { də'lü·shən }

dimedone See 5,5-dimethyl-1,3-cyclohexanedione. { 'dī·məˌdōn }

dimer [CHEM] A molecule that results from a chemical combination of two entities of the same species, for example, the chlorine molecule (Cl₂) or cyanogen (NCCN). { 'dī·mər }

dimeric water [INORG CHEM] Water in which pairs of molecules are joined by hydrogen bonds. { dī'mer·ik 'wòd·ər }

dimerization [CHEM] A chemical reaction in which two identical molecular entities react to form a single dimer. { ˌdī·mər·ə'zā·shən }

dimetan [ORG CHEM] The generic name for 5,5-dimethyldehydroresorcinol dimethylcarbamate, a synthetic carbamate insecticide. { 'dī·məˌtan }

dimethachlon [ORG CHEM] C₁₀H₇Cl₂NO₂ A yellowish, crystalline solid with a melting point of 136.5–138°C; insoluble in water; used as a fungicide. { dī·mə'thaˌklän }

dimethoate [ORG CHEM] C₅H₁₂NO₃PS₂ A crystalline compound, soluble in most organic solvents; used as an insecticide. { dī'meth·əˌwāt }

dimethrin [ORG CHEM] C₁₉H₂₈O₂ An amber liquid with a boiling point of 175°C; soluble in petroleum hydrocarbons, alcohols, and methylene chloride; used as an insecticide for mosquitoes, body lice, stable flies, and cattle flies. { dī'me·thrən }

dimethyl [ORG CHEM] A compound that has two methyl groups. { dī'meth·əl }

dimethylamine [ORG CHEM] (CH₃)₂NH Flammable gas with ammonia aroma, boiling at 7°C; soluble in water, ether, and alcohol; used as an acid-gas absorbent, solvent, and flotation agent, in pharmaceuticals and electroplating, and in dehairing hides. { ˌdīˌmeth·əl'amˌēn }

***para*-dimethylaminobenzalrhodanine** [ORG CHEM] C₁₂H₁₂N₂OS₂ Deep red, needlelike crystals that decompose at 270°C; soluble in strong acids; used in acetone solution for

116

the detection of ions such as silver, mercury, copper, gold, palladium, and platinum. { ¦par·ə ‚dī¦meth·əl¦am·ə‚nō‚ben·zəl'rō·də‚nēn }

2-dimethylaminoethanol [ORG CHEM] $(CH_3)_2NCH_2CH_2OH$ A colorless liquid with a boiling point of 134.6°C; used for the synthesis of dyestuffs, pharmaceuticals, and corrosion inhibitors, in medicine, and as an emulsifier. { ¦tü ‚dī¦meth·əl¦am·ə‚nō'eth· ə‚nól }

N,N-dimethylaniline [ORG CHEM] $C_6H_5N(CH_3)_2$ A yellowish liquid slightly soluble in water; used in dyes and solvent and in the manufacture of vanillin. Also known as aniline N,N-dimethyl. { ¦en ¦en ‚dī¦meth·əl'an·ə‚lēn }

dimethylbenzene *See* xylene. { ‚dī¦meth·əl'ben‚zēn }

2,3-dimethylbutane [ORG CHEM] $(CH_3)_2CHCH(CH_3)_2$ A colorless liquid with a boiling point of 57.9°C; used as a high-octane fuel. Also known as diisopropyl. { ¦tü ¦thrē ‚dī¦meth·əl'byü‚tān }

dimethyl carbate [ORG CHEM] $C_{11}H_{14}O_4$ A colorless liquid with a boiling point of 114–115°C; used as an insect repellent. { ‚dī'meth·əl 'kär‚bāt }

5,5-dimethyl-1,3-cyclohexanedione [ORG CHEM] $C_8H_{12}O_2$ Crystals that decompose at 148–150°C; soluble in water and inorganic solvents such as methanol and ethanol; used as a reagent for the identification of aldehydes. Also known as dimedone. { ¦fīv ¦fīv ‚dī¦meth·əl ¦wən ¦thrē ‚sī·klō¦hek‚sān 'dī‚ōn }

dimethyl diaminophenazine chloride *See* neutral red. { ‚dī'meth·əl dī¦am·ə‚nō'fen· ə‚zēn 'klór‚īd }

2,2-dimethyl-1,3-dioxolane-4-methanol [ORG CHEM] $C_6H_{12}O_3$ The acetone ketal of glycerin; a liquid miscible with water and many organic solvents; used as a plasticizer and a solvent. { ¦tü ¦tü ‚dī¦meth·əl ¦wən ¦thrē dī¦äk·sə‚lān ¦fór 'meth·ə‚nól }

dimethyl ether [ORG CHEM] CH_3OCH_3 A flammable, colorless liquid, boiling at −25°C; soluble in water and alcohol; used as a solvent, extractant, reaction medium, and refrigerant. Also known as methyl ether; wood ether. { ‚dī'meth·əl 'ē·thər }

dimethylethylene *See* butylene. { ‚dī¦meth·əl'eth·ə‚lēn }

N,N-dimethylformamide [ORG CHEM] $HCON(CH_3)_2$ A liquid that boils at 152.8°C; extensively used as a solvent for organic compounds. Abbreviated DMF. { ¦en ¦en ‚dī¦meth·əl'fór·mə‚mīd }

dimethylglyoxime [ORG CHEM] $(CH_3)_2C_2(NOH)_2$ White, crystalline or powdered solid, used in analytical chemistry as a reagent for nickel. { ‚dī¦meth·əl·glī'äk‚sīm }

uns-dimethylhydrazine [ORG CHEM] $(CH_3)_2NNH_2$ A flammable, highly toxic, colorless liquid; used as a component of rocket and jet fuels and as a stabilizer for organic peroxide fuel additives. Abbreviated UDMH. { ¦əns dī¦meth·əl'hī·drə‚zēn }

dimethylisopropanolamine [ORG CHEM] $(CH_3)_2NCN_2CH(OH)CH_3$ A colorless liquid with a boiling point of 125.8°C; soluble in water; used in methadone synthesis. { ‚dī¦meth·əl‚ī·sə‚prō·pə'näl·ə‚mēn }

dimethylolurea [ORG CHEM] $CO(NHCH_2OH)_2$ Colorless crystals melting at 126°C, soluble in water; used to increase fire resistance and hardness of wood, and in textiles to prevent wrinkles. Also known as 1,3-bis-hydroxymethylurea; DMU. { ‚dī¦meth· əl‚ól·yü'rē·ə }

dimethyl phthalate [ORG CHEM] $C_6H_4(COOCH_3)_2$ Odorless, colorless liquid, boiling at 282°C; soluble in organic solvents, slightly soluble in water; used as a plasticizer, in resins, lacquers, and perfumes; also as an insect repellent. { ‚dī'meth·əl 'tha‚lāt }

dimethyl sebacate [ORG CHEM] $[(CH_2)_4COOCH_3]_2$ Clear, colorless liquid, boiling at 294°C; used as a vinyl resin, nitrocellulose solvent, or plasticizer. { ‚dī'meth·əl 'seb·ə‚kāt }

dimethyl sulfate [ORG CHEM] $(CH_3)_2SO_4$ Poisonous, corrosive, colorless liquid, boiling at 188°C; slightly soluble in water, soluble in ether and alcohol; used to methylate amines and phenols. Also known as methyl sulfate. { ‚dī'meth·əl 'səl‚fāt }

dimethyl sulfide *See* methyl sulfide. { ‚dī'meth·əl 'səl‚fīd }

2,4-dimethylsulfolane [ORG CHEM] $C_6H_{12}O_2S$ A yellow to colorless liquid miscible with lower aromatic hydrocarbons; used as a solvent in liquid-liquid and vapor-liquid extraction processes. { ¦tü ¦fór ‚dī¦meth·əl'səl·fə‚lān }

dimethyl sulfoxide [ORG CHEM] $(CH_3)_2SO$ A colorless liquid used as a local analgesic

dimethyl terephthalate

and anti-inflammatory agent, as a solvent in industry, and in laboratories as a medium for carrying out chemical reactions. Abbreviated DMSO. { ‚dī'meth·əl səl'fäk‚sīd }

dimethyl terephthalate [ORG CHEM] $C_6H_4(COOCH_3)_2$ Colorless crystals, melting at 140°C and subliming above 300°; slightly soluble in water, soluble in hot alcohol and ether; used to make polyester fibers and film. Abbreviated DMT. { ‚dī'meth· əl ‚ter·ə'tha‚lāt }

dimethyl-2,3,5,6-tetrachloroterephthalate [ORG CHEM] $C_{10}H_6Cl_4O_4$ A colorless, crystalline compound with a melting point of 156°C; used as a herbicide for turf, ornamental flowers, and certain vegetables and berries. Abbreviated DCPA. { ‚dī'meth·əl ¦tü ¦thrē ¦fīv ¦siks ‚te·trə·klȯr·ō‚ter·ə'tha‚lāt }

dimorphism [CHEM] Having crystallization in two forms with the same chemical composition. { dī'mȯr‚fiz·əm }

dineric [PHYS CHEM] **1.** Having two liquid phases. **2.** Pertaining to the interface between two liquids. { dī'ner·ik }

dinitramine [ORG CHEM] $C_{11}H_{13}N_3O_4F_3$ A yellow solid with a melting point of 98–99°C; used as a preemergence herbicide for annual grass and broadleaf weeds in cotton and soybeans. { dī'nī·trə‚mēn }

dinitrate [CHEM] A molecule that contains two nitrate groups. { dī'nī‚trāt }

dinitrite [CHEM] A molecule that has two nitrite groups. { dī'nī‚trīt }

2,4-dinitroaniline [ORG CHEM] $(NO_2)C_6H_3NH_2$ A compound which crystallizes as yellow needles or greenish-yellow plates, melting at 187.5–188°C; soluble in alcohol; used in the manufacture of azo dyes. { ¦tü ¦fȯr dī¦nī·trō'an·ə‚lēn }

2,4-dinitrobenzaldehyde [ORG CHEM] $(NO_2)_2C_6H_3CHO$ Yellow to light brown crystals with a melting point of 72°C; soluble in alcohol, ether, and benzene; used to make Schiff bases. { ¦tü ¦fȯr dī¦nī·tro‚ben'zal·də‚hīd }

dinitrobenzene [ORG CHEM] Any one of three isomeric substitution products of benzene having the empirical formula $C_6H_4(NO_2)_2$. { dī‚nī·trō'ben‚zēn }

2,4-dinitrobenzenesulfenyl chloride [ORG CHEM] $(NO_2)_2C_6H_3SCl$ Crystals soluble in glacial acetic acid, with a melting point of 96°C; used as a reagent for separation and identification of naturally occurring indoles. { ¦tü ¦fȯr dī¦nī·trō‚ben‚zēn'səl· fə‚nil 'klȯr‚īd }

3,4-dinitrobenzoic acid [ORG CHEM] $C_7H_4N_2O_6$ Crystals with a bitter taste and a melting point of 166°C; used in quantitative sugar analysis. { ¦thrē ¦fȯr dī¦nī·trō·ben'zō·ik 'as·əd }

dinitrogen [CHEM] N_2 The diatomic molecule of nitrogen. { dī'nī·trə·jən }

dinitrogen tetroxide See nitrogen dioxide. { dī'nī·trə·jən te'träk‚sīd }

dinitrophenol [ORG CHEM] Any one of six isomeric substituent products of benzene having the empirical formula $(NO_2)_2C_6H_3OH$. { dī¦nī·trō'fē‚nȯl }

2,4-dinitrophenylhydrazine [ORG CHEM] $(NO_2)_2C_6H_3NHNH_2$ A red, crystalline powder with a melting point of approximately 200°C; soluble in dilute inorganic acids; used as a reagent for determination of ketones and aldehydes. { ¦tü ¦fȯr dī¦nī·trō‚fen· əl'hī·drə‚zēn }

dinitrotoluene [ORG CHEM] Any one of six isomeric substitution products of benzene having the empirical formula $CH_3C_6H_3(NO_2)_2$; they are high explosives formed by nitration of toluene. Abbreviated DNT. { dī¦nī·trō'täl·yə‚wēn }

dinoseb [ORG CHEM] $C_{10}H_{12}O_5N_2$ A reddish-brown liquid with a melting point of 32°C; used as an insecticide and herbicide for numerous crops and in fruit and nut orchards. { 'dī·nə‚seb }

dinoterb acetate [ORG CHEM] $C_{12}H_{14}N_2O_6$ A yellow, crystalline compound with a melting point of 133–134°C; used as a preemergence herbicide for sugarbeets, legumes, and cereals; and as a postemergence herbicide for maize, sorghum, and alfalfa. { 'dī· nə‚tərb 'as·ə‚tāt }

dioctyl [ORG CHEM] A compound that has two octyl groups. { dī'äkt·əl }

dioctyl phthalate [ORG CHEM] $(C_8H_{17}OOC)_2C_6H_4$ Pale, viscous liquid, boiling at 384°C; insoluble in water; used as a plasticizer for acrylate, vinyl, and cellulosic resins, and as a miticide in orchards. Abbreviated DOP. { dī¦äkt·əl 'tha‚lāt }

dioctyl sebacate [ORG CHEM] $(CH_2)_8(COOC_8H_{17})_2$ Water-insoluble, straw-colored liquid,

118

boiling at 248°C; used as a plasticizer for vinyl, cellulosic, and styrene resins.   { dī'äkt·əl 'seb·ə,kāt }

diodide   [CHEM]   A molecule that contains two iodine atoms bonded to an element or radical.   { 'dī·ə,dīd }

diolefin   *See* diene.   { dī'ō·lə,fən }

-dione   [ORG CHEM]   Suffix indicating the presence of two keto groups.   { 'dī,ōn }

1,4-dioxane   [ORG CHEM] $C_4H_8O_2$ The cyclic ether of ethylene glycol; it is soluble in water in all proportions and is used as a solvent.   { ¦wən ¦fòr dī'äk,sān }

dioxide   [CHEM]   A compound containing two atoms of oxygen.   { dī'äk,sīd }

dioxin   [ORG CHEM]   A member of a family of highly toxic chlorinated aromatic hydrocarbons; found in a number of chemical products as lipophilic contaminants. Also known as polychlorinated dibenzo-*para*-dioxin.   { dī'äk·sən }

dioxolane   [ORG CHEM] $C_3H_6O_2$ A cyclic acetal that is a liquid; used as a solvent and extractant.   { dī'äk·sə,lān }

dioxopurine   *See* xanthine.   { dī¦äk·sō'pyùr,ēn }

dioxygen   [CHEM] O_2 Molecular oxygen.   { dī'äk·si·jən }

DIPA   *See* diisopropanolamine.   { 'dip·ə *or* ¦dē¦ī¦pē¦ā }

dipentene   [ORG CHEM]   A racemate of limonene.   { dī'pen,tēn }

dipentene glycol   *See* terpin hydrate.   { dī'pen,tēn 'glī,kòl }

dipentene hydrochloride   *See* terpene hydrochloride.   { dī'pen,tēn,hī·drə'klòr,īd }

diphacinone   [ORG CHEM] $C_{23}H_{16}O_3$ A yellow powder with a melting point of 145–147°C; used to control rats, mice, and other rodents; acts as an anticoagulant.   { də'fas·ə,nōn }

diphenamid   [ORG CHEM] $C_{16}H_{17}ON$ An off-white, crystalline compound with a melting point of 134–135°C; used as a preemergence herbicide for food crops, fruits, and ornamentals.   { dī'fen·ə·məd }

diphenatrile   [ORG CHEM] $C_{14}H_{11}N$ A yellow, crystalline compound with a melting point of 73–73.5°C; used as a preemergence herbicide for turf.   { dī'fen·ə·trəl }

diphenol   [ORG CHEM]   A compound that has two phenol groups, for example, resorcinol.   { dī'fē,nól }

diphenyl   *See* biphenyl.   { dī'fen·əl }

diphenylamine   [ORG CHEM] $(C_6H_5)_2NH$ Colorless leaflets, sparingly soluble in water; melting point 54°C; used as an additive in propellants to increase the storage life by neutralizing the acid products formed upon decomposition of the nitrocellulose. Also known as phenylaniline.   { dī¦fen·əl'am,ēn }

diphenylcarbazide   [ORG CHEM] $CO(NHNHC_6H_5)_2$ A white powder, melting point 170°C; used as an indicator, pink for alkalies, colorless for acids.   { dī¦fen·əl'kär·bə,zīd }

diphenyl carbonate   [ORG CHEM] $(C_6H_5O)_2CO$ Easily hydrolyzed, white crystals, melting at 78°C; soluble in organic solvents, insoluble in water; used as a solvent, plasticizer, and chemical intermediate.   { dī¦fen·əl 'kär·bə,nāt }

diphenylchloroarsine   [ORG CHEM] $(C_6H_5)_2AsCl$ Colorless crystals used during World War I as an antipersonnel device to generate a smoke causing sneezing and vomiting.   { dī¦fen·əl,klòr·ō'är,sēn }

diphenylene oxide   [ORG CHEM] $C_{12}H_8O$ A crystalline solid derived from coal tar; melting point is 87°C; used as an insecticide.   { dī¦fen·əl,ēn 'äk,sīd }

diphenyl ether   *See* diphenyl oxide.   { dī¦fen·əl 'e,thər }

diphenylethylene   *See* stilbene.   { dī¦fen·əl'eth·ə,lēn }

diphenylguanidine   [ORG CHEM] $HNC(NHC_6H_5)_2$ A white powder, melting at 147°C; used as a rubber accelerator. Also known as DPG; melaniline.   { dī¦fen·əl'gwän·ə,dēn }

diphenyl ketone   *See* benzophenone.   { dī¦fen·əl 'kē,tōn }

diphenylmethane   [ORG CHEM] $(C_6H_5)_2CH_2$ Combustible, colorless crystals melting at 26.5°C; used in perfumery, dyes, and organic synthesis.   { dī¦fen·əl'me,thān }

diphenyl oxide   [ORG CHEM] $(C_6H_5)_2O$ A colorless liquid or crystals with a melting point of 27°C and a boiling point of 259°C; soluble in alcohol and ether; used in perfumery, soaps, and resins for laminated electrical insulation. Also known as diphenyl ether; phenyl ether.   { dī¦fen·əl 'äk,sīd }

119

diphenyl phthalate

diphenyl phthalate [ORG CHEM] $C_6H_4(COOC_6H_5)_2$ White powder, melting at 80°C; soluble in chlorinated hydrocarbons, esters, and ketones, insoluble in water; used as a plasticizer for cellulosic and other resins. { dī¦fen·əl 'tha,lāt }

diphosgene *See* trichloromethyl chloroformate. { dī'fäz,jēn }

diphosphate [CHEM] A salt that has two phosphate groups. { dī'fäs,fāt }

diphosphoglyceric acid [ORG CHEM] $C_3H_8O_9P_2$ An ester of glyceric acid, with two molecules of phosphoric acid, characterized by a high-energy phosphate bond. { dī,fäs·fə·glə'ser·ik 'as·əd }

dipicrylamine [ORG CHEM] [(NO₂)₃C₆H₂]₂NH Yellow, prismlike crystals used in the gravimetric determination of potassium. { dī·pə'kril·ə,mēn }

dipnone [ORG CHEM] $C_{16}H_{14}O$ A liquid ketone, formed by condensation of two acetophenone molecules; used as a plasticizer. { 'dip,nōn }

dipolar gas [PHYS CHEM] A gas whose molecules have a permanent electric dipole moment. { dī¦pōl·ər 'gas }

dipolar ion [CHEM] An ion carrying both a positive and a negative charge. Also known as zwitterion. { 'dī,pō·lər 'ī,än }

dipolar aprotic solvent [ORG CHEM] A solvent with characteristically high polarity and low reactivity, that is, a solvent having a sizable permanent dipole moment that cannot donate labile hydrogen atoms to form strong hydrogen bonds; examples include acetonitrile, dimethyl sulfoxides, and hexamethylphosphoramide. { dī,pō·lər ā,präd·ik 'säl·vənt }

dipole-dipole force *See* orientation force. { ¦dī,pōl ¦dī,pōl ,fórs }

dipole moment [PHYS CHEM] The vector sum of the bond moments in a molecule, a measure of the polarity of the molecule. { 'dī,pōl ,mō·mənt }

dipping acid *See* sulfuric acid. { 'dip·iŋ ,as·əd }

dipropyl [ORG CHEM] A compound containing two propyl groups. { dī'prō·pəl }

dipropylene glycol [ORG CHEM] (CH₃CHOHCH₂)₂O A colorless, slightly viscous liquid with a boiling point of 233°C; soluble in toluene and in water; used as a solvent and for lacquers and printing inks. { dī'prō·pə,lēn 'glī,kȯl }

diprotic [CHEM] Pertaining to a chemical structure that has two ionizable hydrogen atoms. { dī'präd·ik }

diprotic acid [CHEM] An acid that has two ionizable hydrogen atoms in each molecule. { di'präd·ik 'as·əd }

2,2′-dipyridyl [ORG CHEM] $C_{10}H_8N_2$ A crystalline substance soluble in organic solvents; melting point is 69.7°C; used as a reagent for the determination of iron. Also known as 2,2′-bipyridine. { ¦tü ¦tü,prīm dī'pir·ə·dəl }

diquat [ORG CHEM] $C_{12}H_{12}N_2Br_2$ A yellow water-soluble solid used as a herbicide. { 'dī,kwät }

direct effect [PHYS CHEM] A chemical effect caused by the direct transfer of energy from ionizing radiation to an atom or molecule in a medium. { də'rekt i'fekt }

direct-vision spectroscope [SPECT] A spectroscope that allows the observer to look in the direction of the light source by means of an Amici prism. { də¦rekt ¦vizh·ən 'spek·trə,skōp }

discontinuous phase *See* disperse phase. { ,dis·kən'tin·yə·wəs 'fāz }

discrete spectrum [SPECT] A spectrum in which the component wavelengths constitute a discrete sequence of values rather than a continuum of values. { di'skrēt 'spek·trəm }

disilane [INORG CHEM] Si_2H_6 A spontaneously flammable compound of silicon and hydrogen; it exists as a liquid at room temperature. { dī'si,lān }

disilicate [CHEM] A silicate compound that has two silicon atoms in the molecule. { dī'sil·ə,kāt }

disilicide [CHEM] A compound that has two silicon atoms joined to a radical or another element. { dī'sil·ə,sīd }

disk colorimeter [ANALY CHEM] A device for comparing standard and sample colors by means of rotating color disks. { 'disk kə·lə'rim·əd·ər }

disodium hydrogen phosphate *See* disodium phosphate. { dī'sōd·ē·əm 'hī·drə·jən 'fäs,fāt }

disodium methylarsonate [ORG CHEM] $CH_3AsO(ONa)_2$ A colorless, hygroscopic, crystalline solid; soluble in water and methanol; used in pharmaceuticals and as a herbicide. Abbreviated DMA. { dī'sōd·ē·əm ¦meth·əl'ärs·ən,āt }

disodium phosphate [INORG CHEM] Na_2HPO_4 Transparent crystals, soluble in water; used in the textile processing and other industries to control pH in the range 4–9, as an additive in processed cheese to maintain spreadability, and as a laxative and antacid. Also known as disodium hydrogen phosphate. { dī'sōd·ē·əm 'fäs,fāt }

disodium tartrate See sodium tartrate. { dī'sōd·ē·əm 'tär,trāt }

disperse phase [CHEM] The phase of a disperse system consisting of particles or droplets of one substance distributed through another system. Also known as discontinuous phase; internal phase. { də'spərs ,fāz }

disperse system [CHEM] A two-phase system consisting of a dispersion medium and a disperse phase. { də'spərs ,sis·təm }

dispersible inhibitor [CHEM] An additive that can be dispersed in a liquid with only moderate agitation to retard undesirable chemical action. { di'spər·sə·bəl in'hib· əd·ər }

dispersion [CHEM] A distribution of finely divided particles in a medium. { də'spər·zhən }

dispersion force [PHYS CHEM] The force of attraction that exists between molecules that have no permanent dipole. Also known as London force; van der Waals force. { də'spər·zhən ,fōrs }

dispersion medium See continuous phase. { də'spər·zhən ,mēd·ē·əm }

dispersoid [CHEM] Matter in a form produced by a disperse system. { də'spər,sȯid }

displacement [CHEM] A chemical reaction in which an atom, radical, or molecule displaces and sets free an element of a compound. { dis'plās·mənt }

displacement chromatography [ANALY CHEM] Variation of column-development or elution chromatography in which the solvent is sorbed more strongly than the sample components; the freed sample migrates down the column, pushed by the solvent. { dis'plās·mənt ,krō·mə'täg·rə·fē }

displacement series [CHEM] The elements in decreasing order of their negative potentials. Also known as constant series; electromotive series; Volta series. { dis'plās· mənt ,sir·ēz }

disproportionation [CHEM] The changing of a substance, usually by simultaneous oxidation and reduction, into two or more dissimilar substances. { ,dis·prə,pȯr· shə'nā·shən }

dissociation [PHYS CHEM] Separation of a molecule into two or more fragments (atoms, ions, radicals) by collision with a second body or by the absorption of electromagnetic radiation. { də,sō·sē'ā·shən }

dissociation constant [PHYS CHEM] A constant whose numerical value depends on the equilibrium between the undissociated and dissociated forms of a molecule; a higher value indicates greater dissociation. { də,sō·sē'ā·shən ,kän·stənt }

dissociation energy [PHYS CHEM] The energy required for complete separation of the atoms of a molecule. { də,sō·sē'ā·shən ,en·ər·jē }

dissociation limit [SPECT] The wavelength, in a series of vibrational bands in a molecular spectrum, corresponding to the point at which the molecule dissociates into its constituent atoms; it corresponds to the convergence limit. { də'sō·sē'ā·shən ,lim·ət }

dissociation pressure [PHYS CHEM] The pressure, for a given temperature, at which a chemical compound dissociates. { də,sō·sē'ā·shən ,presh·ər }

dissociation-voltage effect [PHYS CHEM] A change in the dissociation of a weak electrolyte produced by a strong electric field. { də,sō·sē¦ā·shən 'vōl·tij i,fekt }

dissolution [CHEM] Dissolving of a material. { ,dis·ə'lü·shən }

dissolve [CHEM] **1.** To cause to disperse. **2.** To cause to pass into solution. { də'zälv }

dissymmetry coefficient [ANALY CHEM] Ratio of the intensities of scattered light at 45 and 135°, used to correct for destructive interference encountered in light-scattering-photometric analyses of liquid samples. { di'sim·ə·trē ,kō·i'fish·ənt }

121

distillate

distillate [CHEM] The products of distillation formed by condensing vapors. { 'dis·tə,lāt }

distillation [CHEM] The process of producing a gas or vapor from a liquid by heating the liquid in a vessel and collecting and condensing the vapors into liquids. { ,dis·tə'lā·shən }

distillation column [CHEM] A still for fractional distillation. { ,dis·tə'lā·shən ,käl·əm }

distillation curve [CHEM] The graphical plot of temperature versus overhead product (distillate) volume or weight for a distillation operation. { ,dis·tə'lā·shən ,kərv }

distillation loss [CHEM] In a laboratory distillation, the difference between the volume of liquid introduced into the distilling flask and the sum of the residue and condensate received. { ,dis·tə'lā·shən ,lós }

distillation range [CHEM] The difference between the temperature at the initial boiling point and at the end point of a distillation test. { ,dis·tə'lā·shən ,rānj }

distilled mustard gas [ORG CHEM] A delayed-action casualty gas (mustard gas) that has been distilled, or purified, to greatly reduce the odor and thereby increase its difficulty of detection. { də'stild 'məs·tərd ,gas }

distilled water [CHEM] Water that has been freed of dissolved or suspended solids and organisms by distillation. { də'stild 'wòd·ər }

distilling flask [CHEM] A round-bottomed glass flask that is capable of holding a liquid to be distilled. { də'stil·iŋ ,flask }

distribution coefficient [PHYS CHEM] The ratio of the amounts of solute dissolved in two immiscible liquids at equilibrium. { ,dis·trə'byü·shən ,ko·i'fish·ənt }

distribution law [ANALY CHEM] The law stating that if a substance is dissolved in two immiscible liquids, the ratio of its concentration in each is constant. { ,dis·trə'byü·shən ,lò }

distribution ratio [ANALY CHEM] The ratio of the concentrations of a given solute in equal volumes of two immiscible solvents after the mixture has been shaken and equilibrium established. { ,di·strə'byü·shən ,rā·shō }

disubstituted alkene [ORG CHEM] An alkene with the general formula $R_2C=CH_2$ or $RHC=CHR$, where R is any organic group; a carbon atom is bonded directly to each end of the double bond. { dī'səb·stə,tüd·əd 'al,kēn }

disulfate [CHEM] A compound that has two sulfate radicals. { dī'səl,fāt }

disulfide [CHEM] **1.** A compound that has two sulfur atoms bonded to a radical or element. **2.** One of a group of organosulfur compounds RSSR' that may be symmetrical (R=R') or unsymmetrical (R and R', different). { dī'səl,fīd }

disulfide bond *See* disulfide bridge. { dī'səl,fīd 'bänd }

disulfide bridge [ORG CHEM] A sulfur-to-sulfur bond linking the sulfur atoms of two polypeptide chains. Also known as disulfide bond. { dī'səl,fīd 'brij }

disulfonate [CHEM] A molecule that has two sulfonate groups. { dī'səl·fə,nāt }

disulfonic acid [CHEM] A molecule that has two sulfonic acid groups. { ,dī·səl'fän·ik 'as·əd }

diterpene [ORG CHEM] $C_{20}H_{32}$ A group of terpenes that have twice as many atoms in the molecule as monoterpenes. Any derivative of diterpene. { dī'tər,pēn }

dithiocarbamate [ORG CHEM] **1.** A salt of dithiocarbamic acid. **2.** Any other derivative of dithiocarbamic acid. { ¦dī,thī·ō'kär·bə,māt }

dithiocarbamic acid [ORG CHEM] NH_2CS_2H A colorless, unstable powder; various metal salts are readily obtained, and used as strong accelerators for rubber. Also known as aminodithioformic acid. { ¦dī,thī·ō,kär'bam·ik 'as·əd }

dithioic acid [ORG CHEM] An organic acid in which sulfur atoms have replaced both oxygen atoms of the carboxyl group. { ¦dī,thī¦ō·ik 'as·əd }

dithionate [CHEM] Any salt formed from dithionic acid. { dī'thi·ə,nāt }

dithionic acid [INORG CHEM] $H_2S_2O_6$ A strong acid formed by the oxidation of sulfurous acid, and known only by its salts and in solution. { ¦dī,thī'än·ik 'as·əd }

dithiooxamide [ORG CHEM] $NH_2CSCSNH_2$ Red crystals soluble in alcohol; used as a reagent for copper, cobalt, and nickel, and for the determination of osmium. { ¦dī,thī·ō'äk·sə,mīd }

1,4-dithiothreitol [ORG CHEM] $C_4H_{10}O_2S_2$ Needlelike crystals soluble in water, ethanol acetone, ethylacetate; used as a protective agent for thiol (SH) groups. { ¦wən ¦fór ¦dī,thī·ō'thrē·ə,tȯl }

ditungsten carbide [INORG CHEM] W_2C A gray powder having hardness approaching that of diamond; forms hexagonal crystals with specific gravity 17.2; melting point 2850°C. { ¦dī,təŋ·stən 'kär,bīd }

divalent carbon [ORG CHEM] A charged or uncharged carbon atom that has formed only two covalent bonds. Also known as dicovalent carbon. { dī'vā·lənt 'kär·bən }

divalent metal [CHEM] A metal whose atoms are each capable of chemically combining with two atoms of hydrogen. { dī'vā·lənt 'med·əl }

diver method [PHYS CHEM] Measure of the size of suspended solid particles; small glass divers of known density sink to the level where the liquid-suspension density is equal to that of the diver, allowing calculation of particle size. Also known as Berg's diver method. { 'dī·vər ,meth·əd }

divinyl [ORG CHEM] **1.** A molecule that has two vinyl groups. **2.** See 1,3-butadiene. { dī'vīn·əl }

divinyl acetylene [ORG CHEM] C_6H_6 A linear trimer of acetylene, made by passing acetylene into a hydrochloric acid solution that has metallic catalysts; used as an intermediate in neoprene manufacture. { dī'vīn·əl ə'sed·əl,ēn }

divinylbenzene [ORG CHEM] $C_6H_4(CHCH_2)_2$ Polymerizable, water-white liquid used to make rubbers, drying oils, and ion-exchange resins and other polymers; forms include ortho, meta, and para isomers. Also known as vinylstyrene. { dī¦vīn·əl'ben,zēn }

divinyl ether See vinyl ether. { dī'vīn·əl 'ē·thər }

divinyl oxide See vinyl ether. { dī'vīn·əl 'äk,sīd }

D line [SPECT] The yellow line that is the first line of the major series of the sodium spectrum; the doublet in the Fraunhofer lines whose almost equal components have wavelengths of 5895.93 and 5889.96 angstroms respectively. { 'dē ,līn }

DMA See disodium methylarsonate.

DMB See hydroquinone dimethyl ether.

DMC See p,p'-dichlorodiphenylmethyl carbinol.

DMDT See methoxychlor.

DMF See N,N-dimethylformamide.

DMSO See dimethyl sulfoxide.

DMT See dimethyl terephthalate.

DMU See dimethylolurea.

Dobbin's reagent [ANALY CHEM] A mercuric chloride-potassium iodide reagent used to test for caustic alkalies in soap. { 'däb·ənz rē'ā·jənt }

Dobson spectrophotometer [SPECT] A photoelectric spectrophotometer used in the determination of the ozone content of the atmosphere; compares the solar energy at two wavelengths in the absorption band of ozone by permitting the radiation of each to fall alternately upon a photocell. { 'däb·sən ,spek·trō·fə'täm·əd·ər }

docosane [ORG CHEM] $C_{22}H_{46}$ A paraffin hydrocarbon, especially the normal isomer $CH_3(CH_2)_{20}CH_3$. { 'däk·ə,sān }

docosanoic acid [ORG CHEM] $CH_3(CH_2)_{20}CO_2H$ A crystalline fatty acid, melting at 80°C, slightly soluble in water and alcohol, and found in the fats and oils of some seeds such as peanuts. Also known as behenic acid. { ¦dak·ə·sə¦nō·ik 'as·əd }

1-docosanol See behenyl alcohol. { ¦wən də'käs·ə,nȯl }

docosapentanoic acid [ORG CHEM] $C_{21}H_{33}CO_2H$ A pale-yellow liquid, boils at 236°C (5 mmHg), insoluble in water, soluble in ether, and found in fish blubber. { ¦däk· ə·sə,pen·tə¦nō·ik 'as·əd }

dodecahedrane [ORG CHEM] $C_{20}H_{20}$ A highly strained saturated hydrocarbon cage structure in the shape of a dodecahedron (12 faces). { dō,dek·ə'hē·drān }

dodecahydrate [CHEM] A hydrated compound that has a total of 12 water molecules associated with it. { dō,dek·ə'hī,drāt }

dodecane [ORG CHEM] $CH_3(CH_2)_{10}CH_3C_{12}H_{26}$ An oily paraffin compound, a colorless liquid, boiling at 214.5°C, insoluble in water; used as a solvent and in jet fuel research. Also known as dihexy; propylene tetramer; tetrapropylene. { 'dō·də,kān }

1-dodecene

1-dodecene [ORG CHEM] $CH_2CH(CH_2)_9CH_3$ A colorless liquid, boiling at 213°C, insoluble in water; used in flavors, dyes, perfumes, and medicines. { ¦wən 'dō·də,sēn }

dodecyl [ORG CHEM] $C_{12}H_{25}$ A radical derived from dodecane by removing one hydrogen atom; in particular, the normal radical, $CH_3(CH_2)_{10}CH_2-$. { 'dō·də,sil }

dodecylbenzene [ORG CHEM] Blend of isomeric (mostly monoalkyl) benzenes with saturated side chains averaging 12 carbon atoms; used in the alkyl amyl sulfonate type of detergents. Also known as detergent alkylate. { 'dō·də,sil'ben,zēn }

dodecyl sodium sulfate See sodium lauryl sulfate. { 'dō·də,sil 'sōd·ē·əm 'səl,fāt }

dolomol See magnesium stearate. { 'dä·lə,mȯl }

Donnan distribution coefficient [PHYS CHEM] A coefficient in an expression giving the distribution, on two sides of a boundary between electrolyte solutions in Donnan equilibrium, of ions which can diffuse across the boundary. { ¦dän·ən ,dis·trə'byü· shən ,kō·ə,fish·ənt }

Donnan equilibrium [PHYS CHEM] The particular equilibrium set up when two coexisting phases are subject to the restriction that one or more of the ionic components cannot pass from one phase into the other; commonly, this restriction is caused by a membrane which is permeable to the solvent and small ions but impermeable to colloidal ions or charged particles of colloidal size. Also known as Gibbs-Donnan equilibrium. { 'dō·nən ē·kwə'lib·rē·əm }

Donnan potential [PHYS CHEM] The potential difference across a boundary between two electrolytic solutions in Donnan equilibrium. { 'dän·ən pə,ten·chəl }

DOP See dioctyl phthalate.

Doppler broadening [SPECT] Frequency spreading that occurs in single-frequency radiation when the radiating atoms, molecules, or nuclei do not all have the same velocity and may each give rise to a different Doppler shift. { 'däp·lər ,brȯd·ən·iŋ }

Doppler-free spectroscopy [SPECT] Any of several techniques which make use of the intensity and monochromatic nature of a laser beam to overcome the Doppler broadening of spectral lines and measure their wavelengths with extremely high accuracy. { däp·lər ,frē spek'träs·kə·pē }

Doppler-free two-photon spectroscopy [SPECT] A version of Doppler free spectroscopy in which the wavelength of a transition induced by the simultaneous absorption of two photons is measured by placing a sample in the path of a laser beam reflected on itself, so that the Doppler shifts of the incident and reflected beams cancel. { däp·lər ,frē ¦tü ¦fō,tän spek'träs·kə·pē }

Doppler spectroscopy [SPECT] A technique for measuring the speed with which an object is moving toward or away from the observer by measuring the amount that light from the object is shifted to a higher or lower frequency by the Doppler effect. { ,däp·lər spek'träs·kə·pē }

Dorn effect [PHYS CHEM] A difference in a potential resulting from the motions of particles through water; the potential exists between the particles and the water. { 'dȯrn i,fekt }

dotriacontane [ORG CHEM] $C_{32}H_{66}$ A paraffin hydrocarbon, in particular, the normal isomer $CH_3(CH_2)_{30}CH_3$, which is crystalline. { ¦dō,trī·ə'kän,tān }

double-beam spectrophotometer [SPECT] An instrument that uses a photoelectric circuit to measure the difference in absorption when two closely related wavelengths of light are passed through the same medium. { ¦dəb·əl ¦bēm spek·trō·fə'täm·əd·ər }

double-blind sample [ANALY CHEM] In chemical analysis, a sample submitted in such a way that neither its composition nor its identification as a check sample is known to the analyst. { ¦dəb·əl ¦blīnd 'sam·pəl }

double bond [PHYS CHEM] A type of linkage between atoms in which two pair of electrons are shared equally. { ¦dəb·əl 'bänd }

double-bond isomerism [PHYS CHEM] Isomerism in which two or more substances possess the same elementary composition but differ in having double bonds in different positions. { ¦dəb·əl ¦bänd ī'säm·ə,riz·əm }

double-bond shift [ORG CHEM] In an organic molecular structure, the occurrence when a pair of valence bonds that join a pair of carbons (or other atoms) shifts, via

124

chemical reaction, to a new position, for example, $H_2C=C-C-CH_2$ (butene-1) to $H_2C-C=C-CH_2$ (butene-2).   { ¦dəb·əl ¦bänd 'shift }

double decomposition   [CHEM] The simple exchange of elements of two substances to form two new substances; for example, $CaSO_4 + 2NaCl \rightarrow CaCl_2 + Na_2SO_4$.   { ¦dəb·əl dē,käm·pə'zish·ən }

double layer   See electric double layer.   { ¦dəb·əl 'lā·ər }

double nickel salt   See nickel ammonium sulfate.   { ¦dəb·əl ¦nik·əl 'sȯlt }

double-replacement reaction   [CHEM] A chemical reaction between compounds in which the elements in the reactants recombine to form two different compounds, each of the products having one element from each of the reactants.   { ¦dəb·əl ri'plās·mənt rē,ak·shən }

double salt   [INORG CHEM] **1.** A salt that upon hydrolysis forms two different anions and cations.   **2.** A salt that is a molecular combination of two other salts.   { ¦dəb·əl 'sȯlt }

doublet   [PHYS CHEM] Two electrons which are shared between two atoms and give rise to a nonpolar valence bond.   [SPECT] Two closely separated spectral lines arising from a transition between a single state and a pair of states forming a doublet as described in the atomic physics definition.   { 'dəb·lət }

downflow   [CHEM] In an ion-exchange system, the direction of the flow of the solution being processed.   { 'daȯn,flō }

D.P.   See degree of polymerization.

2,4-DP   See dichlorprop.

DPG   See diphenylguanidine.

Drew number   [PHYS CHEM] A dimensionless group used in the study of diffusion of a solid material A into a stream of vapor initially composed of substance B, equal to $\dfrac{Z_A(M_A - M_B) + M_B}{(Z_A - Y_{AW})(M_B - M_A)} \cdot \ln \dfrac{M_V}{M_W}$ where M_A and M_B are the molecular weights of components A and B, M_V and M_W are the molecular weights of the mixture in the vapor and at the wall, and Y_{AW} and Z_A are the mole fractions of A at the wall and in the diffusing stream, respectively. Symbolized N_D.   { 'drü ,nəm·bər }

driving force   [CHEM] In a chemical reaction, the formation of products such as an insoluble compound, a gas, a nonelectrolyte, or a weak electrolyte that enable the reaction to go to completion as a metathesis.   { 'drīv·iŋ ,fȯrs }

dropping-mercury electrode   [PHYS CHEM] An electrode consisting of a fine-bore capillary tube above which a constant head of mercury is maintained; the mercury emerges from the tip of the capillary at the rate of a few milligrams per second and forms a spherical drop which falls into the solution at the rate of one every 2–10 seconds.   { ¦dräp·iŋ ¦mər·kyə·rē i'lek·trōd }

dropping point   [CHEM] The temperature at which grease changes from a semisolid to a liquid state under standardized conditions.   { 'dräp·iŋ ,pȯint }

dry acid   [CHEM] Nonaqueous acetic acid used for oil-well reservoir acidizing treatment.   { ¦drī 'as·əd }

dry ashing   [ORG CHEM] The conversion of an organic compound into ash (decomposition) by a burner or in a muffle furnace.   { ¦drī 'ash·iŋ }

dry box   [CHEM] A container or chamber filled with argon, or sometimes dry air or air with no carbon dioxide (CO_2), to provide an inert atmosphere in which manipulation of very reactive chemicals is carried out in the laboratory.   { 'drī ,bäks }

dry distillation   [CHEM] A process in which a solid is heated in the absence of liquid to release vapors or liquids from the solid, for example, heating a hydrate to produce the anhydrous salt.   { ¦drī dis·tə'lā·shən }

dry ice   [INORG CHEM] Carbon dioxide in the solid form, usually made in blocks to be used as a coolant; changes directly to a gas at $-78.5°C$ as heat is absorbed.   { ¦drī 'īs }

drying   [CHEM] **1.** An operation in which a liquid, usually water, is removed from a wet solid in equipment termed a dryer.   **2.** A process of oxidation whereby a liquid such as linseed oil changes into a solid film.   { 'drī·iŋ }

drying agent |CHEM| Soluble or insoluble chemical substance that has such a great affinity for water that it will abstract water from a great many fluid materials; soluble chemicals are calcium chloride and glycerol, and insoluble chemicals are bauxite and silica gel. Also known as desiccant. { 'drī·iŋ ,ā·jənt }

dry point [ANALY CHEM] The temperature at which the last drop of liquid evaporates from the bottom of the flask. { 'drī ,pȯint }

dual-function catalyst See bifunctional catalyst. { ¦dül ¦fəŋk·shən 'kad·ə·list }

dubnium |CHEM| A chemical element, symbolized Db, atomic number 105, a synthetic element; the thirteenth transuranium element. { 'düb·nē·əm }

Duhem's equation See Gibbs-Duhem equation. { dü'emz i,kwā·zhən }

Dühring's rule [PHYS CHEM] The rule that a plot of the temperature at which a liquid exerts a particular vapor pressure against the temperature at which a similar reference liquid exerts the same vapor pressure produces a straight or nearly straight line. { 'dir·iŋz ,rül }

dulcitol [ORG CHEM] $C_6H_8(OH)_6$ A sugar with a slightly sweet taste; white, crystalline powder with a melting point of 188.5°C; soluble in hot water; used in medicine and bacteriology. { 'dəl·sə,tȯl }

dulcose See dulcitol. { 'dəl,kōs }

Dumas method [ANALY CHEM] A procedure for the determination of nitrogen in organic substances by combustion of the substance. { 'dü·mä ,meth·əd }

dunnite See ammonium picrate. { 'də,nīt }

duplicate measurement [ANALY CHEM] An additional measurement made on the same (identical) sample of material to evaluate the variance in the measurement. { ¦düp·lə·kət 'mezh·ər·mənt }

duplicate sample [ANALY CHEM] A second sample randomly selected from a material being analyzed in order to evaluate sample variance. { ¦düp·lə·kət 'sam·pəl }

durable-press resin See permanent-press resin. { ¦dür·ə·bəl ¦pres 'rez·ən }

durene [ORG CHEM] $C_6H_2(CH_3)_4$ Colorless crystals with camphor aroma; boiling point 190°C; soluble in organic solvents, insoluble in water; used as a chemical intermediate. Also known as durol. { 'dü,rēn }

durol See durene. { 'dü,rōl }

Dutch liquid See ethylene chloride. { ¦dəch ¦lik·wəd }

Dy See dysprosium.

dye |CHEM| A colored substance which imparts more or less permanent color to other materials. Also known as dyestuff. { dī }

dyeing assistant |CHEM| Material such as sodium sulfate added to a dye bath to control or promote the action of a textile dye. { 'dī·iŋ ə,sis·tənt }

dyestuff See dye. { 'dī,stəf }

dynamic allotropy |CHEM| A phenomenon in which the allotropes of an element exist in dynamic equilibrium. { dī¦nam·ik ə'lä·trə·pē }

dypnone [ORG CHEM] $C_6H_5COCHC(CH_3)C_6H_5$ A light-colored liquid with a boiling point of 246°C at 50 mmHg; used as a plasticizer and perfume base and in light-stable coatings. { 'dip,nōn }

Dyson notation [ORG CHEM] A notation system for representing organic chemicals developed by G. Malcolm Dyson; the compound is described on a single line, symbols are used for the chemical elements involved as well as for the functional groups and various ring systems; for example, methyl alcohol is C.Q and phenol is B6.Q. { 'dī·sən nō,tā·shən }

dysprosium |CHEM| A metallic rare-earth element, symbol Dy, atomic number 66, atomic weight 162.50. { dis'prō·zē·əm }

dystetic mixture [PHYS CHEM] A mixture of two or more substances that has the highest possible melting point of all mixtures of these substances. { di'sted·ik 'miks·chər }

E

eagle mounting [SPECT] A mounting for a diffraction grating, based on the principle of the Rowland circle, in which the diffracted ray is returned along nearly the same direction as the incident beam. { 'ē·gəl ˌmaún·tiŋ }

easin [ORG CHEM] $C_{20}H_6O_5I_4Na_2$ The sodium salt of tetraiodofluorescein; a brown powder, insoluble in water; used as a dye and a pH indicator (hydrogen ion) at pH 2.0. Also known as iodoeasin; sodium tetrafluorescein. { 'ē·ə·zən }

ebulliometer [PHYS CHEM] The instrument used for ebulliometry. Also known as ebullioscope. { əˌbü·lē'äm·əd·ər }

ebulliometry [PHYS CHEM] The precise measurement of the absolute or differential boiling points of solutions. { əˌbü·lē'äm·ə·trē }

ebullioscope See ebulliometer. { ə'bü·lē·əˌskōp }

ebullioscopic constant [PHYS CHEM] The ratio of the elevation of the boiling point of a solvent caused by dissolving a solute to the molality of the solution, taken at extremely low concentrations. Also known as molal elevation of the boiling point. { e'bü·lē·əˌskōp·ik ˌkän·stənt }

ecgonine [ORG CHEM] $C_9H_{15}NO_3$ An alkaloid obtained in crystalline form by the hydrolysis of cocaine. { 'ek·gəˌnēn }

echelette grating [SPECT] A diffraction grating with coarse groove spacing, designed for the infrared region; has grooves with comparatively flat sides and concentrates most of the radiation by reflection into a small angular coverage. { ˈesh·ə¦let *or* ¦äsh¦let ˌgrād·iŋ }

echelle grating [SPECT] A diffraction grating designed for use in high orders and at angles of illumination greater than 45° to obtain high dispersion and resolving power by the use of high orders of interference. { ā'shel ˌgrād·iŋ }

echelle spectrograph [SPECT] A spectrograph that employs gratings intended to be used in very high orders (greater than 10), and is equipped with a second dispersal element (another grating or a prism) at right angles to the first in order to separate the successive spectral strips from each other. { eˌshel 'spek·trəˌgraf }

echelon grating [SPECT] A diffraction grating which consists of about 20 plane-parallel plates about 1 centimeter thick, cut from one sheet, each plate extending beyond the next by about 1 millimeter, and which has a resolving power on the order of 10^6. { 'esh·əˌlän ˌgrād·iŋ }

echinopsine [ORG CHEM] $C_{10}H_9O$ An alkaloid obtained from *Echinops* species; crystallizes as needles from benzene solution, melts at 152°C; physiological action is similar to that of brucine and strychnine. { ˌek·ə'näpˌsēn }

eclipsed conformation [PHYS CHEM] A particular arrangement of constituent atoms that may rotate about a single bond in a molecule; for ethane it is such that when viewed along the axis of the carbon-carbon bond the hydrogen atoms of one methyl group are exactly in line with those of the other methyl group. { i'klipst ˌkän·fər'mā·shən }

edge-bridging ligand [ORG CHEM] A ligand that forms a bridge over one edge of the polyhedron of a metal cluster structure. { 'ej ˌbrij·iŋ 'lī·gənd }

EDTA See ethylenediaminetetraacetic acid.

EDTC See S-ethyl-N,N-dipropylthiocarbamate.

EELS See electron energy loss spectroscopy.

eff *See* efficiency.

effective molecular diameter [PHYS CHEM] The general extent of the electron cloud surrounding a gas molecule as calculated in any of several ways. { ə¦fek·tiv mə‚lek·yə·lər dī'am·əd·ər }

effective permeability [PHYS CHEM] The observed permeability exhibited by a porous medium to one fluid phase when there is physical interaction between this phase and other fluid phases present. { ə¦fek·tiv pər·mē·ə'bil·əd·ē }

effervescence [CHEM] The bubbling of a solution of an element or chemical compound as the result of the emission of gas without the application of heat; for example, the escape of carbon dioxide from carbonated water. { ‚ef·ər'ves·əns }

efflorescence [CHEM] The property of hydrated crystals to lose water of hydration and crumble when exposed to air. { ‚ef·lə'res·əns }

effusion [PHYS CHEM] The movement of a gas through an opening which is small as compared with the average distance which the gas molecules travel between collisions. { e'fyü·zhən }

EGA *See* evolved gas analysis.

EGT *See* ethylene glycol bis(trichloroacetate).

Ehrlich's reagent [ORG CHEM] $(CH_3)_2NC_6H_4CHO$ Granular or leafletlike crystals that are soluble in many organic solvents; melting point is 74°C; used in the preparation of dyes, as a reagent for arsphenamine, anthranilic acid, antipyrine, indole, and skatole, and as a differentiating agent between true scarlet fever and serum eruptions. { 'er·liks rē‚ā·jənt }

eicosanoic acid [ORG CHEM] $CH_3(CH_2)_{18}COOH$ A white, crystalline, saturated fatty acid, melting at 75.4°C; a constituent of butter. Also known as arachic acid; arachidic acid. { ¦ī·kə·sə¦nō·ik 'as·əd }

Einschluss thermometer [ANALY CHEM] All-glass, liquid-filled thermometer, temperature range −201 to +360°C, used for laboratory test work. { 'īn‚shlús thər‚mäm·əd·ər }

einsteinium [CHEM] Synthetic radioactive element, symbol Es, atomic number 99; discovered in debris of 1952 hydrogen bomb explosion; now made in cyclotrons. { īn'stīn·ē·əm }

Einstein photochemical equivalence law [PHYS CHEM] The law that each molecule taking part in a chemical reaction caused by electromagnetic radiation absorbs one photon of the radiation. Also known as Stark-Einstein law. { 'īn‚stīn ¦fōd·ō¦kem·ə·kəl i'kwiv·ə·ləns ‚lȯ }

Einstein viscosity equation [PHYS CHEM] An equation which gives the viscosity of a sol in terms of the volume of dissolved particles divided by total volume. { 'īn‚stīn vis'käs·əd·ē i‚kwä·zhən }

elaidic acid [ORG CHEM] $CH_3(CH_2)_7CH:CH(CH_2)_7COOH$ A transisomer of an unsaturated fatty acid, oleic acid; crystallizes as colorless leaflets, melts at 44°C, boils at 288°C (100 mmHg), insoluble in water, soluble in alcohol and ether; used in chromatography as a reference standard. { ¦el·ə¦id·ik 'as·əd }

elaidinization [ORG CHEM] The process of changing the geometric cis form of an unsaturated fatty acid or a compound related to it into the trans form, resulting in an acid that is more resistant to oxidation. { ə¦lā·ə‚din·ə'zā·shən }

elaidin reaction [ANALY CHEM] A test that differentiates nondrying oils such as olein from semidrying oils and drying oils; nitrous acid converts olein into its solid isomer, while semidrying oils in contact with nitrous acid thicken slowly, and drying oils such as tung oil become hard and resinous. { ə'lā·əd·ən rē‚ak·shən }

Elbs reaction [ORG CHEM] The formation of anthracene derivatives by dehydration and cyclization of diaryl ketone compounds which have a methyl group or methylene group; heating to an elevated temperature is usually required. { 'elbs rē‚ak·shən }

ELDOR *See* electron electron double resonance. { 'el‚dȯr *or* ¦e¦el¦de¦ō'är }

electrical calorimeter [ANALY CHEM] Device to measure heat evolved (from fusion or vaporization, for example); measured quantities of heat are added electrically to the sample, and the temperature rise is noted. { ə'lek·trə·kəl kal·ə'rim·əd·ər }

electrical equivalent [ANALY CHEM] In conductometric analyses of electrolyte solutions, an outside, calibrated current source as compared to (equivalent to) the current passing through the sample under analysis; for example, a Wheatstone-bridge balanced reading. { i'lek·trə·kəl i'kwiv·ə·lənt }

electrically active fluid [PHYS CHEM] A fluid whose properties are altered by either an electric field (electrorheological fluid) or a magnetic field (ferrofluid). { i'lek·trə·klē ¦ak·tiv 'flü·əd }

electric double layer [PHYS CHEM] A phenomenon found at a solid-liquid interface; it is made up of ions of one charge type which are fixed to the surface of the solid and an equal number of mobile ions of the opposite charge which are distributed through the neighboring region of the liquid; in such a system the movement of liquid causes a displacement of the mobile counterions with respect to the fixed charges on the solid surface. Also known as double layer. { i'lek·trik ¦dəb·əl 'lā·ər }

electric-field effect See Stark effect. { i¦lek·trik ¦fēld i'fekt }

electride [INORG CHEM] A member of a class of ionic compounds in which the anion is believed to be an electron. { i'lek,trīd }

electrobalance [ANALY CHEM] Analytical microbalance utilizing electromagnetic weighing; the sample weight is balanced by the torque produced by current in a coil in a magnetic field, with torque proportional to the current. { i,lek·trō'bal·əns }

electrocatalysis [CHEM] Any one of the mechanisms which produce a speeding up of half-cell reactions at electrode surfaces. { i,lek·trō·kə'tal·ə·səs }

electrochemical cell [PHYS CHEM] A combination of two electrodes arranged so that an overall oxidation-reduction reaction produces an electromotive force; includes dry cells, wet cells, standard cells, fuel cells, solid-electrolyte cells, and reserve cells. { i,lek·trō'kem·ə·kəl 'sel }

electrochemical effect [PHYS CHEM] Conversion of chemical to electric energy, as in electrochemical cells; or the reverse process, used to produce elemental aluminum, magnesium, and bromine from compounds of these elements. { i,lek·trō'kem·ə·kəl i'fekt }

electrochemical emf [PHYS CHEM] Electrical force generated by means of chemical action, in manufactured cells (such as dry batteries) or by natural means (galvanic reaction). { i,lek·trō'kem·ə·kəl ¦ē¦em'ef }

electrochemical equivalent [PHYS CHEM] The weight in grams of a substance produced or consumed by electrolysis with 100% current efficiency during the flow of a quantity of electricity equal to 1 faraday (96,485.34 coulombs). { i,lek·trō'kem·ə·kəl i'kwiv·ə·lənt }

electrochemical potential [PHYS CHEM] The difference in potential that exists when two dissimilar electrodes are connected through an external conducting circuit and the two electrodes are placed in a conducting solution so that electrochemical reactions occur. { i,lek·trō'kem·ə·kəl pə'ten·chəl }

electrochemical process [PHYS CHEM] **1.** A chemical change accompanying the passage of an electric current, especially as used in the preparation of commercially important quantities of certain chemical substances. **2.** The reverse change, in which a chemical reaction is used as the source of energy to produce an electric current, as in a battery. { i,lek·trō'kem·ə·kəl 'präs·əs }

electrochemical reduction cell [PHYS CHEM] The cathode component of an electrochemical cell, at which chemical reduction occurs (while at the anode, chemical oxidation occurs). { i,lek·trō'kem·ə·kəl ri'dək·shən ,sel }

electrochemical series [PHYS CHEM] A series in which the metals and other substances are listed in the order of their chemical reactivity or electrode potentials, the most reactive at the top and the less reactive at the bottom. Also known as electromotive series. { i,lek·trō'kem·ə·kəl 'sir·ēz }

electrochemical techniques [PHYS CHEM] The experimental methods developed to study the physical and chemical phenomena associated with electron transfer at the interface of an electrode and solution. { i,lek·trō'kem·ə·kəl tek'nēks }

electrochemiluminescence

electrochemiluminescence [PHYS CHEM] Emission of light produced by an electrochemical reaction. Also known as electrogenerated chemiluminescence. { i‚lek·trō‚kem·ē·ə‚lüm·ə'nes·əns }

electrochemistry [PHYS CHEM] A branch of chemistry dealing with chemical changes accompanying the passage of an electric current; or with the reverse process, in which a chemical reaction is used to produce an electric current. { i¦lek·trō¦kem·ə·strē }

electrochromatography [ANALY CHEM] Type of chromatography that utilizes application of an electric potential to produce an electric differential. Also known as electropherography. { i¦lek·trō‚krō·mə'täg·rə·fē }

electrocratic [CHEM] Referring to the repulsion exhibited by soap films and other colloids in solutions; such repulsion involves a strong osmotic contribution but is largely controlled by electrical forces. { i‚lek·trō'krad·ik }

electrocyclic reaction [PHYS CHEM] The interconversion of a linear π-system containing n π-electrons and a cyclic molecule containing $(n-2)$ π-electrons which is formed by joining the ends of the linear molecule. { i‚lek·trō¦sī·klik rē'ak·shən }

electrodecantation [PHYS CHEM] A modification of electrodialysis in which a cell is divided into three sections by two membranes and electrodes are placed in the end sections; colloidal matter is concentrated at the sides and bottom of the middle section, and the liquid that floats to the top is drawn off. { i¦lek·trō‚dē‚kan'tā·shən }

electrode efficiency [PHYS CHEM] The ratio of the amount of metal actually deposited in an electrolytic cell to the amount that could theoretically be deposited as a result of electricity passing through the cell. { i'lek‚trōd ə‚fish·ən·sē }

electrodeposition analysis [ANALY CHEM] An electroanalytical technique in which an element is quantitatively deposited on an electrode. { i¦lek·trō‚dep·ə'zish·ən ə'nal·ə·səs }

electrode potential [PHYS CHEM] The voltage existing between an electrode and the solution or electrolyte in which it is immersed; usually, electrode potentials are referred to a standard electrode, such as the hydrogen electrode. Also known as electrode voltage. { i'lek‚trōd pə'ten·chəl }

electrodialysis [PHYS CHEM] Dialysis that is conducted with the aid of an electromotive force applied to electrodes adjacent to both sides of the membrane. { i¦lek·trō·dī'al·ə·səs }

electrodialyzer [PHYS CHEM] An instrument used to conduct electrodialysis. { i‚lek·trō'dī·ə‚līz·ər }

electrofocusing See isoelectric focusing. { i‚lek·trō'fō·kəs·iŋ }

electrogenerated chemiluminescence See electrochemiluminescence. { i‚lek·trō¦jen·ə‚rād·əd ‚kem·ē·‚lüm·ə'nes·əns }

electrogravimetry [ANALY CHEM] Electrodeposition analysis in which the quantities of metals deposited may be determined by weighing a suitable electrode before and after deposition. { i‚lek·trə·grə'vim·ə·trē }

electrohydraulic effect [PHYS CHEM] Generation of shock waves and highly reactive species in a liquid as the result of application of very brief but powerful electrical pulses. { i¦lek·trō·hī¦dról·ik i'fekt }

electrohydrodynamic ionization mass spectroscopy [SPECT] A technique for analysis of nonvolatile molecules in which the nonvolatile material is dissolved in a volatile solvent with a high dielectric constant such as glycerol, and high electric-field gradients at the surface of droplets of the liquid solution induce ion emission. { i¦lek·trō¦hī·drō·dī'nam·ik ‚ī·ə·nə'zā·shən ¦mas spek'träs·kə·pē }

electrokinetic phenomena [PHYS CHEM] The phenomena associated with movement of charged particles through a continuous medium or with the movement of a continuous medium over a charged surface. { i¦lek·trō·kə'ned·ik fə'näm·ə·nə }

electrolysis [PHYS CHEM] A method by which chemical reactions are carried out by passage of electric current through a solution of an electrolyte or through a molten salt. { i‚lek'trä·lə·səs }

electrolyte [PHYS CHEM] A chemical compound which when molten or dissolved in certain solvents, usually water, will conduct an electric current. { i'lek·trə‚līt }

130

electrolytic analysis [ANALY CHEM] Basic electrochemical technique for quantitative analysis of conducting solutions containing oxidizable or reducible material; measurement is based on the weight of material plated out onto the electrode. { i'lek·trə,lid·ik ə'nal·ə·səs }

electrolytic cell [PHYS CHEM] A cell consisting of electrodes immersed in an electrolyte solution, for carrying out electrolysis. { i'lek·trə,lid·ik 'sel }

electrolytic conductance [PHYS CHEM] The transport of electric charges, under electric potential differences, by charged particles (called ions) of atomic or larger size. { i'lek·trə,lid·ik kən'dək·təns }

electrolytic conductivity [PHYS CHEM] The conductivity of a medium in which the transport of electric charges, under electric potential differences, is by particles of atomic or larger size. { i'lek·trə,lid·ik ,kän·dək'tiv·əd·ē }

electrolytic dissociation [CHEM] The ionization of a compound in a solution. { i'lek·trə,lid·ik di,sō·sē'ā·shən }

electrolytic migration [PHYS CHEM] The motions of ions in a liquid under the action of an electric field. { i¦lek·trə,lid·ik mī'grā·shən }

electrolytic polarization [PHYS CHEM] The existence of a minimum potential difference necessary to cause a steady current to flow through an electrolytic cell, resulting from the tendency of the products of electrolysis to recombine. { i¦lek·trə,lid·ik pō·lər·ə'zā·shən }

electrolytic potential [PHYS CHEM] Difference in potential between an electrode and the immediately adjacent electrolyte, expressed in terms of some standard electrode difference. { i'lek·trə,lid·ik pə'ten·chəl }

electrolytic process [PHYS CHEM] An electrochemical process involving the principles of electrolysis, especially as relating to the separation and deposition of metals. { i'lek·trə,lid·ik 'präs·əs }

electrolytic separation [PHYS CHEM] Separation of isotopes by electrolysis, based on differing rates of discharge at the electrode of ions of different isotopes. { i'lek·trə,lid·ik ,sep·ə'rā·shən }

electrolytic solution [PHYS CHEM] A solution made up of a solvent and an ionically dissociated solute; it will conduct electricity, and ions can be separated from the solution by deposition on an electrically charged electrode. { i'lek·trə,lid·ik sə'lü·shən }

electromigration [ANALY CHEM] A process used to separate isotopes or ionic species by the differences in their ionic mobilities in an electric field. [PHYS CHEM] The movement of ions under the influence of an electrical potential difference. { i¦lek·trō·mī'grā·shən }

electromodulation [SPECT] Modulation spectroscopy in which changes in transmission or reflection spectra induced by a perturbing electric field are measured. { i¦lek·trō,mäj·ə'lā·shən }

electromotance See electromotive force. { i¦lek·trō'mōt·əns }

electromotive force [PHYS CHEM] **1.** The difference in electric potential that exists between two dissimilar electrodes immersed in the same electrolyte or otherwise connected by ionic conductors. **2.** The resultant of the relative electrode potential of the two dissimilar electrodes at which electrochemical reactions occur. Abbreviated emf. Also known as electromotance. { i¦lek·trə'mōd·iv 'fōrs }

electromotive series See electrochemical series. { i¦lek·trə'mōd·iv 'sir·ēz }

electron acceptor [PHYS CHEM] **1.** An atom or part of a molecule joined by a covalent bond to an electron donor. **2.** See electrophile. { i'lek,trän ak'sep·tər }

electron-capture detector [ANALY CHEM] Extremely sensitive gas chromatography detector that is a modification of the argon ionization detector, with conditions adjusted to favor the formation of negative ions. { i'lek,trän ,kap·chər di'tek·tər }

electron distribution curve [PHYS CHEM] A curve indicating the electron distribution among the different available energy levels of a solid substance. { i'lek,trän dis·trə'byü·shən ,kərv }

electron donor [PHYS CHEM] An atom or part of a molecule which supplies both electrons of a duplet forming a covalent bond. See nucleophile. { i'lek,trän ,dō·nər }

electron-dot formula

electron-dot formula *See* Lewis structure. { i¦lek,trän ¦dät ¸fȯr·myə·lə }

electronegative [PHYS CHEM] Pertaining to an atom or group of atoms that has a relatively great tendency to attract electrons to itself. { i¦lek·trō'neg·əd·iv }

electronegative potential [PHYS CHEM] Potential of an electrode expressed as negative with respect to the hydrogen electrode. { i¦lek·trō'neg·əd·iv pə'ten·chəl }

electron-electron double resonance [SPECT] A type of electron paramagnetic resonance (EPR) spectroscopy in which a material is irradiated at two different microwave frequencies, and the changes in the EPR spectrum resulting from sweeping either the second frequency or the magnetic field are monitored through detection at the first frequency. Abbreviated ELDOR. { i¸lek,trän i¦lek,tran ¸dəb·əl 'rez·ən·əns }

electron energy loss spectroscopy [SPECT] A technique for studying atoms, molecules, or solids in which a substance is bombarded with monochromatic electrons, and the energies of scattered electrons are measured to determine the distribution of energy loss. Abbreviated EELS. { i'lek,trän 'en·ər·jē ¸lȯs spek'träs·kə·pē }

electroneutrality principle [PHYS CHEM] The principle that in an electrolytic solution the concentrations of all the ionic species are such that the solution as a whole is neutral. { i¦lek·trō·nü'tral·əd·ē ¸prin·sə·pəl }

electron exchanger *See* redox polymer. { i'lek,trän iks,chān·jər }

electronic absorption spectrum [SPECT] Spectrum resulting from absorption of electromagnetic radiation by atoms, ions, and molecules due to excitations of their electrons. { i¸lek'trän·ik əb'sȯrp·shən ¸spek·trəm }

electronic band spectrum [SPECT] Bands of spectral lines associated with a change of electronic state of a molecule; each band corresponds to certain vibrational energies in the initial and final states and consists of numerous rotational lines. { i¸lek'trän·ik 'band ¸spek·trəm }

electronic emission spectrum [SPECT] Spectrum resulting from emission of electromagnetic radiation by atoms, ions, and molecules following excitations of their electrons. { i¸lek'trän·ik i'mish·ən ¸spek·trəm }

electronic energy curve [PHYS CHEM] A graph of the energy of a diatomic molecule in a given electronic state as a function of the distance between the nuclei of the atoms. { i¸lek'trän·ik 'en·ər·jē ¸kərv }

electronic spectrum [SPECT] Spectrum resulting from emission or absorption of electromagnetic radiation during changes in the electron configuration of atoms, ions, or molecules, as opposed to vibrational, rotational, fine-structure, or hyperfine spectra. { i¸lek'trän·ik 'spek·trəm }

electron nuclear double resonance [SPECT] A type of electron paramagnetic resonance (EPR) spectroscopy permitting greatly enhanced resolution, in which a material is simultaneously irradiated at one of its EPR frequencies and by a second oscillatory field whose frequency is swept over the range of nuclear frequencies. Abbreviated ENDOR. { i'lek,trän ¦nü·klē·ər ¦dəb·əl 'rez·ən·əns }

electron pair [PHYS CHEM] A pair of valence electrons which form a nonpolar bond between two neighboring atoms. { i'lek,trän 'per }

electron pair bond *See* covalent bond. { i'lek,trän 'per ¸bänd }

electron probe x-ray microanalysis [ANALY CHEM] An analytical technique that uses a narrow electron beam, usually with a diameter less than 1 millimeter, focused on a solid specimen to excite an x-ray spectrum that provides qualitative and quantitative information characteristic of the elements in the sample. Abbreviated EPXMA. { i¦lek,trän ¸prōb ¦eks,rā ¸mī·krō·ə'nal·ə·səs }

electron spectroscopy [SPECT] The study of the energy spectra of photoelectrons or Auger electrons emitted from a substance upon bombardment by electromagnetic radiation, electrons, or ions; used to investigate atomic, molecular, or solid-state structure, and in chemical analysis. { i'lek,trän spek'träs·kə·pē }

electron spectroscopy for chemical analysis *See* x-ray photoelectron spectroscopy. { i'lek,trän spek'träs·kə·pē fər 'kem·i·kəl ə'nal·ə·səs }

electron spectrum [SPECT] Visual display, photograph, or graphical plot of the intensity of electrons emitted from a substance bombarded by x-rays or other radiation as a function of the kinetic energy of the electrons. { i'lek,trän 'spek·trəm }

electron spin echo envelope modulation [SPECT] **1.** The variation in the intensity of an electron spin echo as the time interval between the two microwave pulses producing the echo is incremented in small steps in the case of a two-pulse echo, or time intervals between suitable pulses are incremented for multiple-pulse echoes. **2.** A type of electron paramagnetic resonance spectroscopy in which this variation is mathematically transformed, using the Fourier transform, to yield the spectrum of nuclear frequencies. Abbreviated ESEEM. { i‚lek‚trän ¦spin ‚ek·ō ¦en·və‚lōp ‚mäj· ə'lā·shən }

electroosmosis [PHYS CHEM] The movement in an electric field of liquid with respect to colloidal particles immobilized in a porous diaphragm or a single capillary tube. { i¦lek·trō·äs'mō·səs }

electropherography See electrochromatography. { i¦lek·trō·fə'räg·rə·fē }

electrophile [PHYS CHEM] An electron-deficient ion or molecule that takes part in an electrophilic process. Also known as electron acceptor. { i'lek·trō‚fīl }

electrophilic [PHYS CHEM] **1.** Pertaining to any chemical process in which electrons are acquired from or shared with other molecules or ions. **2.** Referring to an electron-deficient species. { i¦lek·trō'fil·ik }

electrophilic reagent [PHYS CHEM] A reactant which accepts an electron pair from a molecule, with which it forms a covalent bond. { i¦lek·trō¦fil·ik rē'a·jənt }

electrophoresis [PHYS CHEM] An electrochemical process in which colloidal particles or macromolecules with a net electric charge migrate in a solution under the influence of an electric current. Also known as cataphoresis. { i‚lek·trō·fə'rē·səs }

electrophoretic effect [PHYS CHEM] Retarding effect on the characteristic motion of an ion in an electrolytic solution subjected to a potential gradient, which results from motion in the opposite direction by the ion atmosphere. { i¦lek·trō·fə'red·ik i'fekt }

electropositive [PHYS CHEM] Pertaining to elements, ions, or radicals that tend to give up or lose electrons. { i‚lek·trə'päz·əd·iv }

electropositive potential [PHYS CHEM] Potential of an electrode expressed as positive with respect to the hydrogen electrode. { i¦lek·trə¦päz·əd·iv pə'ten·chəl }

electroreflectance [SPECT] Electromodulation in which reflection spectra are studied. Abbreviated ER. { i¦lek·trō·ri'flek·təns }

electrorheological fluid [PHYS CHEM] A colloidal suspension of finely divided particles in a carrier liquid, usually an insulating oil, whose rheological properties are changed through an increase in resistance when an electric field is applied. { i¦lek·trō‚rē· ə¦läj·ə·kəl 'flü·əd }

electrostatic bond [PHYS CHEM] A valence bond in which two atoms are kept together by electrostatic forces caused by transferring one or more electrons from one atom to the other. { i‚lek·trə'stad·ik 'bänd }

electrostatic valence rule [PHYS CHEM] The postulate that in a stable ionic structure the valence of each anion, with changed sign, equals the sum of the strengths of its electrostatic bonds to the adjacent cations. { i'lek·trə‚stad·ik 'vā·ləns ‚rül }

electrosynthesis [CHEM] A reaction in which synthesis occurs as the result of an electric current. { i¦lek·trō'sin·thə·səs }

electrovalence [PHYS CHEM] The valence of an atom that has formed an ionic bond. { i¦lek·trō'vā·ləns }

electrovalent bond See ionic bond. { i¦lek·trō¦vā·lənt 'bänd }

element [CHEM] A substance made up of atoms with the same atomic number; common examples are hydrogen, gold, and iron. Also known as chemical element. { 'el· ə·mənt }

element 110 [CHEM] A synthetic chemical element, atomic number 110; the eighteenth transuranium element. { ¦el·ə·mənt ‚wən'ten }

element 111 [CHEM] A synthetic chemical element, atomic number 111; the nineteenth transuranium element. { ¦el·ə·mənt wən·i'lev·ən }

element 112 [CHEM] A synthetic chemical element, atomic number 112; the twentieth transuranium element. { ¦el·ə·mənt wən'twelv }

elementary process [PHYS CHEM] In chemical kinetics, the particular events at the atomic or molecular level which make up an overall reaction. { ‚el·ə'men·trē 'präs·əs }

elementary reaction [ORG CHEM] A reaction which involves only a single transition state with no intermediates. Also known as step. { ‚el·ə'men·trē rē'ak·shən }

eleostearic acid [ORG CHEM] $CH_3(CH_2)_7(CH:CH)_3-(CH_2)_3COOH$ A colorless, water-insoluble, crystalline, unsaturated fatty acid; the glycerol ester is a chief component of tung oil. { ‚el·ē·ō'stir·ik 'as·əd }

elimination reaction [ORG CHEM] A chemical reaction involving elimination of some portion of a reactant compound, with the production of a second compound. { ə‚lim· ə'nā·shən rē‚ak·shən }

ellagic acid [ORG CHEM] $C_{14}H_6O_8$ A compound isolated from tannins as yellow crystals that are minimally soluble in hot water. Also known as gallogen. { e'laj·ik 'as·əd }

eluant [CHEM] A liquid used to extract one material from another, as in chromatography. { 'el·yə·wənt }

eluant gas See carrier gas. { el'yü·ənt ‚gas }

eluate [CHEM] The solution that results from the elution process. { 'el·yə‚wāt }

elution [CHEM] The removal of adsorbed species from a porous bed or chromatographic column by means of a stream of liquid or gas. { ē'lü·shən }

emf See electromotive force.

emission flame photometry [ANALY CHEM] A form of flame photometry in which a sample solution to be analyzed is aspirated into a hydrogen-oxygen or acetylene-oxygen flame; the line emission spectrum is formed, and the line or band of interest is isolated with a monochromator and its intensity measured photoelectrically. { i'mish·ən ‚flām fō'täm·ə·trē }

emission lines [SPECT] Spectral lines resulting from emission of electromagnetic radiation by atoms, ions, or molecules during changes from excited states to states of lower energy. { i'mish·ən ‚līnz }

emission spectrometer [SPECT] A spectrometer that measures percent concentrations of preselected elements in samples of metals and other materials; when the sample is vaporized by an electric spark or arc, the characteristic wavelengths of light emitted by each element are measured with a diffraction grating and an array of photodetectors. { i'mish·ən spek'träm·əd·ər }

emission spectrum [SPECT] Electromagnetic spectrum produced when radiations from any emitting source, excited by any of various forms of energy, are dispersed. { i'mish·ən ‚spek·trəm }

emodin [ORG CHEM] $C_{14}H_4O_2(OH)_3CH_3$ Orange needles crystallizing from alcohol solution, melting point 256–257°C, practically insoluble in water, soluble in alcohol and aqueous alkali hydroxide solutions, occurs as the rhamnoside in plants such as rhubarb root and alder buckthorn; used as a laxative. { 'em·ə·dən }

empirical formula [CHEM] A chemical formula that indicates the composition of a compound in terms of the relative numbers and kinds of atoms in the simplest ratio; for example, the empirical formula for fluorobenzene is C_6H_5F. { em'pir·ə·kəl 'fòr·myə·mlə }

emulsification [CHEM] The process of dispersing one liquid in a second immiscible liquid; the largest group of emulsifying agents are soaps, detergents, and other compounds, whose basic structure is a paraffin chain terminating in a polar group. { ə‚məl·sə·fə'kā·shən }

emulsion [CHEM] A stable dispersion of one liquid in a second immiscible liquid, such as milk (oil dispersed in water). { ə'məl·shən }

emulsion breaking [CHEM] In an emulsion, the combined sedimentation and coalescence of emulsified drops of the dispersed phase so that they will settle out of the carrier liquid; can be accomplished mechanically (in settlers, cyclones, or centrifuges) with or without the aid of chemical additives to increase the surface tension of the droplets. { ə'məl·shən ‚brāk·iŋ }

emulsion polymerization [ORG CHEM] A polymerization reaction that occurs in one phase of an emulsion. { ə'məl·shən pə‚lim·ə·rə'zā·shən }

enamine [ORG CHEM] An amine in which there is a carbon-to-carbon double bond adjacent to the nitrogen, $-C=C-N-$; considered to be the nitrogen analog of an enol. { 'en·ə,mēn }

enantiomer *See* enantiomorph. { ə¦nan·tē¦ō·mər }

enantiomerically pure [ORG CHEM] Referring to a sample of molecules having the same chirality. IUPAC discourages use of homchiral as a synonym. { ə¦nan·tē·ō¦mer·ə· klē 'pyür }

enantiomeric excess [ORG CHEM] In an asymmetric synthesis, a chemical yield that contains more of the desired enantiomer than other products. { ə¦nan·tē·ō¦mer· ik ek'ses }

enantiomorph [CHEM] One of an isomeric pair of either crystalline forms or chemical compounds whose molecules are nonsuperimposable mirror images. Also known as enantiomer; optical antipode; optical isomer. { ə'nan·tē·ə,mòrf }

enantiomorphism [CHEM] A phenomenon of mirror-image relationship exhibited by right-handed and left-handed crystals or by the molecular structures of two stereoisomers. { ə¦nan·tē·ə'mòr,fiz·əm }

enantioselective reaction *See* stereoselective reaction. { ə¦nan·tē·ə·si¦lek·tiv rē'ak· shən }

enantiotopic ligand [ORG CHEM] A ligand whose replacement or addition gives rise to enantiomers. { ə¦nan·tē·ə¦täp·ik 'līg·ənd }

enantiotropy [CHEM] The relation of crystal forms of the same substance in which one form is stable above the transition-point temperature and the other stable below it, so that the forms can change reversibly one into the other. { ə,nan·te'ä·trə·pē }

encounter [PHYS CHEM] A group of collisions, each of which consists of two molecules that collide without reacting and do not separate immediately because of the cage of surrounding molecules. { en'kaùn·tər }

endo- [ORG CHEM] Prefix that denotes inward-directed valence bonds of a six-membered ring in its boat form. { 'en·dō }

endocyclic double bond [ORG CHEM] In a molecular structure, a double bond that is part of the ring system. { ¦en·dō'sī·klik ¦dəb·əl 'bänd }

endoergic *See* endothermic.

ENDOR *See* electron nuclear double resonance. { 'en,dòr }

endosulfan [ORG CHEM] $C_9H_6Cl_6O_3S$ A tan solid that melts between -10 and $100°C$; used as an insecticide and miticide on vegetable and forage crops, on ornamental flowers, and in controlling termites and tsetse flies. { ¦en·dō'səl,fan }

endotherm [PHYS CHEM] In differential thermal analysis, a graph of the temperature difference between a sample compound and a thermally inert reference compound (commonly aluminum oxide) as the substances are simultaneously heated to elevated temperatures at a predetermined rate, and the sample compound undergoes endothermal or exothermal processes. { 'en·də,thərm }

endothermic [PHYS CHEM] Pertaining to a chemical reaction which absorbs heat. Also known as endoergic. { ,en·də'thər·mik }

end point [ANALY CHEM] That stage in the titration at which an effect, such as a color change, occurs, indicating that a desired point in the titration has been reached. { 'end ,pòint }

end radiation *See* quantum limit. { 'end ,rād·ē,ā·shən }

endrin [ORG CHEM] $C_{12}H_8OCl_6$ Poisonous, white crystals that are insoluble in water; it is used as a pesticide and is a stereoisomer of dieldrin, another pesticide. { 'en·drən }

ene reaction [ORG CHEM] The addition of a compound with a double bond having an allylic hydrogen (ene, such as propene) to a compound with a multiple bond (enophile, such as ethene). { 'ēn rē,ak·shən }

energy of activation *See* activation energy. { 'en·ər·jē əv ak·tə'vā·shən }

energy profile [PHYS CHEM] A diagram of the energy changes that take place during a reaction in a chemical system. { 'en·ər·jē ,prō,fīl }

engineering plastics [ORG CHEM] A class of polymers, based on aromatic backbones, having high strength, stiffness, and toughness together with high thermal and oxidative stability, low creep, and the ability to be processed by standard techniques for thermoplastics; examples include polyacetal, polyamide, polycarbonate, and polysulfone resins. { ‚en·jə'nir·iŋ 'plas·tiks }

English degree [CHEM] A unit of water hardness, equal to 1 part calcium carbonate to 70,000 parts water; equivalent to 1 grain of calcium carbonate per gallon of water. Also known as Clark degree. { 'iŋ·glish di‚grē }

English vermilion [INORG CHEM] Bright vermilion pigment of precipitated mercury sulfide; in paints, it tends to darken when exposed to light. { 'iŋ·glish vər'mil·yən }

enhanced line See enhanced spectral line. { en'hanst 'līn }

enhanced spectral line [SPECT] A spectral line of a very hot source, such as a spark, whose intensity is much greater than that of a line in a flame or arc spectrum. Also known as enhanced line. { en'hanst 'spek·trəl ‚līn }

enium ion [ORG CHEM] A cationic portion of an ionic species in which the valence shell of a positively charged nonmetallic atom has two electrons less than normal, and the charged entity has one covalent bond less than the corresponding uncharged species; used as a suffix with the root name. Also known as ylium ion. { 'en·ē·əm ‚ī·ən }

enol [ORG CHEM] An organic compound with a hydroxide group adjacent to a double bond; varies with a ketone form in the effect known as enol-keto tautomerism; an example is the compound $CH_3COH=CHCO_2C_2H_5$. { 'ē‚nȯl }

enolate anion [ORG CHEM] The delocalized anion which is left after the removal of a proton from an enol, or of the carbonyl compound in equilibrium with the enol. { 'ē·nə‚lāt 'an‚ī·ən }

enol-keto tautomerism [ORG CHEM] The tautomeric migration of a hydrogen atom from an adjacent carbon atom to a carbonyl group of a keto compound to produce the enol form of the compound; the reverse process of hydrogen atom migration also occurs. { ¦ē·nȯl ¦kēd·ō tȯ'tä·mə‚riz·əm }

entering group [ORG CHEM] An atom or group that becomes bonded to the main portion of the substrate during a chemical reaction. { 'en·tər·iŋ ‚grüp }

enthalpimetric analysis [ANALY CHEM] Generic designation for a group of modern thermochemical methodologies such as thermometric enthalpy titrations which rely on monitoring the temperature changes produced in adiabatic calorimeters by heats of reaction occurring in solution; in contradistinction, classical methods of thermoanalysis such as thermogravimetry focus primarily on changes occurring in solid samples in response to externally imposed programmed alterations in temperature. { en‚thal·pə'me·trik ə'nal·ə·səs }

enthalpy of reaction [PHYS CHEM] The change in enthalpy accompanying a chemical reaction. { en'thal·pē əv rē'ak·shən }

enthalpy of transition [PHYS CHEM] The change of enthalpy accompanying a phase transition. { en'thal·pē əv tran'zish·ən }

enthalpy titration See thermometric titration. { en'thal·pē tī'trā·shən }

entrance slit [SPECT] Narrow slit through which passes the light entering a spectrometer. { 'en·trəns ‚slit }

entropy of activation [PHYS CHEM] The difference in entropy between the activated complex in a chemical reaction and the reactants. { 'en·trə·pē əv‚ak·tə'vā·shən }

entropy of mixing [PHYS CHEM] After mixing substances, the difference between the entropy of the mixture and the sum of the entropies of the components of the mixture. { 'en·trə·pē əv 'mik·siŋ }

entropy of transition [PHYS CHEM] The heat absorbed or liberated in a phase change divided by the absolute temperature at which the change occurs. { 'en·trə·pē əv tran'zish·ən }

eosin [ORG CHEM] $C_{20}H_8O_5Br_4$ **1.** A red fluorescent dye in the form of triclinic crystals that are insoluble in water; used chiefly in cosmetics and as a toner. Also known as bromeosin; bromo acid; eosine; tetrabromofluorescein. **2.** The red to brown crystalline sodium or potassium salt of this dye; used in organic pigments, as a biological stain, and in pharmaceuticals. { 'ē·ə·sən }

ephedrine [ORG CHEM] $C_{10}H_{15}NO$ A white, crystalline, water-soluble alkaloid present in several Ephedra species and also produced synthetically; a sympathomimetic amine, it is used for its action on the bronchi, blood pressure, blood vessels, and central nervous system. { ə'fed·rən }

epi- [ORG CHEM] A prefix used in naming compounds to indicate the presence of a bridge or intramolecular connection. { 'ep·ē }

epichlorohydrin [ORG CHEM] C_3H_5OCl A colorless, unstable liquid, insoluble in water; used as a solvent for resins. { ¦ep·ə‚klȯr·ə'hī·drən }

epihydrin alcohol See glycidol. { ¦ep·ə¦hī·drən 'al·kə‚hȯl }

epimer [ORG CHEM] A type of isomer in which the difference between the two compounds is the relative position of the H (hydrogen) group and OH (hydroxyl) group on the last asymmetric C (carbon) atom of the chain, as in the sugars D-glucose and D-mannose. { 'ep·ə·mər }

epimerization [ORG CHEM] In an optically active compound that contains two or more asymmetric centers, a process in which only one of these centers is altered by some reaction to form an epimer. { ‚e·pim·ə·rə'zā·shən }

EPN See O-ethyl-O-para-nitrophenyl phenylphosphonothioate.

epoxidation [ORG CHEM] Reaction yielding an epoxy compound, such as the conversion of ethylene to ethylene oxide. { e‚päk·sə'dā·shən }

epoxide [ORG CHEM] **1.** A reactive group in which an oxygen atom is joined to each of two carbon atoms which are already bonded. **2.** A three-membered cyclic ether. Also known as oxirane. See ethylene oxide. { e'päk‚sīd }

epoxy- [ORG CHEM] A prefix indicating presence of an epoxide group in a molecule. { ə'päk·sē }

1,2-epoxyethane See ethylene oxide. { ¦wən ¦tü ə¦päk·sē'e‚thān }

epoxy resin [ORG CHEM] A polyether resin formed originally by the polymerization of bisphenol A and epichlorohydrin, having high strength, and low shrinkage during curing; used as a coating, adhesive, casting, or foam. { ə'päk·sē 'rez·ən }

EPXMA See electron probe x-ray microanalysis.

equation [CHEM] A symbolic expression that represents in an abbreviated form the laboratory observations of a chemical change; an equation (such as $2H_2 + O_2 \rightarrow 2H_2O$) indicates what reactants are consumed (H_2 and O_2) and what products are formed (H_2O), the correct formula of each reactant and product, and satisfies the law of conservation of atoms in that the symbols for the number of atoms reacting equals the number of atoms in the products. { i'kwā·zhən }

equation of state [PHYS CHEM] A mathematical expression which defines the physical state of a homogeneous substance (gas, liquid, or solid) by relating volume to pressure and absolute temperature for a given mass of the material. { i'kwā·zhən əv 'stāt }

equidensity technique [ANALY CHEM] Interference microscopy technique utilizing the Sabattier effect in photographic emulsions; the equidensities (lines of equal density in a photographic emulsion) are produced by exactly superimposing a positive and a negative of the same interferogram, and making a copy; used to measure photographic film emulsion density. { ¦ē·kwə¦den·səd·ē ‚tek‚nēk }

equilibrium See chemical equilibrium. { ‚ē·kwə'lib·rē·əm }

equilibrium constant [CHEM] A constant at a given temperature such that when a reversible chemical reaction $cC + bB = gG + hH$ has reached equilibrium, the value of this constant K^0 is equal to $\dfrac{a_G^g a_H^h}{a_C^c a_B^b}$ where a_G, a_H, a_C, and a_B represent chemical activities of the species G, H, C, and B at equilibrium. { ‚ē·kwə'lib·rē·əm ‚kän·stənt }

equilibrium diagram [PHYS CHEM] A phase diagram of the equilibrium relationship between temperature, pressure, and composition in any system. { ‚ē·kwə'lib·rē·əm 'dī·ə‚gram }

equilibrium dialysis [ANALY CHEM] A technique used to determine the degree of ion bonding by protein; the protein solution, placed in a bag impermeable to protein but permeable to small ions, is immersed in a solution containing the diffusible ion whose binding is being studied; after equilibration of the ion across the membrane, the concentration of ion in the protein-free solution is determined; the concentration of ion in the protein solution is determined by subtraction; if binding has occurred, the concentration of ion in the protein solution must be greater. { ˌē·kwə'lib·rē·əm dī'al·ə·səs }

equilibrium film [PHYS CHEM] A liquid film that is stable or metastable at a certain thickness with respect to small changes in the thickness. { ˌē·kwə'lib·rē·əm ˌfilm }

equilibrium moisture content [PHYS CHEM] The moisture content in a hydroscopic material that is being dried by contact with air at constant temperature and humidity when a definite, fixed (equilibrium) moisture content in the solid is reached. { ˌē·kwə'lib·rē·əm 'mȯis·chər ˌkän·tent }

equilibrium potential [PHYS CHEM] A point in which forward and reverse reaction rates are equal in an electrolytic solution, thereby establishing the potential of an electrode. { ˌē·kwə'lib·rē·əm pə'ten·chəl }

equilibrium prism [PHYS CHEM] Three-dimensional (solid) diagram for multicomponent mixtures to show the effects of composition changes on some key property, such as freezing point. { ˌē·kwə'lib·rē·əm ˌpriz·əm }

equilibrium ratio [PHYS CHEM] In any system, relation of the proportions of the various components (gas, liquid) at equilibrium conditions. See equilibrium vaporization ratio. { ˌē·kwə'lib·rē·əm ˌrā·shō }

equilibrium solubility [PHYS CHEM] The maximum solubility of one material in another (for example, water in hydrocarbons) for specified conditions of temperature and pressure. { ˌē·kwə'lib·rē·əm ˌsäl·yə'bil·əd·ē }

equilibrium still [ANALY CHEM] Recirculating distillation apparatus (no product withdrawal) used to determine vapor-liquid equilibria data. { ˌē·kwə'lib·rē·əm ˌstil }

equilibrium vaporization ratio [PHYS CHEM] In a liquid-vapor equilibrium mixture, the ratio of the mole fraction of a component in the vapor phase (y) to the mole fraction of the same component in the liquid phase (x), or $y/x = K$ (the K factor). Also known as equilibrium ratio. { ˌē·kwə'lib·rē·əm ˌvā·pə·rə'zā·shən ˌrā·shō }

equipartition [CHEM] 1. The condition in a gas where under equal pressure the molecules of the gas maintain the same average distance between each other. 2. The equal distribution of a compound between two solvents. 3. The distribution of the atoms in an orderly fashion, such as in a crystal. { ¦e·kwə·pär'tish·ən }

equivalence point [CHEM] The point in a titration where the amounts of titrant and material being titrated are equivalent chemically. { i'kwiv·ə·ləns ˌpȯint }

equivalent conductance [PHYS CHEM] Property of an electrolyte, equal to the specific conductance divided by the number of gram equivalents of solute per cubic centimeter of solvent. { i'kwiv·ə·lənt kən'dək·təns }

equivalent nuclei [PHYS CHEM] A set of nuclei in a molecule which are transformed into each other by rotations, reflections, or combinations of these operations, leaving the molecule invariant. { i'kwiv·ə·lənt 'nü·klē·ī }

equivalent weight [CHEM] The number of parts by weight of an element or compound which will combine with or replace, directly or indirectly, 1.008 parts by weight of hydrogen, 8.00 parts of oxygen, or the equivalent weight of any other element or compound. { i'kwiv·ə·lənt 'wāt }

Er See erbium.

ER See electroreflectance.

erbia See erbium oxide. { 'ər·bē·ə }

erbium [CHEM] A trivalent metallic rare-earth element, symbol Er, of the yttrium subgroup, found in euxenite, gadolinite, fergusonite, and xenotine; atomic number 68, atomic weight 167.26, specific gravity 9.051; insoluble in water, soluble in acids; melts at 1400–1500°C. { 'ər·bē·əm }

erbium halide [INORG CHEM] A compound of erbium and one of the halide elements. { 'ər·bē·əm 'hal,īd }

erbium nitrate [INORG CHEM] $Er(NO_3)_3 \cdot 5H_2O$ Pink crystals that are soluble in water, alcohol, and acetone; may explode if it is heated or shocked. { 'ər·bē·əm 'nī,trāt }

erbium oxalate [ORG CHEM] $Er_2(C_2O_4)_3 \cdot 10H_2O$ A red powder that decomposes at 575°C; used to separate erbium from common metals. { 'ər·bē·əm 'äk·sə,lāt }

erbium oxide [INORG CHEM] Er_2O_3 Pink powder that is insoluble in water; used as an actuator for phosphors and in manufacture of glass that absorbs in the infrared. Also known as erbia.　{ 'ər·bē·əm 'äk,sīd }

erbium sulfate　[INORG CHEM]　$Er_2(SO_4)_3 \cdot 8H_2O$ Red crystals that are soluble in water.　{ 'ər·bē·əm 'səl,fāt }

erbon　[ORG CHEM]　$C_{11}H_9Cl_5O_3$ A white solid with a melting point of 49–50°C; insoluble in water; used as a herbicide for perennial broadleaf weeds.　{ 'ər,bän }

ergot　[ORG CHEM]　Any of the five optically isomeric pairs of alkaloids obtained from this fungus; only the levorotatory isomers are physiologically active.　{ 'ər·gət }

ergotamine　[ORG CHEM]　$C_{33}H_{35}N_5O_5$ An alkaloid found in the fungal parasite ergot; causes smooth muscles in peripheral blood vessels to constrict, limiting blood flow; used to treat migraine headaches.　{ ər'gäd·ə,mēn }

ergotinine　[ORG CHEM]　An alkaloid and an isomer of ergotoxine that is a 1:1:1 mixture of ergocornine, ergocristine, and ergocryptine; crystallizes in long needles from acetone solutions, melting point 229°C, and soluble in chloroform, alcohol, and absolute ether.　{ ər'gät·ən,ēn }

ergotoxine　[ORG CHEM]　An alkaloid and an isomer of ergotinine that is a 1:1:1 mixture of ergocornine, ergocristine, and ergocryptine; crystallizes in orthorhombic crystals, melts at 190°C, and is soluble in methyl alcohol, ethyl alcohol, acetone, and chloroform.　{ ,ər·gə'täk·sēn }

eriodictyol　[ORG CHEM]　$C_{15}H_{22}O_6$ A compound isolated from *Eriodictyon californicum* as needlelike crystals from a dilute alcohol solution, sparingly soluble in boiling water, hot alcohol, and glacial acetic acid; used in medicine as an expectorant.　{ ,er·ē·ō'dik·tē,ól }

Erlenmeyer flask　[CHEM]　A conical glass laboratory flask, with a broad bottom and a narrow neck.　{ 'ər·lən,mī·ər 'flask }　■

Erlenmeyer synthesis　[ORG CHEM]　Preparation of cyclic ethers by the condensation of an aldehyde with an α-acylamino acid in the presence of acetic anhydride and sodium acetate.　{ 'ər·lən,mī·ər 'sin·thə·səs }

erucic acid　[ORG CHEM]　$C_{22}H_{42}O_2$ A monoethenoid acid that is the cis isomer of brassidic acid and makes up 40 to 50% of the total fatty acid in rapeseed, wallflower seed, and mustard seed; crystallizes as needles from alcohol solution, insoluble in water, soluble in ethanol and methanol.　{ ə'rüs·ik 'as·əd }

erythrite *See* erythritol.　{ 'er·ə,thrīt }

erythritol　[ORG CHEM]　$H(CHOH)_4H$ A tetrahydric alcohol; occurs as tetragonal prisms, melting at 121°C, soluble in water; used in medicine as a vasodilator.　Also known as erythrite; erythrol.　{ ə'rith·rə,tól }

erythroidine　[ORG CHEM]　$C_{16}H_{19}NO_3$ An alkaloid existing in two forms: α-erythroidine and β-erythroidine, isolated from *Erythrina* species; β-erythroidine has an action similar to that of curare as a skeletal muscle relaxant.　{ ,er·ə'thrō·ə,dēn }

erythrol *See* erythritol.　{ 'er·ə,thról }

erythrophleine　[ORG CHEM]　$C_{24}H_{39}NO_5$ An alkaloid isolated from the bark of *Erythrophleum guineense*; used in medicine experimentally for its digitalislike action.　{ ə,rith·rə'flē·ən }

erythrose　[ORG CHEM]　$HOCH_2(CHOH)_2CHO$ A tetrose sugar obtained from erythrol; a syrupy liquid at room temperature.　{ 'er·ə,thrōs }

erythrosin　[ORG CHEM]　$C_{13}H_{18}O_6N_2$ A red compound obtained by reacting tyrosine with nitric acid.　{ ə'rith·rə·sən }

Es *See* einsteinium.

ESCA *See* x-ray photoelectron spectroscopy.

escaping tendency　[PHYS CHEM]　The tendency of a solute species to escape from solution; related to the chemical potential of the solute.　{ ə'skap·iŋ ,ten·dən·sē }

Eschka mixture　[ANALY CHEM]　A mixture of two parts magnesium oxide and one part anhydrous sodium carbonate; used as a fusion mixture for determining sulfur in coal.　{ 'esh·kə ,miks·chər }

Eschweiler-Clarke modification　[ORG CHEM]　A modification of the Leuckart reaction, involving reductive alkylation of ammonia or amines (except tertiary amines) with formaldehyde and formic acid.　{ ¦esh,vīl·ər ¦klärk ,mäd·ə·fə'kā·shən }

ESEEM

ESEEM *See* electron spin echo envelope modulation. { 'ē,sēm *or* |ē|es|ē|ē'em }

eserine *See* physostigmine. { 'es·ə,rēn }

ester [ORG CHEM] The compound formed by the elimination of water and the bonding of an alcohol and an organic acid. { 'es·tər }

ester gum [ORG CHEM] A compound obtained by forming an ester of a natural resin with a polyhydric alcohol; used in varnishes, paints, and cellulosic lacquers. Also known as rosin ester. { 'es·tər ,gəm }

ester hydrolysis [ORG CHEM] A reaction in which an ester is converted into its alcohol and acid moieties. Also known as esterolysis. { |e·stər hī'dräl·ə·səs }

esterification [ORG CHEM] A chemical reaction whereby esters are formed. { e,ster·ə· fə'kā·shən }

esterolysis *See* ester hydrolysis. { ,e·stər'äl·ə·səs }

estersil [ORG CHEM] Hydrophobic silica powder, an ester of -SiOH with a monohydric alcohol; used as a filler in silicone rubbers, plastics, and printing inks. { 'es·tər,sil }

estragole [ORG CHEM] $C_6H_4(C_3H_5)(OCH_3)$ A colorless liquid with the odor of anise, found in basil oil, estragon oil, and anise bark oil; used in perfumes and flavorings. { 'es·tra,gōl }

Etard reaction [ORG CHEM] Direct oxidation of an aromatic or heterocyclic bound methyl group to an aldehyde by utilizing chromyl chloride or certain metallic oxides. { ā'tär rē,ak·shən }

ethamine *See* ethyl amine. { 'eth·ə·mēn }

ethane [ORG CHEM] CH_3CH_3 A colorless, odorless gas belonging to the alkane series of hydrocarbons, with freezing point of $-183.3°C$ and boiling point of $-88.6°C$; used as a fuel and refrigerant and for organic synthesis. { 'eth,ān }

1,2-ethanedithiol [ORG CHEM] $HSCH_2CH_2SH$ A liquid, freely soluble in alcohol and in alkalies; used as a metal complexing agent. { |wən |tü |eth,ān'dī·ə,mēn }

ethanoic acid *See* acetic acid. { |eth·ə|nō·ik,as·əd }

ethanol [ORG CHEM] C_2H_5OH A colorless liquid, miscible with water, boiling point 78.32°C; used as a reagent and solvent. Also known as ethyl alcohol; grain alcohol. { 'eth·ə,nól }

ethanolamine [ORG CHEM] $NH_2(CH_2)_2OH$ A colorless liquid, miscible in water; used in scrubbing hydrogen sulfide (H_2S) and carbon dioxide (CO_2) from petroleum gas streams, for dry cleaning, in paints, and in pharmaceuticals. { ,eth·ə'näl·ə,mēn }

ethanolurea [ORG CHEM] $NH_2CONHCH_2CH_2OH$ A white solid; its formaldehyde condensation products are thermoplastic and water-soluble. { |eth·ə,nól·yú'rē·ə }

ethene *See* ethylene. { 'e,thēn }

ethenol *See* vinyl alcohol. { 'eth·ə,nól }

ethephon [ORG CHEM] $C_2H_6ClO_3P$ A white solid with a melting point of 74.75°C; very soluble in water; used as a growth regulator for tomatoes, apples, cherries, and walnuts. Also known as CEPHA. { 'eth·ə,fän }

ether [ORG CHEM] **1.** One of a class of organic compounds characterized by the structural feature of an oxygen linking two hydrocarbon groups (such as $R-O-R$). **2.** $(C_2H_5)_2O$ A colorless liquid, slightly soluble in water; used as a reagent, intermediate, anesthetic, and solvent. Also known as ethyl ether. { 'e·thər }

etherification [ORG CHEM] The process of making an ether from an alcohol. { ē,thir· ə·fə'kā·shən }

ethidine *See* ethylidene. { 'eth·ə,dēn }

ethidium bromide [ORG CHEM] $C_{21}H_{20}BrN_3$ Dark red crystals with a melting point of 238–240°C; used in treating trypanosomiasis in animals and as an inhibitor of deoxyribonucleic and ribonucleic acid synthesis. Also known as homidium bromide. { e'thid·ē·əm 'brō,mīd }

ethinyl [ORG CHEM] The $HC≡C-$ radical from acetylene. Also known as acetenyl; acetylenyl; ethynyl. { e'thīn·əl }

ethiolate [ORG CHEM] $C_7H_{15}ONS$ A yellow liquid with a boiling point of 206°C; used as a preemergence herbicide for corn. { ə'thī·ə,lāt }

ethionic acid [ORG CHEM] $HO·SO_2·CH_2·CH_2·SO_2OH$ An unstable diacid, known only in solution. Also known as ethylene sulfonic acid. { 'eth·ē,än·ik 'as·əd }

ethohexadiol [ORG CHEM] $C_8H_{18}O_2$ A slightly oily liquid, used as an insect repellent. { ¦eth·ō,hek·sə'dī·ȯl }

ethoprop [ORG CHEM] $C_8H_{19}O_2PS_2$ A pale yellow liquid compound, insoluble in water; used as an insecticide for soil insects and as a nematicide for plant parasitic nematodes. { 'ē·thō,präp }

ethoxide [ORG CHEM] A compound formed from ethanol by replacing the hydrogen of the hydroxy group by a monovalent metal. Also known as ethylate. { e'thäk,sīd }

ethoxy [ORG CHEM] The C_2H_5O- radical from ethyl alcohol. Also known as ethyoxyl. { e'thäk·sē }

2-ethoxyethanol See cellosolve. { ¦tü e¦thäk·sē'eth·ə,nȯl }

ethoxyquin [ORG CHEM] $C_{14}H_{19}NO$ A dark liquid, used as a growth regulator to protect apples and pears in storage.

ethyl [ORG CHEM] **1.** The hydrocarbon radical C_2H_5. **2.** Trade name for the tetraethyllead antiknock compound in gasoline. { 'eth·əl }

ethyl acetate [ORG CHEM] $CH_3COOC_2H_5$ A colorless liquid, slightly soluble in water; boils at 77°C; a medicine, reagent, and solvent. Also known as acetic ester; acetic ether; acetidin. { 'eth·əl 'as·ə,tāt }

ethyl acetoacetate [ORG CHEM] $CH_3COCH_2COOC_2H_5$ A colorless liquid, boiling at 181°C; used as a reagent, intermediate, and solvent. Also known as acetoacetic ester; diacetic ether. { 'eth·əl¦as·ə·to¦as·ə,tāt }

ethyl acetylene [ORG CHEM] Compound with boiling point 8.1°C; insoluble in water, soluble in alcohol; used in organic synthesis. { ə'sed·əl,ēn }

ethyl acrylate [ORG CHEM] $C_5H_8O_2$ A colorless liquid, boiling at 99°C; used to manufacture chemicals and resins. { 'eth·əl 'ak·rə,lāt }

ethyl alcohol See ethanol. { 'eth·əl 'al·kə,hȯl }

ethyl amine [ORG CHEM] A colorless liquid, boiling at 15°C, water-soluble; used as a solvent, as a dye intermediate, and in organic synthesis. Also known as aminoethane; ethamine. { 'eth·əl 'am,ēn }

ethyl-_para_-aminobenzoate [ORG CHEM] $C_6H_4NH_2CO_2C_2H_5$ A white powder, melting point 88–92°C, slightly soluble in ethanol and ether, very slightly soluble in water; used as a local anesthetic. Also known as benzocaine. { 'eth·əl ¦par·ə ¦am·ə·nō'ben·zə,wāt }

ethyl amyl ketone [ORG CHEM] $C_8H_{16}O$ A colorless liquid, almost insoluble in water; used in perfumery. { ¦eth·əl ¦am·əl 'ke,tōn }

ethylate See ethoxide. { 'eth·ə,lāt }

ethylation [ORG CHEM] Formation of a new compound by introducing the ethyl functional group (C_2H_5). { ,eth·ə'lā·shən }

ethyl benzene [ORG CHEM] $C_6H_5C_2H_5$ A colorless liquid that boils at 136°C, insoluble in water; used in organic synthesis, as a solvent, and in making styrene. { 'eth·əl 'ben,zēn }

ethyl benzoate [ORG CHEM] $C_6H_5COOCH_2CH_3$ Colorless, aromatic liquid, boiling at 213°C, insoluble in water; used as a solvent, in flavoring extracts, and in perfumery. { 'eth·əl 'ben·zə,wāt }

ethyl borate [ORG CHEM] $B(OC_2H_5)_3$ A salt of ethanol and boric acid; colorless, flammable liquid; used in antiseptics, disinfectants, and fireproofing. Also known as boron triethoxide; triethylic borate. { 'eth·əl 'bȯr,āt }

ethyl bromide [ORG CHEM] C_2H_5Br A colorless liquid, boiling at 39°C; used as a refrigerant and in organic synthesis. { 'eth·əl 'brō,mīd }

2-ethylbutene [ORG CHEM] $CH_3CH_2(C_2H_5)CCH_2$ Colorless liquid, soluble in alcohol and organic solvents, insoluble in water; used in organic synthesis. { ¦tü ,eth·əl'byü,tān }

2-ethylbutyl acetate [ORG CHEM] $C_2H_5CH(C_2H_5)CH_2O_2CCH_3$ Colorless liquid with mild odor; used as a solvent for resins, lacquers, and nitrocellulose. { ¦tü ¦eth·əl¦byüd·əl 'as·ə,tāt }

2-ethylbutyl alcohol [ORG CHEM] $(C_2H_5)_2CHCH_2OH$ A stable, colorless liquid, miscible in most organic solvents, slightly water-soluble; used as a solvent for resins, waxes, and dyes, and in the synthesis of perfumes, drugs, and flavorings. { ¦tü ¦eth·əl¦byüd·əl 'al·kə,hȯl }

ethyl butyl ketone

ethyl butyl ketone [ORG CHEM] $C_2H_5COC_4H_9$ A colorless liquid, boiling at 147°C; used in solvent mixtures. Also known as 3-heptanone. { ¦eth·əl ¦byüd·əl 'kē₁tōn }

ethyl butyrate [ORG CHEM] $C_3H_7COOC_2H_5$ A colorless liquid, boiling at 121°C; used in flavoring extracts and perfumery. { ¦eth·əl 'byüd·ə₁rāt }

ethyl caprate [ORG CHEM] $CH_3(CH_2)_8COOC_2H_5$ A colorless liquid, used in the manufacture of wine bouquets and cognac essence. { ¦eth·əl 'ka₁prāt }

ethyl caproate [ORG CHEM] $C_5H_{11}COOC_2H_5$ A colorless to yellow liquid, boiling at 167°C, soluble in ether and alcohol, and having a pleasant odor; used as a chemical intermediate and in the food industry as an artificial fruit essence. Also known as ethyl hexanoate; ethyl hexoate. { ¦eth·əl kə'prō₁āt }

ethyl caprylate [ORG CHEM] $CH_3(CH_2)_6COOC_2H_5$ A clear, colorless liquid with a pineapple odor; used to make fruit ethers. Also known as ethyl octanoate. { ¦eth·əl 'kap·rə₁lāt }

ethyl carbamate See urethane. { ¦eth·əl 'kär·bə₁māt }

ethyl carbinol See propyl alcohol. { ¦eth·əl'kär·bə₁nól }

ethyl carbonate See diethyl carbonate. { ¦eth·əl 'kär·bə₁nāt }

ethyl cellulose [ORG CHEM] The ethyl ester of cellulose; it has film-forming properties and is inert to alkalies and dilute acids; used in adhesives, lacquers, and coatings. { ¦eth·əl'sel·yə₁lōs }

ethyl chloride [ORG CHEM] C_2H_5Cl A colorless gas, liquefying at 12.2°C, slightly soluble in water; used as a solvent, in medicine, and as an intermediate. Also known as chloroethane. { ¦eth·əl 'klór₁īd }

ethyl chloroacetate [ORG CHEM] $CH_2ClCOOC_2H_5$ A colorless liquid, boiling at 145°C; used as a poison gas, solvent, and chemical intermediate. { ¦eth·əl ₁klór·ō'as·ə₁tāt }

ethyl cinnamate [ORG CHEM] $C_6H_5CH=CHCOOC_2H_5$ An oily liquid with a faint cinnamon odor; used as a fixative for perfumes. Also known as ethyl phenylacrylate. { ¦eth·əl 'sin·ə₁māt }

ethyl crotonate [ORG CHEM] $CH_3CHCHCO_2C_2H_5$ A compound with a pungent aroma; boiling point of 143–147°C, soluble in water, soluble in ether; one of two isomeric forms used as an organic intermediate, a solvent for cellulose esters, and as a plasticizer for acrylic resins. { ¦eth·əl 'krōt·ən₁āt }

ethyl crotonic acid [ORG CHEM] $CH_3CHCC_2H_5COOH$ Colorless monoclinic crystals, subliming at 40°C; used as a peppermint flavoring. { ¦eth·əl krə'tän·ik 'as·əd }

ethyl cyanide [ORG CHEM] C_2H_5CN A colorless liquid that boils at 97.1°C; poisonous. { ¦eth·əl 'sī·ə₁nīd }

S-ethyl-N,N-dipropylthiocarbamate [ORG CHEM] $C_9H_{19}NOS$ An amber liquid soluble in water at 370 parts per million; used as a pre- and postemergence herbicide on vegetable crops. Abbreviated EDTC. { ¦es ¦eth·əl ¦en ¦en dī¦prō·pəl₁thī·ō'kär·bə₁māt }

ethyl enanthate [ORG CHEM] $CH_3(CH_2)_5COOC_2H_5$ A clear, colorless oil with a boiling point of 187°C; soluble in alcohol, chloroform, and ether; taste and odor are fruity; used as a flavor for liqueurs and soft drinks. Also known as cognac oil; ethyl heptanoate; ethyl oenanthate. { ¦eth·əl ə'nan₁thāt }

ethylene [ORG CHEM] C_2H_4 A colorless, flammable gas, boiling at −102.7°C; used as an agricultural chemical, in medicine, and for the manufacture of organic chemicals and polyethylene. Also known as ethene; olefiant gas. { 'eth·ə₁lēn }

ethylene bromide See ethylene dibromide. { 'eth·ə₁lēn 'bro₁mīd }

ethylene carbonate [ORG CHEM] $(CH_2O)_2CO$ Odorless, colorless solid with low melting point; soluble in water and organic solvents; used as a polymer and resin solvent, in solvent extraction, and in organic syntheses. { 'eth·ə₁lēn 'kär·bə₁nāt }

ethylene chloride [ORG CHEM] $ClCH_2CH_2Cl$ A colorless, oily liquid, boiling at 83.7°C; used as a solvent and fumigant, for organic synthesis, and for ore flotation. Also known as Dutch liquid; ethylene dichloride. { 'eth·ə₁lēn 'klór₁īd }

ethylene chlorobromide [ORG CHEM] CH_2BrCH_2Cl Volatile, colorless liquid with chloroformlike odor; soluble in ether and alcohol but not in water; general-purpose solvent for cellulosics; used in organic synthesis. { 'eth·ə₁lēn klór·ə'brō₁mīd }

ethylene chlorohydrin [ORG CHEM] $ClCH_2CH_2OH$ A colorless, poisonous liquid, boiling

142

at 129°C; used as a solvent and in organic synthesis. Also known as chloroethyl alcohol. { 'eth·ə,lēn klȯr·ə'hī·drən }

ethylene cyanide [ORG CHEM] $C_2H_4(CN)_2$ Colorless crystals, melting at 57°C; used in organic synthesis. Also known as succinonitrite. { 'eth·ə,lēn 'sī·ə,nīd }

ethylene cyanohydrin [ORG CHEM] C_3H_5ON A colorless liquid that is miscible with water and boils at 221°C. { 'eth·ə,lēn ,sī·ə·nō'hī·drən }

ethylene diacetate See ethylene glycol diacetate. { 'eth·ə,lēn dī'as·ə,tāt }

ethylenediamine [ORG CHEM] $NH_2CH_2CH_2NH_2$ Colorless liquid, melting at 8.5°C, soluble in water; used as a solvent, corrosion inhibitor, and resin and in adhesive manufacture. { ¦eth·ə·,lēn'dī·ə,mēn }

ethylenediaminetetraacetic acid [ORG CHEM] $(HOOCCH_2)_2NCH_2CH_2N(CH_2COOH)$ White crystals, slightly soluble in water and decomposing above 160°C; the sodium salt is a strong chelating agent, reacting with many metallic ions to form soluble nonionic chelate. Abbreviated EDTA. { ¦eth·ə·lēn¦dī·ə,mēn,te·trə·ə'sēd·ik 'as·əd }

ethylene dibromide [ORG CHEM] $BrCH_2CH_2Br$ A colorless, poisonous liquid, boiling at 131°C; insoluble in water; used in medicine, as a solvent in organic synthesis, and in antiknock gasoline. Also known as ethylene bromide. { 'eth·ə·lēn dī'brō,mīd }

ethylene dichloride See ethylene chloride. { 'eth·ə·lēn dī'klȯr,īd }

ethylene glycol See glycol. { 'eth·ə·lēn 'glī,kȯl }

ethylene glycol bis(trichloroacetate) [ORG CHEM] $C_4H_4Cl_6O_4$ A white solid with a melting point of 40.3°C; used as a herbicide for cotton and soybeans. Abbreviated EGT. { 'eth·ə·lēn 'glī,kȯl ,bis·trī,klȯr·ō'as·ə,tāt }

ethylene glycol diacetate [ORG CHEM] $CH_3COOCH_2CH_2OOCCH_3$ A liquid used as a solvent for oils, cellulose esters, and explosives. Also known as ethylene diacetate; glycol diacetate. { 'eth·ə·lēn 'glī,kȯl dī'as·ə,tāt }

ethyleneimine [ORG CHEM] C_2H_4NH Highly corrosive liquid, colorless and clear; miscible with organic solvents and water; used as an intermediate in fuel oil production, refining lubricants, textiles, and pharmaceuticals. Also known as aziridine. { ,eth· ə'lēn·ə,mīn }

ethylene nitrate [ORG CHEM] $(CH_2NO_3)_2$ An explosive yellow liquid, insoluble in water. Also known as glycol dinitrate. { 'eth·ə,lēn 'nī,trāt }

ethylene oxide [ORG CHEM] **1.** $(CH_2)_2O$ A colorless gas, soluble in organic solvents and miscible in water, boiling point 11°C; used in organic synthesis, for sterilizing, and for fumigating. **2.** Also known as 1,2-epoxyethane; epoxide; oxirane { 'eth·ə,lēn 'äk,sīd }

ethylene resin [ORG CHEM] A thermoplastic material composed of polymers of ethylene; the resin is synthesized by polymerization of ethylene at elevated temperatures and pressures in the presence of catalysts. Also known as polyethylene; polyethylene resin. { 'eth·ə,lēn 'rez·ən }

ethylene sulfonic acid See ethionic acid. { 'eth·ə,lēn səl¦fän·ik 'as·əd }

ethylethanolamine [ORG CHEM] $C_2H_5NHCH_2CH_2OH$ Water-white liquid with amine odor; soluble in alcohol, ether, and water; used in dyes, insecticides, fungicides, and surface-active agents. { ¦eth·əl,eth·ə'näl·ə,mēn }

ethyl ether See ether.

ethyl formate [ORG CHEM] $HCOOC_2H_5$ A colorless liquid, boiling at 54.4°C; used as a solvent, fumigant, and larvicide and in flavors, resins, and medicines. { ¦eth· əl 'fȯr,māt }

ethyl hexanoate See ethyl caproate. { ¦eth·əl hek'san·ə,wāt }

ethyl hexoate See ethyl caproate. { ¦eth·əl 'hek·sə,wāt }

2-ethyl hexoic acid [ORG CHEM] $C_4H_9CH(C_2H_5)COOH$ A liquid that is slightly soluble in water, boils at 226.9°C, and has a mild odor; used as an intermediate to make metallic salts for paint and varnish driers, esters for plasticizers, and light metal salts for conversion of some oils to grease. { ¦tü ¦eth·əl hek'sō·ik 'as·əd }

2-ethylhexyl acetate [ORG CHEM] $CH_3COOCH_2CHC_2H_5C_4H_9$ Water-white, stable liquid; used as a solvent for nitrocellulose, resins, and lacquers. Also known as octyl acetate. { ¦tü ¦eth·əl ¦hek·səl'as·ə,tāt }

143

2-ethylhexyl acrylate

2-ethylhexyl acrylate [ORG CHEM] $CH_2CHCOOCH_2CH(C_2H_5)C_4H_9$ Pleasant-smelling liquid; used as monomer for plastics, protective coatings, and paper finishes. { ¦tü ¦eth·əl 'ak·rə‚lāt }

2-ethylhexyl alcohol [ORG CHEM] $C_4H_9CH(C_2H_5)CH_2OH$ Colorless, slightly viscous liquid; used as a defoaming or wetting agent, as a solvent for protective coatings, waxes, and oils, and as a raw material for plasticizers. Also known as octyl alcohol. { ¦tü ¦eth·əl 'al·kə‚hól }

2-ethylhexylamine [ORG CHEM] $C_4H_9CH(C_2H_5)CH_2NH_2$ Water-white liquid with slight ammonia odor; slightly water-soluble; used to synthesize detergents, rubber chemicals, and oil additives. { ¦tü¦eth·əl·hek'sil·ə‚mēn }

2-ethylhexyl bromide [ORG CHEM] $C_4H_9CH(C_2H_5)CH_2Br$ Water-white, water-insoluble liquid; used to prepare pharmaceuticals and disinfectants. { ¦tü ¦eth·əl¦hek·səl 'brō‚mīd }

2-ethylhexyl chloride [ORG CHEM] $C_4H_9CH(C_2H_5)CH_2Cl$ Colorless liquid; used to synthesize cellulose derivatives, pharmaceuticals, resins, insecticides, and dyestuffs. { ¦tü ¦eth·əl¦hek·səl 'klór‚īd }

ethyl-*para*-hydroxybenzoate [ORG CHEM] $HOC_6H_4COOC_2H_5$ Crystals with a melting point of 116°C that are soluble in water, alcohol, and ether; used as a preservative for pharmaceuticals. Also known as ethylparaben. { ¦eth·əl ¦par·ə hī‚dräk·sē'ben‚zə‚wāt }

ethyl-2-hydroxypropionate See ethyl lactate. { ¦eth·əl ¦tü hī‚dräk·sē'prō·pē·ə‚nāt }

ethylic compound [ORG CHEM] Generic term for ethyl compounds. { e'thil·ik 'käm‚paünd }

ethylic ether See diethyl ether. { e'thil·ik 'ē·thər }

ethylidine [ORG CHEM] The $CH_3·CH=$ radical from ethane, C_2H_5. Also known as ethidine. { e'thil·ə‚dēn }

ethyl iodide [ORG CHEM] C_2H_5I A colorless liquid, boiling at 72.3°C; used in medicine and in organic synthesis. Also known as hydroiodic ether; iodoethane. { ¦eth·əl 'ī·ə‚dīd }

ethyl isobutylmethane See 2-methylhexane. { ¦eth·əl ‚ī·sō‚byüd·əl'me‚thān }

ethyl isovalerate [ORG CHEM] $(CH_3)_2CHCH_2COOC_2H_5$ A colorless, oily liquid with an apple odor, soluble in water and miscible with alcohol, benzene, and ether; used for flavoring beverages and confectioneries. { ¦eth·əl ī·sō'val·ə‚rāt }

ethyl lactate [ORG CHEM] $CH_3CHOHCOOC_2H_5$ A colorless liquid that boils at 154°C, has a mild odor, and is miscible with water and organic solvents such as alcohols, ketones, esters, and hydrocarbons; used as a flavoring and as a solvent for cellulose compounds such as nitrocellulose, cellulose acetate, and cellulose ethers. Also known as ethyl-2-hydroxypropionate. { ¦eth·əl 'lak‚tāt }

ethyl malonate [ORG CHEM] $CH_2(COOC_2H_5)_2$ A colorless liquid, boiling at 198°C; used as an intermediate and a plasticizer. Also known as malonic ester. { ¦eth·əl 'mal·ə‚nāt }

ethyl mercaptan [ORG CHEM] C_2H_5SH A colorless liquid, boiling at 36°C. Also known as ethyl sulfhydrate; thioethyl alcohol. { ¦eth·əl mər'kap·tan }

ethyl methacrylate [ORG CHEM] $CH_2CCH_3COOC_2H_5$ Colorless, easily polymerized liquid, water-insoluble; used to produce polymers and chemical intermediates. { ¦eth·əl me'thak·rə‚lāt }

ethyl methyl ketone See methyl ethyl ketone. { ¦eth·əl ¦meth·əl 'kē‚tōn }

ethyl nitrate [ORG CHEM] $C_2H_5NO_3$ A colorless, flammable liquid, boiling at 87.6°C; used in perfumes, drugs, and dyes and in organic synthesis. { ¦eth·əl 'nī‚trāt }

ethyl nitrite [ORG CHEM] $C_2H_5NO_2$ A colorless liquid, boiling at 16.4°C; used in medicine and in organic synthesis. Also known as sweet spirits of niter. { ¦eth·əl 'nī‚trīt }

ethyl octanoate See ethyl caprylate. { ¦eth·əl äk'tan·ə‚wāt }

ethyl oenanthate See ethyl enanthate. { ¦eth·əl ē'nan‚thāt }

ethyl oleate [ORG CHEM] $C_{20}H_{38}O_2$ A yellow oil, insoluble in water; used as a solvent, plasticizer, and lubricant. { ¦eth·əl 'ō·lē‚āt }

ethyl orthosilicate See ethyl silicate. { ¦eth·əl ‚ór·thō'sil·ə‚kāt }

144

ethyl oxalate [ORG CHEM] (COOC$_2$H$_5$)$_2$ Oily, unstable, colorless liquid that is combustible; miscible with organic solvents, very slightly soluble in water; used as a solvent for cellulosics and resins, and as an intermediate for dyes and pharmaceuticals. { ¦eth·əl 'äk·sə‚lāt }

ethyl oxide See diethyl ether. { ¦eth·əl 'äk‚sīd }

ethylparaben See ethyl-*para*-hydroxybenzoate. { ¦eth·əl'par·ə·bən }

O-ethyl-O-*para*-nitrophenyl phenylphosphonothioate [ORG CHEM] C$_2$H$_5$O$_4$NPS A yellow, crystalline compound with a melting point of 36°C; used as an insecticide and miticide on fruit crops. Abbreviated EPN. { ¦ō ¦eth·əl ¦ō ¦par·ə ‚nī·trō'fen·əl‚fen·əl·fäs¦fä·nō 'thī·ə‚wāt }

ethyl phenylacrylate See ethyl cinnamate. { ¦eth·əl ‚fen·əl'ak·rə‚lāt }

N-ethyl-5-phenylisoxazolium-3′-sulfonate [ORG CHEM] C$_{11}$H$_{11}$NO$_4$S Crystals that decompose at 207–208°C; used to form peptide bonds. Also known as Woodward's Reagent K. { ¦en ¦eth·əl ¦fīv ¦fen·əl‚ī·säk·sə'zō·lē·əm ¦thrē‚prīm 'səl·fə‚nāt }

1-ethyl-3-piperidinol See 1-ethyl-3-hydroxypiperidine. { ¦wən ¦eth·əl ¦thrē ‚pi'per·ə·də‚nȯl }

ethyl propionate [ORG CHEM] C$_2$H$_5$COOC$_2$H$_5$ A colorless liquid, slightly soluble in water, boiling at 99°C; used as solvent and pyroxylin cutting agent. Also known as propionic ether. { ¦eth·əl 'prō·pē·ə‚nāt }

ethyl salicylate [ORG CHEM] (HO)C$_6$H$_4$COOC$_2$H$_5$ A clear liquid with a pleasant odor; used in commercial preparation of artificial perfumes. Also known as sal ethyl; salicylic acid ethyl ether; salicylic ether. { ¦eth·əl sə'lis·əl‚āt }

ethyl silicate [ORG CHEM] (C$_2$H$_5$)$_4$SiO$_4$ A colorless, flammable liquid, hydrolyzed by water; used as a preservative for stone, brick, and masonry, in lacquers, and as a bonding agent. Also known as ethyl orthosilicate. { ¦eth·əl 'sil·ə‚kāt }

ethyl sulfate See diethyl sulfate. { ¦eth·əl 'səl‚fāt }

ethyl sulfhydrate See ethyl mercaptan. { ¦eth·əl ‚səlf'hī‚drāt }

ethyl sulfide [ORG CHEM] (C$_2$H$_5$)$_2$S A colorless, oily liquid, boiling at 92°C; used as a solvent and in organic synthesis. Also known as diethyl sulfide; ethylthioethane. { ¦eth·əl 'səl‚fīd }

ethylthioethane See ethyl sulfide. { ¦eth·əl‚thī·ō'e‚thān }

ortho-ethyl(O-2,4,5-trichlorophenyl)ethylphosphonothioate [ORG CHEM] C$_{10}$H$_{12}$-OPSCl$_2$ An amber liquid with a boiling point of 108°C at 0.01 mmHg; solubility in water is 50 parts per million; used as an insecticide for vegetable crops and soil pests on meadows. Also known as trichloronate. { ¦ȯr·thō ¦eth·əl ¦ō ¦tü ¦fȯr ¦fiv ¦trī‚klȯr·ō¦fen·əl¦eth·əl‚fäs¦fan·ō'thī·ə‚wāt }

ethyl urethane See urethane. { ¦eth·əl 'yùr·ə‚thān }

ethyl vanillin [ORG CHEM] C$_2$H$_5$O(OH)C$_6$H$_3$CHO A compound, crystallizing in fine white crystals that melt at 76.5°C, has a strong vanilla odor and four times the flavor of vanilla, soluble in organic solvents such as alcohol, chloroform, and ether; used in the food industry as a flavoring agent to replace or fortify vanilla. { ¦eth·əl və'nil·ən }

ethyne See acetylene. { 'e‚thēn }

ethynyl See ethinyl. { 'eth·ə‚nil }

ethynylation [ORG CHEM] Production of an acetylenic derivative by the condensation of acetylene with a compound such as an aldehyde; for example, production of butynediol from the union of formaldehyde withacetylene. { ‚eth·ən·əl'ā·shən }

ethyoxyl See ethoxy. { ‚eth·əl'äk·səl }

etioporphyrin [ORG CHEM] C$_{31}$H$_{34}$N$_4$ A synthetic porphyrin that has four ethyl and four methyl groups in a red-pigmented compound whose crystals melt at 280°C. { ‚ēd·ē·ō'pȯr·fə·rən }

Eu See europium.

eucalyptol [ORG CHEM] C$_{10}$H$_{18}$O A colorless oil with a camphorlike odor; boiling point is 174–177°C; used in pharmaceuticals, perfumery, and flavoring. Also known as cajeputol; cineol. { ‚yü·kə'lip‚tȯl }

eugenol [ORG CHEM] CH$_2$CHCH$_2$C$_6$H$_3$(OCH$_3$)OH A colorless or yellowish aromatic liquid with spicy odor and taste, soluble in organic solvents, and extracted from clove oil; used in flavors, perfumes, medicines, and the manufacture of vanilla. { 'yü·jə‚nȯl }

europium [CHEM] A member of the rare-earth elements in the cerium subgroup, symbol Eu, atomic number 63, atomic weight 151.96, steel gray and malleable, melting at 1100–1200°C.   { yu̇'rō·pē·əm }

europium halide [INORG CHEM] Any of the compounds of the element europium and the halogen elements; for example, europium chloride, $EuCl_3·xH_2O$.   { yu̇'rō·pē·əm 'ha‚līd }

europium oxide [INORG CHEM] Eu_2O_3 A white powder, insoluble in water; used in red- and infrared-sensitive phosphors.   { yu̇'rō·pē·əm 'äk‚sīd }

eutectic [PHYS CHEM] An alloy or solution that has the lowest possible constant melting point.   { yü'tek·tik }

eutectic mixture See eutectic system.   { yü¦tek·tik 'miks·chər }

eutectic point [PHYS CHEM] The point in the constitutional diagram indicating the composition and temperature of the lowest melting point of a eutectic.   { yü'tek·tik 'pȯint }

eutectic system [PHYS CHEM] The particular composition and temperature of materials at the eutectic point.   Also known as eutectic mixture.   { yü'tek·tik 'sis·təm }

eutectic temperature [PHYS CHEM] The temperature at the lowest melting point of a eutectic.   { yü'tek·tik 'tem·prə·chər }

eutectogenic system [PHYS CHEM] A multicomponent liquid-solid mixture in which pure solid phases of each component are in equilibrium with the remaining liquid mixture at a specific (usually minimum) temperature for a given composition, that is, the eutectic point.   { yü¦tek·tə¦jen·ik 'sis·təm }

eutectoid [PHYS CHEM] The point in an equilibrium diagram for a solid solution at which the solution on cooling is converted to a mixture of solids.   { yü'tek‚tȯid }

evolved gas analysis [ANALY CHEM] An analytical technique in which the characteristics or the amount of volatile products released by a substance and its reaction products are determined as a function of temperature while the sample is subjected to a series of controlled temperature changes.   Abbreviated EGA.   { ē¦välvd 'gas ə‚nal·ə·səs }

exchange broadening [SPECT] The broadening of a spectral line by some type of chemical or spin exchange process which limits the lifetime of the absorbing or emitting species and produces the broadening via the Heisenberg uncertainty principle.   { iks'chānj 'brȯd·ən·iŋ }

exchange narrowing [SPECT] The phenomenon in which, when a spectral line is split and thereby broadened by some variable perturbation, the broadening may be narrowed by a dynamic process that exchanges different values of the perturbation.   { iks'chānj 'nar·ə·wiŋ }

exchange reaction [CHEM] Reaction in which two atoms or ions exchange places either in two different molecules or in the same molecule.   { iks'chānj rē‚ak·shən }

exchange-repulsion [PHYS CHEM] A force that arises between neighboring molecules when they are close enough that their electron clouds overlap and, as a consequence of the Pauli exclusion principle, electrons are squeezed out from the region between the nuclei, which them repel each other.   { iks'chānj ri‚pəl·shən }

exchange velocity [CHEM] In an ion-exchange process, the speed with which one ion is displaced from an exchanger in favor of another ion.   { iks'chānj və'läs·əd·ē }

excimer [CHEM] An excited diatomic molecule where both atoms are of the same species and are dissociated in the ground state.   { 'ek·sə·mər }

exciplex [CHEM] An excited electron donor-acceptor complex which is dissociated in the ground state.   { 'ek·sə‚pleks }

excitation index [SPECT] In emission spectroscopy, the ratio of intensities of a pair of extremely nonhomologous spectra lines; used to provide a sensitive indication of variation in excitation conditions.   { ‚ek‚sī'tā·shən ‚in‚deks }

excitation purity [ANALY CHEM] The ratio of the departure of the chromaticity of a specified color to that of the reference source, measured on a chromaticity diagram; used as a guide of the wavelength of spectrum color needed to be mixed with a reference color to give the specified color.   { ‚ek‚sī'tā·shən ‚pyu̇r·əd·ē }

excitation spectrum |SPECT| The graph of luminous efficiency per unit energy of the exciting light absorbed by a photoluminescent body versus the frequency of the exciting light. { ˌekˌsī'tā·shən ˌspek·trəm }

exciting line |SPECT| The frequency of electromagnetic radiation, that is, the spectral line from a noncontinuous source, which is absorbed by a system in connection with some particular process. { ek'sīd·iŋ ˌlīn }

exhaustion point |CHEM| In an ion-exchange process, the state of an adsorbent at which it no longer can produce a useful ion exchange. { ig'zȯs·chən ˌpȯint }

exo- |ORG CHEM| A conformation of carbon bonds in a six-membered ring such that the molecule is boat-shaped with one or more substituents directed outward from the ring. { 'ek·sō }

exocyclic double bond |ORG CHEM| A double bond that is connected to and external to a ring structure. { ˌek·sōˌsī·klik ˌdəb·əl 'bänd }

explosion |CHEM| A chemical reaction or change of state which is effected in an exceedingly short space of time with the generation of a high temperature and generally a large quantity of gas. { ik'splō·zhən }

extender |CHEM| A material used to dilute or extend or change the properties of resins, ceramics, paints, rubber, and so on. { ik'sten·dər }

extensive property |PHYS CHEM| A noninherent property of a system, such as volume or internal energy, that changes with the quantity of material in the system; the quantitative value equals the sum of the values of the property for the individual constituents. { ik'sten·siv 'präp·ərd·ē }

external circuit |PHYS CHEM| All connecting wires, devices, and current sources which achieve desired conditions within an electrolytic cell. { ek'stərn·əl 'sər·kət }

external phase See continuous phase. { ek'stərn·əl 'fāz }

extinction See absorbance. { ek'stiŋk·shən }

extinction coefficient See absorptivity. { ek'stiŋk·shən ˌkō·i,fish·ənt }

extract |CHEM| Material separated from liquid or solid mixture by a solvent. { 'ek ˌstrakt (noun) or ik'strakt (verb) }

extractant |CHEM| The liquid used to remove a solute from another liquid. { ik 'strak·tənt }

extracting agent |CHEM| In a liquid-liquid distribution, the reagent forming a complex or other adduct that has different solubilities in the two immiscible liquids of the extraction system. { ik'strak·tiŋ ˌā·jənt }

extraction |CHEM| A method of separation in which a solid or solution is contacted with a liquid solvent (the two being essential mutually insoluble) to transfer one or more components into the solvent. { ik'strak·shən }

extreme narrowing approximation |SPECT| A mathematical approximation in the theory of spectral-line shapes to the effect that the exchange narrowing of a perturbation is complete. { ek'strēm 'nar·ə·wiŋ ə,präk·sə'mā·shən }

extrinsic sol |PHYS CHEM| A colloid whose stability is attributed to electric charge on the surface of the colloidal particles. { ekˌstrinz·ik 'säl }

Eyring equation |PHYS CHEM| An equation, based on statistical mechanics, which gives the specific reaction rate for a chemical reaction in terms of the heat of activation, entropy of activation, the temperature, and various constants. { 'ī·riŋ iˌkwā·zhən }

147

F

F *See* fluorine.

face-bridging ligand [ORG CHEM] A ligand that forms a bridge over one triangular face of the polyhedron of a metal cluster structure. { 'fās ,brij·iŋ 'līg·ənd }

family [CHEM] A group of elements whose chemical properties, such as valence, solubility of salts, and behavior toward reagents, are similar. { 'fam·lē }

famphur [ORG CHEM] $C_{10}H_{16}NO_5PS_2$ A crystalline compound with a melting point of 55°C; slightly soluble in water; used as an insecticide for lice and grubs of reindeer and cattle. { 'fam·fər }

faradaic current *See* faradic current. { ,far·ə¦dā·ik ¦kər·ənt }

Faraday's laws of electrolysis [PHYS CHEM] **1.** The amount of any substance dissolved or deposited in electrolysis is proportional to the total electric charge passed. **2.** The amounts of different substances dissolved or desposited by the passage of the same electric charge are proportional to their equivalent weights. { 'far·ə,dāz ¦lȯz əv i,lek'träl·ə·səs }

fast chemical reaction [PHYS CHEM] A reaction with a half-life of milliseconds or less; such reactions occur so rapidly that special experimental techniques are required to observe their rate. { 'fast ¦kem·ə·kəl rē'ak·shən }

fatty acid [ORG CHEM] An organic monobasic acid of the general formula $C_nH_{2n+1}COOH$ derived from the saturated series of aliphatic hydrocarbons; examples are palmitic acid, stearic acid, and oleic acid; used as a lubricant in cosmetics and nutrition, and for soaps and detergents. { ¦fad·ē 'as·əd }

fatty alcohol [ORG CHEM] A high-molecular-weight, straight-chain primary alcohol derived from natural fats and oils; includes lauryl, stearyl, oleyl, and linoleyl alcohols; used in pharmaceuticals, cosmetics, detergents, plastics, and lube oils and in textile manufacture. { 'fad·ē 'al·kə,hȯl }

fatty amine [ORG CHEM] RCH_2NH_2 A normal aliphatic amine from oils and fats; used as a plasticizer, in medicine, as a chemical intermediate, and in rubber manufacture. { 'fad·ē 'am,ēn }

fatty ester [ORG CHEM] RCOOR′ A fatty acid in which the alkyl group (R′) of a monohydric alcohol replaces the active hydrogen; for example, $RCOOCH_3$ from reaction of RCOOH with methane. { 'fad·ē 'es·tər }

fatty nitrile [ORG CHEM] RCN An ester of hydrogen cyanide derived from fatty acid; used in lube oil additives and plasticizers, and as a chemical intermediate. { 'fad·ē 'nī,trəl }

Favorskii rearrangement [ORG CHEM] A reaction in which α-halogenated ketones undergo rearrangement in the presence of bases, with loss of the halogen and formation of carboxylic acids or their derivatives with the same number of carbon atoms. { fa'vȯr·skē ,rē·ə'rānj·mənt }

FCC *See* chlorofluorocarbon.

Fe *See* iron.

feedback [CHEM] In a stepwise reaction, the formation of a substance in one step that affects the rate of a previous step. { 'fēd,bak }

Fehling's reagent [ANALY CHEM] A solution of cupric sulfate, sodium potassium tartrate, and sodium hydroxide, used to test for the presence of reducing compounds such as sugars. { 'fāl·iŋz rē,ā·jənt }

fenaminosulf

fenaminosulf [ORG CHEM] $C_8H_{10}N_3SO_3Na$ A yellow-brown powder, decomposing at 200°C; used as a fungicide for seeds and seedlings in crops. { ‚fen'am·ə·nō‚səlf }

fenazaflor [ORG CHEM] $C_{15}H_7Cl_2F_3N_2O_2$ A greenish-yellow, crystalline compound with a melting point of 103°C; used as an insecticide and miticide for spider mites and eggs. { fə'naz·ə‚flōr }

fenbutatin oxide [ORG CHEM] $C_{60}H_{78}OSn_2$ A white, crystalline compound, insoluble in water; used to control mites in deciduous and citrus fruits. { fen'byüd·əd·ən 'äk‚sīd }

fenchol See fenchyl alcohol. { 'fen·chöl }

fenchone [ORG CHEM] $C_{10}H_{16}O$ An isomer of camphor; a colorless oil that boils at 193°C and is soluble in ether; a constituent of fennel oil; used as a flavoring. { 'fen‚chōn }

fenchyl alcohol [ORG CHEM] $C_{10}H_{18}O$ A colorless solid or oily liquid, boiling at 198–204°C, isolated from pine oil and turpentine and also made synthetically; used as a solvent, an intermediate in organic synthesis, and as a flavoring. Also known as fenchol. { 'fen·chəl 'al·kə‚höl }

fenitrothion [ORG CHEM] $C_9H_{12}NO_5PS$ A yellow-brown liquid, insoluble in water; used as a miticide and insecticide for rice, orchards, vegetables, cereals, and cotton, and for fly and mosquito control. { ‚fen·ə·trō'thī‚än }

fensulfothion [ORG CHEM] $C_{11}H_{17}S_2O_2P$ A brown liquid with a boiling point of 138–141°C; used as an insecticide and nematicide in soils. { ‚fen‚səl·fō'thī‚än }

fentinacetate [ORG CHEM] $C_{20}H_{18}O_2Sn$ A yellow to brown, crystalline solid that melts at 124–125°C; used as a fungicide, molluscicide, and algicide for early and late blight on potatoes, sugarbeets, peanuts, and coffee. Also known as triphenyltinacetate. { ‚fent·ən'as·ə‚tāt }

fenuron [ORG CHEM] $C_9H_{12}N_2O$ A white, crystalline compound with a melting point of 133–134°C; soluble in water; used as a herbicide to kill weeds and bushes. { ‚fen'yü‚rän }

fenuron-TCA [ORG CHEM] $C_{11}H_{13}Cl_3N_2O_3$ A white, crystalline compound with a melting point of 65–68°C; moderately soluble in water; used as a herbicide for noncrop areas. { ‚fen'yü‚rän ‚tē‚sē‚ā }

FEP resin See fluorinated ethylene propylene resin. { ‚ef‚ē‚pē 'rez·ən }

ferbam [ORG CHEM] $C_9H_{18}FeN_3S_6$ [iron(III) dimethyldithiocarbamate] A fungicide for protecting fruits, vegetables, melons, and ornamental plants. { 'fər·bəm }

Fermi resonance [PHYS CHEM] In a polyatomic molecule, the relationship of two vibrational levels that have in zero approximation nearly the same energy; they repel each other, and the eigenfunctions of the two states mix. { 'fer·mē ‚rez·ən·əns }

fermium [CHEM] A synthetic radioactive element, symbol Fm, with atomic number 100; discovered in debris of the 1952 hydrogen bomb explosion, and now made in nuclear reactors. { 'fer·mē·əm }

ferrate [INORG CHEM] A multiple iron oxide with another oxide, for example, Na_2FeO_4. { 'fe‚rāt }

ferric [INORG CHEM] The term for a compound of trivalent iron, for example, ferric bromide, $FeBr_3$. { 'fer·ik }

ferric acetate [ORG CHEM] $Fe_2(C_2H_3O_2)_3$ A brown compound, soluble in water; used as a tonic and dye mordant. { 'fer·ik 'as·ə‚tāt }

ferric ammonium alum See ferric ammonium sulfate. { 'fer·ik ə'mōn·ē·əm 'al·əm }

ferric ammonium citrate [ORG CHEM] $Fe(NH_4)_3(C_6H_5O_7)_2$ Red, deliquescent scales or granules; odorless, water soluble, and affected by light; used in medicine and blueprint photography. { 'fer·ik ə'mōn·ē·əm 'sī‚trāt }

ferric ammonium oxalate [ORG CHEM] $(NH_4)_3Fe(C_2O_4)_3 \cdot 3H_2O$ Green, crystalline material, soluble in water and alcohol, sensitive to light; used in blueprint photography. { 'fer·ik ə'mōn·ē·əm 'äk·sə‚lāt }

ferric ammonium sulfate [INORG CHEM] $FeNH_4(SO_4)_2 \cdot 12H_2O$ Efflorescent, water-soluble crystals; used in medicine, in analytical chemistry, and as a mordant in textile dyeing. Also known as ferric ammonium alum; iron ammonium sulfate. { 'fer·ik ə'mōn·ē·əm 'səl‚fāt }

ferric arsenate [INORG CHEM] $FeAsO_4 \cdot 2H_2O$ A green or brown powder, insoluble in water, soluble in dilute mineral acids; used as an insecticide. { 'fer·ik 'ärs·ən‚āt }

ferric bromide [INORG CHEM] $FeBr_3$ Red, deliquescent crystals that decompose upon heating; soluble in water, ether, and alcohol; used in medicine and analytical chemistry. Also known as ferric sesquibromide; ferric tribromide; iron bromide. { 'fer·ik 'brō,mīd }

ferric chloride [INORG CHEM] $FeCl_3$ Brown crystals, melting at 300°C, that are soluble in water, alcohol, and glycerol; used as a coagulant for sewage and industrial wastes, as an oxidizing and chlorinating agent, as a disinfectant, in copper etching, and as a mordant. Also known as anhydrous ferric chloride; ferric trichloride; flores martis; iron chloride. { 'fer·ik 'klȯr,īd }

ferric citrate [ORG CHEM] $FeC_6H_5O_7·3H_2O$ Red scales that react to light; soluble in water, insoluble in alcohol; used as a medicine for certain blood disorders, and for blueprint paper. Also known as iron citrate. { 'fer·ik 'sī,trāt }

ferric dichromate [INORG CHEM] $Fe_2(CrO_4)_3$ A red-brown, granular powder, miscible in water; used as a mordant. { 'fer·ik dī'krō,māt }

ferric ferrocyanide [INORG CHEM] $Fe_4[Fe(CN)_6]_3$ Dark-blue crystals, used as a pigment, and with oxalic acid in blue ink. Also known as iron ferrocyanide. { 'fer·ik ,fer·ə'sī·ə,nīd }

ferric fluoride [INORG CHEM] FeF_3 Green, rhombohedral crystals, soluble in water and acids; used in porcelain and pottery manufacture. Also known as iron fluoride. { 'fer·ik 'flur,īd }

ferric hydrate See ferric hydroxide. { 'fer·ik 'hī,drāt }

ferric hydroxide [INORG CHEM] $Fe(OH)_3$ A brown powder, insoluble in water; used as arsenic poisoning antidote, in pigments, and in pharmaceutical preparations. Also known as ferric hydrate; iron hydroxide. { 'fer·ik hī'dräk,sīd }

ferric nitrate [INORG CHEM] $Fe(NO_3)_3·9H_2O$ Colorless crystals, soluble in water and decomposed by heat; used as a dyeing mordant, in tanning, and in analytical chemistry. Also known as iron nitrate. { 'fer·ik 'nī,trāt }

ferric oxalate [ORG CHEM] $Fe_2(COO)_3$ Yellow scales, soluble in water, decomposing when heated at about 100°C; used as a catalyst and in photographic printing papers. { 'fer·ik 'äk·sə,lāt }

ferric oxide [INORG CHEM] Fe_2O_3 Red, hexagonal crystals or powder, insoluble in water and soluble in acids, melting at 1565°C; used as a catalyst and pigment for metal polishing, in metallurgy, and in medicine. Also known as ferric oxide red; jeweler's rouge; red ocher. { 'fer·ik 'äk,sīd }

ferric oxide red See ferric oxide. { 'fer·ik ¦ak,sīd 'red }

ferric phosphate [INORG CHEM] $FePO_4·2H_2O$ Yellow, rhombohedral crystals, insoluble in water, soluble in acids; used in medicines and fertilizers. Also known as iron phosphate. { 'fer·ik 'fäs,fāt }

ferric resinate [ORG CHEM] Reddish-brown, water-insoluble powder; used as a drier for paints and varnishes. Also known as iron resinate. { 'fer·ik 'rez·ən,āt }

ferric sesquibromide See ferric bromide. { 'fer·ik ,ses·kwə'brō,mīd }

ferric stearate [ORG CHEM] $Fe(C_{18}H_{35}O_2)_3$ A light-brown, water-insoluble powder; used as a varnish drier. Also known as iron stearate. { 'fer·ik 'stir,āt }

ferric sulfate [INORG CHEM] $Fe_2(SO_4)_3·9H_2O$ Yellow, water-soluble, rhombohedral crystals, decomposing when heated; used as a chemical intermediate, disinfectant, soil conditioner, pigment, and analytical reagent, and in medicine. Also known as iron sulfate. { 'fer·ik 'səl,fāt }

ferric tribromide See ferric bromide. { 'fer·ik trī'brō,mīd }

ferric trichloride See ferric chloride. { 'fer·ik trī'klȯr,īd }

ferric vanadate [INORG CHEM] $Fe(VO_3)_3$ Grayish-brown powder, insoluble in water and alcohol; used in metallurgy. Also known as iron metavanadate. { 'fer·ik 'van·ə,dāt }

ferricyanic acid [INORG CHEM] $H_3Fe(CN)_6$ A red-brown unstable solid. { ,fer·i·sī'an·ik 'as·əd }

ferricyanide [INORG CHEM] A salt containing the radical $Fe(CN)_6^{3-}$. { fer·i'sī·ə,nīd }

ferrisulphas See ferrous sulfate. { ¦fe·ri'səl·fəs }

ferrite [INORG CHEM] An unstable compound of a strong base and ferric oxide which exists in alkaline solution, such as $NaFeO_2$. { 'fe,rīt }

ferrocene

ferrocene [ORG CHEM] $(CH_2)_5Fe(CH_2)_5$ Orange crystals that are soluble in ether, melting point 174°C; used as a combustion control additive in fuels, and for heat stabilization in greases and plastics. { 'fer·ə,sēn }

ferrocyanic acid [INORG CHEM] $H_4Fe(CN)_6$ A white solid obtained by treating ferrocyanides with acid. { ¦fe·rō·sī'an·ik 'as·əd }

ferrocyanide [INORG CHEM] A salt containing the radical $Fe(CN)_6^{4-}$. { ¦fe·rō'sī·ə,nīd }

ferrofluid [PHYS CHEM] A colloidal suspension that becomes magnetized in a magnetic field because of a disperse phase consisting of ferromagnetic or ferrimagnetic particles. { 'fe·rō,flü·əd }

ferrous [CHEM] The term or prefix used to denote compounds of iron in which iron is in the divalent (2+) state. { 'fer·əs }

ferrous acetate [ORG CHEM] $Fe(CH_3COO)_2·4H_2O$ Soluble green crystals, soluble in water and alcohol, that are combustible and that oxidize to basic ferric acetate in air; used as textile dyeing mordant, as wood preservative, and in medicine. Also known as iron acetate. { 'fer·əs 'as·ə,tāt }

ferrous ammonium sulfate [INORG CHEM] $Fe(SO_4)·(NH_4)_2SO_4·6H_2O$ Light-green, water-soluble crystals; used in medicine, analytical chemistry, and metallurgy. Also known as iron ammonium sulfate; Mohr's salt. { 'fer·əs ə'mōn·ē·əm 'səl,fāt }

ferrous arsenate [INORG CHEM] $Fe_3(AsO_4)_2·6H_2O$ Water-insoluble, toxic green amorphous powder, soluble in acids; used in medicine and as an insecticide. Also known as iron arsenate. { 'fer·əs 'ärs·ən,āt }

ferrous carbonate [INORG CHEM] $FeCO_3$ Green rhombohedral crystals that are soluble in carbonated water and decompose when heated; used in medicine. { 'fer·əs 'kär·bə,nāt }

ferrous chloride [INORG CHEM] $FeCl_2·4H_2O$ Green, monoclinic crystals, soluble in water; used as a mordant in dyeing, for sewage treatment, in metallurgy, and in pharmaceutical preparations. Also known as iron chloride; iron dichloride. { 'fer·əs 'klòr,īd }

ferrous hydroxide [INORG CHEM] $Fe(OH)_2$ A white, water-insoluble, gelatinous solid that turns reddish-brown as it oxidizes to ferric hydroxide. { 'fer·əs hī'dräk,sīd }

ferrous oxalate [ORG CHEM] $Fe(COO)_2$ A water-soluble, yellow powder; used in photography and medicine. Also known as iron oxalate. { 'fer·əs 'äk·sə,lāt }

ferrous oxide [INORG CHEM] FeO A black powder, soluble in water, melting at 1419°C. Also known as black iron oxide; iron monoxide. { 'fer·əs 'äk,sīd }

ferrous sulfate [INORG CHEM] $FeSO_4·7H_2O$ Blue-green, water-soluble, monoclinic crystals; used as a mordant in dyeing wool, in the manufacture of ink, and as a disinfectant. Also known as copperas; ferrisulphas; green copperas; green vitriol; iron sulfate. { 'fer·əs 'səl,fāt }

ferrous sulfide [INORG CHEM] FeS Black crystals, insoluble in water, soluble in acids, melting point 1195°C; used to generate hydrogen sulfide in ceramics manufacture. Also known as iron sulfide. { 'fer·əs 'səl,fīd }

ferrum [CHEM] Latin term for iron; derivation of the symbol Fe. { 'fer·əm }

ferulic acid [ORG CHEM] $C_{10}H_{10}O_4$ A compound widely distributed in small amounts in plants, having two isomers: the cis form is a yellow oil, and the trans form is obtained from water solutions as orthorhombic crystals. { fə'rül·ik 'as·əd }

Féry spectrograph [SPECT] A spectrograph whose only optical element consists of a back-reflecting prism with cylindrically curved faces. { ¦fār·ē 'spek·trə,graf }

Feulgen reaction [ANALY CHEM] An aldehyde specific reaction based on the formation of a purple-colored compound when aldehydes react with fuchsin-sulfuric acid; deoxyribonucleic acid gives this reaction after removal of its purine bases by acid hydrolysis; used as a nuclear stain. { 'fòil·gən rē,ak·shən }

ficin [ORG CHEM] A proteolytic enzyme obtained from fig latex or sap; hydrolyzes casein, meat, fibrin, and other proteinlike materials; used in the food industry and as a diagnostic aid in medicine. { 'fī·sən }

FID See free induction decay.

field-desorption mass spectroscopy [SPECT] A technique for analysis of nonvolatile molecules in which a sample is deposited on a thin tungsten wire containing sharp microneedles of carbon on the surface; a voltage is applied to the wire, thus producing high electric-field gradients at the points of the needles, and moderate heating then causes desorption from the surface or molecular ions, which are focused into a mass spectrometer. { ¦fēld dē'sȯrp·shən ¦mas spek'trä·skə·pē }

figure of merit [ANALY CHEM] A performance characteristic of an analytical chemical method that influences its choice for a specific type of determination, such as selectivity, sensitivity, detection limit, precision, and bias. { 'fig·yər əv 'mer·ət }

film boiling [PHYS CHEM] A stage in the boiling process in which the heater surface is totally covered by a film of vapor and the liquid does not contact the solid. { 'film ¸bȯil·iŋ }

film-development chromatography [ANALY CHEM] Liquid-analysis chromatographic technique in which the stationary phase (adsorbent) is a strip or layer, as in paper or thin-layer chromatography. { film di¸vel·əp·mənt ¸krō·mə'täg·rə·fē }

film tension [PHYS CHEM] The contractile force per unit length that is exerted by an equilibrium film in contact with a supporting substrate. { 'film ¸ten·chən }

filter flask [CHEM] A flask with a side arm to which a vacuum can be applied; usually filter flasks have heavy side walls to withstand high vacuum. { 'fil·tər ¸flask }

filter photometry [ANALY CHEM] **1.** Colorimetric analysis of solution colors with a filter applied to the eyepiece of a conventional colorimeter. **2.** Inspection of a pair of Nessler tubes through a filter. { 'fil·tər fə'täm·ə·trē }

filter-press cell [PHYS CHEM] An electrolytic cell consisting of several units in series, as in a filter press, in which each electrode, except the two end ones, acts as an anode on one side and a cathode on the other, and the space between electrodes is divided by porous asbestos diaphragms. { 'fil·tər ¸pres ¸sel }

filter spectrophotometer [SPECT] Spectrophotographic analyzer of spectral radiations in which a filter is used to isolate narrow portions of the spectrum. { 'fil·tər spek·trə·fə'täm·əd·ər }

fingerprint [ANALY CHEM] Evidence for the presence or the identity of a substance that is obtained by techniques such as spectroscopy, chromatography, or electrophoresis. { 'fiŋ·gər¸print }

fire [CHEM] The manifestation of rapid combustion, or combination of materials with oxygen. { fīr }

fire point [CHEM] The lowest temperature at which a volatile combustible substance vaporizes rapidly enough to form above its surface an air-vapor mixture which burns continuously when ignited by a small flame. { 'fīr ¸pȯint }

first-order reaction [PHYS CHEM] A chemical reaction in which the rate of decrease of concentration of component A with time is proportional to the concentration of A. { ¦fərst ¸ȯrd·ər rē'ak·shən }

first-order spectrum [SPECT] A spectrum, produced by a diffraction grating, in which the difference in path length of light from adjacent slits is one wavelength. { ¦fərst ¸ȯrd·ər 'spek·trəm }

Fischer-Hepp rearrangement [ORG CHEM] The rearrangement of a nitroso derivative of a secondary aromatic amine to a p-nitrosoarylamine; the reaction is brought about by an alcoholic solution of hydrogen chloride. { ¦fish·ər ¦hep rē·ə'rānj·mənt }

Fischer indole synthesis [ORG CHEM] A reaction to form indole derivatives by means of a ring closure of aromatic hydrazones. { ¸fish·ər 'in¸dōl ¸sin·thə·səs }

Fischer polypeptide synthesis [ORG CHEM] A synthesis of peptides in which α-amino acids or those peptides with a free amino group react with acid halides of α-haloacids, followed by amination with ammonia. { 'fish·ər ¦päl·ē'pep¸tīd ¸sin·thə·səs }

Fischer projection [ORG CHEM] A method for representing the spatial arrangement of groups around chiral carbon atoms; the four bonds to the chiral carbon are represented by a cross, with the assumption that the horizontal bonds project toward the viewer and the vertical bonds away from the viewer. { 'fish·ər prə¸jek·shən }

Fischer's salt See cobalt potassium nitrite. { ¦fish·ərz 'sȯlt }

fissiochemistry [CHEM] The process of producing chemical change by means of nuclear energy. { ¦fish·ō¦kem·ə·strē }

Fittig's synthesis

Fittig's synthesis [ORG CHEM] The synthesis of aromatic hydrocarbons by the condensation of aryl halides with alkyl halides, using sodium as a catalyst. { 'fid·iks ,sin·thə·səs }

fixed carbon [CHEM] Solid, combustible residue remaining after removal of moisture, ash, and volatile materials from coal, coke, and bituminous materials; expressed as a percentage. { ¦fikst 'kär·bən }

fixed ion [ANALY CHEM] An ion in the lattice of a solid ion exchanger. [PHYS CHEM] One of a group of nonexchangeable ions in an ion exchanger that have a charge opposite to that of the counterions. { ¦fikst 'ī,än }

flame [CHEM] A hot, luminous reaction front (or wave) in a gaseous medium into which the reactants flow and out of which the products flow. { flām }

flame emission spectroscopy [SPECT] A flame photometry technique in which the solution containing the sample to be analyzed is optically excited in an oxyhydrogen or oxyacetylene flame. { 'flām i,mish·ən spek'träs·kə·pē }

flame excitation [SPECT] Use of a high-temperature flame (such as oxyacetylene) to excite spectra emission lines from alkali and alkaline-earth elements and metals. { ¦flām ,ek·sī'tā·shən }

flame ionization detector [ANALY CHEM] A device in which the measured change in conductivity of a standard flame (usually hydrogen) due to the insertion of another gas or vapor is used to detect the gas or vapor. { ¦flām ,ī·ə·nə'zā·shən di,tek·tər }

flame photometer [SPECT] One of several types of instruments used in flame photometry, such as the emission flame photometer and the atomic absorption spectrophotometer, in each of which a solution of the chemical being analyzed is vaporized; the spectral lines resulting from the light source going through the vapors enters a monochromator that selects the band or bands of interest. { 'flām ,fə'täm·əd·ər }

flame photometry [SPECT] A branch of spectrochemical analysis in which samples in solution are excited to produce line emission spectra by introduction into a flame. { 'flām fə'täm·ə·tre }

flame propagation [CHEM] The spread of a flame in a combustible environment outward from the point at which the combustion started. { ¦flām ,präp·ə'gā·shən }

flame spectrometry [SPECT] A procedure used to measure the spectra or to determine wavelengths emitted by flame-excited substances. { ¦flām spek'träm·ə·trē }

flame spectrophotometry [SPECT] A method used to determine the intensity of radiations of various wavelengths in a spectrum emitted by a chemical inserted into a flame. { ¦flām ¦spek·trə·fə'täm·ə·trē }

flame spectrum [SPECT] An emission spectrum obtained by evaporating substances in a nonluminous flame. { 'flām ,spek·trəm }

flame speed [CHEM] The rate at which combustion moves through an explosive mixture. { 'flām ,spēd }

flammability [CHEM] A measure of the extent to which a material will support combustion. Also known as inflammability. { ,flam·ə'bil·əd·ē }

flammability limits [CHEM] The stoichiometric composition limits (maximum and minimum) of an ignited oxidizer-fuel mixture what will burn indefinitely at given conditions of temperature and pressure without further ignition. { ,flam·ə'bil·əd·ē ,lim·əts }

flash photolysis [PHYS CHEM] A method of studying fast photochemical reactions in gas molecules; a powerful lamp is discharged in microsecond flashes near a reaction vessel holding the gas, and the products formed by the flash are observed spectroscopically. { 'flash fə,täl·ə·səs }

flash point [CHEM] The lowest temperature at which vapors from a volatile liquid will ignite momentarily upon the application of a small flame under specified conditions; test conditions can be either open- or closed-cup. { 'flash ,póint }

flash spectroscopy [SPECT] The study of the electronic states of molecules after they absorb energy from an intense, brief light flash. { ¦flash spek'träs·kə·pē }

flask [CHEM] A long-necked vessel, frequently of glass, used for holding liquids. { flask }

F line [SPECT] A green-blue line in the spectrum of hydrogen, at a wavelength of 486.133 nanometers. { 'ef ,līn }

floc [CHEM] Small masses formed in a fluid through coagulation, agglomeration, or biochemical reaction of fine suspended particles. { fläk }

flocculant *See* flocculating agent. { 'fläk·yə·lənt }

flocculate [CHEM] To cause to aggregate or coalesce into a flocculent mass. { 'fläk· yə‚lāt }

flocculating agent [CHEM] A reagent added to a dispersion of solids in a liquid to bring together the fine particles to form flocs. Also known as flocculant. { 'fläk· yə‚lād·iŋ ‚ā·jənt }

flocculent [CHEM] Pertaining to a material that is cloudlike and noncrystalline. { 'fläk· yə·lənt }

floc point [ANALY CHEM] The temperature at which wax or solids separate from kerosine and other illuminating oils as a definite floc. { 'fläk ‚póint }

floc test [ANALY CHEM] A quantitative test applied to kerosine and other illuminating oils to detect substances rendered insoluble by heat. { 'fläk ‚test }

Flood's equation [PHYS CHEM] A relation used to determine the liquidus temperature in a binary fused salt system. { 'flədz i‚kwā·zhən }

flores [CHEM] A form of a chemical compound made by the process of sublimation. { 'flòr·ēz }

flores martis *See* ferric chloride. { 'flòr·ēz 'märd·əs }

flotation agent [CHEM] A chemical which alters the surface tension of water or which makes it froth easily. { flō'tā·shən ‚ā·jənt }

flow birefringence [PHYS CHEM] Orientation of long, thin asymmetric molecules in the direction of flow of a solution forced to flow through a capillary tube. { ‚flō ‚bī· rə'frin·jəns }

flowers of tin *See* stannic oxide. { 'flaú·ərz əv 'tin }

flow-programmed chromatography [ANALY CHEM] A chromatographic procedure in which the rate of flow of the mobile phase is periodically changed. { ‚flō ‚prō‚gramd ‚krō·mə'täg·rə·fē }

fluoborate *See* fluoroborate. { ‚flü·ə'bòr‚āt }

fluometuron [ORG CHEM] $C_{10}H_{11}F_3N_2O$ A white, crystalline solid with a melting point of 163–164.5°C; used as a herbicide for cotton and sugarcane. Also known as 1,1-dimethyl-3-(α,α,α-trifluoro-*meta*-tolyl)urea. { ‚flü·ō'me·chə ‚rän }

fluoranthene [ORG CHEM] $C_{16}H_{10}$ A tetracyclic hydrocarbon found in coal tar fractions and petroleum, forming needlelike crystals, boiling point 250°C, and soluble in organic solvents such as ether and benzene. { flü'ran‚thēn }

fluorene [ORG CHEM] $C_{13}H_{10}$ A hydrocarbon chemical present in the middle oil fraction of coal tar; insoluble in water, soluble in ether and acetone, melting point 116–117°C; used as the basis for a group of dyes. Also known as 2,3-benzindene; diphenylenemethane. { 'flü‚rēn }

fluorescein [ORG CHEM] $C_{20}H_{12}O_5$ A yellowish to red powder, melts and decomposes at 290°C, insoluble in water, benzene, and chloroform, soluble in glacial acetic acid, boiling alcohol, ether, dilute acids, and dilute alkali; used in medicine, in oceanography as a marker in sea water, and in textiles to dye silk and wool. { ‚flü're· sē·ən }

fluorescence analysis *See* fluorometric analysis. { flü'res·əns ə‚nal·ə·səs }

fluorescence spectra [SPECT] Emission spectra of fluorescence in which an atom or molecule is excited by absorbing light and then emits light of characteristic frequencies. { flü'res·əns ‚spek·trə }

fluorescent dye [CHEM] A highly reflective dye that serves to intensify color and add to the brilliance of a fabric. { flü¦res·ənt 'dī }

fluorescent pigment [CHEM] A pigment capable of absorbing both visible and nonvisible electromagnetic radiations and releasing them quickly as energy of desired wavelength; examples are zinc sulfide or cadmium sulfide. { flü¦res·ənt 'pig·mənt }

fluoride [INORG CHEM] A salt of hydrofluoric acid, HF, in which the fluorine atom is in the −1 oxidation state. { 'flùr‚īd }

fluorinated ethylene propylene resin [ORG CHEM] Copolymers of tetrafluoroethylene and hexafluoropropylene. Abbreviated FEP resin. { 'flúr·ə,nād·əd 'eth·ə,lēn 'prō·pə,lēn 'rez·ən }

fluorination [CHEM] A chemical reaction in which fluorine is introduced into a chemical compound. { ,flúr·ə'nā·shən }

fluorine [CHEM] A gaseous or liquid chemical element, symbol F, atomic number 9, atomic weight 18.998403; a member of the halide family, it is the most electronegative element and the most chemically energetic of the nonmetallic elements; highly toxic, corrosive, and flammable; used in rocket fuels and as a chemical intermediate. { 'flúr,ēn }

fluoroacetate [ORG CHEM] Acetate in which carbon-connected hydrogen atoms are replaced by fluorine atoms. { ¦flúr·ō'as·ə,tāt }

fluoroacetic acid [ORG CHEM] CH_2FCOOH A poisonous, crystalline compound obtained from plants, such as those of the Dichapetalaceae family, South Africa, soluble in water and alcohol, and burns with a green flame; the sodium salt is used as a water-soluble rodent poison. Also known as gifblaar poison. { ¦flúr·ō·ə'sēd·ik 'as·əd }

fluoroalkane [ORG CHEM] Straight-chain, saturated hydrocarbon compound (or analog thereof) in which some of the hydrogen atoms are replaced by fluorine atoms. { ¦flúr·ō'al,kān }

para-fluoroaniline [ORG CHEM] $FC_6H_4NH_2$ A liquid that is an intermediate in the manufacture of herbicides and plant growth regulators. { ¦par·ə ,flúr·ō'an·ə,lēn }

fluorobenzene [ORG CHEM] C_6H_5F A colorless liquid with a boiling point of 84.9°C; used as an insecticide intermediate. Also known as phenyl fluoride. { ¦flúr·ō¦ben,zēn }

fluoroborate [INORG CHEM] **1.** Any of a group of compounds related to the borates in which one or more oxygens have been replaced by fluorine atoms. **2.** The BF_4^- ion, which is derived from fluoroboric acid, HBF_4. Also known as fluoborate. { ,flúr·ə'bòr,āt }

fluoroboric acid [INORG CHEM] HBF_4 Colorless, clear, water-miscible acid; used for electrolytic brightening of aluminum and for forming stabilized diazo salts. { ¦flúr·ə,bòr·ik 'as·əd }

fluorocarbon [ORG CHEM] A hydrocarbon in which part or all hydrogen atoms have been replaced by fluorine atoms, including chlorinated and brominated fluorocarbons. Also known as fluorohydrocarbon. { ¦flúr·ō'kär·bən }

fluorocarbon-11 See trichlorofluoromethane. { ¦flúr·ō'kär·bən ə'lev·ən }

fluorocarbon-21 See dichlorofluoromethane. { ¦flúr·ō'kär·bən ,twen·tē'wən }

fluorocarbon fiber [ORG CHEM] Fiber made from a fluorocarbon resin, such as polytetrafluoroethylene resin. { ¦flúr·ō'kär·bən 'fī·bər }

fluorocarbon resin [ORG CHEM] Polymeric material made up of carbon and fluorine with or without other halogens (such as chlorine) or hydrogen; the resin is extremely inert and more dense than corresponding fluorocarbons such as Teflon. { ¦flúr·ō'kär·bən 'rez·ən }

fluorochemical [CHEM] Any chemical compound containing fluorine; usually refers to the fluorocarbons. { ¦flúr·ō'kem·ə·kəl }

fluorochlorocarbon See chlorofluorocarbon. { ¦flúr·ō¦klòr·ō'kär·bən }

fluorodichloromethane See dichlorofluoromethane. { ¦flúr·ō·dī,klòr·ō'meth,ān }

fluorodifen [ORG CHEM] $C_{13}H_7F_3N_2O_4$ A yellow, crystalline compound with a melting point of 93°C; used as a pre- and postemergence herbicide for food crops. { flú'räd·ə·fen }

1-fluoro-2,4-dinitrobenzene [ORG CHEM] $(NO_2)_2C_6H_3F$ Crystals that are soluble in benzene, propylene glycol, and ether; used as a reagent for labeling terminal amino acid groups and in the detection of phenols. Also known as Sanger's reagent. { ¦wən ¦flúr·ō ¦tü ¦fòr dī,nī·trō'ben,zēn }

fluoroform [ORG CHEM] CHF_3 A colorless, nonflammable gas, boiling point 84°C at 1 atmosphere (101,325 pascals), freezing point 160°C at 1 atmosphere; used in refrigeration and as an intermediate in organic synthesis. Also known as propellant 23; refrigerant 23; trifluoromethane. { 'flúr·ə,fòrm }

fluorogenic substrate [CHEM] A nonfluorescent material that is acted upon by an enzyme to produce a fluorescent compound. { 'flúr·ə,jen·ik 'səb,strāt }

fluorohydrocarbon *See* fluorocarbon. { ¦flür·ō,hī·drə'kär·bən }

fluorometric analysis [ANALY CHEM] A method of chemical analysis in which a sample, exposed to radiation of one wavelength, absorbs this radiation and reemits radiation of the same or longer wavelength in about 10^{-9} second; the intensity of reemited radiation is almost directly proportional to the concentration of the fluorescing material. Also known as fluorescence analysis; fluorometry. { ¦flür·ə¦me·trik ə'nal· ə·səs }

fluorometry *See* fluorometric analysis. { flü'räm·ə·trē }

para-fluorophenylacetic acid [ORG CHEM] $FC_6H_4CH_2COOH$ Crystals with a melting point of 86°C; used as an intermediate in the manufacture of fluorinated anesthetics. { ¦par·ə ¦flü·rə,fen·əl·ə'sēd·ik 'as·əd }

fluorophosphoric acid [INORG CHEM] H_2PO_3F A colorless, viscous liquid that is miscible with water; used in metal cleaners and as a catalyst. { ¦flür·ō,fäs'fòr·ik 'as·əd }

fluorothene *See* chlorotrifluoroethylene polymer. { 'flür·ə,thēn }

fluorotrichloromethane *See* trichlorofluoromethane. { ¦flür·ō·trī¦klòr·ō'meth,ān }

fluosilicate [INORG CHEM] A salt derived from fluosilicic acid, H_2SiF_6, and containing the SiF_6^{-2} ion. { ¦flü·ə'sil·ə,kāt }

fluosilicic acid [INORG CHEM] H_2SiF_6 A colorless acid, soluble in water, which attacks glass and stoneware; highly corrosive and toxic; used in water fluoridation and electroplating. Also known as hydrofluorosilicic acid; hydrofluosilicic acid. { ¦flü· ə·sə'lis·ik 'as·əd }

fluosulfonic acid [INORG CHEM] HSO_3F Colorless, corrosive, fuming liquid; soluble in water with partial decomposition; used as organic synthesis catalyst and in electroplating. { ¦flü·ə·səl'fän·ik 'as·əd }

flurenol [ORG CHEM] $C_{18}H_{18}O_3$ A solid, crystalline compound with a melting point of 70–71°C; used as an herbicide for vegetables, cereals, and ornamental flowers. { 'flür·ə,nòl }

fluxional compound [ORG CHEM] **1.** Any of a group of molecules which undergo rapid intramolecular rearrangements in which the component atoms are interchanged among equivalent structures. **2.** Molecules in which bonds are broken and reformed in the rearrangement process. { 'flək·shən·əl ,käm,paúnd }

Fm *See* fermium.

foam [CHEM] An emulsionlike two-phase system where the dispersed phase is gas or air. { fōm }

folic acid sodium salt *See* sodium folate. { ¦fō·lik ¦as·əd ¦sōd·ē·əm 'sòlt }

folimat [ORG CHEM] $C_5H_{12}NO_4PS$ An oily liquid that decomposes at 135°C; soluble in water; used as an insecticide and miticide on fruit and vegetable crops and on ornamental flowers. Also known as omethioate. { 'fä·lə,mat }

Folin solution [ANALY CHEM] An aqueous solution of 500 grams of ammonium sulfate, 5 grams of uranium acetate, and 6 grams of acetic acid in a volume of 1 liter; used to test for uric acid. { 'fō·lən sə,lü·shən }

folpet [ORG CHEM] $C_9H_4Cl_3NO_2S$ A buff or white, crystalline compound with a melting point of 177–178°C; insoluble in water; used as a fungicide on fruits, vegetables, and ornamental flowers. { 'fäl·pet }

foot's oil [CHEM] The oil sweated out of slack wax; it takes its name from the fact that it goes to the bottom, or foot, of the pan when sweated. { 'fúts ,òil }

force constant [PHYS CHEM] An expression for the force acting to restrain the relative displacement of the nuclei in a molecule. { 'fòrs ,kän·stənt }

forced-flow boiling [PHYS CHEM] Boiling of a liquid whose flow over a heater surface is imposed by external means. { ¦fòrst ¦flō 'bòil·iŋ }

force field method *See* molecular mechanics. { 'fòrs ¦fēld ,meth·əd }

formal charge [PHYS CHEM] The apparent charge of an element in a compound; for example, magnesium has a formal charge of +2 in MgO and oxygen has a charge of −2. { ¦fòr·məl ¦chärj }

formaldehyde [ORG CHEM] HCHO The simplest aldehyde; a gas at room temperature, and a poisonous, clear, colorless liquid solution with pungent odor; used to make synthetic resins by reaction with phenols, urea, and melamine, as a chemical interme-diate, as an embalming fluid, and as a disinfectant.   Also known as formol; methanal; methylene oxide.   { fȯr′mal·də‚hīd }

formaldehyde sodium bisulfite   [ORG CHEM] CH_3NaO_4S A compound used as a fixing agent for fibers containing keratin, in metallurgy for flotation of lead-zinc ores, and in photography.   { fȯr′mal·də‚hīd ¦sōd·ē·əm bī′səl‚fīt }

formality   [CHEM] A concentration scale that gives the number of formula weights of solute per liter of solution; designated by F preceded by a number to show solute concentration.   { fȯr′mal·əd·ē }

formamide   [ORG CHEM] **1.** A compound containing the radical HCONH. **2.** $HCONH_2$ A clear, colorless hygroscopic liquid, boiling at 200–212°C; soluble in water and alcohol; used as a solvent, softener, and chemical intermediate.   Also known as formylamine; methanamide.   { fȯrm′am·əd }

formamidinesulfinic acid   [ORG CHEM] $H_2NC(NH)SO_2H$ A reagent for the reduction of ketones to secondary alcohols.   { fȯr¦mam·ə‚dēn·səl′fin·ik ′as·əd }

formate   [ORG CHEM] A compound containing the HCOO−functional group.   { ′fȯr‚māt }

formic acid   [ORG CHEM] HCOOH A colorless, pungent, toxic, corrosive liquid melting at 8.4°C; soluble in water, ether, and alcohol; used as a chemical intermediate and solvent, in dyeing and electroplating processes, and in fumigants.   Also known as methanoic acid.   { ¦fȯr·mik ′as·əd }

formic ether See ethyl formate.   { ′fȯr·mik ′ē·thər }

formol See formaldehyde.   { ′fȯr‚mȯl }

formonitrile See hydrocyanic acid.   { ¦fȯr·mō¦nī·trəl }

formula   [CHEM] **1.** A combination of chemical symbols that expresses a molecule's composition. **2.** A reaction formula showing the interrelationship between reactants and products.   { ′fȯr·myə·lə }

formulation   [CHEM] The particular mixture of base chemicals and additives required for a product.   { ‚fȯr·myə′lā·shən }

formula weight   [CHEM] **1.** The gram-molecular weight of a substance. **2.** In the case of a substance of uncertain molecular weight such as certain proteins, the molecular weight calculated from the composition, assuming that the element present in the smallest proportion is represented by only one atom.   { ′fȯr·myə·lə ‚wāt }

formyl   [ORG CHEM] The formic acid radical, HCO−; it is characteristic of aldehydes.   { ′fȯr‚mil }

formylamine See formamide.   { ‚fȯr·məl′am‚ēn }

Fortrat parabola   [SPECT] Graph of wave numbers of lines in a molecular spectral band versus the serial number of the successive lines.   { ‚fȯrträ pə′rab·ə·lə }

Foulger's test   [ANALY CHEM] A test for fructose in which urea, sulfuric acid, and stan-nous chloride are added to the solution to be tested, the solution is boiled, and in the presence of fructose a blue coloration forms.   { fül‚jāz ‚test }

four-degree calorie   [CHEM] The heat needed to change the temperature of 1 gram of water from 3.5 to 4.5°C.   { ′fȯr di¦grē ′kal·ə·rē }

Fourier transform spectroscopy   [SPECT] A spectroscopic technique in which all perti-nent wavelengths simultaneously irradiate the sample for a short period of time, and the absorption spectrum is found by mathematical manipulation of the Fourier transform so obtained.   { ¦für·ē¦ā ′tranz‚fȯrm spek′träs·kə·pē }

fp See freezing point.

Fr See francium.

fraction   [CHEM] One of the portions of a volatile liquid within certain boiling point ranges, such as petroleum naphtha fractions or gas-oil fractions.   { ′frak·shən }

fractional condensation   [CHEM] Separation of components of vaporized liquid mix-tures by condensing the vapors in stages (partial condensation); highest-boiling-point components condense in the first condenser stage, allowing the remainder of the vapor to pass on to subsequent condenser stages.   { ¦frak·shən·əl ‚kän·den′sā·shən }

fractional distillation [CHEM] A method to separate a mixture of several volatile components of different boiling points; the mixture is distilled at the lowest boiling point, and the distillate is collected as one fraction until the temperature of the vapor rises, showing that the next higher boiling component of the mixture is beginning to distill; this component is then collected as a separate fraction. { ¦frak·shən·əl dis·tə'lā·shən }

fractional precipitation [ANALY CHEM] Method for separating elements or compounds with similar solubilities by a series of analytical precipitations, each one improving the purity of the desired element. { ¦frak·shən·əl prə,sip·ə'tā·shən }

fractionating column [CHEM] An apparatus used widely for separation of fluid (gaseous or liquid) components by vapor-liquid fractionation or liquid-liquid extraction or liquid-solid adsorption. { 'frak·shə,nād·iŋ ,käl·əm }

fractionation [CHEM] Separation of a mixture in successive stages, each stage removing from the mixture some proportion of one of the substances, as by differential solubility in water-solvent mixtures. { ,frak·shə'nā·shən }

francium [CHEM] A radioactive alkali-metal element, symbol Fr, atomic number 87, atomic weight distinguished by nuclear instability; exists in short-lived radioactive forms, the chief isotope being francium-223. { 'fran·sē·əm }

Franck-Condon principle [PHYS CHEM] The principle that in any molecular system the transition from one energy state to another is so rapid that the nuclei of the atoms involved can be considered to be stationary during the transition. { ¦fräŋk 'kän·dən ,prin·sə·pəl }

Franck-Rabinowitch hypothesis [PHYS CHEM] The hypothesis that the decreased quantum efficiencies of certain photochemical reactions observed in the dissolved or liquid state are due to the formation of a cage of solvent molecules around the molecule which has been excited by absorption of a photon. { ¦fräŋk rə'bin·ə,wich hī,päth·ə·səs }

frangula emodin See emodin. { 'fraŋ·gyə·lə 'em·ə·dən }

frangulic acid See emodin. { fraŋ'gyü·lik 'as·əd }

Frankland's method [ORG CHEM] Reaction of dialkyl zinc compounds with alkyl halides to form hydrocarbons; may be used to form paraffins containing a quaternary carbon atom. { 'fraŋk·lənz ,meth·əd }

Fraude's reagent See perchloric acid. { 'frōdz rē,ā·jənt }

fraunhofer [SPECT] A unit for measurement of the reduced width of a spectrum line such that a spectrum line's reduced width in fraunhofers equals 10^6 times its equivalent width divided by its wavelength. { 'fraún,hōf·ər }

Fraunhofer lines [SPECT] The dark lines constituting the Fraunhofer spectrum. { 'fraún,hōf·ər ,līnz }

Fraunhofer spectrum [SPECT] The absorption lines in sunlight, due to the cooler outer layers of the sun's atmosphere. { 'fraún,hōf·ər ,spek·trəm }

freeboard [ANALY CHEM] The space provided above the resin bed in an ion-exchange column to allow for expansion of the bed during backwashing. { ¦frē,bórd }

free cyanide [CHEM] Cyanide not combined as part of an ionic complex. { ¦frē 'sī·ə,nīd }

free induction decay [SPECT] A type of electron paramagnetic resonance spectroscopy in which a material is exposed to a short high-power pulse (as short as 2 nanoseconds) of microwave radiation, and the response of the material is Fourier transformed into the normal spectrum. Abbreviated FID. { ¦frē in'dək·shən di,kā }

free ion [PHYS CHEM] An ion, such as found in an ionized gas, whose properties, such as spectrum and magnetic moment, are not significantly affected by other atoms, ions, or molecules nearby. { ¦frē 'ī,än }

free molecule [PHYS CHEM] A molecule, as in a gas, whose properties, such as spectrum and magnetic moment, are not affected by other atoms, ions, and molecules nearby. { ¦frē 'mäl·ə,kyül }

free radical [CHEM] An atom or a diatomic or polyatomic molecule which possesses one unpaired electron. Also known as a radical. { ¦frē 'rad·ə·kəl }

free-radical reaction See homolytic cleavage. { ¦frē ¦rad·ə·kəl rē'ak·shən }

free water [CHEM] The volume of water that is not contained in suspension in a vessel containing both water and a suspension of water and another liquid. { 'frē 'wȯd·ər }

freeze [PHYS CHEM] To solidify a liquid by removal of heat. { frēz }

freezing mixture [PHYS CHEM] A mixture of substances whose freezing point is lower than that of its constituents. { 'frēz·iŋ ‚miks·chər }

freezing point [PHYS CHEM] The temperature at which a liquid and a solid may be in equilibrium. Abbreviated fp. { 'frēz·iŋ ‚pȯint }

freezing-point depression [PHYS CHEM] The lowering of the freezing point of a solution compared to the pure solvent; the depression is proportional to the active mass of the solute in a given amount of solvent. { 'frēz·iŋ ‚pȯint di‚presh·ən }

frequency factor [PHYS CHEM] The constant A (or v) in the Arrhenius equation, which is the relation between reaction rate and absolute temperature T; the equation is $k = A e^{-(\Delta H_{act}/RT)}$, where k is the specific rate constant, ΔH_{act} is the heat of activation, and R is the gas constant. { 'frē·kwən·sē ‚fak·tər }

Freund method [ORG CHEM] A method for preparation of cycloparaffins in which dihalo derivatives of the paraffins are treated with zinc to produce the cycloparaffin. { 'frȯind ‚meth·əd }

Friedel-Crafts reaction [ORG CHEM] A substitution reaction, catalyzed by aluminum chloride in which an alkyl (R—) or an acyl (RCO—) group replaces a hydrogen atom of an aromatic nucleus to produce hydrocarbon or a ketone. { frē|del 'krafs rē‚ak·shən }

Friedlander synthesis [ORG CHEM] A synthesis of quinolines; the method is usually catalyzed by bases and consists of condensation of an aromatic o-amino-carbonyl derivative with a compound containing a methylene group in the alpha position to the carbonyl. { 'frēd‚lan·dər ‚sin·thə·səs }

Fries rearrangement [ORG CHEM] The conversion of a phenolic ester into the corresponding o- and p-hydroxyketone by treatment with catalysts of the type of aluminum chloride. { 'frēz rē·ə'rānj·mənt }

Fries' rule [ORG CHEM] The rule that the most stable form of the bonds of a polynuclear compound is that arrangement which has the maximum number of rings in the benzenoid form, that is, three double bonds in each ring. { 'frēz ‚rül }

frontier orbitals [PHYS CHEM] Orbitals of two molecules that are spatially arranged so that a significant amount of overlap occurs between them. { frən|tir 'ȯr·bə·təlz }

frother [CHEM] Substance used in flotation processes to make air bubbles sufficiently permanent, principally by reducing surface tension. { 'frȯ·thər }

froth promoter [CHEM] A chemical compound used with a frothing agent. { 'frȯth prə‚mōd·ər }

frustrated internal reflectance See attenuated total reflectance. { 'frəs‚trād·əd in‚tərn·əl ri'flek·təns }

fuchsin [ORG CHEM] $C_{20}H_{19}N_3$ Brownish-red crystals, used as a dye or in the commercial preparation of other dyes, and as an antifungal drug. Also known as magenta; rosaniline. { 'fyük·sən }

fuel cell [PHYS CHEM] An electrochemical device in which the reaction between a fuel, such as hydrogen, and an oxidant, such as oxygen or air, converts the chemical energy of the fuel directly into electrical energy without combustion. { 'fyül ‚sel }

fuel-cell catalyst [CHEM] A substance, such as platinum, silver, or nickel, from which the electrodes of a fuel cell are made, and which speeds the reaction of the cell; it is especially important in a fuel cell which does not operate at high temperatures. { 'fyül ‚sel 'kad·ə‚list }

fuel-cell electrolyte [CHEM] The substance which conducts electricity between the electrodes of a fuel cell. { 'fyül ‚sel i'lek·trə‚līt }

fuel-cell fuel [CHEM] A substance, such as hydrogen, carbon monoxide, sodium, alcohol, or a hydrocarbon, which reacts with oxygen to generate energy in a fuel cell. { 'fyül ‚sel 'fyül }

fugitive dye [CHEM] A dye that is unstable, that is, not fast; used in the textile processing for purposes of identity. { |fyü·jəd·iv 'dī }

160

Fulcher bands |SPECT| A group of bands in the spectrum of molecular hydrogen that are preferentially excited by a low-voltage discharge.   { 'fəl·chər ˌbanz }

fullerene |CHEM| A large molecule composed entirely of carbon, with the chemical formula C_n, where n is any even number from 32 to over 100; believed to have the structure of a hollow spheroidal cage with a surface network of carbon atoms connected in hexagonal and pentagonal rings.   { 'fúl·ə₁rēn }

fully developed nucleate boiling |PHYS CHEM| A stage in the boiling process in which vapor bubbles from neighboring sites on the heater surface merge, and the vapor appears to leave the heater in the form of jets, so that the vapor structures resemble mushrooms with multiple stems.   { ¦fúl·ē di'vel·əpt ¦nük·lē₁āt 'bòil·iŋ }

fulminate |ORG CHEM| **1.** A salt of fulminic acid. **2.** $HgC_2N_2O_2$ An explosive mercury compound derived from the fulminic acid; used for the caps or exploders by means of which charges of gunpowder, dynamite, and other explosives are fired. Also known as mercury fulminate.   { 'fúl·mə₁nāt }

fulminic acid |ORG CHEM| CNOH An unstable isomer of cyanic acid, whose salts are known for their explosive characteristics.   { fúl'min·ik 'as·əd }

fulminuric acid |ORG CHEM| $CN·CH(NO_2)·CONH_2$ A trimer of cyanuric acid; a water-soluble compound, crystallizing in colorless needles, melting at 138°C, and exploding at 145°C. Also known as isocyanuric acid.   { ¦fúl·mə¦núr·ik 'as·əd }

fulvene |ORG CHEM| C_6H_6 A yellow oil, an isomer of benzene.   { 'fúl·vēn }

fumaric acid |ORG CHEM| $C_4H_4O_4$ A dicarboxylic organic acid produced commercially by synthesis and fermentation; the trans isomer of maleic acid; colorless crystals, melting point 287°C; used to make resins, paints, varnishes, and inks, in foods, as a mordant, and as a chemical intermediate. Also known as boletic acid.   { fyü'mar·ik 'as·əd }

fume hood |CHEM| A fume-collection device over an enclosed shelf or table, so that experiments involving poisonous or unpleasant fumes or gases may be conducted away from the experimental area.   { 'fyüm ˌhúd }

fumes |CHEM| Particulate matter consisting of the solid particles generated by condensation from the gaseous state, generally after volatilization from melted substances, and often accompanied by a chemical reaction, such as oxidation.   { fyümz }

fumigant |CHEM| A chemical compound which acts in the gaseous state to destroy insects and their larvae and other pests; examples are dichlorethyl ether, p-dichlorobenzene, and ethylene oxide.   { 'fyü·mə·gənt }

fuming nitric acid |INORG CHEM| Concentrated nitric acid containing dissolved nitrogen dioxide; may be prepared by adding formaldehyde to concentrated nitric acid.   { ¦fyüm·iŋ 'nī₁trik ˌas·əd }

fuming sulfuric acid |INORG CHEM| Concentrated sulfuric acid containing dissolved sulfur trioxide. Also known as oleum.   { ¦fyüm·iŋ səl'fyúr·ik ˌas·əd }

functional group |ORG CHEM| An atom or group of atoms, acting as a unit, that has replaced a hydrogen atom in a hydrocarbon molecule and whose presence imparts characteristic properties to this molecule; frequently represented as R−. Also known as functionality.   { ¦fəŋk·shən·əl 'grüp }

functionality See functional group.   { ˌfəŋk·shə'nal·əd·ē }

fundamental series |SPECT| A series occurring in the line spectra of many atoms and ions having one, two, or three electrons in the outer shell, in which the total orbital angular momentum quantum number changes from 3 to 2.   { ¦fən·də¦ment·əl 'sir·ēz }

funicular distribution |CHEM| The distribution of a two-phase, immiscible liquid mixture (such as oil and water, one a wetting phase, the other nonwetting) in a porous system when the wetting phase is continuous over the surface of the solids.   { fə'nik·yə·lər dis·trə'byü·shən }

2-furaldehyde See furfural.   { ¦tü fə'ral·də₁hīd }

furan |ORG CHEM| **1.** One of a group of organic heterocyclic compounds containing a diunsaturated ring of four carbon atoms and one oxygen atom. **2.** $C_4H_4O_2$ The simplest furan type of molecule; a colorless, mildly toxic liquid, boiling at 32°C, insoluble in water, soluble in alcohol and ether; used as a chemical intermediate. Also known as furfuran; tetrol.   { 'fyúr₁an }

furancarboxylic acid

furancarboxylic acid *See* furoic acid. { ¦fyür·ən¸kar¸bäk'sil·ik 'as·əd }

2,5-furandione *See* maleic anhydride. { ¦tü ¦fīv ¦fyür·ən'dī¸ōn }

furanoside [ORG CHEM] A glycoside whose cyclic sugar component resembles that of furan. { fyə'ran·ə¸sīd }

furan resin [ORG CHEM] A liquid, thermosetting resin in which the furan ring is an integral part of the polymer chain, made by the condensation of furfuryl alcohol; used as a cement and adhesive, casting resin, coating, and impregnant. { 'fyür¸an ¸rez·ən }

furfural [ORG CHEM] C_4H_3OCHO When pure, a colorless liquid, soluble in organic solvents, slightly soluble in water; used as a lube oil-refining solvent, in cellulosic formulations, in making resins, as a weed killer, as a fungicide, and as a chemical intermediate. Also known as 2-furaldehyde; furfuraldehyde; furfurol; furol. { 'fər·fə¸ral }

furfuraldehyde *See* furfural. { 'fər·fə'ral·də¸hīd }

furfuran *See* furan. { 'fər·fə¸ran }

furfurol *See* furfural. { 'fər·fə¸rōl }

furfuryl [ORG CHEM] The functional group C_5H_6O- from furfural. { 'fər·fə¸ril }

furfuryl alcohol [ORG CHEM] $C_5H_6O_2$ A liquid with a faint burning odor and bitter taste, soluble in alcohol and ether, usually prepared from furfural; used as a solvent in the manufacturing of wetting agents and resins. { 'fər·fə¸ril 'al·kə¸hȯl }

furnace black [CHEM] A carbon black formed by partial combustion of liquid and gaseous hydrocarbons in a closed furnace with a deficiency of oxygen; used as a reinforcing filler for synthetic rubber. { 'fər·nəs ¸blak }

furoic acid [ORG CHEM] $C_5H_4O_3$ Long monoclinic prisms crystallized from the water solution, soluble in ether and alcohol; used as a preservative and bactericide. Also known as furancarboxylic acid; pyromucic acid. { fyü'rō·ik 'as·əd }

furol *See* furfural. { 'fyü¸rōl }

fused aromatic ring [ORG CHEM] A molecular structure in which two or more aromatic rings have two carbon atoms in common. { ¦fyüzd ar·ə¦mad·ik 'riŋ }

fused potassium sulfide *See* potassium sulfide. { 'fyüzd pə'tas·ē·əm 'səl¸fīd }

fused-salt electrolysis [PHYS CHEM] Electrolysis with use of purified fused salts as raw material and as an electrolyte. { ¦fyüzd ¦sȯlt i¸lek'trä·lə·səs }

fusion [PHYS CHEM] A change of the state of a substance from the solid phase to the liquid phase. Also known as melting. { 'fyü·zhən }

fusion tube [ANALY CHEM] Device used for the analysis of the elements in a compound by fusing them with another compound; for example, analysis of nitrogen in organic compounds by fusing the compound with sodium and analyzing for sodium cyanide. { 'fyü·zhən ¸tüb }

G

Ga *See* gallium.

GABA *See* γ-aminobutyric acid. { 'ga·bə *or* ¦jē,ä,bē'ā }

Gabriel's synthesis [ORG CHEM] A synthesis of primary amines by the hydrolysis of N-alkylphthalimides; the latter are obtained from potassium phthalimide and alkyl halides. { 'gā·brē·əlz ¸sin·thə·səs }

gadoleic acid [ORG CHEM] $C_{20}H_{38}O_2$ A fatty acid derived from cod liver oil, and melting at 20°C. { ¦gad·ə¦lē·ik 'as·əd }

gadolinium [CHEM] A rare-earth element, symbol Gd, atomic number 64, atomic weight 157.25; highly magnetic, especially at low temperatures. { ¸gad·əl'in·ē·əm }

galactaric acid *See* mucic acid. { ¦gal·ak¦tar·ik 'as·əd }

galipol [ORG CHEM] $C_{15}H_{26}O$ A terpene alcohol derived from the oil of the angostura bark; colorless crystals that melt at 89°C. { 'gal·ə¸pȯl }

gallacetophenone [ORG CHEM] $C_8H_8O_4$ A white to brownish-gray, crystalline powder, melting at 173°C, soluble in water, alcohol, and ether; used as an antiseptic. { ¦gȯl¦as·ə·tä·fə'nōn }

gallein [ORG CHEM] $C_{20}H_{10}O_7$ A brown powder or green scales, broken down by heat; used as a pH indicator in the analysis of phosphates in urine and as an intermediate in the manufacture of dyes. Also known as anthracene violet; gallin; pyrogallolphthalein. { 'gal·ē·ən }

gallic acid [ORG CHEM] $C_7H_6O_5$ A crystalline compound that forms needles from solutions of absolute methanol or chloroform, dissolves in water, alcohol, ether, and glycerol; obtained from nutgall tannins or from *Penicillium notatum* fermentation; used to make antioxidants and ink dyes and in photography. { 'gal·ik 'as·əd }

gallin *See* gallein. { 'gal·ən }

gallium [CHEM] A chemical element, symbol Ga, atomic number 31, atomic weight 69.72. { 'gal·ē·əm }

gallium arsenide [INORG CHEM] GaAs A crystalline material, melting point 1238°C; frequently alloys of this material are formed with gallium phosphide or indium arsenide. { 'gal·ē·əm 'ärs·ən¸īd }

gallium halide [INORG CHEM] A compound formed by bonding of gallium to either chlorine, bromine, iodine, fluorine, or astatine. { 'gal·ē·əm 'ha¸līd }

gallium phosphide [INORG CHEM] GaP Transparent crystals made by reacting phosphorus and gallium suboxide at low temperature. { 'gal·ē·əm 'fäs¸fīd }

gallocyanine [ORG CHEM] $C_{15}H_{13}ClN_2O_5$ Green crystals soluble in alcohol, glacial acetic acid, alkali carbonates, and concentrated hydrochloric acid; used as a dye and as a reagent for the determination of lead. { ¸ga·lō'sī·ə¸nēn }

gallogen *See* ellagic acid. { 'gal·ə·jən }

gallotannic acid *See* tannic acid. { ¦ga·lō¦tan·ik 'as·əd }

gallotannin *See* tannic acid. { ¦ga·lō¦tan·ən }

galvanic series [CHEM] The relative hierarchy of metals arranged in order from magnesium (least noble) at the anodic, corroded end through platinum (most noble) at the cathodic, protected end. { gal'van·ik 'sir·ēz }

gamma [CHEM] The gamma position (the third carbon atom in an aliphatic carbon chain) on a chemical compound. { 'gam·ə }

gamma acid [ORG CHEM] $C_{10}H_5NH_2OHSO_3H$ White crystals, slightly soluble in water;

an intermediate in dyestuff manufacture.   Also known as 2-amino-8-naphthol-6-sulfonic acid; 7-amino-1-naphthol-3-sulfonic acid; 2,5-naphthylamine sulfonic acid; 3-sulfonic acid; 6-sulfonic acid.   { 'gam·ə 'as·əd }

gamma-ray spectrum [SPECT] The set of wavelengths or energies of gamma rays emitted by a given source.   { 'gam·ə ,rā ,spek·trəm }

gammil [CHEM] A unit of concentration, equal to a concentration of 1 milligram of solute in 1 liter of solvent.   Also known as micril; microgammil.   { 'gam·əl }

gas adsorption [PHYS CHEM] The concentration of a gas upon the surface of a solid substance by attractive forces between the surface and the gas molecules.   { ¦gas ad'sorp·shən }

gas analysis [ANALY CHEM] Analysis of the constituents or properties of a gas (either pure or mixed); composition can be measured by chemical adsorption, combustion, electrochemical cells, indicator papers, chromatography, mass spectroscopy, and so on; properties analyzed for include heating value, molecular weight, density, and viscosity.   { 'gas ə,nal·ə·səs }

gas black [CHEM] Fine particles of carbon formed by partial combustion or thermal decomposition of natural gas; used to reinforce rubber products such as tires.   Also known as carbon black; channel black.   { 'gas ,blak }

gas chromatograph [ANALY CHEM] The instrument used in gas chromatography to detect volatile compounds present; also used to determine certain physical properties such as distribution or partition coefficients and adsorption isotherms, and as a preparative technique for isolating pure components or certain fractions from complex mixtures.   { ¦gas krō'mad·ə,graf }

gas chromatography [ANALY CHEM] A separation technique involving passage of a gaseous moving phase through a column containing a fixed adsorbent phase; it is used principally as a quantitative analytical technique for volatile compounds.   { ¦gas ,krō·mə'täg·rə·fē }

gas-condensate liquid [ORG CHEM] A hydrocarbon, such as propane, butane, and pentane, obtained as condensate when wet natural gas is compressed or refrigerated.   { ¦gas 'känd·ən,sāt ,lik·wəd }

gas generator [CHEM] A device used to generate gases in the laboratory.   { 'gas ,jen·ə,rād·ər }

gas-liquid chromatography [ANALY CHEM] A form of gas chromatography in which the fixed phase (column packing) is a liquid solvent distributed on an inert solid support.   Abbreviated GLC.   Also known as gas-liquid partition chromatography.   { 'gas ,lik·wəd ,krō·mə'täg·rə·fē }

gas-liquid partition chromatography See gas-liquid chromatography.   { 'gas ,lik·wəd pär'tish·ən ,krō·mə'täg·rə·fē }

gasometric method [ANALY CHEM] An analytical technique for gases; the gas may be measured by instrumental methods or through chemical reactions with specific reagents.   { ,gas·ə¦me·trik 'meth·əd }

gas-solid chromatography [ANALY CHEM] A form of gas chromatography in which the moving phase is a gas and the stationary phase is a surface-active sorbent (charcoal, silica gel, or activated alumina).   Abbreviated GSC.   { ¦gas ¦säl·əd ,krō·mə'täg·rə·fē }

gas solubility [PHYS CHEM] The extent that a gas dissolves in a liquid to produce a homogeneous system.   { ¦gas ,säl·yə'bil·əd·ē }

Gatterman-Koch synthesis [ORG CHEM] A synthesis of aldehydes; aldehydes form when an aromatic hydrocarbon is heated in the presence of hydrogen chloride, certain metallic chloride catalysts, and either carbon monoxide or hydrogen cyanide.   { ¦gäd·ər·män 'kȯk ,sin·thə·səs }

Gatterman reaction [ORG CHEM] **1.** Reaction of a phenol or phenol ester, and hydrogen chloride or hydrogen cyanide, in the presence of a metallic chloride such as aluminum chloride to form, after hydrolysis, an aldehyde. **2.** Reaction of an aqueous ethanolic solution of diazonium salts with precipitated copper powder or other reducing agent to form diaryl compounds.   { 'gäd·ər·män rē,ak·shən }

gaultheria oil See methyl salicylate.   { gȯl'thir·ē·ə ,ȯil }

geraniol

Gay-Lussac's law of volumes See combining-volumes principle. { ¸gäˈlüˌsäks ¦lȯ əv ˈvälˌyəmz }

Gd See gadolinium.

Ge See germanium.

gel [CHEM] A two-phase colloidal system consisting of a solid and a liquid in more solid form than a sol. { jel }

gelatin [ORG CHEM] A protein derived from the skin, white connective tissue, and bones of animals; used as a food and in photography, the plastics industry, metallurgy, and pharmaceuticals. { ˈjelˈətˈən }

gelation [CHEM] **1.** The act or process of freezing. **2.** Formation of a gel from a sol. { jəˈläˈshən }

gel electrophoresis [CHEM] Electrophoresis performed in silica gel, which is a porous, inert medium. { ¦jel iˌlekˈtrōˈfəˈrēˈsəs }

gel filtration [ANALY CHEM] A type of column chromatography which separates molecules on the basis of size; higher-molecular-weight substances pass through the column first. Also known as molecular exclusion chromatography; molecular sieve chromatography. { ¦jel filˈträˈshən }

gel permeation chromatography [ANALY CHEM] Analysis by chromatography in which the stationary phase consists of beads of porous polymeric material such as a cross-linked dextran carbohydrate derivative sold under the trade name Sephadex; the moving phase is a liquid. { ¦jel ˌpərˈmēˈäˈshən ˌkrōˈməˈtägˈrəˈfē }

gel point [PHYS CHEM] Stage at which a liquid begins to exhibit elastic properties and increased viscosity. { ¦jel ˌpȯint }

geminal [ORG CHEM] Referring to like atoms or groups attached to the same atom in a molecule. { ˈjemˈəˈnəl }

general formula [CHEM] A formula that can apply not only to one specific compound but to a series of related compounds; for example, the general formula for an aldehyde RCHO, where R is hydrogen in formaldehyde (the simplest aldehyde) and is a hydrocarbon radical for other aldehydes in the series such as CH_3 for acetaldehyde and C_2H_5 for proprionaldehyde. { ¦jenˈrəl ˈfȯrˈmyəˈlə }

Geneva system [ORG CHEM] An international system of nomenclature for organic compounds based on hydrocarbon derivatives; names correspond to the longest straight carbon chain in the molecule. { jə¦nēˈvə ˈsisˈtəm }

genicide [ORG CHEM] $C_{13}H_8O_2$ A compound with needlelike crystals and a melting point of 174°C; insoluble in water; used as an insecticide, miticide, and ovicide. Also known as oxoxanthone; 9-xanthenone; xanthone. { ˈjenˈəˌsīd }

genistin [ORG CHEM] $C_{21}H_{20}O_{10}$ A pale-yellow glucoside derived from soybean meal, crystallizes from 80% methanol solution, melting point 256°C, soluble in hot 80% ethanol, hot 80% methanol, and hot acetone. Also known as 7-D-glucoside. { jəˈnisˈtən }

gentianic acid See gentisic acid. { ¦jenˈchē¦anˈik ˈasˈəd }

gentian violet See methyl violet. { ˈjenˈchən ˈvīˈlət }

gentisic acid [ORG CHEM] $C_7H_6O_4$ A crystalline compound that forms monoclinic prisms from a water solution, sublimes at 200°C, melts at 250°C, and is soluble in water, alcohol, ether, sodium, and salt; used in medicine. Also known as gentianic acid. { jenˈtisˈik ˈasˈəd }

geometrical isomerism [PHYS CHEM] The phenomenon in which isomers contain atoms attached to each other in the same order and with the same bonds but with different spatial, or geometrical, relationships; the explicit geometry imposed upon a molecule by, say, a double bond between carbon atoms makes possible the existence of these isomers. { ¦jēˈəˌmeˈtrəˈkəl ītˈsäˈməˌrizˈəm }

geranial See citral. { jəˈräˈnēˈəl }

geranialdehyde See citral. { jəˌräˈnēˈalˈdəˌhīd }

geraniol [ORG CHEM] $(CH_3)_2CCH(CH_2)_2C(CH_3)CHCH_2OH$ A colorless to pale-yellow liquid, an alcohol and a terpene, boiling point 230°C; soluble in alcohol and ether, insoluble in water; used in perfumery and flavoring. { jəˈräˈnēˌȯl }

165

geranyl

geranyl |ORG CHEM| $C_{10}H_{17}$ The functional group from geraniol, $(CH_3)_2:CHCH_2CH_2 \cdot CHCH_3:CH \cdot CH_2OH$. { jə'ran·əl }

Gerard reagent |CHEM| The quaternary ammonium compounds, acethydrazide-pyridinium chloride and trimethylacethydrazide ammonium chloride; used to separate aldehydes and ketones from oily or fatty natural materials and to extract sex hormones from urine. { jə'rärd rē,ā·jənt }

germane |INORG CHEM| **1.** A hydride of germanium whose general formula is Ge_nH_{2n+2}. **2.** The compound GeH_4, a hydride of germanium, a colorless gas that is combustible in air and burns with a blue flame. { ¦jər¦mān }

germanide |INORG CHEM| A compound of an alkaline earth or alkali metal with germanium; an example is magnesium germanide, Mg_2Ge; the germanides are reactive with water. { 'jər·mə,nīd }

germanium |CHEM| A brittle, water-insoluble, silvery-gray metallic element in the carbon family, symbol Ge, atomic number 32, atomic weight 72.59, melting at 959°C. { jər'mān·ē·əm }

germanium halide |INORG CHEM| A dihalide or tetrahalide of fluorine, chlorine, bromine, or iodine with germanium. { jər'mān·ē·əm 'ha,līd }

germanium oxide |INORG CHEM| The monoxide GeO or dioxide GeO_2; a study of GeO indicates it exists in polymeric form; GeO_2 is a white powder, soluble in alkalies; used in special glass and in medicine. { jər'mān·ē·əm 'äk,sīd }

getter |CHEM| *See* scavenger. |PHYS CHEM| **1.** A substance, such as thallium, that binds gases on its surface and is used to maintain a high vacuum in a vacuum tube. **2.** A special metal alloy that is placed in a vacuum tube during manufacture and vaporized after the tube has been evacuated; when the vaporized metal condenses, it absorbs residual gases. Also known as degasser. { 'ged·ər }

ghost image |SPECT| A false image of a spectral line produced by irregularities in the ruling of a diffraction grating. { 'gōst ,im·ij }

Gibbs adsorption equation |PHYS CHEM| A formula for a system involving a solvent and a solute, according to which there is an excess surface concentration of solute if the solute decreases the surface tension, and a deficient surface concentration of solute if the solute increases the surface tension. { 'gibz ad'sȯrp·shən i,kwā·zhən }

Gibbs adsorption isotherm |PHYS CHEM| An equation for the surface pressure of surface monolayers, $\phi = RT \int_0^p \Gamma d(\ln p)$, where ϕ is surface pressure, T is absolute temperature, R is the gas constant, Γ is the number of molecules adsorbed per gram per unit surface area, and p is the pressure of the gas. { 'gibz ad'sȯrp·shən 'ī·sō,thərm }

Gibbs-Donnan equilibrium *See* Donnan equilibrium. { ¦gibz 'dän·ən ē·kwə'lib·rē·əm }

Gibbs-Duhem equation |PHYS CHEM| A relation that imposes a condition on the composition variation of the set of chemical potentials of a system of two or more components, $SdT - VdP + \sum_{i=1}^{r} n_i d\mu_i = 0$, where S is entropy, T absolute temperature, P pressure, n_i the number of moles of the ith component, and μ_i is the chemical potential of the ith component. Also known as Duhem's equation. { ¦gibz 'dü·əm i,kwā·zhən }

Gibbs-Helmholtz equation |PHYS CHEM| An expression for the influence of temperature upon the equilibrium constant of a chemical reaction, $(d \ln K^0/dT)_P = \Delta H^0/RT^2$, where K^0 is the equilibrium constant, ΔH^0 the standard heat of the reaction at the absolute temperature T, and R the gas constant. { 'gibz 'helm,hōlts i,kwā·zhən }

Gibbs phase rule |PHYS CHEM| A relationship used to determine the number of state variables F, usually chosen from among temperature, pressure, and species composition in each phase, which must be specified to fix the thermodynamic state of a system in equilibrium: $F = C - P - M + 2$, where C is the number of chemical species presented at equilibrium, P is the number of phases, and M is the number of independent chemical reactions. Also known as Gibbs rule; phase rule. { 'gibz 'fāz ,rül }

Gibbs-Poynting equation |PHYS CHEM| An expression relating the effect of the total applied pressure P upon the vapor pressure p of a liquid, $(dp/dP):yT = V_l/V_g$, where V_l and V_g are molar volumes of the liquid and vapor. { ¦gibz 'pȯint·iŋ i,kwä·zhən }

Gibbs rule See Gibbs phase rule. { 'gibz ,rül }

Giemsa stain |CHEM| A stain for hemopoietic tissue and hemoprotozoa consisting of a stock glycerol methanol solution of eosinates of Azure B and methylene blue with some excess of the basic dyes. { 'gēm·sə ,stān }

gifblaar poison See fluoroacetic acid. { 'gif,blär ,pȯiz·ən }

Gillespie equilibrium still |ANALY CHEM| A recirculating equilibrium distillation apparatus used to establish azeotropic properties of liquid mixtures. { gə'les·pē ,ē·kwə 'librē·əm ,stil }

gitonin |ORG CHEM| The gitogenin tetraglycoside in Digitalis purpurea seed; resembles digitonin. { jə'tōn·ən }

glacial acetic acid |ORG CHEM| CH_3COOH Pure acetic acid (containing less than 1% water); a clear, colorless, caustic hygroscopic liquid, boiling at 118°C, soluble in water, alcohol, and ether, and crystallizing readily; used as a solvent for oils and resins. { ¦glā·shəl ə¦sēd·ik 'as·əd }

glass electrode |PHYS CHEM| An electrode or half cell in which potential measurements are made through a glass membrane, which acts as a cation-exchange membrane; thus, the potential arises from phase-boundary and diffusion potentials which, depending on the composition of the glass, are logarithmic functions of the activity of the cations such as H^+, Na^+, or K^+ of the solutions in which the electrode is immersed. { ¦glas i'lek,trōd }

glass transition |PHYS CHEM| The change in an amorphous region of a partially crystalline polymer from a viscous or rubbery condition to a hard and relatively brittle one; usually brought about by changing the temperature. Also known as gamma transition; glassy transition. { 'glas tran,zish·ən }

glass transition temperature |PHYS CHEM| The temperature at which a liquid changes to an amorphous or glassy solid. { ¦glas ,tran'zish·ən ,tem·prə·chər }

Glauber's salt |INORG CHEM| $Na_2SO_4·10H_2O$ Crystalline hydrated sodium sulfate; loses water when exposed to air; water soluble, alcohol insoluble; used in textile dyeing and medicine. { 'glau̇·bərz ,sȯlt }

glaze stain |INORG CHEM| Colorant for ceramic glazes; made of a finely ground calcined oxide, such as of cobalt, copper, manganese, or iron. { 'glāz ,stān }

GLC See gas-liquid chromatography.

glucinium |CHEM| The former name for the element beryllium, coined because the salts of beryllium are sweet-tasting. { glü'sin·ē·əm }

glucochloral See chloralose. { ¦glü·kō¦klȯr·əl }

glucochloralose See chloralose. { ¦glü·kō¦klȯr·ə,lōs }

α-D-glucochloralose See chloralose. { ¦al·fə ¦dē ¦glü·kō¦klȯr·ə,lōs }

gluconate |ORG CHEM| A salt of gluconic acid. { glü·kə,nāt }

gluconic acid |ORG CHEM| $C_6H_{12}O_7$ A crystalline acid obtained from glucose by oxidation; used in cleaning metals. { glü'kän·ik 'as·əd }

gluconic acid sodium salt See sodium gluconate. { glü'kän·ik 'as·əd 'sōd·ē·əm 'sȯlt }

7-D-glucoside See genistin. { ¦sev·ən ¦dē 'glü·kə,sīd }

gluside See saccharin. { 'glü,sīd }

glutaraldehyde |ORG CHEM| $OHC(CH_2)_3CHO$ A liquid with a boiling point of 188°C; soluble in water and alcohol; used as a biological solution (50) and for leather tanning. { ,glüd·ə'ral·də,hīd }

glycerin See glycerol. { 'glis·ə·rən }

glycerine See glycerol. { 'glis·ə·rən }

glycerol |ORG CHEM| $CH_2OHCHOHCH_2OH$ The simplest trihedric alcohol; when pure, it is a colorless, odorless, viscous liquid with a sweet taste; it is completely soluble in water and alcohol but only partially soluble in common solvents such as ether and ethyl acetate; used in manufacture of alkyd resins, explosives, antifreezes, medicines, inks, perfumes, cosmetics, soaps, and finishes. Also known as glycerin; Also known as glycerin; glycerine; glycyl alcohol. glycyl alcohol. { 'glis·ə,rȯl }

glyceryl [ORG CHEM] $OCH_2OCHOCH_2\equiv$ The functional group from glycerol, $(CH_2OH)_2$-CHOH. { 'glis·ə·rəl }

glyceryl diacetate See diacetin. { 'glis·ə·rəl dī'as·ə,tāt }

glyceryl tristearate See stearin. { 'glis·ə·rəl trī'stir,āt }

glycidic acid [ORG CHEM] $C_2H_3O\cdot CO_2H$ A volatile liquid. Also known as epoxy-propionic acid. { glə'sid·ik 'as·əd }

glycidol [ORG CHEM] $C_3H_6O_2$ A colorless, liquid epoxide that boils at 162°C and is miscible with water; used in organic synthesis. Also known as epihydrin alcohol. { 'glis·ə,dȯl }

glycin [ORG CHEM] $C_8H_9NO_3$ A crystalline compound that forms shiny leaflets from water solution, melts at 245–247°C, and is soluble in alkalies and mineral acids; used as a photographic developer and in the analytical determination of iron, phosphorus, and silicon. Also known as photoglycine. { 'glī·sən }

glyco- [ORG CHEM] Chemical prefix indicating sweetness, or relating to sugar or glycine. { 'glī·kō }

glycol [ORG CHEM] **1.** $C_nH_{2n}(OH)_2$ An organic chemical with two hydroxyl groups on an essentially aliphatic carbon chain. Also known as dihydroxy alcohol. **2.** $HOCH_2$-CH_2OH A colorless dihydroxy alcohol used as an antifreeze, in hydraulic fluids, and in the manufacture of dynamites and resins. Also known as ethlene glycol. { 'glī,kȯl }

glycol diacetate See ethylene glycol diacetate. { 'glī,kȯl dī'as·ə,tāt }

glycoldinitrate See ethylene nitrate. { 'glī,kȯl dī'nī,trāt }

glycol ester [ORG CHEM] Chemical compound composed of the reaction products of a glycol, $C_nH_{2n}(OH)_2$, and an organic acid; an example is ethylene glycol diacetate, the product of ethylene glycol and acetic acid. { 'glī,kȯl 'es·tər }

glycol ether [ORG CHEM] A colorless liquid used as a solvent, in detergents, and as a diluent; a typical example is ethylene glycol diethyl ether, $C_2H_5OCH_2CH_2OC_2H_5$. { 'glī,kȯl 'ē·thər }

glycolic acid [ORG CHEM] $CH_2OHCOOH$ Colorless, deliquescent leaflets, decomposing about 78°C; soluble in water, alcohol, and ether; used as a chemical intermediate in fabric dyeing. Also known as hydroxyacetic acid. { glī'käl·ik 'as·əd }

glycolythiourea See 2-thiohydantoin. { ¦gli·kȯl¦thī·ō·yü'rē·ə }

glycolyurea See hydantoin. { ¦glī,kȯl·yü'rē·ə }

glycyl [ORG CHEM] NH_2CH_2COO- or $NHCH_2COO=$ The radical from glycine, NH_2CH_2COOH; found in peptides. { 'glī·səl }

glycyl alcohol See glycerol. { 'glī·səl 'al·kə,hȯl }

glyoxal [ORG CHEM] $(CHO)_2$ Colorless, deliquescent powder or liquid with mild odor, melting point 15°C, boiling point 51°C; used to insolubilize starches, cellulosic materials, and proteins, in embalming fluids, for leather tanning, and for rayon shrinkproofing. { glī'äk,sal }

glyoxalic acid [ORG CHEM] $CHOCOOH$ Colorless crystals that are soluble in water, forming glyoxylic acid. { ¦glī,äk¦sal·ik 'as·əd }

glyphosate [ORG CHEM] $C_3H_8NO_5P$ A white solid with a melting point of 200°C; slight solubility in water; used as a herbicide in postharvest treatment of crops. { 'glif·ə,sāt }

glyphosine [ORG CHEM] $C_4H_{11}NO_8P_2$ A white solid with a melting point of 203°C; quite soluble in water; used as a growth regulator in sugarcane. { 'glif·ə,sēn }

glyptal resin [ORG CHEM] A phthalic anhytxride glycerol made from an emulsion of an alkyd resin; used in lacquers and insulation. { 'glipt·əl 'rez·ən }

gold [CHEM] A chemical element, symbol Au, atomic number 79, atomic weight 196.96765; soluble in aqua regia; melts at 1065°C. { gōld }

gold chloride [INORG CHEM] $AuCl_3$ A red, soluble compound made by reaction of gold and chlorine or by reaction of $HAuCl_4$ with chlorine; decomposes by heat; soluble in water, alcohol, and ether; used in photography, plating, inks, medicine, and ceramics. { ¦gōld 'klȯr,īd }

golden antimony sulfide See antimony pentasulfide. { 'gōl·dən 'ant·ə,mō·nē 'səl,fīd }

gold hydroxide [INORG CHEM] Au(OH)₃ A yellow-brown, light-sensitive, water-insoluble powder; dissolves in most acids; easily reduced to metallic gold; used in medicine, porcelain, gold plating, and daguerreotypes. { ¦gōld hī'dräk,sīd }

gold number [ANALY CHEM] A measure of the amount of protective colloid which must be added to a standard red gold sol mixed with sodium chloride solution to prevent the solution from causing the sol to coagulate, as manifested by a change in color from red to blue. { 'gōld ,nəm·bər }

gold oxide [INORG CHEM] Au₂O₃ Water-insoluble, heat-decomposable, brownish-black powder; soluble in hydrochloric acid; used to gild, in medicine and porcelain, and for daguerreotypes. Also known as auric oxide; gold trioxide. { 'gōld 'äk,sīd }

gold potassium chloride See potassium gold chloride. { ¦gōld pə'tas·ē·əm 'klōr,īd }

gold potassium cyanide See potassium gold cyanide. { ¦gōld pə'tas·ē·əm 'sī·ə,nīd }

gold salt See sodium gold chloride. { 'gōld ,sòlt }

gold size [CHEM] A solution of white and red lead and yellow ocher in linseed oil; used to seal permanently microscopical preparations. { ¦gōld ¦sīz }

gold sodium chloride See sodium gold chloride. { ¦gōld 'sōd·ē·əm 'klōr,īd }

gold sodium cyanide See sodium gold cyanide. { ¦gōld 'sōd·ē·əm 'sī·ə,nīd }

gold tin precipitate See gold tin purple. { ¦gōld ¦tin prə'sip·ə,tāt }

gold tin purple [ORG CHEM] A brown powder which is a mixture of gold chloride and brown tin oxide, soluble in ammonia; used in coloring enamels, manufacturing ruby glass, and painting porcelain. Also known as gold tin precipitate; purple of Cassius. { ¦gōld ¦tin 'pər·pəl }

gold trioxide See gold oxide. { ¦gōld trī'äk,sīd }

Gomberg-Bachmann-Hey reaction [ORG CHEM] Production of diaryl compounds by adding alkali to a mixture of a diazonium salt and a liquid aromatic hydrocarbon or a derivative. { 'gòm,berk 'bäk̲,män 'hī rē,ak·shən }

Gomberg reaction [ORG CHEM] The production of free radicals by reaction of metals with triarylmethyl halides. { 'gòm,berk rē,ak·shən }

Gooch crucible [ANALY CHEM] A ceramic crucible with a perforated base; in analysis it is used for filtration through asbestos or glass. { 'güch ,krü·sə·bəl }

gorlic acid [ORG CHEM] C₅H₇(C₁₂H₂₂)COOH An unsaturated acid derived from sapucainha oil, obtained from the seeds of a tree in the Amazon Valley. { 'gòr·lik 'as·əd }

gouy [PHYS CHEM] An electrokinetic unit equal to the product of the electrokinetic potential and the electric displacement divided by 4π times the polarization of the electrolyte. { 'gō·ē }

Gouy balance [ANALY CHEM] Device for measurement of diamagnetic and paramagnetic susceptibilities of samples (solid, liquid, solution). { 'gō·ē ,bal·əns }

gradient elution analysis [ANALY CHEM] A form of gas-liquid chromatography in which the eluting solvent is changed with time, either by gradually mixing a second solvent of greater eluting power with the first, less powerful solvent, or by a gradual change in pH or other property. { 'grād·ē·ənt i'lü·shən ə,nal·ə·səs }

graduate [CHEM] A cylindrical vessel that is calibrated in fluid ounces or milliliters or both; used to measure the volume of liquids. { 'graj·ə·wət }

Graebe-Ullman reaction [ORG CHEM] **1.** Production of fluorenone by boiling 2-benzoyl-benzenediazonium salts in dilute acid solution. **2.** Reaction of 2-aminodiphenylamines with nitrous acid to form a benzotriazole which on heating loses nitrogen to form a carbazole. { ¦gre·bə 'ùl·mən rē,ak·shən }

graft copolymer [ORG CHEM] Any high polymer composed of two or more different polymeric entities chemically united. { ¦graft kō'päl·ə·mər }

grain alcohol See ethanol. { 'grān 'al·kə,hòl }

gram-atomic weight [CHEM] The atomic weight of an element expressed in grams, that is, the atomic weight on a scale on which the atomic weight of carbon-12 isotope is taken as 12 exactly. { 'gram ə¦tam·ik 'wāt }

gram-equivalent weight [CHEM] The equivalent weight of an element or compound expressed in grams on a scale in which carbon-12 has an equivalent weight of

gram-molecular volume

3 grams in those compounds in which its formal valence is 4. { ¦gram i¦kwiv·ə·lənt 'wāt }

gram-molecular volume [CHEM] The volume occupied by a gram-molecular weight of a chemical in the gaseous state at 0°C and 760 millimeters of pressure (101,325 pascals). { ¦gram mə¦lek·yə·lər 'väl·yəm }

gram-molecular weight [CHEM] The molecular weight of compound expressed in grams, that is, the molecular weight on a scale on which the atomic weight of carbon-12 isotope is taken as 12 exactly. { ¦gram mə¦lek·yə·lər 'wāt }

granulate [CHEM] To form or crystallize into grains, granules, or small masses. { 'gran·yə,lāt }

graphical formula [CHEM] A chemical formula that suggests a three-dimensional representation of the structure of a molecule by rendering chemical bonds within the plane of the paper as straight lines, those above the plane of the paper as wedge-shaped bonds, and those below the plane of the paper either as broken lines or broken-line wedges. { ¦graf·ə·kəl 'fòr·myə·lə }

graphics-based molecular modeling See molecular graphics. { 'graf·iks ¦bāst mə'lek·yə·lər 'mäd·əl·iŋ }

graphitization [ORG CHEM] The formation of graphitelike material from organic compounds. { ,graf·əd·ə'zā·shən }

Grassmann's laws [ANALY CHEM] Seven laws of color identification and mixing that form the basis of modern analytical colorimetry. { 'gräs·mənz ,lòz }

grating See diffraction grating. { 'grād·iŋ }

grating constant [SPECT] The distance between consecutive grooves of a diffraction grating. { 'grād·iŋ ,kän·stənt }

grating spectrograph [SPECT] A grating spectroscope provided with a photographic camera or other device for recording the spectrum. { 'grād·iŋ 'spek·trə,graf }

grating spectroscope [SPECT] A spectroscope which employs a transmission or reflection grating to disperse light, and usually also has a slit, a mirror or lenses to collimate the light sent through the slit and to focus the light dispersed by the grating into spectrum lines, and an eyepiece for viewing the spectrum. { 'grād·iŋ 'spek·trə,skōp }

gravimetric absorption method [ANALY CHEM] A method of measuring the moisture content of a gas in which a known volume of gas is passed through a suitable desiccant, such as phosphorus pentoxide or silica gel, and the change in weight of the desiccant is observed. { grav·ə'me·trik əb'sòrp·shən ,meth·əd }

gravimetric analysis [ANALY CHEM] That branch of quantitative analytical chemistry in which a desired constituent is converted, usually by precipitation or combustion, to a pure compound or element, of definite known composition, and is weighed; in a few cases a compound or element is formed which does not contain the constituent but bears a definite mathematical relationship to it. { grav·ə'me·trik ə'nal·ə·səs }

gravity cell [PHYS CHEM] An electrolytic cell in which two ionic solutions are separated by means of gravity. { 'grav·əd·ē ,sel }

green chemistry [CHEM] The use of chemical products and processes that reduce or eliminate substances hazardous to human health or the environment. { ¦grēn 'kem·ə·strē }

green copperas See ferrous sulfate. { ¦grēn 'käp·ə·rəs }

green nickel oxide See nickel oxide. { 'grēn ¦nik·əl 'äk,sīd }

green salt See uranium tetrafluoride. { 'grēn ¦sòlt }

green vitriol See ferrous sulfate. { 'grēn ¦vi·trē,ól }

grid spectrometer [SPECT] A grating spectrometer in which a large increase in light flux without loss of resolution is achieved by replacing entrance and exit slits with grids consisting of opaque and transparent areas, patterned to have large transmittance only when the entrance grid image coincides with that of the exit grid. { 'grid spek'träm·əd·ər }

Griess reagent [ANALY CHEM] A reagent used to test for nitrous acid; it is a solution of sulfanilic acid, α-naphthylamine and acetic acid in water. { 'grēs rē,ā·jənt }

Grignard reaction [ORG CHEM] A reaction between an alkyl or aryl halide and magnesium metal in a suitable solvent, usually absolute ether, to form an organometallic halide. { grin'yär rē,ak·shən }

Grignard reagent [ORG CHEM] RMgX The organometallic halide formed in the Grignard reaction; an example is C_2H_5MgCl; it is useful in organic synthesis. { grin'yär rē,ā·jənt }

Grignard synthesis [ORG CHEM] Use of the Grignard reagent in any one of a vast number of reactions, usually condensations; typical syntheses involve formation of a hydrocarbon, acid, ketone, or secondary or tertiary alcohol. { grin'yär ,sin·thə·səs }

grinding-type resin [ORG CHEM] Vinyl or other resin that requires grinding before dispersal into plastisols or organosols. { 'grīn·diŋ ,tīp ,rez·ən }

grism [SPECT] A combination of a diffraction grating and a prism, wherein the grating spreads light into colors and the prism moves the spectrum's position to the point in an image where the observed object appears. { 'griz·əm }

gross sample [ANALY CHEM] One or more increments taken from a larger quantity of a material that is to be analyzed. Also known as bulk sample; lot sample. { ¦grōs 'sam·pəl }

Grotthus' chain theory [PHYS CHEM] An early theory used to explain the conductivity of an electrolyte, in which it was assumed that the cathode and anode attract hydrogen and oxygen respectively, and the molecules of the electrolyte are stretched out in chains between the electrodes, with decomposition occurring in molecules closest to the electrodes. { 'grōt·hùs 'chān ,thē·ə·rē }

group [CHEM] **1.** A family of elements with similar chemical properties. **2.** A combination of bonded atoms that behave as a unit under certain conditions, for example, the sulfate group, SO_4^{2-}. { grüp }

Grove's synthesis [ORG CHEM] Production of alkyl chlorides by passing hydrochloric acid into an alcohol in the presence of anhydrous zinc chloride. { 'grōvz 'sin·thə·səs }

GR-S rubber [ORG CHEM] Former designation for general-purpose synthetic rubbers formed by copolymerization of emulsions of styrene and butadiene; used in tires and other rubber products; previously also known as Buna-S, currently known as SBR (styrene-butadiene rubber). { ¦jē¦är'es ,rəb·ər }

GSC *See* gas-solid chromatography.

guaiacol [ORG CHEM] $C_6H_4(OH)OCH_3$ A colorless, crystalline compound, soluble in water; used as a reagent to determine the presence of such substances as lignin, narceine, and nitrous acid. { 'gwī·ə,kòl }

Guerbet reaction [ORG CHEM] A condensation of alcohols at high temperatures through the action of sodium alkoxides. { ,ger'bā rē,ak·shən }

guest [CHEM] Cationic, anionic, or neutral organic, inorganic, or biological substance, bound by means of various interactions (electrostatic, hydrogen bonding, van der Waals, donor-acceptor) within a crystalline or molecular structure. Also known as guest molecule; guest substance. { gest }

guest molecule *See* guest. { 'gest ,mäl·ə,kyül }

guest substance *See* guest. { 'gest ,səb·stəns }

Guldberg and Waage law *See* mass action law. { 'gùlt·berk and 'väg·ə ,lò }

gum accroides *See* acaroid resin. { 'gəm ə'kròi·dēz }

Günzberg reagent [ANALY CHEM] A solution of 2 grams of vanillin and 4 grams of phloroglucinol in 80 milliliters of 95% alcohol; used as a test reagent for determining free hydrochloric acid in gastric juice. { 'gints·berk rē,ā·jənt }

Gurney-Mott theory [CHEM] A theory of the photographic process that proposes a two-stage mechanism; in the first stage, a light quantum is absorbed at a point within the silver halide gelatin, releasing a mobile electron and a positive hole; these mobile defects diffuse to trapping sites (sensitivity centers) within the volume or on the surface of the grain; in the second stage, trapped (negatively charged) electron is neutralized by an interstitial (positively charged) silver ion, which combines with the electron to form a silver atom; the silver atom is capable of trapping a second electron, after which the process repeats itself, causing the silver speck to grow. { 'gər·nē 'mät ,thē·ə·rē }

Gutzeit test

Gutzeit test [ANALY CHEM] A test for arsenic; zinc and dilute sulfuric acid are added to the substance, which is then covered with a filter paper moistened with mercuric chloride solution; a yellow spot forms on the paper if arsenic is in the sample. { 'güt,sīt ,test }

H

H *See* hydrogen.

Ha *See* hahnium.

H acid [ORG CHEM] $H_2NC_{10}H_4(OH)(SO_3H)_2$ A gray powder or crystalline substance that is soluble in water, ether, and alcohol; used as a dye intermediate. { ¦āch 'as·əd }

hafnium [CHEM] A metallic element, symbol Hf, atomic number 72, atomic weight 178.49; melting point 2000°C, boiling point above 5400°C. { 'haf·nē·əm }

hafnium carbide [INORG CHEM] HfC Gray powder, melting at 3887°C; used in the control rods of nuclear reactors. { 'haf·nē·əm 'kär,bīd }

Haggenmacher equation [CHEM] Equation to calculate latent heats of vaporizations of pure compounds by using critical conditions with Antoine constants. { ¦häg·ən¦mäk·ər i,kwā·zhən }

halazone [ORG CHEM] $COOHC_6H_4SO_2NCl_2$ White crystals, with strong chlorine aroma; slightly soluble in water and chloroform; used as water disinfectant. { 'hal·ə,zōn }

half-cell [PHYS CHEM] A single electrode immersed in an electrolyte. { 'haf ¦sel }

half-cell potential [PHYS CHEM] In electrochemical cells, the electrical potential developed by the overall cell reaction; can be considered, for calculation purposes, as the sum of the potential developed at the anode and the potential developed at the cathode, each being a half-cell. { 'haf ¦sel pə'ten·chəl }

half-life [CHEM] The time required for one-half of a given material to undergo chemical reactions. { 'haf ,līf }

halide [CHEM] A compound of the type MX, where X is fluorine, chlorine, iodine, bromine, or astatine, and M is another element or organic radical. { 'ha,līd }

haloalkane [ORG CHEM] Halogenated aliphatic hydrocarbon. { ¦ha·lō¦al,kān }

halocarbon [ORG CHEM] A compound of carbon and a halogen, sometimes with hydrogen. { ¦ha·lō¦kär·bən }

halocarbon plastic [ORG CHEM] Plastic made from halocarbon resins. { ¦ha·lō¦kär·bən 'plas·tik }

halocarbon resin [ORG CHEM] Resin produced by the polymerization of monomers made of halogenated hydrocarbons, such as tetrafluoroethylene, C_2F_4, and trifluorochloroethylene, C_2F_3Cl. { ¦ha·lō¦kär·bən 'rez·ən }

haloform [ORG CHEM] CHX_3 A compound made by reaction of acetaldehyde or methyl ketones with NaOX, where X is a halogen; an example is iodoform, HCl_3, or bromoform, $HCBr_3$ or chloroform, $HCCl_3$. { 'hal·ə,fórm }

haloform reaction [ORG CHEM] Halogenation of acetaldehyde or a methyl ketone in aqueous basic solution; the reaction is characteristic of compounds containing a CH_3CO group linked to a hydrogen or to another carbon. { 'hal·ə,fórm rē,ak·shən }

halogen [CHEM] Any of the elements of the halogen family, consisting of fluorine, chlorine, bromine, iodine, and astatine. { 'hal·ə·jən }

halogen acid [INORG CHEM] A compound composed of hydrogen bonded to a halogen element, for example, hydrochloric acid. { 'hal·ə·jən ,as·əd }

halogenated hydrocarbon [ORG CHEM] One of a group of halogen derivatives of organic hydrogen- and carbon-containing compounds; the group includes monohalogen compounds (alkyl or aryl halides) and polyhalogen compounds that contain the same or different halogen atoms. { 'hal·ə·jə,nād·əd ,hī·drə'kär·bən }

halogenation [ORG CHEM] A chemical process or reaction in which a halogen element

halohydrin

is introduced into a substance, generally by the use of the element itself. { ‚hal·ə·jə'nā·shən }

halohydrin [ORG CHEM] A compound with the general formula X-R-OH where X is a halide such as Cl^-; an example is chlorohydrin. { ‚hal·ə'hī·drən }

halon [ORG CHEM] A fluorocarbon that has one or more bromine atoms in its molecule. { 'ha‚län }

Hammett acidity function [CHEM] An expression for the acidity of a medium, defined as $h_0 = K_{BH^+}[BH^+]/[B]$, where K_{BH^+} is the dissociation constant of the acid form of the indicator, and $[BH^+]$ and $[B]$ are the concentrations of the protonated base and the unprotonated base respectively. { 'ham·ət ə'sid·əd·ē ‚faŋk·shən }

hand sugar refractometer [ANALY CHEM] Portable device to read refractive indices of sugar solutions. Also known as proteinometer. { 'hand 'shúg·ər ‚rē,frak'täm·əd·ər }

Hansa yellow [ORG CHEM] Group of organic azo pigments with strong tinting power, but poor opacity in paints; used where nontoxicity is important. { 'hän·sə 'yel·ō }

Hantzsch synthesis [ORG CHEM] The reaction whereby a pyrrole compound is formed when a β-ketoester, chloroacetone, and a primary amine condense. { 'hänsh ‚sin·thə·səs }

Hanus solution [ANALY CHEM] Iodine monobromide in glacial acetic acid; used to determine iodine values in oils containing unsaturated organic compounds. { 'han·əs sə‚lü·shən }

hard acid [CHEM] A Lewis acid of low polarizability, small size, and high positive oxidation state; it does not have easily excitable outer electrons; some examples are H^+, Li^+, and Al^+. { 'härd 'as·əd }

hard base [CHEM] A Lewis base (electron donor) that has high polarizability and low electronegativity, is easily oxidized, or possesses low-lying empty orbitals; some examples are H_2O, HO^-, OCH_3^-, and F^-. { 'härd ‚bās }

hard detergent [CHEM] A nonbiodegradable detergent. { 'härd di'tər·jənt }

hardener [ORG CHEM] Compound reacted with a resin polymer to harden it, such as the amines or anhydrides that react with epoxides to cure or harden them into plastic materials. Also known as curing agent. { 'härd·ən·ər }

hardness [CHEM] The amount of calcium carbonate dissolved in water, usually expressed as parts of calcium carbonate per million parts of water. { 'härd·nəs }

hardness test [ANALY CHEM] A test to determine the calcium and magnesium content of water. { 'härd·nəs ‚test }

hard-sphere collision theory [PHYS CHEM] A theory for calculating reaction rate constants for biomolecular gas-phase reactions in which the molecules are considered to be colliding, hard spheres. { 'härd ‚sfir kə'lizh·ən ‚thē·ə·rē }

hard water [CHEM] Water that contains certain salts, such as those of calcium or magnesium, which form insoluble deposits in boilers and form precipitates with soap. { 'härd ‚wòd·ər }

Hardy-Schulz rule [PHYS CHEM] An increase in the charge of ions results in a large increase in their flocculating power. { 'härd·ē 'shúlts ‚rül }

Haring cell [PHYS CHEM] An electrolytic cell with four electrodes used to measure electrolyte resistance and polarization of electrodes. { 'her·iŋ ‚sel }

harman [ORG CHEM] $C_{12}H_{10}N_2$ Crystals that melt at 237–238°C; inhibits growth of molds and certain bacteria. Also known as arabine; loturine; passiflorin. { 'här·mən }

harmonic vibration-rotation band [SPECT] A vibration-rotation band of a molecule in which the harmonic oscillator approximation holds for the vibrational levels, so that the vibrational levels are equally spaced. { här'män·ik vī'brā·shən rō'tā·shən ‚band }

Hartmann diaphragm [ANALY CHEM] Comparison device for positive-element-identification readings from emission spectra. { 'härt·män ‚dī·ə‚fram }

Hartmann test [SPECT] A test for spectrometers in which light is passed through different parts of the entrance slit; any resulting changes of the spectrum indicate a fault in the instrument. { 'härt·män ‚test }

Hartman's solution [ANALY CHEM] Solution of thymol, ethyl alcohol, and sulfuric ether; used for selective dentin analysis. { 'härt·mənz sə‚lü·shən }

hassium [CHEM] A chemical element, symbolized Hs, atomic number 108, a synthetic element; the sixteenth transuranium element. { 'hä·sē·əm }
HBFC *See* hydrobromofluorocarbon.
HCB *See* hexachlorobenzene.
HCFC *See* hydrochlorofluorocarbon.
HDPE *See* high-density polyethylene.
He *See* helium.
heating value *See* heat of combustion. { 'hēd·iŋ ‚val·yü }
heat of activation [PHYS CHEM] The increase in enthalpy when a substance is transformed from a less active to a more reactive form at constant pressure. { 'hēt əv ‚ak·tə'vā·shən }
heat of association [PHYS CHEM] Increase in enthalpy accompanying the formation of 1 mole of a coordination compound from its constituent molecules or other particles at constant pressure. { 'hēt əv ə‚sō·sē'ā·shən }
heat of atomization [PHYS CHEM] The change in enthalpy accompanying the conversion of 1 mole of an element or a compound at 298 K (77°F) and 1 atmosphere (10⁵ pascals) into free atoms. { ‚hēt əv ‚ad·ə·mə'zā·shən }
heat of combustion [PHYS CHEM] The amount of heat released in the oxidation of 1 mole of a substance at constant pressure, or constant volume. Also known as heat value; heating value. { 'hēt əv kəm'bəs·chən }
heat of decomposition [PHYS CHEM] The change in enthalpy accompanying the decomposition of 1 mole of a compound into its elements at constant pressure. { 'hēt əv dē‚käm·pə'zish·ən }
heat of dilution [PHYS CHEM] **1.** The increase in enthalpy accompanying the addition of a specified amount of solvent to a solution of constant pressure. Also known as integral heat of dilution; total heat of dilution. **2.** The increase in enthalpy when an infinitesimal amount of solvent is added to a solution at constant pressure. Also known as differential heat of dilution. { 'hēt əv də'lü·shən }
heat of dissociation [PHYS CHEM] The increase in enthalpy at constant pressure, when molecules break apart or valence linkages rupture. { 'hēt əv di‚sō·sē'ā·shən }
heat of formation [PHYS CHEM] The increase in enthalpy resulting from the formation of 1 mole of a substance from its elements at constant pressure. { 'hēt əv fór'mā·shən }
heat of hydration [PHYS CHEM] The increase in enthalpy accompanying the formation of 1 mole of a hydrate from the anhydrous form of the compound and from water at constant pressure. { 'hēt əv hī'drā·shən }
heat of ionization [PHYS CHEM] The increase in enthalpy when 1 mole of a substance is completely ionized at constant pressure. { 'hēt əv ‚ī·ən·ə'zā·shən }
heat of linkage [PHYS CHEM] The bond energy of a particular type of valence linkage between atoms in a molecule, as determined by the energy required to dissociate all bonds of the type in 1 mole of the compound divided by the number of such bonds in a compound. { 'hēt əv 'liŋk·ij }
heat of reaction [PHYS CHEM] **1.** The negative of the change in enthalpy accompanying a chemical reaction at constant pressure. **2.** The negative of the change in internal energy accompanying a chemical reaction at constant volume. { 'hēt əv rē'ak·shən }
heat of solution [PHYS CHEM] The enthalpy of a solution minus the sum of the enthalpies of its components. Also known as integral heat of solution; total heat of solution. { 'hēt əv sə'lü·shən }
heat value *See* heat of combustion. { 'hēt ‚val·yü }
heavy acid *See* phosphotungstic acid. { ‚hev·ē 'as·əd }
heavy water [INORG CHEM] A compound of hydrogen and oxygen containing a higher proportion of the hydrogen isotope deuterium than does naturally occurring water. Also known as deuterium oxide. { 'hev·ē 'wȯd·ər }
Hefner lamp [CHEM] A flame lamp that burns amyl acetate. { 'hef·nər ‚lamp }
Hehner number [ANALY CHEM] Weight percent of water-insoluble fatty acids in fats and oils. { 'hān·ər ‚nəm·bər }
Heitler-London covalence theory [PHYS CHEM] A calculation of the binding energy and the distance between the atoms of a diatomic hydrogen molecule, which assumes

that the two electrons are in atomic orbitals about each of the nuclei, and then combines these orbitals into a symmetric or antisymmetric function. { 'hīt·lər 'lən·dən kō'vā·ləns ˌthē·ə·rē }

helicate [ORG CHEM] Any member of a group of synthetic, helical arrays of molecules formed by the chemical recognition and organization of metals and organic bases. { 'hel·i͟ˌkāt }

helicin See salicylaldehyde. { 'hel·ə·sən }

heliotropin See piperonal. { ¦hē·lē·ə¦trō·pən }

helium [CHEM] A gaseous chemical element, symbol He, atomic number 2, and atomic weight 4.0026; one of the noble gases in group 0 of the periodic table. { 'hē·lē·əm }

helium spectrometer [SPECT] A small mass spectrometer used to detect the presence of helium in a vacuum system; for leak detection, a jet of helium is applied to suspected leaks in the outer surface of the system. { 'hē·lē·əm spek'träm·əd·ər }

Hell-Volhard-Zelinsky reaction [ORG CHEM] Preparation of an ester or α-halo substituted acid (chloro or bromo) by reacting the halogen on the acid in the presence of phosphorus or phosphorus halide, and then followed by hydrolysis or alcoholysis of the haloacyl halide resulting. { ¦hel ¦fōl¦härt zə'lins·kē rē͟ˌak͟ˌshən }

Helmholtz equation [PHYS CHEM] The relationship stating that the emf (electromotive force) of a reversible electrolytic cell equals the work equivalent of the chemical reaction when charge passes through the cell plus the product of the temperature and the derivative of the emf with respect to temperature. { 'helm͟ˌhōlts i͟ˌkwā·zhən }

hematin [ORG CHEM] $C_{34}H_{33}O_5N_4Fe$ The hydroxide of ferriheme derived from oxidized heme. { 'hē·məd·ən }

hematoxylin [ORG CHEM] $C_{16}H_{14}O_6$ A colorless, crystalline compound occurring in hematoxylon; upon oxidation, it is converted to hematein which forms deeply colored lakes with various metals; used as a stain in microscopy. { ˌhē·mə'täk·sə·lən }

hemiacetal [ORG CHEM] A class of compounds that have the grouping C(OH)−(OR) and that result from the reaction of an aldehyde and alcohol. { ¦he·mē'as·ə͟ˌtal }

hemihydrate [ORG CHEM] A hydrate with a 2:1 molecular ratio of anhydrous compound to water, plaster of paris is the hemihydrate of calcium sulfate, composition $CaSO_4 \cdot {}^1\!/_2H_2O$. { ˌhem·ē'hī͟ˌdrāt }

hemiketal [ORG CHEM] A carbonyl compound that results from the addition of an alcohol to the carbonyl group of a ketone, with the general formula (R)(R′)C(OH)(OR). { ¦he·mē'ked·əl }

hemimellitic acid [ORG CHEM] $C_6H_3(COOH)_3$ A compound crystallizing in colorless needles; melting point 196°C; slightly soluble in water. { ¦he·mē·mə¦lid·ik 'as·əd }

hendecanal See undecanal. { hen'dek·ə·nəl }

hendecane See undecane. { 'hen·də͟ˌkān or hen'de͟ˌkān }

hendecyl See undecyl. { hen'des·əl }

Henderson equation for pH [PHYS CHEM] An equation for the pH of an acid during its neutralization: $pH = pK_a + \log \lfloor salt \rfloor / \lfloor acid \rfloor$, where pK_a is the logarithm to base 10 of the reciprocal of the dissociation constant of the acid; the equation is found to be useful for the pH range 4–10, providing the solutions are not too dilute. { 'hen·dər·sən i¦kwā·zhən fər ¦pē'ach }

heneicosane [ORG CHEM] $C_{21}H_{44}$ Saturated hydrocarbon of the methane series; the crystals melt at 40°C and boil at 215°C (at 15 mm Hg). { hen'ī·kə͟ˌsān }

Henry's law [PHYS CHEM] The law that at sufficiently high dilution in a liquid solution, the fugacity of a nondissociating solute becomes proportional to its concentration. { 'hen·rēz ˌlò }

hentriacontane [ORG CHEM] $C_{31}H_{64}$ A hydrocarbon; a crystalline material melting at 68°C and boiling at 302°C (at 15 mmHg); derived from roots of *Oenanthe crocata* and found in beeswax. { ˌhen͟ˌtrī·ə'kän͟ˌtän }

hepar calcies See calcium sulfide. { 'hē͟ˌpär 'kal͟ˌsēz }

hepar sulfuris See potassium sulfide. { 'hē͟ˌpär səl'fyùr·əs }

heptachlor [ORG CHEM] $C_{10}H_7Cl_7$ An insecticide; a white to tan, waxy solid; insoluble in water, soluble in alcohol and xylene; melts at 95–96°C. { 'hep·tə͟ˌklòr }

heptacosane [ORG CHEM] $C_{27}H_{56}$ A hydrocarbon; water-insoluble crystals melting at

60°C and boiling at 270°C (at 15 mmHg); soluble in alcohol; found in beeswax. { hep'täk·ə,sān *or* ‚hep·tə'kō,sän }

heptadecane [ORG CHEM] $C_{17}H_{36}$ A hydrocarbon; water-insoluble, alcohol-soluble solid melting at 23°C and boiling at 303°C; used as a chemical intermediate. { ‚hep·tə'de,kān }

***n*-heptadecanoic acid** [ORG CHEM] $CH_3(CH_2)_{15}COOH$ A fatty acid that is saturated; soluble in ether and alcohol, insoluble in water; colorless crystals melt at 61°C. Also known as margaric acid. { ¦en ¦hep·tə¦dek·ə¦nō·ik 'as·əd }

heptadecanol [ORG CHEM] $C_{17}H_{35}OH$ An alcohol; colorless liquid boiling at 309°C; slightly soluble in water; used as a chemical intermediate, as a perfume fixative, in cosmetics and soaps, and to manufacture surfactants. { ¦hep·tə'dek·ə,nȯl }

heptadione-2,3 *See* acetyl valeryl. { ‚hep·tə'dī,ōn ¦tü ¦thrē }

heptaldehyde [ORG CHEM] $C_6H_{13}CHO$ An aldehyde; ether-soluble, colorless oil with fruity aroma; slightly soluble in water; boils at 153°C; used as a chemical intermediate and for perfumes and pharmaceuticals. Also known as heptanal. { ‚hep·'tal·də,hīd }

heptanal *See* heptaldehyde. { 'hep·tə,nal }

heptane [ORG CHEM] $CH_3(CH_2)_5CH_3$ A hydrocarbon; water-insoluble, flammable, colorless liquid boiling at 98°C; soluble in alcohol, chloroform, and ether; used as an anesthetic, solvent, and chemical intermediate, and in standard octane-rating tests. { 'hep,tān }

heptanoic acid [ORG CHEM] $CH_3(CH_2)_5COOH$ Clear oil boiling at 223°C; soluble in alcohol and ether, insoluble in water; used as a chemical intermediate. { ¦hep·tə¦nō·ik 'as·əd }

1-heptanol [ORG CHEM] $C_7H_{15}OH$ An alcohol; a fragrant, colorless liquid boiling at 174°C; soluble in water, ether, or alcohol; used as a chemical intermediate, as a solvent, and in cosmetics. Also known as heptyl alcohol. { ¦wən 'hep·tə,nȯl }

3-heptanol [ORG CHEM] $CH_3CH_2CH(OH)C_4H_9$ An alcohol; a liquid boiling at 156°C; used as a coating, solvent, and diluent, as a chemical intermediate, and as a flotation frother. { ¦thrē 'hep·tə,nȯl }

2-heptanone *See* methyl-*n*-amyl ketone. { ¦tü 'hep·tə,nōn }

3-heptanone *See* ethyl butyl ketone. { ¦thrē 'hep·tə,nōn }

4-heptanone [ORG CHEM] $(CH_3CH_2CH_2)_2CO$ A colorless liquid that is stable and has a pleasant odor; boils at approximately 98°C; used to put nitrocellulose and raw and blown oils into solution, and used in lacquers and as a flavoring in foods. { ¦fȯr 'hep·tə,nōn }

heptene [ORG CHEM] $C_{17}H_{14}$ A liquid that is a mixture of isomers; boils at 189.5°C; used as an additive in lubricants, as a catalyst, and as a surface active agent. Also known as heptylene. { 'hep,tēn }

heptoxide [CHEM] An oxide whose molecule contains seven atoms of oxygen. { hep'täk,sīd }

heptyl [ORG CHEM] $CH_3(CH_2)_6-$ The functional group from heptane, $CH_3(CH_2)_5CH_3$. { 'hep·təl }

heptyl alcohol *See* 1-heptanol. { 'hep·təl 'al·kə,hȯl }

heptylene *See* heptene. { 'hep·tə,lēn }

Hercules trap [ANALY CHEM] Water-measuring liquid trap used in aquametry when the material collected is heavier than water. { 'hər·kyə,lēz ‚trap }

Hess's law [PHYS CHEM] The law that the evolved or absorbed heat in a chemical reaction is the same whether the reaction takes one step or several steps. Also known as the law of constant heat summation. { 'hes·əz ‚lȯ }

hetero- [CHEM] Prefix meaning different; for example, a heterocyclic compound is one in which the ring is made of more than one kind of atom. { 'hed·ə·rō }

heteroatom [ORG CHEM] In an organic compound, any atom other than carbon or hydrogen. { 'hed·ə·rō,ad·əm }

heteroazeotrope [CHEM] Liquid mixture that is not completely miscible in all proportions in the liquid phase, yet does not form an azeotrope. Also known as heterogeneous zeotrope. { ¦hed·ə·rō·ā'zē·ə,trōp }

177

heterocyclic compound

heterocyclic compound [ORG CHEM] Compound in which the ring structure is a combination of more than one kind of atom; for example, pyridine, C_5H_5N. { ‚hed·ə·rō'sī·klik ¦käm‚paůnd }

heterogeneous [CHEM] Pertaining to a mixture of phases such as liquid-vapor, or liquid-vapor-solid. { ‚hed·ə'räj·ə·nəs }

heterogeneous catalysis [CHEM] A chemical process in which the catalyst is in a separate phase; usually the reactants and products are in gaseous or liquid phases and the catalyst is a solid, and the catalytic reaction occurs on the surface of the solid. { ‚hed·ə·rə'jē·nē·əs kə'tal·ə·səs }

heterogeneous chemical reaction [CHEM] Chemical reaction system in which the reactants are of different phases; for example, gas with liquid, liquid with solid, or a solid catalyst with liquid or gaseous reactants. { ‚hed·ə·rə'jē·nē·əs ¦kem·ə·kəl rē'ak·shən }

heterogeneous nucleation [PHYS CHEM] The formation of vapor bubbles on cavities or scratches of a surface bounding a superheated liquid. { ‚hed·ə·rə¦jē·nē·əs ‚nü·klē'ā·shən }

heterolysis See heterolytic cleavage. { ‚hed·ə'räl·ə·səs }

heterolytic bond dissociation energy [PHYS CHEM] The change in enthalpy that occurs when a chemical bond undergoes heterolytic cleavage. { ‚hed·ə·rō¦lid·ik ‚bänd ‚dis·ə‚sō·sē'ā·shən ‚en·ər·jē }

heterolytic cleavage [ORG CHEM] The breaking of a single (two-electron) chemical bond in which both electrons remain on one of the atoms. Also known as heterolysis. { ‚hed·ə·rō¦lid·ik 'klēv·ij }

heteronuclear molecule [CHEM] A diatomic molecule having atoms of different elements. { ¦hed·ə·rə‚nü·klē·ər 'mäl·ə‚kyül }

heteropolar bond [PHYS CHEM] A covalent bond whose total dipole moment is not 0. { ‚hed·ə·rə¦pō·lər 'bänd }

heteropoly acid [INORG CHEM] Complex acids of metals, whose specific gravity is greater than 4, with phosphoric acid; an example is phosphomolybdic acid. { ‚hed·ə'räp·ə·lē 'as·əd }

heteropoly compound [INORG CHEM] Polymeric compounds of molybdates with anhydrides of other elements such as phosphorus; the yellow precipitate $(NH_4)_3P(Mo_3O_{10})_4$ is such a compound. { ‚hed·ə'räp·ə·lē 'käm‚paůnd }

heteropolymer [CHEM] A compound comprising two or more molecules that are different from one another. { ‚hed·ə·rə'päl·ə·mər }

heterotopic faces [ORG CHEM] On molecules, faces of double bonds where addition gives rise to isomeric structures. { ¦hed·ə·rō¦täp·ik 'fäs·əz }

heterotopic ligands [CHEM] Constitutionally identical ligands whose separate replacement by a different ligand gives rise to isomeric structures. { ¦hed·ə·rō¦täp·ik 'līg·ənz }

hexachlorobenzene [ORG CHEM] C_6Cl_6 Colorless, needlelike crystals with a melting point of 231°C; used in organic synthesis and as a fungicide. Abbreviated HCB. { ¦hek·sə¦klȯr·ō'ben‚zēn }

hexachlorobutadiene [ORG CHEM] $Cl_2C:CClCCl:CCl_2$ A colorless liquid with mild aroma, boiling at 210–220°C; soluble in alcohol and ether, insoluble in water; used as solvent, heat-transfer liquid, and hydraulic fluid. { ¦hek·sə¦klȯr·ō‚byüd·ə'dī‚ēn }

1,2,3,4,5,6-hexachlorocyclohexane [ORG CHEM] $C_6H_6Cl_6$ A white or yellow powder or flakes with a musty odor; a systemic insecticide toxic to flies, cockroaches, aphids, and boll weevils. Abbreviated TBH. { ¦wən ¦tü ¦thrē ¦fȯr ¦fīv ¦siks ¦hek·sə¦klȯr·ō‚sī·klō'hek‚sān }

hexachloroethane [ORG CHEM] Cl_3CCCl_3 Colorless crystals with a camphorlike odor, melting point 185°C, toxic; used in organic synthesis, as a retarding agent in fermentation, and as a rubber accelerator. { ¦hek·sə¦klȯr·ō'e‚thān }

hexachlorophene [ORG CHEM] $(C_6HCl_3OH)_2CH_2$ A white powder melting at 161°C; soluble in alcohol, ether, acetone, and chloroform, insoluble in water; bacteriostat used in antiseptic soaps, cosmetics, and dermatologicals. { ‚hek·sə'klȯr·ə‚fēn }

hexachloropropylene [ORG CHEM] $CCl_3CCl:CCl_2$ A water-white liquid boiling at 210°C, soluble in alcohol, ether, and chlorinated solvents, insoluble in water; used as a solvent, plasticizer, and hydraulic fluid. { ¦hek·sə¦klȯr·ō'prō·pə‚lēn }

hexacontane [ORG CHEM] $C_{60}H_{122}$ Solid, saturated hydrocarbon of the methane series; melts at 101°C. { ‚hek·sə'kän‚tān }

hexacosane [ORG CHEM] $C_{26}H_{54}$ Saturated hydrocarbon of the methane series; colorless crystals melting at 57°C. { ‚hek·sə'kō‚sān }

hexacosanoic acid See cerotic acid. { ¦hek·sə‚kō·sə'nō·ik 'as·əd }

n-hexadecane [ORG CHEM] $C_{16}H_{34}$ A colorless, solid hydrocarbon, melting point 20°C; a standard reference fuel in determining the ignition quality (cetane number) of diesel fuels. Also known as cetane. { ¦en¦hek·sə'de‚kān }

1-hexadecene [ORG CHEM] $CH_3(CH_2)_{13}CH:CH_2$ A colorless liquid made by treating cetyl alcohol with phosphorus pentoxide; boils at 274°C; soluble in organic solvents such as alcohol, ether, and petroleum; used as an intermediate in organic synthesis. { ¦wən ‚hek·sə'de‚sēn }

hexadentate ligand [INORG CHEM] A chelating agent having six groups capable of attachment to a metal ion. Also known as sexadentate ligand. { ‚hek·sə'den‚tāt 'līg·ənd }

hexadiene [ORG CHEM] C_6H_{10} A group of unsaturated hydrocarbons with two double bonds; some members of the group are 1,4-hexadiene, 1,5-hexadiene, and 2,4-hexadiene. { ‚hek·sə'dī‚ēn }

hexahydric alcohol [ORG CHEM] A member of the mannitol-sorbitol-dulcitol sugar group; isomer of $C_6H_8(OH)_6$. { ¦hek·sə¦hī·drik 'al·kə‚hȯl }

hexahydrotoluene See methyl cyclohexane. { ¦hek·sə¦hī·drō'täl·yə‚wēn }

n-hexaldehyde [ORG CHEM] $CH_3(CH_2)_4CHO$ Colorless liquid with sharp aroma, boiling at 128.6°C; used as an intermediate for plasticizers, dyes, insecticides, resins, and rubber chemicals. { ¦en ‚heks'al·də‚hīd }

hexametapol [ORG CHEM] $C_6H_{18}N_3OP$ A liquid used as a solvent in organic synthesis, as a deicing additive for jet engine fuel, and as an insect pest chemosterilant and chemical mutagen. { ‚hek·sə'med·ə‚pȯl }

hexamethylene See cyclohexane. { ‚hek·sə'meth·ə‚lēn }

hexamethylenediamine [ORG CHEM] $H_2N(CH_2)_6NH_2$ Colorless solid boiling at 205°C; slightly soluble in water, alcohol, and ether; used to make nylon and other high polymers. { ‚hek·sə'meth·ə‚lēn‚dī·ə‚mēn }

hexamethylene tetramine See cystamine. { ‚hek·sə'meth·ə‚lēn 'te·trə‚mēn }

hexamethylphosphoric triamide See bempa. { ‚hek·sə‚meth·əl'fäs'fȯr·ik trī'am·əd }

hexane [ORG CHEM] C_6H_{14} Water-insoluble, toxic, flammable, colorless liquid with faint aroma; forms include: n-hexane, a straight-chain compound boiling at 68.7°C and used as a solvent, paint diluent, alcohol denaturant, and polymerization-reaction medium; isohexane, a mixture of hexane isomers boiling at 54–61°C and used as a solvent and freezing-point depressant; and neohexane. { 'hek‚sān }

1,6-hexanediol [ORG CHEM] $HO(CH_2)_6OH$ A crystalline substance, soluble in water and alcohol; used in gasoline refining, as an intermediate in nylon manufacturing, and in making polyesters and polyurethanes. { ¦wən ¦siks ‚hek‚sān'dī‚ȯl }

hexanitrodiphenyl amine [ORG CHEM] $(NO_2)_3C_6H_2NHC_6H_2(NO_2)_3$ Explosive, yellow solid melting at 238–244°C; insoluble in water, ether, alcohol, or benzene; soluble in alkalies and acetic and nitric acids; used as an explosive and in potassium analysis. { ¦hek·sə¦nī·trō·dī'fen·əl 'am‚ēn }

hexaphenylethane [ORG CHEM] $(C_6H_5)_3CC(C_6H_5)_3$ The dimer of triphenylmethyl radical. { ¦hek·sə¦fen·əl'eth‚ān }

hexaprismane [CHEM] $C_{12}H_{12}$ A highly strained saturated hydrocarbon cage structure which consists of two flat cyclohexanes fused by six cyclobutanes. { ‚hek·sə'riz‚mān }

1-hexene [ORG CHEM] $CH_3(CH_2)_3HC:CH_2$ Colorless, olefinic hydrocarbon boiling at 64°C; soluble in alcohol, acetone, ether, and hydrocarbons, insoluble in water; used as a chemical intermediate and for resins, drugs, and insecticides. Also known as hexylene. { ¦wən 'hek‚sēn }

hexone See methyl isobutyl ketone. { 'hek‚sōn }

n-hexyl acetate [ORG CHEM] $CH_3COOC_6H_{13}$ Colorless liquid boiling at 169°C; soluble

hexyl alcohol

in alcohol and ether, insoluble in water; used as a solvent for resins and cellulosic esters.   { ¦en 'hek·səl 'as·ə,tāt }

hexyl alcohol   [ORG CHEM] $CH_3(CH_2)_4CH_2OH$ Colorless liquid boiling at 156°C; soluble in alcohol and ether, slightly soluble in water; used as a chemical intermediate for pharmaceuticals, perfume esters, and antiseptics.   { 'hek·səl 'al·kə,hȯl }

hexylamine   [ORG CHEM] $CH_3(CH_2)_5NH_2$ Poisonous, water-white liquid with amine aroma; boils at 129°C; a ptomaine base from the autolysis of protoplasm.   { hek 'sil·ə,mēn }

hexylene See 1-hexene.   { 'hek·sə,lēn }

hexylene glycol   [ORG CHEM] $C_6H_{14}O_2$ Water-miscible, colorless liquid boiling at 198°C; used in hydraulic brake fluids, in printing inks, and in textile processing.   { 'hek· sə,lēn 'glī,kȯl }

n-hexyl ether   [ORG CHEM] $C_6H_{13}OC_6H_{13}$ Faintly colored liquid with a characteristic odor, only slightly water-soluble; used in solvent extraction and in the manufacture of collodion and various cellulosic products.   { ¦en 'hek·səl 'ē·thər }

hexylresorcinol   [ORG CHEM] $C_6H_{13}C_6H_3(OH)_2$ Sharp-tasting, white to yellowish crystals melting at 64°C; slightly soluble in water, soluble in glycerin, vegetable oils, and organic solvents; used in medicine.   { ¦hek·səl·ri'sȯrs·ən,ȯl }

1-hexyne   [ORG CHEM] C_4H_9CCH A colorless, water-white liquid, either n-butylacetylene, boiling at 71.5°C, or methylpropylacetylene, boiling at 84°C.   { ¦wən 'hek,sēn }

Hf See hafnium.

HFC See hydrofluorocarbon.

hfs See hyperfine structure.

Hg See mercury.

high-density polyethylene   [ORG CHEM] A thermoplastic polyolefin with a density of 0.941–0.960 gram per cubic centimeter (0.543–0.555 ounce per cubic inch).   Abbreviated HDPE.   { ,hī ¦den·səd·ē ,päl·ē'eth·ə,lēn }

high-energy bond   [PHYS CHEM] Any chemical bond yielding a decrease in free energy of at least 5 kilocalories per mole.   { 'hī ,en·ər·jē 'bänd }

highest occupied molecular orbital   [PHYS CHEM] The highest-energy molecular orbital that is occupied by electrons.   Abbreviated HOMO.   { ,hī·əst ¦äk·yə·pīd mə¦lek·yə· lər 'ȯr·bə·təl }

high-frequency titration   [ANALY CHEM] A conductimetric titration in which two electrodes are mounted on the outside of the beaker or vessel containing the solution to be analyzed and an alternating current source in the megahertz range is used to measure the course of a titration.   { 'hī ¦frē·kwən·sē tī'trā·shən }

high-performance liquid chromatography   [ANALY CHEM] A type of column chromatography in which the solvent is conveyed through the column under pressure.   Abbreviated HPLC.   { ¦hī pər'fȯrm·əns ,lik·wəd ,krō·mə'täg·rə·fē }

high polymer   [ORG CHEM] A large molecule (of molecular weight greater than 10,000) usually composed of repeat units of low-molecular-weight species; for example, ethylene or propylene.   { 'hī 'päl·ə·mər }

high-pressure chemistry   [PHYS CHEM] The study of chemical reactions and phenomena that occur at pressures exceeding 10,000 bars (a bar is nearly equivalent to a kilogram per square centimeter), mainly concerned with the properties of the solid state.   { 'hī ¦presh·ər 'kem·ə·strē }

high-resolution electron energy loss spectroscopy   [SPECT] A type of electron energy loss spectroscopy in which electron scattering is performed by using a monoenergetic beam and electron energy analyzers to achieve a resolution of 5 to 10 millielectron-volts.   Abbreviated HREELS.   { 'hī ,rez·ə'lü·shən i'lek,trän 'en·ər·jē ,lȯs spek'träs· kə·pē }

high-temperature chemistry   [PHYS CHEM] The study of chemical phenomena occurring above about 500 K.   { 'hī ,tem·prə·chər 'kem·ə·strē }

Hill reaction   [ORG CHEM] Production of substituted phenylacetic acids by the oxidation of the corresponding alkylbenzene by potassium permanganate in the presence of acetic acid.   { 'hil rē,ak·shən }

Hinsberg test   [ANALY CHEM] A test to distinguish between primary and secondary amines; it involves reaction of an amine with benzene disulforyl chloride in alkaline solution; primary amines give sulfonamides that are soluble in basic solution; secondary amines give insoluble derivatives; tertiary amines do not react with the reagent. { 'hinz·bərg ‚test }

hippuric acid [ORG CHEM] $C_6H_5CONHCH_2 \cdot COOH$ Colorless crystals melting at 188°C; soluble in hot water, alcohol, and ether; used in medicine and as a chemical intermediate. { hi'pyür·ik 'as·əd }

Hittorf method [PHYS CHEM] A procedure for determining transference numbers in which one measures changes in the composition of the solution near the cathode and near the anode of an electrolytic cell, due to passage of a known amount of electricity. { 'hi·dȯrf ‚meth·əd }

HMPA *See* hexametapol.

Ho *See* holmium.

Hofmann amine separation [ORG CHEM] A technique to separate a mixture of primary, secondary, and tertiary amines; they are heated with ethyl oxalate; there is no reaction with tertiary amines, primary amines form a diamide, and the secondary amines form a monoamide; when the reaction mixture is distilled, the mixture is separated into components. { 'häf·mən 'am‚ēn ‚sep·ə‚rā·shən }

Hofmann degradation [ORG CHEM] The action of bromine and an alkali on an amide so that it is converted into a primary amine with one less carbon atom. { 'häf·mən deg·rə'dā·shən }

Hofmann exhaustive methylation reaction [ORG CHEM] The thermal decomposition of quaternary ammonium hydroxide compounds to yield an olefin and water; an exception is tetramethylammonium hydroxide, which decomposes to give an alcohol. { 'häf·mən ig‚zȯs·tiv ‚meth·ə'lā·shən rē‚ak·shən }

Hofmann mustard-oil reaction [ORG CHEM] Preparation of alkylisothiocyanates by heating together a primary amine, mercuric chloride, and carbon disulfide. { 'häf·mən 'məs·tərd ‚ȯil rē‚ak·shən }

Hofmann reaction [ORG CHEM] A reaction in which amides are degraded by treatment with bromine and alkali (caustic soda) to amines containing one less carbon; used commercially in the production of nylon. { 'häf·mən rē‚ak·shən }

Hofmann rearrangement [ORG CHEM] A chemical rearrangement of the hydrohalides of N-alkylanilines upon heating to give aminoalkyl benzenes. { 'häf·mən ‚rē·ə'rānj·mənt }

Hofmeister series [CHEM] An arrangement of anions or cations in order of decreasing ability to produce coagulation when their salts are added to lyophilic sols. Also known as lyotopic series. { 'hȯf‚mīs·tər ‚sir·ēz }

hole-burning spectroscopy [SPECT] A method of observing extremely narrow line widths in certain ions and molecules embedded in crystalline solids, in which broadening produced by crystal-site-dependent statistical field variations is overcome by having a monochromatic laser temporarily remove ions or molecules at selected crystal sites from their absorption levels, and observing the resulting dip in the absorption profile with a second laser beam. { 'hōl ‚bərn·iŋ spek'träs·kə·pē }

holmium [CHEM] A rare-earth element belonging to the yttrium subgroup, symbol Ho, atomic number 67, atomic weight 164.9304, melting point 1400–1525°C. { 'hōl·mē·əm }

homatropine [ORG CHEM] $C_{16}H_{21}O_3N$ An alkaloid that causes pupil dilation and paralysis of accommodation. { hōm'a·trə‚pēn }

homidium bromide *See* ethidium bromide. { hə'mid·ē·əm 'brō‚mīd }

homo- [ORG CHEM] **1.** Indicating the homolog of a compound differing in formula from the latter by an increase of one CH_2 group. **2.** Indicating a homopolymer made up of a single type of monomer, such as polyethylene from ethylene. **3.** Indicating that a skeletal atom has been added to a well-known structure. { 'hō·mō }

HOMO *See* Highest Occupied Molecular Orbital. { ‚āch‚ō‚em‚ō or 'hō‚mō }

homochiral *See* enantiomerically pure. { ‚hō·mə'kī·rəl }

homocyclic compound [ORG CHEM] A ring compound that has one type of atom in its structure; an example is benzene. { ‚hä·mə'sī·klik 'käm‚paůnd }

homogeneous [CHEM] Pertaining to a substance having uniform composition or structure. { ˌhä·mə'jē·nē·əs }

homogeneous catalysis [CHEM] Catalysis occurring within a single phase, usually a gas or liquid. { ˌhä·mə'jē·nē·əs kə'tal·ə·səs }

homogeneous chemical reaction [CHEM] Chemical reaction system in which all constitutents (reactants and catalyst) are of the same phase. { ˌhä·mə'jē·nē·əs ¦kem·i·kəl rē'ak·shən }

homogeneous nucleation [PHYS CHEM] The process of creation of vapor bubble nuclei in a superheated liquid away from bounding walls and in the absence of any foreign material. { ˌhō·mə¦jē·nē·əs ˌnü·klē'a·shən }

homologation [ORG CHEM] A type of hydroformylation in which carbon monoxide reacts with certain saturated alcohols to yield either aldehydes or alcohols (or a mixture of both) containing one more carbon atom than the parent. { hə,mäl·ə'gā·shən }

homology [CHEM] The relation among elements of the same group, or family, in the periodic table. [ORG CHEM] That state, in a series of organic compounds that differ from each other by a CH_2 such as the methane series C_{2n+2}, in which there is a similarity between the compounds in the series and a graded change of their properties. { hə'mäl·ə·jē }

homolysis See homolytic cleavage. { hə'mäl·ə·sis }

homolytic cleavage [ORG CHEM] The breaking of a single (two-electron) bond in which one electron remains on each of the atoms. Also known as free-radical reaction; homolysis. { ¦häm·ə,lid·ik 'klēv·ij }

homomorphs [CHEM] Chemical molecules that are similar in size and shape, but not necessarily having any other characteristics in common. { 'hä·mə,mȯrfs }

homonuclear molecule [CHEM] A diatomic molecule, both of whose atoms are of the same element. { ˌhō·mō¦nü·klē·ər 'mäl·ə,kyül }

homopolar bond [PHYS CHEM] A covalent bond whose total dipole moment is zero. { ¦hä·mə'pō·lər 'bänd }

homopolymer [ORG CHEM] A polymer formed from a single monomer; an example is polyethylene, formed by polymerization of ethylene. { ˌhä·mō'päl·ə·mər }

homozeotrope [CHEM] Mixture in which the liquid components are miscible in all proportions in the liquid phase, and may be separated by ordinary distillation. { ¦hä·mə'zē·ə,trōp }

Hopkins-Cole reaction [ANALY CHEM] The appearance of a violet ring when concentrated sulfuric acid is added to a mixture that includes a protein and glyoxylic acid; however, gelatin and zein do not show the reaction. { 'häp·kənz'kōl rē,ak·shən }

horizontal chromatography [ANALY CHEM] Paper chromatography in which the chromatogram is horizontal instead of vertical. { ˌhär·ə'zänt·əl ˌkrō·mə'täg·rə·fē }

Hortvet sublimator [ANALY CHEM] Device for the determination of the condensation temperature (sublimation point) of sublimed solids. { ¦hȯrt¦vet 'səb·lə,mād·ər }

host [CHEM] A crystalline lattice or receptor molecule for the strong and selective binding of a cationic, anionic, or neutral organic, inorganic, or biological substance (guest) by means of electrostatic, hydrogen-bonding, van der Waals, or donor-acceptor interactions. Examples include clathrates, crown ethers, cryptands, cyclodextrins, calixarenes, cavitands, cyclophanes, and cryptophanes. Also known as host substance. { hōst }

host-guest complexation chemistry [ORG CHEM] The design, synthesis, and study of highly structured organic molecular complexes that mimic biological complexes. { 'hōst 'gest ˌkäm·plek'sā·shən ˌkem·ə·strē }

host substance See host. { ¦hōst 'səb·stəns }

Houben-Hoesch synthesis [ORG CHEM] Condensation of cyanides with polyhydric phenols in the presence of hydrogen chloride and zinc chloride to yield phenolic ketones. { 'hü·bən 'hərsh 'sin·thə·səs }

HPLC See high-performance liquid chromatography.

HREELS See high-resolution electron energy loss spectroscopy. { 'äch ˌrēlz }

Hs See hassium.

Huber's reagent [ANALY CHEM] Aqueous solution of ammonium molybdate and potassium ferrocyanide used as a reagent to detect free mineral acid. { 'hyü·bərz rē,ā·jənt }

Hubl's reagent [ANALY CHEM] Solution of iodine and mercuric chloride in alcohol; used to determine the iodine content of oils and fats. { 'həb·əlz rē,ā·jənt }

Hückel's 4n + 2 rule [ORG CHEM] Aromatic (ring) compounds must have $4n + 2$ pi-bonding electrons, where n is a whole number and generally limited to $n = 0$ to 5. When $n = 1$, for example, there are six pi-electrons, as for benzene. Also known as Hückel's rule. { 'húk·əlz ¦fór ,en pləs 'tü ,rül }

Hückel's rule *See* Hückel's 4n + 2 rule. { 'húk·əlz ,rül }

Hull cell [PHYS CHEM] An electrodeposition cell that operates within a simultaneous range of known current densities. { 'həl ,sel }

humectant [CHEM] A substance which absorbs or retains moisture; examples are glycerol, propylene glycol, and sorbitol; used in preparing confectioneries and dried fruit. { hyü'mek·tənt }

humic acid [ORG CHEM] Any of various complex organic acids obtained from humus; insoluble in acids and organic solvents. { 'hyü·mik 'as·əd }

humidity indicator [INORG CHEM] Cobalt salt (for example, cobaltous chloride) that changes color as the surrounding humidity changes; changes from pink when hydrated, to greenish-blue when anhydrous. { hyü'mid·əd·ē ,in·də,kād·ər }

humin [ORG CHEM] An insoluble pigment formed in the acid hydrolysis of a protein that contains tryptophan. { 'hyü·mən }

Humphreys series [SPECT] A series of lines in the infrared spectrum of atomic hydrogen whose wave numbers are given by $R_H(1/36)l − (1/n^2)$, where R_H is the Rydberg constant for hydrogen, and n is any number greater than 6. { 'həm·frēz ,sir·ēz }

Hundsdieke reaction [ORG CHEM] Production of an alkyl halide by boiling a silver carboxylate with an equivalent weight of bromine in carbon tetrachloride. { 'hənz,dēk·ə rē,ak·shən }

hybridization [PHYS CHEM] The mixing together on the same atom of two or more orbitals that have similar energies, forming a hybrid orbital. { ,hī·brəd·ə'zā·shən }

hybridized orbital [PHYS CHEM] A molecular orbital which is a linear combination of two or more orbitals of comparable energy (such as 2s and 2p orbitals), is concentrated along a certain direction in space, and participates in formation of a directed valence bond. { 'hī·brəd,īzd 'ór·bad·əl }

hybrid orbital [PHYS CHEM] An orbital formed by the combination of two or more atomic orbitals on a single atom. { ¦hī·brəd 'ór·bad·əl }

hydantoin [ORG CHEM] $C_3N_2O_2H$ A white, crystalline compound, melting point 220°C; used as an intermediate in certain pharmaceutical manufacturing and as a textile softener and lubricant. Also known as glycolyurea. { hī'dant·ə·wən }

hydnocarpic acid [ORG CHEM] $C_{16}H_{28}O_2$ A nonedible fat and oil isolated from chaulmoogra oil, forming white crystals that melt at 60°C; used to treat Hansen's disease. { ¦hīd·nə¦kär·pik 'as·əd }

hydracrylic acid [ORG CHEM] $CH_2OH·CH_2COOH$ An oily liquid that is an isomer of lactic acid and that breaks down on heating to acrylic acid. { ¦hī·drə¦kril·ik 'as·əd }

hydrargyrum *See* mercury. { hī'drär·jə·rəm }

hydrastine [ORG CHEM] $C_{21}H_{21}NO_6$ An alkaloid isolated from species of the family Ranunculaceae and from *Hydrastis canadensis*; orthorhombic prisms crystallize from alcohol solution, melting point 132°C; highly soluble in acetone and benzene, soluble in chloroform, less soluble in ether and alcohol. { 'hī·dra,stēn }

hydrastinine [ORG CHEM] $C_{11}H_{13}O_3N$ A compound formed by the decomposition of hydrastine; crystallizes as needles from petroleum-ether solution, soluble in organic solvents such as alcohol, chloroform, and ether; used in medicine as a stimulant in coronary disease and as a hemostatic in uterine hemorrhage. { hī'dras·tə,nēn }

hydrate [CHEM] **1.** A form of a solid compound which has water in the form of H_2O molecules associated with it; for example, anhydrous copper sulfate is a white solid with the formula $CuSO_4$, but when crystallized from water a blue crystalline solid with formula $CuSO_4·5H_2O$ results, and the water molecules are an integral part of

hydrate aluminum oxide

the crystal. **2.** A crystalline compound resulting from the combination of water and a gas; frequently a constituent of natural gas that is under pressure. { 'hī,drāt }

hydrate aluminum oxide *See* alumina trihydrate. { 'hī,drāt ə'lü·mə·nəm 'äk,sīd }

hydrated alumina *See* alumina trihydrate. { 'hī,drād·əd ə'lü·mə·nə }

hydrated chloral *See* chloral hydrate. { 'hī,drād·əd 'klȯr·əl }

hydrated electron [PHYS CHEM] An electron released during ionization of a water molecule by water and surrounded by water molecules oriented so that the electron cannot escape. Also known as aqueous electron. { 'hī,drād·əd i'lek,trän }

hydrated lime *See* calcium hydroxide. { 'hī,drād·əd 'līm }

hydrated manganic hydroxide *See* manganic hydroxide. { 'hī,drād·əd maŋ'gan·ik hī'dräk,sīd }

hydrated mercurous nitrate [INORG CHEM] $Hg_2(NO_3)_2 \cdot 2H_2O$ Poisonous, light-sensitive crystals, soluble in warm water, decomposes at 70°C; used as an analytical reagent and in cosmetics and medicine. { 'hī,drād·əd mər'kyūr·əs 'nī,trāt }

hydrated silica *See* silicic acid. { 'hī,drād·əd 'sil·ə·kə }

hydrate inhibitor [CHEM] A material (such as alcohol or glycol) added to a gas stream to prevent the formation and freezing of gas hydrates in low-temperature systems. { 'hī,drāt in,hib·əd·ər }

hydration [CHEM] The incorporation of molecular water into a complex molecule with the molecules or units of another species; the complex may be held together by relatively weak forces or may exist as a definite compound. { hī'drā·shən }

hydrazide [INORG CHEM] An acyl hydrazine; a compound of the formula

where R may be an alkyl group. { 'hī·drə,zīd }

hydrazine [INORG CHEM] H_2NNH_2 A colorless, hygroscopic liquid, boiling point 114°C, with an ammonialike odor; it is reducing, decomposable, basic, and bifunctional; used as a rocket fuel, in corrosion inhibition in boilers, and in the synthesis of biologically active materials, explosives, antioxidants, and photographic chemicals. { 'hī·drə,zēn }

hydrazine hydrate [ORG CHEM] $H_2NNH_2OH_2O$ A colorless, fuming liquid that boils at 119.4°C; used as a component in jet fuels and as an intermediate in organic synthesis. { 'hī·drə,zēn 'hī,drāt }

hydrazinobenzene *See* phenylhydrazine. { hi¦draz·ə·nō'ben,zēn }

2-hydrazinoethanol *See* 2-hydroxyethylhydrazine. { ¦tü hī¦draz·ə·nō'e·thə,nȯl }

hydrazobenzene [ORG CHEM] $C_{12}H_{12}N_2$ A colorless, crystalline compound, melts at 132°C, slightly soluble in water, soluble in alcohol; used as an intermediate in the synthesis of benzidine. { ¦hī·drə·zō'ben,zēn }

hydrazoic acid [INORG CHEM] NHN:N Explosive liquid, a strong protoplasmic poison boiling at 37°C. { ¦hī·drə¦zō·ik 'as·əd }

hydrazone [ORG CHEM] A compound containing the grouping −NH·N:C−, and obtained from a condensation reaction involving hydrazines with aldehydes or ketones; has been used as an exotic fuel. { 'hī·drə,zōn }

hydride [INORG CHEM] A compound containing hydrogen and another element; examples are H_2S, which is a hydride although it may be properly called hydrogen sulfide, and lithium hydride, LiH. { 'hī,drīd }

hydrindantin [ORG CHEM] $C_{18}H_{10}O_6$ A compound used as a reagent for the photometric determination of amino acids. { ,hī·drən'dant·ən }

hydriodic acid [INORG CHEM] A yellow liquid that is a water solution of the gas hydrogen iodide; a solution of 59% hydrogen iodide produces a liquid that is constant-boiling; it is a strong acid used in organic synthesis and as a reagent in analytical chemistry. { 'hī·drē,äd·ik 'as·əd }

hydriodic acid gas *See* hydrogen iodide. { 'hī·drē,äd·ik ¦as·əd 'gas }

hydrobenzoin [ORG CHEM] $C_{14}H_{14}O_2$ A colorless, crystalline compound formed by action of sodium amalgam on benzaldehyde, melts at 136°C, and is slightly soluble in water. { ¦hī·drō'ben·zə·wən }

hydroboration [ORG CHEM] The process of producing organoboranes by the addition of a compound with a B-H bond to an unsaturated hydrocarbon; for example, the reaction of diborane ion with a carbonyl compound. Also known as boration. { ¦hī·drō·bə'rā·shən }

hydrobromic acid [INORG CHEM] HBr A solution of hydrogen bromide in water, usually 40%; a clear, colorless liquid; used in medicine, analytical chemistry, and synthesis of organic compounds. { ¦hī·drə'brō·mik 'as·əd }

hydrobromofluorocarbon [ORG CHEM] A compound consisting of hydrogen, bromine, fluorine, and carbon. Abbreviated HBFC. { ¦hī·drə¦brō·mō¦flúr·ō'kär·bən }

hydrocarbon [ORG CHEM] One of a very large group of chemical compounds composed only of carbon and hydrogen; the largest source of hydrocarbons is from petroleum crude oil. { ¦hī·drə'kär·bən }

hydrocarbon resins [ORG CHEM] Brittle or gummy materials prepared by the polymerization of several unsaturated constituents of coal tar, rosin, or petroleum; they are inexpensive and find uses in rubber and asphalt formulations and in coating and caulking compositions. { ¦hī·drə'kär·bən 'rez·ənz }

hydrochinone See hydroquinone. { ¦hī·drə·kə'nōn }

hydrochloric acid [INORG CHEM] HCl A solution of hydrogen chloride gas in water; a poisonous, pungent liquid forming a constant-boiling mixture at 20% concentration in water; widely used as a reagent, in organic synthesis, in acidizing oil wells, ore reduction, food processing, and metal cleaning and pickling. Also known as muriatic acid. { ¦hī·drə'klór·ik 'as·əd }

hydrochlorofluorocarbon [ORG CHEM] A compound composed of hydrogen, chlorine, fluorine, and carbon atoms. Also known as HCFC. { ¦hī·drə¦klór·ə'flúr·ə¦kär·bən }

hydrocinnamic acid [ORG CHEM] $C_6H_5CH_2CH_2COOH$ A compound whose crystals have a floral odor (hyacinth-rose) and melt at 46°C; used in perfumes and flavoring. { ¦hī·drō·si'nam·ik 'as·əd }

hydrocinnamic alcohol See phenylpropyl alcohol. { ¦hī·drō·si'nam·ik 'al·kə¦hòl }

hydrocinnamic aldehyde See phenylpropyl aldehyde. { ¦hī·drō·si'nam·ik 'al·də¦hīd }

hydrocrackate [ORG CHEM] The product of a hydrocracker. { ¦hī·drō'kra¦kāt }

hydrocyanic acid [INORG CHEM] HCN A highly toxic liquid that has the odor of bitter almonds and boils at 25.6°C; used to manufacture cyanide salts, acrylonitrile, and dyes, and as a fumigant in agriculture. Also known as formonitrile; hydrogen cyanide; prussic acid. { ¦hī·drō·sī'an·ik 'as·əd }

hydrofluoric acid [INORG CHEM] An aqueous solution of hydrogen fluoride, HF; colorless, fuming, poisonous liquid; extremely corrosive, it is a weak acid as compared to hydrochloric acid, but will attack glass and other silica materials; used to polish, frost, and etch glass, to pickle copper, brass, and alloy steels, to clean stone and brick, to acidize oil wells, and to dissolve ores. { ¦hī·drə'flúr·ik 'as·əd }

hydrofluorocarbon [ORG CHEM] A compound consisting of hydrogen, fluorine, and carbon. Abbreviated HFC. { ¦hī·drə¦flúr·ə'kär·bən }

hydrofluorosilicic acid See fluosilicic acid. { ¦hī·drō¦flúr·ō·sə'lis·ik 'as·əd }

hydrogel [CHEM] The formation of a colloid in which the disperse phase (colloid) has combined with the continuous phase (water) to produce a viscous jellylike product; for example, coagulated silicic acid. { 'hī·drə¦jel }

hydrogen [CHEM] The first chemical element, symbol H, in the periodic table, atomic number 1, atomic weight 1.00797; under ordinary conditions it is a colorless, odorless, tasteless gas composed of diatomic molecules, H_2; used in manufacture of ammonia and methanol, for hydrofining, for desulfurization of petroleum products, and to reduce metallic oxide ores. { 'hī·drə·jən }

hydrogenated oil [ORG CHEM] Unsaturated liquid vegetable oil that has had hydrogen catalytically added so as to convert the oil to a hydrogen-saturated solid. { 'hī·drə·jə¦nād·əd ¦òil }

hydrogenation [ORG CHEM] Catalytic reaction of hydrogen with other compounds, usually unsaturated; for example, unsaturated cottonseed oil is hydrogenated to form solid fats. { hī¦dräj·ə'nā·shən }

hydrogen bond [PHYS CHEM] A type of bond formed when a hydrogen atom bonded

hydrogen bromide

to atom A in one molecule makes an additional bond to atom B either in the same or another molecule; the strongest hydrogen bonds are formed when A and B are highly electronegative atoms, such as fluorine, oxygen, or nitrogen.   { 'hī·drə·jən 'bänd }

hydrogen bromide   [INORG CHEM]   HBr A hazardous, toxic gas used as a chemical intermediate and as an alkylation catalyst; forms hydrobromic acid in aqueous solution.   { 'hī·drə·jən 'brō͵mīd }

hydrogen chloride   [INORG CHEM]   HCl A fuming, highly toxic, colorless gas soluble in water, alcohol, and ether; used in the production of vinyl chloride and alkyl chloride, and in polymerization, isomerization, and other reactions.   { 'hī·drə·jən 'klȯr͵īd }

hydrogen cyanide   See hydrocyanic acid.   { 'hī·drə·jən 'sī·ə͵nīd }

hydrogen cycle   [CHEM]   The complete process of a cation-exchange operation in which the adsorbent is used in the hydrogen or free acid form.   { 'hī·drə·jən ͵sī·kəl }

hydrogen disulfide   See hydrogen sulfide.   { 'hī·drə·jən dī'səl͵fīd }

hydrogen electrode   [PHYS CHEM]   A noble metal (such as platinum) of large surface area covered with hydrogen gas in a solution of hydrogen ion saturated with hydrogen gas; metal is used in a foil form and is welded to a wire sealed in the bottom of a hollow glass tube, which is partially filled with mercury; used as a standard electrode with a potential of zero to measure hydrogen ion activity.   { 'hī·drə·jən i'lek͵trōd }

hydrogen equivalent   [CHEM]   The number of replaceable hydrogen atoms or hydroxyl groups in a molecule of an acid or a base.   { 'hī·drə·jən i'kwiv·ə·lənt }

hydrogen fluoride   [INORG CHEM]   HF The hydride of fluoride; anhydrous HF is a mobile, colorless, liquid that fumes in air, melts at $-83°C$, boils at $19.8°C$; used to make fluorine-containing refrigerants (such as Freon) and organic fluorocarbon compounds, as a catalyst in alkylate gasoline manufacture, as a fluorinating agent, and in preparation of hydrofluoric acid.   { 'hī·drə·jən 'flu̇r͵īd }

hydrogen iodide   [INORG CHEM]   HI A water-soluble, colorless gas that may be used in organic synthesis and as a reagent.   Also known as hydriodic acid gas.   { 'hī·drə·jən 'ī·ə͵dīd }

hydrogen ion   See hydronium ion.   { 'hī·drə·jən 'ī͵än }

hydrogen ion concentration   [CHEM]   The normality of a solution with respect to hydrogen ions, H^+; it is related to acidity measurements in most cases by $pH = \log \frac{1}{2} [1/(H^+)]$, where (H^+) is the hydrogen ion concentration in gram equivalents per liter of solution.   { 'hī·drə·jən 'ī͵än ͵käns·ən͵trā·shən }

hydrogen ion exponent   [CHEM]   A way of expressing pH; namely, $pH = -\log c_H$, where c_H = hydrogen ion concentration.   { 'hī·drə·jən 'ī͵än ik'spō·nənt }

hydrogen line   [SPECT]   A spectral line emitted by neutral hydrogen having a frequency of 1420 megahertz and a wavelength of 21 centimeters; radiation from this line is used in radio astronomy to study the amount and velocity of hydrogen in the Galaxy.   { 'hī·drə·jən ͵līn }

hydrogenolysis   [CHEM]   A reaction in which hydrogen gas causes a chemical change that is similar to the role of water in hydrolysis.   { 'hī·drə·jə'näl·ə·səs }

hydrogenous   [CHEM]   Of, pertaining to, or containing hydrogen.   { hī'dräj·ə·nəs }

hydrogen oxide   See water.   { 'hī·drə·jən 'äk͵sīd }

hydrogen peroxide   [INORG CHEM] H_2O_2 Unstable, colorless, heavy liquid boiling at 158°C; soluble in water and alcohol; used as a bleach, chemical intermediate, rocket fuel, and antiseptic.   Also known as peroxide.   { 'hī·drə·jən pə'räk͵sīd }

hydrogen phosphide   See phosphine.   { 'hī·drə·jən 'fäs͵fīd }

hydrogen selenide   [INORG CHEM] H_2Se A toxic, colorless gas, soluble in water, carbon disulfide, and phosgene; used to make metallic selenides and organoselenium compounds and in the preparation of semiconductor materials.   { 'hī·drə·jən 'sel·ə͵nīd }

hydrogen sulfide   [INORG CHEM] H_2S Flammable, toxic, colorless gas with offensive odor, boiling at $-60°C$; soluble in water and alcohol; used as an analytical reagent, as a sulfur source, and for purification of hydrochloric and sulfuric acids.   Also known as hydrogen disulfide.   { 'hī·drə·jən 'səl͵fīd }

hydrogen tellurate   See telluric acid.   { 'hī·drə·jən 'tel·yə͵rāt }

hydroiodic ether   See ethyl iodide.   { ͵hī·drȯi'äd·ik 'ē·thər }

186

hydrolysis [CHEM] **1.** Decomposition or alteration of a chemical substance by water. **2.** In aqueous solutions of electrolytes, the reactions of cations with water to produce a weak base or of anions to produce a weak acid. { hī'dräl·ə·səs }

hydrolytic process [CHEM] A reaction of both organic and inorganic chemistry wherein water effects a double decomposition with another compound, hydrogen going to one compound and hydroxyl to another. { ¦hī·drə¦lid·ik 'prä·səs }

hydronium ion [INORG CHEM] H_3O^+ An oxonium ion consisting of a proton combined with a molecule of water; found in pure water and in all aqueous solutions. Also known as hydrogen ion. { hī'drō·nē·əm ¦ī,än }

hydrophile-lipophile balance [ORG CHEM] The relative simultaneous attraction of an emulsifier for two phases of an emulsion system; for example, water and oil. { 'hī·drə,fīl 'lip·ə,fīl ,bal·əns }

hydrophilic [CHEM] Having an affinity for, attracting, adsorbing, or absorbing water. { ,hī·drə'fil·ik }

hydrophobic [CHEM] Lacking an affinity for, repelling, or failing to adsorb or absorb water. { ¦hī·drə'fō·bik }

hydroquinol See hydroquinone. { ¦hī·drō'kwi,nȯl }

hydroquinone [ORG CHEM] $C_6H_4(OH)_2$ White crystals melting at 170°C and boiling at 285°C; soluble in alcohol, ether, and water; used in photographic dye chemicals, in medicine, as an antioxidant and inhibitor, and in paints, varnishes, and motor fuels and oils. Also known as hydrochinone; hydroquinol; quinol. { ¦hī·drə·kwə'nōn }

hydroquinone dimethyl ether [ORG CHEM] $C_6H_4(OCH_3)_2$ White flakes with a melting point of 56°C; used as a weathering agent in paint, as a flavoring, and in dyes and cosmetics. { ¦hī·drə·kwə'nōn dī¦meth·əl 'ē·thər }

hydroquinone monomethyl ether [ORG CHEM] $CH_3OC_6H_4OH$ A white, waxy solid with a melting point of 52.5°C; soluble in benzene, acetone, and alcohol; used for antioxidants, pharmaceuticals, and dyestuffs. { ¦hī·drə·kwə'nōn ,män·ō¦meth·əl 'ē·thər }

hydrosilylation [ORG CHEM] The addition of a Si-H bond to a C-C double bond of an olefin. { ¦hī·drō,sil·ē'ā·shən }

hydrosol [CHEM] A colloidal system in which the dispersion medium is water, and the dispersed phase may be a solid, a gas, or another liquid. Also known as aquasol. { 'hī·drə,sȯl }

hydrosulfide [CHEM] A compound that has the SH— radical; for example, sulfhydrates, sulfhydryls, thioalcohols, thiols, sulfur alcohols, and mercaptans. { ¦hī·drə'səl,fīd }

hydrotrope [CHEM] Compound with the ability to increase the solubilities of certain slightly soluble organic compounds. { 'hī·drə,trōp }

hydrous [CHEM] Indicating the presence of an indefinite amount of water. { 'hī·drəs }

hydroxamic acid [ORG CHEM] An organic compound that contains the group $-C(=O)$-NHOH. { ¦hī,dräk¦sam·ik 'as·əd }

hydroxide [CHEM] Compound containing the OH^- group; the hydroxides of metals are usually bases and those of nonmetals are usually acids; a hydroxide can be organic or inorganic. { hī'dräk,sīd }

hydroximino See nitroso. { ¦hī,dräk'sim·ə·nō }

hydroxisoxazole [ORG CHEM] $C_4H_5NO_2$ A colorless, crystalline compound with a melting point of 86–87°C; used as a fungicide in soil and as a growth regulator for seeds. Also known as 3-hydroxy-5-methylisoxazole; hymexazol. { hī¦dräk·sə¦säk·sə,zōl }

hydroxy- [ORG CHEM] Chemical prefix indicating the OH^- group in an organic compound, such as hydroxybenzene for phenol, C_6H_5OH; the use of just oxy- for the prefix is incorrect. Also spelled hydroxyl-. { hī'dräk·sē }

hydroxyacetic acid See glycolic acid. { hī¦dräk·sə·ə'sēd·ik 'as·əd }

hydroxy acid [ORG CHEM] Any organic acid, with an OH^- group, such as hydroxyacetic acid. { hī'dräk·sē 'as·əd }

hydroxybenzoic acid [ORG CHEM] $C_7H_6O_3$ Any one of three crystalline derivatives of benzoic acid: ortho, meta, and para forms; the ester of the para compound is used as a bacteriostatic agent. { hī¦dräk·sē·ben'zō·ik 'as·əd }

187

para-hydroxybenzoic acid

para-hydroxybenzoic acid [ORG CHEM] $C_6H_4(OH)COOH \cdot 2H_2O$ Colorless crystals melting at 210°C; soluble in alcohol, water, and ether; used as a chemical intermediate and for synthetic drugs. { ¦par·ə hī¦dräk·sē·ben'zō·ik 'as·əd }

2-hydroxybiphenyl *See* phenylphenol. { ¦tü hī¦dräk·sē·bī'fen·əl }

hydroxycarbonyl compound [ORG CHEM] A compound possessing one or more hydroxy (—OH) groups and one or more carbonyl (=C=O) groups. { hī¦dräk·sē'kär·bə·nəl 'käm,paúnd }

hydroxycholine *See* muscarine. { hī¦dräk·sē'kō,lēn }

hydroxycinchonine *See* cupreine. { hī¦dräk·sē'siŋ·kə,nēn }

hydroxycitronellal [ORG CHEM] $C_{10}H_{20}O_2$ A colorless or light yellow, viscous liquid with a boiling range of 94–96°C; soluble in 50% alcohol and fixed oils; used in perfumery and flavoring. Also known as citronellal hydrate. { hī¦dräk·sē,sī·trə'nel·əl }

2-(hydroxydiphenyl)methane [ORG CHEM] $C_6H_5CH_2C_6H_4OH$ A crystalline substance with a melting point of 20.2–20.9°C, or a liquid; used as a germicide, preservative, and antiseptic. { ¦tü hī¦dräk·sē·dī'fen·əl 'meth,ān }

2-hydroxyethylhydrazine [ORG CHEM] $HOCH_2CH_2NHNH_2$ A colorless, slightly viscous liquid with a melting point of-70°C; soluble in lower alcohols; used as an abscission agent in fruit. Also known as 2-hydrazinoethanol. { ¦tü hī¦dräk·sē¦eth·əl'hī·drə,zēn }

hydroxyl- *See* hydroxy-. { hī'dräk·səl }

hydroxylamine [INORG CHEM] NH_2OH A colorless, crystalline compound produced commercially by acid hydrolysis of nitroparaffins, decomposes on heating, melts at 33°C; used in organic synthesis and as a reducing agent. { ,hī,dräk'sil·ə,mēn }

hydroxylamine hydrochloride [ORG CHEM] $(NH_2OH)Cl$ A crystalline substance with a melting point of 151°C; soluble in glycerol and propylene glycol; used as a reducing agent in photography and in synthetic and analytic chemistry, as an antioxidant in fatty acids and soaps, and as a reagent for enzyme reactivation. { ,hī,dräk'sil·ə,mēn ,hī·drə'klòr,īd }

ortho-hydroxylaniline [ORG CHEM] $C_6H_4NH_2OH$ White crystals that turn brownish upon standing for some time; melts at 172–173°C, and will sublime upon more heating; soluble in cold water and benzene; used as a dye for hair and furs, and as a dye intermediate. Also known as *ortho*-aminophenol; oxammonium. { ¦òr·thō ,hī¦dräk·səl'an·əl·ən }

hydroxylation reaction [ORG CHEM] One of several types of reactions used to introduce one or more hydroxyl groups into organic compounds; an oxidation reaction as opposed to hydrolysis. { hī,dräk·sə'lā·shən rē,ak·shən }

β-hydroxynaphthoic acid [ORG CHEM] $C_{10}H_6OHCOOH$ A yellow solid that is soluble in ether and alcohol and melts at about 218°C; used as a dye and a pigment. { ¦bād·ə hī¦dräk·sē·naf'thō·ik 'as·əd }

4-hydroxy-3-nitrobenzenearsonic acid [ORG CHEM] $HOC_6H_3(NO_2)AsO(OH)_2$ Crystals used as a reagent for zirconium; also used to control enteric infections and to improve growth and feed efficiency in animals. Also known as roxarsone. { ¦fòr hī¦dräk·sē ¦thrē ,nī·trō¦ben,zēn·är'sän·ik 'as·əd }

8-hydroxyquinoline [ORG CHEM] C_9H_6NOH White crystals or powder that darken on exposure to light, slightly soluble in water, soluble in benzene, melting at 73–75°C; used in preparing fungicides and in the separation of metals by acting as a precipitating agent. Also known as oxine; oxyquinoline; 8-quinolinol. { ¦āt hī¦dräk·sē'kwin·ə·lən }

3-hydroxytyramine hydrobromide [ORG CHEM] $(HO)_2C_6H_3CH_2CH_2NH_2 \cdot HBr$ A source of dopamine for the synthesis of catecholamine analogs. { ¦thrē hī¦dräk·sē'tī·rə,mēn ,hī·drə'brō,mīd }

hygroscopic [CHEM] **1.** Possessing a marked ability to accelerate the condensation of water vapor; applied to condensation nuclei composed of salts which yield aqueous solutions of a very low equilibrium vapor pressure compared with that of pure water at the same temperature. **2.** Pertaining to a substance whose physical characteristics are appreciably altered by effects of water vapor. **3.** Pertaining to water absorbed by dry soil minerals from the atmosphere; the amounts depend on the physicochemical character of the surfaces, and increase with rising relative humidity. { ¦hī·grə ¦skäp·ik }

hygroscopic depression [CHEM] The measure of a desiccant's capacity to take on water. { ¦hī·grə¦skäp·ik di'presh·ən }

hymecromone [ORG CHEM] $C_{10}H_8O_3$ A crystalline substance with a melting point of 194–195°C; soluble in methanol and glacial acetic acid; used as choleretic and antispasmodic drugs and as a standard for the fluorometric determination of enzyme activity. { hī'mek·rə,mōn }

hymexazol See hydroxisoxazole. { hī'mek·sə,zȯl }

hyoscyamine [ORG CHEM] $C_{17}H_{23}O_3N$ A white, crystalline alkaloid isolated from henbane, belladonna, and other plants of the family Solanaceae, which is freely soluble in alcohol and dilute acids; used in medicine as an anticholinergic. { ,hī·ə'sī·ə,mēn }

hyperchromicity [PHYS CHEM] An increase in the absorption of ultraviolet light by polynucleotide solutions due to a loss of the ordered secondary structure. { ,hī·pər·krō'mis·əd·ē }

hyperconjugation [PHYS CHEM] An arrangement of bonds in a molecule that is similar to conjugation in its formulation and manifestations, but the effects are weaker; it occurs when a CH_2 or CH_3 group (or in general, an AR_2 or AR_3 group where A may be any polyvalent atom and R any atom or radical) is adjacent to a multiple bond or to a group containing an atom with a lone π-electron, π-electron pair or quartet, or π-electron vacancy; it can be sacrificial (relatively weak) or isovalent (stronger). { ,hī·pər,kän·jə'gā·shən }

hyperfine structure [SPECT] A splitting of spectral lines due to the spin of the atomic nucleus or to the occurrence of a mixture of isotopes in the element. Abbreviated hfs. { 'hī·pər,fīn 'strək·chər }

hypergolic [CHEM] Capable of igniting spontaneously upon contact. { ¦hī·pər¦gäl·ik }

hypervalent atom [CHEM] A central atom in a single-bonded structure that imparts more than eight valence electrons in forming covalent bonds. { ,hī·pər'vā·lənt 'ad·əm }

hypervalent compounds [CHEM] Stable compounds of the main group elements in the third row of the periodic table, such as silicon, phosphorus, and sulfur, that have more than eight valence shell electrons, for example, PF_5 and SF_4. { ,hī·pər¦vā·lənt 'käm,paúnz }

hypo See sodium thiosulfate. { 'hī·pō }

hypochlorite [INORG CHEM] ClO_3sw A negative ion derived from hypochlorous acid, $HClO$; the ion is an oxidizing agent and a constituent of bleaching agents. { ,hī· pə'klȯr,īt }

hypochlorous acid [INORG CHEM] $HOCl$ Weak, unstable acid existing in solution only; its salts (such as calcium hypochlorite) are used as bleaching agents. { ¦hī·pə'klȯr· əs 'as·əd }

hypochromicity [PHYS CHEM] A decrease in the absorption of ultraviolet light by polynucleotide solutions due to the formation of an ordered secondary structure. { ,hī· pə·krə'mis·əd·ē }

hypohalous acid [INORG CHEM] An oxyacid of a halogen (fluorine, chlorine, bromine, iodine, or astatine) possessing the general chemical formula HOX, where X is the halogen atom. { hī¦pöt·ə·ləs 'as·əd }

hypoiodous acid [INORG CHEM] HIO A very weak unstable acid that occurs as the result of the weak hydrolysis of iodine in water. { ¦hī·pō,ī'ōd·əs 'as·əd }

189

I *See* iodine.

IBA *See* indolebutyric acid.

IBIB *See* isobutyl isobutyrate.

ibogaine [ORG CHEM] $C_{26}H_{32}O_2N_2$ An alkaloid isolated from the stems and leaves of the shrub *Tabernanthe iboga*, crystallizing from absolute ethanol as prismatic needles, melting at 152–153°C, soluble in ethanol, ether, and chloroform; used in medicine. { ə'bō·gə,ēn }

ice [PHYS CHEM] **1.** The dense substance formed by the freezing of water to the solid state; has a melting point of 32°F (0°C) and commonly occurs in the form of hexagonal crystals. **2.** A layer or mass of frozen water. { īs }

ice color *See* azoic dye. { 'īs ,kəl·ər }

ice crystal [PHYS CHEM] Any one of a number of macroscopic crystalline forms in which ice appears, including hexagonal columns, hexagonal platelets, dendritic crystals, ice needles, and combinations of these forms; although the crystal lattice of ice is hexagonal in its symmetry, varying conditions of temperature and vapor pressure can lead to growth of crystalline forms in which the simple hexagonal pattern is almost undiscernible. { 'īs ,krist·əl }

ice needle [PHYS CHEM] A long, thin ice crystal whose cross section perpendicular to its long dimension is typically hexagonal. Also called ice spicule. { 'īs ,nēd·əl }

ice point [PHYS CHEM] The true freezing point of water; the temperature at which a mixture of air-saturated pure water and pure ice may exist in equilibrium at a pressure of 1 standard atmosphere (101,325 pascals). { 'īs ,pòint }

ice spicule *See* ice needle. { 'īs ¦spik·yəl }

ice splinters [PHYS CHEM] Minute, electrically charged fragments of ice which have been observed under laboratory conditions to be torn away from dendritic crystals or spatial aggregates exposed to moving air. { 'īs ¦splin·tərz }

ICP-AES *See* inductively coupled plasma-atomic emission spectroscopy.

IDA *See* iminodiacetic acid.

ideal solution [CHEM] A solution that conforms to Raoult's law over all ranges of temperature and concentration and shows no internal energy change on mixing and no attractive force between components. { ī'dēl sə'lü·shən }

ignite [CHEM] To start a fuel burning. { ig'nīt }

ignition [CHEM] The process of starting a fuel mixture burning, or the means for such a process. { ig'nish·ən }

ignition point *See* ignition temperature. { ig'nish·ən ,pòint }

ignition temperature [CHEM] The lowest temperature at which combustion begins and continues in a substance when it is heated in air. Also known as autogenous ignition temperature; ignition point. { ig'nish·ən ,tem·prə·chər }

Ilkovič equation [ANALY CHEM] Mathematical relationship between diffusion current, diffusion coefficient, and active-substance concentration; used for polarographic analysis calculations. { 'il·kə,vich i,kwā·zhən }

imbibition [PHYS CHEM] Absorption of liquid by a solid or a semisolid material. { ,im·bə'bish·ən }

imidazole [ORG CHEM] $C_3H_4N_2$ One of a group of organic heterocyclic compounds containing a five-membered diunsaturated ring with two nonadjacent nitrogen atoms

imidazolyl

as part of the ring; the particular compound imidazole is a member of the group. { ¦im·ə'da¦zōl }

imidazolyl [ORG CHEM] $C_3H_3N_2$· A free radical derived from imidazole. { ¦im·ə'da· zə¦lil }

imide [ORG CHEM] **1.** A compound derived from acid anhydrides by replacing the oxygen (O) with the =NH group. **2.** A compound that has either the =NH group or a secondary amine in which R is an acyl functional group, as R_2NH. { 'i¦mīd }

imine [ORG CHEM] A class of compounds that are the product of condensation reactions of aldehydes or ketones with ammonia or amines; they have the NH radical attached to the carbon with the double bond, as R−HC=NH; an example is benzaldimine. { 'i¦mēn }

imino acid [ORG CHEM] Organic acid in which the =NH group is attached to one or two carbons; for example, acetic acid, $NH(CH_2COOH)_2$. { 'im·ə¦nō 'as·əd }

imino compound [ORG CHEM] A compound that has the =NH radical attached to one or two carbon atoms. { 'im·ə¦nō ¦käm¦paùnd }

iminodiacetic acid [ORG CHEM] $C_4H_7NO_4$ A crystalline substance used as an intermediate in the manufacture of chelating agents, surface-active agents, and complex salts. Abbreviated IDA. Also known as diglycine; iminodiethanoic acid. { ¦im·ə·nō·dī· ə'sēd·ik 'as·əd }

imino nitrogen [ORG CHEM] Nitrogen combined with hydrogen in the imino group. { 'im·ə¦nō 'nī·trə·jən }

immersion sampling [ANALY CHEM] Collection of a liquid sample for laboratory or other analysis by immersing a container in the liquid and filling it. { ə'mər·zhən ¦sam·pliŋ }

immiscible [CHEM] Pertaining to liquids that will not mix with each other. { i'mis· ə·bəl }

immobilized catalyst [CHEM] A molecular catalyst that is bound without substantial change in its structure to an insoluble solid to prevent solution of the catalyst in the contacting liquid. Also known as anchored catalyst. { i¦mō·bə¦līzd 'kad·ə¦list }

imperial red [INORG CHEM] Any of the red varieties of ferric oxide used as pigment. { im'pir·ē·əl 'red }

implosion [CHEM] The sudden reduction of pressure by chemical reaction or change of state which causes an inrushing of the surrounding medium. { im'plō·zhən }

In See indium.

inactive tartaric acid See racemic acid. { in'ak·tiv tär'tär·ik 'as·əd }

incineration [CHEM] The process of burning a material so that only ashes remain. { in¦sin·ə'rā·shən }

inclusion complex [CHEM] An unbonded association in which the molecules of one component are contained wholly or partially within the crystal lattice of the other component. { in'klü·zhən 'käm¦pleks }

incomplete combustion [CHEM] Combustion in which oxidation of the fuel is incomplete. { ¦in·kəm'plēt kəm'bəs·shən }

increment [ANALY CHEM] An individual portion of material of a group of samples collected by a single operation of a sampling device from parts of a lot that are separated in time or space. { 'iŋ·krə·mənt }

incubation [CHEM] Maintenance of chemical mixtures at specified temperatures for varying time periods to study chemical reactions, such as enzyme activity. { ¦iŋ· kyə'bā·shən }

indamine [ORG CHEM] $HN:C_6H_4:N·C_6H_4NH_2$ An unstable dye obtained by the reaction of *para*-phenylenediamine and aniline. Also known as phenylene blue. { 'in·də¦mēn }

indan [ORG CHEM] $C_6H_4(CH_2)_3$ Colorless liquid boiling at 177°C; soluble in alcohol and ether, insoluble in water; derived from coal tar. { 'in¦dan }

indanthrone [ORG CHEM] $C_{28}H_{14}N_2O_4$ A blue pigment or vat dye soluble in dilute base solutions; used in cotton dyeing and as a pigment in paints and enamels. { in'dan¦thrōn }

indene [ORG CHEM] C_9H_8 A colorless, liquid, polynuclear hydrocarbon; boils at 181°C and freezes at −2°C; derived from coal tar distillates; copolymers with benzofuran have been manufactured on a small scale for use in coatings and floor coverings. { 'in₁dēn }

independent migration law [ANALY CHEM] The law that each ion in a conductiometric titration contributes a definite amount to the total conductance, irrespective of the nature of the other ions in the electrolyte. { ,in·də'pen·dənt mī'grā·shən ,lò }

index of unsaturation [ORG CHEM] A numerical value that represents the number of rings or double bonds in a molecule; a triple bond is considered to have the numerical value of 2. { 'in₁deks əv ¦ən,sach·ə'ra·shən }

indican [ORG CHEM] $C_{14}H_{17}O_6N$ A glucoside of indoxyl occurring in the indigo plant; on hydrolysis indican gives rise to indoxyl, which is oxidized to indigo by air. { 'in·də,kan }

indicator See chemical indicator. { 'in·də,kād·ər }

indigo [ORG CHEM] **1.** A blue dye extracted from species of the Indigofera bush. **2.** See indigo blue. { 'in·də·gō }

indigo blue [ORG CHEM] $C_{16}H_{10}O_2N_2$ A component of the dye indigo, crystallizing as dark-blue rhomboids that break down at 30°C, that are soluble in hot aniline and hot chloroform, and that are also made synthetically; used as a reagent and a dye. Also known as indigo. { 'in·də·gō 'blü }

indigo carmine [ORG CHEM] $C_{16}H_8N_2Na_2O_8S_2$ A dark blue powder with coppery luster; used as a dye in testing kidney function and as a reagent in detecting chlorate and nitrate. Also known as soluble indigo blue. { 'in·də·gō 'kär·mən }

indigoid dye [ORG CHEM] Any of the vat dyes with $C_{16}H_{10}O_2N_2$ (indigo) or $C_{16}H_8S_2O_2$ (thioindigo) groupings; used to dye cotton and rayon, sometimes silk. { 'in·də,gòid ,dī }

indigo red [ORG CHEM] $C_{16}H_{10}O_2N_2$ A red isomer of indigo obtained in the manufacture of indigo. Also known as indirubin. { 'in·də·gō 'red }

indirect effect [PHYS CHEM] A chemical effect of ionizing radiation on a dilute solution caused by the interaction of solute molecules with highly reactive transient molecules or ions formed by reaction of the radiation with the solvent. { ,in·də'rekt i'fekt }

indirubin See indigo red. { ,in·də'rü·bən }

indium [CHEM] A metallic element, symbol In, atomic number 49, atomic weight 114.82; soluble in acids; melts at 156°C, boils at 1450°C. { 'in·dē·əm }

indium antimonide [INORG CHEM] InSb Crystals that melt at 535°C; an intermetallic compound having semiconductor properties and the highest room-temperature electron mobility of any known material; used in Hall-effect and magnetoresistive devices and as an infrared detector. { 'in·dē·əm ,an'tim·ə,nīd }

indium arsenide [INORG CHEM] InAs Metallic crystals that melt at 943°C; an intermetallic compound having semiconductor properties; used in Hall-effect devices. { 'in·dē·əm ¦ärs·ən,īd }

indium chloride [INORG CHEM] InCl₃ Hygroscopic white powder, soluble in water and alcohol. { 'in·dē·əm 'klòr,īd }

indium phosphide [INORG CHEM] InP A metallic mass that is brittle and melts at 1070°C; an intermetallic compound having semiconductor properties. { 'in·dē·əm 'fäs,fīd }

indium sulfate [INORG CHEM] In₂(SO₄)₃ Deliquescent, water-soluble, grayish powder; decomposes when heated. { 'in·dē·əm 'səl,fāt }

indogen [ORG CHEM] The functional group $C_6H_4(NH)COC=$; it occurs, for example, in the molecule indigo. { 'in·də·jən }

indogenide [ORG CHEM] A compound containing the function group $C_6H_4(NH)·CO·C=$ from indogen. { 'in·də·jə,nīd }

indole [ORG CHEM] Carcinogenic, white to yellowish scales with unpleasant aroma; soluble in alcohol, ether, hot water, and fixed oils; melt at 52°C; used as a chemical reagent and in perfumery and medicine. Also known as 2,3-benzopyrrole. { 'in ,dōl }

indolebutyric acid [ORG CHEM] $C_{12}H_{13}O_2N$ A crystalline acid similar to indoleacetic acid in auxin activity. Abbreviated IBA. { ¦in,dōl·byü'tir·ik 'as·əd }

indoxyl [ORG CHEM] $(C_8H_6N)OH$ A yellow crystalline glycoside, used as an intermediate in the manufacture of indigo. { in'däk·səl }

induction force [PHYS CHEM] A type of van der Waals force resulting from the interaction of the dipole moment of a polar molecule and the induced dipole moment of a nonpolar molecule. Also known as Debye force. { in'dək·shən ‚förs }

induction period [PHYS CHEM] A time of acceleration of a chemical reaction from zero to a maximum rate. { in'dək·shən ‚pir·ē·əd }

inductive effect [PHYS CHEM] In a molecule, a shift of electron density due to the polarization of a bond by a nearby electronegative or electropositive atom. { in'dək·tiv ə'fekt }

inductively coupled plasma-atomic emission spectroscopy [SPECT] A type of atomic spectroscopy in which the light emitted by atoms and ions in an inductively coupled plasma is observed. Abbreviated ICP-AES. { in'dək·tiv·lē ¦kəp·əld ¦plaz·mə ə¦täm·ik i¦mish·ən spek'träs·kə·pē }

industrial alcohol [ORG CHEM] Ethyl alcohol that has been denatured by acetates, ketones, gasoline, or other additives to make it unfit for beverage purposes. { in'dəs·trē·əl 'al·kə‚hòl }

inert gas See noble gas. { i'nərt 'gas }

inflammability See flammability. { in‚flam·ə'bil·əd·ē }

infrared reflectography [ANALY CHEM] In art conservation, a nondestructive digital imaging technique used to investigate underdrawings (below the painted surface) of paintings. { ¦in·frə‚red ‚rē‚flek'täg·rə·fē }

infrared spectrometer [SPECT] An instrument used to identify and measure the concentration of chemical compounds (gases, nonaqueous liquids, and solids) with electromagnetic radiation from 800 nanometers to 1 millimeter. { ¦in·frə¦red spek'träm·əd·ər }

infrared spectrophotometry [SPECT] Spectrophotometry in the infrared region, usually for the purpose of chemical analysis through measurement of absorption spectra associated with rotational and vibrational energy levels of molecules. { ¦in·frə¦red ¦spek·trə·fə'täm·ə·trē }

infrared spectroscopy [SPECT] The study of the properties of material systems by means of their interaction with infrared radiation; ordinarily the radiation is dispersed into a spectrum after passing through the material. { ¦in·frə¦red spek'träs·kə·pē }

infusion [CHEM] The aqueous solution of a soluble constituent of a substance as the result of the substance's steeping in the solvent for a period of time. { in'fyü·zhən }

ingrain color See azoic dye. { 'in‚grān ‚kəl·ər }

inhibitor [CHEM] A substance which is capable of stopping or retarding a chemical reaction; to be technically useful, it must be effective in low concentration. { in'hib·əd·ər }

initiation step [CHEM] The reaction that causes a chain reaction to begin but is not itself the principal source of products. { i‚nish·ē'ā·shən ‚step }

initiator [CHEM] The substance or molecule (other than reactant) that initiates a chain reaction, as in polymerization; an example is acetyl peroxide. { i'nish·ē‚ād·ər }

inorganic [INORG CHEM] Pertaining to or composed of chemical compounds that do not contain carbon as the principal element (excepting carbonates, cyanides, and cyanates), that is, matter other than plant or animal. { ¦in·òr¦gan·ik }

inorganic acid [INORG CHEM] A compound composed of hydrogen and a nonmetal element or radical; examples are hydrochloric acid, HCl, sulfuric acid, H_2SO_4, and carbonic acid, H_2CO_3. { ¦in·òr¦gan·ik 'as·əd }

inorganic chemistry [CHEM] The study of chemical reactions and properties of all the elements and their compounds, with the exception of hydrocarbons, and usually including carbides, oxides of carbon, metallic carbonates, carbon-sulfur compounds, and carbon-nitrogen compounds. { ¦in·òr¦gan·ik 'kem·ə·strē }

inorganic peroxide [INORG CHEM] An inorganic compound containing an element at its highest state of oxidation (such as perchloric acid, $HClO_4$), or having the peroxy group, $-O-O-$ (such as perchromic acid, $H_3CrO_8·2H_2O$). { ¦in·òr¦gan·ik pə'räk‚sīd }

inorganic pigment [INORG CHEM] A natural or synthetic metal oxide, sulfide, or other salt used as a coloring agent for paints, plastics, and inks. { ¦in·òr¦gan·ik 'pig·mənt }

inorganic polymer [INORG CHEM] Large molecules, usually linear or branched chains with atoms other than carbon in their backbone; an example is glass, an inorganic polymer made up of rings and chains of repeating silicate units. { |in·ȯr|gan·ik 'päl· ə·mər }

inositol [ORG CHEM] $C_6H_6(OH)_6\cdot2H_2O$ A water-soluble alcohol often grouped with the vitamins; there are nine stereoisomers of hexahydroxycyclohexane, and the only one of biological importance is optically inactive *meso*-inositol, comprising white crystals, widely distributed in animals and plants; it serves as a growth factor for animals and microorganisms. { i'näs·ə,tȯl }

insol *See* insoluble. { 'in,säl }

insoluble [CHEM] Incapable of being dissolved in another material; usually refers to solid-liquid or liquid-liquid systems. Abbreviated insol. { in'säl·yə·bəl }

insoluble anode [CHEM] An anode that resists dissolution during electrolysis. { in'säl·yə·bəl 'an,ōd }

inspissation [CHEM] The process of thickening a liquid by evaporation. { ,in·spi'sā· shən }

integral heat of dilution *See* heat of dilution. { 'int·ə·grəl |hēt əv də'lü·shən }

integral heat of solution *See* heat of solution. { 'int·ə·grəl |hed əv sə'lü·shən }

integral procedure decomposition temperature [PHYS CHEM] Decomposition temperatures derived from graphical integration of the thermogravimetric analysis of a polymer. { 'int·ə·grəl prə|sē·jər dē,käm·pə'zish·ən ,tem·prə·chər }

intensive properties [CHEM] Properties independent of the quantity or shape of the substance under consideration; for example, temperature, pressure, or composition. { in'ten·siv 'präp·ərd·ēz }

intercalibration [ANALY CHEM] A state achieved by a group of laboratories engaged in a monitoring program in which they produce and maintain compatible data outputs. { ,in·tər,kal·ə'brā·shən }

interdiffusion [PHYS CHEM] The self-mixing of two fluids, initially separated by a diaphragm. { |in·tər·də'fyü·zhən }

interface [PHYS CHEM] The boundary between any two phases: among the three phases (gas, liquid, and solid), there are five types of interfaces: gas-liquid, gas-solid, liquid-liquid, liquid-solid, and solid-solid. { 'in·tər,fās }

interface mixing [PHYS CHEM] The mixing of two immiscible or partially miscible liquids at the plane of contact (interface). { 'in·tər,fās |mik·siŋ }

interfacial layer [PHYS CHEM] A one- or two-molecules-thick boundary between any two bulk phases (gas, liquid, or solid) in contact where the properties differ from the properties of the bulk phases. { ,in·tər|fā·shəl 'lā·ər }

interference [ANALY CHEM] A systematic error in measurement that occurs when concomitants are present in the sample being analyzed. { ,in·tər'fir·əns }

interference spectrum [SPECT] A spectrum that results from interference of light, as in a very thin film. { ,in·tər'fir·əns |spek·trəm }

interferogram [SPECT] A graph of the variation of the output signal from an interferometer as the condition for interference within the interferometer is varied. { ,in· tə'fir·ə,gram }

interhalogen [INORG CHEM] Any of the compounds formed from the elements of the halogen family that react with each other to form a series of binary compounds; for example, iodine monofluoride. { |in·tər'hal·ə·jən }

interionic attraction [PHYS CHEM] The Coulomb attraction between ions of opposite sign in a solution. { ,in·tir·ē'än·ik ə|trak·shən }

intermediate [CHEM] A precursor to a desired product; ethylene is an intermediate for polyethylene, and ethane is an intermediate for ethylene. { ,in·tər'mēd·ē·ət }

intermolecular force [PHYS CHEM] The force between two molecules; it is that negative gradient of the potential energy between the interacting molecules, if energy is a function of the distance between the centers of the molecules. { ,in·tər·mə'lek·yə· lər 'fȯrs }

internal phase *See* disperse phase. { in'tərn·əl |fāz }

internal reflectance spectroscopy *See* attenuated total reflectance. { in'tərn·əl ri|flek· təns spek'träs·kə·pē }

internal standard

internal standard [SPECT] The principal line in spectrum analysis by the logarithmic sector method, a quantitative spectroscopy procedure. { in'tərn·əl 'stan·dərd }

International Union of Pure and Applied Chemistry [CHEM] An international scientific (nongovernmental) organization, recognized as the world authority on chemical nomenclature, terminology, standardized methods for measurement, atomic weights, and many other critically evaluated data. Abbreviated IUPAC.

internuclear distance [PHYS CHEM] The distance between two nuclei in a molecule. { ¦in·tər¦nü·klē·ər ‚dis·təns }

interpenetrating polymer network [ORG CHEM] Two or more polymer components, each of which is a crosslinked three-dimensional network, one of which is formed (crosslinked) in the presence of the other. The polymer networks are physically entangled with, but not covalently bonded to, each other. Characteristically, these networks do not dissolve in solvent or flow when heated. Abbreviated IPN. { ‚in·tər‚pen·ə‚trād·iŋ ‚päl·ə·mər 'net‚wərk }

interphase [CHEM] A region between the two phases of a newly created interface that contains particles of both phases. { 'in·tər‚fāz }

interpolymer [ORG CHEM] A mixed polymer made from two or more starting materials. { ¦in·tər'päl·ə·mər }

interstitial compound [CHEM] A compound of a transition metal and hydrogen, boron, carbon, or nitrogen whose crystals have a close-packed structure of the metal ions, with the nonmetal atoms being located in the interstices. { ¦in·tər¦stish·əl 'käm‚paund }

intimate ion pair See contact ion pair. { 'in·tə·mət 'ī‚än ‚per }

intracavity absorption spectroscopy [SPECT] A highly sensitive technique in which an absorbing sample is placed inside the resonator of a broad-band dye laser, and absorption lines are detected as dips in the laser emission spectrum. { ¦in·trə'kav·əd·ē əb¦sòrp·shən spek'träs·kə·pē }

intrinsic viscosity [PHYS CHEM] The ratio of a solution's specific viscosity to the concentration of the solute, extrapolated to zero concentration. Also known as limiting viscosity number. { in'trin·sik vi'skäs·əd·ē }

introfaction [CHEM] Change in fluidity and specific wetting properties (for impregnation acceleration) of an impregnating compound, caused by an introfier (impregnation accelerator). { ¦in·trə¦fak·shən }

inverse micelle See inverted micelle. { 'in‚vərs mī'sel }

inverse Stark effect [SPECT] The Stark effect as observed with absorption lines, in contrast to emission lines. { 'in‚vərs 'stärk i‚fekt }

inverse Zeeman effect [SPECT] A splitting of the absorption lines of atoms or molecules in a static magnetic field; it is the Zeeman effect observed with absorption lines. { 'in‚vərs 'zē·mən i‚fekt }

inversion [CHEM] Change of a compound into an isomeric form. { in'vər·zhən }

inversion spectrum [SPECT] Lines in the microwave spectra of certain molecules (such as ammonia) which result from the quantum-mechanical analog of an oscillation of the molecule between two configurations which are mirror images of each other. { in'vər·zhən ‚spek·trəm }

inverted micelle [PHYS CHEM] An aggregate of colloidal dimension in which the polar groups are concentrated in the interior and the lipophilic groups extend outward into the solvent. Also known as inverse micelle. { in¦vərd·əd mī'sel }

iodate [INORG CHEM] A salt of iodic acid containing the IO_3sw radical; sodium and potassium iodates are the most important salts and are used in medicine. { 'ī·ə‚dāt }

iodcyanin See cyanine dye. { ¦ī·əd¦sī·ə·nən }

iodic acid [INORG CHEM] HIO_3 Water-soluble, moderately strong acid; colorless or white powder or crystals; decomposes at 110°C; used in analytical chemistry and medicine. { ī'äd·ik 'as·əd }

iodic acid anhydride See iodine pentoxide. { ī'äd·ik 'as·əd an'hī‚drīd }

iodide [CHEM] **1.** A compound which contains the iodine atom in the −1 oxidation state and which may be considered to be derived from hydriodic acid (HI); examples

196

are KI and NaI. **2.** A compound of iodine, such as CH_3CH_2I, in which the iodine has combined with a more electropositive group. { 'ī·ə,dīd }

iodine [CHEM] A nonmetallic halogen element, symbol I, atomic number 53, atomic weight 126.9045; melts at 114°C, boils at 184°C; the poisonous, corrosive, dark plates or granules are readily sublimed; insoluble in water, soluble in common solvents; used as germicide and antiseptic, in dyes, tinctures, and pharmaceuticals, in engraving lithography, and as a catalyst and analytical reagent. { 'ī·ə,dīn }

iodine bisulfide See sulfur iodine. { 'ī·ə,dīn bī'səl,fīd }

iodine cyanide [INORG CHEM] ICN Poisonous, colorless needles with pungent aroma and acrid taste; melts at 147°C; soluble in water, alcohol, and ether; used in taxidermy as a preservative. Also known as cyanogen iodide. { 'ī·ə,dīn 'sī·ə,nīd }

iodine disulfide See sulfur iodine. { 'ī·ə,dīn dī'səl,fīd }

iodine number [ANALY CHEM] A measure of the iodine absorbed in a given time by a chemically unsaturated material, such as a vegetable oil or a rubber; used to measure the unsaturation of a compound or mixture. Also known as iodine value. { 'ī· ə,dīn ,nəm·bər }

iodine pentoxide [INORG CHEM] I_2O_5 White crystals, decomposing at 275°C, very soluble in water, insoluble in absolute alcohol, ether, and chloroform; used as an oxidizing agent to oxidize carbon monoxide to dioxide at ordinary temperatures, and in organic synthesis. Also known as iodic acid anhydride. { 'ī·ə,dīn pen'täk,sīd }

iodine test [ANALY CHEM] Placing a few drops of potassium iodide solution on a sample to detect the presence of starch; test is positive if sample turns blue. { 'ī·ə,dīn ,test }

iodine value See iodine number. { 'ī·ə,dīn ,val·yü }

iodoacetic acid [ORG CHEM] CH_2ICOOH White or colorless crystals that are soluble in water and alcohol, and melt at 82–83°C; used in biological research for its inhibitive effect on enzymes. { ī|ō·dō·ə|sēd·ik 'as·əd }

iodoalkane [ORG CHEM] An alkane hydrocarbon in which an iodine atom replaces one or more hydrogen atoms in the molecule; an example is iodomethane, CH_3I, better known as methyl iodide. { ī|ō·dō·al'kān }

iodoeasin See easin. { ī|ō·dō'ē·ə·sən }

iodoethane See ethyl iodide. { ī|ō·dō'eth,ān }

iodoethylene See tetraiodoethylene. { ī|ō·dō'eth·ə,lēn }

iodoform [ORG CHEM] CHI_3 A yellow, hexagonal solid; melting point 119°C; soluble in chloroform, ether, and water; has weak bactericidal qualities and is used in ointments for minor skin diseases. Also known as triiodomethane. { ī'ō·də,fȯrm }

iodohydrocarbon [ORG CHEM] A hydrocarbon in which an iodine atom replaces one or more hydrogen atoms in the molecule, as in an alkane, aromatic, or olefin. { ī|ō· də,hī·drə'kär·bən }

iodomethane See methyl iodide. { ī|ō·də'meth,ān }

iodometry [ANALY CHEM] An application of iodine chemistry to oxidation-reduction titrations for the quantitative analysis in certain chemical compounds, in which iodine is used as a reductant and the iodine freed in the associated reaction is titrated, usually in neutral or slightly acid mediums with a standard solution of a reductant such as sodium thiosulfate or sodium arsenite; examples of chemicals analyzed are copper(III), gold(VI), arsenic(V), antimony(V), chlorine, and bromine. { ,ī·ō'däm·ə·trē }

iodonium [INORG CHEM] A halonium ion such as H_2I^+ or R_2I^+; it may be open-chain or cyclic. { ī·ə'dōn·ē·əm }

iodophor [CHEM] Any compound that is a carrier of iodine. { i'äd·ə,fȯr }

iodosobenzene [ORG CHEM] C_6H_5IO A yellowish-white amorphous solid that explodes at 200°C, soluble in hot water and alcohol; a strong oxidizing agent. { ,ī·ə'dō· sō'ben,zēn }

iodoxybenzene [ORG CHEM] $C_6H_5IO_2$ Clear white crystals that explode at 227–228°C, slightly soluble in water, insoluble in chloroform, acetone, and benzene; a strong oxidizing agent. { ī·ə|däk·sē'ben,zēn }

ion [CHEM] An isolated electron or positron or an atom or molecule which by loss or gain of one or more electrons has acquired a net electric charge. { 'ī,än }

ion cloud

ion cloud [PHYS CHEM] A slight preponderance of negative ions around a positive ion in an electrolyte, and vice versa, according to the Debye-Hückel theory. Also known as ion atmosphere. { 'ī,än ,klaúd }

ion-cyclotron-resonance mass spectrometer [SPECT] A device for detecting and measuring the mass distribution of ions orbiting in an applied magnetic field, either by applying a constant radio-frequency signal and varying the magnetic field to bring ion frequencies equal to the applied radio frequency sequentially into resonance, or by rapidly varying the radio frequency and applying Fourier transform techniques. { 'ī,än 'sī·klə,trän 'rez·ən·əns 'mas spek'träm·əd·ər }

ion detector [ANALY CHEM] Device for detection of presence or concentration of liquid solution ions, such as with a pH meter or by conductimetric techniques. { 'ī,än di,tek·tər }

ion exchange [PHYS CHEM] A chemical reaction in which mobile hydrated ions of a solid are exchanged, equivalent for equivalent, for ions of like charge in solution; the solid has an open, fishnetlike structure, and the mobile ions neutralize the charged, or potentially charged, groups attached to the solid matrix; the solid matrix is termed the ion exchanger. { 'ī,än iks,chānj }

ion-exchange chromatography [ANALY CHEM] A chromatographic procedure in which the stationary phase consists of ion-exchange resins which may be acidic or basic. { 'ī,än iks,chānj ,krō·mə'täg·rə·fē }

ion exchanger [PHYS CHEM] A solid or liquid material containing ions that are exchangeable with other ions with a like charge that are present in a solution in which the material is insoluble. { 'ī,än iks,chānj·ər }

ion exclusion [CHEM] Ion-exchange resin system in which the mobile ions in the resin-gel phase electrically neutralize the immobilized charged functional groups attached to the resin, thus preventing penetration of solvent electrolyte into the resin-gel phase; used in separations where electrolyte is to be excluded from the resin, but not nonpolar materials, as the separation of salt from nonpolar glycerin. { 'ī,än iks,klü·zhən }

ion-exclusion chromatography [ANALY CHEM] Chromatography in which the adsorbent material is saturated with the same mobile ions (cationic or anionic) as are present in the sample-carrying solvent (solvent), thus repelling the similar sample ions. { 'ī,än iks,klü·zhən ,krō·mə'täg·rə·fē }

ionic bond [PHYS CHEM] A type of chemical bonding in which one or more electrons are transferred completely from one atom to another, thus converting the neutral atoms into electrically charged ions; these ions are approximately spherical and attract one another because of their opposite charge. Also known as electrovalent bond. { ī'än·ik 'bänd }

ionic conductance [PHYS CHEM] The contribution of a given type of ion to the total equivalent conductance in the limit of infinite dilution. { ī'än·ik kən'dək·təns }

ionic dissociation [PHYS CHEM] Dissociation that results in the production of ions. { i'än·ik di,sō·sē'ā·shən }

ionic equilibrium [PHYS CHEM] The condition in which the rate of dissociation of nonionized molecules is equal to the rate of combination of the ions. { ī'än·ik ,ē·kwə'lib·rē·əm }

ionic equivalent conductance [PHYS CHEM] The contribution made by each ion species of a salt toward an electrolyte's equiviconductance. { ī'än·ik i¦kwiv·ə·lənt kən'dək·təns }

ionic gel [CHEM] A gel with ionic groups attached to the structure of the gel; the groups cannot diffuse out into the surrounding solution. { ī'än·ik 'jel }

ionicity [CHEM] The ionic character of a solid. { ,ī·ə'nis·əd·ē }

ionic polymerization [ORG CHEM] Polymerization that proceeds via ionic intermediates (carbonium ions or carbanions) than through neutral species (olefins or acetylenes). { ī'än·ik pə,lim·ə·rə'zā·shən }

ionic radii [PHYS CHEM] Radii which can be assigned to ions because the rapid variation of their repulsive interaction with distance makes them repel like hard spheres; these radii determine the dimensions of ionic crystals. { ī'än·ik 'rād·ē,ī }

198

iridium chloride

ionic strength [PHYS CHEM] A measure of the average electrostatic interactions among ions in an electrolyte; it is equal to one-half the sum of the terms obtained by multiplying the molality of each ion by its valence squared. { ī'än·ik 'streŋkth }

ionization [CHEM] A process by which a neutral atom or molecule loses or gains electrons, thereby acquiring a net charge and becoming an ion; occurs as the result of the dissociation of the atoms of a molecule in solution (NaCl → Na$^+$ + Cl$^-$) or of a gas in an electric field (H$_2$ → 2H$^+$). { ˌī·ə·nə'zā·shən }

ionization constant [PHYS CHEM] Analog of the dissociation constant, where k = |H$^+$||A$^-$|/|HA|; used for the application of the law of mass action to ionization; in the equation HA represents the acid, such as acetic acid. { ˌī·ə·nə'zā·shən ˌkän·stənt }

ionization degree [PHYS CHEM] The proportion of potential ionization that has taken place for an ionizable material in a solution or reaction mixture. { ˌī·ə·nə'zā·shən di,grē }

ionization isomer [CHEM] One of two or more compounds that have identical molecular formulas but different ionic forms. { ˌī·ə·nə'zā·shən 'ī·sə·mər }

ionized atom [CHEM] An atom with an excess or deficiency of electrons, so that it has a net charge. { 'ī·ə,nīzd 'ad·əm }

ion kinetic energy spectrometry [SPECT] A spectrometric technique that uses a beam of ions of high kinetic energy passing through a field-free reaction chamber from which ionic products are collected and energy analyzed; it is a generalization of metastable ion studies in which both unimolecular and bimolecular reactions are considered. { 'ī,än ki,ned·ik 'en·ər·jē spek'träm·ə·trē }

ion mean life [PHYS CHEM] The average time between the ionization of an atom or molecule and its recombination with one or more electrons, or its loss of the excess electrons. { 'ī,än ˌmēn 'līf }

ionogenic group [PHYS CHEM] A fixed group of atoms in an ion exchanger that is either ionized or capable of dissociation into fixed ions and mobile counterions. { ˌī·ə·nə'jen·ik ˌgrüp }

ionography [ANALY CHEM] A type of electrochromatography involving migration of ions. { ˌī·ə'näg·rə·fē }

ionomer [ORG CHEM] Polymer with covalent bonds between the elements of the chain, and ionic bonds between the chains. { ī'än·ə·mər }

ionomer resin [ORG CHEM] A polymer which has ethylene as the major component, but which contains both covalent and ionic bonds. { ī'än·ə·mər 'rez·ən }

ionone [ORG CHEM] C$_{13}$H$_{20}$O A colorless to light yellow liquid with a boiling point of 126–128°C at 12 mmHg (1600 pascals); soluble in alcohol, ether, and mineral oil; used in perfumery, flavoring, and vitamin A production. Also known as irisone. { 'ī·ə,nōn }

ion scattering spectroscopy [SPECT] A spectroscopic technique in which a low-energy (about 1000 electronvolts) beam of inert-gas ions is directed at a surface, and the energies and scattering angles of the scattered ions are used to identify surface atoms. Abbreviated ISS. { 'ī,än ˌskad·ə·riŋ spek'träs·kə·pē }

ioxynil [ORG CHEM] C$_7$H$_3$I$_2$NO A colorless solid with a melting point of 212–213°C; used for postemergence control of seedling weeds in cereals and sports turf. { ī'äk·sə,nil }

ioxynil octanoate [ORG CHEM] C$_{15}$H$_{17}$I$_2$NO$_2$ A waxy solid with a melting point of 59–60°C; insoluble in water; used as an insecticide for cereals and sugarcane. { ī'äk·sə,nil ˌäk·tə 'nō·ət }

IPC See propham.

IPN See interpenetrating polymer network.

Ir See iridium.

iridescent layer See schiller layer. { ˌi·ri'des·ənt ˌlā·ər }

iridic chloride [INORG CHEM] IrCl$_4$ A hygroscopic brownish-black mass, soluble in water and alcohol; used to analyze for nitric acid, HNO$_3$, and in analytical microscopic work. Also known as iridium chloride; iridium tetrachloride. { ī'rid·ik 'klȯr,īd }

iridium [CHEM] A metallic element, symbol Ir, atomic number 77, atomic weight 192.2, in the platinum group; insoluble in acids, melting at 2454°C. { i'rid·ē·əm }

iridium chloride See iridic chloride. { i'rid·ē·əm 'klȯr,īd }

iridium tetrachloride

iridium tetrachloride See iridic chloride. { i'rid·ē·əm ˌte·trə'klȯrˌīd }
irisone See ionone. { 'ī·rəˌsōn }
iron [CHEM] A silvery-white metallic element, symbol Fe, atomic number 26, atomic weight 55.847, melting at 1530°C. { 'ī·ərn }
iron acetate See ferrous acetate. { 'ī·ərn 'as·əˌtāt }
iron ammonium sulfate See ferric ammonium sulfate; ferrous ammonium sulfate. { 'ī·ərn ə'mō·nē·əm 'səlˌfāt }
iron arsenate See ferrous arsenate. { 'ī·ərn 'ärs·ənˌāt }
iron black [CHEM] Fine black antimony powder used to give a polished-steel look to papier-maché and plaster of paris; made by reaction of zinc with acid solution of an antimony salt and precipitation of black antimony powder. { 'ī·ərn 'blak }
iron blue [INORG CHEM] Ferric ferrocyanide used as blue pigment by the paint industry for permanent body and trim paints; also used in blue ink, in paper dyeing, and as a fertilizer ingredient. { 'ī·ərn 'blü }
iron bromide See ferric bromide. { 'ī·ərn 'brōˌmīd }
iron carbonyl See iron pentacarbonyl. { 'ī·ərn 'kär·bəˌnil }
iron chloride See ferric chloride; ferrous chloride. { 'ī·ərn 'klȯrˌīd }
iron citrate See ferric citrate. { 'ī·ərn 'sīˌtrāt }
iron dichloride See ferrous chloride. { 'ī·ərn dī'klȯrˌīd }
irone [ORG CHEM] $C_{14}H_{22}O$ A colorless liquid terpene; a component of essential oil from the orrisroot; used in perfumes. { 'īˌrōn }
iron ferrocyanide See ferric ferrocyanide. { 'ī·ərn ˌfer·ə'sī·əˌnīd }
iron fluoride See ferric fluoride. { 'ī·ərn 'flúrˌīd }
iron hydroxide See ferric hydroxide. { 'ī·ərn hī'dräkˌsīd }
iron metavanadate See ferric vanadate. { 'ī·ərn ˌmed·ə'van·əˌdāt }
iron monoxide See ferrous oxide. { 'ī·ərn mə'näkˌsīd }
iron nitrate See ferric nitrate. { 'ī·ərn 'nīˌtrāt }
iron nonacarbonyl [INORG CHEM] $Fe_2(CO)_9$ Orange-yellow crystals that break down at 100°C to yield tetracarbonyl, slightly soluble in alcohol and acetone, almost insoluble in water, ether, and benzene. { 'ī·ərn ˌnō·nə'kär·bəˌnil }
iron oxalate See ferrous oxalate. { 'ī·ərn 'äk·səˌlāt }
iron oxide [INORG CHEM] Any of the hydrated, synthetic, or natural oxides of iron: ferrous oxide, ferric oxide, ferriferous oxide. { 'ī·ərn 'äkˌsīd }
iron pentacarbonyl [INORG CHEM] $Fe(CO)_5$ An oily liquid that decomposes upon exposure to light, soluble in most organic solvents; used as a source of a pure iron catalyst and for magnet cores. Also known as iron carbonyl. { 'ī·ərn ˌpen·tə'kär·bəˌnil }
iron phosphate See ferric phosphate. { 'ī·ərn 'fäsˌfāt }
iron resinate See ferric resinate. { 'ī·ərn 'rez·ənˌāt }
iron stearate See ferric stearate. { 'ī·ərn 'stirˌāt }
iron sulfate See ferric sulfate; ferrous sulfate. { 'ī·ərn 'səlˌfāt }
iron sulfide See ferrous sulfide. { 'ī·ərn 'səlˌfīd }
iron tetracarbonyl [INORG CHEM] $Fe_3(CO)_{12}$ Dark-green lustrous crystals that break down at 140–50°C; soluble in organic solvents. Also known as tri-iron dodecacarbonyl. { 'ī·ərn ˌte·trə'kär·bəˌnil }
irregular polymer [CHEM] A polymer whose molecular structure does not consist of only one species of constitutional unit in a single sequential arrangement. { i̇'reg·yə·lər 'päl·i·mər }
isatin [ORG CHEM] $C_6H_5NO_2$ An indole substituted with oxygen at carbon position 2 and 3; crystallizes as red needles that are soluble in hot water; used in dye manufacture. { 'ī·sə·tən }
isethionic acid [ORG CHEM] $CH_2OH·CH·SO_2OH$ A water-soluble liquid, boiling at 100°C; used in the manufacture of detergents. { ˌīs·ə·thī̵än·ik 'as·əd }
iso- [CHEM] A prefix indicating an isomer of an element in which there is a difference in the nucleus when compared to the most prevalent form of the element. [ORG CHEM] A prefix indicating a single branching at the end of the carbon chain. { 'ī·sō }
isoactyl thioglycolate [ORG CHEM] $HSCH_2COOCH_2 C_7H_{15}$ A colorless liquid with a slight

fruity odor and a boiling point of 125°C; used in antioxidants, insecticides, oil additives, and plasticizers.   { ¦ī·sō'akt·əl ˌthī·ə'glī·kəˌlāt }

isoalkane   [ORG CHEM] An alkane with a branched chain whose next-to-last carbon atom is bonded to a single methyl group.   { ¦ī·sō'alˌkān }

isoalkyl group   [ORG CHEM] A group of atoms resulting from the removal of a hydrogen atom from a methyl group situated at the end of the straight-chain segment of an isoalkane.   { ¦ī·so'al·kəl ˌgrüp }

isoamyl acetate See amyl acetate.   { ¦ī·sō¦am·əl 'as·əˌtāt }

isoamyl alcohol See isobutyl carbinol.   { ¦ī·sō¦am·əl 'al·kəˌhȯl }

isoamyl benzoate   [ORG CHEM] C₆H₅COOC₅H₁₁ Colorless liquid with fruity aroma; boils at 260°C; soluble in alcohol, insoluble in water; used in flavors and perfumes.   Also known as amyl benzoate.   { ¦ī·sō¦am·əl 'ben·zəˌwāt }

isoamyl bromide   [ORG CHEM] (CH₃)₂CHCH₂CH₂Br A colorless liquid with a boiling point of 120–121°C; miscible with alcohol and with ether; used in organic synthesis.   { ¦ī·sō¦am·əl 'brōˌmīd }

isoamyl butyrate   [ORG CHEM] C₅H₁₁COOC₃H₇ A water-white liquid boiling at 150–180°C; soluble in alcohol and ether; used as a solvent and plasticizer for cellulose acetate and in flavor extracts.   { ¦ī·sō¦am·əl 'byüd·əˌrāt }

isoamyl chloride   [ORG CHEM] C₅H₁₁Cl Water-insoluble, colorless liquid boiling at 100°C; it can be any one of several compounds, such as 1-chloro-3-methylbutane, (CH₃)₂CH(CH₂)₂Cl, or mixtures thereof; used as a solvent, in inks, for soil fumigation, and as a chemical intermediate.   { ¦ī·sō¦am·əl 'klȯrˌīd }

isoamyl nitrite See amyl nitrite.   { ¦ī·sō¦am·əl 'nīˌtrīt }

isoamyl salicylate See amyl salicylate.   { ¦ī·sō¦am·əl sə'lis·əˌlāt }

isoamyl valerate   [ORG CHEM] C₄H₉CO₂C₅H₁₁ Clear liquid with apple aroma; boils at 204°C; soluble in alcohol and ether, insoluble in water; used in medicine and fruit flavors.   { ¦ī·sō¦am·əl 'val·əˌrāt }

isobornyl acetate   [ORG CHEM] C₁₀H₁₇OOCCH₃ A colorless liquid with an odor of pine needles and a boiling point of 220–224°C; soluble in fixed oils and mineral oil; used in toiletries and soaps and antiseptics, and as a flavoring agent.   { ¦ī·sə'bȯrn·əl 'as·əˌtāt }

isobornyl thiocyanoacetate   [ORG CHEM] C₁₀H₁₇OOCCH₂SCN An oily, yellow liquid; soluble in alcohol, benzene, chloroform, and ether; used in medicine and as an insecticide.   { ¦ī·sə'bȯrn·əl ¦thī·ə¦sī·ə·nō'as·əˌtāt }

isobutane   [ORG CHEM] (CH₃)₂CHCH₃ A colorless, stable gas, noncorrosive to metals, nonreactive with water; boils at −11.7°C; used as a chemical intermediate, refrigerant, and fuel.   { ¦ī·sō'byüˌtān }

isobutanol See isobutyl alcohol.   { ¦ī·sō'byüt·ənˌȯl }

isobutene See isobutylene.   { ¦ī·sō'byüˌtēn }

isobutyl   [ORG CHEM] The radical (CH₃)₂CHCH₂−, occurring, for example, in isobutanol (isobutyl alcohol), (CH₃)₂CHCH₂OH.   { ¦ī·sō'byüd·əl }

isobutyl acetate   [ORG CHEM] C₄H₉OOCCH₃ Colorless liquid with fruitlike aroma; soluble in alcohols, ether, and hydrocarbons, insoluble in water; boils at 116°C; used as a solvent for lacquer and nitrocellulose.   { ¦ī·sō'byüd·əl 'as·əˌtāt }

isobutyl alcohol   [ORG CHEM] (CH₃)₂CHCH₂OH A colorless liquid that is a by-product of the synthetic production of methanol, boils at 107°C; soluble in water, ether, and alcohol; used as a solvent in paints and lacquers, in organic synthesis, and in resin coatings.   Also known as isobutanol; isopropylcarbinol; 2-methyl-l-propanol.   { ¦ī·sō'byüd·əl 'al·kəˌhȯl }

isobutyl aldehyde   [ORG CHEM] (CH₃)₂CHCHO Colorless, transparent liquid with pungent aroma; soluble in alcohol, insoluble in water; boils at 64°C; used as a chemical intermediate.   Also known as isobutyraldehyde.   { ¦ī·sō'byüd·əl 'al·dəˌhīd }

isobutyl carbinol   [ORG CHEM] (CH₃)₂CH(CH₂)₂OH Colorless liquid with pungent taste and disagreeable aroma; soluble in alcohol and ether, slightly soluble in water; boils at 132°C; used as a chemical intermediate and solvent, and in pharmaceutical products and medicines.   Also known as isoamyl alcohol.   { ¦ī·sō'byüd·əl 'kär·bəˌnȯl }

201

isobutylene [ORG CHEM] $(CH_3)_2CCH_2$ Flammable, colorless, volatile liquid boiling at $-7°C$; easily polymerized; used in gasolines, as a chemical intermediate, and to make butyl rubber. Also known as isobutene. { ¦ī·sō'byüd·əl¦ēn }

isobutyl isobutyrate [ORG CHEM] $(CH_3)_2CHCOOCH_2CH(CH_3)_2$ A colorless liquid with a fruity odor and a boiling point of 148.7°C; soluble in alcohol and ether; used for flavoring and as an insect repellent. Abbreviated IBIB. { ¦ī·sō'byüd·əl ¦ī·sō'byüd·ə,rāt }

isobutyraldehyde See isobutyl aldehyde. { ¦ī·sō,byüd·ə'ral·də,hīd }

isobutyric acid [ORG CHEM] $(CH_3)_2CHCOOH$ Colorless liquid boiling at 154°C; soluble in water, alcohol, and ether; used as a chemical intermediate and disinfectant, in flavor and perfume bases, and for leather treating. { ¦ī·sō·byü'tir·ik 'as·əd }

isobutyryl [ORG CHEM] $(CH_3)_2C\cdot CHO$ The radical group from isobutyric acid, $(CH_3)_2CHCOOH$. { ¦ī·sō'byüd·ə·rəl }

isocetyl laurate [ORG CHEM] $C_{11}H_{23}COOC_{16}H_{33}$ An oily, combustible liquid, soluble in most organic solvents; used in cosmetics and pharmaceuticals and as a plasticizer and textile softener. { ¦ī·sə¦sēd·əl 'lȯ,rāt }

isocyanate [ORG CHEM] **1.** One of a group of neutral derivatives of primary amines; its formula is $R-N=C=O$, where R may be an alkyl or aryl group; an example is 2,4-toluene diisocyanate. **2.** Any compound containing the isocyanato functional group. { ¦ī·sō'sī·ə,nāt }

isocyanate resin [ORG CHEM] A linear alkyd resin lengthened by reaction with isocyanates, then treated with a glycol or diamine to cross-link the molecular chain; the product has good abrasion resistance. { ¦ī·sō'sī·ə,nāt 'rez·ən }

isocyanato group [ORG CHEM] A functional group $(-N=C-O)$ which forms isocyanates by replacing the hydrogen atom of a hydrocarbon. { ¦ī·sō·sī'an·ə,tō ,grüp }

isocyanic acid [ORG CHEM] $HN-C-O$ One of two forms of cyanic acid; a gas used as an intermediate in the preparation of polyurethane and other resins. { ¦ī·sō·sī'an·ik 'as·əd }

isocyanide [ORG CHEM] A compound with the general formula $RN\equiv C$ in which the hydrogen of a hydrocarbon has been replaced by the $-N\equiv C$ group. { ¦ī·sō'sī·ə,nīd }

isocyanine [ORG CHEM] Any one of a series of dyes whose structure has two heterocyclic or quinoline rings connected by an odd number chain of carbon atoms containing conjugated double bonds; for example, cyanine blue. { ¦ī·sō'sī·ə,nēn }

isocyanuric acid See fulminuric acid. { ¦ī·sō,sī·ə¦núr·ik 'as·əd }

isocyclic compound [ORG CHEM] A compound in which the ring structure is made up of one kind of atom. { ¦ī·sō'sī·klik 'käm,paúnd }

isodecyl chloride [ORG CHEM] $C_{10}H_{21}Cl$ A colorless liquid with a boiling point of 210.6°C; used as a solvent and in extractants, cleaning compounds, pharmaceuticals, insecticides, and plasticizers. { ,ī·sə'des·əl 'klȯr,īd }

isodisperse [CHEM] **1.** Having dispersed particles, of colloidal dimensions, that are all of the same size. **2.** Dispersible in solutions with the same pH value. { ,īs·ə·di'spərs }

isoelectric focusing [PHYS CHEM] Protein separation technique in which a mixture of protein molecules is resolved into its components by subjecting the mixture to an electric field in a supporting gel having a previously established pH gradient. Also known as electrofocusing. { ¦ī·sō·i'lek·trik 'fō·kəs·iŋ }

isoelectric point [PHYS CHEM] The pH value of the dispersion medium of a colloidal suspension at which the colloidal particles do not move in an electric field. { ¦ī·sō·i'lek·trik 'pȯint }

isoelectric precipitation [CHEM] Precipitation of materials at the isoelectric point (the pH at which the net charge on a molecule in solution is zero); proteins coagulate best at this point. { ¦ī·sō·i'lek·trik prə,sip·ə'tā·shən }

isoelectronic principle [CHEM] The concept that molecules having the same number of electrons and the same number of atoms whose atomic masses are greater than that of hydrogen (heavy atoms) tend to have similar electronic structures, similar chemical properties, and heavy-atom geometries. { ¦ī·sō·i,lek'trän·ik 'prin·sə·pəl }

isoelectronic sequence [SPECT] A set of spectra produced by different chemical elements ionized so that their atoms or ions contain the same number of electrons. { ¦ī·sō·i,lek'trän·ik 'sē·kwəns }

isoeugenol [ORG CHEM] C₁₀H₁₂O₂ An oily liquid prepared from eugenol by heating, slightly soluble in water; used in the manufacture of vanillin. { ¦ī·sō'yü·jə,nȯl }

isohexane [ORG CHEM] C₆H₁₄ A liquid mixture of isomeric hydrocarbons, flammable and explosive, insoluble in water, soluble in most organic solvents, boils at 54–61°C; used as a solvent, freezing-point depressant, and chemical intermediate. { ¦ī·sō'hek,sān }

isohydric [CHEM] Referring to a set of solutions with the same hydrogen ion concentration and not affecting the conductivity of each of the various solutions on mixing. { ¦ī·sə¦hī·drik }

isokinetic relationship [PHYS CHEM] A linear relationship that exists between the enthalpies and entropies of activation of a series of related reactions. { ,i·sə·ki¦ned·ik ri'lā·shən,ship }

isokinetic temperature [PHYS CHEM] The actual or virtual temperature at which rates of all members of a series of related reactions are equal. { ,ī·sə·ki¦ned·ik 'tem·prə·chər }

isolation [CHEM] Separation of a pure chemical substance from a compound or mixture; as in distillation, precipitation, or absorption. { ,ī·sə'lā·shən }

isomer [CHEM] One of two or more chemical substances having the same elementary percentage composition and molecular weight but differing in structure, and therefore in properties; there are many ways in which such structural differences occur; one example is provided by the compounds *n*-butane, CH₃(CH₂)₂CH₃, and isobutane, CH₃CH(CH₃)₂. { 'ī·sə·mər }

isomeric shift [PHYS CHEM] Shift in the Mössbauer resonance caused by the effect of the valence of the atom on the interaction of the electron density at the nucleus with the nuclear charge. Also known as chemical shift. { ¦ī·sə¦mer·ik 'shift }

isomerism [CHEM] The phenomenon whereby certain chemical compounds have structures that are different although the compounds possess the same elemental composition. { ī'säm·ə,riz·əm }

isomerization [CHEM] A process whereby a compound is changed into an isomer; for example, conversion of butane into isobutane. { ī,säm·ə·rə'zā·shən }

isomolecule See nonlinear molecule. { ¦ī·sō'mäl·ə,kyül }

isomorphism [PHYS CHEM] A condition present when an ion at high dilution is incorporated by mixed crystal formation into a precipitate, even though such formation would not be predicted on the basis of crystallographic and ionic radii; an example is coprecipitation of lead with potassium chloride. { ¦ī·sə¦mȯr,fiz·əm }

isonicotinic acid [ORG CHEM] C₆H₅NO₂ White platelets or powder, slightly soluble in water, sublimes at 260°C; used in the manufacture of isonicotinic acid hydrazide, an antitubercular agent. { ¦ī·sə,nik·ə'tin·ik 'as·əd }

isonitrosoacetophenone [ORG CHEM] C₈H₇NO₂ Platelike crystals with a melting point of 126–128°C; soluble in alkalies and alkali carbonates; used to detect ferrous ions and palladium. { ¦ī·sə¦nī·trə·sō,as·ə'täf·ə,nōn }

isooctane [ORG CHEM] (CH₃)₂CHCH₂C(CH₃)₃ Flammable, colorless liquid boiling at 99°C; slightly soluble in alcohol and ether, insoluble in water; used in motor fuels and as a chemical intermediate. { ¦ī·sō'äk,tān }

isooctyl alcohol [ORG CHEM] C₇H₁₅CH₂OH Mixture of isomers from oxo-process synthesis; boils at 182–195°C; used as a chemical intermediate, resin solvent, emulsifier, and antifoaming agent. { ¦ī·sō'äkt·əl 'al·kə,hȯl }

isoparaffin [ORG CHEM] A branched-chain version of a straight-chain (normal) saturated hydrocarbon; for example, isooctane, or 2,2,4-trimethyl pentane, (CH₃)₃C₅H₉, is the branched-chain version of *n*-octane, CH₃(CH₂)₆CH₃. { ¦ī·sō'par·ə·fən }

isopentane [ORG CHEM] CH₃CHCH₃CH₂CH₃ Flammable, colorless liquid with pleasant aroma; boils at 28°C; soluble in oils, ether, and hydrocarbons, insoluble in water; used as a solvent and chemical intermediate. Also known as 2-methylbutane. { ¦ī·sō'pen,tān }

isopentanoic acid

isopentanoic acid [ORG CHEM] C₄H₉COOH A colorless, combustible liquid with a boiling point of 183.2°C; used for manufacture of plasticizers, pharmaceuticals, and synthetic lubricants. { ¦ī·sə‚pen·tə'nō·ik 'as·əd }

isopentyl unit *See* isoprene unit. { ‚ī·sə'pent·əl ‚yü·nət }

isophorone [ORG CHEM] COCHC(CH₃)CH₂C(CH₃)₂CH₂ A water-white liquid boiling at 215°C; used as a solvent for lacquers and polyvinyl and nitrocellulose resins. { ¦ī·sə'fó‚rōn }

isophthalic acid [ORG CHEM] C₆H₄(COOH)₂ Colorless crystals subliming at 345°C; slightly soluble in water, soluble in alcohol and acetic acid, and insoluble in benzene; used as an intermediate for polyester and polyurethane resins, and as a plasticizer. Also known as *meta*-phthalic acid. { ¦ī·sō¦thal·ik 'as·əd }

isopolymolybdate [INORG CHEM] A class of compounds formed by the acidification of a molybdate solution, or in some cases by heating normal molybdates. { ¦ī·sō‚päl·i·mə'lib‚dāt }

isopolytungstate [INORG CHEM] A compound formed by the condensation of tungstate compounds, usually classified into metatungstates, such as Na₆W₁₂O₄₀·xH₂O, and paratungstates, such as Na₁₀W₁₂O₄₁·xH₂O. { ¦ī·sō‚päl·i'təŋ‚ stāt }

isoprene [ORG CHEM] C₅H₈ A conjugated diolefin; a mobile, colorless liquid having a boiling point of 34.1°C; insoluble in water, soluble in alcohol and ether; polymerizes readily to form dimers and high-molecular-weight elastomer resins. { 'ī·sə‚prēn }

isoprene unit [ORG CHEM] The five-carbon structural unit characteristic of terpenes. Also known as isopentyl unit. { ¦ī·sə‚prēn ‚yü·nət }

isoprenoid *See* terpene. { ‚ī·sə·'prē‚nóid }

isopropaline [ORG CHEM] C₁₅H₂₃N₃O₄ An orange liquid with limited solubility in water; used as a preemergence herbicide for control of grass and broadleaf weeds on tobacco. { ¦ī·sə'prō·pə‚lēn }

isopropanol *See* isopropyl alcohol. { ¦ī·sə'prō·pə‚nól }

isopropanolamine [ORG CHEM] CH₃CH(OH)CH₂NH₂ A combustible liquid with a faint ammonia odor and a boiling point of 159.9°C; soluble in water; used as an emulsifying agent and for dry-cleaning soaps, wax removers, cosmetics, plasticizers, and insecticides. { ¦ī·sə‚prō·pə'nal·ə‚mēn }

isopropenyl acetate [ORG CHEM] CH₃CO₂C(CH₃)=CH₂ A liquid with a boiling point of 97°C; used for acylation of potential enols. { ‚ī·sə'prō·pə·nəl 'as·ə‚tāt }

2-isopropoxyphenyl N-methylcarbamate [ORG CHEM] C₁₁H₁₅O₃N A colorless solid with a melting point of 91°C; used as an insecticide for cockroaches, flies, mosquitoes, and lawn insects. { ¦tü ¦ī·sō·prə¦päk·sē'fen·əl ¦en ‚meth·əl'kär·bə‚mät }

isopropyl [ORG CHEM] The radical (CH₃)₂CH, from isopropane; an example of its occurrence is in isopropyl alcohol, (CH₃)₂CHOH. { ¦ī·sə'prō·pəl }

isopropyl acetate [ORG CHEM] CH₃COOCH(CH₃)₂ A colorless, aromatic liquid with a boiling point of 89.4°C; used as a solvent and for paints and printing inks. { ¦ī·sə'prō·pəl 'as·ə‚tāt }

isopropyl alcohol [ORG CHEM] (CH₃)₂CHOH A colorless liquid that boils at 82.4°C; soluble in water, ether, and ethanol; used in manufacturing of acetone and its derivatives, of glycerol, and as a solvent. Also known as isopropanol; 2-propanol; *sec*-propyl alcohol. { ¦ī·sə'prō·pəl 'al·kə‚hól }

isopropylamine [ORG CHEM] (CH₃)₂CHNH₂ A volatile, colorless liquid with a boiling point of 32.4°C; used as a solvent and in the manufacture of pharmaceuticals, dyes, insecticides, and bactericides. Also known as 2-aminopropane. { ¦ī·sə·prō'pil·ə‚mēn }

isopropyl-2-(N-benzoyl-3-chloro-4-fluoroanilino)propionate [ORG CHEM] C₁₉H₁₉O₃NClF Off-white crystals with a melting point of 56–57°C; used as a postemergence herbicide for wild oats and barley. { ¦ī·sə¦prō·pəl 'tü ¦en ¦ben·zə·wəl ¦thrē ¦klór·ō ¦fór ‚flür·ō¦an·ə·lō 'prō·pē·ə‚nāt }

isopropyl 4,4′-dibromobenzilate [ORG CHEM] C₁₇H₁₆O₃Br₂ A brownish solid with a melting point of 77°C; solubility in water is less than 0.5 part per million at 20°C; used as a miticide for deciduous fruit and citrus. { ¦ī·sə¦prō·pəl ¦fór ¦fór‚prīm dī‚brō·mō'ben·zə‚lāt }

isopropyl 4,4'-dichlorobenzilate [ORG CHEM] $C_{17}H_{16}O_3Cl_2$ A white powder with a melting point of 70–72°C; solubility in water is less than 10 parts per million at 20°C; used as a miticide for spider mites on apple and pear trees. { ¦ī·sə¦prō·pəl ¦fór ¦fór,prīm dī,klór·ō'ben·zə,lāt }

isopropyl ether [ORG CHEM] $(CH_3)_2CHOCH(CH_3)_2$ Water-soluble, flammable, colorless liquid with etherlike aroma; boils at 68°C; used as a solvent and extractant, in paint and varnish removers, and in spotting formulas. Also known as diisopropyl ether. { ¦ī·sə¦prō·pəl 'ē·thər }

N-4-isopropylphenyl-N′,N′-dimethylurea [ORG CHEM] $(CH_3)_2CHC_6H_4NHCON(CH_3)_2$ A crystalline solid with a melting point of 151–153°C; solubility in water is 170 parts per million; used as an herbicide for wheat, barley, and rye. { ¦en ¦fór ¦ī·sə¦prō·pəl¦fen·əl ¦en,prīm ¦en,prīm dī,meth·əl·yú'rē·ə }

ortho-isopropylphenyl-methylcarbamate [ORG CHEM] $C_{11}H_{15}O_2N$ A white, crystalline compound with a melting point of 88–89°C; used as an insecticide for rice and cacao crops. Also known as MIPC. { ¦ór·thō ¦ī·sə¦prō·pəl¦fen·əl ,meth·əl'kär·bə,māt }

isopulegol [ORG CHEM] $C_{10}H_{17}OH$ An alcohol derived from terpene as a water-white liquid that has a mintlike odor; used in making perfumes. { ¦ī·sō'pyü·lə,gól }

isoquinoline [ORG CHEM] $C_6H_4CHNCHCH$ Colorless liquid boiling at 243°C; soluble in most organic solvents and dilute mineral acids, insoluble in water; derived from coal tar or made synthetically; used to make dyes, insecticides, pharmaceuticals, and rubber accelerators, and as a chemical intermediate. { ¦ī·sə'kwin·ə,lēn }

isosafrole [ORG CHEM] $C_{10}H_{10}O_2$ A liquid with the odor of anise that is obtained from safrole, and that boils at 253°C; used to make perfumes and flavors. { ¦ī·sō'sa,frōl }

isosbestic point [PHYS CHEM] During a chemical reaction, a point in the absorption spectrum (that is, a wavelength) where at least two chemical species (for example, reactant and product) have identical molar absorption coefficients, which remain constant as the reaction proceeds. A stable isosbestic point is evidence that a reaction is proceeding without forming an intermediate or multiple products. { ,ī·səs,bes·tik 'point }

isosteric [CHEM] Referring to similar electronic arrangements in chemical compounds. { ¦ī·sə¦ster·ik }

isosterism [PHYS CHEM] A similarity in the physical properties of ions, compounds, or elements, as a result of electron arrangements that are identical or similar. { ī'säs·tə,riz·əm }

isosynthesis [ORG CHEM] A process in which mixtures of hydrogen and carbon monoxide are reacted over a thorium oxide catalyst (sometimes mixed with additional substances) to produce branched hydrocarbons. { ¦ī·sō'sin·thə·səs }

isotachophoresis [PHYS CHEM] A variant of electrophoresis in which ionic species move with equal velocity in the presence of an electric field. { ¦ī·sə,tak·ə·fə'rē·səs }

isotactic [ORG CHEM] Designating crystalline polymers in which substituents in the asymmetric carbon atoms have the same (rather than random) configuration in relation to the main chain. { ¦ī·sə¦tak·tik }

isothiocyanate [ORG CHEM] A compound of the type R—N=C=S, where R may be an alkyl or aryl group; an example is mustard oil. Also known as sulfocarbimide. { ,ī·sə,thī·ō'sī·ə,nāt }

isotope-dilution analysis [ANALY CHEM] Variation on paper-chromatography analysis; a labeled radioisotope of the same type as the one being quantitated is added to the solution, then quantitatively analyzed afterward via radioactivity measurement. { 'ī·sə,tōp də¦lü·shən ə,nal·ə·səs }

isotope effect [PHYS CHEM] The effect of difference of mass between isotopes of the same element on nonnuclear physical and chemical properties, such as the rate of reaction or position of equilibrium, of chemical reactions involving the isotopes. { 'ī·sə,tōp i,fekt }

isotope-exchange reaction [CHEM] A chemical reaction in which interchange of the atoms of a given element between two or more chemical forms of the element occurs, the atoms in one form being isotopically labeled so as to distinguish them from atoms in the other form. { 'ī·sə,tōp iks¦chānj rē,ak·shən }

isotope shift

isotope shift [SPECT] A displacement in the spectral lines due to the different isotopes of an element. { 'ī·sə,tōp ,shift }

isotopic carrier [CHEM] A carrier that differs from the trace it is carrying only in isotopic composition. { ¦ī·sə¦täp·ik 'kar·ē·ər }

isotopic exchange [PHYS CHEM] A process in which two atoms belonging to different isotopes of the same element exchange valency states or locations in the same molecule or different molecules. { ¦ī·sə¦täp·ik iks'chānj }

isotopic indicator *See* isotopic tracer. { ¦ī·sə¦täp·ik 'in·də,kād·ər }

isotopic label *See* isotopic tracer. { ¦ī·sə¦täp·ik 'lā·bəl }

isotopic tracer [CHEM] An isotope of an element, either radioactive or stable, a small amount of which may be incorporated into a sample material (the carrier) in order to follow the course of that element through a chemical, biological, or physical process, and also follow the larger sample. Also known as isotopic indicator; isotopic label; label; tag. { ¦ī·sə¦täp·ik 'trā·sər }

isovalent conjugation [PHYS CHEM] An arrangement of bonds in a conjugated molecule such that alternative structures with an equal number of bonds can be written; an example occurs in benzene. { ¦ī·sə¦vā·lənt kən'jəŋk·shən }

isovalent hyperconjugation [PHYS CHEM] An arrangement of bonds in a hyperconjugated molecule such that the number of bonds is the same in the two resonance structures but the second structure is energetically less favorable than the first structure; examples are $H_3 \equiv C - C^+H_2$ and $H_3 \equiv C - CH_2$. { ¦ī·sə¦vā·lənt ,hī·pər,känjə'gā·shən }

isovaleral *See* isovaleraldehyde. { ¦ī·sō¦val·ə·rəl }

isovaleraldehyde [ORG CHEM] $(CH_3)_2CHCH_2CHO$ A colorless liquid with an applelike odor and a boiling point of 92°C; soluble in alcohol and ether; used in perfumes and pharmaceuticals and for flavoring. { ¦ī·sō¦val·ər'al·də,hīd }

isovaleric acid [ORG CHEM] $(CH_3)_2CHCH_2COOH$ Color-less liquid with disagreeable taste and aroma; boils at 176°C; soluble in alcohol and ether; found in valeriana, hop, tobacco, and other plants; used in flavors, perfumes, and medicines. { ¦ī·sō·və'ler·ik 'as·əd }

2-isovaleryl-1,3-indandione [ORG CHEM] $C_{14}H_{14}O_3$ A yellow, crystalline compound with a melting point of 67–68°C; insoluble in water; used as a rodenticide. { tü ¦ī·sō'val·ə,ril ¦wən ¦thrē ,in·dən'dī,ōn }

ISS *See* ion scattering spectroscopy.

itaconic acid [ORG CHEM] $CH_2:C(COOH)CH_2COOH$ A colorless crystalline compound that decomposes at 165°C, prepared by fermentation with *Aspergillus terreus*; used as an intermediate in organic synthesis and in resins and plasticizers. { ¦id·ə¦kän·ik 'as·əd }

itatartaric acid [ORG CHEM] $C_5H_8O_6$ A compound produced experimentally by fermentation; formed as a minor product, 5.8% of total acidity produced, of an itaconic-acid producing strain of *Aspergillus niger*. { ¦id·ə¦tär·də·rik 'as·əd }

ium ion [ORG CHEM] A positively charged group of atoms in which a charged nonmetallic ion other than carbon or silicon possesses a closed-shell electron configuration; often joined to a root word, as in carbonium ion. { 'ī·əm 'ī,än }

IUPAC *See* International Union of Pure and Applied Chemistry. { ī'yü,pak *or* ¦ī¦yü¦pē¦ā¦sē }

Ivanov reagent [ORG CHEM] A reagent that is similar to a Grignard reagent, and that is formed by reacting an arylacetic acid or its sodium salt with isopropyl magnesium halide. { ē·və·nóf rē,ā·jənt }

J

Jacquemart's reagent [ANALY CHEM] Analytical reagent used to test for ethyl alcohol; consists of an aqueous solution of mercuric nitrate and nitric acid. { zhak'märz rē,ä·jənt }

Jahn-Teller effect [PHYS CHEM] The effect whereby, except for linear molecules, degenerate orbital states in molecules are unstable. { 'yän 'tel·ər i,fekt }

jasmone [ORG CHEM] $C_{11}H_{16}O$ A liquid ketone found in jasmine oil and other essential oils from plants. { 'jaz,mōn }

jellium model [PHYS CHEM] A model describing the delocalized valence electrons in a metallic atom cluster in which the positive charge is regarded as being smeared out over the entire volume of the cluster while the valence electrons are free to move within this homogeneously distributed, positively charged background. { 'jel·ē·əm ,mäd·əl }

jeweler's rouge See ferric oxide. { 'jü·lərz 'rüzh }

jodfenphos [ORG CHEM] $C_8H_8O_3Cl_2IPS$ A crystalline compound with a melting point of 76°C; slight solubility in water; used as an insecticide in homes, farm buildings, and industrial sites. { 'yȯd·fən,fäs }

Jones reductor [CHEM] A device used to chemically reduce solutions, such as ferric salt solutions, consisting of a vertical tube containing granular zinc into which the solution is poured. { 'jōnz ri,dək·tər }

juglone [ORG CHEM] $C_{10}H_6O_3$ A naphthoquinone derivative that occurs naturally in black walnuts and is toxic to plants. { 'jəg,lōn }

juniperic acid [ORG CHEM] $C_{16}H_{32}O_3$ A crystalline hydroxy acid that melts at 95°C, obtained from waxy exudations from conifers. { ¦jü·nə¦per·ik 'as·əd }

K

K *See* potassium.
K acid [ORG CHEM] $C_{10}H_4NH_2OH(SO_3H)_2$ An acid derived from naphthylamine trisulfonic acid; used in dye manufacture. { 'kā ,as·əd }
kalium *See* potassium. { 'kāl·ēl·əm }
karbutilate [ORG CHEM] $C_{14}H_{21}N_3O_3$ An off-white solid with a melting point of 176–177°C; used as a herbicide on noncroplands, railroad rights-of-way, and plant sites. { kär'byüd·əl,āt }
Karl Fischer reagent [ANALY CHEM] A solution of 8 moles pyridine to 2 moles sulfur dioxide, with the addition of about 15 moles methanol and then 1 mole iodine; used to determine trace quantities of water by titration. { 'kärl 'fish·ər rē'ā·jənt }
Karl Fischer technique [ANALY CHEM] A method of determining trace quantities of water by titration; the Karl Fischer reagent is added in small increments to a glass flask containing the sample until the color changes from yellow to brown or a change in potential is observed at the end point. { 'kärl 'fish·ər tek'nēk }
kauri-butanol value [ANALY CHEM] The measure of milliliters of paint or varnish petroleum thinner needed to cause cloudiness in a solution of kauri gum in butyl alcohol. { 'kaú·rē 'byüt·ən,ȯl ,val·yü }
kayser [SPECT] A unit of reciprocal length, especially wave number, equal to the reciprocal of 1 centimeter. Also known as rydberg. { 'kī·zər }
Keesom force *See* orientation force. { 'kā·səm ,fȯrs }
Keesom relationship [PHYS CHEM] An equation for the potential energy associated with the interaction of the dipole moments of two polar molecules. { 'kā·səm ri'lā·shən,ship }
Kekulé structure [ORG CHEM] A molecular structure of a cyclic conjugated system that is depicted with alternating single and double bonds. { 'kā·kə,lā ,strək·chər }
ketal [ORG CHEM] **1.** Former term for the =CO group, as in dimethyl ketal (acetone). **2.** Any of the ketone acetates from condensation of alkyl orthoformates with ketones in the presence of alcohols. { 'kē,tal }
ketene [ORG CHEM] C_2H_2O A colorless, toxic, highly reactive gas, with disagreeable taste; boils at −56°C; soluble in ether and acetone, and decomposes in water and alcohol; used as an acetylating agent in organic synthesis. { 'kē,tēn }
ketimide [ORG CHEM] A compound that is represented by R_2:C:NX, where X is an acyl radical. { 'ked·ə,mīd }
ketimine [ORG CHEM] An organic compound that contains the divalent group ⟩C=NH; a Schiff base is an example. { 'ked·ə,mēn }
keto- [ORG CHEM] Organic chemical prefix for the keto or carbonyl group, C:O, as in a ketone. { 'kēd·ō }
keto acid [ORG CHEM] A compound that is both an acid and a ketone; an example is β-acetoacetic acid. { 'kēd·ō ,as·əd }
ketoglutarate [ORG CHEM] A salt or ester of ketoglutaric acid. { ,kēd·ə'glüd·ə,rāt }
ketone [ORG CHEM] One of a class of chemical compounds of the general formula RR'CO, where R and R' are alkyl, aryl, or heterocyclic radicals; the groups R and R' may be the same or different, or incorporated into a ring; the ketones, acetone, and methyl ethyl ketone are used as solvents, and ketones in general are important intermediates in the synthesis of organic compounds. { 'kē,tōn }

Kiliani reaction

Kiliani reaction [ORG CHEM] A method of synthesizing a higher aldose from a lower aldose; monosaccharides, such as aldehydes and ketones, react with hydrogen cyanide to form cyanohydrins, which are hydrolyzed to hydroxy acids, converted to lactones, and reduced to aldoses with sodium amalgams. { ˌkil·ē'an·ē rē,ak·shən }

kilogram-equivalent weight [CHEM] A unit of mass 1000 times the gram-equivalent weight. { 'kil·ə,gram i'kwiv·ə·lənt 'wāt }

king's blue *See* cobalt blue. { 'kiŋz 'blü }

kinic acid *See* quinic acid. { 'kin·ik 'as·əd }

Kistiakowsky-Fishtine equation [PHYS CHEM] An equation to calculate latent heats of vaporization of pure compounds; useful when vapor pressure and critical data are not available. { ˌkis·tē·ə'kȯf·skē fə'shtīn i,kwā·zhən }

kitol [ORG CHEM] $C_{40}H_{60}O_2$ One of the provitamins of vitamin A derived from whale liver oil; crystallizes from methanol solution. { 'kē,tȯl }

Kjeldahl method [ANALY CHEM] Quantitative analysis of organic compounds to determine nitrogen content by interaction with concentrated sulfuric acid; ammonia is distilled from the NH_4SO_4 formed. { 'kel,däl ˌmeth·əd }

Klein-Rydberg method [PHYS CHEM] A method for determining the potential energy function of the distance between the nuclei of a diatomic molecule from the molecule's vibrational and rotational levels. { 'klīn 'rid,berg ˌmeth·əd }

Klein's reagent [CHEM] Saturated solution of borotungstate; used to separate minerals by specific gravity. { 'klīnz rē,ā·jənt }

Knoevenagel reaction [ORG CHEM] The condensation of aldehydes with compounds containing an activated methylene (=CH_2) group. { kə'nē·və,näg·əl rē,ak·shən }

Knorr synthesis [ORG CHEM] A condensation reaction carried out in either glacial acetic acid or an aqueous alkali in which an α-aminoketone combines with an α-carbonyl compound to form a pyrrole; possibly the most versatile pyrrole synthesis. { 'nȯr ˌsin·thə·səs }

knot [ORG CHEM] A chiral structure in which rings containing 50 or more members have a knotlike configuration. { nät }

Knudsen cell [PHYS CHEM] A vessel used to measure very low vapor pressures by measuring the mass of vapor which escapes when the vessel contains a liquid in equilibrium with its vapor. { kə'nüd·sən ˌcel }

Kohlrausch law [PHYS CHEM] **1.** The law that every ion contributes a definite amount to the equivalent conductance of an electrolyte in the limit of infinite dilution, regardless of the presence of other ions. **2.** The law that the equivalent conductance of a very dilute solution of a strong electrolyte is a linear function of the concentration. { 'kōl,raúsh ˌlȯ }

Kohlrausch method [PHYS CHEM] A method of measuring the electrolytic conductance of a solution using a Wheatstone bridge. { 'kōl,raúsh ˌmeth·əd }

Kojic acid [ORG CHEM] $C_6H_6O_4$ A crystalline antibiotic with a melting point of 152–154°C; soluble in water, acetone, and alcohol; used in insecticides and as an antifungal and antimicrobial agent. { 'kō·jik ˌas·əd }

Kolbe hydrocarbon synthesis [ORG CHEM] The production of an alkane by the electrolysis of a water-soluble salt of a carboxylic acid. { 'kōl·bə ˌhī·drə'kär·bən ˌsin·thə·səs }

Kolbe-Schmitt synthesis [ORG CHEM] The reaction of carbon dioxide with sodium phenoxide at 125°C to give salicyclic acid. { 'kōl·bə 'shmit ˌsin·thə·səs }

Konowaloff rule [PHYS CHEM] An empirical rule which states that in the vapor over a liquid mixture there is a higher proportion of that component which, when added to the liquid, raises its vapor pressure, than of other components. { ˌkò·nə'vä·lȯf ˌrül }

Kopp's law [PHYS CHEM] The law that for solids the molal heat capacity of a compound at room temperature and pressure approximately equals the sum of heat capacities of the elements in the compound. { 'käps ˌlȯ }

Korner's method [ORG CHEM] A method for determining the absolute position of substituents for positional isomers in benzene by the experimental production of positional isomers from a given disubstituted benzene. { 'kȯr·nərz ˌmeth·əd }

210

Korshun method [ANALY CHEM] Microdetermination of carbon and hydrogen in organic compounds; the sample is prepyrolyzed (cracked) in a shortage of oxygen, then oxidized in an excess of oxygen. { 'kòr·shən ‚meth·əd }

Kossel-Sommerfeld law [SPECT] The law that the arc spectra of the atom and ions belonging to an isoelectronic sequence resemble each other, especially in their multiplet structure. { 'käs·əl 'zòm·ər‚felt ‚lò }

Kovat's retention indexes [ANALY CHEM] Procedure to identify compounds in gas chromatography; the behavior of a compound is indicated by its position on a scale of normal alkane values (for example, methane = 100, ethane = 200). { 'kō·vats ri'ten·chən ‚in‚dek·səs }

Kr *See* krypton.

krypton [CHEM] A colorless, inert gaseous element, symbol Kr, atomic number 36, atomic weight 83.80; it is odorless and tasteless; used to fill luminescent electric tubes. { 'krip·tän }

Ku *See* kurchatovium.

Kundt rule [SPECT] The rule that the optical absorption bands of a solution are displaced toward the red when its refractive index increases because of changes in composition or other causes. { 'kùnt ‚rül }

kurchatovium [CHEM] The name suggested by workers in the Soviet Union for element 104. Symbolized Ku. { ‚kər·chə'tō·vē·əm }

211

L

La *See* lanthanum.

laboratory sample [ANALY CHEM] A sample of a material to be tested or analyzed that is prepared from a gross sample and retains the latter's composition. { ˈlab·rə,tȯr· ē ˌsam·pəl }

lachesne [ORG CHEM] $C_{20}H_{26}ClNO_3$ A compound that crystallizes from a solution of ethanol and acetone, and whose melting point is 213°C; used in ophthalmology. Also known as chloride benzilate. { lə'shēn }

lactam [ORG CHEM] An internal (cyclic) amide formed by heating gamma (γ) and delta (δ) amino acids; thus γ-aminobutyric acid readily forms γ-butyrolactam (pyrrolidone); many lactams have physiological activity. { 'lak,tam }

lactate [ORG CHEM] A salt or ester of lactic acid in which the acidic hydrogen of the carboxyl group has been replaced by a metal or an organic radical. { 'lak,tāt }

lactide [ORG CHEM] A cyclic, intermolecular, double ester formed from α-hydroxy acids; most lactides are relatively low melting solids and are easily hydrolyzed by base to form salts of the parent acid, such as sodium lactate. { 'lak,tīd }

lactim [ORG CHEM] A tautomeric enol form of a lactam with which it forms an equilibrium whenever the lactam nitrogen carries a free hydrogen. { 'lak·təm }

lactone [ORG CHEM] An internal cyclic mono ester formed by gamma (γ) or delta (δ) hydroxy acids spontaneously; thus γ-hydroxybutyric acid forms γ-butyrolactone. { 'lak,tōn }

lactonitrile [ORG CHEM] $CH_3CHOHCN$ A straw-colored liquid boiling at 183°C; soluble in water, insoluble in carbon disulfide and petroleum ether; used as a solvent, and as a chemical intermediate in making esters of lactic acid. Also known as acetaldehyde cyanohydrin. { ˌlak·tō'nī,tril }

lactonization [ORG CHEM] The process in which a lactone is formed by intramolecular attack of a hydroxyl group on an activated carbonyl group. { ,lak·tə·nə'zā·shən }

lambda sulfur [CHEM] One of the two components of plastic (or gamma) sulfur; soluble in carbon disulfide. { 'lam·də ,səl·fər }

Lambert-Beer law *See* Bouguer-Lambert-Beer law. { 'lam·bərt 'bir ,lȯ }

Lambert's law *See* Bouguer-Lambert law. { 'lam·bərts ,lȯ }

Langelier index [CHEM] A measure, based on pH, of the degree of calcium carbonate saturation in water, where negative values indicate that corrosion may result (pH below 7, dissolves calcium carbonate), and positive values indicate that scale deposition may result (pH above 7, precipitates calcium carbonate). { ,länzh·əl,yā 'in,deks }

Langmuir-Blodgett film [PHYS CHEM] A highly ordered monomolecular film that results from compressing a surface layer of amphiphilic molecules into a floating monolayer and transferring it to a substrate by dipping. { 'laŋ,myür 'bläj·ət ,film }

Langmuir isotherm equation [PHYS CHEM] An equation, useful chiefly for gaseous systems, for the amount of material adsorbed on a surface as a function of pressure, while the temperature is held constant, assuming that a single layer of molecules is adsorbed; it is $f = ap/(1 + ap)$, where f is the fraction of surface covered, p is the pressure, and a is a constant. { 'laŋ,myür 'īs·ə,thərm i,kwā·zhən }

lanthana *See* lanthanum oxide. { 'lan·thə·nə }

lanthanide series [CHEM] Rare-earth elements of atomic numbers 57 through 71; their

lanthanum

chemical properties are similar to those of lanthanum, atomic number 57. { 'lan·thə,nīd ,sir·ēz }

lanthanum [CHEM] A chemical element, symbol La, atomic number 57, atomic weight 138.91; it is the second most abundant element in the rare-earth group. { 'lan·thə·nəm }

lanthanum nitrate [INORG CHEM] $La(NO_3)_3 \cdot 6H_2O$ Hygroscopic white crystals melting at 40°C; soluble in alcohol and water; used as an antiseptic and in gas mantles. { 'lan·thə·nəm 'nī,trāt }

lanthanum oxide [INORG CHEM] La_2O_3 A white powder melting at about 2000°C; soluble in acid, insoluble in water; used to replace lime in calcium lights and in optical glass. Also known as lanthana; lanthanum sesquioxide; lanthanum trioxide. { 'lan·thə·nəm 'äk,sīd }

lanthanum sesquioxide See lanthanum oxide. { 'lan·thə·nəm ,ses·kwē'äk,sīd }

lanthanum sulfate [INORG CHEM] $La_2(SO_4)_3 \cdot 9H_2O$ White crystals; slightly soluble in water, soluble in alcohol; used for atomic weight determinations for lanthanum. { 'lan·thə·nəm 'səl,fāt }

lanthanum trioxide See lanthanum oxide. { 'lan·thə·nəm trī'äk,sīd }

larixinic acid See maltol. { ¦lar·ik¦sin·ik 'as·əd }

laser heterodyne spectroscopy [SPECT] A high-resolution spectroscopic technique, used in astronomical and atmospheric observations, in which the signal to be measured is mixed with a laser signal in a solid-state diode, producing a difference-frequency signal in the radio-frequency range. { 'lā·zər ¦hed·ə·rə,dīn spek'träs·kə·pē }

laser spectroscopy [SPECT] A branch of spectroscopy in which a laser is used as an intense, monochromatic light source; in particular, it includes saturation spectroscopy, as well as the application of laser sources to Raman spectroscopy and other techniques. { 'lā·zər spek'träs·kə,pē }

laudanidine [ORG CHEM] $C_{20}H_{25}NO_4$ An optically active alkaloid found in opium that crystallizes as prisms from an alcohol solution, and melts at 185°C. Also known as l-laudanine; tritopine. { lò'dan·ə,dēn }

laudanine [ORG CHEM] $C_{20}H_{25}NO_4$ An optically inactive alkaloid derived from alkaline mother liquors from morphine extraction; it crystallizes in orthorhombic prisms from alcohol and chloroform; the prisms melt at 167°C, and are soluble in hot alcohol, benzene, and chloroform. Also known as dl-laudanidine. { lò'dan·ə,nēn }

laudanosine [ORG CHEM] $C_{21}H_{27}NO_4$ An alkaloid that is the methyl ether of laudanine; the optically inactive form crystallizes from dilute alcohol and melts at about 115°C; the levorotatory active form crystallizes from light petroleum solution and melts at 89°C. { lò'dan·ə,sēn }

laughing gas See nitrous oxide. { 'laf·iŋ ,gas }

lauric acid [ORG CHEM] $CH_3(CH_2)_{10}COOH$ A fatty acid melting at 44°C, boiling at 225°C (100 mmHg; 13,332 pascals); colorless needles soluble in alcohol and ether, insoluble in water; found as the glyceride in vegetable fats, such as coconut and laurel oils; used for wetting agents, in cosmetics, soaps, resins, and insecticides, and as a chemical intermediate. { 'lòr·ik 'as·əd }

lauryl alcohol [ORG CHEM] $CH_3(CH_2)_{11}OH$ A colorless solid which is obtained from coconut oil fatty acids, has a floral odor, and boils at 259°C; used in detergents, lubricating oils, and pharmaceuticals. { 'lòr·əl 'al·kə,hòl }

lauryl aldehyde [ORG CHEM] $CH_3(CH_2)_{10}CHO$ A constituent of an essential oil from the silver fir; a colorless solid or a liquid, with a floral odor, that is soluble in 90% alcohol; used in perfumes. { 'lòr·əl 'al·də,hīd }

lauryl mercaptan [ORG CHEM] $C_{12}H_{25}SH$ Pale-yellow or water-white liquid with mild odor; insoluble in water, soluble in organic solvents; used to manufacture plastics, pharmaceuticals, insecticides, fungicides, and elastomers. { 'lòr·əl mər'kap,tan }

law of constant heat summation See Hess's law. { 'lò əv ¦kän·stənt 'hēt sə,mā·shən }

law of corresponding states [CHEM] The law that when, for two substances, any two ratios of pressure, temperature, or volume to their respective critical properties are equal, the third ratio must equal the other two. { 'lò əv ¦kär·ə¦spän·diŋ 'stāts }

law of definite composition *See* law of definite proportion. { 'lȯ əv ¦def·ə·nət ˌkäm·pə'zish·ən }

law of definite proportion [CHEM] The law that a given chemical compound always contains the same elements in the same fixed proportion by weight. Also known as law of definite composition. { 'lȯ əv ¦def·ə·nət prə'pȯr·shən }

law of mass action [CHEM] The law stating that the rate at which a chemical reaction proceeds is directly proportional to the molecular concentrations of the reacting compounds. { 'lȯ əv ¦mas 'ak·shən }

lawrencium [CHEM] A chemical element, symbol Lr, atomic number 103; isotopes with mass numbers 251–263 have been discovered, all unilable; mass number 262 has the longest half-life (3.6 hours). { 'lȯ'ren·sē·əm }

LDPE *See* low-density polyethylene.

leachate [CHEM] A solution formed by leaching. { 'lē,chāt }

lead [CHEM] A chemical element, symbol Pb, atomic number 82, atomic weight 207.19. { led }

lead acetate [ORG CHEM] $Pb(C_2H_3O_2)_2 \cdot 3H_2O$ Poisonous, water-soluble white crystals decomposing at 280°C; loses water at 75°C; used in hair dyes, medicines, and textile mordants, for waterproofing, for manufacture of varnishes and pigments, and as an analytical reagent. Also known as sugar of lead. { 'led 'as·ə,tāt }

lead antimonite [INORG CHEM] $Pb_3(SbO_4)_2$ Poisonous, water-insoluble orange-yellow powder; used as a paint pigment and to stain glass and ceramics. Also known as antimony yellow; Naples yellow. { 'led an'tim·ə,nīt }

lead arsenate [INORG CHEM] $Pb_3(AsO_4)_2$ Poisonous, water-insoluble white crystals; soluble in nitric acid; used as an insecticide. { 'led 'ärs·ən,āt }

lead azide [INORG CHEM] $Pb(N_3)_2$ Unstable, colorless needles that explode at 350°C; lead azide is shipped submerged in water to reduce sensitivity; used as a detonator for high explosives. { 'led 'a,zīd }

lead borate [INORG CHEM] $Pb(BO_2)_2 \cdot H_2O$ Poisonous, water-insoluble white powder; soluble in dilute nitric acid; used as varnish and paint drier, for galvanoplastic work, in lead glass, and in waterproofing paints. { 'led 'bȯr,āt }

lead bromide [INORG CHEM] $PbBr_2$ An alcohol-insoluble white powder melting at 373°C, boiling at 916°C; slightly soluble in hot water. { 'led 'brō,mīd }

lead carbonate [INORG CHEM] $PbCO_3$ Poisonous, acid-soluble white crystals decomposing at 315°C; insoluble in alcohol and water; used as a paint pigment. { 'led 'kär·bə,nāt }

lead chloride [INORG CHEM] $PbCl_2$ Poisonous white crystals melting at 498°C, boiling at 950°C; slightly soluble in hot water, insoluble in alcohol and cold water; used to make lead salts and lead chromate pigments and as an analytical reagent. { 'led 'klȯr,īd }

lead chromate [INORG CHEM] $PbCrO_4$ Poisonous, water-insoluble yellow crystals melting at 844°C; soluble in acids; used as a paint pigment. { 'led 'krō,māt }

lead cyanide [INORG CHEM] $Pb(CN)_2$ Poisonous white to yellow powder; slightly soluble in water, decomposed by acids; used in metallurgy. { 'led 'sī·ə,nīd }

lead dioxide [INORG CHEM] PbO_2 Poisonous brown crystals that decompose when heated; insoluble in water and alcohol, soluble in glacial acetic acid; used as an oxidizing agent, in electrodes, batteries, matches, and explosives, as a textile mordant, in dye manufacture, and as an analytical reagent. Also known as anhydrous plumbic acid; brown lead oxide; lead peroxide. { 'led dī'äk,sīd }

lead fluoride [INORG CHEM] PbF_2 A crystalline solid with a melting point of 824°C; used for laser crystals and electronic and optical applications. { 'led 'flu̇r,īd }

lead formate [ORG CHEM] $Pb(CHO_2)_2$ Poisonous, water-soluble brownish-white crystals that decompose at 190°C; used as an analytical reagent. { 'led 'fȯr,māt }

lead halide [INORG CHEM] PbX_2, where X is a halogen (such as F, Br, Cl, or I). { 'led 'ha,līd }

lead hexafluorosilicate [INORG CHEM] $PbSiF_6 \cdot 2H_2O$ Poisonous, colorless, water-soluble crystals; used in the electrolytic method for refining lead. { 'led ¦hek·sə,flu̇r·ə'sil·ə,kāt }

lead iodide

lead iodide [INORG CHEM] PbI$_2$ Poisonous, water- and alcohol-insoluble golden-yellow crystals melting at 402°C, boiling at 954°C; used in photography, medicine, printing, mosaic gold, and bronzing. { 'led 'ī·ə,dīd }

lead metasilicate *See* lead silicate. { 'led ,med·ə'sil·ə,kāt }

lead molybdate [INORG CHEM] PbMoO$_4$ Poisonous, acid-soluble yellow powder; insoluble in water and alcohol; used in pigments and as an analytical reagent. { 'led mə'lib,dāt }

lead monoxide [INORG CHEM] PbO Yellow, tetragonal crystals that melt at 888°C and are soluble in alkalies and acids; used in storage batteries, ceramics, pigments, and paints. Also known as litharge; plumbous oxide; yellow lead oxide. { 'led mə'näk,sīd }

lead nitrate [INORG CHEM] Pb(NO$_3$)$_2$ Strongly oxidizing, poisonous, water- and alcohol-soluble white crystals that decompose at 205–223°C; used as a textile mordant, paint pigment, and photographic sensitizer and in medicines, matches, explosives, tanning, and engraving. { 'led 'nī,trāt }

lead oleate [ORG CHEM] Pb(C$_{18}$H$_{33}$O$_2$)$_2$ Poisonous, water-insoluble, white, ointmentlike material; soluble in alcohol, benzene, and ether; used in varnishes, lacquers, and high-pressure lubricants, and as a paint drier. { 'led 'ō·lē,āt }

lead orthoplumbate *See* lead tetroxide. { 'led ,òr·thō'pləm,bāt }

lead oxide red *See* lead tetroxide. { 'led 'äk,sīd 'red }

lead peroxide *See* lead dioxide. { 'led pə'räk,sīd }

lead phosphate [INORG CHEM] Pb$_3$PO$_4$ A poisonous, white powder that melts at 1014°C; soluble in nitric acid and in fixed alkali hydroxide; used as a stabilizer in plastics. { 'led 'fäs,fāt }

lead pigments [CHEM] Chemical compounds of lead used in paints to give color; examples are white lead; basic lead carbonate; lead carbonate; lead thiosulfate; lead sulfide; basic lead sulfate (sublimed white lead); silicate white lead; basic lead silicate; lead chromate; basic lead chromate; lead oxychloride; and lead oxide (monoxide and dioxide). { 'led 'pig·məns }

lead resinate [ORG CHEM] Pb(C$_{20}$H$_{29}$O$_2$)$_2$ Poisonous, insoluble, brown, lustrous, translucent lumps; used as a paint and varnish drier and for textile waterproofing. { 'led 'rez·ən,āt }

lead silicate [INORG CHEM] PbSiO$_3$ Toxic, insoluble white crystals; used in ceramics, paints, and enamels, and to fireproof fabrics. Also known as lead metasilicate. { 'led 'sil·ə,kāt }

lead sodium hyposulfate *See* lead sodium thiosulfate. { 'led 'sōd·ē·əm ,hī·pō'səl,fāt }

lead sodium thiosulfate [INORG CHEM] Na$_4$Pb(S$_2$O$_3$)$_3$ Poisonous, small, white, heavy crystals that are soluble in thiosulfate solutions; used in the manufacture of matches. Also known as lead sodium hyposulfate; sodium lead hyposulfate; sodium lead thiosulfate. { 'led 'sōd·ē·əm ,thī·ə'səl,fāt }

lead stearate [ORG CHEM] Pb(C$_{18}$H$_{35}$O$_2$)$_2$ Poisonous white powder; soluble in alcohol and ether, insoluble in water; used as a lacquer and varnish drier and in high-pressure lubricants. { 'led 'stir,āt }

lead sulfate [INORG CHEM] PbSO$_4$ Poisonous white crystals melting at 1170°C; slightly soluble in hot water, insoluble in alcohol; used in storage batteries and as a paint pigment. { 'led 'səl,fāt }

lead sulfide [INORG CHEM] PbS Blue, metallic, cubic crystals that melt at 1120°C, derived from the mineral galena or by reacting hydrogen sulfide gas with a solution of lead nitrate; used in semiconductors and ceramics. Also known as plumbous sulfide. { 'led 'səl,fīd }

lead telluride [INORG CHEM] PbTe A crystalline solid that is very toxic if inhaled or ingested; melts at 902°C; used as a semiconductor and photoconductor in the form of single crystals. { 'led 'tel·yə,rīd }

lead tetraacetate [ORG CHEM] Pb(CH$_3$COO)$_4$ Crystals that are faintly pink or colorless; melts at 175°C; used as an oxidizing agent in organic chemistry, cleaving 1,2-diols to form aldehydes or ketones. { 'led ,te·trə'as·ə,tāt }

lead tetroxide [INORG CHEM] Pb$_3$O$_4$ A poisonous, bright-red powder, soluble in excess glacial acetic acid and dilute hydrochloric acid; used in medicine, in cement for special applications, in manufacture of colorless glass, and in ship paint. Also known as lead orthoplumbate; lead oxide red; red lead. { 'led ,te'träk,sīd }

lead thiocyanate [INORG CHEM] $Pb(SCN)_2$ Yellow, monoclinic crystals, soluble in potassium thiocyanate and slightly soluble in water; used in the powder mixture that primes small arm cartridges, in dyes, and in safety matches. { 'led ,thī·ō'sī·ə,nāt }

lead titanate [INORG CHEM] $PbTiO_3$ A water-insoluble, pale-yellow solid; used as coloring matter in paints. { 'led 'tīt·ən,āt }

lead tungstate [INORG CHEM] $PbWO_4$ A yellowish powder, melting at 1130°C; insoluble in water, soluble in acid; used as a pigment. Also known as lead wolframate. { 'led 'təŋ·stə,nāt }

lead vanadate [INORG CHEM] $Pb(VO_3)_2$ A water-insoluble, yellow powder; used as a pigment and for the preparation of other vanadium compounds. { 'led 'van·ə,dāt }

lead wolframate *See* lead tungstate. { 'led 'wúl·frə,māt }

leakage [PHYS CHEM] A phenomenon occurring in an ion-exchange process in which some influent ions are not adsorbed by the ion-exchange bed and appear in the effluent. { 'lēk·ij }

leaving group [ORG CHEM] The group of charged or uncharged atoms that departs during a substitution or displacement reaction. Also known as nucleofuge. { 'lēv·iŋ ,grüp }

Lennard-Jones potential [PHYS CHEM] A semiempirical approximation to the potential of the force between two molecules, given by $v = (A/r^{12}) - (B/r^6)$, where r is the distance between the centers of the molecules, and A and B are constants. { 'len·ərd 'jōnz pə,ten·chəl }

lepidine [ORG CHEM] $C_9H_6NCH_3$ An alkaloid derived as an oily liquid from cinchona bark; boils at 266°C; soluble in ether, benzene, and alcohol; used in organic synthesis. { 'lep·ə,dēn }

leptophos [ORG CHEM] $C_{13}H_{10}BrCl_2O_2PS$ A white solid with a melting point of 70.2–70.6°C; slight solubility in water; used as an insecticide on vegetables, fruit, turf, and ornamentals. Also known as O-(4-bromo-2,5-dichlorophenyl) O-methyl phenylphosphorothioate. { 'lep·tə,fäs }

leucaenine *See* mimosine. { 'lü·sə,nēn }

leucaenol *See* mimosine. { 'lü·sə,nòl }

leucenine *See* mimosine. { 'lü·sə,nēn }

leucenol *See* mimosine. { 'lü·sə,nòl }

leuco base [ORG CHEM] Any group of colorless derivatives of triphenylmethane dyes that are produced by reducing the dye and are capable of being reconverted to the original dye by oxidation. Also known as leuco compound. { 'lü·kō ,bās }

leuco compound *See* leuco base. { 'lü·ko ,käm,paúnd }

leucoline *See* quinoline. { 'lü·kə,lēn }

leukol *See* quinoline. { 'lü,kòl }

levigate [CHEM] **1.** To separate a finely divided powder from a coarser material by suspending in a liquid in which both substances are insoluble. Also known as elutriation. **2.** To grind a moist solid to a fine powder. { 'lev·ə,gāt }

levorotatory enantiomer [ORG CHEM] An optically active substance that rotates the plane of plane-polarized light counterclockwise. Symbolized l. Abbreviated levo. { ‚lē·vō'rōt·ə,tòr·e ə,nan·tē,ō·mər }

levulinic acid [ORG CHEM] $CH_3COCH_2CH_2COOH$ Crystalline compound forming plates or leaflets that melt at 37°C; freely soluble in alcohol, ether, and chloroform; used in the manufacture of pharmaceuticals, plastics, rubber, and synthetic fibers. { ‚lev·yə,lin·ik 'as·əd }

Lewis acid [CHEM] A substance that can accept an electron pair from a base; thus, $AlCl_3$, BF_3, and SO_3 are acids. { 'lü·əs ,as·əd }

Lewis base [CHEM] A substance that can donate an electron pair; examples are the hydroxide ion, OH^-, and ammonia, NH_3. { 'lü·əs ,bās }

Lewis formula *See* Lewis structure. { 'lü·is ,fòr·myə·lə }

217

lewisite

lewisite [ORG CHEM] $C_2H_2AsCl_3$ An oily liquid, colorless to brown or violet; forms a toxic gas, used in World War I. { 'lü·ə,sīt }

Lewis structure [CHEM] A structural formula in which electrons are represented by dots; two dots between atoms represent a covalent bond. Also known as electron-dot formula; Lewis formula. { 'lü·is ,strək·chər }

Li See lithium.

lidocaine [ORG CHEM] $C_{14}H_{22}N_2O$ A crystalline compound, used as a local anesthetic. Also known as lignocaine. { 'līd·ə,kān }

ligand [CHEM] The molecule, ion, or group bound to the central atom in a chelate or a coordination compound; an example is the ammonia molecules in $[Co(NH_3)_6]^{3+}$. { 'lī·gənd }

ligand membrane [CHEM] A solvent immiscible with water and a reagent and acting as an extractant and complexing agent for an ion. { 'lī·gənd 'mem,brān }

light-emitting polymer See polymer light-emitting diode. { ¦līt·ə,mid·iŋ 'pöl·ə·mər }

light-scattering photometry [ANALY CHEM] Use of optical methods to measure the extent of scattering of light by particles suspended in fluids or by macromolecules in solution. { ¦līt ,skad·ər·iŋ fə'täm·ə·trē }

lignin plastic [ORG CHEM] A plastic based on resins derived from lignin; used as a binder or extender. { 'lig·nən 'plas·tik }

lignocaine See lidocaine. { 'lī·nə,kān }

lignosulfonate [ORG CHEM] Any of several substances manufactured from waste liquor of the sulfate pulping process of soft wood; used in the petroleum industry to reduce the viscosity of oil well muds and slurries, and as extenders in glues, synthetic resins, and cements. { ¦lig·nō'səl·fə,nāt }

limiting current density [PHYS CHEM] The maximum current density to achieve a desired electrode reaction before hydrogen or other extraneous ions are discharged simultaneously. { 'lim·əd·iŋ ¦kə·rənt ,den·səd·ē }

limiting density [PHYS CHEM] The density of a gas when the ratio of density per unit pressure is extrapolated to zero pressure, the point at which a gas exhibits ideal-gas behavior. { 'lim·ət·iŋ ,den·səd·ē }

limiting mean [ANALY CHEM] The value that the average approaches as the number of measurements made in a stable chemical measurement process increases indefinitely. { ¦lim·əd·iŋ 'mēn }

limiting reagent [CHEM] In a chemical reaction, the reagent that controls the quantity of product which can be formed. { 'lim·əd·iŋ rē,ā·jənt }

limiting viscosity number See intrinsic viscosity. { 'lim·əd·iŋ vi'skäs·əd·ē ,nəm·bər }

limit of detection [ANALY CHEM] The quantity or concentration that represents the smallest measure of an analyte that can be detected with reasonable certainty by a given analytical procedure. { ¦lim·ət əv di'tek·shən }

limonene [ORG CHEM] $C_{10}H_{16}$ A terpene with a lemon odor that is optically active and is found in oils from citrus fruits and in oils from peppermint and spearmint; a colorless, water-insoluble liquid that boils at 176°C. { 'lim·nə,lēn }

linalool [ORG CHEM] $(CH_3)_2C:CH(CH_2)_2CCH_3OHCH:CH_2$ A terpene that is a colorless liquid, has a bergamot odor, boils at 195–196°C, and is found in many essential oils, particularly bergamot and rosewood; used as a flavoring agent and in perfumes. Also known as coriandrol. { lə'näl·ə,wöl }

linalyl acetate [ORG CHEM] $(CH_3)_2C:CH(CH_2)_2CCH_3(OCOCH_3)CH:CH_2$ The acetic acid ester of linalool, a colorless oily liquid with a bergamot odor that boils at 108–110°C; used in perfumes and as a flavoring agent. { 'lin·ə,lil 'as·ə,tāt }

lindane [INORG CHEM] The gamma isomer of 1,2,3,4,5,6-hexachlorocyclohexane, constituting a persistent, bioaccumulative pesticide and a neurotoxin. { 'lin,dān }

linear molecule [PHYS CHEM] A molecule whose atoms are arranged so that the bond angle between each is 180°; an example is carbon dioxide, CO_2. { 'lin·ē·ər 'mäl·ə,kyül }

linear polymer [ORG CHEM] A polymer whose molecule is arranged in a chainlike fashion with few branches or bridges between the chains. { 'lin·ē·ər 'päl·ə·mər }

line broadening [SPECT] An increase in the range of wavelengths over which the characteristic absorption or emission of a spectral line takes place, due to a number of causes such as collision broadening and Doppler broadening. { 'līn ˌbròd·ən·iŋ }

line-formula method [ORG CHEM] A system of notation for hydrocarbons showing the chemical elements, functional groups, and ring systems in linear form; an example is acetone, CH_3COCH_3. { 'līn ˌfòr·myə·lə ˌmeth·əd }

line pair [SPECT] In spectrographic analysis, a particular spectral line and the internal standard line with which it is compared to determine the concentration of a substance. { 'līn ˌper }

line-segment formula See bond-line formula. { 'līn ˌseg·mənt ˌfòr·myə·lə }

line spectrum [SPECT] **1.** A spectrum of radiation in which the quantity being studied, such as frequency or energy, takes on discrete values. **2.** Conventionally, the spectra of atoms, ions, and certain molecules in the gaseous phase at low pressures; distinguished from band spectra of molecules, which consist of a pattern of closely spaced spectral lines which could not be resolved by early spectroscopes. { 'līn ˌspek·trəm }

linolenyl alcohol [ORG CHEM] $C_{18}H_{32}O$ A colorless, combustible solid used for paints, paper, leather, and flotation processes. Also known as octadecatrienol. { ˌlin·əˌlēn·əl 'al·kəˌhòl }

lipophilic [CHEM] **1.** Having a strong affinity for fats. **2.** Promoting the solubilization of lipids. { ˌlip·əˌfil·ik }

lipophobic [CHEM] Lacking an affinity for, repelling, or failing to absorb or adsorb fats. { ˌlip·ə'fōb·ik }

liquid chromatography [ANALY CHEM] A form of chromatography employing a liquid as the moving phase and a solid or a liquid on a solid support as the stationary phase; techniques include column chromatography, gel permeation chromatography, and partition chromatography. { 'lik·wəd ˌkrō·mə'täg·rə·fē }

liquid crystal [PHYS CHEM] A liquid which is not isotropic; it is birefringent and exhibits interference patterns in polarized light; this behavior results from the orientation of molecules parallel to each other in large clusters. { 'lik·wəd 'krist·əl }

liquid crystal polymers [ORG CHEM] Aromatic polyester copolymers that have characteristically high-temperature resistance, yet can be melted and molded. Upon melting, the polymer chains undergo parallel ordering in the direction of the flow, resulting in superior mechanical properties in that direction. { ˌlik·wəd ˌkrist·əl 'päl·ə·mərs }

liquid dioxide See nitrogen dioxide. { 'lik·wəd dī'äkˌsīd }

liquid glass See sodium silicate. { 'lik·wəd 'glas }

liquid hydrocarbon [ORG CHEM] A hydrocarbon that has been converted from a gas to a liquid by pressure or by reduction in temperature; usually limited to butanes, propane, ethane, and methane. { 'lik·wəd 'hī·drəˌkär·bən }

liquid junction emf [PHYS CHEM] The emf (electromotive force) generated at the area of contact between the salt bridge and the test solution in a pH cell electrode. { 'lik·wəd ˌjəŋk·shən ˌēˌem'ef }

liquid junction potential See diffusion potential. { 'lik·wəd ˌjəŋk·shən pə'ten·chəl }

liquid-liquid chemical reaction [CHEM] Chemical reaction in which the reactants, two or more, are liquids. { 'lik·wəd 'lik·wəd ˌkem·ə·kəl rē'ak·shən }

liquid-liquid distribution [CHEM] The process in which a dissolved substance is transferred from one liquid phase to another, immiscible liquid phase. { 'lik·wəd 'lik·wəd ˌdis·trə'byü·shən }

liquid-solid chemical reaction [CHEM] Chemical reaction in which at least one of the reactants is a liquid, and another of the reactants is a solid. { 'lik·wəd 'säl·əd ˌkem·ə·kəl rē'ak·shən }

liquid-solid equilibrium See solid-liquid equilibrium. { 'lik·wəd 'säl·əd ˌē·kwə'lib·rē·əm }

liquid-vapor chemical reaction [CHEM] Chemical reaction in which at least one of the reactants is a liquid, and another of the reactants is a vapor. { 'lik·wəd ˌvā·pər ˌkem·ə·kəl rē'ak·shən }

219

liquid-vapor equilibrium [PHYS CHEM] The equilibrium relationship between the liquid and its vapor phase for a partially vaporized compound or mixture at specified conditions of pressure and temperature; for mixtures, it is expressed by $K = x/y$, where K is the equilibrium constant, x the mole fraction of a key component in the vapor, and y the mole fraction of the same key component in the liquid. Also known as vapor-liquid equilibrium. { 'lik·wəd |vā·pər ‚ē·kwə'lib·rē·əm }

lithamide See lithium amide. { 'lith·ə‚mīd }

litharge See lead monoxide. { 'li‚thärj }

lithium [CHEM] A chemical element, symbol Li, atomic number 3, atomic weight 6.939; an alkali metal. { 'lith·ē·əm }

lithium aluminum hydride [INORG CHEM] LiAlH₄ A compound made by the reaction of lithium hydride and aluminum chloride; a powerful reducing agent for specific linkages in complex molecules; used in organic synthesis. { 'lith·ē·əm ə'lü·mə·nəm 'hī‚drīd }

lithium amide [INORG CHEM] LiNH₂ A compound crystallizing in the cubic form, and melting at 380–400°C; used in organic synthesis. Also known as lithamide. { 'lith·ē·əm 'am‚īd }

lithium bromide [INORG CHEM] LiBr·H₂O A white, deliquescent, granular powder with a bitter taste, melting at 547°C; soluble in alcohol and glycol; used to add moisture to air-conditioning systems and as a sedative and hypnotic in medicine. { 'lith·ē·əm 'brō‚mīd }

lithium carbonate [INORG CHEM] Li₂CO₃ A colorless, crystalline compound that melts at 700°C and has slight solubility in water; used in ceramic industries in the manufacture of powdered glass for porcelain enamel formulation. { 'lith·ē·əm 'kär·bə‚nāt }

lithium cell [CHEM] An electrolytic cell for the production of metallic lithium. { 'lith·ē·əm ‚sel }

lithium chloride [INORG CHEM] LiCl·2H₂O A colorless, water-soluble compound, forming octahedral crystals and melting at 614°C; used to form concentrated brine in commercial air-conditioning systems and as a pyrotechnic in welding and brazing fluxes. { 'lith·ē·əm 'klȯr‚īd }

lithium citrate [ORG CHEM] Li₃C₆H₅O₇·4H₂O White powder that decomposes when heated; slightly soluble in alcohol; soluble in water; used in beverages and pharmaceuticals. { 'lith·ē·əm 'sī‚trāt }

lithium fluoride [INORG CHEM] LiF Poisonous, white powder melting at 870°C, boiling at 1670°C; insoluble in alcohol, slightly soluble in water, and soluble in acids; used as a heat-exchange medium, as a welding and soldering flux, in ceramics, and as crystals in infrared instruments. { 'lith·ē·əm 'flu̇r‚īd }

lithium halide [INORG CHEM] A binary compound of lithium, LiX, where X is a halide; examples are lithium chloride, LiCl, and lithium fluoride, LiF. { 'lith·ē·əm 'hal‚īd }

lithium hydride [INORG CHEM] LiH Flammable, brittle, white, translucent crystals; decomposes in water; insoluble in ether, benzene, and toluene; used as a hydrogen source and desiccant, and to prepare lithium amide and double hydrides. { 'lith·ē·əm 'hī‚drīd }

lithium hydroxide [INORG CHEM] LiOH; LiOH·H₂O Colorless crystals; used as a storage-battery electrolyte, as a carbon dioxide absorbent, and in lubricating greases and ceramics. { 'lith·ē·əm hī'dräk‚sīd }

lithium iodide [INORG CHEM] LiI; LiI·3H₂O White, water- and alcohol-soluble crystals; LiI melts at 446°C; LiI·3H₂O loses water at 72°C; used in medicine, photography, and mineral waters. { 'lith·ē·əm 'ī·ə‚dīd }

lithium molybdate [INORG CHEM] Li₂MoO₄ Water-soluble white crystals melting at 705°C; used as a catalytic cracking (petroleum) catalyst and as a mill additive for steel. { 'lith·ē·əm mə'lib‚dāt }

lithium nitrate [INORG CHEM] LiNO₃ Water- and alcohol-soluble colorless powder melting at 261°C; used as a heat-exchange medium and in ceramics, pyrotechnics, salt baths, and refrigeration systems. { 'lith·ē·əm 'nī‚trāt }

lithium perchlorate [INORG CHEM] $LiClO_4 \cdot 3H_2O$ A compound with high oxygen content (60% available oxygen), used as a source of oxygen in rockets and missiles. { 'lith·ē·əm pər'klȯr,āt }

lithium stearate [ORG CHEM] $LiC_{18}H_{35}O_2$ A white, crystalline compound with a melting point of 220°C; used in cosmetics, plastics, and greases, and as a corrosion inhibitor in petroleum. { 'lith·ē·əm 'stir,āt }

lithium tetraborate [INORG CHEM] $Li_2B_4O_7 \cdot 5H_2O$ White crystals that lose water at 200°C; insoluble in alcohol, soluble in water; used in ceramics. { 'lith·ē·əm ,te·trə'bȯr,āt }

lithium titanate [INORG CHEM] Li_2TiO_3 A water-insoluble white powder with strong fluxing ability when used in titanium-containing enamels; also used as a mill additive in vitreous and semivitreous glazes. { 'lith·ē·əm 'tī·tən,āt }

Littrow grating spectrograph [SPECT] A spectrograph having a plane grating at an angle to the axis of the instrument, and a lens in front of the grating which both collimates and focuses the light. { 'li,trō ¦grād·iŋ 'spek·trə,graf }

Littrow mounting [SPECT] The arrangement of the grating and other components of a Littrow grating spectrograph, which is analogous to that of a Littrow quartz spectrograph. { 'li,trō ,maúnt·iŋ }

Littrow quartz spectrograph [SPECT] A spectrograph in which dispersion is accomplished by a Littrow quartz prism with a rear reflecting surface that reverses the light; a lens in front of the prism acts as both collimator and focusing lens. { 'li,trō ¦kwȯrts 'spek·trə,graf }

Lobry de Bruyn-Ekenstein transformation [ORG CHEM] The change in which an aldose sugar treated with dilute alkali results in a mixture of an epimeric pair and 2-ketohexose due to the production of enolic forms in the presence of hydroxyl ions, followed by a rearrangement. { lō¦brē·də¦brīn 'ā·kən,shtīn ,trans·fər,mā·shən }

locant [CHEM] The portion of a chemical name, usually a number or a letter, that designates the position of an atom or group of atoms in a formula unit. { 'lō,kant }

London force See dispersion force; van der Waals force. { 'lun·dən ,fȯrs }

lone-pair electrons [PHYS CHEM] A nonbonding pair of electrons in the valence shell of an atom. { ¦lōn ¦per i'lek,tränz }

Loomis-Wood diagram [SPECT] A graph used to assign lines in a molecular spectrum to the various branches of rotational bands when these branches overlap, in which the difference between observed wave numbers and wave numbers extrapolated from a few lines that apparently belong to one branch are plotted against arbitrary running numbers for that branch. { 'lü·məs 'wúd ,dī·ə,gram }

lophine [ORG CHEM] $C_{21}H_{16}O_2$ A colorless, crystalline, water-insoluble compound that melts at 275°C; used as an indicator in fluorescent neutralization tests. { 'lō,fēn }

Lorentz unit [SPECT] A unit of reciprocal length used to measure the difference, in wave numbers, between a (zero field) spectrum line and its Zeeman components; equal to $eH/4\pi mc^2$, where H is the magnetic field strength, c is the speed of light, and e and m are the charge and mass of the electron respectively (gaussian units). { 'lȯr,ens ,yü·nət }

lot [ANALY CHEM] A specimen of bulk material that is to undergo chemical analysis. { lät }

lot sample See gross sample. { 'lät ,sam·pəl }

loturine See harman. { ,lä·chə,rēn }

low-boiling butene-2 See butene-2. { 'lō ,bȯil·iŋ ¦byü,tēn 'tü }

low-density polyethylene [ORG CHEM] A thermoplastic polymer with a density of 0.910–0.940 gram per cubic centimeter (0.526–0.543 ounce per cubic inch). Abbreviated LDPE. { 'lō ,den·səd·ē ,päl·ē'eth·ə,lēn }

lowest unoccupied molecular orbital [PHYS CHEM] The lowest-energy molecular orbital that is occupied by electrons. Abbreviated LUMO. { ¦lō·əst ən¦äk·yə,pīd mə¦lek·yə·lər 'ȯr·bəd·əl }

low-frequency spectrum [SPECT] Spectrum of atoms and molecules in the microwave region, arising from such causes as the coupling of electronic and nuclear angular momenta, and the Lamb shift. { 'lō ,frē·kwən·sē 'spek·trəm }

221

Lr

Lr *See* lawrencium.

LSD *See* lysergic acid diethylamide.

LSD-25 *See* lysergic acid diethylamide.

Lu *See* lutetium.

Luggin probe [PHYS CHEM] A device which transmits a significant current density on the surface of an electrode to measure its potential. { 'ləg·ən ˌprōb }

Lugol solution [CHEM] A solution of 5 grams of iodine and 10 grams of potassium iodide per 100 milliliters of water; used in medicine. { 'lüˌgȯl sə'lü·shən }

luminol [ORG CHEM] $C_8H_7N_3O_2$ A white, water-soluble, crystalline compound that melts at 320°C; used in an alkaline solution for analytical testing in chemistry. Also known as 3-aminophthalic hydrazide. { 'lü·məˌnȯl }

LUMO *See* lowest unoccupied molecular orbital. { ¦lü¦mō *or* ¦el¦yü¦em'ō }

Lundegardh vaporizer [ANALY CHEM] A device used for emission flame photometry in which a compressed air aspirator vaporizes the solution within a chamber; smaller droplets are carried into the fuel-gas stream and to the burner orifice where the solvent is evaporated, dissociated, and optically excited. { 'lün·dəˌgard 'vā·pəˌrīz·ər }

lupinidine *See* sparteine. { lü'pin·əˌdēn }

lutetium [CHEM] A chemical element, symbol Lu, atomic number 71, atomic weight 174.967; a very rare metal and the heaviest member of the rare-earth group. { lü'tē·shəm }

lyate ion [CHEM] The anion that is produced when a solvent molecule loses a proton (hydrogen nucleus), for example, the hydroxide ion is the lyate ion of water. { 'lī,āt ˌī·ən }

lycine *See* betaine. { 'lī,sēn }

lye [INORG CHEM] **1.** A solution of potassium hydroxide or sodium hydroxide used as a strong alkaline solution in industry. **2.** The alkaline solution that is obtained from the leaching of wood ashes. { lī }

Lyman-alpha radiation [SPECT] Radiation emitted by hydrogen associated with the spectral line in the Lyman series whose wavelength is 121.5 nanometers. { 'lī·mən 'al·fə ˌrād·ē'ā·shən }

Lyman band [SPECT] A band in the ultraviolet spectrum of molecular hydrogen, extending from 125 to 161 nanometers. { 'lī·mən ˌband }

Lyman continuum [SPECT] A continuous range of wavelengths (or wave numbers or frequencies) in the spectrum of hydrogen at wavelengths less than the Lyman limit, resulting from transitions between the ground state of hydrogen and states in which the single electron is freed from the atom. { 'lī·mən kən'tin·yə·wəm }

Lyman ghost [SPECT] A false line observed in a spectroscope as a result of a combination of periodicities in the ruling. { 'lī·mən ˌgōst }

Lyman limit [SPECT] The lower limit of wavelengths of spectral lines in the Lyman series (912 angstrom units), or the corresponding upper limit in frequency, energy of quanta, or wave number (equal to the Rydberg constant for hydrogen). { 'lī·mən ˌlim·ət }

Lyman series [SPECT] A group of lines in the ultraviolet spectrum of hydrogen covering the wavelengths of 121.5–91.2 nanometers. { 'lī·mən ˌsir·ēz }

lyonium ion [CHEM] The cation that is produced when a solvent molecule is protonated. { lī'än·ē·əm ˌī·ən }

lyophilic [CHEM] Referring to a substance which will readily go into colloidal suspension in a liquid. { ¦lī·ə¦fil·ik }

lyophobic [CHEM] Referring to a substance in a colloidal state that has a tendency to repel liquids. { ¦lī·ə¦fō·bik }

lyotopic series *See* Hofmeister series. { ¦lī·ə¦täp·ik 'sir·ēz }

lyotropic liquid crystal [PHYS CHEM] A liquid crystal prepared by mixing two or more components, one of which is polar in character (for example, water). { ¦lī·ə¦träp·ik ¦lik·wəd 'krist·əl }

lysergic acid [ORG CHEM] $C_{16}H_{16}N_2O_2$ A compound that crystallizes in the form of hexagonal plates that melt and decompose at 240°C; derived from ergot alkaloids; used as a psychotomimetic agent. { lə'sər·jīk 'as·əd }

lysergic acid diethylamide [ORG CHEM] $C_{15}H_{15}N_2CON(C_2H_5)_2$ A psychotomimetic drug synthesized from compounds derived from ergot. Abbreviated LSD; LSD-25. { lə'sərjik ¦as·əd dī,eth·əl'am·əd }

M

M *See* molarity.

MAA *See* methanearsonic acid.

M acid [ORG CHEM] $NH_2C_{10}H_5(OH)SO_3H$ A sulfonic acid formed by alkaline fusion of a disulfonic acid of α-naphthylamine; used as a dye intermediate. { 'em ˌas·əd }

Macquer's salt *See* potassium arsenate. { mə'kerz ˌsȯlt }

macroanalysis [ANALY CHEM] Qualitative or quantitative analysis of chemicals that are in quantities of the order of grams. { 'mak·rō·ə'nal·ə·səs }

macrocycle *See* macrocyclic compound. { 'mak·rō₁sī·kəl }

macrocyclic compound [ORG CHEM] An organic compound containing a large ring, that is, a closed chain of 12 or more carbon atoms; examples include crown ethers, cryptands, spherands, carcerands, cyclodextrins, cyclophanes, and calixarenes. Also known as macrocycle. { ¦mak·rō₁sī 'käm₁paúnd }

macrolide [ORG CHEM] A large ring molecule with many functional groups bonded to it. { 'mak·rə₁līd }

macromolecular [ORG CHEM] Composed of or characterized by large molecules. { ¦mak·rō·mə'lek·yə·lər }

macromolecule [ORG CHEM] A large molecule in which there is a large number of one or several relatively simple structural units, each consisting of several atoms bonded together. { ¦mak·rō'mäl·ə₁kyül }

macropore [CHEM] A pore in a catalytic material whose width is greater than 0.05 micrometer. { 'mak·rə₁pȯr }

macroporous resin [ORG CHEM] A member of a class of very small, highly cross-linked polymer particles penetrated by channels through which solutions can flow; used as ion exchanger. Also known as macroreticular resin. { ¦mak·rə¦pȯr·əs 'rez·ən }

macroreticular resin *See* macroporous resin. { ¦mak·rō·rə'tik·yə·lər 'rez·ən }

magenta *See* fuchsin. { mə'jen·tə }

magic acid [INORG CHEM] A superacid consisting of equal molar quantities of fluorosulfonic acid (HSO_3F) and antimony pentafluoride (SbF_5). { 'maj·ik ˌas·əd }

magic numbers [PHYS CHEM] Numbers of atoms or molecules for which certain atom or molecular clusters have an unusually high abundance. { 'maj·ik 'nəm·bərz }

magister of sulfur [CHEM] Amorphous sulfur produced by acid precipitation from solutions of hyposulfites or polysulfides. { mə'jis·tər əv 'səl·fər }

magnesia [INORG CHEM] Magnesium oxide that is processed for a particular purpose. { mag'nē·zhə }

magnesia mixture [ANALY CHEM] Reagent used to analyze for phosphorus; consists of the filtered liquor from an aqueous mixture of ammonium chloride, magnesium sulfate, and ammonia. { mag'nē·zhə ˌmiks·chər }

magnesium [CHEM] A metallic element, symbol Mg, atomic number 12, atomic weight 24.305. { mag'nē·zē·əm }

magnesium acetate [ORG CHEM] $Mg(OOCCH_3)_2 \cdot 4H_2O$ or $Mg(OOCCH_3)_2$ A compound forming colorless crystals that are soluble in water and melt at 80°C; used in textile printing, in medicine as an antiseptic, and as a deodorant. { mag'nē·zē·əm 'as·ə₁tāt }

magnesium arsenate [INORG CHEM] $Mg_3(AsO_4)_2 \cdot xH_2O$ A white, poisonous, water-insoluble powder used as an insecticide. { mag'nē·zē·əm 'ärs·ən₁āt }

magnesium benzoate

magnesium benzoate   [ORG CHEM] $Mg(C_7H_5O_2)_2 \cdot 3H_2O$ A crystalline white powder melting at 200°C; soluble in alcohol and hot water; used in medicine.   { mag'nē·zē·əm 'ben·zə,wāt }

magnesium borate   [INORG CHEM] $3MgO \cdot B_2O_3$ Crystals that are white or colorless and transparent; soluble in alcohol and acids, slightly soluble in water; used as a fungicide, antiseptic, and preservative.   { mag'nē·zē·əm 'bȯr,āt }

magnesium boride *See* magnesium diboride.   { mag'nē·zē·əm 'bȯr,īd }

magnesium bromate   [INORG CHEM] $Mg(BrO_3)_2 \cdot 6H_2O$ A white crystalline compound, insoluble in alcohol, soluble in water; a fire hazard; used as an analytical reagent.   { mag'nē·zē·əm 'brō,māt }

magnesium bromide   [INORG CHEM] $MgBr_2 \cdot 6H_2O$ Deliquescent, colorless, bitter-tasting crystals, melting at 172°C; soluble in water, slightly soluble in alcohol; used in medicine and in the synthesis of organic chemicals.   { mag'nē·zē·əm 'brō,mīd }

magnesium carbonate   [INORG CHEM] $MgCO_3$ A water-insoluble, white powder, decomposing at about 350°C; used as a refractory material.   { mag'nē·zē·əm 'kär·bə,nāt }

magnesium chlorate   [INORG CHEM] $Mg(ClO_3)_2 \cdot 6H_2O$ A white powder, bitter-tasting and hygroscopic; slightly soluble in alcohols, soluble in water; used in medicine.   { mag'nē·zē·əm 'klȯr,āt }

magnesium chloride   [INORG CHEM] $MgCl_2 \cdot 6H_2O$ Deliquescent white crystals; soluble in water and alcohol; used in disinfectants and fire extinguishers, and in ceramics, textiles, and paper manufacture.   { mag'nē·zē·əm 'klȯr,īd }

magnesium diboride   [INORG CHEM] MgB_2 A crystalline intermetallic compound, produced as a black powder, that becomes superconducting at the unusually high temperature of 39 K (−389°F; −234°C); melts at 800 °C. Also known as magnesium boride.   { mag'nē·zē·əm dī'bȯr,īd }

magnesium fluoride   [INORG CHEM] MgF_2 White, fluorescent crystals; insoluble in water and alcohol, soluble in nitric acid; melts at 1263°C; used in ceramics and glass. Also known as magnesium flux.   { mag'nē·zē·əm 'flür,īd }

magnesium fluosilicate   [INORG CHEM] $MgSiF_6 \cdot 6H_2O$ Water-soluble, efflorescent white crystals; used in ceramics, in mothproofing and waterproofing, and as a concrete hardener. Also known as magnesium silicofluoride.   { mag'nē·zē·əm 'flü·ə'sil·ə,kāt }

magnesium flux *See* magnesium fluoride.   { mag'nē·zē·əm 'fləks }

magnesium formate   [ORG CHEM] $Mg(CHO_2)_2 \cdot 2H_2O$ Colorless, water-soluble crystals; insoluble in alcohol and ether; used in analytical chemistry and medicine.   { mag'nē·zē·əm 'fȯr,māt }

magnesium gluconate   [ORG CHEM] $Mg(C_6H_{11}O_7)_2 \cdot 2H_2O$ An odorless, tasteless, water-soluble powder; used in medicine.   { mag'nē·zē·əm 'glü·kə,nāt }

magnesium halide   [INORG CHEM] A compound formed from the metal magnesium and any of the halide elements; an example is magnesium bromide.   { mag'nē·zē·əm 'ha,līd }

magnesium hydrate *See* magnesium hydroxide.   { mag'nē·zē·əm 'hī,drāt }

magnesium hydride   [INORG CHEM] MgH_2 A hydride compound formed from the metal magnesium; it decomposes violently in water, and in a vacuum at about 280°C.   { mag'nē·zē·əm 'hī,drīd }

magnesium hydroxide   [INORG CHEM] $Mg(OH)_2$ A white powder, very slightly soluble in water, decomposing at 350°C; used as an intermediate in extraction of magnesium metal, and as a reagent in the sulfite wood pulp process. Also known as magnesium hydrate.   { mag'nē·zē·əm hī'dräk,sīd }

magnesium hyposulfite *See* magnesium thiosulfate.   { mag'nē·zē·əm ,hī·pō'səl,fīt }

magnesium iodide   [INORG CHEM] $MgI_2 \cdot 8H_2O$ Crystalline powder, white and deliquescent, discoloring in air; soluble in water, alcohol, and ether; used in medicine.   { mag'nē·zē·əm 'ī·ə,dīd }

magnesium lactate   [ORG CHEM] $Mg(C_3H_5O_3)_2 \cdot 3H_2O$ Bitter-tasting, water-soluble white crystals; slightly soluble in alcohol; used in medicine.   { mag'nē·zē·əm 'lak,tāt }

magnesium methoxide   [ORG CHEM] $(CH_3O)_2Mg$ Colorless crystals that decompose

226

when heated; used as a catalyst, dielectric coating, and cross-linking agent, and to form gels. Also known as magnesium methylate. { mag'nē·zē·əm me'thäk‚sīd }

magnesium methylate See magnesium methoxide. { mag'nē·zē·əm 'meth·ə‚lāt }

magnesium nitrate [INORG CHEM] $Mg(NO_3)_2 \cdot 6H_2O$ Deliquescent white crystals; soluble in alcohol and water; a fire hazard; used as an oxidizing material in pyrotechnics. { mag'nē·zē·əm 'nī‚trāt }

magnesium oleate [ORG CHEM] $Mg(C_{18}H_{33}O_2)_2$ Water-insoluble, yellowish mass; soluble in hydrocarbons, alcohol, and ether; used as a plasticizer lubricant and emulsifying agent, and in varnish driers and dry-cleaning solutions. { mag'nē·zē·əm 'ō·lē‚āt }

magnesium oxide [INORG CHEM] MgO A white powder that (depending on the method of preparation) may be light and fluffy, or dense; melting point 2800°C; insoluble in acids, slightly soluble in water; used in making refractories, and in cosmetics, pharmaceuticals, insulation, and medicine. { mag'nē·zē·əm 'äk‚sīd }

magnesium perchlorate [INORG CHEM] $Mg(ClO_4)_2 \cdot 6H_2O$ White, deliquescent crystals; soluble in water and alcohol; explosive when in contact with reducing materials; used as a drying agent for gases. { mag'nē·zē·əm pər'klȯr‚āt }

magnesium peroxide [INORG CHEM] MgO_2 A tasteless, odorless white powder; soluble in dilute acids, insoluble in water; a fire hazard; used as a bleaching and oxidizing agent, and in medicine. { mag'nē·zē·əm pə'räk‚sīd }

magnesium phosphate [INORG CHEM] A compound with three forms: monobasic, $MgH_4(PO_4)_2 \cdot 2H_2O$, used in medicine and wood fireproofing; dibasic, $MgHPO_4 \cdot 3H_2O$, used in medicine and as a plastics stabilizer; tribasic, $Mg_3(PO_4)_2 \cdot 8H_2O$, used in dentifrices, as an adsorbent, and in pharmaceuticals. { mag'nē·zē·əm 'fäs‚fāt }

magnesium salicylate [ORG CHEM] $Mg(C_7H_5O_3) \cdot 4H_2O$ Efflorescent colorless crystals; soluble in water and alcohol; used in medicine. { mag'nē·zē·əm sə'lis·ə‚lāt }

magnesium silicate [INORG CHEM] $3MgSiO_3 \cdot 5H_2O$ White, water-insoluble powder, containing variable proportions of water of hydration; used as a filler for rubber and in medicine. { mag'nē·zē·əm 'sil·ə‚kāt }

magnesium silicofluoride See magnesium fluosilicate. { mag'nē·zē·əm ‚sil·ə·kō'flür‚īd }

magnesium stearate [ORG CHEM] $Mg(C_{18}H_{35}O_2)_2$ Tasteless, odorless white powder; soluble in hot alcohol, insoluble in water; melts at 89°C; used in paints and medicine, and as a plastics stabilizer and lubricant. Also known as dolomol. { mag'nē·zē·əm 'stir‚āt }

magnesium sulfate [INORG CHEM] $MgSO_4$ Colorless crystals with a bitter, saline taste; soluble in glycerol; used in fireproofing, textile processes, ceramics, cosmetics, and fertilizers. { mag'nē·zē·əm 'səl‚fāt }

magnesium sulfite [INORG CHEM] $MgSO_3 \cdot 6H_2O$ A white, crystalline powder; insoluble in alcohol, slightly soluble in water; used in medicine and paper pulp. { mag'nē·zē·əm 'səl‚fīt }

magnesium thiosulfate [INORG CHEM] $MgS_2O_3 \cdot 6H_2O$ Colorless crystals that lose water at 170°C; used in medicine. Also known as magnesium hyposulfite. { mag'nē·zē·əm ‚thī·ə'səl‚fāt }

magnesium trisilicate [INORG CHEM] $Mg_2Si_3O_8 \cdot 5H_2O$ A white, odorless, tasteless powder; insoluble in water and alcohol; used as an industrial odor absorbent and in medicine. { mag'nē·zē·əm ‚trī'sil·ə‚kāt }

magnesium tungstate [INORG CHEM] $MgWoO_4$ White crystals, insoluble in alcohol and water, soluble in acid; used in luminescent paint and for fluorescent x-ray screens. { mag'nē·zē·əm 'təŋ‚stāt }

magneson [ORG CHEM] $C_{12}H_9N_3O_4$ A brownish-red powder, soluble in dilute aqueous sodium hydroxide; used in the detection of magnesium and molybdenum. { 'mag·nə‚sän }

magnetic scanning [SPECT] The magnetic field sorting of ions into their respective spectrums for analysis by mass spectroscopy; accomplished by varying the magnetic field strength while the electrostatic field is held constant. { mag'ned·ik 'skan·iŋ }

magnetochemistry [PHYS CHEM] A branch of chemistry which studies the interrelationship between the bulk magnetic properties of a substance and its atomic and molecular structure. { mag¦nēd·ō'kem·ə·strē }

magnetofluid [PHYS CHEM] A Newtonian or shear-thinning fluid whose flow properties become viscoplastic when it is modulated by a magnetic field. { ¦mag·nəd·ō'flü·əd }

malathion [ORG CHEM] $C_{10}H_{19}O_6PS_2$ A yellow liquid, slightly soluble in water; malathion is the generic name for S-1,2-bis(ethoxycarbonyl)ethyl O,O-dimethylphosphorodithioate; used as an insecticide. { ˌmal·ə'thī,än }

maleate [ORG CHEM] An ester or salt of maleic acid. { mə'lē,āt }

maleic acid [ORG CHEM] HOOCCH:CHCOOH A colorless, crystalline dibasic acid; soluble in water, acetone, and alcohol; melting point 130–131°C; used in textile processing, and as an oil and fat preservative. { mə'lā·ik 'as·əd }

maleic anhydride [ORG CHEM] $C_4H_2O_3$ Colorless crystals, soluble in acetone, hydrolyzing in water; used to form polyester resins. Also known as 2,5-furandione. { mə'lā·ik an'hī,drīd }

maleic hydrazide [ORG CHEM] $C_4N_2H_4O_2$ Solid material, decomposing at 260°C; slightly soluble in alcohol and water; used as a weed killer and growth inhibitor. { mə'lā·ik 'hī·drə,zīd }

malonamide nitrile See cyanoacetamide. { mə'län·ə·məd 'nī·trəl }

malonic acid [ORG CHEM] $CH_2(COOH)_2$ A white, crystalline dicarboxylic acid, melting at 132–134°C; used to manufacture pharmaceuticals. { mə'län·ik 'as·əd }

malonic ester See ethyl malonate. { mə'län·ik 'es·tər }

malonic mononitrile See cyanoacetic acid. { mə'län·ik ¦män·ō'nī·trəl }

malonyl [ORG CHEM] $CH_2(COO)_2$ A bivalent functional group formed from malonic acid. { 'mal·ə,nil }

maltol [ORG CHEM] $C_6H_6O_3$ Crystalline substance with a melting point of 161–162°C and a fragrant caramellike odor; used as a flavoring agent in bread and cakes. Also known as larixinic acid. { 'mȯl,tȯl }

mandelic acid [ORG CHEM] $C_6H_5CHOHCOOH$ A white, crystalline compound, melting at 117–119°C, darkening upon exposure to light; used in organic synthesis. { man'del·ik 'as·əd }

mandelic acid nitrile See mandelonitrile. { man'del·ik 'as·əd 'nī·trəl }

mandelonitrile [ORG CHEM] $C_6H_5CH(OH)CN$ A liquid used to prepare bitter almond water. Also known as mandelic acid nitrile. { man¦del·ō'nī·trəl }

maneb [ORG CHEM] $Mn[SSCH(CH_2)_2NHCSS]$ A generic term for manganese ethylene-1,2-bisdithiocarbamate; irritating to eyes, nose, skin, and throat; used as a fungicide. { 'ma,neb }

manganate [INORG CHEM] **1.** Salts that have manganese in the anion. **2.** In particular, a salt of manganic acid formed by fusion of manganese dioxide with an alkali. { 'maŋ·gə,nāt }

manganese [CHEM] A metallic element, symbol Mn, atomic weight 54.938, atomic number 25; a transition element whose properties fall between those of chromium and iron. { 'maŋ·gə,nēs }

manganese acetate [ORG CHEM] $Mn(C_2H_3O_2)_2·4H_2O$ A pale-red crystalline compound melting at 80°C; soluble in water and alcohol; used in textile dyeing, as a catalyst, and for leather tanning. { 'maŋ·gə,nēs 'as·ə,tāt }

manganese binoxide See manganese dioxide. { 'maŋ·gə,nēs bi'näk,sīd }

manganese black See manganese dioxide. { 'maŋ·gə,nēs 'blak }

manganese borate [INORG CHEM] MnB_4O_7 Water-insoluble, reddish-white powder; used as a varnish and oil drier. { 'maŋ·gə,nēs 'bȯr,āt }

manganese bromide See manganous bromide. { 'maŋ·gə,nēs 'brō,mīd }

manganese carbonate [INORG CHEM] $MnCO_3$ Rose-colored crystals found in nature as rhodocrosite; soluble in dilute acids, insoluble in water; used in medicine, in fertilizer, and as a paint pigment. { 'maŋ·gə,nēs 'kär·bə,nāt }

manganese citrate [ORG CHEM] $Mn_3(C_6H_5O_7)_2$ A white powder, water-insoluble in the presence of sodium citrate; used in medicine. { 'maŋ·gə,nēs 'sī,trāt }

manganese dioxide [INORG CHEM] MnO_2 A black, crystalline, water-insoluble compound, decomposing to manganese sesquioxide, Mn_2O_3, and oxygen when heated to 535°C; used as a depolarizer in certain dry-cell batteries, as a catalyst, and in dyeing of textiles. Also known as battery manganese; manganese binoxide; manganese black; manganese peroxide. { 'maŋ·gə,nēs dī'äk,sīd }

manganese fluoride See manganous fluoride. { 'maŋ·gə,nēs 'flúr,īd }

manganese gluconate [ORG CHEM] $Mn(C_6H_{11}O_7)_2·2H_2O$ A pinkish powder, insoluble in benzene and alcohol, soluble in water; used in medicine, in vitamin tablets, and as a feed additive and dietary supplement. { 'maŋ·gə,nēs 'glü·kə,nāt }

manganese green See barium manganate. { 'maŋ·gə,nēs 'grēn }

manganese halide [INORG CHEM] Compound of manganese with a halide, such as chlorine, bromine, fluorine, or iodine. { 'maŋ·gə,nēs 'ha,līd }

manganese heptoxide [INORG CHEM] Mn_2O_7 A compound formed as an explosive dark-green oil by the action of concentrated sulfuric acid on permanganate compounds. { 'maŋ·gə,nēs hep'täk,sīd }

manganese hydroxide See manganous hydroxide. { 'maŋ·gə,nēs hī'dräk,sīd }

manganese hypophosphite [INORG CHEM] $Mn(H_2PO_2)_2·H_2O$ Odorless, tasteless pink crystals which explode if heated with oxidants; used in medicine. { 'maŋ·gə,nēs ,hī·pō'fäs,fīt }

manganese iodide See manganous iodide. { 'maŋ·gə,nēs 'ī·ə,dīd }

manganese lactate [ORG CHEM] $Mn(C_3H_5O_3)_2·3H_2O$ Pale-red crystals; insoluble in water and alcohol; used in medicine. { 'maŋ·gə,nēs 'lak,tāt }

manganese linoleate [ORG CHEM] $Mn(C_{18}H_{31}O_2)_2$ A dark-brown mass, soluble in linseed oil; used in pharmaceutical preparations and as a varnish and paint drier. { 'maŋ·gə,nēs lə'nō·lē,āt }

manganese monoxide See manganese oxide. { 'maŋ·gə,nēs mə'näk,sīd }

manganese naphthenate [ORG CHEM] Hard brown resinous mass, soluble in mineral spirits; melts at 135°C; contains 6% manganese in commercial solutions; used as a paint and varnish drier. { 'maŋ·gə,nēs 'naf·thə,nāt }

manganese oleate [ORG CHEM] $Mn(C_{18}H_{33}O_2)_2$ Granular brown mass, soluble in oleic acid and ether, insoluble in water; used in medicine and as a varnish drier. { 'maŋ·gə,nēs 'ō·lē,āt }

manganese oxalate [ORG CHEM] $MnC_2O_4·2H_2O$ A white crystalline compound, soluble in dilute acids, only slightly soluble in water; used as a paint and varnish drier. { 'maŋ·gə,nēs 'äk·sə,lāt }

manganese oxide [INORG CHEM] MnO Green powder, soluble in acids, insoluble in water; melts at 1650°C; used in medicine, in textile printing, as a catalyst, in ceramics, and in dry batteries. Also known as manganese monoxide; manganous oxide. { 'maŋ·gə,nēs 'äk,sīd }

manganese peroxide See manganese dioxide. { 'maŋ·gə,nēs pə'räk,sīd }

manganese resinate [ORG CHEM] $Mn(C_{20}H_{29}O_2)_2$ Water-insoluble mass, flesh-colored or brownish black; used as a varnish and oil drier. { 'maŋ·gə,nēs 'rez·ən,āt }

manganese silicate See manganous silicate. { 'maŋ·gə,nēs 'sil·ə,kāt }

manganese sulfate See manganous sulfate. { 'maŋ·gə,nēs 'səl,fāt }

manganese sulfide See manganous sulfide. { 'maŋ·gə,nēs 'səl,fīd }

manganic fluoride [INORG CHEM] MnF_3 Poisonous red crystals, decomposed by heat and water; used as a fluorinating agent. { man'gan·ik 'flúr,īd }

manganic hydroxide [INORG CHEM] $Mn(OH)_3$ A brown powder that rapidly loses water to form MnO(OH); used in ceramics and as a fabric pigment. Also known as hydrated manganic hydroxide. { man'gan·ik hī'dräk,sīd }

manganic oxide [INORG CHEM] Mn_2O_3 Hard black powder, insoluble in water, soluble in cold hydrochloric acid, hot nitric acid, and sulfuric acid; occurs in nature as manganite. Also known as manganese sesquioxide. { man'gan·ik 'äk,sīd }

manganous bromide [INORG CHEM] $MnBr_2·4H_2O$ Water-soluble, deliquescent red crystals. Also known as manganese bromide. { 'maŋ·gə·nəs 'brō,mīd }

manganous chloride [INORG CHEM] $MnCl_2·4H_2O$ Water-soluble, deliquescent rose-colored crystals melting at 88°C; used as a catalyst and in paints, dyeing, and pharmaceutical preparations. { 'maŋ·gə·nəs 'klór,īd }

manganous fluoride [INORG CHEM] MnF_2 Reddish powder, insoluble in water, soluble in acid. Also known as manganese fluoride. { 'maŋ·gə·nəs 'flúr,īd }

manganous hydroxide

manganous hydroxide [INORG CHEM] $Mn(OH)_2$ Heat-decomposable white-pink crystals; insoluble in water and alkali, soluble in acids; occurs in nature as pyrochroite. Also known as manganese hydroxide. { 'maŋ·gə·nəs hī'dräk,sīd }

manganous iodide [INORG CHEM] $MnI_2·4H_2O$ Water-soluble, deliquescent yellowish-brown crystals. Also known as manganese iodide. { 'maŋ·gə·nəs 'ī·ə,dīd }

manganous silicate [INORG CHEM] $MnSiO_3$ Water-insoluble red crystals or yellowish-red powder; occurs in nature as rhodonite. Also known as manganese silicate. { 'maŋ·gə·nəs 'sil·ə,kāt }

manganous sulfate [INORG CHEM] $MnSO_4·4H_2O$ Water-soluble, translucent, efflorescent rose-red prisms; melts at 30°C; used in medicine, textile printing, and ceramics, as a fungicide and fertilizer, and in paint manufacture. Also known as manganese sulfate. { 'maŋ·gə·nəs 'səl,fāt }

manganous sulfide [INORG CHEM] MnS An almost water-insoluble powder that decomposes on heating; used as a pigment and as an additive in making steel. Also known as manganese sulfide. { 'maŋ·gə·nəs 'səl,fīd }

manganous sulfite [INORG CHEM] $MnSO_3$ Grayish-black or brownish-red powder, soluble in sulfur dioxide, insoluble in water. { 'maŋ·gə·nəs 'səl,fīt }

manna sugar See mannitol. { 'man·ə ,shúg·ər }

Mannich condensation reaction See Mannich reaction. { 'män·i<u>k</u> ,kän·dən'sā·shən rē,ak·shən }

Mannich reaction [ORG CHEM] Condensation of a primary or secondary amine or ammonia (usually as the hydrochloride) with formaldehyde and a compound containing at least one reactive hydrogen atom, for example, acetophenone. Also known as Mannich condensation reaction. { 'män·i<u>k</u> rē,ak·shən }

mannite See mannitol. { 'ma,nīt }

mannitol [ORG CHEM] $C_6H_8(OH)_6$ A straight-chain alcohol with six hydroxyl groups; a white, water-soluble, crystalline powder; used in medicine and as a dietary supplement. Also known as manna sugar; mannite. { 'man·ə,tòl }

mannitol hexanitrate [ORG CHEM] $C_6H_8(ONO_2)_6$ Explosive colorless crystals; soluble in alcohol, acetone, and ether, insoluble in water; melts at 112°C; used in explosives and medicine. { 'man·ə,tòl ,hek·sə'nī,trāt }

manure salts [INORG CHEM] Potash salts that have a high proportion of chloride and 20–30% potash; used in fertilizers. { mə'nùr ,sòlts }

margaric acid See n-heptadecanoic acid. { mär'gär·ik 'as·əd }

Mark-Houwink equation [PHYS CHEM] The relationship between intrinsic viscosity and molecular weight for homogeneous linear polymers. { 'märk 'haù,wiŋk i,kwā·zhən }

Markovnikoff's rule [ORG CHEM] In an addition reaction, the additive molecule RH adds as H and R, with the R going to the carbon atom with the lesser number of hydrogen atoms bonded to it. { mär'kòv·nə,kòfs ,rül }

Marsh-Berzelius test See Marsh test. { 'märsh ber'zā·lē·əs ,test }

Marsh test [ANALY CHEM] A test for the presence of arsenic in a compound; the substance to be tested is mixed with granular zinc, and dilute hydrochloric acid is added to the mixture; gaseous arsine forms, which decomposes to a black deposit of arsenic, when the gas is passed through a heated glass tube. Also known as Marsh-Berzelius test. { 'märsh ,test }

Mars pigments [INORG CHEM] A group of five pigments produced when milk of lime is added to a ferrous sulfate solution, and the precipitate is calcined; color is controlled by calcination temperature to give yellow, orange, brown, red, or violet. { 'märz ,pig·məns }

masking agent See masking reagent. { 'mask·iŋ ,ā·jənt }

masking reagent [ANALY CHEM] A substance that decreases the concentration of a free metal ion or ligand by conversion into an essentially unreactive form, thus preventing undesirable chemical reactions that would interfere with the determination. Also known as masking agent. { 'mask·iŋ rē,ā·jənt }

mass action law [PHYS CHEM] The law that the rate of a chemical reaction for a uniform system at constant temperature is proportional to the concentrations of the substances reacting. Also known as Guldberg and Waage law. { 'mas ¦ak·shən ,lò }

mass-analyzed ion kinetic energy spectrometry [SPECT] A type of ion kinetic energy spectrometry in which the ionic products undergo mass analysis followed by energy analysis. Abbreviated MIKES. { 'mas ¦an·ə‚līzd 'ī‚än kə¦ned·ik 'en·ər·jē spek'träm· ə·trē }

Massenfilter See quadrupole spectrometer. { 'mäs·ən‚fil·tər }

mass spectrometry [ANALY CHEM] An analytical technique for identification of chemical structures, determination of mixtures, and quantitative elemental analysis, based on application of the mass spectrometer. { 'mas spek'träm·ə·trē }

mass susceptibility [PHYS CHEM] Magnetic susceptibility of a compound per gram. Also known as specific susceptibility. { 'mas sə‚sep·tə'bil·əd·ē }

mass-to-charge ratio [ANALY CHEM] In analysis by mass spectroscopy, the measurement of the sample mass as a ratio to its ionic charge. { ¦mas tə 'chärj ‚rā·shō }

matrix [ANALY CHEM] The analyte as considered in terms of its being an assemblage of constituents, each with its own properties. { 'mā·triks }

matrix effects [ANALY CHEM] **1.** The enhancement or suppression of minor element spectral lines from metallic oxides during emission spectroscopy by the matrix element (such as graphite) used to hold the sample. **2.** The combined effect exerted by the various constituents of the matrix on the measurements of the analysis. { 'mā·triks i‚feks }

matrix isolation [SPECT] A spectroscopic technique in which reactive species can be characterized by maintaining them in a very cold, inert environment while they are examined by an absorption, electron-spin resonance, or laser excitation spectroscope. { 'mā·triks ‚i·sə'lā·shən }

matrix spectrophotometry [SPECT] Spectrophotometric analysis in which the specimen is irradiated in sequence at more than one wavelength, with the visible spectrum evaluated for the energy leaving for each wavelength of irradiation. { 'mā·triks ‚spek·trō·fə'täm·ə·trē }

MBT See mercaptobenzothiazole.

Md See mendelevium.

measured spectrum See spectrogram. { 'mezh·ərd 'spek·trəm }

mechanism [CHEM] A detailed description of the course of a chemical reaction as it proceeds from the reactants to the products, with as complete a characterization as possible of the reaction steps and intermediate species. Also known as reaction path. { 'mek·ə‚niz·əm }

mechanochemical effect [PHYS CHEM] Changes in the dimensions of certain polymers, particularly photoelectrolytic gels and crystalline polymers, in response to changes in their chemical environment. { ¦mek·ə·nō'kem·ə·kəl i‚fekt }

mechanochemistry [PHYS CHEM] The study of the conversion of mechanical energy into chemical energy in polymers. { ¦mek·ə·nō'kem·ə·strē }

mechanophotochemistry [PHYS CHEM] The study of changes in the dimensions of certain photoresponsive polymers upon exposure to light. { ¦mek·ə·nō‚fō·dō'kem· ə·strē }

meconin [ORG CHEM] $C_{10}H_{10}O_4$ A neutral principle of opium; white crystals, soluble in hot water and alcohol and melting at 102–103°C. Also known as opianyl. { 'mek· ə·nən }

meitnerium [CHEM] A chemical element, symbolized Mt, atomic number 109, a synthetic element; the seventeenth transuranium element. { mīt'nir·ē·əm }

MEK See methyl ethyl ketone.

melamine [ORG CHEM] $C_3H_6N_4$ A white crystalline compound that is slightly soluble in water, melts at 354°C and is a cyclic trimer of cyanamide; used to make melamine resins and in tanning of leather. { 'mel·ə‚mēn }

melaniline See diphenylguanidine. { mel'an·ə·lən }

melissic acid [ORG CHEM] $CH_3(CH_2)_{28}COOH$ Fatty acid found in beeswax; soluble in benzene and hot alcohol; melts at 90°C; used in biochemical research. { mə'lis·ik 'as·əd }

mellitate [ORG CHEM] An ester or salt of mellitic acid. { 'mel·ə‚tāt }

mellitic acid [ORG CHEM] $C_6(COOH)_6$ A water-soluble compound forming colorless needles that melt at 287°C. { mə'lid·ik 'as·əd }

melt [CHEM] **1.** To change a solid to a liquid by the application of heat. **2.** A melted material. { melt }

melting See fusion. { 'melt·iŋ }

membrane mimetic chemistry [ORG CHEM] The study of processes and reactions that have been developed by using information obtained from biological membrane systems. { ¦mem,brān mi¦med·ik 'kem·ə·strē }

MEMC See methoxyethylmercury chloride.

menadione [ORG CHEM] $C_{11}H_8O_2$ (2-methyl-1,4-naphthoquinone) A compound used as a vitamin K supplement; important in blood clotting. { ˌmen·ə'dī·ōn }

mendelevium [CHEM] Synthetic radioactive element, symbol Md, with atomic number 101; made by bombarding lighter elements with light nuclei accelerated in cyclotrons. { ˌmen·də'lē·vē·əm }

menthane [ORG CHEM] $C_{10}H_{20}$ A colorless, water-insoluble liquid hydrocarbon; used in organic synthesis. { 'men,thān }

menthene [ORG CHEM] $C_{10}H_{18}$ A colorless, water-insoluble, liquid hydrocarbon; used in organic synthesis. { 'men,thēn }

menthol [ORG CHEM] $CH_3C_6H_9(C_3H_7)OH$ An alcohol-soluble, white crystalline compound that may exist in levo form or a mixture of dextro and levo isomers; used in medicines and perfumes, and as a flavoring agent. Also known as peppermint camphor. { 'men,thȯl }

menthone [ORG CHEM] $C_{10}H_{18}O$ Oily, colorless ketonic liquid with slight peppermint odor; slightly soluble in water, soluble in organic solvents. { 'men,thōn }

menthyl [ORG CHEM] $C_{10}H_{19}$ A univalent radical that is derived from menthol by removal of the hydroxyl group. { 'men·thəl }

meq See milliequivalent. { mek }

-mer [ORG CHEM] A combining form denoting the repeating structure unit of any high polymer. { mər }

merbromin [ORG CHEM] $C_{20}H_8O_6Na_2Br_2Hg$ A green crystalline powder that gives a deep-red solution in water; used as an antiseptic. { mər'brō·mən }

mercapt-, mercapto- [CHEM] A combining form denoting the presence of the thiol (SH) group. { mər'kap·tō }

mercaptal [ORG CHEM] A group of organosulfur compounds that contain the group $=C(SR)_2$. { mər'kap,tal }

mercaptan [ORG CHEM] A group of organosulfur compounds that are derivatives of hydrogen sulfide in the same way that alcohols are derivatives of water; have a characteristically disagreeable odor, and are found with other sulfur compounds in crude petroleum; an example is methyl mercaptan. Also known as thiol. { mər'kap,tan }

mercaptide [ORG CHEM] A compound consisting of a metal and a mercaptan. { mər'kap,tīd }

mercaptoacetic acid See thioglycolic acid. { mər¦kap·tō·ə¦sēd·ik 'as·əd }

2-mercaptobenzoic acid See thiosalicylic acid. { ¦tü mər¦kap·tō·ben'zō·ik 'as·əd }

mercaptobenzothiazole [ORG CHEM] C_7H_5NS A yellow powder, melting at 164–174°C; used in rubber as a vulcanization accelerator with stearic acid. Abbreviated MBT. { mər¦kap·tō,ben·zō'thī·ə,zōl }

mercapto compound See sulfhydryl compound. { mər'kap·tō ,käm,pau̇nd }

mercaptoethanol [ORG CHEM] $HSCH_2CH_2OH$ Mobile liquid, water-white; soluble in water, benzene, ether, and most organic solvents; boils at 157°C; used as a solvent, chemical intermediate, and reducing agent. { mər,kap·tō'eth·ə,nȯl }

mercaptol [ORG CHEM] A compound formed by combining a mercaptal and a ketone. { mər'kap,tȯl }

mercaptosuccinic acid See thiomalic acid. { mər,kap·tō·sək'sin·ik 'as·əd }

mercuric [INORG CHEM] The mercury ion with a 2+ oxidation state, for example $Hg(NO_3)_2$. { mər'kyu̇r·ik }

mercuric acetate [ORG CHEM] $Hg(C_2H_3O_2)_2$ Poisonous, light-sensitive white crystals;

soluble in alcohol and water; used in medicine and as a catalyst in organic synthesis. Also known as mercury acetate.   { mər'kyùr·ik 'as·ə,tāt }

mercuric arsenate   [INORG CHEM] $HgHAsO_4$ A poisonous yellow powder; soluble in hydrochloric acid, insoluble in water; used in antifouling and waterproof paints and in medicine.   Also known as mercury arsenate; mercury arseniate.   { mər'kyùr·ik 'ärs·ən,āt }

mercuric barium iodide   [INORG CHEM] $HgI_2·BaI_2·5H_2O$ Crystals that are yellow or reddish and deliquescent; soluble in alcohol and water; used in aqueous solution as Rohrbach's solution for mineral separation on the basis of density.   Also known as barium mercury iodide; mercury barium iodide.   { mər'kyùr·ik 'bar·ē·əm 'ī·ə,dīd }

mercuric benzoate   [ORG CHEM] $Hg(C_7H_5O_2)_2·H_2O$ Poisonous white crystals, sensitive to light, melting at 165°C; slightly soluble in alcohol and water; used in medicine.   Also known as mercury benzoate.   { mər'kyùr·ik 'ben·zə,wāt }

mercuric bromide   [INORG CHEM] $HgBr_2$ Poisonous white crystals, sensitive to light, melting at 235°C; soluble in alcohol and ether; used in medicine.   Also known as mercury bromide.   { mər'kyùr·ik 'brō,mīd }

mercuric chloride   [INORG CHEM] $HgCl_2$ An extremely toxic compound that forms white, rhombic crystals which sublime at 300°C and are soluble in alcohol or benzene; used for the manufacture of other mercuric compounds, as a fungicide, and in medicine and photography.   Also known as bichloride of mercury; corrosive sublimate.   { mər'kyùr·ik 'klòr,īd }

mercuric cyanate   See mercury fulminate.   { mər'kyùr·ik 'sī·ə,nāt }

mercuric cyanide   [INORG CHEM] $Hg(CN)_2$ Poisonous, colorless, transparent crystals that darken in light, decompose when heated; soluble in water and alcohol; used in photography, medicine, and germicidal soaps.   Also known as mercury cyanide.   { mər'kyùr·ik 'sī·ə,nīd }

mercuric fluoride   [INORG CHEM] HgF_2 Poisonous, transparent crystals that decompose when heated; moderately soluble in alcohol and water; used to synthesize organic fluorides.   { mər'kyùr·ik 'flùr,īd }

mercuric iodide   [INORG CHEM] HgI_2 Poisonous red crystals that turn yellow when heated to 150°C; soluble in boiling alcohol; used in medicine and in Nessler's and Mayer's reagents.   { mər'kyùr·ik 'ī·ə,dīd }

mercuric lactate   [ORG CHEM] $Hg(C_3H_5O_3)_2$ A poisonous white powder that decomposes when heated; soluble in water; used in medicine.   { mər'kyùr·ik 'lak,tāt }

mercuric nitrate   [INORG CHEM] $Hg(NO_3)_2·H_2O$ Poisonous, colorless crystals that decompose when heated; soluble in water and nitric acid, insoluble in alcohol; a fire hazard; used in medicine, in nitrating organic aromatics, and in felt manufacture.   Also known as mercury nitrate; mercury pernitrate.   { mər'kyùr·ik 'nī,trāt }

mercuric oleate   [ORG CHEM] $Hg(C_{18}H_{33}O_2)_2$ A poisonous yellowish-to-red liquid or solid mass; insoluble in water; used in medicine and antifouling paints, and as an antiseptic.   Also known as mercury oleate.   { mər'kyùr·ik 'ōl·ē,āt }

mercuric oxide   [INORG CHEM] HgO A compound of mercury that exists in two forms, red mercuric oxide and yellow mercuric oxide; the red form decomposes upon heating, is insoluble in water, and is used in pigments and paints, and in ceramics; the yellow form is insoluble in water, decomposes upon heating, and is used in medicine.   Also known as mercury oxide; red precipitate; yellow precipitate.   { mər'kyùr·ik 'äk,sīd }

mercuric phosphate   [INORG CHEM] $Hg_3(PO_4)_2$ Poisonous yellowish or white powder; insoluble in alcohol and water, soluble in acids; used in medicine.   Also known as mercury phosphate; trimercuric orthophosphate.   { mər'kyùr·ik 'fäs,fāt }

mercuric salicylate   [ORG CHEM] $Hg(C_7H_5O_3)_2$ Poisonous, white powder; odorless and tasteless; almost insoluble in water and alcohol; variable composition; used in medicine.   Also known as salicylated mercury.   { mər'kyùr·ik sə'lis·ə,lāt }

mercuric stearate   [ORG CHEM] $Hg(C_{17}H_{35}CO_2)_2$ Poisonous yellow powder; soluble in fatty acids, slightly soluble in alcohol; used as a germicide and in medicine.   Also known as mercury stearate.   { mər'kyùr·ik 'stir,āt }

mercuric sulfate   [INORG CHEM] $HgSO_4$ A toxic, white, crystalline powder, soluble in

mercuric sulfide

acids; used in medicine, as a catalyst, and for galvanic batteries.   Also known as mercury persulfate; mercury sulfate.   { mər'kyür·ik 'səl,fāt }

mercuric sulfide   [INORG CHEM] **1.** HgS The black variety is a poisonous powder; insoluble in water, alcohol, and nitric acid, soluble in sodium sulfide solution; sublimes at 583°C; used as a pigment.   Also known as black mercury sulfide; ethiops mineral.   **2.** The red variety is a poisonous powder; insoluble in water and alcohol; sublimes at 446°C; used as a medicine and pigment.   Also known as Chinese vermilion; quicksilver vermilion; red mercury sulfide; vermilion.   { mər'kyür·ik 'səl,fīd }

mercuric thiocyanate   [INORG CHEM] $Hg(SCN)_2$ Poisonous white powder; soluble in alcohol, slightly soluble in water; decomposes when heated; used in photography.   Also known as mercury thiocyanate.   { mər'kyür·ik ,thī·ə'sī·ə,nāt }

mercurous   [INORG CHEM] Referring to mercury with a valence of 1; for example, mercurous chloride, Hg_2Cl_2, where the mercury is covalently bonded, as $Cl-Hg-Hg-Cl$.   { mər'kyür·əs }

mercurous acetate   [ORG CHEM] $HgC_2H_3O_2$ Poisonous colorless plates or scales; decomposed by boiling water and by light; soluble in dilute nitric acids, slightly soluble in water.   Also known as mercury acetate; mercury protoacetate.   { mər 'kyür·əs 'as·ə,tāt }

mercurous bromide   [INORG CHEM] HgBr Poisonous white powder, crystals, or fibrous mass; odorless and tasteless; darkens in light; soluble in hot sulfuric acid and fuming nitric acid, insoluble in alcohol and ether; used in medicine.   Also known as mercury bromide.   { mər'kyür·əs 'brō,mīd }

mercurous chlorate   [INORG CHEM] $Hg_2(ClO_3)_2$ Poisonous white crystals that decompose at 250°C; soluble in alcohol and water; explodes in contact with combustible substances.   Also known as mercury chlorate.   { mər'kyür·əs 'klȯr,āt }

mercurous chloride   [INORG CHEM] Hg_2Cl_2 Odorless, nonpoisonous white crystals that darken in light; insoluble in water, alcohol, and ether; melts at 302°C; used in medicine and pyrotechnics.   Also known as mercury monochloride; mercury protochloride; mild mercury chloride.   { mər'kyür·əs 'klȯr,īd }

mercurous chromate   [INORG CHEM] Hg_2CrO_4 Red powder with variable composition; decomposes when heated; soluble in nitric acid, insoluble in water and alcohol; used to color ceramics green.   Also known as mercury chromate.   { mər'kyür·əs 'krō,māt }

mercurous iodide   [INORG CHEM] Hg_2I_2 Odorless, tasteless, poisonous yellow powder; darkens when heated; insoluble in water, alcohol, and ether; sublimes at 140°C; used as external medicine.   Also known as mercury protoiodide.   { mər'kyür·əs 'ī·ə,dīd }

mercurous oxide   [INORG CHEM] Hg_2O A poisonous black powder; insoluble in water, soluble in acids; decomposes at 100°C.   { mər'kyür·əs 'äk,sīd }

mercurous phosphate   [INORG CHEM] Hg_3PO_4 Light-sensitive white powder with variable composition; insoluble in alcohol and water, soluble in nitric acids; used in medicine.   Also known as mercury phosphate; trimercurous orthophosphate.   { mər'kyür·əs 'fäs,fāt }

mercurous sulfate   [INORG CHEM] Hg_2SO_4 Poisonous yellow-to-white powder; soluble in hot sulfuric acid or dilute nitric acid, insoluble in water; used as a catalyst and in laboratory batteries.   { mər'kyür·əs 'səl,fāt }

mercury   [CHEM] A metallic element, symbol Hg, atomic number 80, atomic weight 200.59, existing at room temperature as a silvery, heavy liquid.   Also known as quicksilver.   { 'mər·kyə·rē }

mercury acetate See mercuric acetate; mercurous acetate.   { 'mər·kyə·rē 'as·ə,tāt }
mercury arsenate See mercuric arsenate.   { 'mər·kyə·rē 'ärs·ən,āt }
mercury arseniate See mercuric arsenate.   { 'mər·kyə·rē är'sē·nē,āt }
mercury barium iodide See mercuric barium iodide.   { 'mər·kyə·rē ba·rē·əm 'ī·ō·dīd }
mercury benzoate See mercuric benzoate.   { 'mər·kyə·rē 'ben·zə,wāt }
mercury bromide See mercuric bromide; mercurous bromide.   { 'mər·kyə·rē 'brō,mīd }
mercury chlorate See mercurous chlorate.   { 'mər·kyə·rē 'klȯr,āt }
mercury chromate See mercurous chromate.   { 'mər·kyə·rē 'krō,māt }
mercury cyanide See mercuric cyanide.   { 'mər·kyə·rē 'sī·ə,nīd }
mercury fulminate   [ORG CHEM] $Hg(CNO)_2$ A gray, crystalline powder; explodes at the

melting point; soluble in alcohol, ammonium hydroxide, and hot water; used for explosive caps and detonators.   Also known as mercuric cyanate.   { 'mǝr·kyǝ·rē 'fül·mǝ͵nāt }

mercury monochloride See mercurous chloride.   { 'mǝr·kyǝ·rē ͵män·ǝ'klȯr͵īd }

mercury naphthenate   [ORG CHEM] Poisonous dark-amber liquid; soluble in mineral oils; used in gasoline antiknock compounds and as a paint antimildew promoter.   { 'mǝr·kyǝ·rē 'naf·thǝ͵nāt }

mercury nitrate See mercuric nitrate.   { 'mǝr·kyǝ·rē 'nī͵trāt }

mercury oleate See mercuric oleate.   { 'mǝr·kyǝ·rē 'ōl·ē͵āt }

mercury oxide See mercuric oxide.   { 'mǝr·kyǝ·rē 'äk͵sīd }

mercury pernitrate See mercuric nitrate.   { 'mǝr·kyǝ·rē pǝr'nī͵trāt }

mercury persulfate See mercuric sulfate.   { 'mǝr·kyǝ·rē pǝr'sǝl͵fāt }

mercury phosphate See mercuric phosphate; mercurous phosphate.   { 'mǝr·kyǝ·rē 'fäs͵fāt }

mercury protoacetate See mercurous acetate.   { 'mǝr·kyǝ·rē ͵prō·dō'as·ǝ͵tāt }

mercury protochloride See mercurous chloride.   { 'mǝr·kyǝ·rē ͵prō·dō'klȯr͵īd }

mercury protoiodide See mercurous iodide.   { 'mǝr·kyǝ·rē ͵prō·dō'ī·ǝ͵dīd }

mercury stearate See mercuric stearate.   { 'mǝr·kyǝ·rē 'stir͵āt }

mercury sulfate See mercuric sulfate.   { 'mǝr·kyǝ·rē 'sǝl͵fāt }

mercury thiocyanate See mercuric thiocyanate.   { 'mǝr·kyǝ·rē ͵thī·ǝ'sī·ǝ͵nāt }

merthiolate See thimerosal.   { mǝr'thī·ǝ͵lāt }

mesaconic acid   [ORG CHEM] $C_5H_6O_4$ An unsaturated dibasic acid, an isomer of citraconic acid, that melts at 202°C.   Also known as methyl fumaric acid.   { ͵mes·ǝ'kän·ik 'as·ǝd }

mescaline   [ORG CHEM] $C_{11}H_{17}NO_3$ The alkaloid 3,4,5-trimethoxyphenethylamine, found in mescal buttons; produces unusual psychic effects and visual hallucinations.   { 'mes·kǝ͵lēn }

mesitylene   [ORG CHEM] C_9H_{12} A colorless fragrant liquid that boils at 164.7°C (328.6°F); it is an aromatic hydrocarbon that is part of the benzene series, and occurs naturally in coal tar or is synthesized from acetone.   { mǝ'sid·ǝl·ēn }

mesityl oxide   [ORG CHEM] $(CH_3)_2C=CHCOCH_3$ A colorless, oily liquid with a honeylike odor; solidifies at −41.5°C; used as a solvent for resins, particularly vinyl resins, many gums, and nitrocellulose; also used in lacquers, paints, and varnishes.   { 'mez·ǝ͵til 'äk͵sīd }

meso-   [CHEM] A prefix meaning intermediate or middle, as in denoting inactive optical isomers, the form of intermediate inorganic acid, the middle position in cyclic organic compounds, or a ring system with middle ring positions.   { 'me·zō }

mesogen See mesogenic unit.   { 'mez·ǝ͵jen }

mesogenic unit   [PHYS CHEM] A component of a molecule that induces a mesomorphic or liquid crystalline phase.   Also known as mesagen.   { ¦mez·ǝ¦jen·ik 'yü·nǝt }

meso-ionic compound   [ORG CHEM] Any of a class of five-membered ring heterocycles and their benzo derivatives which possess a sextet of pi electrons in association with the atoms composing the ring but which cannot be represented satisfactorily by any one covalent or polar structure.   { ¦mez·ō·ī¦än·ik 'käm͵paúnd }

mesomorphism   [PHYS CHEM] A state of matter intermediate between a crystalline solid and a normal isotropic liquid, in which long rod-shaped organic molecules contain dipolar and polarizable groups.   { ¦mez·ǝ¦mȯr͵fiz·ǝm }

mesopore   [CHEM] A pore in a catalytic material whose width ranges from 2 nanometers to 0.05 micrometer.   { 'mez·ǝ͵pȯr }

mesyl See methylsulfonyl.   { 'mes·ǝl }

meta-   [ORG CHEM] A prefix for benzene-ring compounds when two side chains are connected to carbon atoms with an unsubstituted carbon atom between them.   { 'med·ǝ }

metachromasia   [CHEM] **1.** The property exhibited by certain pure dyestuffs, chiefly basic dyes, of coloring certain tissue elements in a different color, usually of a shorter wavelength absorption maximum, than most other tissue elements.   **2.** The

metachromatism

assumption of different colors or shades by different substances when stained by the same dye. Also known as metachromatism. { ,med,ə·krō'mā·zhə }

metachromatism *See* metachromasia. { ¦med·ə'krō·mə ,tiz·əm }

metachrome yellow [ORG CHEM] $C_{13}H_8N_3NaO_5$ A yellow dye that is slightly soluble in water. { ¦med·ə,krōm 'yel·ō }

metahydrate sodium carbonate [INORG CHEM] $Na_2CO_3 \cdot H_2O$ Water-soluble, white crystals with an alkaline taste, loses water at 109°C; melts at 851°C; used in medicine, photography, and water pH control; and as a food additive. Also known as crystal carbonate; soda crystals. { ,med·ə'hī,drāt ¦sōd·ē·əm 'kär·bə,nāt }

metal alkyl [ORG CHEM] One of the family of organometallic compounds, a combination of an alkyl organic radical with a metal atom or atoms. { 'med·əl 'al·kəl }

metal cluster compound [CHEM] A compound in which two or more metal atoms aggregate so as to be within bonding distance of one another and each metal atom is bonded to at least two other metal atoms; some nonmetal atoms may be associated with the cluster. { 'med·əl ¦kləs·tər 'käm,paúnd }

metaldehyde [ORG CHEM] $(CH_3CHO)_n$ White acetaldehyde-polymer prisms; soluble in organic solvents, insoluble in water; used as a pesticide or fuel. { me'tal·də,hīd }

metal ion indicator [ANALY CHEM] A substance, usually a dyestuff, that changes color after forming a metal ion complex with a color different from that of the uncomplexed indicator. Also known as complexation indicator. { ¦med·əl ¦ī,än ,in·də,kād·ər }

metallation [ORG CHEM] The direct replacement of a hydrogen atom by a metal atom in an organic molecule to form a carbon-metal bond. { ,med·ə'lā·shən }

metallic bond [PHYS CHEM] The type of chemical bond that is present in all metals, and may be thought of as resulting from a sea of valence electrons which are free to move throughout the metal lattice. { mə'tal·ik 'bänd }

metallic element [CHEM] An element generally distinguished (from a nonmetallic one) by its luster, electrical conductivity, malleability, and ability to form positive ions. { mə'tal·ik 'el·ə·mənt }

metallic hydrogen [PHYS CHEM] **1.** A phase of hydrogen believed to occur at extremely high pressures, in which the material transforms to a conducting molecular solid. **2.** A phase of hydrogen believed to occur at still higher pressures, in which the molecular bonds that exist at lower pressures are broken and an atomic solid with the structure of an alkali metal is formed. { mə'tal·ik 'hī·drə·jən }

metallic soap [ORG CHEM] A salt of stearic, oleic, palmitic, lauric, or erucic acid with a heavy metal such as cobalt or copper; used as a drier in paints and inks, in fungicides, decolorizing varnish, and waterproofing. { mə'tal·ik 'sōp }

metallo-carbohedrene [CHEM] A member of a class of molecular clusters in which atoms of an early transition metal (scandium through nickel in the third period of the periodic table) are bonded with carbon atoms in a cagelike network. { mə¦ta·lō ,kär·bə'hed,rēn }

metallocene [ORG CHEM] Organometallic coordination compound which is obtained as a cyclopentadienyl derivative of a transition metal or a metal halide. { mə'tal·ə,sēn }

metallocene catalyst [ORG CHEM] A molecular structure with a well-defined single catalytic site, consisting of an organometallic coordination compound in which one or two cyclopentadienyl rings (with or without substituents) are bonded to a central transition-metal atom; used to produce uniform polyolefins with unique structures and physical properties. { mə¦tal·ə,sēn 'kad·ə,list }

metallocycle [ORG CHEM] A compound whose structure consists of a cyclic array of atoms of which one is a metal atom; frequently the ring contains three or four carbon atoms and one transition-metal atom. { mə'tal·ə,sī·kəl }

metalloid [CHEM] An element whose properties are intermediate between those of metals and nonmetals. Also known as semimetal. { 'med·ə,lóid }

metamer [ORG CHEM] One of two or more chemical compounds that exhibits isomerism with the others. { 'med·ə·mər }

metanillic acid [ORG CHEM] $C_6H_4(NH_2)SO_3H$ A water-soluble, crystalline compound, isomeric with sulfanilic acid; used in medicines and dyes. { ¦med·ə¦nil·ik 'as·əd }

metastable equilibrium [PHYS CHEM] A state of pseudo-equilibrium having higher free energy than the true equilibrium state. { ¦med·ə'stā·bəl ¸ē·kwə'lib·rē·əm }

metastable ion [ANALY CHEM] In mass spectroscopy, an ion formed by a secondary dissociation process in the analyzer tube (formed after the parent or initial ion has passed through the accelerating field). { ¦med·ə'stā·bəl 'ī¸än }

metastable phase [PHYS CHEM] Existence of a substance as either a liquid, solid, or vapor under conditions in which it is normally unstable in that state. { ¦med·ə'stā·bəl ¦fāz }

metathesis [CHEM] A reaction involving the exchange of elements or groups as in the general equation AX + BY → AY + BX. { mə'tath·ə·səs }

metathetical salts [CHEM] Salts that form a four-component, ternary equilibrium system in which there are four possible binary systems, resulting in two quadruple points. { ¦med·ə¦thed·ə·kəl 'sȯls }

metatitanic acid See titanic acid. { ¦med·ə·tī'tan·ik 'as·əd }

methacrolein [ORG CHEM] CH₂C(CH₃)CHO Liquid with 68°C boiling point; slightly soluble in water; used to make resins and copolymers. { mə'thak·rə·lən }

methacrylate ester [ORG CHEM] CH₂:C(CH₃)COOR Methacrylic acid ester in which R can be methyl, ethyl, isobutyl, or 50-50 n-butyl-isobutyl groups; used to make thermoplastic polymers or copolymers. { meth'ak·rə¸lāt 'es·tər }

methacrylic acid [ORG CHEM] CH₂C(CH₃)COOH Easily polymerized, colorless liquid melting at 15–16°C; soluble in water and most organic solvents; used to make water-soluble polymers and as a chemical intermediate. { ¦meth·ə¦kril·ik 'as·əd }

methacrylic polymer [ORG CHEM] A polymer whose monomer is a methacrylic ester with the general formula H₂C=C(CH₃)COOR. { 'meth·ə¸kril·ik 'päl·ə·mər }

methacrylonitrile [ORG CHEM] CH₂:C(CH₃)CN Clear, colorless liquid boiling at 90°C; used to make solvent-resistant thermoplastic polymers and copolymers. { ¸meth·ə·krə'län·ə¸tril }

methallyl alcohol [ORG CHEM] H₂C:C(CH₃)CH₂OH Flammable, toxic, water-soluble, colorless liquid boiling at 115°C; has pungent aroma; soluble in most organic solvents; used as a chemical intermediate. Also spelled methyl allyl alcohol. { meth'al·əl 'al·kə¸hȯl }

methanal See formaldehyde. { 'meth·ə¸nal }

methane [ORG CHEM] CH₄ A colorless, odorless, and tasteless gas, lighter than air and reacting violently with chlorine and bromine in sunlight, a chief component of natural gas; used as a source of methanol, acetylene, and carbon monoxide. Also known as methyl hydride. { 'meth¸ān }

methanearsonic acid [ORG CHEM] CH₃AsO(OH)₂ A white solid with a melting point of 161°C; very soluble in water; used as an herbicide for cotton crops and for noncrop areas. Abbreviated MAA. { ¦meth¸ān·är'sän·ik 'as·əd }

methane hydrate [CHEM] Methane gas trapped or dissolved in ice formed in deep-sea sediments. { 'mēth¸ān 'hī¸drāt }

methanesulfonic acid [ORG CHEM] CH₃SO₂OH A solid with a melting point of 20°C; used as a catalyst in polymerization, esterification, and alkylation reactions, and as a solvent. Also known as methysulfonic acid. { ¦meth¸ān·səl'fän·ik 'as·əd }

methanethiol See methyl mercaptan. { mə¸than·ə'thī¸ȯl }

methanoic acid See formic acid. { ¸meth·ə'nō·ik 'as·əd }

methanol See methyl alcohol. { 'meth·ə¸nȯl }

methenyl See methine group. { 'meth·ə¸nil }

methidathion [ORG CHEM] C₆H₁₁O₄N₂PS₃ A colorless, crystalline compound with a melting point of 39–40°C; used as an insecticide and miticide for pests on alfalfa, citrus, and cotton. { mə¸thid·ə'thī¸än }

methide [ORG CHEM] A binary compound consisting of methyl and, most commonly, a metal, such as sodium (sodium methide, NaCH₃). { 'me¸thīd }

methine group [ORG CHEM] HC≡ A radical consisting of a single carbon and a single hydrogen. Also known as methenyl; methylidyne. { 'me¸thēn 'grüp }

methionic acid [ORG CHEM] CH₂(SO₃H)₂ An acid that exists as hygroscopic crystals; used in organic synthesis. { ¦meth·ē¦än·ik 'as·əd }

237

methoxide

methoxide [ORG CHEM] A compound formed from a metal and the methoxy radical; an example is sodium methoxide. Also known as methylate. { mə'thäk₁sīd }

methoxy- [ORG CHEM] OCH₃⁻ A combining form indicating the oxygen-containing methane radical, found in many organic solvents, insecticides, and plasticizer intermediates. { mə'thäk·sē }

methoxychlor [ORG CHEM] Cl₃CCH(C₆H₄OCH₃)₂ White, water-insoluble crystals melting at 89°C; used as an insecticide. Also known as DMDT; methoxy DDT. { me'thäk·si₁klȯr }

methoxy DDT *See* methoxychlor. { me'thäk·sē ₁de̱₁dē'tē }

2-methoxyethanol [ORG CHEM] CH₃OCH₂CH₂OH A poisonous liquid, used as a solvent for low-viscosity cellulose acetate, natural and some synthetic resins, and alcohol-soluble dyes, and also used in dyeing leather. { ₁tü mə₁thäk·sē'eth·ə₁nȯl }

methoxyethylmercury chloride [ORG CHEM] CH₃OCH₂CH₂HgCl A white, crystalline compound with a melting point of 65°C; used as a fungicide in diseases of sugarcane, pineapples, seed potatoes, and flower bulbs, and as seed dressings for cereals, legumes, and root crops. Abbreviated MEMC. { ma₁thäk·sē₁eth·əl₁mər·kyə·rē 'klȯr₁īd }

4-methoxy-2-hydroxybenzophenone *See* oxybenzone. { ₁fȯr mə₁thäk·sē ₁tü ₁hī₁dräk·sē·ben'zä·fə₁nōn }

methoxyl [ORG CHEM] CH₃O⁻ A functional group which is univalent. { mə'thäk·səl }

methyl [ORG CHEM] The alkyl group derived from methane and usually written CH₃⁻. Also known as carbinyl. { 'meth·əl }

methyl abietate [ORG CHEM] C₁₉H₂₉COOCH₃ Colorless to yellow liquid boiling at 365°C; miscible with most organic solvents; used as a solvent and plasticizer for lacquers, varnishes, and coatings. { 'meth·əl 'ab·ē·ə₁tāt }

methyl acetate [ORG CHEM] CH₃CO₂CH₃ Flammable, colorless liquid with fragrant odor; boils at 54°C; partially soluble in water, miscible with hydrocarbon solvents; used as a solvent and extractant. { 'meth·əl 'as·ə₁tāt }

methylacetic acid *See* propionic acid. { ₁meth·əl·ə₁sēd·ik ₁as·əd }

methyl acetoacetate [ORG CHEM] CH₃COCH₂CO₂CH₃ Alcohol-soluble, colorless liquid boiling at 172°C; used as a chemical intermediate and as a solvent for cellulosics. { 'meth·əl ₁as·əd·ō₁as·ə₁tāt }

methyl acetophenone [ORG CHEM] CH₃C₆H₄COCH₃ Fragrant (coumarin aroma), colorless or pale-yellow liquid, soluble in alcohol; used in perfumery. { 'meth·əl ₁as·ə'täf·ə₁nōn }

methyl acrylate [ORG CHEM] CH₂:CHCOOCH₃ A readily polymerized, volatile, colorless liquid boiling at 80°C; slightly soluble in water; used as a chemical intermediate and in making polymers. { 'meth·əl 'ak·rə₁lāt }

methylal [ORG CHEM] CH₃OCH₂OCH₃ Flammable, volatile, colorless liquid boiling at 42°C; soluble in ether, hydrocarbons, and alcohol, partially soluble in water; used as a solvent and chemical intermediate, and in perfumes, adhesives, coatings. Also known as formal. { 'meth·ə₁lal }

methyl alcohol [ORG CHEM] CH₃OH A colorless, toxic, flammable liquid, boiling at 64.5°C; miscible with water, ether, alcohol; used in manufacture of formaldehyde, chemical synthesis, antifreeze for autos, and as a solvent. Also known as methanol; wood alcohol. { 'meth·əl 'al·kə₁hȯl }

methyl allyl alcohol *See* methallyl alcohol. { 'meth·əl 'al·əl 'al·kə₁hȯl }

methyl allyl chloride [ORG CHEM] CH₂:C(CH₃)CH₂Cl Volatile, flammable, colorless liquid boiling at 72°C; has disagreeable odor; used as an insecticide and fumigant, and for chemical synthesis. { 'meth·əl 'al·əl 'klȯr₁īd }

methylamine [ORG CHEM] CH₃NH₂ A colorless gas that is highly toxic and flammable; used to prepare dyes, and as a chemical intermediate. { ₁meth·ə·lə₁mēn }

N-methyl-*para*-aminophenol [ORG CHEM] CH₃NHC₆H₄OH Colorless, combustible needles with a melting point of 87°C; soluble in water, alcohol, and ether; used as a photographic developer. { ₁en ₁meth·əl ₁par·ə ₁am·ə·nō'fē₁nȯl }

methyl amyl acetate [ORG CHEM] CH₃COOCH(CH₃)CH₂CH(CH₃)₂ Toxic, flammable, colorless liquid with mild, agreeable odor; boils at 146°C; used as nitrocellulose lacquer

238

solvent. Also known as methyl isobutyl carbinol acetate. { 'meth·əl 'am·əl 'as· ə,tāt }

methyl amyl alcohol [ORG CHEM] $(CH_3)_2CHCH_2CHOHCH_3$ Toxic, flammable, colorless liquid; boils at 132°C; miscible with water and most organic solvents; used as a solvent and as a chemical intermediate. Also known as methyl isobutyl carbinol (MIBC). { 'meth·əl 'am·əl 'al·kə ,hòl }

methyl amyl carbinol [ORG CHEM] $CH_3(CH_2)_4CHOHCH_3$ Colorless liquid with mild aroma; boils at 160°C; miscible with most organic liquids; used as an ore-flotation frothing agent and as a synthetic-resin solvent. { 'meth·əl 'am·əl 'kär·bə,nòl }

methyl-n-amyl ketone [ORG CHEM] $CH_3(CH_2)_4COCH_3$ Stable, water-white liquid; miscible with organic lacquer solvents, slightly soluble in water; used as an inert reaction medium and as a solvent for nitrocellulose lacquers. Also known as 2-heptanone. { 'meth·əl ¦en ¦am·əl 'kē,tōn }

N-methylaniline [ORG CHEM] $C_6H_5NH(CH_3)$ Oily liquid, colorless to reddish-brown; soluble in water and organic solvents; boils at 190°C; used as an acid acceptor, solvent, and chemical intermediate. { ¦en ¦meth·əl'an·ə·lən }

α-methylanisalacetone [ORG CHEM] $CH_3OC_6H_4CH:CHCOCH_2CH_3$ A white to pale yellow, combustible solid with a melting point of 60°C; used as a flavoring. { ¦al·fə ¦meth·əl,an·ə·sə'las·ə,tōn }

methyl anisole See methyl para-cresol. { 'meth·əl 'an·ə,sōl }

methyl anthranilate [ORG CHEM] $H_2NC_6H_4CO_2CH_3$ A yellowish to colorless liquid, slightly soluble in water; used in flavoring and in perfumery. Also known as artificial neroli oil. { 'meth·əl an'thran·ə,lāt }

2-methyl anthraquinone See tectoquinone. { ¦tü ¦meth·əl ,an·thrə,kwē'nōn }

methyl arachidate [ORG CHEM] $CH_3(CH_2)_{18}COOCH_3$ A waxlike solid with a melting point of 45.8°C; soluble in alcohol and ether; used in medical research and as a reference standard for gas chromatography. Also known as methyl eicosanoate. { 'meth·əl ə'rak·ə,dāt }

methylarsinic sulfide [ORG CHEM] CH_3AsS A colorless compound whose flakes melt at 110°C; insoluble in water; used as a fungicide in treating cotton seeds. Also known as rhizoctol. { ¦meth·əl·är¦sin·ik 'səl,fīd }

methylate See methoxide. { 'meth·ə,lāt }

methylation [ORG CHEM] A chemical process for introducing a methyl group (CH_3-) into an organic compound. { ,meth·ə'lā·shən }

methyl behenate [ORG CHEM] $CH_3(CH_2)_{20}COOCH_3$ A combustible, waxlike solid with a melting point of 53.2°C; soluble in alcohol and ether; used in medical and biochemical research and as a reference standard for gas chromatography. Also known as methyl docosanoate. { 'meth·əl bə'he,nāt }

methylbenzene See toluene. { ¦meth·əl¦ben,zēn }

methylbenzethonium chloride [ORG CHEM] $C_{27}H_{44}O_2Cl·H_2O$ Colorless crystals with a melting point of 161–163°C; soluble in alcohol, hot benzene, chloroform, and water; used as a bactericide. { ,meth·əl,ben·zə'thō·nē·əm 'klòr,īd }

methyl benzoate [ORG CHEM] $C_6H_5CO_2CH_3$ Colorless, fragrant liquid boiling at 199°C; slightly soluble in alcohol and water, soluble in ether; used in perfumery and as a solvent. Also known as niobe oil. { 'meth·əl 'ben·zə,wāt }

methyl ortho-benzoylbenzoate [ORG CHEM] $C_6H_5COC_6H_4COOCH_3$ A colorless, combustible liquid with a boiling point of 351°C; slightly soluble in water; used as a plasticizer. { 'meth·əl ¦or·thō,ben·zə,wil'ben·zə,wāt }

α-methylbenzyl acetate [ORG CHEM] $C_6H_5CH(CH_3)OOCCH_3$ A colorless, combustible liquid with a strong floral odor; soluble in glycerin, mineral oil, and 70% alcohol; used in perfumes and as a flavoring. { ¦al·fə ¦meth·əl¦ben·zəl 'as·ə,tāt }

α-methylbenzyl alcohol [ORG CHEM] $C_6H_5CH(CH_3)OH$ A colorless, combustible liquid with a mild floral odor and a boiling point of 204°C; soluble in water; used in perfumes and dyes and as a flavoring agent. { ¦al·fə ¦meth·əl¦ben·zəl 'al·kə,hòl }

α-methylbenzylamine [ORG CHEM] $C_6H_5CH(CH_3)NH_2$ A colorless, combustible liquid with a boiling point of 188.5°C; soluble in most organic solvents; used as an emulsifying agent. { ¦al·fə ,meth·əl·ben'zal·ə,mēn }

α-methylbenzyl ether

α-methylbenzyl ether [ORG CHEM] $C_6H_5CH(CH_3)OCH(CH_3)C_6H_5$ A straw-colored, combustible liquid with a boiling point of 286.3°C; at 760 mmHg (101,325 pascals); slightly soluble in water; used as a solvent and as a synthetic rubber softener. { ¦al·fə ¦meth·əl¦ben·zəl 'ē·thər }

methyl blue [ORG CHEM] Dark-blue powder or dye; sodium triphenyl *para*-rosaniline sulfonate; used as a biological and bacteriological stain and as an antiseptic. { 'meth·əl 'blü }

methyl borate *See* trimethyl borate. { 'meth·əl 'bȯr,āt }

methyl bromide [ORG CHEM] CH_3Br A toxic, colorless gas that forms a crystalline hydrate with cold water; used in synthesis of organic compounds, and as a fumigant. { 'meth·əl 'brō,mīd }

2-methyl-1,3-butadiene *See* isoprene. { ¦tü 'meth·əl ¦wən ¦thrē ,byüd·ə'dī,ēn }

2-methylbutanal *See* 2-methylbutyraldehyde. { ¦tü ,meth·əl'byüt·ən,al }

2-methylbutane *See* isopentane. { ¦tü ,meth·əl'byü,tān }

2-methyl-1-butanol [ORG CHEM] $C_5H_{12}O$ A liquid with a boiling point of 128°C, miscible with alcohol and with ether, slightly soluble in water; used as a solvent, in organic synthesis, and as an additive in oils and paints. { ¦tü ¦meth·əl ¦wən 'byüt·ən,ȯl }

methyl butene [ORG CHEM] C_5H_{10} Either of two colorless, flammable, volatile liquid isomers; soluble in alcohol, insoluble in water: 3-methyl-1-butene boils at 20°C, is used as a chemical intermediate and in the manufacture of high-octane fuel, and is also known as isopropylethylene; 3-methyl-2-butene boils at 38°C, is used as an anesthetic and high-octane fuel and as a chemical intermediate, and is also known as trimethylethylene. { 'meth·əl 'byü,tēn }

2-methyl-2-butene *See* amylene. { ¦tü 'meth·əl ¦tü 'byü,tēn }

methyl butyl ketone [ORG CHEM] $CH_3COC_4H_9$ A liquid boiling at 127°C; soluble in water, alcohol, and ether; used as a solvent. Also known as propylacetone. { 'meth·əl 'byüd·əl 'kē,tōn }

methylbutynol [ORG CHEM] $HC:CCOH(CH_3)_2$ Water-miscible, colorless liquid boiling at 104°C; soluble in most organic solvents; used as a stabilizer for chlorinated organic compounds, as a solvent, and as a chemical intermediate. { ,meth·əl'byüt·ən,ȯl }

2-methylbutyraldehyde [ORG CHEM] $CH_3CH_2CH(CH_3)CHO$ A combustible liquid with a boiling point of 92.93°C; soluble in alcohol and ether; used as a brightener in electroplating. Also known as 2-methylbutanal. { ¦tü ,meth·əl,byüd·ə'ral·də,hīd }

methyl butyrate [ORG CHEM] $CH_3CH_2CH_2COOCH_3$ Liquid boiling at 102°C; used as a solvent for cellulosic materials. { 'meth·əl 'byüd·ə,rāt }

methyl caprate [ORG CHEM] $CH_3(CH_2)_8COOCH_3$ A colorless, combustible liquid with a boiling point of 244°C; soluble in alcohol and ether; used in the manufacture of detergents, stabilizers, plasticizers, textiles, and lubricants. Also known as methyl decanoate. { 'meth·əl 'ka,prāt }

methyl caproate [ORG CHEM] $CH_3(CH_2)_4COOCH_3$ Colorless liquid boiling at 150°C; soluble in alcohol and ether, insoluble in water; used as an intermediate to make caproic acid. Also known as methyl hexanoate. { 'meth·əl 'kap·rə,wāt }

methyl caprylate [ORG CHEM] $CH_3(CH_2)_6COOCH_3$ Colorless liquid boiling at 193°C; soluble in ether and alcohol, insoluble in water; used as an intermediate to make caprylic acid. { 'meth·əl 'kap·rə,lāt }

methyl carbonate [ORG CHEM] $CO(OCH_3)_2$ Water-insoluble, colorless liquid boiling at 91°C; has pleasant odor; miscible with acids and alkalies; used as a chemical intermediate. { 'meth·əl 'kär·bə,nāt }

methylcellulose [ORG CHEM] A grayish-white powder derived from cellulose; swells in water to a colloidal solution; soluble in glacial acetic acid; used in water-based paints and ceramic glazes, for leather tanning, and as a thickening and sizing agent, adhesive, and food additive. Also known as cellulose methyl ether. { 'meth·əl'sel·yə,lōs }

methyl chloride *See* chloromethane. { 'meth·əl 'klȯr,īd }

methyl chloroacetate [ORG CHEM] $ClHC_2COOCH_3$ Colorless liquid boiling at 131°C; miscible with ether and alcohol, slightly soluble in water; used as a solvent. { 'meth·əl ,klȯr·ō'as·ə,tāt }

240

methyl chlorocarbonate *See* methyl chloroformate. { 'meth·əl ,klȯr·ō'kär·bə,nāt }
methyl chloroform *See* trichloroethane. { 'meth·əl 'klȯr·ə,fȯrm }
methyl chloroformate [ORG CHEM] ClCOOCH₃ A toxic, corrosive, colorless liquid with a boiling point of 71.4°C; soluble in benzene, ether, and methanol; used as a lacrimator in military poison gas and for insecticides. Also known as methyl chlorocarbonate. { 'meth·əl ¦klȯrə¦fȯr,māt }
methyl cinnamate [ORG CHEM] C₆H₅CH:CHCO₂CH₃ A white crystalline compound with strawberry aroma; soluble in ether and alcohol, insoluble in water; boils at 260°C; used to flavor confectioneries and in perfumes. { 'meth·əl 'sin·ə,māt }
methyl *para*-cresol [ORG CHEM] CH₃C₆H₄OCH₃ Colorless liquid with floral aroma; used in perfumery. Also known as methyl anisole. { 'meth·əl ¦par·ə 'krē,sȯl }
methyl cyanoacetate [ORG CHEM] CNCH₂COOCH₃ A toxic, combustible, colorless liquid with a boiling point of 203°C; soluble in water, ether, and alcohol; used in pharmaceuticals and dyes. { 'meth·əl ¦sī·ō,nō'as·ə,tāt }
methyl cyclohexane [ORG CHEM] C₇H₁₄ Colorless liquid boiling at 101°C; used as a cellulosic solvent and as a chemical intermediate. Also known as hexahydrotoluene. { 'meth·əl ¦sī·klō'hek,sān }
methyl cyclohexanol [ORG CHEM] CH₃C₆H₁₀OH A toxic, colorless liquid with menthol aroma; a mixture of three isomers; used as a solvent for lacquer and cellulosics, as a lubricant antioxidant, and in detergents and textile soaps. { 'meth·əl ¦sī·klō 'hek·sə,nȯl }
methyl cyclohexanone [ORG CHEM] CH₃C₅H₉CO A toxic, clear to pale-yellow liquid with acetonelike aroma; a mixture of cyclic ketones; used as a solvent and in lacquers. { 'meth·əl ¦sī·klō'hek·sə,nōn }
methylcyclopentadiene dimer [ORG CHEM] C₁₂H₁₆ A flammable, colorless liquid with a boiling range of 78–183°C; soluble in alcohol, benzene, and ether; used in high-energy fuels, plasticizers, dyes, and pharmaceuticals. { ¦meth·əl,sī·klō,pen·tə'dī,ēn 'dī·mər }
methyl cyclopentane [ORG CHEM] C₅H₉CH₃ Flammable, colorless liquid boiling at 72°C; used as a chemical intermediate. { 'meth·əl ¦sī·klō'pen,tān }
methyl decanoate *See* methyl caprate. { 'meth·əl də'kan·ə,wāt }
methyldichlorosilane [ORG CHEM] CH₃SiHCl₂ A colorless liquid with a melting point of −91°C (−130°F) and boiling point of 41°C (106°F). Also known as dichloromethylsilane. { 'meth·əl·dī,klȯr·ō 'sī,lān }
methyl-*N*-(3,4-dichlorophenyl)carbamate *See* swep. { 'meth·əl ¦en ¦thrē ¦fȯr dī,klȯr· ō'fen·əl 'kär·bə,māt }
methyl diethanolamine [ORG CHEM] CH₃N(C₂H₄OH)₂ A colorless liquid miscible with water and benzene; has amine aroma; boils at 247°C; used as a chemical intermediate and as an acid-gas absorbent. { 'meth·əl dī,eth·ə'näl·ə,mēn }
methyl dioxolane [ORG CHEM] C₄H₇O₂ Water-soluble, clear liquid boiling at 81°C; used as a solvent and extractant. { 'meth·əl dī'äk·sə,lān }
methyl docosanoate *See* methyl behenate. { 'meth·əl ,dō·kə'san·ə,wāt }
methyl eicosanoate *See* methyl arachidate. { 'meth·əl ,ī·kə'san·ə,wāt }
methylene [ORG CHEM] −CH₂− A radical that contains a bivalent carbon. { 'meth· ə,lēn }
methylene blue [ORG CHEM] Dark green crystals or powder; soluble in water (deep blue solution), alcohol, and chloroform; C₁₆H₁₈N₃SCl·3H₂O used in medicine; (C₁₆H₁₈N₃SCl)₂·ZnCl₂·H₂O used as a textile dye, biological stain, and indicator. Also known as methylthionine chloride. { 'meth·ə,lēn 'blü }
methylene bromide [ORG CHEM] CH₂Br₂ Colorless, clear liquid boiling at 97°C; miscible with organic solvents, slightly soluble in water; used as a solvent and chemical intermediate. Also known as dibromomethane. { 'meth·ə,lēn 'brō,mīd }
methylene chloride [ORG CHEM] CH₂Cl₂ A colorless liquid, practically nonflammable and nonexplosive; used as a refrigerant in centrifugal compressors, a solvent for organic materials, and a component in nonflammable paint-remover mixtures. { 'meth·ə,lēn 'klȯr,īd }
methylene iodide [ORG CHEM] CH₂I₂ Yellow liquid boiling at 180°C; soluble in ether

methylene oxide

and alcohol, insoluble in water; used as a chemical intermediate and to separate mineral mixtures. Also known as diiodomethane. { 'meth·ə₁lēn 'ī·ə₁dīd }

methylene oxide *See* formaldehyde. { 'meth·ə₁lēn 'äk₁sīd }

methylene succinic acid *See* itaconic acid. { 'meth·ə₁lēn sək'sin·ik 'as·əd }

methyl ester [ORG CHEM] An ester that forms methanol when hydrolyzed. { 'meth·əl 'es·tər }

methyl ether *See* dimethyl ether. { 'meth·əl 'ē·thər }

methylethylcellulose [ORG CHEM] A combustible, white to cream-colored, fibrous solid or powder; disperses in cold water, forming solutions which undergo reversible transformation from sol to gel; used as an emulsifier and foaming agent. { ¦meth·əl¦eth·əl'sel·yə₁lōs }

methyl ethylene *See* propylene. { 'meth·əl 'eth·ə₁lēn }

methyl ethyl ketone [ORG CHEM] CH₃COC₂H₅ A water-soluble, colorless liquid that is miscible in oil; used as a solvent in vinyl films and nitrocellulose coatings, and as a reagent in organic synthesis. Also known as ethyl methyl ketone; MEK. { ¦meth·əl ¦eth·əl 'kē₁tōn }

methyl formate [ORG CHEM] HCOOCH₃ A flammable, colorless liquid with a boiling point of 31.8°C; soluble in ether, water, and alcohol; used in military poison gases and larvicides, and as a fumigant. { 'meth·əl 'fȯr₁māt }

methyl fumaric acid *See* mesaconic acid. { 'meth·əl fyü'mar·ik 'as·əd }

2-methylfuran [ORG CHEM] C₄H₃OCH₃ A colorless liquid with ether flike aroma; boils at 64°C; used as a chemical intermediate. { ¦tü ₁meth·ə'fyür₁än }

methyl furoate [ORG CHEM] C₄H₃OCO₂CH₃ Colorless liquid that turns yellow in light; soluble in ether and alcohol, insoluble in water; used as a solvent and chemical intermediate. { 'meth·əl 'fyür·ə₁wāt }

methyl glucoside [ORG CHEM] C₇H₁₄O₆ Odorless, water-soluble white crystals; used to make resins, drying oils, plasticizers, and surfactants. { 'meth·əl 'glü·kə₁sīd }

methyl glycocoll *See* sarcosine. { 'meth·əl 'glī·kə₁kȯl }

methyl heptane [ORG CHEM] C₈H₁₈ Either of two colorless, water-insoluble liquids, soluble in alcohol and ether, used as chemical intermediates: 2-methylheptane boils at 118°C, is flammable; 4-methylheptane boils at 122°C. { ₁meth·əl 'hep₁tān }

methylheptenone [ORG CHEM] (CH₃)₂C:CH(CH₂)₂COCH₃ A combustible, colorless liquid with a boiling point of 173–174°C; a constituent of many essential oils; used in perfumes and for flavoring. { ₁meth·əl'hep·tə₁nōn }

2-methylhexane [ORG CHEM] C₇H₁₆ Colorless liquid boiling at 90°C; insoluble in alcohol and water; used as a chemical intermediate. Also known as ethyl isobutylmethane. { ¦tü ¦meth·əl'hek₁sān }

methyl hexanoate *See* methyl caproate. { 'meth·əl ₁hek'san·ə₁wāt }

methyl hexyl ketone [ORG CHEM] CH₃COC₆H₁₃ A combustible, colorless liquid with a boiling point of 173.5°C; soluble in alcohol, hydrocarbons, ether, and esters; used in perfumes and as a flavoring and odorant. { ¦meth·əl ¦hek·səl 'kē₁tōn }

methyl hydride *See* methane. { 'meth·əl 'hī₁drīd }

methyl hydroxystearate [ORG CHEM] C₁₉H₃₈O₃ A white, waxy material; slightly soluble in organic solvents, insoluble in water; used in cosmetics, inks, and adhesives. { 'meth·əl hī₁dräk·sē'stir₁āt }

methylidyne *See* methine group. { me'thil·ə₁dīn }

3-methylindole *See* skatole. { ¦thrē meth·ə'lin₁dōl }

methyl iodide [ORG CHEM] CH₃I Flammable colorless liquid that turns brown in light; boils at 42°C; soluble in ether and alcohol, insoluble in water; used as a chemical intermediate, in medicine, and in analytical chemistry. Also known as iodomethane. { 'meth·əl 'ī·ə₁dīd }

methyl isobutyl carbinol *See* methyl amyl alcohol. { 'meth·əl ₁ī·sō'byüd·əl 'kär·bə₁nȯl }

methyl isobutyl carbinol acetate *See* methyl amyl acetate. { 'meth·əl ₁ī·sō'byüd·əl 'kär·bə₁nȯl 'as·ə₁tāt }

methyl isobutyl ketone [ORG CHEM] (CH₃)₂CHCH₂COCH₃ Flammable colorless liquid with pleasant aroma; boils at 116°C, miscible with most organic solvents; used as

242

a solvent, extractant, and chemical intermediate. Also known as hexone. { 'meth·əl ˌī·sō'byüd·əl 'kē,tōn }

methylisothiocyanate [ORG CHEM] C_2H_3NS A crystalline compound, with a melting point of 35–36°C; soluble in alcohol and ether; used as a pesticide and in amino acid sequence analysis. Also known as methyl mustard oil. { ¦meth·əl¦ī·sō,thī· ə'sī·ə,nāt }

methyl lactate [ORG CHEM] $CH_3CHCHCOOCH_3$ Liquid boiling at 145°C; miscible with water and most organic liquids; used as a solvent for lacquers, stains, and cellulosic materials. { 'meth·əl 'lak,tāt }

methyl laurate [ORG CHEM] $CH_3(CH_2)_{10}COOCH_3$ Water-insoluble, clear, colorless liquid boiling at 262°C; used as a chemical intermediate to make rust removers, and for leather treatment. { 'meth·əl 'lȯ,rāt }

methyl linoleate [ORG CHEM] $C_{19}H_{34}O_2$ A combustible, colorless liquid with a boiling point of 212°C; soluble in alcohol and ether; used in the manufacture of detergents, emulsifiers, lubricants, and textiles, and in medical research. { 'meth·əl lə'nō·lē,āt }

methyl mercaptan [ORG CHEM] CH_3SH Colorless, toxic, flammable gas with unpleasant odor; boils at 6.2°C; insoluble in water, soluble in organic solvents; used as a chemical intermediate. Also known as methanethiol. { 'meth·əl mər'kap,tan }

methylmercury compound [ORG CHEM] Any member of a class of toxic compounds containing the methyl-mercury group, CH_3Hg. { ¦meth·əl'mər·kyə·rē ,käm,paúnd }

methylmercury cyanide See methylmercury nitrile. { ¦meth·əl'mər·kyə·rē 'sī·ə,nīd }

methylmercury nitrile [ORG CHEM] CH_3HgCN A crystalline solid with a melting point of 95°C; soluble in water; used as a fungicide to treat seeds of cereals, flax, and cotton. Also known as methylmercury cyanide. { ¦meth·əl'mər·kyə·rē 'nī·trəl }

methyl methacrylate [ORG CHEM] $CH_2C(CH_3)COOCH_3$ A flammable, colorless liquid, soluble in most organic solvents but insoluble in water; used as a monomer for polymethacrylate resins. { ¦meth·əl mə'thak·rə,lāt }

methyl mustard oil See methylisothiocyanate. { 'meth·əl 'məs·tərd ,ȯil }

methyl myristate [ORG CHEM] $CH_3(CH_2)_{12}COOCH_3$ A colorless liquid with a boiling point of 186.8°C; used in the manufacture of detergents, plasticizers, resins, textiles, and animal feeds, and as a flavoring. Also known as methyl tetradecanoate. { 'meth·əl mə'ri,stāt }

methylnaphthalene [ORG CHEM] $C_{10}H_7CH_3$ A solid melting at 34°C; used in insecticides and organic synthesis. { ,meth·əl'naf·thə,lēn }

methyl nitrate [ORG CHEM] CH_3NO_3 Explosive liquid boiling at 60°C; slightly soluble in water, soluble in ether and alcohol; used as a rocket propellant. { 'meth·əl 'nī,trāt }

methyl nonanoate [ORG CHEM] $CH_3(CH_2)_7COOCH_3$ A colorless liquid with a fruity odor and a boiling point of 213.5°C; soluble in alcohol and ether; used in perfumes and flavors, and for medical research. Also known as methyl pelargonate. { 'meth·əl nə'nan·ə,wāt }

methyl nonyl ketone [ORG CHEM] $CH_3COC_9H_{19}$ An oily liquid with a boiling point of 225°C; soluble in two parts of 70% alcohol; used in perfumes and flavoring. Also known as 2-undecanone. { ¦meth·əl ¦nō·nəl 'kē,tōn }

methyl oleate [ORG CHEM] $C_{17}H_{33}COOCH_3$ Amber liquid with faint fatty odor; soluble in organic liquids, mineral spirits, and vegetable oil, insoluble in water; used as a plasticizer and softener. { 'meth·əl 'ōl·ē,āt }

methylol riboflavin [ORG CHEM] An orange to yellow powder, soluble in water; used as a nutrient and in medicine. { 'meth·ə,lȯl 'rī·bə,flā·vən }

methylol urea [ORG CHEM] $H_2NCONHCH_2OH$ Water-soluble, colorless crystals melting at 111°C; used to treat textiles and wood, and in the manufacture of resins and adhesives. { 'meth·ə,lȯl yü'rē·ə }

methyl palmitate [ORG CHEM] $CH_3(CH_2)_{14}COOCH_3$ A colorless liquid with a boiling point of 211.5°C; soluble in alcohol and ether; used in the manufacture of detergents, resins, plasticizers, lubricants, and animal feed. { 'meth·əl 'pal·mə,tāt }

methyl pelargonate See methyl nonanoate. { 'meth·əl pə'lär·gə,nāt }

3-methylpentane [ORG CHEM] C_6H_{14} Flammable, colorless liquid; insoluble in water,

2-methylpentanoic acid

soluble in alcohol; boils at 64°C; used as a chemical intermediate. { ¦thrē ¦meth· əl'pen,tān }

2-methylpentanoic acid [ORG CHEM] $(CH_3)_2CH(CH_2)_2COOH$ A colorless liquid with a boiling point of 197°C; soluble in alcohol, benzene, and acetone; used for plasticizers, vinyl stabilizers, and metallic salts. { ¦tü ¦meth·əl,pen·tə¦nō·ik 'as·əd }

methylpentene polymer [ORG CHEM] Thermoplastic material based on 4-methylpentene-1; has low gravity, excellent electrical properties, and 90% optical transmission. { ¦meth·əl'pen,tēn 'päl·ə·mər }

methyl pentose [ORG CHEM] **1.** Any compound that is a methyl derivative of a five carbon sugar. **2.** In particular, the compound $CH_3(CHOH)_4CHO$. { 'meth·əl 'pen,tōs }

methyl phenyl acetate [ORG CHEM] $C_6H_5CH_2COOCH_3$ A colorless liquid with honey odor; used to flavor tobacco and in perfumery. { ¦meth·əl ¦fen·əl 'as·ə,tāt }

methylphosphoric acid [ORG CHEM] $CH_3H_2PO_4$ A straw-colored liquid used for textile- and paper-processing compounds, as a rust remover, and in soldering flux. { ,meth· əl,fäs'för·ik 'as·əd }

methyl propionate [ORG CHEM] $CH_3CH_2COOCH_3$ A flammable, colorless liquid with a boiling range of 78.0–79.5°C; soluble in most organic solvents; used as a solvent for cellulose nitrate, in lacquers, varnishes, and paints, and for flavoring. { 'meth· əl 'prō·pē·ə,nāt }

methyl propyl carbinol [ORG CHEM] $CH_3CHOHC_3H_7$ Colorless liquid boiling at 119°C; miscible with ether and alcohol, slightly soluble in water; used as a pharmaceuticals intermediate and as a paint and lacquer solvent. Also known as *sec-n-*amyl alcohol; 2-pentanol. { ¦meth·əl ¦prō·pəl 'kär·bə,nól }

N-methyl-2-pyrrolidone [ORG CHEM] C_5H_9NO A liquid boiling at 202°C; miscible with water, castor oil, and organic solvents; used as a chemical intermediate and as a solvent for petroleum and resins; and in PVC spinning. { ¦en 'meth·əl ¦tü pə'räl· ə,dōn }

methyl red [ORG CHEM] $(CH_3)_2NC_6H_4NNC_6H_4COOH$ A dark red powder or violet crystals with a melting point of 180°C; soluble in alcohol, ether, and glacial acetic acid; used as an acid-base indicator (pH 4.2–6.2). { 'meth·əl 'red }

methyl ricinoleate [ORG CHEM] $C_{19}H_{36}O_3$ Clear, low-viscosity fluid used as a wetting agent, cutting oil additive, lubricant, and plasticizer. { 'meth·əl ,ris·ən'ōl·ē,āt }

methyl salicylate [ORG CHEM] $C_6H_4OHCOOCH_3$ A colorless, yellow, or reddish liquid, slightly soluble in water, boiling at 222.2°C, with an odor of wintergreen; used in medicine and perfumery, and as a solvent for cellulose derivatives. Also known as betula oil; gaultheria oil; wintergreen oil. { 'meth·əl sə'lis·ə,lāt }

3-methylsalicylic acid [ORG CHEM] $C_8H_8O_3$ A white to reddish, crystalline compound with a melting point of 165–166°C; soluble in chloroform, alcohol, ether, and alkali hydroxides; used to make dyes. { ¦thrē ¦meth·əl¦sal·ə¦sil·ik 'as·əd }

methyl silicone [ORG CHEM] $[(CH_3)_2SiO]_x$, $[C(CH_3)_2Si_2O_3]_y$, etc. The common varieties of silicones with properties of oil, resin, or rubber, depending on molecular size and arrangement. { 'meth·əl 'sil·ə,kōn }

methyl stearate [ORG CHEM] $C_{17}H_{35}COOCH_3$ Colorless crystals melting at 39°C; soluble in alcohol and ether, insoluble in water; used as an intermediate for stearic acid manufacture. { 'meth·əl 'stir,āt }

methyl styrene See vinyltoluene. { 'meth·əl 'stī,rēn }

α-methyl styrene [ORG CHEM] $C_6H_5C(CH_3):CH_2$ Colorless, toxic, polymerizable liquid boiling at 165°C; used to produce polystyrene resins. { ¦al·fə ¦meth·əl 'stī,rēn }

methyl sulfate See dimethyl sulfate. { 'meth·əl 'səl,fāt }

methyl sulfide [ORG CHEM] $(CH_3)_2S$ Flammable, colorless liquid with disagreeable aroma; soluble in ether and alcohol, insoluble in water; boils at 38°C; used as a chemical intermediate. Also known as dimethyl sulfide. { 'meth·əl 'səl,fīd }

methylsulfonic acid See methanesulfonic acid. { ¦meth·əl·səl¦fän·ik 'as·əd }

methylsulfonyl [ORG CHEM] A functional group with the formula CH_3SO_2-. Also known as mesyl. { ¦meth·əl 'səl·fə,nil }

methyl tertiary butyl ether [ORG CHEM] $CH_3OC(CH_3)_3$ A volatile, flammable, colorless

244

liquid, with a boiling point of 55°C (131°F) and a terpene-like odor, originally used in gasoline as an octane enhancer and lead substitute, more recently used to reduce engine exhaust emissions. Abbreviated MTBE. { ¦meth·əl ¦tər·shē,er·ē ¦byüd·əl 'ē·thər }

methyl tetradecanoate *See* methyl myristate. { 'meth·əl ,te·trə·də'kan·ə,wāt }

4-methyl-5-thiazole ethanol [ORG CHEM] C_6H_9NOS A viscous, oily liquid; soluble in alcohol, ether, benzene, chloroform, and water; used as an intermediate in the synthesis of vitamin B_1 and as a sedative and hypnotic. { ¦fór 'meth·əl ¦fīv 'thī·ə· zōl 'eth·ə,nòl }

methylthionine chloride *See* methylene blue. { ¦meth·əl'thī·ə,nēn 'klòr,īd }

methyltrichlorosilane [ORG CHEM] CH_3SiCl_3 A colorless liquid with a pungent odor, boiling point of 66°C (150.8°F), and melting point of −90°C (−130°F). Also known as trichloromethylsilane. { ¦meth·əl ,trī,klòr·ō'sī,lān }

α-methyl-*para*-tyrosine [ORG CHEM] $C_{10}H_{13}NO_3$ A crystalline compound which acts as the inhibitor of the first and rate-limiting reaction in the biosynthesis of catecholamine; used as an inhibitor of tyrosine hydroxylase. { ¦al·fə 'meth·əl ¦par·ə 'tī·rə,sēn }

methyl violet [ORG CHEM] A derivative of pararosaniline, used as an antiallergen and bactericide, acid-base indicator, biological stain, and textile dye. Also known as crystal violet; gentian violet. { 'meth·əl 'vī·lət }

mevalonic acid [ORG CHEM] $HO_2C_5H_9COOH$ A dihydroxy acid used in organic synthesis. { ¦mev·ə¦lan·ik 'as·əd }

mexacarbate [ORG CHEM] $C_{12}H_{18}N_2O_2$ A tan solid with a melting point of 85°C; used to control insect pests of trees, flowers, and shrubs. { ¦mek·sə'kär,bāt }

Mg *See* magnesium.

MIBC *See* methyl amyl alcohol.

micellar catalysis [CHEM] Enhancement of the rate of a chemical reaction in solution by the addition of a surfactant, so that the reaction proceeds in the environment of surfactant aggregates. { mī¦sel·ər kə'tal·ə·səs }

micelle [PHYS CHEM] A colloidal aggregate of a unique number (between 50 and 100) of amphipathic molecules, which occurs at a well-defined concentration known as the critical micelle concentration. { mī'sel }

Michler's ketone *See* tetramethyldiaminobenzophenone. { 'mik·lərz 'kē,tōn }

micril *See* gammil. { 'mī·krəl }

microanalysis [ANALY CHEM] Identification and chemical analysis of material on a small scale so that specialized instruments such as the microscope are needed; the material analyzed may be on the scale of 1 microgram. { ¦mī·krō·ə'nal·ə·səs }

microchemistry [CHEM] The study of chemical reactions, using small quantities of materials, frequently less than 1 milligram or 1 milliliter, and often requiring special small apparatus and microscopical observation. { ¦mī·krō¦kem·ə·strē }

microdensitometer [SPECT] A high-sensitivity densitometer used in spectroscopy to detect spectrum lines too faint on a negative to be seen by the human eye. { ¦mī· krō,den·sə'täm·əd·ər }

microdialysis [ANALY CHEM] A technique for sampling biological systems in which a short length of hollow-fiber dialysis membrane is implanted into any tissue or fluid compartment, and through which compounds in the extracellular fluid are collected for subsequent analysis. { ,mī·krō·dī'al·ə·səs }

microelectrolysis [PHYS CHEM] Electrolysis of small quantities of material. { ¦mī·krō· i,lek'träl·ə·səs }

microelectrophoresis [ANALY CHEM] Direct microscopic observation and measurement of the velocity of migration of ions or other charged bodies through a solution toward oppositely charged electrodes. Also known as optical cytopherometry. { ¦mī·krō· i,lek·trə·fə'rē·səs }

microgammil *See* gammil. { ¦mī·krō'gam·əl }

microheterogeneity [CHEM] A small variation in the chemical structure of a molecule that does not result in a significant change in properties. { ,mī·krō,hed·ə·rə·jə'nē· əd·ē }

microincineration [CHEM] Reduction of small quantities of organic substances to ash by application of heat. { ¦mī·krō·in,sin·ə'rā·shən }

micropore [CHEM] A pore in a catalytic material whose diameter is less than 2 nanometers. { 'mī·krə,pȯr }

microprobe [SPECT] An instrument for chemical microanalysis of a sample, in which a beam of electrons is focused on an area less than a micrometer in diameter, and the characteristic x-rays emitted as a result are dispersed and analyzed in a crystal spectrometer to provide a qualitative and quantitative evaluation of chemical composition. Also known as x-ray microprobe. { 'mī·krə,prōb }

microprobe spectrometry [SPECT] Microanalysis of a sample, using a microprobe. { 'mī·krə,prōb spek'träm·ə·trē }

microradiography [ANALY CHEM] Technique for the study of surfaces of solids by monochromatic-radiation (such as x-ray) contrast effects shown via projection or enlargement of a contact radiograph. { ¦mī·krō,rād·ē'äg·rə·fē }

microspectrograph [SPECT] A microspectroscope provided with a photographic camera or other device for recording the spectrum. { ¦mī·krō'spek·trə,graf }

microspectrophotometer [SPECT] A split-beam or double-beam spectrophotometer including a microscope for the localization of the object under study, and capable of carrying out spectral analyses within the dimensions of a single cell. { ¦mī·kro¦spek·trə·fə'täm·əd·ər }

microspectroscope [SPECT] An instrument for analyzing the spectra of microscopic objects, such as living cells, in which light passing through the sample is focused by a compound microscope system, and both this light and the light which has passed through a reference sample are dispersed by a prism spectroscope, so that the spectra of both can be viewed simultaneously. { ¦mī·krō'spek·trə,skōp }

microthrowing power [PHYS CHEM] Relative ability of an electroplating solution to deposit metal in a small, shallow aperture or crevice not exceeding a few thousandths of an inch in dimensions. { ¦mī·krō'thrō·iŋ ,pau̇·ər }

microwave spectrometer [SPECT] An instrument which makes a graphical record of the intensity of microwave radiation emitted or absorbed by a substance as a function of frequency, wavelength, or some related variable. { 'mī·krə,wāv spek 'träm·əd·ər }

microwave spectroscope [SPECT] An instrument used to observe the intensity of microwave radiation emitted or absorbed by a substance as a function of frequency, wavelength, or some related variable. { 'mī·krə,wāv 'spek·trə,skōp }

microwave spectroscopy [SPECT] The methods and techniques of observing and the theory for interpreting the selective absorption and emission of microwaves at various frequencies by solids, liquids, and gases. { 'mī·krə,wāv spek'träs·kə·pē }

microwave spectrum [SPECT] A display, photograph, or plot of the intensity of microwave radiation emitted or absorbed by a substance as a function of frequency, wavelength, or some related variable. { 'mī·krə,wāv 'spek·trəm }

migration [CHEM] The movement of an atom or group of atoms to new positions during the course of a molecular rearrangement. { mī'grā·shən }

migration current [PHYS CHEM] Additional current produced by electrostatic attraction of cations to the surface of a dropping electrode; an unpredictable and undesirable effect to be avoided during analytical voltammetry. { mī'grā·shən ,kə·rənt }

MIKES See mass-analyzed ion kinetic energy spectrometry. { mīks }

mild mercury chloride See mercurous chloride. { 'mīld 'mər·kyə·rē 'klȯr,īd }

milk [CHEM] A suspension of certain metallic oxides, as milk of magnesia, iron, or bismuth. { milk }

milliequivalent [CHEM] One-thousandth of a compound's or an element's equivalent weight. Abbreviated meq. { ¦mil·ē·ə¦kwiv·ə·lənt }

Millon's reagent [CHEM] Reagent used to test for proteins; made by dissolving mercury in nitric acid, diluting, then decanting the liquid from the precipitate. { mē'lȯnz re,ä·jənt }

mimosine [ORG CHEM] $C_8H_{10}N_2O_4$ A crystalline compound with a melting point of 235–236°C; soluble in dilute acids or bases; used as a depilatory agent. Also known as leucaenine; leucaenol; leucenine; leucenol. { mə'mō,sēn }

mineral acid [INORG CHEM] Any one of the major inorganic acids, such as sulfuric, nitric, or hydrochloric acids. { 'min·rəl ,as·əd }

mineral green See copper carbonate. { 'min·rəl ¦grēn }

mineralize [CHEM] To convert organic compounds to simpler inorganic compounds, namely, carbon dioxide and water (and halogen acids, if the organic substances are halogenated). { 'min·rə,līz }

mineralogy [INORG CHEM] The science which concerns the study of natural inorganic substances called minerals. { ,min·ə'räl·ə·jē }

MIPC See ortho-isopropylphenyl-methylcarbamate.

miscibility [CHEM] The tendency or capacity of two or more liquids to form a uniform blend, that is, to dissolve in each other; degrees are total miscibility, partial miscibility, and immiscibility. { ,mis·ə'bil·əd·ē }

misfire [CHEM] Failure of fuel or an explosive charge to ignite properly. { 'mis,fīr }

Mitscherlich law of isomorphism [CHEM] Substances which have similar chemical properties and crystalline forms usually have similar chemical formulas. { 'mich· ər,lik ¦lȯ əv ¦ī·sō'mȯr,fiz·əm }

mixed acid See nitrating acid. { 'mikst 'as·əd }

mixed aniline point [PHYS CHEM] The minimum temperature at which a mixture of aniline, heptane, and hydrocarbon will form a solution; related to the aromatic character of the hydrocarbon. { 'mikst 'an·ə·lən ,pȯint }

mixed indicator [ANALY CHEM] Color-change indicator for acid-base titration end points in which a mixture of two indicator substances is used to give sharper end-point color changes. { 'mikst 'in·də,kād·ər }

mixed potential [PHYS CHEM] The electrode potential of a material while more than one electrochemical reaction is occurring simultaneously. { 'mikst pə'ten·chəl }

Mn See manganese.

Mo See molybdenum.

mobile electron [PHYS CHEM] An electron that can move readily from one atom to another within a chemical structure in response to changes in the external chemical environment. { ,mō·bəl ə'lek,trän }

mobile phase [ANALY CHEM] **1.** In liquid chromatography, the phase that is moving in the bed, including the fraction of the sample held by this phase. **2.** The carrier gas in a gas chromatography procedure. { 'mō·bəl ,fāz }

mobility coefficient [PHYS CHEM] The average speed of motion of molecules in a solution in the direction of the concentration gradient, at unit concentration and unit osmotic pressure gradient. { mō'bil·əd·ē ,kō·ə,fish·ənt }

modified Lewis acid [PHYS CHEM] An acid that is a halide ion acceptor. { 'mäd·ə,fīd 'lü·əs ¦as·əd }

modulated Raman scattering [SPECT] Application of modulation spectroscopy to the study of Raman scattering; in particular, use of external perturbations to lower the symmetry of certain crystals and permit symmetry-forbidden modes, and the use of wavelength modulation to analyze second-order Raman spectra. { 'mäj·ə,lād·əd 'rä·mən ,skad·ə·riŋ }

modulation spectroscopy [SPECT] A branch of spectroscopy concerned with the measurement and interpretation of changes in transmission or reflection spectra induced (usually) by externally applied perturbation, such as temperature or pressure change, or an electric or magnetic field. { ,mäj·ə'lā·shən spek'träs·kə·pē }

Mohr's salt See ferrous ammonium sulfate. { 'mȯrz ,sȯlt }

Mohr titration [ANALY CHEM] Titration with silver nitrate to determine the concentration of chlorides in a solution; silver chromate precipitation is the end-point indicator. { 'mȯr tī'trā·shən }

moiety [CHEM] A part or portion of a molecule, generally complex, having a characteristic chemical or pharmacological property. { 'mȯi·əd·ē }

moisture [PHYS CHEM] Water that is dispersed through a gas in the form of water vapor or small droplets, dispersed through a solid, or condensed on the surface of a solid. { 'mȯis·chər }

mol See mole. { mōl }

molal average boiling point [PHYS CHEM] A pseudo boiling point for a mixture calculated as the summation of individual mole fraction-boiling point (in degrees Rankine) products. { 'mō·ləl 'av·rij 'bóil·iŋ ,póint }

molal elevation of the boiling point See ebullioscopic constant. { ¦mō·ləl ,el·ə¦vā·shən əv thə 'bóil·iŋ ,póint }

molal heat capacity See molar heat capacity. { 'mō·ləl 'hēt kə,pas·əd·ē }

molality [CHEM] Concentration given as moles per 1000 grams of solvent. { mō'lal·əd·ē }

molal quantity [CHEM] The number of moles (gram-molecular weights) present, expressed with weight in pounds, grams, or such units, numerically equal to the molecular weight; for example, pound-mole, gram-mole. { 'mō·ləl 'kwän·əd·ē }

molal solution [CHEM] Concentration of a solution expressed in moles of solute divided by 1000 grams of solvent. { 'mō·ləl sə,lü·shən }

molal specific heat See molar specific heat. { 'mō·ləl spə¦sif·ik 'hēt }

molal volume See molar volume. { 'mō·ləl 'väl·yəm }

molar [PHYS CHEM] Denoting a physical quantity divided by the amount of substance expressed in moles. { 'mō·lər }

molar conductivity [PHYS CHEM] The ratio of the conductivity of an electrolytic solution to the concentration of electrolyte in moles per unit volume. { 'mō·lər ,kän,dək'tiv·əd·ē }

molar heat capacity [PHYS CHEM] The amount of heat required to raise 1 mole of a substance 1° in temperature. Also known as molal heat capacity; molecular heat capacity. { 'mō·lər 'hēt kə,pas·əd·ē }

molarity [CHEM] Measure of the number of gram-molecular weights of a compound present (dissolved) in 1 liter of solution; it is indicated by M, preceded by a number to show solute concentration. { mō'lar·əd·ē }

molar solution [CHEM] Aqueous solution that contains 1 mole (gram-molecular weight) of solute in 1 liter of the solution. { 'mō·lər sə,lü·shən }

molar specific heat [PHYS CHEM] The ratio of the amount of heat required to raise the temperature of 1 mole of a compound 1°, to the amount of heat required to raise the temperature of 1 mole of a reference substance, such as water, 1° at a specified temperature. Also known as molal specific heat; molecular specific heat. { 'mō·lər spə,sif·ik ,hēt }

molar susceptibility [PHYS CHEM] Magnetic susceptibility of a compound per gram-mole of that compound. { 'mō·lər sə,sep·tə'bil·əd·ē }

molar volume [PHYS CHEM] The volume occupied by one mole of a substance in the form of a solid, liquid, or gas. Also known as molal volume; mole volume. { 'mō·lər 'väl·yəm }

mole [CHEM] An amount of substance of a system which contains as many elementary units as there are atoms of carbon in 0.012 kilogram of the pure nuclide carbon-12; the elementary unit must be specified and may be an atom, molecule, ion, electron, photon, or even a specified group of such units. Symbolized mol. { mōl }

molecular adhesion [PHYS CHEM] A particular manifestation of intermolecular forces which causes solids or liquids to adhere to each other; usually used with reference to adhesion of two different materials, in contrast to cohesion. { mə'lek·yə·lər ad'hē·zhən }

molecular amplitude [ANALY CHEM] The difference between the molecular rotation at the extreme (peak or trough) value caused by the longer light wavelength and the molecular rotation at the extreme value caused by the shorter wavelength. { mə'lek·yə·lər 'am·plə,tüd }

molecular association [PHYS CHEM] The formation of double molecules or polymolecules from a single species as a result of specific and moderately strong intermolecular forces. { mə'lek·yə·lər ə,sō·sē'ā·shən }

molecular asymmetry See asymmetry. { mə'lek·yə·lər ,ā'sim·ə·trē }

molecular attraction [PHYS CHEM] A force which pulls molecules toward each other. { mə'lek·yə·lər ə'trak·shən }

molecular cluster [PHYS CHEM] An assembly of molecules that are weakly bound together and display properties intermediate between those of isolated gas-phase molecules and bulk condensed media. { mə'lek·yə·lər 'kläs·tər }

molecular conductivity [PHYS CHEM] The conductivity of a volume of electrolyte containing 1 mole of dissolved substance. { mə'lek·yə·lər ˌkän,dək'tiv·əd·ē }

molecular device [CHEM] An assemblage of a discrete number of molecular components (that is, a supramolecular structure) designed to achieve a specific function. { mə'lek·yə·lər di'vīs }

molecular diamagnetism [PHYS CHEM] Diamagnetism of compounds, especially organic compounds whose susceptibilities can often be calculated from the atoms and chemical bonds of which they are composed. { mə'lek·yə·lər ˌdī·ə'mag·nə,tiz·əm }

molecular diameter [PHYS CHEM] The diameter of a molecule, assuming it to be spherical; has a numerical value of 10^{-8} centimeter multiplied by a factor dependent on the compound or element. { mə'lek·yə·lər dī'am·əd·ər }

molecular dipole [PHYS CHEM] A molecule having an electric dipole moment, whether it is permanent or produced by an external field. { mə'lek·yə·lər 'dī,pōl }

molecular distillation [CHEM] A process by which substances are distilled in high vacuum at the lowest possible temperature and with least damage to their composition. { mə'lek·yə·lər ˌdis·tə'lā·shən }

molecular dynamics [PHYS CHEM] A branch of physical chemistry concerned with molecular mechanisms of the elementary physical and chemical processes that control rates of reaction. { mə'lek·yə·lər di'nam·iks }

molecular energy level [PHYS CHEM] One of the states of motion of nuclei and electrons in a molecule, having a definite energy, which is allowed by quantum mechanics. { mə'lek·yə·lər 'en·ər·jē ˌlev·əl }

molecular entity [CHEM] A chemically or isotopically distinct atom, molecule, ion, complex, free radical, or similar unit that can be distinguished from other kinds of units. { mə'lek·yə·lər 'en·təd·ē }

molecular exclusion chromatography See gel filtration. { mə'lek·yə·lər ik'sklü·zhən ˌkrō·mə'täg·rə·fē }

molecular formula [CHEM] A chemical formula that indicates the actual numbers and kinds of atoms in a molecule, but not the chemical structure. { mə'lek·yə·lər 'fȯr·myə·lə }

molecular gas [CHEM] A gas composed of a single species, such as oxygen, chlorine, or neon. { mə'lek·yə·lər 'gas }

molecular graphics [PHYS CHEM] The use of computer graphics to display and manipulate chemical structures with sufficient accuracy that bond distances and angles may be displayed and reported and it is possible to dock or fit together two or more molecules. Also known as graphics-based molecular modeling. { mə'lek·yə·lər 'graf·iks }

molecular heat capacity See molar heat capacity. { mə'lek·yə·lər 'hēt kə,pas·əd·ē }

molecular imprinting [PHYS CHEM] A technique for creating receptor structures on a polymer surface that can selectively bind to molecules of interest, molecularly imprinted polymers are used for separations, as catalysts, and in biosensors. { mə'lek·yə·lər 'im,print·iŋ }

molecular ion [ORG CHEM] An ion that results from the loss of an electron by an organic molecule following bombardment with high-energy electrons during mass spectrometry. { mə'lek·yə·lər 'ī,än }

molecularity [PHYS CHEM] In a chemical reaction, the number of molecules which come together and form the activated complex. { mə,lek·yə'lar·əd·ē }

molecular machine [CHEM] A molecular device in which the component parts can display changes (reversible movement) in their relative positions as a result of some external stimulus (such as light, electrical energy, or chemical energy), resulting in a signal (a change in a chemical or physical property of the supramolecular system) that can be used to monitor the operation of the device. { mə'lek·yə·lər mə'shēn }

molecular magnet [PHYS CHEM] A molecule having a nonvanishing magnetic dipole

molecular mechanics

moment, whether it is permanent or produced by an external field. { mə'lek·yə·lər 'mag·nət }

molecular mechanics [PHYS CHEM] An empirical method of calculating the dynamics of molecules, in which bonds between atoms are represented by springs obeying Hooke's law, and additional terms representing bond angle bending, torsional interactions, and van der Waals-type interactions are included. Also known as force-field method. { mə'lek·yə·lər mi'kan·iks }

molecular modeling [CHEM] The use of computers for the simulation of chemical entities and processes. { mə'lek·yə·lər 'mäd·liŋ }

molecular orbital [PHYS CHEM] A wave function describing an electron in a molecule. { mə'lek·yə·lər 'òr·bəd·əl }

molecular paramagnetism [PHYS CHEM] Paramagnetism of molecules, such as oxygen, some other molecules, and a large number of organic compounds. { mə'lek·yə·lər ¦par·ə'mag·nə,tiz·əm }

molecular polarizability [PHYS CHEM] The electric dipole moment induced in a molecule by an external electric field, divided by the magnitude of the field. { mə'lek· yə·lər ,pō·lə,rīz·ə'bil·əd·ē }

molecular rearrangement See rearrangement reaction. { mə'lek·yə·lər ,rē·ə'rānj·mənt }

molecular receptor [ORG CHEM] A species that can select one of many possible binding partners and form a complex that is stabilized by interactions such as hydrogen bonding or changes in solvation. { mə'lek·yər·lər ri'sep·tər }

molecular recognition [CHEM] The (molecular) storage and the (supramolecular) retrieval and processing of molecular structural information and interactions. { mə¦lek·yə·lər ,rek·ig'nish·ən }

molecular relaxation [PHYS CHEM] Transition of a molecule from an excited energy level to another excited level of lower energy or to the ground state. { mə'lek·yə· lər ,rē,lak'sā·shən }

molecular self-assembly [ORG CHEM] The spontaneous aggregation of molecules into well-defined, stable, noncovalently bonded assemblies that are held together by intermolecular forces. { mə'lek·yə·lər ,self ə'sem·blē }

molecular sieve [CHEM] A naturally occurring or synthetic zeolite characterized by the ability to undergo dehydration with little or no change in crystal structure, thereby offering a very high surface area for adsorption of foreign molecules. { mə'lek·yə· lər 'siv }

molecular-sieve chromatography See gel filtration. { mə'lek·yə·lər ¦siv ,krō·mə'täg· rə·fē }

molecular simulation [CHEM] Computational techniques for predicting many useful functional properties of chemicals and materials, including thermodynamic properties, thermochemical properties, spectroscopic properties, mechanical properties, transport properties, and morphological information. { mə¦lek·yə·lər ,sim·yə'lā· shən }

molecular specific heat See molar specific heat. { mə'lek·yə·lər spə¦sif·ik 'hēt }

molecular spectroscopy [SPECT] The production, measurement, and interpretation of molecular spectra. { mə'lek·yə·lər spek'träs·kə·pē }

molecular spectrum [SPECT] The intensity of electromagnetic radiation emitted or absorbed by a collection of molecules as a function of frequency, wave number, or some related quantity. { mə'lek·yə·lər 'spek·trəm }

molecular still [CHEM] An apparatus used to conduct molecular distillation. { mə'lek· yə·lər 'stil }

molecular structure [PHYS CHEM] The manner in which electrons and nuclei interact to form a molecule, as elucidated by quantum mechanics and a study of molecular spectra. { mə'lek·yə·lər 'strək·chər }

molecular velocity [PHYS CHEM] The velocity of an individual molecule in a given sample of gas; the vector quantity is symbolized **u**, and the magnitude is symbolized *u*. { mə'lek·yə·lər və'läs·əd·ē }

molecular vibration [PHYS CHEM] The theory that all atoms within a molecule are in continuous motion, vibrating at definite frequencies specific to the molecular structure as a whole as well as to groups of atoms within the molecule; the basis of spectroscopic analysis. { mə'lek·yə·lər vī'brä·shən }

molecular volume [CHEM] The volume that is occupied by 1 mole (gram-molecular weight) of an element or compound; equals the molecular weight divided by the density. { mə'lek·yə·lər 'väl·yəm }

molecular weight [CHEM] The sum of the atomic weights of all the atoms in a molecule. Also known as relative molecular mass. { mə'lek·yə·lər 'wāt }

molecular-weight distribution [ORG CHEM] Frequency of occurrence of the different molecular-weight chains in a homologous polymeric system. { mə'lek·yə·lər ¦wāt ‚di·strə'byü·shən }

molecule [CHEM] A group of atoms held together by chemical forces; the atoms in the molecule may be identical as in H_2, S_2, and S_8, or different as in H_2O and CO_2; a molecule is the smallest unit of matter which can exist by itself and retain all its chemical properties. { 'mäl·ə‚kyül }

mole fraction [CHEM] The ratio of the number of moles of a substance in a mixture or solution to the total number of moles of all the components in the mixture or solution. { 'mōl ‚frak·shən }

mole percent [CHEM] Percentage calculation expressed in terms of moles rather than weight. { 'mōl pər‚sent }

mole volume See molar volume. { 'mōl 'väl·yəm }

molinate [ORG CHEM] $C_9H_{17}NOS$ A light yellow liquid with limited solubility in water; used as a herbicide to control watergrass in rice. { 'mäl·ə‚nāt }

molybdate [INORG CHEM] A salt derived from a molybdic acid. { mə'lib‚dät }

molybdenum [CHEM] A chemical element, symbol Mo, atomic number 42, and atomic weight 95.94. { mə'lib·de·nəm }

molybdenum dioxide [INORG CHEM] MoO_2 Lead-gray powder; insoluble in hydrochloric and hydrofluoric acids; used in pigment for textiles. { mə'lib·də·nəm dī'äk‚sīd }

molybdenum disilicide [INORG CHEM] $MoSi_2$ A dark gray, crystalline powder with a melting range of 1870–2030°C; soluble in hydrofluoric and nitric acids; used in electrical resistors and for protective coatings for high-temperature conditions. { mə'lib·də·nəm dī'sil·ə‚sīd }

molybdenum disulfide [INORG CHEM] MoS_2 A black lustrous powder, melting at 1185°C, insoluble in water, soluble in aqua regia and concentrated sulfuric acid; used as a dry lubricant and an additive for greases and oils. Also known as molybdenum sulfide; molybdic sulfide. { mə'lib·də·nəm dī'səl‚fīd }

molybdenum pentachloride [INORG CHEM] $MoCl_5$ Hygroscopic gray-black needles melting at 194°C; reacts with water and air; soluble in anhydrous organic solvents; used as a catalyst and as raw material to make molybdenum hexacarbonyl. { mə'lib·də· nəm ‚pen·tə'klór‚īd }

molybdenum sesquioxide [INORG CHEM] MoO_3 Water-insoluble, gray-black powder with slight solubility in acids; used as a catalyst and as a coating for metal articles. { mə'lib·də·nəm ‚ses·kwē'äk‚sīd }

molybdenum sulfide See molybdenum disulfide. { mə'lib·də·nəm 'səl‚fīd }

molybdenum trioxide [INORG CHEM] MoO_3 A white solid at room temperature, with a melting point of 795°C; soluble in concentrated mixtures of nitric and sulfuric acids and nitric and hydrochloric acids; used as a corrosion inhibitor, in enamels and ceramic glazes, in medicine and agriculture, and as a catalyst in the petroleum industry. { mə'lib·də·nəm trī'äk‚sīd }

molybdic acid [INORG CHEM] Any acid derived from molybdenum trioxide, especially the simplest acid H_2MoO_4, obtained as white crystals. { mə'lib·dik 'as·əd }

molybdic sulfide See molybdenum disulfide. { mə'lib·dik 'səl‚fīd }

monatomic [CHEM] Composed of one atom. { ¦män·ə'täm·ik }

monatomic gas [CHEM] A gas whose molecules have only one atom; the inert gases are examples. { ¦män·ə¦täm·ik 'gas }

mono- [CHEM] A prefix for chemical compounds to show a single radical; for example, monoglyceride, a glycol ester on which a single acid group is attached to the glycerol group. { 'män·ō }

monoacetate [ORG CHEM] A compound such as a salt or ester that contains one acetate group. { ¦män·ō'as·ə,tāt }

monoacid [CHEM] **1.** An acid that has only one replaceable hydrogen. **2.** A base or an alcohol that has a single hydroxyl (−OH) group which can be replaced by an atom or a functional group to form a salt or ester. { ¦män·ō'as·əd }

monoamine [ORG CHEM] An amine compound that has only one amino group. { ¦män·ō'am,ēn }

monoammonium tartrate See ammonium bitartrate. { ¦män·ō·ə'mō·nē·əm 'tär,trāt }

monobasic [CHEM] Pertaining to an acid with one displaceable hydrogen atom, such as hydrochloric acid, HCl. { ¦män·ō'bās·ik }

monobasic calcium phosphate See calcium phosphate. { män·ō'bās·ik 'kal·sē·əm 'fäs,fāt }

monobasic sodium phosphate [INORG CHEM] NaH_2PO_4 White crystals that are slightly hygroscopic, soluble in water, insoluble in alcohol; used in baking powders and acid cleansers, and as a cattle-food supplement. { män·ō'bās·ik 'sōd·ē·əm 'fäs,fāt }

monocalcium phosphate See calcium phosphate. { ¦män·ō'kal·sē·əm 'fäs,fāt }

monochromator [SPECT] A spectrograph in which a detector is replaced by a second slit, placed in the focal plane, to isolate a particular narrow band of wavelengths for refocusing on a detector or experimental object. { ¦män·ə¦krō,mād·ər }

monodisperse colloidal system [CHEM] A colloidal system in which the suspended particles have identical size, shape, and interaction. { ¦män·ō·di¦spərs kə'lòid·əl 'sis·təm }

monodispersity [ORG CHEM] Polymer system that is homogeneous in molecular weight, that is, it does not have a distribution of different molecular-weight chains within the total mass. { ¦män·ō·di'spər·səd·ē }

monoester [ORG CHEM] An ester that has only one ester group. { ¦män·ō'es·tər }

monofunctional compound [ORG CHEM] An organic compound whose chemical structure possesses a single highly reactive site. { ,män·ō¦fəŋk·shən·əl 'käm,paùnd }

monoglyceride [ORG CHEM] Any of the fatty-acid glycerol esters where only one acid group is attached to the glycerol group, for example, $RCOOCH_2CHOHCH_2OH$; examples are glycerol monostearate and monolaurate; used as emulsifiers in cosmetics and lubricants. { ¦män·ō'glis·ə,rīd }

monolayer See monomolecular film. { 'män·ō,lā·ər }

monolayer capacity [CHEM] **1.** In chemisorption, the amount of adsorbate required to occupy all adsorption sites on the solid surface. **2.** In physisorption, the amount of material required to cover the solid surface with a complete monolayer of the adsorbate in a close-packed array. { 'män·ə,lā·ər kə,pas·əd·ē }

monomer [ORG CHEM] A molecule which is capable of combining with like or unlike molecules to form a polymer; it is a repeating structure unit within a polymer. Also known as repeating unit. { 'män·ə·mər }

monomethylhydrazine [CHEM] $CH_3N_2H_3$ A volatile toxic liquid that will react with carbon dioxide and oxygen. { ,mä·nō,meth·əl 'hī·drə,zēn }

monomeric unit See repeating unit. { ,män·ə¦mer·ik 'yü·nət }

monomolecular film [PHYS CHEM] A film one molecule thick. Also known as monolayer. { ¦män·ō·mə¦lek·yə·lər 'film }

monopotassium ʟ-glutamate See potassium glutamate. { ¦män·ō·pə'tas·ē·əm ¦el'glüd·ə,māt }

monoprotic acid [CHEM] An acid that has only one ionizable hydrogen atom in each molecule. { ,män·ə¦präd·ik 'as·əd }

monosodium acid methanearsonate [ORG CHEM] CH_4AsNaO_3 A white, crystalline solid; melting point is 132–139°C; soluble in water; used as an herbicide for grassy weeds on rights-of-way, storage areas, and noncrop areas, and as a preplant treatment for cotton, citrus trees, and turf. Abbreviated MSMA. { ¦män·ə'sōd·ē·əm ¦as·əd ¦meth ,än'ärs·ən,āt }

monosodium glutamate *See* sodium glutamate. { ¦män·ə'sōd·ē·əm 'glüd·ə,māt }
monosubstituted alkene [ORG CHEM] An alkene with the general formula RHC=CH₂, where R is any organic group; only one carbon atom is bonded directly to one of the carbons of the carbon-to-carbon double bond. { ,män·ō¦səb·stə,tüd·əd 'al·kēn }
monoterpene [ORG CHEM] **1.** A class of terpenes with molecular formula C₁₀H₁₆; the members of the class contain two isoprene units. **2.** A derivative of a member of such a class. { ¦män·ō'tər,pēn }
monovalent [CHEM] A radical or atom whose valency is 1. { ¦män·ō'vā·lənt }
monoxide [CHEM] A compound that contains a single oxygen atom, such as carbon monoxide, CO. { mə'näk,sīd }
mordant [CHEM] An agent, such as alum, phenol, or aniline, that fixes dyes to tissues, cells, textiles, and other materials by combining with the dye to form an insoluble compound. Also known as dye mordant. { 'mórd·ənt }
morin [ORG CHEM] C₁₅H₁₀O₇·2H₂O Colorless needles soluble in boiling alcohol, slightly soluble in water; used as a mordant dye and analytical reagent. { 'mòr·ən }
morpholine [ORG CHEM] C₄H₈ONH A hygroscopic liquid, soluble in water; used as a solvent and rubber accelerator. { 'mòr·fə,lēn }
morphosan [ORG CHEM] C₁₇H₁₉NO₃·CH₃Br A solid morphine derivative without morphine's disagreeable after effects; used in medicine. { 'mòr·fə,san }
Morse equation [PHYS CHEM] An equation according to which the potential energy of a diatomic molecule in a given electronic state is given by a Morse potential. { 'mórs i,kwā·zhən }
Morse potential [PHYS CHEM] An approximate potential associated with the distance *r* between the nuclei of a diatomic molecule in a given electronic state; it is V(*r*) = D{1 − exp[− *a*(*r* − *r*ₑ)]}², where *r*ₑ is the equilibrium distance, D is the dissociation energy, and *a* is a constant. { 'mórs pə,ten·chəl }
mosaic gold *See* stannic sulfide. { mō'zā·ik ,gōld }
Moseley's law [SPECT] The law that the square-root of the frequency of an x-ray spectral line belonging to a particular series is proportional to the difference between the atomic number and a constant which depends only on the series. { 'mōz·lēz ,lò }
Mössbauer spectroscopy [SPECT] The study of Mössbauer spectra, for example, for nuclear hyperfine structure, chemical shifts, and chemical analysis. { 'mùs,baù·ər spek'träs·kə·pē }
Mössbauer spectrum [SPECT] A plot of the absorption, by nuclei bound in a crystal lattice, of gamma rays emitted by similar nuclei in a second crystal, as a function of the relative velocity of the two crystals. { 'mùs,baù·ər ,spek·trəm }
mountain blue [INORG CHEM] 2CuCO₃·Cu(OH)₂ Ground azurite used as a paint pigment. Also known as copper blue. { 'maùnt·ən 'blü }
moving-boundary electrophoresis [ANALY CHEM] A U-tube variation of electrophoresis analysis that uses buffered solution so that all ions of a given species move at the same rate to maintain a sharp, moving front (boundary). { 'müv·iŋ ¦baùn·drē i¦lek·trə·fə'rē·səs }
MPK *See* pentanone.
MSG *See* sodium glutamate.
MSMA *See* monosodium acid methanearsonate.
MTBE *See* methyl tertiary butyl ether.
mucic acid [ORG CHEM] HOOC(CHOH)₄COOH A white, crystalline powder with a melting point of 210°C; soluble in water; used as a metal ion sequestrant and to retard concrete hardening. Also known as glactaric acid; saccharolactic acid; tetrahydroxyadipic acid. { 'myü·sik 'as·əd }
mull technique [SPECT] Method for obtaining infrared spectra of materials in the solid state; material to be scanned is first pulverized, then mulled with mineral oil. { 'məl tek,nēk }
multident *See* polydent. { 'məl·tə,dent }
multidentate ligand [CHEM] A ligand capable of donating two or more pairs of electrons in a complexation reaction to form coordinate bonds. { ,məl·tē¦den,tāt 'lī·gənd }

multiplet [SPECT] A collection of relatively closely spaced spectral lines resulting from transitions to or from the members of a multiplet. { 'məl·tə·plət }

multiplet intensity rules [SPECT] Rules for the relative intensities of spectral lines in a spin-orbit multiplet, stating that the sum of the intensities of all lines which start from a common initial level, or end on a common final level, is proportional to $2J+1$, where J is the total angular momentum of the initial level or final level respectively. { 'məl·tə·plət in'ten·səd·ē ‚rülz }

multivalent *See* polyvalent. { ¦məl·tə'vā·lənt }

muriatic acid *See* hydrochloric acid. { myür·ē'ad·ik 'as·əd }

muscarine [ORG CHEM] $C_8H_{19}NO_3$ A quaternary ammonium compound, the toxic ingredient of certain mushrooms, as *Amanita muscaria*. Also known as hydroxycholine. { 'məs·kə‚rēn }

musk ambrette [ORG CHEM] $C_{12}H_{16}N_2O_5$ White to yellow powder with heavy musky aroma; soluble in various oils and phthalates, insoluble in water; congeals at 83°C; used as a perfume fixative. Also known as 2,6-dinitro-3-methoxy-4-tert-butyltoluene. { 'məsk ‚am‚bret }

musk ketone [ORG CHEM] $C_{14}H_{18}N_2O_5$ White to yellow crystals with sweet musk aroma; soluble in various oils and phthalates, insoluble in water; used as a perfume fixative. Also known as 3,5-dinitro-2,6-dimethyl-4-tert-butylacetophenone. { 'məsk 'kē‚tōn }

musk xylene *See* musk xylol. { 'məsk 'zī‚lēn }

musk xylol [ORG CHEM] $(NO_2)_3C_6(CH_3)_2C(CH_3)_3$ White to yellow crystals with powerful musk aroma; soluble in various oils and phthalates, insoluble in water; congeals at 105°C; used as a perfume fixative. Also known as musk xylene; 2,4,6-trinitro-1,3-dimethyl-5-tert-butylbenzene. { 'məsk 'zī‚lól }

mustard gas [ORG CHEM] $HS(CH_2ClCH_2)_2S$ An oil with density 1.28, boiling point 215°C; used in chemical warfare. Also known as dichlorodiethylsulfide. { 'məs·tərd ‚gas }

mustard oil *See* allyl isothiocyanate. { 'məs·tərd ‚óil }

mutarotation [CHEM] A change in the optical rotation of light that takes place in the solutions of freshly prepared sugars. { ¦myüd·ə·rō'tā·shən }

mutual exclusion rule [PHYS CHEM] The rule that if a molecule has a center of symmetry, then no transition is allowed in both its Raman scattering and infrared emission (and absorption), but only in one or the other. { 'myü·chə·wəl ik'sklü·zhən ‚rül }

mutuality of phases [CHEM] The rule that if two phases, with respect to a reaction, are in equilibrium with a third phase at a certain temperature, then they are in equilibrium with respect to each other at that temperature. { ‚myü·chə'wal·əd·ē əv 'fāz·əz }

β-myrcene [ORG CHEM] $C_{10}H_{16}$ An oily liquid with a pleasant odor; soluble in alcohol, chloroform, ether, and glacial acetic acid; used as an intermediate in the preparation of perfume chemicals. { ¦bād·ə 'mər‚sēn }

myricetin [ORG CHEM] $C_{15}H_{10}O_8$ A yellow, crystalline compound with a melting point of 357°C; soluble in alcohol; used as an inhibitor of adenosine triphosphatase. Also known as cannabiscetin; delphidenolon. { mə'ris·ə·tən }

myristic acid [ORG CHEM] $CH_3(CH_2)_{12}COOH$ Oily white crystals melting at 58°C; soluble in ether and alcohol, insoluble in water; used to synthesize flavor and perfume esters, and in soaps and cosmetics. { mə'ris·tik 'as·əd }

myristyl alcohol [ORG CHEM] $C_{14}H_{29}OH$ Liquid boiling at 264°C; soluble in ether and alcohol, insoluble in water; used as a chemical intermediate, plasticizer, and perfume fixative. { mə'rist·əl 'al·kə‚hól }

myristyl mercaptan *See* tetradecyl mercaptan. { mə'rist·əl mər'kap‚tan }

N

n- [ORG CHEM] Chemical prefix for normal (straight-carbon-chain) hydrocarbon compounds.

N *See* nitrogen; normality.

Na *See* sodium.

NAA *See* naphthaleneacetic acid.

naled [ORG CHEM] $C_4H_7Br_2Cl_2O_4$ A white solid with a melting point of 27°C; slight solubility in water; used as an insecticide and miticide for crops, farm buildings, and kennels, and for mosquito control.   { 'nal·əd }

nanochemistry [CHEM] The study of the synthesis and analysis of materials in the nanoscale range (1 – 10 nanometers), including large organic molecules, inorganic cluster compounds, and metallic or semiconductor particles.   { ,nan·ō'kem·ə·strē }

nantokite *See* cuprous chloride.   { 'nan·tə,kīt }

naphthacene [ORG CHEM] $C_{18}H_{12}$ A hydrocarbon molecule that may be considered to be four benzene rings fused together; it is explosive when shocked; used in organic synthesis.   Also known as rubene; tetracene.   { 'naf·thə,sēn }

naphthalene [ORG CHEM] $C_{10}H_8$ White, volatile crystals with coal tar aroma; insoluble in water, soluble in organic solvents; structurally it is represented as two benzenoid rings fused together; boiling point 218°C, melting point 80.1°C; used for moth repellents, fungicides, lubricants, and resins, and as a solvent.   Also known as naphthalin; tar camphor.   { 'naf·thə,lēn }

naphthaleneacetamide [ORG CHEM] $C_{12}H_{11}NO$ A colorless solid with a melting point of 183°C; used as a growth regulator for root cuttings and for thinning of apples and pears.   { ¦naf·thə,lēn·ə'sed·ə·məd }

naphthaleneacetic acid [ORG CHEM] $C_{10}H_7CH_2COOH$ White, odorless crystals, melting at 132–135°C; soluble in organic solvents, slightly soluble in water; used as an agricultural spray.   Abbreviated NAA.   Also known as 1-naphthylacetic acid.   { ¦naf·thə,lēn·ə¦sēd·ik 'as·əd }

naphthalene-1,5-disulfonic acid [ORG CHEM] $C_{10}H_6-(SO_3H)_2$ White crystals, decomposing when heated; used to make dyes.   Also known as Armstrong's acid.   { 'naf·thə,lēn ¦wən ¦fīv ,dī·səl'fän·ik 'as·əd }

1-naphthalenesulfonic acid [ORG CHEM] $C_{10}H_8O_3S$ A crystalline compound with a melting point of 90°C (dihydrate); soluble in water or alcohol; used to make α-naphthol.   { ¦wən ¦naf·thə,lēn·səl'fän·ik 'as·əd }

naphthalic acid *See* phthalic acid.   { naf'thal·ik 'as·əd }

naphthalin *See* naphthalene.   { 'naf·thə·lən }

naphthene [ORG CHEM] Any of the cycloparaffin derivatives of cyclopentane (C_5H_{10}) or cyclohexane (C_6H_{12}) found in crude petroleum.   { 'naf,thēn }

naphthenic acid [ORG CHEM] Any of the derivatives of cyclopentane, cyclohexane, cycloheptane, or other naphthenic homologs derived from petroleum; molecular weights 180 to 350; soluble in organic solvents and hydrocarbons, slightly soluble in water; used as a paint drier and wood preservative, and in metals production.   { naf'thēn·ik 'as·əd }

naphthionic acid [ORG CHEM] $C_{10}H_6(NH_2)SO_3H$ White powder or crystals that decompose when heated; used to manufacture dyes.   { ¦naf·thē¦än·ik 'as·əd }

α-naphthol

α-naphthol [ORG CHEM] $C_{10}H_7OH$ Colorless to yellow powder, melting at 96°C; used to make dyes and perfumes, and in synthesis of organic molecules. { ¦al·fə 'naf‚thȯl }

β-naphthol [ORG CHEM] $C_{10}H_7OH$ White crystals that melt at 121.6°C; insoluble in water; used to make pigments, dyes, and antioxidants. { ¦bād·ə 'naf‚thȯl }

1,2-naphthoquinone [ORG CHEM] $C_{10}H_6O_2$ A golden yellow, crystalline compound that decomposes at 145–147°C; soluble in benzene and ether; used as a reagent for resorcinol and thalline. { ¦wən ¦tü ‚naf·thə·kwə'nōn }

1,4-naphthoquinone [ORG CHEM] $C_{10}H_6O_2$ Greenish-yellow powder soluble in organic solvents, slightly soluble in water; melts at 123–126°C; used as an antimycotic agent, in synthesis, and as a rubber polymerization regulator. { ¦wən ¦fȯr ‚naf·thə·kwə'nōn }

naphthoresorcinol [ORG CHEM] $C_{10}H_6(OH)_2$ Crystals with a melting point of 124–125°C; soluble in ether, alcohol, and water; used as a reagent for sugars and oils, and to determine glucuronic acid in urine. { ‚naf·thə·ri'sȯrs·ən‚ȯl }

β-naphthoxyacetic acid [ORG CHEM] $C_{12}H_{10}O_3$ A crystalline compound soluble in water, with a melting point of 156°C; used as a growth regulator to set blossoms and regulate growth for pineapples, strawberries, and tomatoes. Also known as O-(2-naphthyl)glycolic acid. Abbreviated BNOA. { ¦bād·ə naf¦thäk·sē·ə'sēd·ik 'as·əd }

2-(α-naphthoxy)-N,N-diethylpropionamide See devrinol. { ¦tü ¦al·fə naf'thäk·sē ¦en ¦en dī‚eth·əl‚pro·pē'än·ə·məd }

1-naphthylacetic acid See naphthaleneacetic acid. { ¦wən ‚naf·thil·ə'sēd·ik 'as·əd }

naphthylamine [ORG CHEM] $C_{10}H_7NH_2$ White, toxic crystals, soluble in alcohol and ether; used in dyes; the two forms are α-naphthylamine, boiling at 301°C, and β-naphthylamine, boiling at 306°C. { naf'thil·ə‚mēn }

2,5-naphthylamine sulfonic acid See gamma acid. { ¦tü ¦fīv naf'thil·ə‚mēn səl'fän·ik 'as·əd }

β-naphthylmethyl ether [ORG CHEM] $C_{10}H_7OCH_3$ White, crystalline scales with a melting point of 72°C; soluble in alcohol and ether; used for soap perfumes. { ¦bād·ə ¦naf·thil¦meth·əl 'ē·thər }

N-1-naphthylphthalamic acid [ORG CHEM] $C_{10}H_7NHCOC_6H_4COOH$ A crystalline solid with a melting point of 185°C; used as a preemergence herbicide. { ¦en ¦wən ¦naf·thil·thə'lam·ik 'as·əd }

α-naphthylthiocarbamide See 1-(1-naphthyl)-2-thiourea. { ¦al·fə ¦naf·thil‚thī·ō'kär·bə·məd }

1-(1-naphthyl)-2-thiourea [ORG CHEM] $C_{10}H_7NHCSNH_2$ A crystalline compound with a melting point of 198°C; soluble in water, acetone, triethylene glycol, and hot alcohol; used as a poison to control the adult Norway rat. { ¦wən ¦wən 'naf·thil ¦tü ‚thī·ə·yü'rē·ə }

Naples yellow See lead antimonite. { 'nā·pəlz 'yel·ō }

narceine [ORG CHEM] $C_{23}H_{27}O_8N·3H_2O$ White, odorless crystals with bitter taste; soluble in alcohol and water, insoluble in ether; melts at 170°C; used in medicine. { 'när·sē‚ēn }

narcissistic reaction [CHEM] A chemical reaction in which a reactant is converted into a product whose structure is the mirror image of the reactant molecule. { ‚när·sə'sis·tik rē'ak·shən }

naringin [ORG CHEM] $C_{27}H_{32}O_{14}$ A crystalline bioflavonoid with a melting point of 171°C; soluble in acetone and alcohol; used as a food supplement. Also known as aurantiin. { nə'rin·jən }

nascent [CHEM] Pertaining to an atom or simple compound at the moment of its liberation from chemical combination, when it may have greater activity than in its usual state. { 'nā·sənt }

natrium [CHEM] Latin name for sodium; source of the symbol Na. { 'nā·trē·əm }

natural linewidth [SPECT] The part of the linewidth of an absorption or emission line that results from the finite lifetimes of one or both of the energy levels between which the transition takes place. { 'nach·rəl 'līn‚width }

natural red See purpurin. { 'nach·rəl 'red }

Nb See niobium.

Nd See neodymium.

NDGA *See* nordihydroguaiaretic acid.

Ne *See* neon.

near-infrared spectrophotometry [ANALY CHEM] Spectrophotometry at wavelengths in the near-infrared region, generally using instruments with quartz prisms in the monochromators and lead sulfide photoconductor cells as detectors to observe absorption bands which are harmonics of bands at longer wavelengths. { 'nir ˌin·frə'red ˌspek· trō·fə'täm·ə·trē }

nebulium line [SPECT] An optical emission line in the spectrum of oxygen at a wavelength of 500.7 nanometers, prominent in the spectra of H II regions. { nə'bül·ē· əm ˌlīn }

negative catalysis [CHEM] A catalytic reaction such that the reaction is slowed down by the presence of the catalyst. { 'neg·əd·iv kə'tal·ə·səs }

negative ion [CHEM] An atom or group of atoms which by gain of one or more electrons has acquired a negative electric charge. { 'neg·əd·iv 'ī‚än }

neighboring-group participation *See* anchimeric assistance. { 'nā·bər·iŋ ˌgrüp pär‚tis· ə'pā·shən }

nematic phase [PHYS CHEM] A phase of a liquid crystal in the mesomorphic state, in which the liquid has a single optical axis in the direction of the applied magnetic field, appears to be turbid and to have mobile threadlike structures, can flow readily, has low viscosity, and lacks a diffraction pattern. { nə'mad·ik ˌfāz }

nematogenic solid [PHYS CHEM] A solid which will form a nematic liquid crystal when heated. { nə'mad·ə‚jen·ik 'säl·əd }

neo-, ne- [ORG CHEM] Prefix indicating hydrocarbons where a carbon is bonded directly to at least four other carbon atoms, such as neopentane. { 'nē·ō }

neodymium [CHEM] A metallic element, symbol Nd, with atomic weight 144.24, atomic number 60; a member of the rare-earth group of elements. { ˌnē·ō'dim·ē·əm }

neodymium chloride [INORG CHEM] $NdCl_3 \cdot xH_2O$ Water-and acid-soluble, pink lumps; used to prepare metallic neodymium. { ˌnē·ō'dim·ē·əm 'klȯr‚īd }

neodymium oxide [INORG CHEM] Nd_2O_3 A hygroscopic, blue-gray powder; insoluble in water, soluble in acids; used to color glass and in ceramic capacitors. { ˌnē·ō'dim· ē·əm 'äk‚sīd }

neohexane [ORG CHEM] C_6H_{14} Volatile, flammable, colorless liquid boiling at 50°C; used as high-octane component of motor and aviation gasolines. { ¦nē·ō'hek‚sān }

neon [CHEM] A gaseous element, symbol Ne, atomic number 10, atomic weight 20.179; a member of the family of noble gases in the zero group of the periodic table. { 'nē‚än }

neopentane [ORG CHEM] C_5H_{12} Colorless liquid boiling at 10°C; soluble in alcohol, insoluble in water; a hydrocarbon found as a minor component of natural gasoline. { ¦nē·ō'pen‚tān }

neptunium [CHEM] A chemical element, symbol Np, atomic number 93, atomic weight 237.0482; a member of the actinide series of elements. { nep'tü·nē·əm }

neptunium decay series [CHEM] Little-known radioactive elements with short lives; produced as successive series of decreasing atomic weight when uranium-237 and plutonium-241 decay radioactively through neptunium-237 to bismuth-209. { nep'tü·nē·əm di'kā ˌsir·ēz }

Nernst equation [PHYS CHEM] The relationship showing that the electromotive force developed by a dry cell is determined by the activities of the reacting species, the temperature of the reaction, and the standard free-energy change of the overall reaction. { 'nernst i‚kwā·zhən }

Nernst-Thomson rule [PHYS CHEM] The rule that in a solvent having a high dielectric constant the attraction between anions and cations is small so that dissociation is favored, while the reverse is true in solvents with a low dielectric constant. { 'nernst 'täm·sən ˌrül }

Nernst zero of potential [PHYS CHEM] An electrode potential corresponding to the reversible equilibrium between hydrogen gas at a pressure of 1 standard atmosphere and hydrogen ions at unit activity. { 'nernst 'zir·ō əv pə'ten·chəl }

257

nerol

nerol [ORG CHEM] $C_{10}H_{17}OH$ Colorless liquid with rose-neroli odor; derived from geraniol (a trans isomer); used in perfumery.   { 'ne,ròl }

nerolidol [ORG CHEM] $C_{15}H_{26}O$ A straw-colored sesquiterpene alcohol; liquid with rose and apple aroma derived from cabreuva oil, oils of orange flower, and ylang ylang; soluble in alcohol; used in perfumery.   { nə'räl·ə,dòl }

nerve gas [CHEM] Chemical agent which is absorbed into the body by breathing, by ingestion, or through the skin, and affects the nervous and respiratory systems and various body functions; an example is isopropylphosphonofluoridate.   { 'nərv 'gas }

Nessler's reagent [ANALY CHEM] Mercuric iodide-potassium iodide solution, used to analyze for small amounts of ammonia.   { 'nes·lərz rē,ā·jənt }

Nessler tubes [ANALY CHEM] Standardized glass tubes for filling with standard solution colors for visual color comparison with similar tubes filled with solution samples.   { 'nes·lər ,tübz }

network polymer [ORG CHEM] A three-dimensional material made by crosslinking.   { ¦net,wərk 'päl·ə·mər }

neutral [CHEM] Property of a solution which is neither acidic nor basic, having the same concentration of hydrogen ions as water.   { 'nü·trəl }

neutral flame [CHEM] Gas flame produced by a mixture of fuel and oxygen so as to be neither oxidizing nor reducing.   { 'nü·trəl 'flām }

neutral granulation [CHEM] Propellant granulation in which the surface area of a grain remains constant during burning.   { 'nü·trəl ,gran·yə'lā·shən }

neutralization [CHEM] The process of making a solution neutral (pH = 7) by adding a base to an acid solution, or adding an acid to an alkaline (basic) solution.   Also known as neutralization reaction.   { ,nü·trə·lə'zā·shən }

neutralization equivalent [CHEM] For an acid or base, the same as equivalent weight; multiplication of the neutralization equivalent by the number of acidic or basic groups in the molecule gives the molecular weight.   { ,nü·trə·lə'zā·shən i,kwiv·ə·lənt }

neutralization number [ANALY CHEM] Petroleum product test; it is the milligrams of potassium hydroxide required to neutralize the acid in 1 gram of oil; used as an indication of oil acidity.   { ,nü·trə·lə'zā·shən ,nəm·bər }

neutralization reaction See neutralization.   { ,nü·trə·lə'zā·shən rē,ak·shən }

neutralize [CHEM] To make a solution neutral (neither acidic nor basic, pH of 7) by adding a base to an acidic solution, or an acid to a basic solution.   { 'nü·trə,līz }

neutral molecule [PHYS CHEM] A molecule in which the number of electrons surrounding the nuclei is the same as the total number of protons in the nuclei, so that there is no net electric charge.   { 'nü·trəl 'mäl·ə,kyül }

neutral potassium phosphate See potassium phosphate.   { 'nü·trəl pə'tas·ē·əm 'fäs,fāt }

neutral red [ORG CHEM] $(CH_3)_2NC_6H_3N_2C_6H_2CH_3NH_2·ClH$ Water- and alcohol-soluble green powder; used as pH 6.8–8.0 acid-base indicator, and as a dye to test stomach function.   Also known as dimethyl diaminophenazine chloride; toluylene red.   { 'nü·trəl 'red }

neutral species See uncharged species.   { 'nü·trəl 'spē·shēz }

Newland's law of octaves [CHEM] An arrangement of the elements that predated Mendeleev's periodic table; Newland's arrangement was a grouping of the elements in increasing atomic weights (starting with lithium) in horizontal rows of eight elements, with each new row directly beneath the previous one.   { 'nü·lənz 'lò əv 'äk·tivz }

Newman projection [ORG CHEM] A representation of the conformation of a molecule in which the viewer's eye is considered to be sighting down a carbon-carbon bond; the front carbon is represented by a point and the back carbon by a circle.   { 'nü·mən prə,jek·shən }

Ni See nickel.

nickel [CHEM] A chemical element, symbol Ni, atomic number 28, atomic weight 58.69.   { 'nik·əl }

nickel acetate [ORG CHEM] $Ni(OOCCH_3)_2 \cdot 4H_2O$ Efflorescent green crystals that decompose upon heating; soluble in alcohol and water; used as textile dyeing mordant. { 'nik·əl 'as·ə,tāt }

nickel ammonium sulfate [INORG CHEM] $NiSO_4 \cdot (NH_4)_2SO_4 \cdot 6H_2O$ A green, crystalline compound, soluble in water; used as a nickel electrolyte for electroplating. Also known as ammonium nickel sulfate; double nickel salt. { 'nik·əl ə'mō·nē·əm 'səl,fāt }

nickel arsenate [INORG CHEM] $Ni_3(AsO_4)_2 \cdot H_2O$ Poisonous yellow-green powder; soluble in acids, insoluble in water; used as a fat-hardening catalyst in soapmaking. { 'nik· əl 'ars·ən,āt }

nickel carbonate [INORG CHEM] $NiCO_3$ Light-green crystals that decompose upon heating; soluble in acid, insoluble in water; used in electroplating. { 'nik·əl 'kär·bə,nāt }

nickel carbonyl [INORG CHEM] $Ni(CO)_4$ Colorless, flammable, poisonous liquid boiling at 43°C; soluble in alcohol and concentrated nitric acid, insoluble in water; used in gas plating (vapor decomposes at 60°C) and to produce metallic nickel. { 'nik·əl 'kär·bə,nil }

nickel cyanide [INORG CHEM] $Ni(CN)_2 \cdot 4H_2O$ Poisonous, water-insoluble apple-green powder; melts and loses water at 200°C, decomposes at higher temperatures; used for electroplating and metallurgy. { 'nik·əl 'sī·ə,nīd }

nickel formate [ORG CHEM] $Ni(HCOO)_2 \cdot 2H_2O$ Water-soluble green crystals; used in hydrogenation catalysts. { 'nik·əl 'fòr,māt }

nickel iodide [INORG CHEM] NiI_2 or $NiI_2 \cdot 6H_2O$ Hygroscopic black or blue-green solid; soluble in water and alcohol; sublimes when heated. { 'nik·əl 'ī·ə,dīd }

nickel nitrate [INORG CHEM] $Ni(NO_3)_2 \cdot 6H_2O$ Fire-hazardous oxidant; deliquescent, green, water- and alcohol-soluble crystals; used for nickel plating and brown ceramic colors, and in nickel catalysts. { 'nik·əl 'nī,trāt }

nickelocene [ORG CHEM] $(C_5H_5)_2Ni$ Dark green crystals with a melting point of 171–173°C; soluble in most organic solvents; used as an antiknock agent. { nə'kəl·ə,sēn }

nickel oxide [INORG CHEM] NiO Green powder; soluble in acids and ammonium hydroxide; insoluble in water; used to make nickel salts and for porcelain paints. Also known as green nickel oxide. { 'nik·əl 'äk,sīd }

nickel phosphate [INORG CHEM] $Ni_3(PO_4)_2 \cdot 7H_2O$ A light-green powder; soluble in acids and ammonium hydroxide, insoluble in water; used for electroplating and production of yellow nickel. { 'nik·əl 'fäs,fāt }

nicotine [ORG CHEM] $C_{10}H_{14}N_2$ A colorless liquid with a boiling point of 247.3°C; miscible with water; used as a contact insecticide fumigant in closed spaces. { 'nik·ə,tēn }

ninhydrin [ORG CHEM] $C_9H_4O_3 \cdot H_2O$ White crystals or powder with a melting point of 240–245°C; soluble in water and alcohol; used for the detection and assay of peptides, amines, amino acids, and amino sugars. Also known as triketohydrindene hydrate. { nin'hī·drən }

niobe oil See methyl benzoate. { 'nī·ə,bē,òil }

niobic acid [INORG CHEM] $Nb_2O_5 \cdot nH_2O$ Family of hydrates; white precipitate, soluble in inorganic acids and bases, insoluble in water; its formation is part of the analytical determination of niobium. { nī'ō·bik 'as·əd }

niobium [CHEM] A chemical element, symbol Nb, atomic number 41, atomic weight 92.904. { nī'ō·bē·əm }

niobium carbide [INORG CHEM] NbC A lavender gray powder with a melting point of 3500°C; used for carbide-tipped tools and special steels. { nī'ō·bē·əm 'kär,bīd }

niter See potassium nitrate. { 'nīd·ər }

niter cake See sodium bisulfate. { 'nīd·ər ,kāk }

nitrate [CHEM] **1.** A salt or ester of nitric acid. **2.** Any compound containing the ion NO_3sw. { 'nī,trāt }

nitrating acid [INORG CHEM] Sulfuric-nitric acid mix used to nitrate cellulosics and aromatic chemicals. Also known as mixed acid. { 'nī,trād·iŋ 'as·əd }

nitration [ORG CHEM] Introduction of an NO_2sw group into an organic compound. { nī'trā·shən }

nitrene [ORG CHEM] A molecular fragment that is an uncharged, electron-deficient species containing a monocovalent nitrogen. { 'nī,trēn }

nitric acid [INORG CHEM] HNO_3 Strong oxidant that is fire-hazardous; colorless or yellowish liquid, miscible with water; boils at 86°C; used for chemical synthesis, explosives, and fertilizer manufacture, and in metallurgy, etching, engraving, and ore flotation. Also known as aqua fortis. { 'nī·trik 'as·əd }

nitric oxide [INORG CHEM] NO A colorless gas that, at room temperature, reacts with oxygen to form nitrogen dioxide (NO_2, a reddish-brown gas). It may be used to form other compounds. It is a crucial physiological messenger molecule thought to play a role in blood pressure regulation, control of blood clotting, immune defense, digestion, the senses of sight and smell, and possibly learning and memory. { 'nī·trik 'äk,sīd }

nitride [INORG CHEM] Compound of nitrogen and a metal, such as Mg_3N_2. { 'nī,trīd }

nitrile [ORG CHEM] RC≡N Cyanide derived by removal of water from an acid amide. { 'nī,trīl }

nitrile resin [ORG CHEM] Any one of a family of polymers produced from acrylonitrile, various esters, butadiene, and styrene. { 'nī,trīl ,rez·ən }

nitrilotriacetic acid [ORG CHEM] $N(CH_2COOH)_3$ A white powder, melting point 240°C, with some decomposition; soluble in water; it is toxic, and birth abnormalities may result from ingestion; may be used as a chelating agent in the laboratory. Also known as NTA; TGA. { ¦nī·trə·lō,trī·ə'sēd·ik 'as·əd }

nitrite [CHEM] A compound containing the radical NO_2^-; can be organic or inorganic. { 'nī,trīt }

nitro- [CHEM] Chemical prefix showing the presence of the NO_2^- radical. { 'nī·trō }

nitroalkane See nitroparaffin. { ¦nī·trō'al,kān }

***meta*-nitroaniline** [ORG CHEM] $NO_2C_6H_4NH_2$ Yellow crystals that melt at 112.5°C; a toxic material; used as a dye intermediate. { ¦med·ə ,nī·trō'an·ə·lən }

***ortho*-nitroaniline** [ORG CHEM] $NO_2C_6H_4NH_2$ Orange-red crystals that melt at 69.7°C, soluble in ethanol; a toxic material; used to manufacture dyes. { ¦òr·thō ,nī·trō'an·ə·lən }

***para*-nitroaniline** [ORG CHEM] $NO_2C_6H_4NH_2$ Yellow crystals that melt at 148°C; insoluble in water, soluble in ethanol; a toxic material; used to make dyes, and as a corrosion inhibitor. { ¦par·ə ,nī·trō'an·ə·lən }

nitroaromatic [ORG CHEM] A nitrated benzene or benzene derivative, such as nitrobenzene, $C_6H_5NO_2$, or nitrobenzoic acid, $NO_2·C_6H_4·COOH$. { ¦nī·trō,ar·ə'mad·ik }

nitrobarite See barium nitrate. { ¦nī·trō'ba,rīt }

nitrobenzene [ORG CHEM] $C_6H_5NO_2$ Greenish crystals or a yellowish liquid, melting point 5.70°C; a toxic material; used in aniline manufacture. Also known as oil of mirbane. { ¦nī·trō'ben,zēn }

***ortho*-nitrobiphenyl** [ORG CHEM] $C_{12}H_9NO_2$ A crystalline compound with a sweetish odor; melting point is 36.7°C; used as a plasticizer for resins, cellulose acetate and nitrate, and polystyrenes, and as a fungicide for textiles. Abbreviated ONB. { ¦òr·thō ,nī·trō·bī'fen·əl }

nitrobromoform See bromopicrin. { ¦nī·trō'brō·mə,fòrm }

nitrocalcite See calcium nitrate. { ¦nī·trō'kal,sīt }

nitrocellulose See cellulose nitrate. { ¦nī·trō'sel·yə,lōs }

nitrocotton See cellulose nitrate. { ¦nī·trō'kät·ən }

nitro dye [ORG CHEM] A dye with the NO_2 chromophore group in the molecules. { 'nī·trō ,dī }

nitroethane [ORG CHEM] $CH_3CH_2NO_2$ A colorless liquid, slightly soluble in water; boils at 114°C; used as a solvent for cellulosics, resins, waxes, fats, and dyestuffs, and as a chemical intermediate. { ¦nī·trō'eth,ān }

nitro explosive [ORG CHEM] Explosive compound containing one or more NO_2^- groups, such as nitroglycerine, $C_3H_5(ONO_2)_3$, or trinitrotoluene, $C_6H_2(CH_3)(NO_2)_3$. { 'nī·trō ik'splō·siv }

nitrogen [CHEM] A chemical element, symbol N, atomic number 7, atomic weight 14.0067; it is a gas, diatomic (N_2) under normal conditions; about 78% of the atmosphere is N_2; in the combined form the element is a constituent of all proteins. { 'nī·trə·jən }

2-nitropropane

nitrogen acid anhydride *See* nitrogen pentoxide. { 'nī·trə·jən ¦as·əd an'hī,drīd }
nitrogen dioxide [INORG CHEM] NO₂ A reddish-brown gas; it exists in varying degrees of concentration in equilibrium with other nitrogen oxides; used to produce nitric acid. Also known as dinitrogen tetroxide; liquid dioxide; nitrogen peroxide; nitrogen tetroxide. { 'nī·trə·jən dī'äk,sīd }
nitrogen monoxide *See* nitrous oxide. { 'nī·trə·jən mə'näk,sīd }
nitrogen mustard [ORG CHEM] Any of the substituted mustard gases in which the sulfur is replaced by an amino nitrogen, such as for methyl bis(2-chlorethyl)amine, (CH₂ClCH₂)₂NCH₃; useful in cancer research. { 'nī·trə·jən 'məs·tərd }
nitrogen oxides [INORG CHEM] NOₓ Chemical compounds of nitrogen and oxygen; produced primarily from the combustion of fossil fuels, they contribute to the formation of ground-level ozone. { ¦nī·trə·jən 'äk,sīdz }
nitrogen pentoxide [INORG CHEM] N₂O₅ Colorless crystals, soluble in water (forms HNO₃); decomposes at 46°C. Also known as nitrogen acid anhydride. { 'nī·trə·jən pen 'täk,sīd }
nitrogen peroxide *See* nitrogen dioxide. { 'nī·trə·jən pə'räk,sīd }
nitrogen solution [INORG CHEM] Mixture used to neutralize super-phosphate in fertilizer manufacture; consists of 60% ammonium nitrate, and the balance a 50% aqua ammonia solution. { 'nī·trə·jən sə,lü·shən }
nitrogen tetroxide *See* nitrogen dioxide. { 'nī·trə·jən te'träk,sīd }
nitrogen trifluoride [INORG CHEM] NF₃ A colorless gas that has a melting point of −206.6°C and a boiling point of −128.8°C; used as an oxidizer for high-energy fuels. { 'nī·trə·jən trī'flúr,īd }
nitrogen trioxide [INORG CHEM] N₂O₃ Green, water-soluble liquid; boils at 3.5°C. { 'nī· trə·jən trī'äk,sīd }
nitroglycerin [ORG CHEM] CH₂NO₃CHNO₃CH₂NO₃ Highly unstable, explosive, flammable pale-yellow liquid; soluble in alcohol; freezes at 13°C and explodes at 260°C; used as an explosive, to make dynamite, and in medicine. Also spelled nitroglycerine. { ¦nī·trə'glis·ə·rən }
nitroglycerine *See* nitroglycerin. { ¦nī·trə'glis·ə·rən }
nitroguanidine [ORG CHEM] H₂NC(NH)NHNO₂ Explosive yellow solid, soluble in alcohol; melts at 246°C; used in explosives and smokeless powder. { ¦nī·trō'gwän·ə,dēn }
nitrometer [ANALY CHEM] Glass apparatus used to collect and measure nitrogen and other gases evolved by a chemical reaction. Also known as azotometer. { nī'träm· əd·ər }
nitromethane [ORG CHEM] CH₃NO₂ A liquid nitroparaffin compound; oily and colorless; boils at 101°C; used as a monopropellant for rockets, in chemical synthesis, and as an industrial solvent for cellulosics, resins, waxes, fats, and dyestuffs. { ¦nī· trō'meth,ān }
nitron [ORG CHEM] CN₄(C₆H₅)₃CH Yellow crystals, soluble in chloroform and acetone; used as reagent to detect NO₃ ion in dilute solutions. { 'nī,trän }
nitronium [CHEM] Positively charged NO₂ ion, believed to be formed from HNO₃. Also known as nitryl ion. { nī'trō·nē·əm }
nitroparaffin [ORG CHEM] Any organic compound in which one or more hydrogens of an alkane are replaced by a nitro, or NO₂sw, group, such as nitromethane, CH₃NO₂, or nitroethane, C₂H₅NO₂. Also known as nitroalkane. { ¦nī·trō'par·ə·fən }
ortho-nitrophenol [ORG CHEM] C₆H₅NO₃ A yellow, crystalline compound; melting point is 44–45°C; soluble in hot water, alcohol, benzene, ether, carbon disulfide, and alkali hydroxides; used in the commercial preparation of many compounds. { ¦ȯr·thō ¦nī·trō'fe,nȯl }
para-nitrophenylhydrazine [ORG CHEM] C₆H₇N₃O₂ An orange-red, crystalline compound with a melting point of about 157°C; soluble in hot water or hot benzene; used as a reagent for aliphatic aldehydes and ketones. { ¦par·ə ¦nī·trō¦fen·əl'hī·drə,zēn }
1-nitropropane [ORG CHEM] CH₃CH₂CH₂NO₂ A colorless liquid with a boiling point of 132°C; used as a rocket propellant and gasoline additive. { ¦wən ¦nī·trō'prō,pān }
2-nitropropane [ORG CHEM] CH₃CHNO₂CH₃ A colorless liquid with a boiling point of

261

nitroso

120°C; used as a solvent for vinyl coatings, as a rocket propellant, and as a gasoline additive. { ¦tü ¦nī·trō'prō͟,pän }

nitroso [CHEM] The radical NO⁻ with trivalent nitrogen. Also known as hydroximino; oximido. { nī'trō·sō }

nitrostarch [ORG CHEM] $C_{12}H_{12}(NO_2)_8O_{10}$ Orange powder, soluble in ethyl alcohol; used in explosives. Also known as starch nitrate. { 'nī·trə͟,stärch }

meta-nitrotoluene [ORG CHEM] $NO_2C_6H_4CH_3$ Yellow powder that melts at 15°C; insoluble in water; used in organic synthesis. { ¦med·ə ¦nī·trō'täl·yə͟,wēn }

ortho-nitrotoluene [ORG CHEM] $NO_2C_6H_4CH_3$ A yellow liquid boiling at 220.4°C; insoluble in water; used to produce toluidine and dyes. { ¦ór·thō ¦nī·trō'täl·yə͟,wēn }

para-nitrotoluene [ORG CHEM] $NO_2C_6H_4CH_3$ Yellow crystals that melt at 51.7°C; insoluble in water, soluble in ethanol; used to produce toluidine and to manufacture dyes. { ¦par·ə ¦nī·trō'täl·yə͟,wēn }

nitrourea [ORG CHEM] $NH_2CONHNO_2$ Highly explosive white crystals, melting at 159°C; soluble in ether and alcohol, slightly soluble in water; used as a chemical intermediate. { ¦nī·trō·yü'rē·ə }

nitrous acid [INORG CHEM] HNO_2 Aqueous solution of nitrogen trioxide, N_2O_3. { 'nī·trəs 'as·əd }

nitrous oxide [INORG CHEM] N_2O Colorless, sweet-tasting gas, boiling at −90°C; slightly soluble in water, soluble in alcohol; used as a food aerosol, and as an anesthetic in dentistry and surgery. Also known as laughing gas; nitrogen monoxide. { 'nī·trəs 'äk͟,sīd }

nitroxylene [ORG CHEM] $C_6H_3(CH_3)_2NO_2$ Any of three isomers occurring either as a yellow liquid or as crystalline needles with a melting point of 2°C and boiling point of 246°C; soluble in alcohol and ether; used in gelatinizing accelerators for pyroxylin. { nī'träk·sə͟,lēn }

nitryl halide [INORG CHEM] NO_2X Compound containing a halide (X) and a nitro group (NO_2). { 'nī͟,tril 'ha͟,līd }

nitryl ion See nitronium. { 'nī͟,tril ͟,ī͟,än }

N line [SPECT] One of the characteristic lines in an atom's x-ray spectrum, produced by excitation of an N electron. { 'en ͟,līn }

No See nobelium.

nobelium [CHEM] A chemical element, symbol No, atomic number 102; a synthetic element, in the actinium series; isotopes with mass numbers 250–260 and 262 have been produced in the laboratory, with mass number 259 having the longest known half-life, 58 minutes. { nō'bel·ē·əm }

noble gas [CHEM] A gas in group 0 of the periodic table of the elements; it is monatomic and, with limited exceptions, chemically inert. Also known as inert gas; rare gas. { 'nō·bəl 'gas }

noble-gas electron configuration [CHEM] An electron structure of an atom or ion in which the outer electron shell contains eight electrons, corresponding to the electron configuration of a noble gas, such as neon or argon. { ¦nō·bəl ¦gas i'lek͟,trän kən͟,fig·yə͟,rā·shən }

noble potential [PHYS CHEM] A potential equaling or approaching that of the noble elements, such as gold, silver, or copper, of the electromotive series. { 'nō·bəl pə'ten·chəl }

NODA See n-octyl n-decyl adipate. { 'nō·də }

noise [SPECT] Random fluctuations of electronic signals appearing in a recorded spectrum. { nóiz }

nonacosane [ORG CHEM] $C_{29}H_{60}$ Colorless hydrocarbon, melting at 63°C; found in beeswax and the fat of cabbage leaves. { ͟,nō·nə'kō͟,sān }

nonadecane [ORG CHEM] $CH_3(CH_2)_{17}CH_3$ Flammable crystals, soluble in ether and alcohol, insoluble in water; melts at 32°C; used as a chemical intermediate. { ͟,nō·nə'de͟,kān }

nonanal [ORG CHEM] $C_8H_{17}CHO$ A colorless liquid with an orange rose odor; used in perfumes and for flavoring. { 'nän·ə͟,näl }

nonane |ORG CHEM| $CH_3(CH_2)_7CH_3$ Flammable, colorless liquid, boiling at 151°C; soluble in alcohol, insoluble in water; used as a chemical intermediate. Also known as nonyl hydride. { 'nō,nān }

nonaqueous [CHEM] Pertaining to a liquid or solution containing no water. { ¦nän'ā·kwē·əs }

nonbenzenoid aromatic compound |ORG CHEM| A compound exhibiting aromatic character but not containing a benzene nucleus, or having one or more rings in a fused ring system that are not benzene rings. { ¦nän'ben·zə,nȯid ¦ar·ə¦mad·ik 'käm,paúnd }

nonbonded distance [PHYS CHEM] The distance between atoms in a molecule that are not bonded to each other. { ¦nän,bän·dəd 'dis·təns }

noncovalent bonds [CHEM] Weak chemical bonds that are electrostatic and hydrophobic in nature, for example, hydrogen bonds; important in determining complex biological structures. { ¦nän·kō¦vāl·ənt 'bänd }

noncrossing rule [PHYS CHEM] The rule that when the potential energies of two electronic states of a diatomic molecule are plotted as a function of distance between the nuclei, the resulting curves do not cross, unless the states have different symmetry. { ¦nän'krȯs·iŋ 'rül }

nonene See 1-nonylene. { 'nō,nēn }

nonfaradaic path [PHYS CHEM] One of the two available paths for transfer of energy across an electrolyte-metal interface, in which energy is carried by capacitive transfer, that is, by charging and discharging the double-layer capacitance. { ¦nän,far·ə'dā·ik 'path }

nonhypergolic [CHEM] Not capable of igniting spontaneously upon contact; used especially with reference to rocket fuels. { ,nän,hī·pər'gäl·ik }

nonideal solution [PHYS CHEM] A solution whose behavior does not conform to that of an ideal solution; that is, the behavior is not predictable over a wide range of concentrations and temperatures by the use of Raoult's law. { 'nän·ī,dēl sə·lü·shən }

nonine See nonyne. { 'nō,nīn }

nonlinear molecule |ORG CHEM| A branched-chain molecule, that is, one whose atoms do not all lie along a straight line. Also known as isomolecule. { 'nän,lin·ē·ər 'mäl·ə,kyül }

nonlinear spectroscopy [SPECT] The study of energy levels not normally accessible with optical spectroscopy, through the use of nonlinear effects such as multiphoton absorption and ionization. { 'nän,lin·ē·ər spek'träs·kə·pē }

nonlocalized bond See delocalized bond. { ¦nän'lō·kə,līzd 'bänd }

nonoic acid |ORG CHEM| $C_8H_{17}COOH$ Any of a family of acids which are mixed isomers produced in the Fischer-Tropsch process; pelargonic acid is the straight-chain member; used as a chemical intermediate. { nō'nō·ik 'as·əd }

nonpolar [CHEM] Pertaining to an element or compound which has no permanent electric dipole moment. { ¦nän'pō·lər }

nonpolar bond [PHYS CHEM] A type of covalent bond in which both atoms attract the bonding electrons equally or nearly equally. { ¦nän¦pōl·ər 'bänd }

nonpolar covalent bond [PHYS CHEM] A bond in which a pair of electrons is distributed or shared equally between two atoms. { ¦nän,pō·lər ¦kō,vā·lənt 'bänd }

nonpolar molecule [PHYS CHEM] A molecule with equal distribution of electrons among its atoms. { ¦nän,pō·lər 'mäl·ə,kyül }

nonprotic solvent [CHEM] A solvent that does not contain a hydrogen ion source. { ¦nän¦prōd·ik 'säl·vənt }

nonyl acetate |ORG CHEM| $C_9H_{19}OOCCH_3$ Any of a family of isomers, such as *n*-nonyl acetate and diisobutyl carbinyl acetate, which are products of Fischer-Tropsch and oxo syntheses. { 'nä,nil 'as·ə,tāt }

***n*-nonyl acetate** |ORG CHEM| $CH_3COO(CH_2)_8CH_3$ Alcohol-soluble, colorless liquid with pungent odor; boiling point 208–212°C; used in perfumery. { ¦en 'nä,nil 'as·ə,tāt }

***n*-nonyl alcohol** |ORG CHEM| $CH_3(CH_2)_7CH_2OH$ One of a family of $C_9H_{19}OH$ isomers; a colorless liquid with rose aroma; boils at 215°C; insoluble in water, soluble in alcohol; used in perfumery and flavorings. { ¦en 'nä,nil 'al·kə,hȯl }

263

nonyl benzene

nonyl benzene [ORG CHEM] $C_9H_{19}C_6H_5$ Liquid boiling at 245–252°C; straw-colored with aromatic aroma; used to make surface-active agents. { 'nä,nil 'ben,zēn }

1-nonylene [ORG CHEM] C_9H_{18} Colorless liquid boiling at 150°C; soluble in alcohol, insoluble in water; used as a chemical intermediate. Also known as nonene. { ¦wən 'nän·ə,lēn }

nonyl hydride *See* nonane. { 'nän·əl 'hī,drīd }

nonyl phenol [ORG CHEM] $C_9H_{19}C_6H_4OH$ Pale-yellow liquid boiling at 283–302°C; soluble in organic solvents, insoluble in water; a mixture of monoalkyl phenol isomers, mostly para-substituted; used to make surface-active agents, resins, and plasticizers. { 'nä,nil 'fē,nȯl }

nonylphenoxyacetic acid [ORG CHEM] $C_9H_{19}C_6H_4OCH_2COOH$ A viscous, amber-colored liquid, soluble in alkali; used in turbine oils, lubricants, greases, and other materials as a corrosion inhibitor. { ¦nä,nil·fe¦näk·sē·ə'sēd·ik 'as·əd }

nonyne [ORG CHEM] $CH_3(CH_2)_6{\equiv}CCH$ Water-insoluble, colorless liquid boiling at 160°C. { 'nō,nīn }

nopinene *See* pinene. { 'nä·pə,nēn }

nor- [CHEM] Chemical formula prefix for normal; indicates a parent for another compound to be formed by removal of one or more carbons and associated hydrogens. { nȯr }

nordihydroguaiaretic acid [ORG CHEM] $C_{18}H_{22}O_4$ A crystalline compound with a melting point of 184–185°C; soluble in alcohols, ether, acetone, glycerol, and propyleneglycol; used as an antioxidant in fats and oils. Abbreviated NDGA. { ¦nȯr·dī¦hī·drō¦gwī·ə¦red·ik 'as·əd }

normal bonded-phase chromatography [ANALY CHEM] A technique of bonded-phase chromatography in which the stationary phase is polar and the mobile phase is nonpolar. { 'nȯr·məl ¦ban·dəd ¦fāz ,krō·mə'täg·rə·fē }

normality [CHEM] Measure of the number of gram-equivalent weights of a compound per liter of solution. Abbreviated N. { nȯr'mal·əd·ē }

normal potassium pyrophosphate *See* potassium pyrophosphate. { 'nȯr·məl pə'tas·ē·əm ,pī·rō'fäs,fāt }

normal salt [CHEM] A salt in which all of the acid hydrogen atoms have been replaced by a metal, or the hydroxide radicals of a base are replaced by an acid radical; for example, Na_2CO_3. { 'nȯr·məl 'sȯlt }

normal silver sulfate *See* silver sulfate. { 'nȯr·məl 'sil·vər 'səl,fāt }

normal solution [CHEM] An aqueous solution containing one equivalent of the active reagent in grams in 1 liter of the solution. { 'nȯr·məl sə'lü·shən }

normal thorium sulfate *See* thorium sulfate. { 'nȯr·məl 'thȯr·ē·əm 'səl,fāt }

norphytane *See* pristane. { nȯr'fī,tān }

Np *See* neptunium.

NRS *See* nuclear reaction spectrometry.

NTA *See* nitrilotriacetic acid.

nuclear [CHEM] Pertaining to a group of atoms joined directly to the central group of atoms or central ring of a molecule. { 'nü·klē·ər }

nuclear atom [CHEM] An atomic structure consisting of dense, positively charged nucleus (neutrons and protons) surrounded by a corresponding set of negatively charged electrons. { 'nü·klē·ər 'ad·əm }

nuclear magnetic resonance spectrometer [SPECT] A spectrometer in which nuclear magnetic resonance is used for the analysis of protons and nuclei and for the study of changes in chemical and physical quantities over wide frequency ranges. { 'nü·klē·ər mag'ned·ik 'rez·ən·əns spek'träm·əd·ər }

nuclear reaction spectrometry [SPECT] A method of determining the concentration of a given element as a function of depth beneath the surface of a sample, by measuring the yield of characteristic gamma rays from a resonance reaction occurring when the surface is bombarded by a beam of ions. Abbreviated NRS. { 'nü·klē·ər rē'ak·shən spek'träm·ə·trē }

nuclear Zeeman effect [SPECT] A splitting of atomic spectral lines resulting from the interaction of the magnetic moment of the nucleus with an applied magnetic field. { 'nü·klē·ər 'zē·mən i‚fekt }

nucleation [CHEM] In crystallization processes, the formation of new crystal nuclei in supersaturated solutions. [PHYS CHEM] The formation of vapor bubbles in a superheated liquid. { ‚nü·klē'ā·shən }

nucleofuge *See* leaving group. { 'nü·klē·ə‚fyüj }

nucleophile [PHYS CHEM] A species possessing one or more electron-rich sites, such as an unshared pair of electrons, the negative end of a polar bond, or pi electrons. Also known as electron donor. { 'nü·klē·ə‚fīl }

nucleophilic displacement *See* nucleophilic substitution. { ‚nü·klē·ə¦fil·ik di'splās·mənt }

nucleophilic reagent [PHYS CHEM] A reactant that gives up electrons, or a share in electrons, to other molecules or ions in the course of a chemical reaction. { ¦nü·klē·ō¦fil·ik rē'ā·jənt }

nucleophilic substitution [ORG CHEM] A reaction in which a nucleophile bonds to a carbon atom in a molecule, displacing a leaving group. Also known as nucleophilic displacement. { ‚nü·klē·ə¦fil·ik ‚səb·stə'tü·shən }

Nylander reagent [CHEM] A solution of Rochelle salt (potassium sodium tartrate), potassium or sodium hydroxide, and bismuth subnitrate in water; used to test for sugar in urine. { 'nī·lən·dər rē‚ā·jənt }

O *See* oxygen.

Obermayer's reagent [CHEM] A 0.4% solution of ferric chloride in concentrated hydrochloric acid; used to test for indican in urine, with a pale-blue or deep-violet color indicating positive. { 'ō·bər₁mī·ərz rē₁ā·jənt }

n-octadecane [ORG CHEM] $C_{18}H_{38}$ Colorless liquid boiling at 318°C; soluble in alcohol, acetone, ether, and petroleum, insoluble in water; used as a solvent and chemical intermediate. { ¦en ₁äk·tə'de₁kān }

1-octadecene [ORG CHEM] $C_{18}H_{36}$ Colorless liquid boiling at 180°C; soluble in alcohol, acetone, ether, and petroleum, insoluble in water; used as a chemical intermediate. { ¦wən ₁äk·tə'de₁sēn }

octadecenoic acid *See* oleic acid. { ₁äk·tə'des·ə₁nō·ik 'as·əd }

octadecenyl aldehyde [ORG CHEM] $C_{17}H_{35}CHO$ A flammable liquid with a boiling point of 167°C; used in the manufacture of vulcanization accelerators, rubber antioxidants, and pesticides. { ₁äk·tə'des·ə₁nəl 'al·də₁hīd }

octafluorocyclobutane [ORG CHEM] C_4F_8 A colorless gas or liquid with a boiling point of −4°C and a freezing point of −41.4°C; soluble in ether; used as a dielectric, refrigerant, and aerosol propellant. { ¦äk·tə¦flür·ō₁sī·klō'byü₁tān }

octafluoropropane [ORG CHEM] C_3F_8 A colorless gas with a boiling point of −36.7°C and a freezing point of approximately −160°C; used as a refrigerant and gaseous insulator. { ¦äk·tə₁flür·ō'prō₁pān }

octahedral molecule [CHEM] A molecule whose structure forms an octahedron in which a central atom possesses six valence bonds that are directed to the points of the octahedron, for example, sulfur hexafluoride (SF_6). { ¦ak·tə¦hē·drəl 'mäl·ə₁kyül }

octamethylcyclotetrasiloxane [ORG CHEM] $C_8H_{24}O_4Si_4$ An oily colorless liquid with a boiling point of 175°C (315°F) and melting point of 17.5°C (63.5°F); a component of silicone gel. { ₁äk·tə₁meth·əl₁sīk·lə₁te·trə·sə'läk₁sān }

octanal *See* octyl aldehyde. { 'äk·tə₁nal }

n-octane [ORG CHEM] C_8H_{18} Colorless liquid boiling at 126°C; soluble in alcohol, acetone, and ether, insoluble in water; used as a solvent and chemical intermediate. { ¦en 'äk₁tān }

octanedioic acid *See* suberic acid. { ¦äk₁tan·dī'ō·ik 'as·əd }

1-octene [ORG CHEM] $CH_3(CH_2)_5CHCH_2$ A colorless, flammable liquid; used as a plasticizer and in synthesis of organic compounds. Also known as 1-caprylene; 1-octylene. { ¦wən 'äk₁tēn }

2-octene [ORG CHEM] $CH_3(CH_2)_4CHCHCH_3$ A colorless, flammable liquid, with trans and cis forms; used to manufacture lubricants and to synthesize organic materials. { ¦tü 'äk₁tēn }

octet rule [CHEM] A concept of chemical bonding theory based on the assumption that in the formation of compounds, atoms exhibit a tendency for their valence shells either to be empty or to have a full complement of eight electrons (octet); for some elements there are more than the usual eight valence electrons in some of their compounds. { äk₁tet ₁rül }

octomethylene *See* cyclooctane. { ¦äk·tō'meth·ə₁lēn }

octyl- [ORG CHEM] Prefix indicating the eight-carbon hydrocarbon radical ($C_8H_{17}-$). { 'äkt·əl }

***n*-octyl acetate** [ORG CHEM] $CH_3COO(CH_2)_7CH_3$ A colorless liquid with a fruity odor and a boiling point of 199°C; soluble in alcohol and other organic liquids; used for perfumes and flavoring. { ¦en 'äkt·əl 'as·ə,tāt }

octyl alcohol *See* 2-ethylhexyl alcohol. { 'äkt·əl 'al·kə,hòl }

octyl aldehyde [ORG CHEM] $C_8H_{16}O$ A liquid aldehyde boiling at 172°C; found in essential oils of many plants; used in perfume compositions. Also known as octanal. { 'äkt·əl 'al·də,hīd }

***n*-octyl *n*-decyl adipate** [ORG CHEM] A liquid with a boiling range of 250–254°C; used as a low-temperature plasticizer. Abbreviated NODA. { ¦en 'äkt·əl ¦en 'des·əl 'ad·ə,pāt }

***n*-octyl *n*-decyl phthalate** [ORG CHEM] A clear liquid with a boiling range of 232–267°C; used as a plasticizer for vinyl resins. { ¦en 'äkt·əl ¦en 'des·əl 'tha,lāt }

1-octylene *See* 1-octene. { ¦wən 'äk·tə,lēn }

octyl formate [ORG CHEM] $C_8H_{17}OOCH$ A colorless liquid with a fruity odor; soluble in mineral oil; used for flavoring. { 'äkt·əl 'fòr,māt }

***n*-octyl mercaptan** [ORG CHEM] $C_8H_{17}SH$ Clear, colorless liquid boiling at 199°C; used as a chemical intermediate and polymerization conditioner. { ¦en 'äkt·əl mər'kap,tan }

octyl phenol [ORG CHEM] $C_8H_{17}C_6H_4OH$ White flakes, congealing at 73°C; soluble in organic solvents, insoluble in water; used to make surfactants, plasticizers, and antioxidants. { 'äkt·əl 'fē,nòl }

octyne [ORG CHEM] $CHC(CH_2)_5CH_3$ Colorless hydrocarbon liquid, boiling at 125°C. { 'äk,tīn }

ODMR *See* optical detection of magnetic resonance.

-oic [ORG CHEM] A suffix indicating the presence of a −COOH group, as in ethyloic (−CH_2−COOH). { 'ō·ik }

oil blue [INORG CHEM] Violet-blue copper sulfide pigment used in varnishes. { 'òil 'blü }

oil of mirbane *See* nitrobenzene. { 'òil əv 'mər,bān }

oil of vitriol *See* sulfuric acid. { 'òil əv 'vit·rē,òl }

-ol [ORG CHEM] Chemical suffix for an −OH group in organic compounds, such as phenol (C_6H_5OH). { ,òl }

oleate [ORG CHEM] Salt made up of a metal or alkaloid with oleic acid; used for external medicines and in soaps and paints. { 'ō·lē,āt }

olefiant gas *See* ethylene. { 'ō·lə¦fī·ənt 'gas }

olefin [ORG CHEM] C_nH_{2n} A family of unsaturated, chemically active hydrocarbons with one carbon-carbon double bond; includes ethylene and propylene. { 'ō·lə·fən }

olefin copolymer [ORG CHEM] Polymer made by the interreaction of two or more kinds of olefin monomers, such as butylene and propylene. { 'ō·lə·fən kō'päl·ə·mər }

olefin resin [ORG CHEM] Long-chain polymeric material produced by the chain reaction of olefinic monomers, such as polyethylene from ethylene, or polypropylene from propylene. { 'ō·lə·fən 'rez·ən }

oleic acid [ORG CHEM] $C_{17}H_{33}COOH$ Yellowish, unsaturated fatty acid with lardlike aroma; soluble in organic solvents, slightly soluble in water; boils at 286°C (100 mmHg); the main component of olive and cooking oils; used in soaps, ointments, cosmetics, and ore beneficiation. Also known as octadecenoic acid. { ō'lā·ik 'as·əd }

olein [ORG CHEM] $(C_{17}H_{33}COO)_3C_3H_5$ Oleic acid triglyceride; yellow liquid melting at −5°C; slightly soluble in alcohol, soluble in chloroform, ether, and carbon tetrachloride; found in most fats and oils; used in textile lubrication. { 'ō·lē·ən }

oleum [CHEM] **1.** Latin name for oil. **2.** *See* fuming sulfuric acid. { 'ō·lē·əm }

oleyl alcohol [ORG CHEM] $C_{18}H_{35}OH$ Clear liquid, boiling at 282–349°C; fatty alcohol derived from oleic acid; commercial grade 80–90% pure; used to make resins and surface-active agents, and as a chemical intermediate. { ō'lē·əl 'al·kə,hòl }

oligomer [ORG CHEM] A molecule made up of a relatively small number of monomer units. { ə'lig·ə·mər }

oligopeptide [ORG CHEM] A peptide composed of no more than 10 amino acids. { ,äl·ə·gō'pep,tīd } optical spectroscopy

omethioate See folimat. { ,ō·mə'thī·ə,wāt }
ONB See ortho-nitrobiphenyl.
-one [ORG CHEM] Chemical suffix indicating a ketone, a substance related to starches and sugars, or an alkone. { ōn }
-onium [CHEM] Chemical suffix indicating a complex cation, as for hydronium, $(H_3O)^+$. { 'ō·nē·əm }
Onsager equation [PHYS CHEM] An equation which relates the measured equivalent conductance of a solution at a certain concentration to that of the pure solvent. { 'ön,säg·ər i,kwä·zhən }
open-circuit potential [PHYS CHEM] Steady-state or equilibrium potential of an electrode in absence of external current flow to or from the electrode. { 'ō·pən ¦sər·kət pə'ten·chəl }
open tubular column See capillary column. { 'ō·pən 'tü·byə·lər ,käl·əm }
opianyl See meconin. { 'ō·pē·ə,nil }
Oppenauer oxidation [ORG CHEM] The oxidation of a primary or secondary hydroxyl compound to form the corresponding carbonyl compound; aluminum alkoxide and an excess amount of a carbonyl hydrogen acceptor, such as benzophenone or acetone, are required. { 'äp·ə,naú·ər ,äk·sə'dā·shən }
optical anomaly [PHYS CHEM] The phenomenon in which an organic compound has a molar refraction which does not agree with the value calculated from the equivalents of atoms and other structural units composing it. { 'äp·tə·kəl ə'näm·ə·lē }
optical antipode See enantiomorph. { 'äp·tə·kəl 'ant·i,pōd }
optical cytopherometry See microelectrophoresis. { 'äp·tə·kəl ,sī·dō·fə'räm·ə·trē }
optical detection of magnetic resonance [SPECT] A type of electron paramagnetic resonance (EPR) spectroscopy that takes advantage of the sensitivity of electric dipole transitions, in which paramagnetic states are optically excited and the EPR signal is detected through changes in the optical absorption as the magnetic field is swept through one or more resonances. Abbreviated ODMR. { ¦äp·tə·kəl di,tek·shən əv mag,ned·ik 'rez·ən·əns }
optical exaltation [PHYS CHEM] Optical anomaly in which the observed molar refraction exceeds the calculated one; most cases of optical anomaly are in this category. { 'äp·tə·kəl ,ek·səl'tā·shən }
optical isomer See enantiomorph. { 'äp·tə·kəl 'ī·sə·mər }
optical isomerism [PHYS CHEM] Existence of two forms of a molecule such that one is a mirror image of the other; the two molecules differ in that rotation of light is equal but in opposite directions. { 'äp·tə·kəl ī'säm·ə,riz·ən }
optical monochromator [SPECT] A monochromator used to observe the intensity of radiation at wavelengths in the visible, infrared, or ultraviolet regions. { 'äp·tə·kəl ¦man·ə'kräm·əd·ər }
optical null method [SPECT] In infrared spectrometry, the adjustment of a reference beam's energy transmission to match that of a beam that has been passed through a sample being analyzed. { 'äp·tə·kəl ¦nəl ,meth·əd }
optical spectra [SPECT] Electromagnetic spectra for wavelengths in the ultraviolet, visible and infrared regions, ranging from about 10 nanometers to 1 millimeter, associated with excitations of valence electrons of atoms and molecules, and vibrations and rotations of molecules. { 'äp·tə·kəl 'spek·trə }
optical spectrograph [SPECT] An optical spectroscope provided with a photographic camera or other device for recording the spectrum made by the spectroscope. { 'äp·tə·kəl 'spek·trə,graf }
optical spectrometer [SPECT] An optical spectroscope that is provided with a calibrated scale either for measurement of wavelength or for measurement of refractive indices of transparent prism materials. { 'äp·tə·kəl spek'träm·əd·ər }
optical spectroscope [SPECT] An optical instrument, consisting of a slit, collimator lens, prism or grating, and a telescope or objective lens, which produces an optical spectrum arising from emission or absorption of radiant energy by a substance, for visual observation. { 'äp·tə·kəl 'spek·trə,skōp }
optical spectroscopy [SPECT] The production, measurement, and interpretation of

269

optoacoustic detection method

optical spectra arising from either emission or absorption of radiant energy by various substances.   { 'äp·tə·kəl spek'träs·kə·pē }

optoacoustic detection method   [ANALY CHEM]   A method of detecting trace inpurities in a gas, in which the absorption of a sample of the gas at various light frequencies is measured by directing a periodically interrupted laser beam through the sample in a spectrophone and measuring the sound generated by the optoacoustic effect at the frequency of interruption of the beam.   { ¦äp·tō·ə¦küs·tik di'tek·shən ˌmeth·əd }

optoacoustic spectroscopy   *See* photoacoustic spectroscopy.   { ¦äp·tō·ə¦kü·stik spek'träs·kə·pē }

optogalvanic spectroscopy   [SPECT]   A method of obtaining absorption spectra of atomic and molecular species in flames and electrical discharges by measuring voltage and current changes upon laser irradiation.   { ¦äp·tō·gal¦van·ik spek'träs·kə·pē }

orange oxide   *See* uranium trioxide.   { 'är·inj 'äk‚sīd }

orange spectrometer   [SPECT]   A type of beta-ray spectrometer that consists of a number of modified double-focusing spectrometers employing a common source and a common detector, and has exceptionally high transmission.   { 'är·inj spek'träm·əd·ər }

orange toner   [ORG CHEM]   A diazo dyestuff coupled to diacetoacetic acid anhydride; contains no sulfonic or carboxylic groups; used for printing inks.   { 'är·inj 'tōn·ər }

orbital overlap   [PHYS CHEM]   The overlapping of two electron orbitals, one from each of two different atoms, such that each orbital obtains a share in the electron of the other atom, forming a chemical bond.   { 'ȯr·bəd·əl 'ō·vər‚lap }

orbital symmetry   [PHYS CHEM]   The property of certain molecular orbitals of being carried into themselves or into the negative of themselves by certain geometrical operations, such as a rotation of 180° about an axis in the plane of the molecule, or reflection through this plane.   { 'ȯr·bəd·əl 'sim·ə·trē }

orcin   [ORG CHEM] $CH_3C_6H_3(OH)_2 \cdot H_2O$ White crystals with strong, sweet, unpleasant taste; soluble in water, alcohol, and ether; extracted from lichens; used in medicine and as an analytical reagent.   { 'ȯr·sən }

order   [CHEM]   A classification of chemical reactions, in which the order is described as first, second, third, or higher, according to the number of molecules (one, two, three, or more) which appear to enter into the reaction; decomposition of H_2O_2 to form water and oxygen is a first-order reaction.   { 'ȯrd·ər }

organic   [ORG CHEM]   Of chemical compounds, based on carbon chains or rings and also containing hydrogen with or without oxygen, nitrogen, or other elements.   { ȯr'gan·ik }

organic acid   [ORG CHEM]   A chemical compound with one or more carboxyl radicals (COOH) in its structure; examples are butyric acid, $CH_3(CH_2)_2COOH$, maleic acid, HOOCCHCHCOOH, and benzoic acid, C_6H_5COOH.   { ȯr'gan·ik 'as·əd }

organic chemistry   [CHEM]   The study of the structure, preparation, properties, and reactions of carbon compounds.   { ȯr'gan·ik 'kem·ə·strē }

organic pigment   [ORG CHEM]   Any of the materials with organic-chemical bases used to add color to dyes, plastics, linoleum, tones, and lakes.   { ȯr'gan·ik 'pig·mənt }

organic quantitative analysis   [ANALY CHEM]   Quantitative determination of elements, functional groups, or molecules in organic materials.   { ȯr'gan·ik 'kwän·ə‚tād·iv ə‚nal·ə·səs }

organic reaction mechanism   [ORG CHEM]   A pathway of chemical states traversed by an organic chemical system in its passage from reactants to products.   { ȯr'gan·ik rē'ak·shən ‚mek·ə‚niz·əm }

organic salt   [ORG CHEM]   The reaction product of an organic acid and an inorganic base, for example, sodium acetate (CH_3COONa) from the reaction of acetic acid (CH_3COOH) and sodium hydroxide (NaOH).   { ȯr'gan·ik 'sȯlt }

organic solvent   [ORG CHEM]   Liquid organic compound with the power to dissolve solids, gases, or liquids (miscibility); examples are methanol (methyl alcohol), CH_3OH, and benzene, C_6H_6.   { ȯr'gan·ik 'säl·vənt }

organoborane   [ORG CHEM]   A derivative of a borane (boron hydride) in which one or more hydrogen atoms have been replaced by functional groups.   { ȯr¦gan·ə'bȯr‚ān }

organometallic compound [ORG CHEM] Molecules containing carbon-metal linkage; a compound containing an alkyl or aryl radical bonded to a metal, such as tetraethyllead, $Pb(C_2H_5)_4$. { ȯr|gan·ə·mə'tal·ik 'käm,paůnd }

organophosphate [ORG CHEM] A soluble fertilizer material made up of organic phosphate esters such as glucose, glycol, or sorbitol; useful for providing phosphorus to deep-root systems. { ȯr|gan·ə'fäs,fāt }

organophosphorus compound [ORG CHEM] An organic compound that contains phosphorus in its chemical structure. { ȯr|gan·ə|fäs·fə·rəs 'käm,paůnd }

organoselenium compound [ORG CHEM] An organic compound that contains both selenium and carbon, and frequently other elements, such as the halogens, oxygen, sulfur, or nitrogen. { ȯr|gan·ə·sə|lē·nē·əm 'käm,paůnd }

organosilicon compound [ORG CHEM] A compound in which silicon is bonded to an organic functional group, either directly or indirectly via another atom. { ȯr|gan· ō'sil·ə·kən 'käm,paůnd }

organosulfur compound [ORG CHEM] One of a group of substances which contain both carbon and sulfur. { ȯr|gan·ə|səl·fər 'käm,paůnd }

orientation [PHYS CHEM] The arrangement of radicals in an organic compound in relation to each other and to the parent compound. { ,ȯr·ē·ən'tā·shən }

orientation effect [PHYS CHEM] A method of determining attractive forces among molecules, or components of these forces, from the interaction energy associated with the relative orientation of molecular dipoles. { ,ȯr·ē·ən'tā·shən i,fekt }

orientation force [PHYS CHEM] A type of van der Waals force, resulting from interaction of the dipole moments of two polar molecules. Also known as dipole-dipole force; Keesom force. { ,ȯr·ē·ən'tā·shən ,fȯrs }

Orsat analyzer [ANALY CHEM] Gas analysis apparatus in which various gases are absorbed selectively (volumetric basis) by passing them through a series of preselected solvents. { 'ȯr,sat 'an·ə,līz·ər }

ortho acid [ORG CHEM] **1.** Aromatic acid with a carboxyl group in the ortho position (1,2 position). **2.** Organic acid with one added molecule of water in chemical combination; for example, $HC(OH)_3$, orthoformic acid, in contrast to HCOOH, formic acid; $H_3PO_4(P_2O_5·3H_2O)$, orthophosphoric acid, in contrast to the less hydrated form, metaphosphoric acid, $HPO_3(P_2O_5·H_2O)$. { 'ȯr·thō 'as·əd }

orthoarsenic acid See arsenic acid. { |ȯr·thō·är|sen·ik 'as·əd }

orthoboric acid See boric acid. { |ȯr·thȯ|bȯr·ik 'as·əd }

orthophosphate [INORG CHEM] One of the possible salts of orthophosphoric acid; the general formula is M_3PO_4, where M may be potassium as in potassium orthophosphate, K_3PO_4. { |ȯr·thə'fäs,fāt }

orthophosphoric acid See phosphoric acid. { |ȯr·thə·fäs'fȯr·ik 'as·əd }

orthotungstic acid See tungstic acid. { |ȯr·thō'təŋ·stik 'as·əd }

Os See osmium.

oscillatory reaction [CHEM] A chemical reaction in which a variable of a chemical system exhibits regular periodic changes in time or in space. { 'äs·ə·lə,tȯr·ē rē'ak·shən }

oscillographic polarography [PHYS CHEM] A type of voltammetry using a dropping mercury electrode with oscillographic scanning of the applied potential; used to measure the concentration of electroactive species in solutions. { |äs·ə·lə|graf·ik ,pō·lə'räg·rə·fē }

oscillometric titration [PHYS CHEM] Radio-frequency technique used for conductometric and dielectrometric titrations; the changes in conductance or dielectric properties changes the solution capacity and thus the frequency of the connected oscillator circuit. { |äs·ə·lō|me·trik tī'trā·shən }

oscillometry [PHYS CHEM] Electrode measurement of oscillation-frequency changes to detect the progress of a titration of electrolytic solutions. { ,äs·ə'läm·ə·trē }

oscine See scopoline. { 'ä,sīn }

osmate [INORG CHEM] A salt or ester of osmic acid, containing the osmate radical, $OsO_4{}^{2-}$; for example, potassium osmate (K_2OsO_4). { 'äz·mət }

osmic acid anhydride [INORG CHEM] OsO_4 Poisonous yellow crystals with disagreeable

osmium

odor; melts at 40°C; soluble in water, alcohol, and ether; used in medicine, photography, and catalysis. Also known as osmium oxide; osmium tetroxide. { 'äz·mik 'as·əd an'hī‚drīd }

osmium [CHEM] A chemical element, symbol Os, atomic number 76, atomic weight 190.2. { 'äz·mē·əm }

osmium oxide See osmic acid anhydride. { 'äz·mē·əm 'äk‚sīd }

osmium tetroxide See osmic acid anhydride. { 'äz·mē·əm te'träk‚sīd }

osmolality [CHEM] The molality of an ideal solution of a nondissociating substance that exerts the same osmotic pressure as the solution being considered. { ‚äz·mə'lal·əd·ē }

osmolarity [CHEM] The molarity of an ideal solution of a nondissociating substance that exerts the same osmotic pressure as the solution being considered. { ‚äz·mə'lar·əd·ē }

osmole [CHEM] **1.** The unit of osmolarity equal to the osmolarity of a solution that exerts an osmotic pressure equal to that of an ideal solution of a nondissociating substance that has a concentration of 1 mole of solute per liter of solution. **2.** The unit of osmolality equal to the osmolality of a solution that exerts an osmotic pressure equal to that of an ideal solution of a nondissociating substance that has a concentration of 1 mole of solute per kilogram of solvent. { 'äz‚mōl }

osmometer [ANALY CHEM] A device for measuring molecular weights by measuring the osmotic pressure exerted by solvent molecules diffusing through a semipermeable membrane. { äz'mäm·əd·ər }

osmosis [PHYS CHEM] The transport of a solvent through a semipermeable membrane separating two solutions of different solute concentration, from the solution that is dilute in solute to the solution that is concentrated. { ä'smō·səs }

osmotic gradient See osmotic pressure. { äz'mäd·ik 'grād·ē·ənt }

osmotic pressure [PHYS CHEM] **1.** The applied pressure required to prevent the flow of a solvent across a membrane which offers no obstruction to passage of the solvent, but does not allow passage of the solute, and which separates a solution from the pure solvent. **2.** The applied pressure required to prevent passage of a solvent across a membrane which separates solutions of different concentration, and which allows passage of the solute, but may also allow limited passage of the solvent. Also known as osmotic gradient. { äz'mäd·ik 'presh·ər }

Ostwald coefficient [PHYS CHEM] A measure of the solubility of a gas in a liquid, equal to the volume of gas that can be dissolved in a given volume of liquid divided by the volume of liquid. { 'äst‚vält ‚kō·ə‚fish·ənt }

Ostwald dilution law [PHYS CHEM] The law that for a sufficiently dilute solution of univalent electrolyte, the dissociation constant approximates $a^2c/(1l-la)$, where c is the concentration of electrolyte and a is the degree of dissociation. { 'óst‚vält di'lü·shən ‚ló }

Ostwald ripening [CHEM] Solution-crystallizer phenomenon in which small crystals, more soluble than large ones, dissolve and reprecipitate onto larger particles. { 'óst‚vält ¦rīp·ə·niŋ }

Oswald diagram [ANALY CHEM] Diagram used in fuel Orsat analyses by plotting percent by volume CO_2 (carbon dioxide) maximum in the fuel [ordinate] versus percent by volume O_2 (oxygen) in air [abscissa]; O_2 and CO_2 Orsat readings should fall on a line connecting these maximum values if the analysis is proceeding properly. { 'äz‚wóld 'dī·ə‚gram }

ouabain [ORG CHEM] $C_{29}H_{44}O_{12}·8H_2O$ White crystals that melt with decomposition at 190°C, soluble in water and ethanol; used in medicine. { wä'bī·ən }

Oudeman law [PHYS CHEM] The law that the molecular rotations of the various salts of an acid or base tend toward an identical limiting value as the concentration of the solution is reduced to zero. { 'ód·ə·mən ‚ló }

outer orbital complex [PHYS CHEM] A metal coordination compound in which the d orbital used in forming the coordinate bond is at the same energy level as the s and p orbitals. { ‚aúd·ər 'órb·əd·əl 'käm‚pleks }

overall stability constant [ANALY CHEM] Reaction equilibrium constant for the reaction forming soluble complexes during compleximetric titration. { ¦ō·vər¦ól stə'bil·əd·ē ˌkän·stənt }

overpoint [CHEM] The initial boiling point in a distillation process; specifically, the temperature at which the first drop falls from the tip of the condenser into the condensate flask. { 'ō·vər,póint }

overtone band [SPECT] The spectral band associated with transitions of a molecule in which the vibrational quantum number changes by 2 or more. { 'ō·vər,tōn ˌband }

overvoltage [PHYS CHEM] The difference between electrode potential under electrolysis conditions and the thermodynamic value of the electrode potential in the absence of electrolysis for the same experimental conditions. Also known as overpotential. { ¦ō·vər¦vōl·tij }

ovex [ORG CHEM] ClC₆H₄OSO₂C₆H₄Cl A white, crystalline solid with a melting point of 86.5°C; soluble in acetone and aromatic solvents; used as an insecticide and acaricide. { 'ō,veks }

oxadiazon [ORG CHEM] C₁₃H₁₈Cl₂N₂O₃ A white solid with a melting point of 88–90°C; slight solubility in water; used as a pre- and postemergence herbicide to control weeds in rice, turf, soybeans, peanuts, and orchards. { ˌäk·sə'dī·ə,zän }

oxalate [ORG CHEM] Salt of oxalic acid; contains the (COO)₂ radical; examples are sodium oxalate, Na₂C₂O₄, ammonium oxalate, (NH₄)₂C₂O₄·H₂O, and ethyl oxalate, C₂H₅(C₂O₄)C₂H₅. { 'äk·sə,lāt }

oxalic acid [ORG CHEM] HOOCCOOH·2H₂O Poisonous, transparent, colorless crystals melting at 187°C; soluble in water, alcohol, and ether; used as a chemical intermediate and a bleach, and in polishes and rust removers. { äk'sal·ik 'as·əd }

oxalyl chloride [INORG CHEM] (COCl)₂ Toxic, colorless liquid boiling at 64°C; soluble in ether, benzene, and chloroform; used as a chlorinating agent and for military poison gas. { 'äk·sə,lil 'klór,īd }

oxamide [ORG CHEM] NH₂COCONH₂ Water-insoluble white powder, melting at 419°C; used as a stabilizer for nitrocellulose products. { 'äk·sə·məd }

oxamyl [ORG CHEM] C₇H₁₃N₃O₃S A white, crystalline compound with a melting point of 100–102°C; used to control pests of tobacco, ornamentals, fruits, and crops. { 'äk·sə,mil }

oxazole [ORG CHEM] C₃H₃ON A structure that consists of a five-membered ring containing oxygen and nitrogen in the 1 and 3 position; a colorless liquid (boiling point 69–70°C) that is miscible with organic solvents and water; used to prepare other organic compounds. { 'äk·sə,zōl }

oxidant See oxidizing agent. { 'äk·səd·ənt }

oxidation [CHEM] **1.** A chemical reaction that increases the oxygen content of a compound. **2.** A chemical reaction in which a compound or radical loses electrons, that is in which the positive valence is increased. { ˌäk·sə'dā·shən }

oxidation number [CHEM] **1.** Numerical charge on the ions of an element. **2.** See oxidation state. { ˌäk·sə'dā·shən ˌnəm·bər }

oxidation potential [PHYS CHEM] The difference in potential between an atom or ion and the state in which an electron has been removed to an infinite distance from this atom or ion. { ˌäk·sə'dā·shən pə,ten·chəl }

oxidation-reduction indicator [ANALY CHEM] A compound whose color in the oxidized state differs from that in the reduced state. { ˌäk·sə'dā·shən ri'dək·shən ˌin·də,kād·ər }

oxidation-reduction potential See redox potential. { ˌäk·sə'dā·shən ri'dək·shən pə ˌten·chəl }

oxidation-reduction reaction [CHEM] An oxidizing chemical change, where an element's positive valence is increased (electron loss), accompanied by a simultaneous reduction of an associated element (electron gain). { ˌäk·sə'dā·shən ri'dək·shən rē,ak·shən }

oxidation state [CHEM] The number of electrons to be added (or subtracted) from an atom in a combined state to convert it to elemental form. Also known as oxidation number. { ˌäk·sə'dā·shən ˌstāt }

oxide [CHEM] Binary chemical compound in which oxygen is combined with a metal (such as Na₂O; basic) or nonmetal (such as NO₂; acidic). { 'äk,sīd }

oxidizing agent [CHEM] Compound that gives up oxygen easily, removes hydrogen from another compound, or attracts negative electrons. Also known as oxidant. { 'äk·sə,dīz·iŋ ,ā·jənt }

oxidizing atmosphere [CHEM] Gaseous atmosphere in which an oxidation reaction occurs; usually refers to the oxidation of solids. { 'äk·sə,dīz·iŋ 'at·mə,sfir }

oxidizing flame [CHEM] A flame, or the portion of it, that contains an excess of oxygen. { 'äk·sə,dīz·iŋ ,flām }

oxime [ORG CHEM] Compound containing the CH(:NOH) radical; condensation product of hydroxylamine with aldehydes or ketones. { 'äk,sēm }

oximido See nitroso. { äk'sim·ə·dō }

oxine [ORG CHEM] C₉H₆NOH White powder that darkens when exposed to light; slightly soluble in water, dissolves in ethanol, acetone, and benzene; used to prepare fungicides and to separate metals by precipitation. Also known as 8-hydroxyquinoline; oxyquinoline; 8-quinolinol. { 'äk,sēn }

oxirane See epoxide; ethylene oxide. { 'äk·sə·rān }

oxo- [ORG CHEM] Chemical prefix designating the keto group, C:O. { 'äk·sō }

para-oxon [ORG CHEM] (C₂H₅O)₂P(O)C₆H₄NO₂ A reddish-yellow oil with a boiling point of 148–151°C; soluble in most organic solvents; used as an insecticide. Also known as diethyl para-nitrophenyl phosphate. { ¦par·ə 'äk,sän }

oxonium ion [CHEM] R₃O⁺ A cation in which an oxygen atom is covalently bound to three atoms or groups of atoms. { äk'sō·nē·əm 'ī,än }

oxosilane See siloxane. { ¦äk·sō'si,lān }

oxoxanthone See genicide. { ¦äk·sō'zan,thōn }

oxyacanthine [ORG CHEM] C₃₇H₄₀N₂O₆ An alkaloid obtained from the root of Berberis vulgaris; a white, crystalline powder with a melting point of 202–214°C; soluble in water, chloroform, benzene, alcohol, and ether; used in medicine. Also known as vinetine. { ¦äk·sē·ə'kan,thēn }

oxybenzone [ORG CHEM] C₁₄H₁₂O₃ A crystalline substance with a melting point of 66°C; used as a sunscreen agent. Also known as 4-methoxy-2-hydroxybenzophenone. { ¦äk·sē'ben,zōn }

oxycarboxin [ORG CHEM] C₁₂H₁₃NO₄S An off-white, crystalline compound with a melting point of 127.5–130°C; used to control rust disease in greenhouse carnations. Also known as 5,6-dihydro-2-methyl-1,4-oxathiin-3-carboxanilide-4,4-dioxide. { ¦äk·sē·kär'bäk·sən }

oxy compound [CHEM] A compound containing two or more oxygen atoms that are not joined to each other but are covalently bound to other atoms in the structure. { 'äk·sē ,käm,paůnd }

oxygen [CHEM] A gaseous chemical element, symbol O, atomic number 8, and atomic weight 15.9994; an essential element in cellular respiration and in combustion processes; the most abundant element in the earth's crust, and about 20% of the air by volume. { 'äk·sə·jən }

oxygen absorbent [CHEM] Any material that will absorb (dissolve) oxygen into its body without reacting with it. { 'äk·sə·jən əb'sòr·bənt }

oxygenate [CHEM] To treat, infuse, or combine with oxygen. { 'äk·sə·jə,nāt }

oxygen cell See aeration cell. { 'äk·sə·jən ,sel }

oxygen concentration cell See differential aeration cell. { ¦äk·sə·jən ,kän·sən'trā·shən ,sel }

oxygen-flask method [ANALY CHEM] Technique to determine the presence of combustible elements; the sample is burned with oxygen in a closed flask, and combustion products are absorbed in water of dilute alkali with subsequent analysis of the solution. { 'äk·sə·jən ¦flask ,meth·əd }

oxyhydrogen flame [CHEM] A flame obtained from the combustion of a mixture of oxygen and hydrogen. { ¦äk·sē'hī·drə·jən ¦flām }

oxyneurine See betaine. { ,äk·sē'nů,rēn }

oxyquinoline See oxine. { ,äk·sē'kwin·ə·lən }

ozone |CHEM| O_3 Unstable blue gas with pungent odor; an allotropic form of oxygen; a powerful oxidant boiling at $-112°C$; used as an oxidant, bleach, and water purifier, and to treat industrial wastes. { 'ō͵zōn }

ozonide |ORG CHEM| Any of the oily, thick, unstable compounds formed by reaction of ozone with unsaturated compounds; an example is oleic ozonide from the reaction of oleic acid and ozone. { 'äz·ə͵nīd }

ozonization |CHEM| The process of treating, impregnating, or combining with ozone. { ͵ō͵zō·nə'zā·shən }

ozonolysis |ANALY CHEM| The use of ozone to locate double bonds. |ORG CHEM| Oxidation of an organic substance by means of ozone. { ͵ō·zə'näl·ə·səs }

P

p- *See* para-.

P *See* phosphorus.

Pa *See* protactinium.

Paal-Knorr synthesis [ORG CHEM] A method of converting a 1,4-dicarbonyl compound by cyclization with ammonia or a primary amine to a pyrrole. { 'pòl kə'nòr 'sin·thə·səs }

Paar turbidimeter [ANALY CHEM] A visual-extinction device for measurement of solution turbidity; the length of the column of liquid suspension is adjusted until the light filament can no longer be seen. { pär ‚tər·bə'dim·əd·ər }

paired electron [PHYS CHEM] One of two electrons that form a valence bond between two atoms. { ¦perd i'lek‚trän }

palladium [CHEM] A chemical element, symbol Pd, atomic number 46, atomic weight 106.4. { pə'lād·ē·əm }

palladium chloride [INORG CHEM] PdCl₂ or PdCl₂·2H₂O Dark-brown, deliquescent powder that decomposes at 501°C; soluble in water, alcohol, acetone, and hydrochloric acid; used in medicine, analytical chemisty, photographic chemicals, and indelible inks. { pə'lād·ē·əm 'klòr‚īd }

palladium iodide [INORG CHEM] PdI₂ Black powder that decomposes above 100°C; soluble in potassium iodide solution, insoluble in water and alcohol. { pə'lād·ē·əm 'ī·ə‚dīd }

palladium nitrate [INORG CHEM] Pd(NO₃)₂ Brown, water-soluble, deliquescent salt; used as an analytical reagent. { pə'lād·ē·əm 'nī‚trāt }

palladium oxide [INORG CHEM] PdO Amber or black-green powder that decomposes at 750°C; soluble in dilute acids; used in chemical synthesis as a reduction catalyst. { pə'lād·ē·əm 'äk‚sīd }

palmitate [ORG CHEM] A derivative ester or salt of palmitic acid. { 'päm·ə‚tāt }

palmitic acid [ORG CHEM] C₁₅H₃₁COOH A fatty acid; white crystals, soluble in alcohol and ether, insoluble in water; melts at 63.4°C, boils at 271.5°C (100 mmHg); derived from spermaceti; used to make metallic palmitates and in soaps, waterproofing, and lubricating oils. { päl'mid·ik 'as·əd }

palmitoleic acid [ORG CHEM] C₁₆H₃₀O₂ An unsaturated fatty acid, found in marine animal oils; it is a clear liquid used as a standard in chromatography. { ¦päl·məd·ō¦lē·ik 'as·əd }

Paneth's adsorption rule [PHYS CHEM] The rule that an element is strongly absorbed on a precipitate which has a surface charge opposite in sign to that carried by the element, provided that the resulting adsorbed compound is very sparingly soluble in the solvent. { 'pan·əths ad'sòrp·shən ‚rül }

Papanicolaou's stains [CHEM] A group of stains used on exfoliated cells, particularly those from the vagina, for examination and diagnosis. { ‚pä·pə'nēk·ə‚laúz ‚stān }

papaverine [ORG CHEM] C₂₀H₂₁O₄N A white, crystalline alkaloid, melting at 147°C; soluble in acetone and chloroform, insoluble in water; used as a smooth muscle relaxant and weak analgesic, usually as the water-soluble hydrochloride salt. { pə'pav·ə‚rēn }

paper chromatography [ANALY CHEM] Procedure for analysis of complex chemical mixtures by the progressive absorption of the components of the unknown sample (in a solvent) on a special grade of paper. { 'pā·pər ‚krō·mə'täg·rə·fē }

paper electrochromatography

paper electrochromatography [ANALY CHEM] Variation of paper electrophoresis in which the electrolyte-impregnated absorbent paper is suspended vertically and the electrodes are connected to the sides of the paper, producing a current at right angles to the downward movement of the unknown sample. { 'pā·pər i,lek·trō,krō·mə'täg·rə·fē }

paper electrophoresis [ANALY CHEM] A variation of paper chromatography in which an electric current is applied to the ends of the electrolyte-impregnated absorbent paper, thus moving chargeable molecules of the unknown sample toward the appropriate electrode. { 'pā·pər i,lek·trə·fə're·səs }

paper-tape chemical analyzer [ANALY CHEM] Chemically treated paper tape that is continuously unreeled, exposed to the sample, and viewed by a phototube to measure the color change that is empirically related to changes in the sample's chemical composition. { 'pā·pər ¦tāp 'kem·ə·kəl 'an·ə,līz·ər }

para- [ORG CHEM] Chemical prefix designating the positions of substituting radicals on the opposite ends of a benzene nucleus, for example, paraxylene, $CH_3C_6H_4CH_3$. Abbreviated p-. { 'par·ə }

parabanic acid [ORG CHEM] $C_3H_2O_3N_2$ A water-soluble cyclic compound that decomposes when heated to about 227°C; used in organic synthesis. { ¦par·ə¦ban·ik 'as·əd }

paracyanogen [INORG CHEM] $(CN)_x$ A white solid produced by polymerization of cyanogen gas when heated to 400°C. { ¦par·ə·sī'an·ə·jən }

paraffinic hydrocarbon See alkane. { ,par·ə¦fin·ik 'hī·drə,kär·bən }

paraffinicity [ORG CHEM] The paraffinic nature or composition of crude petroleum or its products. { ,par·ə·fə'nis·əd·ē }

paraformaldehyde [ORG CHEM] $(HCHO)_n$ Polymer of formaldehyde where n is greater than 6; white, alkali-soluble solid, insoluble in alcohol, ether, and water; used as a disinfectant, fumigant, and fungicide, and to make resins. { ¦par·ə·fòr'mal·də,hīd }

paraldehyde [ORG CHEM] $C_6H_{12}O_3$ Acetaldehyde polymer; colorless, flammable, toxic liquid, miscible with most organic solvents, soluble in water; melts at 12.6°C, boils at 124.5°C; used as a chemical intermediate, in medicine, and as a solvent. { pə'ral·də,hīd }

paraldol [ORG CHEM] $(CH_3CHOHCH_2CHO)_2$ Water-soluble, white crystals, boiling at 90–100°C; used as a chemical intermediate, to make resins, and in cadmium plating baths. { par'al,dòl }

paramagnetic analytical method [ANALY CHEM] A method for analyzing fluid mixtures by measurement of the paramagnetic (versus diamagnetic) susceptibilities of materials when exposed to a magnetic field. { ¦par·ə·mag'ned·ik ,an·ə'lid·ə·kəl 'meth·əd }

paramagnetic spectra [SPECT] Spectra associated with the coupling of the electronic magnetic moments of atoms or ions in paramagnetic substances, or in paramagnetic centers of diamagnetic substances, to the surrounding liquid or crystal environment, generally at microwave frequencies. { ¦par·ə·mag'ned·ik 'spek·trə }

paranitraniline red See para red. { ¦par·ə·nə'tran·ə·lən 'red }

paraquat [ORG CHEM] $[CH_3(C_5H_4N)_2CH_3]·2CH_3SO_4$ A yellow, water-soluble solid, used as a herbicide. { 'par·ə,kwät }

para red [ORG CHEM] $C_{10}H_6(OH)NNC_6H_4NO_2$ Red pigment derived from the coupling of β-naphthol with diazotized paranitroaniline. Also known as paranitraniline red. { 'par·ə 'red }

pararosaniline [ORG CHEM] $HOC(C_6H_4NH_2)_3$ Red to colorless crystals, melting at 205°C; soluble in ethanol, in the hydrochloride salt; used as a dye. { ¦par·ə·rō'san·ə·lən }

parent compound [CHEM] A chemical compound that is the basis for one or more derivatives; for example, ethane is the parent compound for ethyl alcohol and ethyl acetate. { 'per·ənt ,käm,paúnd }

parent name [CHEM] That part of a chemical compound's name from which the name of a derivative comes; for example, ethane is the parent name for ethanol. { 'per·ənt ,nām }

partially ionic bond [PHYS CHEM] A chemical bond that is neither wholly ionic nor wholly covalent in character. { 'pär·shə·lē ī¦än·ik 'bänd }

partial molal quantity [PHYS CHEM] The molal concentration of one component of a mixture of components as related to total molal concentration for all components in the mixture.   { 'pär·shəl 'mō·ləl 'kwän·əd·ē }

partial molar volume   [PHYS CHEM] That portion of the volume of a solution or mixture related to the molar content of one of the components within the solution or mixture.   { 'pär·shəl 'mō·lər 'väl·yəm }

partial nucleate boiling   [PHYS CHEM] A stage in the boiling process in which isolated vapor bubbles are released from randomly located active sites on the heater surface.   { ¦pär·shəl ¦nü·klē͞‚āt 'bȯil·iŋ }

particle counting   [ANALY CHEM] Microscopic or photomicrographic technique for the visual counting of the numbers of particles in a known quantity of a solid-liquid suspension.   { 'pärd·ə·kəl ‚kau̇nt·iŋ }

particle electrophoresis   [PHYS CHEM] Electrophoresis in which the particles undergoing analysis are of sufficient size to be viewed either with the naked eye or with the assistance of an optical microscope.   { 'pärd·ə·kəl i‚lek·trō·fə'rē·səs }

particle-induced x-ray emission   [ANALY CHEM] A method of trace analysis in which a beam of ions is directed at a thin foil on which the sample to be analyzed has been deposited, and the energy spectrum of the resulting x-rays is measured.   { 'pärd·ə·kəl in¦düst 'eks‚rā i‚mish·ən }

particle-scattering factor   [ANALY CHEM] Factor in light-scattering equations used to compensate for the loss in scattered light intensity caused by destructive interference during the analysis of macromolecular compounds.   { 'pärd·ə·kəl ¦skad·ə·riŋ ‚fak·tər }

particle-thickness technique   [PHYS CHEM] A microscopic technique for visual measurement of the thickness of a fine particle (in the 3–100 micrometer range).   { 'pärd·ə·kəl ¦thik·nəs tek‚nēk }

partition chromatography   [ANALY CHEM] Chromatographic procedure in which the stationary phase is a high-boiling liquid spread as a thin film on an inert support, and the mobile phase is a vaporous mixture of the components to be separated in an inert carrier gas.   { pär'tish·ən ‚krō·mə'täg·rə·fē }

partition coefficient   [ANALY CHEM] In the equilibrium distribution of a solute between two liquid phases, the constant ratio of the solute's concentration in the upper phase to its concentration in the lower phase. Symbolized K.   { pär'tish·ən ‚kō·i‚fish·ənt }

parylene   [ORG CHEM] Polyparaxylylene, used in ultrathin plastic films for capacitor dielectrics, and as a pore-free coating.   { 'par·ə‚lēn }

PAS See photoacoustic spectroscopy.

Pascal rules   [PHYS CHEM] Rules which give the diamagnetic susceptiblity of a complex molecule in terms of the sum of the susceptibilities of its constituent atoms, and a correction factor which depends on the type of bonds linking the atoms.   { pa'skal ‚rülz }

Paschen-Back effect   [SPECT] An effect on spectral lines obtained when the light source is placed in a very strong magnetic field; the anomalous Zeeman effect obtained with weaker fields changes over to what is, in a first approximation, the normal Zeeman effect.   { 'päsh·ən 'bäk i‚fekt }

Paschen-Runge mounting   [SPECT] A diffraction grating mounting in which the slit and grating are fixed, and photographic plates are clamped to a fixed track running along the corresponding Rowland circle.   { 'päsh·ən 'ru̇ŋ·ə ‚mau̇nt·iŋ }

Paschen series   [SPECT] A series of lines in the infrared spectrum of atomic hydrogen whose wave numbers are given by $R_H [(1/9) - (1/n^2)]$, where R_H is the Rydberg constant for hydrogen, and n is any integer greater than 3.   { 'päsh·ən ‚sir·ēz }

passiflorin See harman.   { ¦pas·ə¦flȯr·ən }

passivation potential   [PHYS CHEM] The potential corresponding to the critical anodic current density on an electrode which behaves in an active-passive manner.   { ‚pas·ə'vā·shən pə‚ten·chəl }

passivity   [CHEM] A state of chemical inactivity, especially of a metal that is relatively resistant to corrosion due to loss of chemical activity.   { pə'siv·əd·ē }

Pasteur's salt solution [ANALY CHEM] Laboratory reagent consisting of potassium phosphate and calcium phosphate, magnesium sulfate, and ammonium tartrate in distilled water. { pa'stərz 'sȯlt sə,lü·shən }

Pauling scale [PHYS CHEM] A numerical scale of electronegativities based on bondenergy calculations for different elements joined by covalent bonds. { 'pȯl·iŋ ,skāl }

Pavy's solution [ANALY CHEM] Laboratory reagent used to determine the concentration of sugars in solution by color titration; contains copper sulfate, sodium potassium tartrate, sodium hydroxide, and ammonia in water solution. { 'pā·vēz sə,lü·shən }

Pb See lead.

p-block elements [CHEM] Elements of the main groups III-VII and 0 in the periodic table whose outer electronic configurations have occupied p levels. { ¦pē 'bläk 'el·ə,mənts }

P-branch [SPECT] A series of lines in molecular spectra that correspond, in the case of absorption, to a unit decrease in the rotational quantum number J. { 'pē ,branch }

PCB See polychlorinated biphenyl.

PCNB See pentachloronitrobenzene.

Pd See palladium.

PDMS See plasma desorption mass spectrometry.

peacock blue [ORG CHEM] $HSO_3C_6H_4COH[C_6H_4N(C_2H_5)CH_2C_6H_4SO_3Na]_2$ Blue pigment used in inks for multicolor printing. { 'pē,käk 'blü }

peak analysis [SPECT] Determination of the relevant peak parameters, such as position or area, from a spectogram. { 'pēk ə,nal·ə·səs }

peak area [ANALY CHEM] The area enclosed between the peak and the base line on a spectrogram or chromatogram. { 'pēk ,er·ē·ə }

peak enthalpimetry [ANALY CHEM] A thermochemical analytical procedure applicable to biochemical and chemical analyses; the salient feature is rapid mixing of a reagent stream with an isothermal solvent stream into which discrete samples are intermittently injected; peak enthalpograms result which exhibit the response characteristics of genuine differential detectors. { 'pēk ¦en·thəl'pim·ə·trē }

peak width [ANALY CHEM] In a gas chromatogram (plot of eluent rise and fall versus time), the width of the base (time duration) of a symmetrical peak (rise and fall) of eluent. { 'pēk 'width }

pearl hardening [INORG CHEM] Commerical name for a crystallized grade of calcium sulfate; used as a paper filler. { 'pərl ¦härd·ən·iŋ }

pelargonic acid [ORG CHEM] $CH_3(CH_2)_7CO_2H$ A colorless or yellowish oil, boiling at 254°C; soluble in ether and alcohol, insoluble in water; used as a chemical intermediate and flotation agent, in lacquers, pharmaceuticals, synthetic flavors and aromas, and plastics. { ¦pe,lär'gän·ik 'as·əd }

pellet technique See potassium bromide-disk technique. { 'pel·ət tek,nēk }

pellicular resins [ANALY CHEM] Glass spheres coated with a thin layer of ion-exchange resin, used in liquid chromatography. { pə'lik·yə·lər 'rez·ənz }

pellotine [ORG CHEM] $C_{13}H_{19}O_3N$ A colorless, crystalline alkaloid, derived from the dried cactus pellote, Lophophora williamsi (Mexico), slightly soluble in water; used as a hypnotic. { 'pel·ə,tēn }

pentabasic [CHEM] A description of a molecule that has five hydrogen atoms that may be replaced by metals or bases. { ¦pen·tə'bā·sik }

pentaborane [INORG CHEM] B_5H_9 Flammable liquid boiling at 48°C; ignites spontaneously in air; proposed as high-energy fuel for aircraft and missiles. { ¦pen·tə'bȯr,ān }

pentachloride [CHEM] A molecule containing five atoms of chlorine in its structure. { ¦pen·tə'klȯr,īd }

pentachloroethane [ORG CHEM] $CHCl_2CCl_3$ Colorless, water-insoluble liquid, boiling at 159°C; used as a solvent to degrease metals. Also known as pentalin. { ¦pen·tə,klȯr·ō'eth,ān }

pentachloronitrobenzene [ORG CHEM] $C_6Cl_5NO_2$ Crystals or cream color with a melting point of 142–145°C; slightly soluble in alcohols; used as a fungicide and herbicide. Abbreviated PCNB. Also known as quintozene; terrachlor. { ¦pen·tə¦klȯr·ō,nī·trə 'ben,zēn }

pentachlorophenol [ORG CHEM] C_6Cl_5OH A toxic white powder, decomposing at 310°C, melting at 190°C; soluble in alcohol, acetone, ether, and benzene; used as a fungicide, bactericide, algicide, herbicide, and chemical intermediate. { ¦pen·tə¦klȯr·ō·'fē͵nȯl }

pentacosane [ORG CHEM] $C_{25}H_{52}$ A water-insoluble hydrocarbon derived from beeswax. { ͵pen·tə'kō͵sān }

pentadecane [ORG CHEM] $C_{15}H_{32}$ A colorless, water-insoluble liquid, boiling at 270.5°C; soluble in alcohol; used as a chemical intermediate. { ¦pen·tə¦de͵kān }

pentadecanolide [ORG CHEM] $C_{15}H_{18}O_2$ A colorless liquid with a musky odor extracted from angelica oil; soluble in 90% ethyl alcohol in equal volume; used in perfumes. { ¦pen·tə·də'kan·əl͵īd }

pentadentate ligand [INORG CHEM] A chelating agent having five groups capable of attachment to a metal ion. Also known as quinquidentate ligand. { ¦pen·tə¦den͵tāt 'lī·gənd }

pentadiene [ORG CHEM] C_5H_8 Any of several straight-chain liquid diolefins: $CH_3CH_2CH=C=CH_2$, a colorless liquid boiling at 45°C, also known as ethylallene; $CH_3CH=CHCH=CH_2$, a colorless liquid boiling at 43°C; $CH_2=CHCH_2CH=CH_2$, a colorless liquid boiling at 26°C. { ¦pen·tə'dī͵ēn }

pentaerythritol [ORG CHEM] $(CH_2OH)_4C$ A white crystalline solid, melting at 261–262°C; moderately soluble in cold water, freely soluble in hot water; used to make the explosive pentaerythritol tetranitrate (PETN) and in the manufacture of alkyol resins and other coating compounds. { ¦pen·tə·ə'rith·rə͵tȯl }

pentaerythritol tetranitrate [ORG CHEM] $C(CH_2ONO_2)_4$ A white crystalline compound, melting at 139°C; explodes at 205–215°C; soluble in acetone, insoluble in water; used in medicines and explosives. Also known as penthrite; PETN. { ¦pen·tə·ə'rith·rə͵tȯl ͵te·trə'nī͵trāt }

pentaerythritol tetrastearate [ORG CHEM] $C(CH_2OOCC_{17}H_{35})_4$ A hard, ivory-colored wax with a softening point of 67°C; used in polishes and textile finishes. { ¦pen·tə·ə'rith·rə͵tȯl ͵te·trə'stir͵āt }

pentafluoride [CHEM] A chemical compound onto which five fluoride atoms are bonded. { ͵pen·tə'flu̇r͵īd }

pentalin See pentachloroethane. { 'pent·əl·ən }

n-pentane [ORG CHEM] $CH_3(CH_2)_3CH_3$ A colorless, flammable, water-insoluble hydrocarbon liquid, freezing at −130°C, boiling at 36°C; soluble in hydrocarbons and ethers; used as a chemical intermediate, solvent, and anesthetic. { ¦en 'pen͵tān }

1,5-pentanediol [ORG CHEM] $HOCH_2(CH_2)_3CH_2OH$ Colorless, water-miscible liquid boiling at 242.5°C; used as a hydraulic fluid, lube-oil additive, and antifreeze, and in manufacture of polyester and polyurethane resins. { ¦wən ¦fīv ͵pen͵tān'dī͵ȯl }

pentane insolubles [ANALY CHEM] Insoluble matter that can be separated from used lubricating oil in solution in n-pentane; may include resinous bitumens produced from the oxidation of oil and fuel; used in an American Society for Testing and Material test. { 'pen͵tān in'säl·yə·bəlz }

pentanol [ORG CHEM] $C_5H_{11}OH$ A toxic organic alcohol; 1-pentanol is n-amyl alcohol, primary; 2-pentanol is methylpropylcarbinol; 3-pentanol is diethylcarbinol; tert-pentanol is tert-amyl alcohol; pentanols are used in pharmaceuticals, as chemical intermediates, and as solvents. { 'pen·tə͵nȯl }

pentanone [ORG CHEM] Either of two isomeric ketones derived from pentane: $CH_3COC_3H_7$ is a flammable, colorless, clear liquid, a mixture of methyl propyl and diethyl ketones; insoluble in water, soluble in ether and alcohol; used as a solvent. $C_2H_5COC_2H_5$ is a colorless, flammable liquid with acetone aroma, boiling at 101°C; soluble in alcohol and ether; used in medicine and organic synthesis. { 'pen·tə͵nōn }

pentaprismane [ORG CHEM] $C_{10}H_{10}$ A highly strained, saturated hydrocarbon cage structure. { ͵pen·tə'priz͵mān }

pentavalent [CHEM] An atom or radical that exhibits a valency of 5. { ¦pen·tə'vā·lənt }

pentene [ORG CHEM] C_5H_{12} Colorless, flammable liquids derived from natural gasoline; isomeric forms are α-n-amylene and β-n-amylene. { 'pen͵tēn }

penthrite See pentaerythritol tetranitrate. { 'pen͵thrīt }

pentoglycerine See trimethylolethane. { ¦pen·tō'glis·ə·rən }

281

pentoxide

pentoxide [INORG CHEM] A compound that is binary and has five atoms of oxygen; for example, phosphorus pentoxide, P_2O_5. { pen'täk,sīd }

pentyl *See* amyl. { 'pent·əl }

para-pentyloxyphenol [ORG CHEM] $C_{11}H_{16}O_2$ Compound melting at 49–50°C; used as a bactericide. { ¦par·ə ¦pent·əl¦äk·sē'fē,nól }

pentyne [ORG CHEM] C_5H_8 Either of two normal isometric acetylene hydrocarbons: $HC≡C(CH_2)_2CH_3$, colorless liquid boiling at 40°C, also known as pentine, propylacetylene; $CH_3C≡CC_2H_5$, liquid boiling at 56°C. { 'pen,tēn }

peppermint camphor *See* menthol. { 'pep·ər,mint 'kam·fər }

peptide bond [ORG CHEM] A bond in which the carboxyl group of one amino acid is condensed with the amino group of another to form a −CO·NH− linkage. Also known as peptide linkage. { 'pep,tīd ,bänd }

peptide linkage *See* peptide bond. { 'pep,tīd ,liŋ·kij }

peptization [CHEM] **1.** Aggregation in which a hydrophobic colloidal sol is stabilized by the addition of electrolytes (peptizing agents) which are adsorbed on the particle surfaces. **2.** Liquefaction of a substance by trace amounts of another substance. { ,pep·tə,zā·shən }

per- [CHEM] Prefix meaning: **1.** Complete, as in hydrogen peroxide. **2.** Extreme, or the presence of the peroxy (−O−O−) group. **3.** Exhaustive (complete) substitution, as in perchloroethylene. { pər *or* per }

peracetic acid [ORG CHEM] CH_3COOH A toxic, colorless liquid with strong aroma; boils at 105°C; explodes at 110°C; miscible with water, alcohol, glycerin, and ether; used as an oxidizer, bleach, catalyst, bactericide, fungicide, epoxy-resin precursor, and chemical intermediate. Also known as peroxyacetic acid. { ¦par·ə¦sēd·ik 'as·əd }

peracid [CHEM] Acid containing the peroxy (−O−O−) group, such as peracetic acid or perchloric acid. { ¦pər'as·əd }

peralcohol [ORG CHEM] Chemical compound containing the peroxy group (−O−O−), such as peracetic acid and perchromic acid. { pər'al·kə,hól }

perbenzoic acid [ORG CHEM] $C_6H_5CO_2OH$ A crystalline compound forming leaflets from benzene solution, melting at 41–45°C; freely soluble in organic solvents; used in analysis of unsaturated compounds and to change ethylinic compounds into oxides. { ¦pər·ben¦zō·ik 'as·əd }

perchlorate [INORG CHEM] A salt of perchloric acid containing the ClO_4^- radical; for example, potassium perchlorate, $KClO_4$. { pər'klór,āt }

perchloric acid [INORG CHEM] $HClO_4$ Strongly oxidizing, corrosive, colorless, hygroscopic liquid, boiling at 16°C (8 mmHg, or 1067 pascals); soluble in water; unstable in pure form, but stable when diluted in water; used in medicine, electrolytic baths, electropolishing, explosives, and analytical chemistry, and as a chemical intermediate. Also known as Fraude's reagent. { pər'klór·ik 'as·əd }

perchloroethylene [ORG CHEM] CCl_2CCl_2 Stable, colorless liquid, boiling at 121°C; nonflammable and nonexplosive, with low toxicity; used as a dry-cleaning and industrial solvent, in pharmaceuticals and medicine, and for metal cleaning. Also known as tetrachloroethylene. { pər¦klór·ō'eth·ə,lēn }

perchloromethyl mercaptan [ORG CHEM] $ClSCCl_3$ Poisonous, yellow oil with disagreeable aroma; decomposes at 148°C; used as a chemical intermediate, granary fumigant, and military poison gas. { pər¦klór·ō'meth·əl mər'kap,tan }

perchloryl fluoride [INORG CHEM] $ClFO_3$ A colorless gas with a sweet odor; boiling point is −46.8°C and melting point is −146°C; used as an oxidant in rocket fuels. { pər'klór·əl 'flúr,īd }

perfect fractionation path [PHYS CHEM] On a phase diagram, a line or a path representing a crystallization sequence in which any crystal that has been formed remains inert, that is, its composition is not altered. { 'pər·fikt ,frak·shə'nā·shən ,path }

perfectly mobile component [PHYS CHEM] A component whose quantity in a system is determined by its externally imposed chemical potential rather than by its initial quantity in the system. Also known as boundary value component. { 'pər·fik·lē 'mō·bəl kəm'pō·nənt }

perfect solution [PHYS CHEM] A solution that is ideal throughout its entire compositional range. { 'pər·fikt sə'lü·shən }

perfluorocarbon [ORG CHEM] **1.** A compound consisting of carbon and fluorine. **2.** A compound in which all the hydrogen atoms of a hydrocarbon are replaced with fluorine atoms. Abbreviated PFC. { pər'flür·ə͵kär·bən }

perfluorochemical [ORG CHEM] A hydrocarbon in which all the hydrogen atoms have been replaced by fluorine. { pər¦flür·ō'kem·ə·kəl }

perhydro- [ORG CHEM] Prefix designating a completely saturated aromatic compound, as for decalin ($C_{10}H_{18}$), also known as perhydronaphthalene. { ¦pər'hī·drō }

pericondensed polycyclic [ORG CHEM] Referring to an aromatic compound in which three or more rings share common carbon atoms. { ¦per·ə·kən'denst ¦päl·ē¦sī·klik }

pericyclic reaction [ORG CHEM] Any one of a group of reactions that involve conjugated polyenes and proceed by single-step (concerted) reaction mechanisms. { ͵per·ə¦sīk·lik rē'ak·shən }

period [CHEM] A family of elements with consecutive atomic numbers in the periodic table and with closely related properties; for example, chromium through copper. { 'pir·ē·əd }

periodate [INORG CHEM] A salt of periodic acid, HIO_4, for example, potassium periodate, KIO_4. { ¦pər'ī·ə͵dāt }

periodic acid [INORG CHEM] $HIO_4 \cdot 2H_2O$ Water- and alcohol-soluble white crystals; loses water at 100°C; used as an oxidant. { ¦pir·ē¦äd·ik 'as·əd }

periodic law [CHEM] The law that the properties of the chemical elements and their compounds are a periodic function of their atomic weights. { ¦pir·ē¦äd·ik ¦lò }

periodic table [CHEM] A table of the elements, written in sequence in the order of atomic number or atomic weight and arranged in horizontal rows (periods) and vertical columns (groups) to illustrate the occurrence of similarities in the properties of the elements as a periodic function of the sequence. { ¦pir·ē¦äd·ik 'tā·bəl }

peritectic [PHYS CHEM] An isothermal reversible reaction in which a liquid phase reacts with a solid phase during cooling to produce a second solid phase. { ¦per·ə¦tek·tik }

peritectic point [PHYS CHEM] In a binary two-phase heteroazeotropic system at constant pressure, that point up to which the boiling point has remained constant until one of the phases has boiled away. { ¦per·ə¦tek·tik ͵pòint }

peritectoid [PHYS CHEM] An isothermal reversible reaction in which a solid phase on cooling reacts with another solid phase to form a third solid phase. { ͵per·ə'tek͵tòid }

Perkin reaction [ORG CHEM] The formation of unsaturated cinnamic-type acids by the condensation of aromatic aldehydes with fatty acids in the presence of acetic anhydride. { 'pər·kən rē͵ak·shən }

permanent hardness [CHEM] The hardness of water persisting after boiling. { 'pər·mə·nənt 'härd·nəs }

permanent-press resin [ORG CHEM] A thermosetting resin, based on chemicals such as formaldehyde and maleic anhydride, which is used to impart crease resistance to textiles and fibers. Also known as durable-press resin. { 'pər·mə·nənt ¦pres 'rez·ən }

permanganate [INORG CHEM] A purple salt of permanganic acid containing the MnO_4sw radical; used as an oxidizing agent and a disinfectant. { ¦pər'maŋ·gə͵nāt }

permanganic acid [INORG CHEM] $HMnO_4$ An unstable acid that exists only in dilute solutions; decomposes to manganese dioxide and oxygen. { ¦pər'man'gan·ik 'as·əd }

permeable membrane [CHEM] A thin sheet or membrane of material through which selected liquid or gas molecules or ions will pass, either through capillary pores in the membrane or by ion exchange; used in dialysis, electrodialysis, and reverse osmosis. { 'pər·mē·ə·bəl 'mem͵brān }

permeametry [ANALY CHEM] Determination of the average size of fine particles in a fluid (gas or liquid) by passing the mixture through a powder bed of known dimensions and recording the pressure drop and flow rate through the bed. { ͵pər·mē'äm·ə·trē }

permeant [CHEM] A material that permeates another material. { 'pər·mē·ənt }

permeation [CHEM] The movement of atoms, molecules, or ions into or through a porous or permeable substance (such as zeolite or a membrane). { ˌpər·mē'ā·shən }

permselective membrane [PHYS CHEM] An ion-exchange material that allows ions of one electrical sign to enter and pass through. { ¦pərm·si¦lek·tiv 'mem,brān }

peroxide [CHEM] A compound containing the peroxy (−O−O−) group, as in hydrogen peroxide. *See* hydrogen peroxide. { pə'räk,sīd }

peroxide number [ANALY CHEM] Measure of millimoles of peroxide (or milliequivalents of oxygen) taken up by 1000 grams of fat or oil; used to measure rancidity. Also known as peroxide value. { pə'räk,sīd ,nəm·bər }

peroxide value *See* peroxide number. { pə'räk,sīd ,val·yü }

peroxyacetic acid *See* peracetic acid. { pə¦räk·sē·ə¦sēd·ik 'as·əd }

peroxydol *See* sodium perborate. { pə'räk·sə,dȯl }

peroxynitrite [INORG CHEM] A nitrogen oxyanion containing an O-O peroxo bond that is a structural isomer of the nitrate ion. Species are generally distinguished by writing the chemical formula for peroxynitrite as $ONOO^-$ and nitrate as NO^{3-}. Other names that have been given to peroxynitrite include pernitrite and peroxonitrite; its recommended IUPAC name is oxoperoxonitrate(1-). { pə,räk·sē'nī,trīt }

Persian red [INORG CHEM] Red pigment made from basic lead chromate or ferric oxide. { 'pər·zhən 'red }

persulfate [INORG CHEM] Salt derived from persulfuric acid and containing the radical $S_2O_8^{2-}$; made by electrolysis of sulfate solutions. { ¦pər'səl,fāt }

persulfuric acid [INORG CHEM] $H_2S_2O_8$ Acid formed in lead-cell batteries by electrolyzing sulfuric acid; strong oxidizing agent. { ¦pər·səl'fyür·ik 'as·əd }

pervaporation [CHEM] A chemical separation technique in which a solution is placed in contact with a heated semipermeable membrane that selectively retains one of the components of a solution. { pər,vap·ə'rā·shən }

PET *See* polyethylene terephthalate. { ¦pē¦ē'tē or pet }

PETN *See* pentaerythritol tetranitrate.

petrochemicals [ORG CHEM] Chemicals made from feedstocks derived from petroleum or natural gas; examples are ethylene, butadiene, most large-scale plastics and resins, and petrochemical sulfur. Also known as petroleum chemicals. { ¦pe·trō'kem·ə·kəlz }

petrochemistry [ORG CHEM] The chemistry and reactions of materials derived from petroleum, natural gas, or asphalt deposits. { ¦pe·trō'kem·ə·strē }

petroleum chemicals *See* petrochemicals. { pə'trō·lē·əm ,kem·ə·kəlz }

petroleum resin [ORG CHEM] Any one of a family of polymers produced from mixed unsaturated monomers recovered from petroleum processing streams. { pə'trō·lē·əm ,rez·ən }

PFC *See* perfluorocarbon.

Pfund series [SPECT] A series of lines in the infrared spectrum of atomic hydrogen whose wave numbers are given by $R_H[(1/25) − (1/n^2)]$, where R_H is the Rydberg constant for hydrogen, and n is any integer greater than 5. { 'fünt ,sir·ēz }

pH [CHEM] A term used to describe the hydrogen-ion activity of a system; it is equal to $-\log a_H{}^+$; here $a_H{}^+$ is the activity of the hydrogen ion; in dilute solution, activity is essentially equal to concentration and pH is defined as $-\log_{10}[H^+]$, where H^+ is hydrogen-ion concentration in moles per liter; a solution of pH 0 to 7 is acid, pH of 7 is neutral, pH over 7 to 14 is alkaline. { pē'äch }

pharmacology [CHEM] The science dealing with the nature and properties of drugs, particularly their actions. { ˌfär·mə'käl·ə·jē }

phase [CHEM] Portion of a physical system (liquid, gas, solid) that is homogeneous throughout, has definable boundaries, and can be separated physically from other phases. { fāz }

phase diagram [PHYS CHEM] A graphical representation of the equilibrium relationships between phases (such as vapor-liquid, liquid-solid) of a chemical compound, mixture of compounds, or solution. { 'fāz ,dī·ə,gram }

phase equilibria [PHYS CHEM] The equilibrium relationships between phases (such as vapor, liquid, solid) of a chemical compound or mixture under various conditions of temperature, pressure, and composition. { 'fāz ‚ē·kwə'lib·rē·ə }

phase ratio [ANALY CHEM] In chromatography, the ratio of the volume of the mobile phase to that of the stationary phase in a chromatographic column. { 'fāz ‚rā·shō }

phase rule See Gibbs phase rule. { 'fāz rül }

phase solubility [PHYS CHEM] The different solubilities of a sample's solid constituents (phases) in a selected solvent. { 'fāz ‚säl·yə‚bil·əd·ē }

phase-solubility analysis [ANALY CHEM] Solvent technique used to determine the amount and number of components in a solid substance; the weight of sample added to the solvent is plotted against the weight of sample dissolved, with breakpoints in the curve occurring with each progressive saturation of the solvent with respect to each of the components; can be combined with extraction and recrystallization procedures. { 'fāz ‚säl·yə‚bil·əd·ē ə‚nal·ə·səs }

phase titration [ANALY CHEM] Analysis of a binary mixture of miscible liquids by titrating with a third liquid that is miscible with only one of the components, using the ternary phase diagram to determine the end point. { 'fāz tī‚trā·shən }

phase transfer catalysis [ORG CHEM] Enhancement of the reaction rate of a two-phase organic-water system by addition of a catalyst which alters the rate of transfer of water-soluble reactant across the interface to the organic phase. { 'fāz ‚trans‚fər kə'tal·ə·səs }

pH electrode [ANALY CHEM] Membrane-type glass electrode used as the hydrogen-ion sensor of most pH meters; the pH-response electrode surface is a thin membrane made of a special glass. { ‚pē'āch i'lek‚trōd }

α-phellandrene [ORG CHEM] $C_{10}H_{16}$ A colorless oil soluble in ether; boiling point of *d*-optical isomer is 66–68°C, of *l*-optical isomer is 58–59°C; used in flavoring and perfumes. { ‚al·fə fə'lan‚drēn }

phenanthrene [ORG CHEM] $C_{14}H_{10}$ A colorless, crystalline hydrocarbon; melts at about 100°C; the nucleus is produced by the degradation of certain alkaloids; used in the synthesis of dyes and drugs. { fə'nan‚thrēn }

phenanthroline [ORG CHEM] $C_{12}H_8N_2$ Any of three nitrogen bases related to phenanthrene; the ortho form is an oxidation-reduction indicator, turning faint blue when oxidized. { fə'nan·thrə‚lēn }

phenanthroline indicator [ANALY CHEM] A sensitive, red-colored specific reagent for iron. { fə'nan·thrə‚lēn 'in·də‚kād·ər }

phenarsazine chloride [ORG CHEM] $C_{12}H_9AsClN$ A yellow, crystalline compound obtained as a precipitate from carbon tetrachloride solutions; it sublimes readily, and is slightly soluble in xylene, benzene, and carbon tetrachloride; used as a war gas. Also known as adamsite. { fə'när·sə‚zēn 'klȯr‚īd }

phenazine [ORG CHEM] $C_6H_4N_2C_6H_4$ Yellow crystals, melting at 170°C; slightly soluble in water, soluble in alcohol and ether; used as chemical intermediate and to make dyes. { 'fen·ə‚zēn }

phenethyl acetate [ORG CHEM] $C_6H_5CH_2CH_2OOCCH_3$ A colorless liquid with a peachlike odor and a boiling point of 226°C; soluble in alcohol, ether, and fixed oils; used in perfumes. Also known as phenylethyl acetate. { fen'eth·əl 'as·ə‚tāt }

phenethyl alcohol [ORG CHEM] $C_8H_{10}O$ A liquid with a floral odor found in many natural essential oils; soluble in 50% alcohol; used in perfumes and flavors, and in medicine as an antibacterial agent in diseases of the eye. { fen'eth·əl 'al·kə‚hȯl }

phenethyl isobutyrate [ORG CHEM] $(CH_3)_2CHCOOC_2H_4C_6H_5$ A colorless liquid, soluble in alcohol and ether; used in perfumes and flavoring. { fen'eth·əl ‚ī·sō'byüd·ə‚rāt }

phenetidine [ORG CHEM] $NH_2C_6H_4OC_2H_5$ Either of two toxic, oily liquids that darken when exposed to light and air; soluble in alcohol, insoluble in water; the ortho form boils at 228–230°C, is used to make dyes, and is also known at 2-aminophenetole; the para form boils at 253–255°C, is used to make dyes and in pharmaceuticals. { fə'ned·ə‚dēn }

phenol [ORG CHEM] **1.** C_6H_5OH White, poisonous, corrosive crystals with sharp, burning taste; melts at 43°C, boils at 182°C; soluble in alcohol, water, ether, carbon disulfide,

285

phenol coefficient

and other solvents; used to make resins and weed killers, and as a solvent and chemical intermediate. Also known as carbolic acid; phenylic acid. **2.** A chemical compound based on the substitution product of phenol, for example, ethylphenol ($C_2H_4C_4H_5OH$), the ethyl substitute of phenol. { 'fē,nȯl }

phenol coefficient [ANALY CHEM] Number scale for comparison of antiseptics, using the efficacy of phenol as unity. { 'fē,nȯl ‚kō·i‚fish·ənt }

phenol-coefficient method [CHEM] A method for evaluating water-miscible disinfectants in which a test organism is added to a series of dilutions of the disinfectant; the phenol coefficient is the number obtained by dividing the greatest dilution of the disinfectant killing the test organism by the greatest dilution of phenol showing the same result. { 'fē,nȯl ‚kō·i‚fish·ənt ‚meth·əd }

phenol-formaldehyde resin [ORG CHEM] Thermosetting resin made by the reaction of phenol and formaldehyde; has good strength and chemical resistance and low cost; used as a molding material for mechanical and electrical parts. Originally known as Bakelite. { 'fē,nȯl fər'mal·də‚hīd ‚rez·ən }

phenol-furfural resin [ORG CHEM] A phenolic resin characterized by the ability to be fabricated by injection molding since it hardens after curing conditions are reached. { 'fē,nȯl 'fər·fə‚ral ‚rez·ən }

phenolphthalein [ORG CHEM] ($C_6H_4OH)_2COC_6H_4CO$ Pale-yellow crystals; soluble in alcohol, ether, and alkalies, insoluble in water; used as an acid-base indicator (carmine-colored to alkalies, colorless to acids) for titrations, as a laxative and dye, and in medicine. { ‚fē,nȯl'thal·ē·ən }

phenol red See phenolsulfonphthalein. { 'fē,nȯl 'red }

phenolsulfonic acid [ORG CHEM] $C_6H_5SO_3H$ Water- and alcohol-soluble mixture of ortho- and para-phenolsulfonic acids; yellowish liquid that turns brown when exposed to air; used as a chemical intermediate and in water analysis. Also known as sulfocarbolic acid. { ‚fē,nȯl·səl'fän·ik 'as·əd }

phenolsulfonphthalein [ORG CHEM] $C_{19}H_{14}O_5S$ A bright-red, crystalline compound, soluble in water, alcohol, and acetone; used as a pH indicator, and to test for kidney function in dogs. Also known as phenol red. { ‚fē,nȯl‚səl‚fōn·'thal·ē·ən }

phenothiazine [ORG CHEM] $C_{12}H_9N$ A yellow, crystalline compound, forming rhomboid leaflets or diamond-shaped plates, obtained from toluene or butanol solution; soluble in hot acetic acid, benzene, and ether; used as an insecticide and in pharmaceutical manufacture. { 'fē·nə'thī·ə‚zēn }

phenotole [ORG CHEM] $C_6H_5OC_2H_5$ Combustible, colorless liquid; boiling at 172°C; soluble in alcohol and ether, insoluble in water. { 'fē·nə‚tōl }

phenoxyacetic acid [ORG CHEM] $C_6H_5OCH_2COOH$ A light tan powder with a melting point of 98°C; soluble in ether, water, carbon disulfide, methanol, and glacial acetic acid; used in the manufacture of pharmaceuticals, pesticides, fungicides, and dyes. { fə'näk·sē·ə‚sēd·ik 'as·əd }

phenoxybenzamine hydrochloride [ORG CHEM] $C_{18}H_{22}ONCl·HCl$ White crystals, slightly soluble in water, melting at 139°C; used in medicine. { fə'näk·sē'ben·zə‚mēn ‚hī·drə'klȯr‚īd }

2-phenoxyethanol [ORG CHEM] $C_6H_5OCH_2CH_2OH$ An oily liquid with a faint aromatic odor; melting point is 14°C; soluble in alcohol and ether; used in perfumes as a fixative, in organic synthesis, as an insect repellent, and as a topical anesthetic. { ‚tü fə‚näk·sē'eth·ə‚nȯl }

phenoxypropanediol [ORG CHEM] $C_9H_{12}O_3$ A white, crystalline solid with a melting point of 53°C; soluble in water, alcohol, glycerin, and carbon tetrachloride; used in medicine and as a plasticizer. { fə'näk·sē‚prō·pān'dī‚ȯl }

phenoxy resin [ORG CHEM] A high-molecular-weight thermoplastic polyether resin based on bisphenol-A and epichlorohydrin with bisphenol-A terminal groups; used for injection molding, extrusion, coatings, and adhesives. { fə'näk·sē 'rez·ən }

phentolamine hydrochloride [ORG CHEM] $C_{17}H_{19}ON_2·HCl$ White, water-soluble crystals, melting at 240°C; a sympatholytic; used in medicine. { fen'täl·ə‚mēn ‚hī·drə'klȯr‚īd }

phenyl [ORG CHEM] C_6H_5- A functional group consisting of a benzene ring from which a hydrogen has been removed. { 'fen·əl }

286

phenylacetaldehyde [ORG CHEM] C_8H_8O A colorless liquid with a boiling point of 193–194°C; soluble in ether and fixed oils; used in perfumes and flavoring. Also known as α-toluic aldehyde. { 'fen·əl,as·ə'täl·də,hīd }

phenylacetic acid [ORG CHEM] $C_8H_8O_2$ White crystals with a boiling point of 262°C; soluble in alcohol and ether; used in perfumes, medicine, and flavoring and in the manufacture of penicillin. Also known as α-toluic acid. { ¦fen·əl·ə'sēd·ik 'as·əd }

phenylaniline See diphenylamine. { ¦fen·əl'an·ə·lən }

N-phenylanthranilic acid [ORG CHEM] $(C_6H_5NH)C_6H_4COOH$ A crystalline compound, soluble in hot alcohol; decomposes at 183–184°C; used to detect vanadium in steel. { ¦en ¦fen·əl,an·thrə¦nil·ik 'as·əd }

phenylbenzene See biphenyl. { ¦fen·əl'ben,zēn }

phenylbutazone [ORG CHEM] $C_{19}H_{20}O_2N_2$ White or light-yellow powder with aromatic aroma and bitter taste; melts at 107°C; slightly soluble in water, soluble in acetone; used in medicine as an analgesic and antipyretic. Also known as butazolidine. { ¦fen·əl'byüd·ə,zōn }

phenyl cyanide See benzonitrile. { ¦fen·əl 'sī·ə,nīd }

phenylcyclohexane [ORG CHEM] $C_{12}H_{16}$ A colorless, oily liquid with a boiling point of 237.5°C; soluble in alcohol, benzene, castor oil, carbon tetrachloride, xylene, and hexane; used as a high-boiling solvent and a penetrating agent. { ¦fen·əl,sī·klō'hek,sān }

phenyldichloroarsine [ORG CHEM] $C_6H_5AsCl_2$ A liquid which becomes a microcrystalline mass at −20°C (melting point) and decomposes in water; soluble in alcohol, ether, and benzene; used as a poison gas. { ¦fen·əl·dī,klór·ō'är,sēn }

phenyl diglycol carbonate [ORG CHEM] $C_{18}H_{18}O_7$ A colorless solid with a melting point of 40°C; soluble in organic solvents; used as a plasticizer. { ¦fen·əl dī'glī,kól 'kär·bə,nät }

phenylene blue See indamine. { 'fən·əl,ēn 'blü }

phenylenediamine [ORG CHEM] $C_6H_4(NH_2)_2$ Also known as diaminobenzene. Any of three toxic isomeric crystalline compounds that are diamino derivatives of benzene; the ortho form, toxic colorless crystals melting at 102–104°C and soluble in alcohol, ether, water, and chloroform, is used to manufacture dyes, in photographic developers, and as a chemical intermediate; the meta form, colorless crystals unstable in air, melting at 63°C, and soluble in alcohol, ether, and water, is used to manufacture dyes, in textile dyeing, and as a nitrous acid detector; the para form, white to purple crystals melting at 147°C, soluble in alcohol and ether, and irritating to the skin, is used to manufacture dyes, in chemical analysis, and in photographic developers. { ¦fen·əl,ēn'dī·ə,mēn }

phenyl ether See diphenyl oxide. { 'fen·əl 'eth·ər }

phenylethyl acetate See phenethyl acetate. { ¦fen·əl¦eth·əl 'as·ə,tāt }

phenylethylene See styrene. { ¦fen·əl'eth·ə,lēn }

phenyl fluoride See fluorobenzene. { 'fen·əl 'flúr,īd }

N-phenylglycine [ORG CHEM] $C_6H_5NHCH_2COOH$ A crystalline compound, moderately soluble in water, melting at 127–128°C; used in dye manufacture (indigo). { ¦en ¦fen·əl'glī,sēn }

phenylglyoxylonitriloxime O,O-diethyl phosphorothioate [ORG CHEM] $(H_5C_2O)_2PSO$-NCCNC$_6$H$_5$ A yellow liquid with a boiling point of 102°C at 0.01 mmHg (1.333 pascals); solubility in water is 7 parts per million at 20°C; used as an insecticide for stored products. Also known as phoxim. { ¦fen·əl·glī¦äk·sē¦län·ə·trəl'äk,sēm ¦ō¦ō dī'eth·əl ,fäs·fə·rō'thī·ə,wāt }

phenylhydrazine [ORG CHEM] $C_6H_5NHNH_2$ Poisonous, oily liquid, boiling at 244°C; soluble in alcohol, ether, chloroform, and benzene, slightly soluble in water; used in analytical chemistry to detect sugars and aldehydes, and as a chemical intermediate. Also known as hydrazinobenzene. { ¦fen·əl'hī·drə,zēn }

phenyl ketone See benzophenone. { 'fen·əl 'kē,tōn }

phenyl mercaptan See thiophenol. { 'fen·əl mər'kap,tan }

phenylmercuric acetate [ORG CHEM] $C_8H_8O_2Hg$ White to cream-colored prisms with a melting point of 148–150°C; soluble in alcohol, benzene, and glacial acetic acid;

phenylmercuric chloride

used as an antiseptic, fungicide, herbicide, and mildewcide.   { ¦fen·əl·mər'kyür·ik 'as·ə͵tāt }

phenylmercuric chloride   [ORG CHEM] C_6H_5HgCl White crystals with a melting point of 251°C; soluble in benzene and ether; used as an antiseptic and fungicide.   { ¦fen·əl·mər'kyür·ik 'klȯr͵īd }

phenylmercuric hydroxide   [ORG CHEM] C_6H_5HgOH White to cream-colored crystals with a melting point of 197–205°C; soluble in acetic acid and alcohol; used as a fungicide, germicide, and alcohol denaturant.   { ¦fen·əl·mər'kyür·ik hi'dräk͵sīd }

phenylmercuric oleate   [ORG CHEM] $C_{41}H_{21}O_2Hg$ A white, crystalline powder with a melting point of 45°C; soluble in organic solvents; used in paints as a mildew-proofing agent, and as a fungicide.   { ¦fen·əl·mər'kyür·ik 'ō·lē͵āt }

phenylmercuric propionate   [ORG CHEM] $C_9H_{10}O_2Hg$ A white, waxlike powder with a melting point of 65–70°C; used in paints as a fungicide and bactericide.   { ¦fen·əl·mər'kyür·ik 'prō·pē·ə͵nāt }

phenylmercuriethanolammonium acetate   [ORG CHEM] $C_{10}H_{15}O_3NHg$ A white, water-soluble, crystalline solid; used as an insecticide and fungicide.   { ¦fen·əl·mər¦kyür·ik¦eth·ə͵nȯl·ə'mō·nē·əm 'as·ə͵tāt }

phenylmethane   See toluene.   { ¦fen·əl'meth͵ān }

phenylmethanol   See benzyl alcohol.   { ¦fen·əl'meth·ə͵nȯl }

phenylmethyl acetate   See benzyl acetate.   { ¦fen·əl¦meth·əl 'as·ə͵tāt }

N-phenylmorpholine   [ORG CHEM] $C_{10}H_{13}NO$ A white, water-soluble solid with a melting point of 57°C; used in the manufacture of dyestuffs, corrosion inhibitors, and photographic developers, and as an insecticide.   { ¦en ¦fen·əl'mȯr·fə͵lēn }

phenyl mustard oil   [ORG CHEM] C_6H_5NCS A pale yellow or colorless liquid with a boiling point of 221°C; soluble in alcohol and ether; used in medicine.   { 'fen·əl 'məs·tərd ͵ȯil }

phenylphenol   [ORG CHEM] $C_6H_5C_6H_4OH$ Almost white crystals, soluble in alcohol, insoluble in water; the ortho form, melting at 56–58°C, is used to manufacture dyes, as germicide and fungicide, and in the rubber industry, and is also known as 2-hydroxybiphenyl, *ortho*-xenol; the para form, melting at 164–165°C, is used to manufacture dyes, resins, and rubber chemicals, and as a fungicide.   { ¦fen·əl'fē͵nȯl }

N-phenylpiperazine   [ORG CHEM] $C_{10}H_{14}N_2$ A pale yellow oil with a boiling point of 286.5°C; soluble in alcohol and ether; used for pharmaceuticals and in the manufacture of synthetic fibers.   { ¦en ¦fen·əl·pī'per·ə͵zēn }

phenylpropane   See propyl benzene.   { ¦fen·əl'prō͵pān }

1-phenyl-1-propanol   [ORG CHEM] $C_6H_5CH(OH)CH_2CH_3$ An oily liquid that has a weak esterlike odor; miscible with methanol, ethanol, ether, benzene, and toluene; used in industry as a heat transfer medium, in the manufacture of perfumes, and as a choleretic in medicine.   { ¦wən ¦fen·əl ¦wən 'prō·pə͵nȯl }

phenylpropyl alcohol   [ORG CHEM] $C_9H_{12}O$ A colorless liquid with a floral odor and a boiling point of 219°C; soluble in 70% alcohol; used in perfumes and flavoring.   Also known as hydrocinnamic alcohol.   { ¦fen·əl¦prō·pəl 'al·kə͵hȯl }

phenylpropyl aldehyde   [ORG CHEM] $C_9H_{10}O$ A colorless liquid with a floral odor; soluble in 50% alcohol; used in perfumes and flavoring.   Also known as hydrocinnamic aldehyde.   { ¦fen·əl¦prō·pəl 'al·də͵hīd }

1-phenyl-3-pyrazolidinone   [ORG CHEM] $C_9H_{10}N_2O$ A crystalline compound soluble in dilute aqueous solutions of acids and alkalies; melting point is 121°C; used as a high-contrast photographic developer.   { ¦wən ¦fen·əl ¦thrē ͵pī·rə·zə'lid·ən͵ōn }

phenyl salicylate   See salol.   { ¦fen·əl sə'lis·ə͵lāt }

phenylthiourea   [ORG CHEM] $C_6H_5NHCSNH_2$ A crystalline compound that has either a bitter taste or is tasteless, depending on the heredity of the taster; used in human genetics studies.   { ¦fen·əl¦thī·ō·yú'rē·ə }

***ortho*-phenyl tolyl ketone**   [ORG CHEM] $CH_3C_6H_4COC_6H_5$ An oily liquid with a boiling point of 309–311°C; soluble in alcohol, oils, and organic solvents; used as a fixative in perfumery.   { ¦ȯr·thō ¦fen·əl 'tä͵lil ¦kē͵tōn }

phenyltrichlorosilane   [ORG CHEM] $C_6H_5SiCl_3$ A liquid with a sharp odor [boiling point

201°C (410°F), melting point −31°C (−23.8°F)] used in the manufacture of various silicone oligomers and polymers. { ‚fen·əl‚trī‚klȯr·ō'sī‚lān }

philosopher's wool See zinc oxide. { fə'läs·ə·fərz 'wùl }

phloridzin [ORG CHEM] $C_{21}H_{24}O_{19} \cdot 2H_2O$ A glycoside extracted from the root bark of apple, plum, and pear trees; white needles with a melting point of 109°C; soluble in alcohol and hot water; used in medicine. { flə'rid·zən }

phloroglucinol [ORG CHEM] $C_6H_3(OH)_3 \cdot 2H_2O$ White to yellow crystals with a melting point of 212–217°C when heated rapidly and 200–209°C when heated slowly; soluble in alcohol and ether; used as a bone decalcifying agent, as a floral preservative, and in the manufacture of pharmaceuticals. { ‚flȯr·ə'glüs·ən‚ȯl }

pH measurement [ANALY CHEM] Determination of the hydrogen-ion concentration in an ionized solution by means of an indicator solution (such as phenolphthalein) or a pH meter. { ‚pē'āch ‚mezh·ər·mənt }

phorate [ORG CHEM] $C_7H_{17}O_2PS_2$ A clear liquid with slight solubility in water; used as an insecticide for a wide range of insects on a wide range of crops. { 'fȯr‚āt }

phosgene [ORG CHEM] $COCl_2$ A highly toxic, colorless gas that condenses at 0°C to a fuming liquid; used as a war gas and in manufacture of organic compounds. { 'fäz‚jēn }

phosphate [CHEM] **1.** Generic term for any compound containing a phosphate group (PO_4^{3-}), such as potassium phosphate, K_3PO_4. **2.** Generic term for a phosphate-containing fertilizer material. { 'fä‚sfāt }

phosphate anion [INORG CHEM] PO_4^{3-} The negative ion of phosphoric acid. { 'fä‚sfāt 'an‚ī·ən }

phosphate buffer [ANALY CHEM] Laboratory pH reference solution made of KH_2PO_4 and Na_2HPO_4; when 0.025 molal (equimolal of the potassium and sodium salts), the pH is 6.865 at 25°C. { 'fä‚sfāt 'bəf·ər }

phosphide [INORG CHEM] Binary compound of trivalent phosphorus, as in Na_3P. { 'fä‚sfīd }

phosphine [INORG CHEM] PH_3 Poisonous, colorless, spontaneously flammable gas with garlic aroma; soluble in alcohol, slightly soluble in cold water; boils at −85°C; used in organic reactions. Also known as hydrogen phosphide; phosphoretted hydrogen. { 'fä‚sfēn }

phosphinic acid [ORG CHEM] Organic derivative of hypophosphorous acid; contains the radical $-H_2PO_2$ or $=HPO_2$; examples are methylphosphinic acid, CH_3HPOOH, and dimethyl phosphinic acid, $(CH_3)_2POOH$. { fä'sfin·ik 'as·əd }

phosphite [INORG CHEM] Salt of phosphorous acid; contains the radical PO_3^{3-}; an example is normal sodium phosphite, Na_3PO_3. { 'fä‚sfīt }

phospholan [ORG CHEM] $C_6H_{14}O_3PNS_2$ A colorless to yellow solid with a melting point of 37–45°C; used as an insecticide and miticide for cotton. { 'fä·sfə‚lan }

phosphomolybdic acid [INORG CHEM] $H_3PO_4 \cdot 12MoO_3 \cdot xH_2O$ Yellow crystals; soluble in alcohol, ether, and water; used as an alkaloid reagent and a pigment. Abbreviated PMA. { ‚fä·sfō·mə'lib·dik 'as·əd }

phosphonic acid [ORG CHEM] $ROP(OH)_2$, where R is an organic radical such as C_6H_5sw, as in phenylphosphonic acid. { fä'sfän·ik 'as·əd }

phosphoretted hydrogen See phosphine. { 'fäs·fə‚red·əd 'hī·drə·jən }

phosphoric acid [INORG CHEM] H_3PO_4 Water-soluble, transparent crystals, melting at 42°C; used as a fertilizer, in soft drinks and flavor syrups, pharmaceuticals, water treatment, and animal feeds and to pickle and rust-proof metals. Also known as orthophosphoric acid. { fä'sfȯr·ik 'as·əd }

phosphoric anhydride [INORG CHEM] P_2O_5 A flammable, dangerous, soft-white deliquescent powder; used as a dehydrating agent, in medicine and sugar refining, and as a chemical intermediate and analytical reagent. Also known as anhydrous phosphoric acid; phosphoric oxide; phosphorus pentoxide. { fä'sfȯr·ik an'hī‚drīd }

phosphoric oxide See phosphoric anhydride. { fä'sfȯr·ik 'äk‚sīd }

phosphorimetry [ANALY CHEM] Low-temperature, analytical procedure related to fluorometry; based on the nature and intensity of the phosphorescent light emitted by an appropriately excited molecule. { ‚fäs·fə'rim·ə·trē }

phosphorous acid

phosphorous acid [INORG CHEM] H_3PO_3 Alcohol- and water-soluble deliquescent white or yellowish crystals; decomposes at 200°C; used as an analytical reagent and reducing agent. { 'fäs·fə·rəs 'as·əd }

phosphorus [CHEM] A nonmetallic element, symbol P, atomic number 15, atomic weight 30.97376; used to manufacture phosphoric acid, in phosphor bronzes, incendiaries, pyrotechnics, matches, and rat poisons; the white (or yellow) allotrope is a soft waxy solid melting at 44.5°C, is soluble in carbon disulfide, insoluble in water and alcohol, and is poisonous and self-igniting in air; the red allotrope is an amorphous powder subliming at 416°C, igniting at 260°C, is insoluble in all solvents, and is nonpoisonous; the black allotrope comprises lustrous crystals similar to graphite, and is insoluble in most solvents. { 'fäs·fə·rəs }

phosphorus nitride [INORG CHEM] P_3N_5 Amorphous white solid that decomposes in hot water; insoluble in cold water, soluble in organic solvents; used to dope semiconductors. { 'fäs·fə·rəs 'nī,trīd }

phosphorus oxide [INORG CHEM] An oxygen compound of phosphorus; examples are phosphorus monoxide (P_2O), phosphorus trioxide (P_2O_3), phosphorus suboxide (P_4O). { 'fäs·fə·rəs 'äk,sīd }

phosphorus oxychloride [INORG CHEM] $POCl_3$ Toxic, colorless, fuming liquid with pungent aroma; boils at 107°C; decomposes in water or alcohol; causes skin burns; used as a catalyst, chlorinating agent, and in manufacture of various anhydrides. Also known as phosphoryl chloride. { 'fäs·fə·rəs ¦äk·sē'klȯr,īd }

phosphorus pentabromide [INORG CHEM] PBr_5 Yellow crystals, decomposing at 106°C and in water; used in organic synthesis. { 'fäs·fə·rəs ¦pen·tə'brō,mīd }

phosphorus pentachloride [INORG CHEM] PCl_5 Toxic, yellowish crystals with irritating aroma; an eye irritant; sublimes on heating, but will melt at 148°C under pressure; soluble in carbon disulfide; decomposes in water; used as a catalyst and chlorinating agent. { 'fäs·fə·rəs ¦pen·tə'klȯr,īd }

phosphorus pentasulfide [INORG CHEM] P_2S_5 Flammable, hygroscopic, yellow crystals, melting at 281°C; decomposes in moist air; soluble in alkali hydroxides; used to make lube-oil additives, rubber additives, and flotation agents. { 'fäs·fə·rəs ¦pen·tə'səl,fīd }

phosphorus pentoxide See phosphoric anhydride. { 'fäs·fə·rəs pen'täk,sīd }

phosphorus sesquisulfide [INORG CHEM] P_4S_3 Flammable, yellow crystals, melting at 172°C; decomposed by hot water, insoluble in water, soluble in carbon disulfide; used as chemical intermediate and to make matches. Also known as tetraphosphorus trisulfide. { 'fäs·fə·rəs ¦ses·kwē'səl,fīd }

phosphorus thiochloride [INORG CHEM] $PSCl_3$ Yellow liquid, boiling at 125°C; used to make insecticides and oil additives. { 'fäs·fə·rəs ,thī·ə'klȯr,īd }

phosphorus tribromide [INORG CHEM] PBr_3 A corrosive, fuming, colorless liquid with penetrating aroma; soluble in acetone, alcohol, carbon disulfide, and hydrogen sulfide; decomposes in water; used as an analytical reagent to test for sugar and oxygen. { 'fäs·fə·rəs trī'brō,mīd }

phosphorus trichloride [INORG CHEM] PCl_3 A colorless, fuming liquid that decomposes rapidly in moist air and water; soluble in ether, benzene, carbon disulfide, and carbon tetrachloride; boils at 76°C; used as a chlorinating agent, phosphorus solvent, and in saccharin manufacture. { 'fäs·fə·rəs trī'klȯr,īd }

phosphorus triiodide [INORG CHEM] PI_3 Hygroscopic, red crystals, melting at 61°C; soluble in alcohol and carbon disulfide; decomposes in water; used in organic syntheses. { 'fäs·fə·rəs trī'ī·ə,dīd }

phosphorus trisulfide [INORG CHEM] P_2S_3 or P_4S_6 Grayish-yellow, tasteless, odorless solid that burns in air; soluble in alcohol, carbon disulfide, and ether; melts at 290°C; used as an analytical reagent. { 'fäs·fə·rəs trī'səl,fīd }

phosphorylation [ORG CHEM] The esterification of compounds with phosphoric acid. { ,fäs,fȯr·ə'lā·shən }

phosphoryl chloride See phosphorus oxychloride. { 'fäs·fə·rəl 'klȯr,īd }

phosphotungstic acid [INORG CHEM] $H_3PO_4\cdot12WO_3\cdot xH_2O$ Heavy-greenish, water- and alcohol-soluble crystals; used as an analytical reagent and in the manufacture of

photoelectric absorption analysis

organic pigments. Also known as heavy acid; phosphowolframic acid; PTA. { ¦fä·sfō'təŋ·stik 'as·əd }

phosphotungstic pigment [ORG CHEM] A green or blue pigment prepared by precipitating solutions of phosphotungstic or phosphomolybdic acid with malachite green, Victoria blue, and other basic dyestuffs; used in printing inks, paints, and enamels. Also known as tungsten lake. { ¦fä·sfō'təŋ·stik 'pig·mənt }

phosphowolframic acid See phosphotungstic acid. { ¦fä·sfō·wùl'fram·ik 'as·əd }

photoacoustic spectroscopy [SPECT] A spectroscopic technique for investigating solid and semisolid materials, in which the sample is placed in a closed chamber filled with a gas such as air and illuminated with monochromatic radiation of any desired wavelength, with intensity modulated at some suitable acoustic frequency; absorption of radiation results in a periodic heat flow from the sample, which generates sound that is detected by a sensitive microphone attached to the chamber. Abbreviated PAS. Also known as optoacoustic spectroscopy. { ¦fōd·ō·ə¦kü·stik spek'träs·kə·pē }

photoaddition [PHYS CHEM] A bimolecular photochemical process in which a single product is formed by electronically excited unsaturated molecules. { ¦fōd·ō·ə'dish·ən }

photobleach [PHYS CHEM] Upon exposure to light, to decrease in absorbance intensity or, for fluorescent compounds, to decrease in emission intensity. { ¦fōd·ō,blēch }

photocatalysis [PHYS CHEM] The phenomenon by which a relatively small amount of light-absorbing material, called a photocatalyst, changes the rate of chemical reaction without itself being consumed. { ,fōd·ō·kə'tal·ə·səs }

photocatalyst [PHYS CHEM] A light-absorbing substance which, when added to a reaction, facilitates the reaction, while remaining unchanged at the end of the reaction. { ,fōd·ō'kad·əl·ist }

photochemical oxidant [CHEM] Any of the chemicals which enter into oxidation reactions in the presence of light or other radiant energy. { ¦fōd·ō'kem·ə·kəl 'äk·sə·dənt }

photochemical reaction [PHYS CHEM] A chemical reaction influenced or initiated by light, particularly ultraviolet light, as in the chlorination of benzene to produce benzene hexachloride. { ¦fōd·ō'kem·ə·kəl rē'ak·shən }

photochemical reduction See photoreduction. { ,fōd·ō¦kem·ə·kəl ri'dək·shən }

photochemistry [PHYS CHEM] The study of the effects of light on chemical reactions. { ¦fōd·ō'kem·ə·strē }

photochromic compound [CHEM] A chemical compound that changes in color when exposed to visible or near-visible radiant energy; the effect is reversible; used to produce very-high-density microimages. { ¦fōd·ō¦krō·mik 'käm,paùnd }

photochromic reaction [CHEM] A chemical reaction that produces a color change. { ¦fōd·ō¦krō·mik rē'ak·shən }

photochromism [CHEM] The ability of a chemically treated plastic or other transparent material to darken reversibly in strong light. { ¦fōd·ō'krō,miz·əm }

photocurrent [PHYS CHEM] An electric current induced at an electrode by radiant energy. { ¦fōd·ō'kə·rənt }

photodegradation [ORG CHEM] Chemical changes resulting from the absorption of light that reduce the useful properties of materials, particularly polymers. The chemical changes can include bond scission (especially of the molecular backbone), color formation, crosslinking, and chemical rearrangements. { ¦fōd·ō,deg·rə'dā·shən }

photodetachment [PHYS CHEM] The removal of an electron from a negative ion by absorption of a photon, resulting in a neutral atom or molecule. { ¦fōd·ō·di'tach·mənt }

photodimerization [PHYS CHEM] A bimolecular photochemical process involving an electronically excited unsaturated molecule that undergoes addition with an unexcited molecule of the same species. { ¦fōd·ō,dī·mə·rə'zā·shən }

photodissociation [PHYS CHEM] The removal of one or more atoms from a molecule by the absorption of a quantum of electromagnetic energy. { ¦fōd·ō·di,sō·shē'ā·shən }

photoelectric absorption analysis [ANALY CHEM] Type of activation analysis in which the γ-photon gives all of its energy to an electron in the crystal under analysis,

291

photoelectric color comparator

generating a maximum-sized pulse for that particular γ-energy. { ¦fōd·ō·i'lek·trik əb'sȯrp·shən ə,nal·ə·səs }

photoelectric color comparator *See* color comparator. { ¦fōd·ō·i'lek·trik 'kal·ər kəm,par· əd·ər }

photoelectric colorimetry [ANALY CHEM] Measurement of the colorant concentration in a solution by means of the tristimulus values of three primary light filter-photocell combinations. { ¦fōd·ō·i'lek·trik ,kəl·ə'rim·ə·trē }

photoelectrolysis [PHYS CHEM] The process of using optical energy to assist or effect electrolytic processes that ordinarily require the use of electrical energy. { ¦fōd· ō,i,lek'träl·ə·səs }

photoelectron spectroscopy [SPECT] The branch of electron spectroscopy concerned with the energy analysis of photoelectrons ejected from a substance as the direct result of bombardment by ultraviolet radiation or x-radiation. { ¦fōd·ō·i'lek,trän spek'träs·kə·pē }

photoglycine *See* glycin. { ¦fōd·ō'glī,sēn }

photographic photometry [SPECT] The use of a comparator-densitometer to analyze a photographed spectrograph spectrum by emulsion density measurements. { ¦fōd· ə¦graf·ik fə'täm·ə·trē }

photohomolysis [PHYS CHEM] A homolysis reaction in which bond breaking is caused by radiant energy. { ¦fōd·ō·hə'mäl·ə·səs }

photoinitiated polymerization [PHYS CHEM] A chain reaction of monomer to polymer initiated by a photogenerated radical or ion. { ,fōd·ō·ə¦nish·ē,ād·əd pə,lim·ə· rə'zā·shən }

photoinitiator [PHYS CHEM] A substance (other than reactant) which, on absorption of light, generates a reactive species (ion or radical), initiates a chemical reaction or transformation, and is consumed. { ,fōd·ō·ə'nish·ē,ād·ər }

photoionization [PHYS CHEM] The removal of one or more electrons from an atom or molecule by absorption of a photon of visible or ultraviolet light. Also known as atomic photoelectric effect. { ¦fōd·ō,ī·ə·nə'zā·shən }

photoisomer [PHYS CHEM] An isomer produced by photolysis. { ¦fōd·ō'ī·sə·mər }

photolysis [PHYS CHEM] The use of radiant energy to produce chemical changes. { fō'täl·ə·səs }

photomechanochemistry [PHYS CHEM] A branch of polymer sciences that deals with photochemical conversion of chemical energy into mechanical energy. { ¦fōd·ō· mə,kan·ō'kem·ə·strē }

photometric titration [ANALY CHEM] A titration in which the titrant and solution cause the formation of a metal complex accompanied by an observable change in light absorbance by the titrated solution. { ¦fōd·ə¦me·trik tī'trā·shən }

photooxidation [PHYS CHEM] **1.** The loss of one or more electrons from a photoexcited chemical species. **2.** The reaction of a substance with oxygen and light. When oxygen remains in the product, the reaction is also known as photooxygenation. { ,fōd·ō,äk·sə'dā·shən }

photopolymer [PHYS CHEM] Any polymer which, on exposure to light, undergoes a spontaneous and permanent change in physical properties, such as crosslinking or depolymerization. { ,fōd·ō'päl·ə·mər }

photoreduction [CHEM] A chemical reduction that is produced by electromagnetic radiation. Also known as photochemical reduction. [PHYS CHEM] **1.** Addition of one or more electrons to a photoexcited chemical species. **2.** Photochemical hydrogenation of a substance. { ,fōd·ō·ri'dək·shən }

photosensitizer [PHYS CHEM] A light-absorbing substance that initiates a photochemical or photophysical reaction in another substance (molecule), and is not consumed in the reaction. { ¦fōd·ō'sen·sə,tīz·ər }

photostabilize [ORG CHEM] To incorporate stabilizers in polymers, such as ultraviolet absorbers, to prevent photodegradation. { ,fōd·ō'stā·bə,līz }

phoxim *See* phenylglyoxylonitriloxime O,O-diethyl phosphorothioate. { 'fäk,sim }

pH standard [ANALY CHEM] Five standard laboratory solutions available from the U. S. National Institute of Standards and Technology, each solution having a known pH

value; the standards cover pH ranges from 3.557 to 8.833.   Abbreviated pH(S).
{ ˌpēˈach ˈstan·dərd }
phthalimide   [ORG CHEM] $C_8H_5NO_2$ The product made by heating phthalic anhydride with ammonia; used in Gabriel's synthesis of primary amines, amino acids, and anthranilic acid (o-aminobenzoic acid).   { ′thal·əˌmīd }
phthalate   [ORG CHEM] A salt of phthalic acid; contains the radical $C_6H_4(COO)_2^{2-}$; an example is dibutylphthalate, $C_{16}H_{22}O_4$; used as a plasticizer in plastics, and as a buffer in standard laboratory solutions.   { ′thaˌlāt }
phthalate buffer   [ANALY CHEM] Laboratory pH reference solution made of potassium hydrogen phthalate, $KHC_8H_4O_4$; at 0.05 molal, the pH is 4.008 at 25°C.   { ′thaˌlāt ′bəf·ər }
phthalate ester   [ORG CHEM] Any of a group of plastics plasticizers made by the direct action of alcohol on phthalic anhydride; generally characterized by moderate cost, good stability, and good general properties.   { ′thaˌlāt ′es·tər }
phthalazine   [ORG CHEM] $C_6H_4CHN_2CH$ Colorless crystals, melting at 91°C; soluble in alcohol.   { ′thal·əˌzēn }
phthalic acid   [ORG CHEM] $C_6H_4(CO_2H)_2$ Any of three isomeric benzene dicarboxylic acids; the ortho form is usually called phthalic acid, comprises alcohol-soluble, colorless crystals decomposing at 191°C, slightly soluble in water and ether, is used to make dyes, medicine, and synthetic perfumes, and as a chemical intermediate, and is also known as benzene orthodicarboxylic acid; the para form, known as terephthalic acid, is used to make polyester resins (Dacron) and as poultry feed additives; the meta form is isophthalic acid.   { ′thal·ik ′as·əd }
meta-phthalic acid See isophthalic acid.   { ˈmed·ə ′thal·ik ′as·əd }
ortho-phthalic acid See phthalic acid.   { ˈȯr·thō ′thal·ik ′as·əd }
para-phthalic acid See terephthalic acid.   { ˈpar·ə ′thal·ik ′as·əd }
phthalic anhydride   [ORG CHEM] $C_6H_4(CO)_2O$ White crystals, melting at 131°C; sublimes when heated; slightly soluble in ether and hot water, soluble in alcohol; used to make dyes, resins, plasticizers, and insect repellents.   { ′thal·ik anˈhīˌdrīd }
phthalocyanine pigments   [ORG CHEM] A group of light-fast organic pigments with four isoindole groups, $(C_6H_4)C_2N$, linked by four nitrogen atoms to form a conjugated chain; included are phthalocyanine (blue-green), copper phthalocyanine (blue), chlorinated copper phthalocyanine (green), and sulfonated copper phthalocyanine (green); used in enamels, plastics, linoleum, inks, wallpaper, and rubber goods.   { ˈthal·ō'sī·ə·nən ′pigˌməns }
phthalonitrile   [ORG CHEM] $C_6H_4(CN)_2$ Buff-colored crystals with a melting point of 138°C; soluble in acetone and benzene; used in organic synthesis and as an insecticide.   { ˈthal·ō'nī·trəl }
physical adsorption   [PHYS CHEM] Reversible adsorption in which the adsorbate is held by weak physical forces.   { ′fiz·ə·kəl ad'sȯrp·shən }
physical chemistry   [CHEM] The branch of chemistry that deals with the interpretation of chemical phenomena and properties in terms of the underlying physical processes, and with the development of techniques for their investigation.   { ′fiz·ə·kəl ′kem·ə·strē }
physical organic chemistry   [ORG CHEM] The study of the scope and limitations of the various rules, effects, and generalizations in use in organic chemistry by application of physical and mathematical means.   { ˈfiz·ə·kəl ȯrˈgan·ik ′kem·ə·strē }
physical property   [CHEM] Property of a compound that can change without involving a change in chemical composition; examples are the melting point and boiling point.   { ′fiz·ə·kəl ′präp·ərd·ē }
physisorption   [PHYS CHEM] A physical adsorption process in which there are van der Waals forces of interaction between gas or liquid molecules and a solid surface.   { ˈfiz·ə'sȯrp·shən }
physostigmine   [ORG CHEM] $C_{15}H_{21}O_2N_3$ An alkaloid; poisonous, colorless-to-pinkish crystals; soluble in alcohol and dilute acids; melts at 86°C; used as a source of salicylate and sulfate forms.   Also known as calabarine; eserine.   { ˌfī·sə'stig·mēn }

physostigmine salicylate

physostigmine salicylate [ORG CHEM] $C_{15}H_{21}O_2N_3 \cdot C_7H_6O_3$ Poisonous, colorless-to-yellow crystals; soluble in water, alcohol, and chloroform; melts at 182°C; used for medicines. { ‚fī·sə'stig‚mēn sə'lis·ə‚lāt }

physostigmine sulfate [ORG CHEM] $(C_{15}H_{21}O_2N_3)_2 \cdot H_2SO_4$ Poisonous, white crystals; soluble in water, alcohol, and chloroform; melts at 150°C; used for medicines. { ‚fī· sə'stig‚mēn 'səl‚fāt }

phytane [ORG CHEM] $C_{20}H_{42}$ A hydrocarbon derivative of chlorophyll that is found in rock specimens $2.5-3 \times 10^9$ years old; frequently associated with Precambrian fossil plant matter. { 'fī‚tān }

phytic acid [ORG CHEM] $C_6H_6[OPO(OH)_2]_6$ An acid found in seeds of plants as the insoluble calcium magnesium salt (phytin); derived from corn steep liquor; inhibits calcium absorption in intestine; used to treat hard water, to remove iron and copper from wines, and to inactivate trace-metal contaminants in animal and vegetable oils. { 'fīd·ik 'as·əd }

phytol [ORG CHEM] $C_{20}H_{40}O$ A liquid with a boiling point of 202–204°C; soluble in organic solvents; used in the synthesis of vitamins E and K. { 'fī‚tól }

phytonadione [ORG CHEM] $C_{31}H_{46}O_2$ A yellow, viscous liquid soluble in benzene, chloroform, and vegetable oils; used in medicine and as a food supplement. Also known as vitamin K_1. { ‚fī‚tän·ə'dī‚ōn }

pi bonding [PHYS CHEM] Covalent bonding in which the greatest overlap between atomic orbitals is along a plane perpendicular to the line joining the nuclei of the two atoms. { 'pī ‚bänd·iŋ }

Pickering series [SPECT] A series of spectral lines of singly ionized helium, observed in very hot O-type stars, associated with transitions between the level with principal quantum number $n = 4$ and higher energy levels. { 'pik·riŋ ‚sir·ēz }

pickling acid [CHEM] Any of the acids used in pickling solutions, such as hydrochloric, sulfuric, nitric, phosphoric, or hydrofluoric acid. { 'pik·liŋ ‚as·əd }

picoline [ORG CHEM] $C_5H_4N(CH_3)$ Family of colorless liquid isomers, soluble in water and alcohol; the alpha form, boiling at 129°C, is used as a solvent and chemical intermediate, and is also known as 2-methyl pyridine; the beta form, boiling at 143.5°C, is used as a solvent for chemical synthesis reactions, to make nicotinic acid, and in fabric waterproofing, and is also known as 3-methyl pyridine; the gamma form, boiling at 143.1°C, is used as a solvent for chemical synthesis reactions and in fabric waterproofing. { 'pik·ə‚lēn }

picolinic acid [ORG CHEM] $C_{10}H_8N_4O_5$ An alcohol-soluble crystalline compound, forming yellow leaflets that melt at 116–117°C; used as a reagent in phenylalanine, tryptophan, and alkaloids production, and for the quantitative detection of calcium. { ‚pik·ə‚lin·ik 'as·əd }

picramic acid [ORG CHEM] $C_6H_5N_3O_5$ A crystalline acid, forming dark red needles from alcohol solutions, melting at 169–170°C; used in dye manufacture and as a reagent in tests for albumin. { pi'kram·ik 'as·əd }

picric acid [ORG CHEM] $C_6H_2(NO_2)_3OH$ Poisonous, explosive, highly oxidative yellow crystals with bitter taste; soluble in water, alcohol, chloroform, benzene, and ether; melts at 122°C; used in explosives, in external medicines; to make dyes, matches, and batteries, and to etch copper. { 'pik·rik 'as·əd }

pi electron [PHYS CHEM] An electron which participates in pi bonding. { 'pī i'lek‚trän }

piezochemistry [CHEM] The field of chemical reactions under high pressures. { pē‚ā·zō'kem·ə·strē }

piezoelectric polymer See piezopolymer. { pē‚ā·zō·ə'lek·trik 'päl·ə·mər }

piezopolymer [ORG CHEM] A polymeric film that has the ability to reversibly convert heat and pressure into electricity. Also known as piezoelectric polymer. { pē‚ā·zō'päl·ə·mər }

pilocarpine [ORG CHEM] $C_{11}H_{16}N_2O_2$ An alkaloid, in either oil or crystal form, melting at 34°C; soluble in chloroform, water, and alcohol; used in medicine. { ‚pī·lə'kär‚pēn }

pimaricin [ORG CHEM] $C_{33}H_{47}NO_{13}$ A compound crystallizing from a methanol-water solution, decomposing at about 200°C; soluble in water and organic solvents; used in medicine as an antifungal agent for *Candida albicans* vaginitis. { pə'mar·ə·sən }

pimelic acid |ORG CHEM| $HOOC(CH_2)_5COOH$ Crystals melting at 105°C; slightly soluble in water, soluble in alcohol and ether; used in biochemical research. { pə'mel·ik 'as·əd }

pinene |ORG CHEM| $C_{10}H_{16}$ Either of two colorless isomeric unsaturated bicyclic terpene hydrocarbon liquids derived from sulfate wood turpentine; 95% of the alpha form boils in the range 156–160°C, and of the beta form boils in the range 164–169°C; used as solvents for coatings and wax formulations, as chemical intermediates for resins, and as lube-oil additives. Also known as nopinene. { 'pī̱,nēn }

pinene hydrochloride *See* terpene hydrochloride. { 'pī̱,nēn ,hi·drə'klȯr,īd }

pinic acid |ORG CHEM| $C_9H_{14}O_4$ A crystalline dicarboxylic acid derived from α-pinene; used to make diesters for plasticizers and lubricants. { 'pī·nik 'as·əd }

piperazine |ORG CHEM| $C_4H_{10}N_2$ A cyclic compound; colorless, deliquescent crystals, melting at 104–107°C; soluble in water, alcohol, glycerol, and glycols; absorbs carbon dioxide from air; used in medicine. { pī'par·ə,zēn }

piperazine dihydrochloride |ORG CHEM| $C_4H_{10}N_2·2HCl$ White, water-soluble needles; used for insecticides and pharmaceuticals. { pī'par·ə,zēn dī,hī·drə'klȯr,īd }

piperazine hexahydrate |ORG CHEM| $C_4H_{10}N_2·6H_2O$ White crystals with a melting point of 44°C; soluble in alcohol and water; used for pharmaceuticals and insecticides. { pī'par·ə,zēn ¦hek·sə'hī,drāt }

piperidine |ORG CHEM| $C_5H_{11}N$ A cyclic compound, and strong base; colorless liquid with a pepper aroma; boils at 106°C; soluble in water, alcohol, and ether; used as a chemical intermediate and rubber accelerator, and in medicine. { pī'per·ə,dēn }

piperine |ORG CHEM| $C_{17}H_{19}NO_3$ A crystalline compound that is found in black pepper; melting point is 130°C; soluble in benzene and acetic acid; used to give a pungent taste to brandy and as an insecticide. { 'pip·ə,rēn }

piperocaine hydrochloride |ORG CHEM| $C_{16}H_{23}NO_2·HCl$ A white, crystalline powder with a bitter taste and a melting point of 172–175°C; soluble in water, chloroform, and alcohol; used in medicine. { pī'per·ə,kān ,hī·drə'klȯr,īd }

piperonal |ORG CHEM| $C_8H_6O_3$ White crystals with a floral odor and a melting point of 35.5–37°C; soluble in alcohol and ether; used in medicine, perfumes, suntan preparations, and mosquito repellents. Also known as heliotropin. { pə'per·ə,nal }

pipet |CHEM| Graduated or calibrated tube which may have a center reservoir (bulb); used to transfer known volumes of liquids from one vessel to another; types are volumetric or transfer, graduated, and micro. { pī'pet }

pirimiphosethyl |ORG CHEM| $C_{13}H_{24}N_3O_3PS$ A straw-colored liquid which decomposes at 130°C; used as an insecticide for the control of soil insects in vegetables and other crops. { ,pir·əm·fäs'eth·əl }

Pitzer equation |PHYS CHEM| Equation for the approximation of data for heats of vaporization for organic and simple inorganic compounds; derived from temperature and reduced temperature relationships. { 'pit·sər i,kwā·zhən }

PIXE *See* proton-induced x-ray emission. { 'pik·sē }

pK |CHEM| The logarithm (to base 10) of the reciprocal of the equilibrium constant for a specified reaction under specified conditions. { pē'kā }

plait point |CHEM| Composition conditions in which the three coexisting phases of partially soluble components of a three-phase liquid system approach each other in composition. { 'plāt ,pȯint }

planocaine base *See* procaine base. { 'plan·ə,kān ,bās }

plasma desorption mass spectrometry |SPECT| A technique for analysis of nonvolatile molecules, particularly heavy molecules with atomic weight over 2000, in which heavy ions with energies on the order of 100 MeV penetrate and deposit energy in thin films, giving rise to chemical reactions that result in the formation of molecular ions and shock waves that result in the ejection of these ions from the surface; the ions are then analyzed in a mass spectrometer. Abbreviated PDMS. { 'plaz·mə dē¦sȯrp·shən 'mas spek'träm·ə·trē }

plasma-jet excitation |SPECT| The use of a high-temperature plasma jet to excite an element to provide measurable spectra with many ion lines similar to those from spark-excited spectra. { 'plaz·mə ,jet ,ek·sə'tā·shən }

plaster of paris

plaster of paris [INORG CHEM] White powder consisting essentially of the hemihydrate of calcium sulfate ($CaSO_4 \cdot \frac{1}{2}H_2O$ or $2CaSO_4 \cdot H_2O$), produced by calcining gypsum until it is partially dehydrated; forms with water a paste that quickly sets; used for casts and molds, building materials, and surgical bandages. Also known as calcined gypsum. { 'plas·tər əv 'par·əs }

plastizymes [PHYS CHEM] Artificial enzymes (artificial polymeric materials with molecule-shaped pores) that possess catalytic properties. { 'plas·ti,zīmz }

plate theory [ANALY CHEM] In gas chromatography, the theory that the column operates similarly to a distillation column; for example, chromatographic columns are considered as consisting of a number of theoretical plates, each performing a partial separation of components. { 'plāt ,thē·ə·rē }

platinic chloride See chloroplatinic acid. { plə'tin·ik 'klȯr,īd }

platinic sodium chloride See sodium chloroplatinate. { plə'tin·ik 'sōd·ē·əm 'klȯr,īd }

platinic sulfate See platinum sulfate. { plə'tin·ik 'səl,fāt }

platinochloride See chloroplatinate. { ¦plat·ən·ō'klȯr,īd }

platinocyanide [INORG CHEM] A double salt of platinous cyanide and another cyanide, such as $K_2Pt(CN)_4$; used in photography and fluorescent x-ray screens. Also known as cyanoplatinate. { ¦plat·ən·ō'sī·ə,nīd }

platinous chloride See platinum dichloride. { 'plat·ən·əs 'klȯr,īd }

platinous iodide See platinum iodide. { 'plat·ən·əs 'ī·ə,dīd }

platinum [CHEM] A chemical element, symbol Pt, atomic number 78, atomic weight 195.08. { 'plat·ən·əm }

platinum bichloride See platinum dichloride. { 'plat·ən·əm bī'klȯr,īd }

platinum chloride [INORG CHEM] $PtCl_4$ or $PtCl_4 \cdot 5H_2O$ A brown solid or red crystals; soluble in alcohol and water; decomposes when heated (loses $4H_2O$ at 100°C); used as an analytical reagent. { 'plat·ən·əm 'klȯr,īd }

platinum dichloride [INORG CHEM] $PtCl_2$ Water-insoluble, green-gray powder; decomposes to platinum at red heat; used to make platinum salts. Also known as platinous chloride; platinum bichloride. { 'plat·ən·əm dī'klȯr,īd }

platinum diiodide See platinum iodide. { 'plat·ən·əm dī'ī·ə,dīd }

platinum electrode [PHYS CHEM] A solid platinum wire electrode used during voltammetric analyses of electrolytes. { 'plat·ən·əm i'lek,trōd }

platinum iodide [INORG CHEM] PtI_2 Water- and alkali-insoluble black powder; slightly soluble in hydrochloric acid; decomposes at 300–350°C. Also known as platinous iodide; platinum diiodide. { 'plat·ən·əm 'ī·ə,dīd }

platinum metal [CHEM] A group of transition metals that includes ruthenium, osmium, rhodium, iridium, palladium, and platinum. { 'plat·ən·əm 'med·əl }

platinum oxide [INORG CHEM] An oxide of platinum; examples are platinum monoxide (or platinous oxide), PtO, and platinum dioxide (or platinic oxide), PtO_2. { 'plat·ən·əm 'äk,sīd }

platinum potassium chloride See potassium chloroplatinate. { 'plat·ən·əm pə'tas·ē·əm 'klȯr,īd }

platinum sodium chloride See sodium chloroplatinate. { 'plat·ən·əm 'sōd·ē·əm 'klȯr,īd }

platinum sulfate [INORG CHEM] $Pt(SO_4)_2$ A hygroscopic, dark mass; soluble in alcohol, ether, water, and dilute acids; used in microanalysis for halogens. Also known as platinic sulfate. { 'plat·ən·əm 'səl,fāt }

Plessy's green [INORG CHEM] $CrPO_4 \cdot xH_2O$ Deep-green pigment made of chromium phosphate mixed with chromium oxide and calcium phosphate. { ple'sēz 'grēn }

plumbous oxide See lead monoxide. { 'pləm·bəs 'äk,sīd }

plumbous sulfide See lead sulfide. { 'pləm·bəs 'səl,fīd }

plumbum [CHEM] Latin name for lead; source of the element symbol Pb. { 'pləm·bəm }

plutonium [CHEM] A reactive metallic element, symbol Pu, atomic number 94, in the transuranium series of elements; the first isotope to be identified was plutonium-239; used as a nuclear fuel, to produce radioactive isotopes for research, and as the fissile agent in nuclear weapons. { plü'tō·nē·əm }

plutonium oxide [INORG CHEM] PuO_2 A radioactively poisonous pyrophoric oxide of plutonium; particles may be easily airborne. { plü'tō·nē·əm 'äk,sīd }

Pm *See* promethium.

PMA *See* phosphomolybdic acid; pyromellitic acid.

PMDA *See* pyromellitic dianhydride.

pNa [CHEM] Logarithm of the sodium-ion concentration in a solution; that is, $pNa = -\log a_{Na^+}$, where a_{Na^+} is the sodium-ion concentration.

pnicogen [CHEM] Any member of the nitrogen family of elements, group 15 in the periodic table. { 'nī·kə·jən }

pnictide [CHEM] A simple compound of a pnicogen and an electropositive element. { 'nik,tīd }

Po *See* polonium.

POD analysis [ANALY CHEM] A precision laboratory distillation procedure used to separate low-boiling hydrocarbon fractions quantitatively for analytical purposes. Also known as Podbielniak analysis. { ¦pē¦ō'dē ə,nal·ə·səs }

Podbielniak analysis *See* POD analysis. { päd'bēl·nē,ak ə,nal·ə·səs }

poison [CHEM] A substance that exerts inhibitive effects on catalysts, even when present only in small amounts; for example, traces of sulfur or lead will poison platinum-based catalysts. { 'pȯiz·ən }

polar compound [CHEM] Molecules which contain polar covalent bonds; they can ionize when dissolved or fused; polar compounds include inorganic acids, bases, and salts. { 'pō·lər 'käm,paůnd }

polar covalent bond [PHYS CHEM] A bond in which a pair of electrons is shared in common between two atoms, but the pair is held more closely by one of the atoms. { 'pō·lər kō'vā·lənt 'bänd }

polarimetric analysis [ANALY CHEM] A method of chemical analysis based on the optical activity of the substance being determined; the measurement of the extent of the optical rotation of the substance is used to identify the substance or determine its quantity. { pō¦lar·ə¦me·trik ə'nal·ə·səs }

polarization potential [PHYS CHEM] The reverse potential of an electrolytic cell which opposes the direct electrolytic potential of the cell. { ,pō·lə·rə'zā·shən pə,ten·chəl }

polarization spectroscopy [SPECT] A type of saturation spectroscopy in which a circularly polarized saturating laser beam depletes molecules with a certain orientation preferentially, leaving the remaining ones polarized; the latter are detected through their induction of elliptical polarization in a probe beam, allowing the beam to pass through crossed linear polarizers. { ,pō·lə·rə'zā·shən spek'träs·kə·pē }

polarized scattering [PHYS CHEM] In a quasi-elastic light scattering experiment performed with polarizers, the type of scattering produced when the polarizers select both the incident and final polarizations perpendicular to the scattering plane. { 'pō·lə,rīzd 'skad·ə·riŋ }

polar molecule [PHYS CHEM] A molecule having a permanent electric dipole moment. { 'pō·lər 'mäl·ə,kyül }

polarogram [ANALY CHEM] Plotted output (current versus electrode voltage) for polarographic analysis of an electrolyte. { pə'lar·ə,gram }

polarographic analysis [ANALY CHEM] An electroanalytical technique in which the current through an electrolysis cell is measured as a function of the applied potential; the apparatus consists of a potentiometer for adjusting the potential, a galvanometer for measuring current, and a cell which contains two electrodes, a reference electrode whose potential is constant and an indicator electrode which is commonly the dropping mercury electrode. Also known as polarography. { pō¦lar·ə¦graf·ik ə'nal·ə·səs }

polarographic cell [ANALY CHEM] Device for polarographic (voltammetric) analysis of an electrolyte solution; a known voltage is applied to the solution, and the ensuing current that passes through the cell (to an electrode) is measured. { pō¦lar·ə¦graf·ik 'sel }

297

polarographic maximum [ANALY CHEM] A deceptively high voltage buildup on an electrode during polarographic analysis of an electrolyte; caused by a reduction or oxidation process at the electrode. { pō¦lar·ə¦graf·ik 'mak·sə·məm }

polarography See polarographic analysis. { ˌpō·lə'räg·rə·fē }

polonium [CHEM] A chemical element, symbol Po, atomic number 84; all polonium isotopes are radioactive; polonium-210 is the naturally occurring isotope found in pitchblende. { pə'lō·nē·əm }

poly- [ORG CHEM] A chemical prefix meaning many; for example, a polymer is made of a number of single molecules known as monomers, as polyethylene is made from ethylene. { 'päl·ē, 'päl·ə, 'päl·i }

polyacetals See acetal resins. { ¦päl·ē'as·ə,talz }

polyacrylamide [ORG CHEM] $(CH_2CHCONH_2)_x$ A white, water-soluble high polymer based on acrylamide; used as a thickening or suspending agent in water-base formulations. { ¦päl·ē·ə'kril·ə·məd }

polyacrylate [ORG CHEM] A polymer of an ester or salt of acrylic acid. { ¦päl·ē'ak·rə,lāt }

polyacrylic acid [ORG CHEM] $(CH_2CHCOOH)_x$ An acrylic or acrylate resin formed by the polymerization of acrylic acid; water-soluble; used as a suspending and textile-sizing agent, and in adhesives, paints, and hydraulic fluids. { ¦päl·ē·ə¦kril·ik 'as·əd }

polyacrylic fiber [ORG CHEM] Continuous-strand fiber extruded from an acrylate resin. { ¦päl·ē·ə¦kril·ik 'fī·bər }

polyacrylonitrile [ORG CHEM] Polymer of acrylonitrile; semiconductive; used like an inorganic oxide catalyst to dehydrogenate tert-butyl alcohol to produce isobutylene and water. { ¦päl·ē¦ak·rə·lō'nī·trəl }

polyalcohol See polyhydric alcohol. { ¦päl·ē'al·kə,hól }

polyallomer [ORG CHEM] A copolymer of propylene with other olefins. { ¦päl·ē'al·ə·mər }

polyamide [ORG CHEM] Any member of a class of polymers in which individual structural units are joined by amide bonds. { ¦päl·ē'am·əd }

polyamide resin [ORG CHEM] Product of polymerization of amino acid or the condensation of a polyamine with a polycarboxylic acid; an example is the nylons. { ¦päl·ē'am·əd 'rez·ən }

polyatomic ion [CHEM] An electrically charged species formed by covalent bonding of atoms of two or more different elements, usually nonmetals, for example, the ammonium ion (NH_4^+). { ˌpäl·ē·ə¦täm·ik 'ī·ən }

polyatomic molecule [CHEM] A chemical molecule with three or more atoms. { ¦päl·ē·ə¦täm·ik 'mäl·ə,kyül }

polybasic [CHEM] A chemical compound in solution that yields two or more H^- ions per molecule, such as sulfuric acid, H_2SO_4. { ¦päl·i,bā·sik }

polybutadiene [ORG CHEM] Oil-extendable synthetic elastomer polymer made from butadiene; resilience is similar to natural rubber; it is blended with natural rubber for use in tire and other rubber products. Also known as butadiene rubber. { ¦päl·i,byüd·ə'dī,ēn }

polybutene [ORG CHEM] A polymer of isobutene, $(CH_3)_2CCH_2$; made in varying chain lengths to give a wide range of properties from oily to solid; used as a lube-oil additive, in adhesives, and in rubber products. { ¦päl·i'byü,tēn }

polybutylene [ORG CHEM] A polymer of one or more butylenes whose consistency ranges from a viscous liquid to a rubbery solid. { ¦päl·i'byüd·ə,lēn }

polycarbonate [ORG CHEM] $[OC_6H_4C(CH_3)_2C_6H_4OCO]_x$ A linear polymer of carbonic acid which is a thermoplastic synthetic resin made from bisphenol and phosgene; used in emulsion coatings with glass fiber reinforcement. { ¦päl·i'kär·bə·nət }

polycarboxylic [ORG CHEM] Prefix for a compound containing two or more carboxyl (−COOH) groups. { ¦päl·i¦kär,bäk¦sil·ik }

polychlorinated biphenyl [ORG CHEM] Any member of the group of chlorinated isomers of biphenyl. Abbreviated PCB. { ¦päl·i'klór·ə,nād·əd bī'fen·əl }

polycondensation [ORG CHEM] A chemical condensation leading to the formation of a polymer by the linking together of molecules of a monomer and the releasing of water or a similar simple substance. { ¦päl·i,kän·dən'sā·shən }

polycyclic [ORG CHEM] A molecule that contains two or more closed atomic rings; can be aromatic (such as DDT), aliphatic (bianthryl), or mixed (dicarbazyl). { ¦päl·i'sī·klik }

polycyclic aromatic hydrocarbon [ORG CHEM] A compound containing two or more fused benzene rings such as naphthalene or anthracene. { ˌpäl·ē¸sī·klik ˌar·ə¸mad·ik 'hī·drə¸kär·bən }

polycyclic hydrocarbon *See* polynuclear hydrocarbon. { ¦päl·i'sī·klik 'hī·drə¸kär·bən }

polydent [ORG CHEM] Pertaining to a chemical species whose molecules possess more than two reactive sites. Also known as multident. { 'päl·ə¸dent }

polydisperse colloidal system [CHEM] A colloidal system in which the suspended particles have various sizes and shapes. { ¦päl·i·di'spərs kə¦lóid·əl 'sis·təm }

polydispersity [CHEM] Molecular-weight nonhomogeneity in a polymer system; that is, there is some molecular-weight distribution throughout the body of the polymer. { ¦päl·i·di'spər·səd·ē }

polyelectrolyte [ORG CHEM] A natural or synthetic electrolyte with high molecular weight, such as proteins, polysaccharides, and alkyl addition products of polyvinyl pyridine; can be a weak or strong electrolyte; when dissociated in solution, it does not give uniform distribution of positive and negative ions (the ions of one sign are bound to the polymer chain while the ions of the other sign diffuse through the solution). { ¦päl·ē·ə'lek·trə¸līt }

polyene [ORG CHEM] Compound containing many double bonds, such as the carotenoids. { 'päl·ē¸ēn }

polyester resin [ORG CHEM] A thermosetting or thermoplastic synthetic resin made by esterification of polybasic organic acids with polyhydric acids; examples are Dacron and Mylar; the resin has high strength and excellent resistance to moisture and chemicals when cured. { 'päl·ē¸es·tər 'rez·ən }

polyester rubber *See* polyurethane rubber. { 'päl·ē¸es·tər 'rəb·ər }

polyether [ORG CHEM] Any compound whose molecular structure contains linked ethers, R−O−R', where R and R' represent functional groups. { 'päl·ē¸ē·thər }

polyether resin [ORG CHEM] Any member of a large group of thermoplastic or thermosetting polymers that contain the typical polyether linkages in the polymer chain. { 'päl·ē¸ē·thər 'rez·ən }

polyethylene *See* ethylene resin. { ¦päl·ē'eth·ə¸lēn }

polyethylene glycol [ORG CHEM] Any of a family of colorless, water-soluble liquids with molecular weights from 200 to 6000; soluble also in aromatic hydrocarbons (not aliphatics) and many organic solvents; used to make emulsifying agents and detergents, and as plasticizers, humectants, and water-soluble textile lubricants. { ¦päl·ē'eth·ə¸lēn 'glī¸kól }

polyethylene glycol distearate *See* polyglycol distearate. { ¦päl·ē'eth·ə¸lēn 'glī¸kól dī'stir¸āt }

polyethylene resin *See* ethylene resin. { ¦päl·ē'eth·ə¸lēn 'rez·ən }

polyethylene terephthalate [ORG CHEM] A thermoplastic polyester resin made from ethylene glycol and terephthalic acid; melts at 265°C; used to make films or fibers. Abbreviated PET. { ¦päl·ē'eth·ə¸lēn ¸ter·ə'tha¸lāt }

polyformaldehyde *See* polyoxymethylene. { ˌpäl·ē·fór'mal·də¸hīd }

polygen *See* polyvalent. { 'päl·i·jən }

polyglycol [ORG CHEM] A dihydroxy ether derived from the dehydration (removal of a water molecule) of two or more glycol molecules; an example is diethylene glycol, $CH_2OHCH_2OCH_2CH_2OH$. { 'päl·i¸glī¸kól }

polyglycol distearate [ORG CHEM] $(C_{17}H_{35})_2CO_2CO(CH_2CH_2O)_x$ An off-white, soft solid with a melting point of 43°C; soluble in chlorinated solvents, acetone, and light esters; used as a resin plasticizer. Also known as polyethylene glycol distearate. { 'päl·i¸glī¸kól dī'stir¸āt }

polyhaloalkane [ORG CHEM] An alkane derivative in which two or more hydrogen atoms have been replaced by halogen atoms. { ˌpäl·ē¸ha·lō'al¸kān }

polyhalogeno compound [ORG CHEM] An organic compound containing more than one halogen atom. { ˌpäl·ē·hə'läj·ə·nō ¸käm¸paúnd }

299

polyhydric alcohol [ORG CHEM] An alcohol with many hydroxyl ($-OH$) radicals, such as glycerol, $C_3H_5(OH)_3$. Also known as polyalcohol; polyol. { ¦päl·i¦hī·drək 'al·kə,hòl }

polyhydric phenol [ORG CHEM] A phenolic compound containing two or more hydroxyl groups, such as diphenol, $C_6H_4(OH)_2$. { ¦päl·i¦hī·drək 'fē,nòl }

polyimide resin [ORG CHEM] An aromatic polyimide made by reacting pyromellitic dianhydride with an aromatic diamine; has high resistance to thermal stresses; used to make components of internal combustion engines. { ¦päl·ē'i,mīd ,rez·ən }

polyisoprene [ORG CHEM] $(C_5H_8)_x$ The basis of natural rubber, balata, gutta-percha, and other rubberlike materials; can also be made synthetically; the stereospecific forms are cis-1,4- and trans-1,4-polyisoprene; the polymer is thermoplastic. { ¦päl·ē'īs·ə,prēn }

polylactic resin [ORG CHEM] A soft, elastic resin made by the heat reaction of lactic acid with castor oil or other fatty oils; used to produce tough, water-resistant coatings. { ¦päl·i¦lak·tik ,rez·ən }

polyLED See polymer light-emitting diode. { ,päl·ē¦el¦ē'dē }

polyligated atom [PHYS CHEM] An atom that is bonded to more than one other atom. { ,päl·ē¦lī,gād·əd 'ad·əm }

polymer [ORG CHEM] Substance made of giant molecules formed by the union of simple molecules (monomers); for example polymerization of ethylene forms a polyethylene chain, or condensation of phenol and formaldehyde (with production of water) forms phenol-formaldehyde resins. { 'päl·ə·mər }

polymer blend [ORG CHEM] A homogeneous mixture of two or more different polymers. { 'päl·ə·mər ,blend }

polymeric [CHEM] Made of repeating subunits. { ,päl·ə'mer·ik }

polymerization [CHEM] **1.** The bonding of two or more monomers to produce a polymer. **2.** Any chemical reaction that produces such a bonding. { pə,lim·ə·rə'zā·shən }

polymer light-emitting diode [ORG CHEM] An organic polymeric material that emits light in response to the application of an electric field. It may be an organic semiconductor sandwiched between metals of high and low work functions or a heterostructure made of two polymers, which increases the likelihood of radiative electron-hole recombination because of the energy-band structure. Also known as light-emitting polymer; polyLED. { ¦päl·ə·mər ,līt·i,mid·iŋ 'dī,ōd }

polymethyl methacrylate [ORG CHEM] A thermoplastic polymer that is derived from methyl methacrylate, $CH_2=C(CH_3)COOCH_3$; transparent solid with excellent optical qualities and water resistance; used for aircraft domes, lighting fixtures, optical instruments, and surgical appliances. { 'päl·i,meth·əl mə'thak·rə,lāt }

polymolecular assembly [CHEM] The spontaneous association of a large number of components into a specific phase (films, layers, membranes, vesicles, micelles, mesophases, surfaces, solids, and so on). { ,päl·ē·mə,lek·yə·lər ə'sem·blē }

polynuclear hydrocarbon [ORG CHEM] Hydrocarbon molecule with two or more closed rings; examples are naphthalene, $C_{10}H_8$, with two benzene rings side by side, or diphenyl, $(C_6H_5)_2$, with two bond-connected benzene rings. Also known as polycyclic hydrocarbon. { ¦päl·ə'nü·klē·ər 'hī·drə,kär·bən }

polyol See polyhydric alcohol. { 'päl·ē,ól }

polyolefin [ORG CHEM] A resinous material made by the polymerization of olefins, such as polyethylene from ethylene, polypropylene from propylene, or polybutene from butylene. { ¦päl·ē'ōl·ə·fən }

polyoxyalkylene resin [ORG CHEM] Condensation polymer produced from an oxyalkene, such as polyethylene glycol from oxyethylene or ethylene glycol. { ¦päl·ē¦äk·sē'al·kə,lēn 'rez·ən }

polyoxyethylene (8) stearate See polyoxyl (8) stearate. { ¦päl·ē,äk·sē'eth·ə,lēn ¦āt 'stir,āt }

polyoxyl (8) stearate [ORG CHEM] A cream-colored, soft, waxy solid at 25°C; soluble in toluene, acetone, ether, and ethanol; used in bakery products as an emulsifier. Also known as polyoxyethylene (8) stearate. { ,päl·ē'äk·səl ¦āt 'stir,āt }

polyoxymethylene [ORG CHEM] $(OCH_2)_n$ A polymer of formaldehyde that has excellent mechanical and high-temperature properties. Also know as polyacetal; polyformaldehyde. { ‚päl·ē‚äk·sē'meth·ə‚lēn }

polyphenyl [ORG CHEM] Any of a group of direct colors used to dye cotton and wool. { 'päl·i‚fen·əl }

polyphenylene oxide [ORG CHEM] A polyether resin of 2,6-dimethylphenol, $(CH_3)_2$-C_6H_3OH; useful temperature range is −275 to 375°F (−168 to 191°C), with intermittent use possible up to 400°F (204°C). { ‚päl·i'fen·əl‚ēn 'äk‚sīd }

polyphosphazene [ORG CHEM] A high-molecular-weight, essentially linear polymer with alternating phosphorus and nitrogen atoms in the skeleton and two side groups attached to each phosphorus. { ‚päl·i'fä·sfə‚zēn }

polyphosphoric acid [INORG CHEM] $H_6P_4O_{13}$ Viscous, water-soluble, hygroscopic, waterwhite liquid; used wherever concentrated phosphoric acid is needed. { ‚päl·i·fä'sfȯr·ik 'as·əd }

polypropylene [ORG CHEM] $(C_3H_6)_x$ A crystalline, thermoplastic resin made by the polymerization of propylene, C_3H_6; the product is hard and tough, resists moisture, oils, and solvents, and withstands temperatures up to 170°C; used to make molded articles, fibers, film, rope, printing plates, and toys. { ‚päl·ə'prō·pə‚lēn }

polypropylene glycol [ORG CHEM] $CH_3CHOH(CH_2OCH-CH_3)_xCH_2OH$ Polymeric material similar to polyethylene glycol, but with greater oil solubility and less water solubility; used as a solvent for vegetable oils, waxes, and resins, in hydraulic fluids and as a chemical intermediate. { ‚päl·ə'prō·pə‚lēn 'glī‚kȯl }

polysiloxane [ORG CHEM] $(R_2SiO)_n$ A polymer in which the chain contains alternate silicon and oxygen atoms; in the formula, R can be H or an alkyl or aryl group; commercially, the R is usually CH_3 (the methylsiloxanes); properties vary with molecular weight, from oils to greases to rubbers to plastics. { ‚päl·i·si'läk‚sän }

polysorbate [ORG CHEM] Any compound that is an ester of sorbitol. { ‚päl·ē'sȯr‚bāt }

polystyrene [ORG CHEM] $(C_6H_5CHCH_2)_x$ A water-white, tough synthetic resin made by polymerization of styrene; soluble in aromatic and chlorinated hydrocarbon solvents; used for injection molding, extrusion or casting for electrical insulation, fabric lamination, and molding of plastic objects. { ‚päl·i'stī‚rēn }

polysulfide rubber [ORG CHEM] A synthetic polymer made by the reaction of sodium polysulfide with an organic dichloride; resistant to light, oxygen, oils, and solvents; impermeable to gases; poor tensile strength and abrasion resistance. { ‚päl·i'səl‚fīd 'rəb·ər }

polyterpene resin [ORG CHEM] A thermoplastic resin or viscous liquid from polymerization of turpentine; used in paints, polishes, and rubber plasticizers, and to cure concrete and impregnate paper. { ‚päl·i'tər‚pēn ‚rez·ən }

polytetrafluoroethylene [ORG CHEM] $(CF_2CF_2)_n$ A highly crystalline perfluorinated polymer that is characteristically resistant to heat and chemicals. { ‚päl·ē‚te·trə‚flür·ō'eth·ə‚lēn }

polythene [ORG CHEM] Common name for polyethlylene in the United Kingdom. { 'päl·i‚thēn }

polytrifluorochloroethylene resin See chlorotrifluoroethylene polymer. { ‚päl·i·trī‚flür·ō‚klȯr·ō'eth mbə‚lēn ‚rez·ən }

polyunsaturated acid [ORG CHEM] A fatty acid with two or more double bonds per molecule, such as linoleic or linolenic acid. { ‚päl·ē‚ən'sach·ə‚rād·əd 'as·əd }

polyurethane resin [ORG CHEM] Any resin resulting from the reaction of diisocyanates (such as toluene diisocyanate) with a phenol, amine, or hydroxylic or carboxylic compound to produce a polymer with free isocyanate groups; used as protective coatings, potting or casting resins, adhesives, rubbers, and foams, and in paints, varnishes, and adhesives. { ‚päl·ē'yür·ə‚thän 'rez·ən }

polyurethane rubber [ORG CHEM] A synthetic polyurethane-resin elastomer made by the reaction of a diisocyanate to a polyester (such as the glycol-adipic acid ester); has high resistance to abrasion, oil, ozone, and high temperatures. Also known as polyester rubber. { ‚päl·ē'yür·ə‚thän 'rəb·ər }

polyvalent [CHEM] Pertaining to an ion with more than one valency, such as the sulfate ion, SO_4^{2-}. Also known as multivalent; polygen. { ‚päl·i'vā·lənt }

301

polyvinyl acetal resin

polyvinyl acetal resin *See* vinyl acetal resin. { ¦päl·i'vīn·əl 'as·ə₁tal 'rez·ən }
polyvinyl acetate [ORG CHEM] (H₂CCHOOCCH₃)ₓ A thermoplastic polymer; insoluble in water, gasoline, oils, and fats, soluble in ketones, alcohols, benzene, esters, and chlorinated hydrocarbons; used in adhesives, films, lacquers, inks, latex paints, and paper sizes. Abbreviated PVA; PVAc. { ¦päl·i'vīn·əl 'as·ə₁tāt }
polyvinyl alcohol [ORG CHEM] Water-soluble polymer made by hydrolysis of a polyvinyl ester (such as polyvinyl acetate); used in adhesives, as textile and paper sizes, and for emulsifying, suspending, and thickening of solutions. Abbreviated PVA. { ¦päl· i'vīn·əl 'al·kə₁hȯl }
polyvinyl carbazole [ORG CHEM] Thermoplastic resin made by reaction of acetylene with carbazole; softens at 150°C; has good electrical properties and heat and chemical stabilities; used as a paper-capacitor impregnant and as a substitute for electrical mica. { ¦päl·i'vīn·əl 'kär·bə₁zōl }
polyvinyl chloride [ORG CHEM] (H₂CCHCl)ₓ Polymer of vinyl chloride; tasteless, odorless; insoluble in most organic solvents; a member of the family of vinyl resins; used in soft flexible films for food packaging and in molded rigid products such as pipes, fibers, upholstery, and bristles. Abbreviated PVC. { ¦päl·i'vīn·əl 'klȯr₁īd }
polyvinyl chloride acetate [ORG CHEM] Thermoplastic copolymer of vinyl chloride, CH₂CHCl, and vinyl acetate, CH₃COOCH=CH₂; colorless solid with good resistance to water, concentrated acids, and alkalies; compounded with plasticizers, it yields a flexible material superior to rubber in aging properties; used for cable and wire coverings and protective garments. { ¦päl·i'vīn·əl 'klȯr₁īd 'as·ə₁tāt }
polyvinyl dichloride [ORG CHEM] A high-strength polymer of chlorinated polyvinyl chloride; it is self-extinguishing and has superior chemical resistance; used for pipes carrying hot, corrosive materials. Abbreviated PVDC. { ¦päl·i'vīn·əl dī'klȯr₁īd }
polyvinyl ether *See* polyvinyl ethyl ether. { ¦päl·i'vīn·əl 'ē·thər }
polyvinyl ethyl ether [ORG CHEM] [−CH(OC₂H₅)CH₂−]ₓ A viscous gum to rubbery solid, soluble in organic solvents; used for pressure-sensitive tape. Also known as polyvinyl ether. { ¦päl·i'vīn·əl 'eth·əl 'ē·thər }
polyvinyl fluoride [ORG CHEM] (−H₂CCHF−)ₓ Vinyl fluoride polymer; has superior resistance to weather, chemicals, oils, and stains; has high strength; used for packaging (but not of food) and electrical equipment. { ¦päl·i'vīn·əl 'flúr₁īd }
polyvinyl formate resin [ORG CHEM] (CH₂=CHOOCH)ₓ Clear-colored resin that is hard and solvent-resistant; used to make clear, hard plastics. { ¦päl·i'vīn·əl 'fȯr₁māt ₁rez·ən }
polyvinylidene chloride [ORG CHEM] Thermoplastic polymer of vinylidene chloride, H₂C=CCl₂; white powder softening at 185–200°C; used to make soft-flexible to rigid products. { ¦päl·i·vī'nil·ə₁dēn 'klȯr₁īd }
polyvinylidene fluoride [ORG CHEM] Fluorocarbon polymer made from vinylidene fluoride, (H₂C=CF₂); has good tensile and compressive strength and high impact strength; used in chemical equipment for gaskets, impellers, and other pump parts, and for drum linings and protective coatings. { ¦päl·i·vī'nil·ə₁dēn 'flúr₁īd }
polyvinylidene resin *See* vinylidene resin. { ¦päl·i·vī'nil·ə₁dēn 'rez·ən }
polyvinyl isobutyl ether [ORG CHEM] [−CH₂CHOCH₂CH(CH₃)₂−]ₓ An odorless synthetic resin; elastomer to viscous liquid depending on molecular weight; soluble in hydrocarbons, esters, ethers, and ketones, insoluble in water; used in adhesives, waxes, plasticizers, lubricating oils, and surface coatings. Abbreviated PVI. { ¦päl·i'vīn·əl ¦ī·sə¦byüd·əl 'e·thər }
polyvinyl methyl ether [ORG CHEM] (−CH₂CHOCH₃−)ₓ A colorless, tacky liquid, soluble in organic solvents, except aliphatic hydrocarbons, and in water below 32°C; used for pressure-sensitive adhesives, as a heat sensitizer for rubber latex, and as a pigment binder in inks and textile finishing. Abbreviated PVM. { ¦päl·i'vīn·əl 'meth·əl 'ē·thər }
polyvinyl pyrrolidone [ORG CHEM] (C₆H₉NO)ₓ A water-soluble, white, resinous solid; used in pharmaceuticals, cosmetics, detergents, and foods, and as a synthetic blood plasma. Abbreviated PVP. { ¦päl·i'vīn·əl pə'räl·ə₁dōn }

polyvinyl resin [ORG CHEM] Any resin or polymer derived from vinyl monomers. Also known as vinyl plastic. { ¦päl·i'vīn·əl 'rez·ən }

pool boiling [PHYS CHEM] Boiling of a liquid whose flow results from natural convection. { 'pül ¸bȯil·iŋ }

porous alum *See* aluminum sodium sulfate. { 'pȯr·əs 'al·əm }

positional isomer [CHEM] One of a set of structural isomers which differ only in the point at which a side-chain group is attached. [ORG CHEM] Constitutional isomer having the same functional group located in different positions along a chain or in a ring. { pə'zish·ən·əl 'ī·sə·mər }

positive ion [CHEM] An atom or group of atoms which by loss of one or more electrons has acquired a positive electric charge; occurs on ionization of chemical compounds as H$^+$ from ionization of hydrochloric acid, HCl. { 'päz·əd·iv 'ī¸än }

positron emission spectroscopy [SPECT] A technique in which a solid surface is bombarded with a low-energy monoenergetic positron beam and the energies of positrons emitted from the surface are measured to determine the amounts of energy lost to molecules adsorbed on the surface. { 'päz·ə¸trän i¦mish·ən spek'träs·kə·pē }

positronium velocity spectroscopy [SPECT] A technique in which a solid surface is bombarded with a low-energy monoenergetic positron beam and the velocities of the emitted positronium atoms are measured to determine the energy and momentum spectrum of the density of electron states near the surface. { ¸päz·ə'trō·nē·əm və'läs·əd·ē spek'träs·kə·pē }

postignition [CHEM] Surface ignition after the passage of the normal spark. { ¸pōst·ig'nish·ən }

postprecipitation [CHEM] Precipitation of an impurity from a supersaturated solution onto the surface of an already present precipitate; used for analytical laboratory separations. { ¦pōs·pri¸sip·ə'tā·shən }

potash *See* potassium carbonate. { 'päd¸ash }

potash blue [INORG CHEM] A pigment made by oxidizing ferrous ferrocyanide; used in making carbon paper. { 'päd¸ash 'blü }

potassium [CHEM] A chemical element, symbol K, atomic number 19, atomic weight 39.0983; an alkali metal. Also known as kalium. { pə'tas·ē·əm }

potassium acetate [ORG CHEM] KC$_2$H$_3$O$_2$ White, deliquescent solid; soluble in water and alcohol, insoluble in ether; melts at 292°C; used as analytical reagent, dehydrating agent, in medicine, and in crystal glass manufacture. { pə'tas·ē·əm 'as·ə¸tāt }

potassium acid carbonate *See* potassium bicarbonate. { pə'tas·ē·əm 'as·əd 'kär·bə¸nāt }

potassium acid fluoride *See* potassium bifluoride. { pə'tas·ē·əm 'as·əd 'flùr¸īd }

potassium acid oxalate *See* potassium binoxalate. { pə'tas·ē·əm 'as·əd 'äk·sə¸lāt }

potassium acid phosphate *See* potassium phosphate. { pə'tas·ē·əm 'as·əd 'fäs¸fāt }

potassium acid phthalate *See* potassium biphthalate. { pə'tas·ē·əm 'as·əd 'tha¸lāt }

potassium acid saccharate [ORG CHEM] HOOC(CHOH)$_4$COOK An off-white powder, soluble in hot water, acid, or alkaline solutions; used in rubber formulations, soaps, and detergents, and for metal plating. { pə'tas·ē·əm 'as·əd 'sak·ə¸rāt }

potassium acid sulfate *See* potassium bisulfate. { pə'tas·ē·əm 'as·əd 'səl¸fāt }

potassium acid sulfite *See* potassium bisulfite. { pə'tas·ē·əm 'as·əd 'səl¸fīt }

potassium acid tartrate *See* potassium bitartrate. { pə'tas·ē·əm 'as·əd 'tär¸trāt }

potassium alginate [ORG CHEM] (C$_6$H$_7$O$_6$K)$_n$ A hydrophilic colloid occurring as filaments, grains, granules, and powder; used in food processing as a thickener and stabilizer. Also known as potassium polymannuronate. { pə'tas·ē·əm 'al·jə¸nāt }

potassium alum *See* potassium aluminum sulfate. { pə'tas·ē·əm 'al·əm }

potassium aluminate [INORG CHEM] K$_2$Al$_2$O$_4$·3H$_2$O Water-soluble, alcohol-insoluble, lustrous crystals; used as a dyeing and printing mordant, and as a paper sizing. { pə'tas·ē·əm ə'lüm·ə¸nāt }

potassium aluminum fluoride [INORG CHEM] K$_3$AlF$_6$ A toxic, white powder used as an insecticide. { pə'tas·ē·əm ə'lüm·ə·nəm 'flùr¸īd }

potassium aluminum sulfate [INORG CHEM] KAl(SO$_4$)$_2$·12H$_2$O White, odorless crystals

potassium antimonate

that are soluble in water; used in medicines and baking powder, in dyeing, papermaking, and tanning.   Also known as alum; aluminum potassium sulfate; potassium alum.   { pə'tas·ē·əm ə'lüm·ə·nəm 'səl,fāt }

potassium antimonate [INORG CHEM] KSbO₃ White, water-soluble crystals.   Also known as potassium stibnate.   { pə'tas·ē·əm 'ant·ə·mə,nāt }

potassium antimonyl tartrate See tartar emetic.   { pə'tas·ē·əm 'ant·ə·mə,nil 'tär,trāt }

potassium argentocyanide See silver potassium cyanide.   { pə'tas·ē·əm ¦ar·jən·tō'sī·ə,nīd }

potassium arsenate [INORG CHEM] K₃AsO₄ Poisonous, colorless crystals; soluble in water, insoluble in alcohol; used as an insecticide, analytical reagent, and in hide preservation and textile printing.   Also known as Macquer's salt.   { pə'tas·ē·əm 'ärs·ən,āt }

potassium arsenite [INORG CHEM] KH(AsO₂)₂ Poisonous, hygroscopic, white powder; soluble in alcohol; decomposes slowly in air; used in medicine, on mirrors, and as an analytical reagent.   Also known as potassium metarsenite.   { pə'tas·ē·əm 'ärs·ən,īt }

potassium aurichloride See potassium gold chloride.   { pə'tas·ē·əm ¦or·ə'klor,īd }

potassium bicarbonate [INORG CHEM] KHCO₃ A white powder or granules, or transparent colorless crystals; used in baking powder and in medicine as an antacid.   Also known as potassium acid carbonate.   { pə'tas·ē·əm bī'kär·bə,nāt }

potassium bichromate See potassium dichromate.   { pə'tas·ē·əm bī'krō,māt }

potassium bifluoride [INORG CHEM] KHF₂ Colorless, corrosive, poisonous crystals; soluble in water and dilute alcohol; used to etch glass and as a metallurgy flux.   Also known as Fremy's salt; potassium acid fluoride.   { pə'tas·ē·əm bī'flur,īd }

potassium binoxalate [ORG CHEM] KHC₂O₄·H₂O A poisonous, white, odorless, crystalline compound; used to clean wood and remove ink stains, as a mordant in dyeing, and in photography.   Also known as potassium acid oxalate; sal acetosella; salt of sorrel.   { pə'tas·ē·əm bə'näk·sə,lāt }

potassium biphthalate [ORG CHEM] HOOCC₆H₄COOH A crystalline compound, soluble in 12 parts of water; used as a buffer in pH determinations and as a primary standard for preparation of volumetric alkali solutions.   Also known as acid potassium phthalate; potassium acid phthalate; potassium hydrogen phthalate.   { pə'tas·ē·əm bī'tha,lāt }

potassium bismuth tartrate [ORG CHEM] A white, granular powder with a sweet taste; soluble in water; used in medicine.   Also known as bismuth potassium tartrate.   { pə'tas·ē·əm 'biz·məth 'tär,trāt }

potassium bisulfate [INORG CHEM] KHSO₄ Water-soluble, colorless crystals, melting at 214°C; used in winemaking, fertilizer manufacture, and as a flux and food preservative.   Also known as acid potassium sulfate; potassium acid sulfate.   { pə'tas·ē·əm bī'səl,fāt }

potassium bisulfite [INORG CHEM] KHSO₃ White, water-soluble powder with sulfur dioxide aroma; insoluble in alcohol; decomposes when heated; used as an antiseptic and reducing chemical, and in analytical chemistry, tanning, and bleaching.   Also known as potassium acid sulfite.   { pə'tas·ē·əm bī'səl,fīt }

potassium bitartrate [ORG CHEM] KHC₄H₄O₆ White, water-soluble crystals or powder; used in baking powder, for medicine, and as an acid and buffer in foods.   Also known as cream of tartar; potassium acid tartrate.   { pə'tas·ē·əm bī'tär,trāt }

potassium borohydride [INORG CHEM] KBH₄ A white, crystalline powder, soluble in water, alcohol, and ammonia; used as a hydrogen source and a reducing agent for aldehydes and ketones.   { pə'tas·ē·əm ¦bor·ō'hī,drīd }

potassium bromate [INORG CHEM] KBrO₃ Water-soluble, white crystals, melting at 434°C; insoluble in alcohol; strong oxidizer and a fire hazard; used in analytical chemistry and as an additive for permanent-wave compounds.   { pə'tas·ē·əm 'brō,māt }

potassium bromide [INORG CHEM] KBr White, hygroscopic crystals with bitter taste; soluble in water and glycerin, slightly soluble in alcohol and ether; melts at 730°C; used in medicine, soaps, photography, and lithography.   { pə'tas·ē·əm 'brō,mīd }

304

potassium bromide-disk technique [ANALY CHEM] Method of preparing an infrared spectrometry sample by grinding it and mixing it with a dry powdered alkali halide (such as KBr), then compressing the mixture into a tablet or pellet. Also known as pellet technique; pressed-disk technique. { pə'tas·ē·əm 'brō,mīd ¦disk tek,nēk }

potassium cadmium iodide See potassium tetraiodocadmate. { pə'tas·ē·əm 'kad·mē· əm 'ī·ə,dīd }

potassium carbonate [INORG CHEM] K_2CO_3 White, water-soluble, deliquescent powder, melting at 891°C; insoluble in alcohol; used in brewing, ceramics, explosives, fertilizers, and as a chemical intermediate. Also known as potash; salt of tartar. { pə'tas· ē·əm 'kär·bə,nāt }

potassium chlorate [INORG CHEM] $KClO_3$ Transparent, colorless crystals or a white powder with a melting point of 356°C; soluble in water, alcohol, and alkalies; used as an oxidizing agent, for explosives and matches, and in textile printing and paper manufacture. { pə'tas·ē·əm 'klȯr,āt }

potassium chloride [INORG CHEM] KCl Colorless crystals with saline taste; soluble in water, insoluble in alcohol; melts at 776°C; used as a fertilizer and in photography and pharmaceutical preparations. Also known as potassium muriate. { pə'tas·ē· əm 'klȯr,īd }

potassium chloroaurate See potassium gold chloride. { pə'tas·ē·əm ¦klȯr·ō'ȯr,āt }

potassium chloroplatinate [INORG CHEM] K_2PtCl_6 Orange-yellow crystals or powder which decomposes when heated (250°C); used in photography. Also known as platinum potassium chloride; potassium platinichloride. { pə'tas·ē·əm ¦klȯr· ō'plat·ən,āt }

potassium chromate [INORG CHEM] K_2CrO_4 Yellow crystals, melting at 971°C; soluble in water, insoluble in alcohol; used as an analytical reagent and textile mordant, in enamels, inks, and medicines, and as a chemical intermediate. { pə'tas·ē·əm 'krō,māt }

potassium chromium sulfate See chrome alum. { pə'tas·ē·əm 'krō·mē·əm 'səl,fāt }

potassium citrate [ORG CHEM] $K_3C_6H_5O_7·H_2O$ Odorless crystals with saline taste; soluble in water and glycerol, deliquescent and insoluble in alcohol; decomposes about 230°C; used in medicine. { pə'tas·ē·əm 'sī,trāt }

potassium cobaltinitrite See cobalt potassium nitrite. { pə'tas·ē·əm kō¦bȯl·tə'nī,trāt }

potassium cyanate [INORG CHEM] KOCN Colorless, water-soluble crystals; used as an herbicide and for the manufacture of drugs and organic chemicals. { pə'tas·ē·əm 'sī·ə,nāt }

potassium cyanide [INORG CHEM] KCN Poisonous, white, deliquescent crystals with bitter almond taste; soluble in water, alcohol, and glycerol; used for metal extraction, for electroplating, for heat-treating steel, and as an analytical reagent and insecticide. { pə'tas·ē·əm 'sī·ə,nīd }

potassium cyanoargentate See silver potassium cyanide. { pə'tas·ē·əm ¦sī·ə·nō'är· jən,tāt }

potassium cyanoaurite See potassium gold cyanide. { pə'tas·ē·əm ¦sī·ə·nō'ȯr,īt }

potassium dichloroisocyanurate [INORG CHEM] White, crystalline powder or granules; strong oxidant used in dry household bleaches, detergents, and scouring powders. { pə'tas·ē·əm dī¦klȯr·ō,ī'sō,sī'an·yúr,āt }

potassium dichromate [INORG CHEM] $K_2Cr_2O_7$ Poisonous, yellowish-red crystals with metallic taste; soluble in water, insoluble in alcohol; melts at 396°C, decomposes at 500°C; used as an oxidizing agent and analytical reagent, and in explosives, matches, and electroplating. Also known as potassium bichromate; red potassium chromate. { pə'tas·ē·əm dī'krō,māt }

potassium dihydrogen phosphate See potassium phosphate. { pə'tas·ē·əm dī'hī·drə· jən 'fäs,fāt }

potassium diphosphate See potassium phosphate. { pə'tas·ē·əm dī'fäs,fāt }

potassium ferric oxalate [INORG CHEM] $K_3Fe(C_2O_4)_3·3H_2O$ Green crystals decomposing at 230°C, soluble in water and acetic acid; used in photography and blueprinting. { pə'tas·ē·əm 'fer·ik 'äk·sə,lāt }

potassium ferricyanide [INORG CHEM] $K_3Fe(CN)_6$ Poisonous, water-soluble, bright-red

potassium ferrocyanide

crystals; decomposes when heated; used in calico printing and wool dyeing.   Also known as red potassium prussiate; red prussiate of potash.   { pə'tas·ē·əm ˌfer· ə'sī·əˌnīd }

potassium ferrocyanide   [INORG CHEM] $K_4Fe(CN)_6 \cdot 3H_2O$ Yellow crystals with saline taste; soluble in water, insoluble in alcohol; loses water at 60°C; used in medicine, dry colors, explosives, and as an analytical reagent.   Also known as yellow prussiate of potash.   { pə'tas·ē·əm ˌfer·ō'sī·əˌnīd }

potassium fluoborate   [INORG CHEM] KBF_4 White powder or gelatinous crystals that decompose at high temperatures; slightly soluble in water and hot alcohol; used as a sand agent to cast magnesium and aluminum, and in electrochemical processes.   { pə'tas·ē·əm ˌflü·ə'bȯrˌāt }

potassium fluoride   [INORG CHEM]   KF or $KF \cdot 2H_2O$ Poisonous, white, deliquescent crystals with saline taste; soluble in water and hydrofluoric acid, insoluble in alcohol; melts at 846°C; used to etch glass and as a preservative and insecticide.   { pə'tas· ē·əm 'flùrˌīd }

potassium fluosilicate   [INORG CHEM] K_2SiF_6 An odorless, white crystalline compound; slightly soluble in water; used in vitreous frits, synthetic mica, metallurgy, and ceramics.   Also known as potassium silicofluoride.   { pə'tas·ē·əm ˌflü·ə'sil·ə·kət }

potassium gluconate   [ORG CHEM] $KC_6H_{11}O_7$ An odorless, white crystalline compound with salty taste; soluble in water, insoluble in alcohol and benzene; used in medicine.   { pə'tas·ē·əm 'glü·kəˌnāt }

potassium glutamate   [ORG CHEM] $KOOC(CH_2)_2CH(NH_2)COOH \cdot H_2O$ White, hygroscopic, water-soluble powder; used as a flavor enhancer and salt substitute.   Also known as monopotassium L-glutamate.   { pə'tas·ē·əm 'glüdˌəˌmāt }

potassium glycerinophosphate   See potassium glycerophosphate.   { pə'tas·ē·əm ˌglis· ə·rə·nō'fäsˌfāt }

potassium glycerophosphate   [ORG CHEM] $K_2C_3H_5O_2 \cdot H_2PO_4 \cdot 3H_2O$ Pale yellow, syrupy liquid, soluble in alcohol; used in medicine and as a dietary supplement.   Also known as potassium glycerinophosphate.   { pə'tas·ē·əm ˌglis·ə·rō'fäsˌfāt }

potassium gold chloride   [INORG CHEM] $KAuCl_4 \cdot 2H_2O$ Yellow crystals, soluble in water, ether, and alcohol; used in photography and medicine.   Also known as gold potassium chloride; potassium aurichloride; potassium chloroaurate.   { pə'tas·ē·əm 'gōld 'klȯrˌīd }

potassium gold cyanide   [INORG CHEM] $KAu(CN)_2$ A white, water-soluble, crystalline powder; used in medicine and for gold plating.   Also known as gold potassium cyanide; potassium cyanoaurite.   { pə'tas·ē·əm 'gōld 'sī·əˌnīd }

potassium hydrate   See potassium hydroxide.   { pə'tas·ē·əm 'hīˌdrāt }

potassium hydrogen phosphate   See potassium phosphate.   { pə'tas·ē·əm 'hī·drə·jən 'fäsˌfāt }

potassium hydrogen phthalate   See potassium biphthalate.   { pə'tas·ē·əm 'hī·drə·jən 'thaˌlāt }

potassium hydroxide   [INORG CHEM]   KOH Toxic, corrosive, water-soluble, white solid, melting at 360°C; used to make soap and matches, and as an analytical reagent and chemical intermediate.   Also known as caustic potash; potassium hydrate.   { pə'tas·ē·əm hī'dräkˌsīd }

potassium hyperchlorate   See potassium perchlorate.   { pə'tas·ē·əm ˌhī·pər'klȯrˌāt }

potassium hypophosphite   [INORG CHEM] KH_2PO_2 White, opaque crystals or powder, soluble in water and alcohol; used in medicine.   { pə'tas·ē·əm ˌhī·pō'fäsˌfīt }

potassium iodate   [INORG CHEM] KIO_3 Odorless, white crystals; soluble in water, insoluble in alcohol; melts at 560°C; used as an analytical reagent and in medicine.   { pə'tas·ē·əm 'ī·əˌdāt }

potassium iodide   [INORG CHEM]   KI Water- and alcohol-soluble, white crystals with saline taste; melts at 686°C; used in medicine and photography, and as an analytical reagent.   { pə'tas·ē·əm 'ī·əˌdīd }

potassium linoleate   [ORG CHEM] $C_{17}H_{31}COOK$ Light-tan, water-soluble paste; used as an emulsifying agent.   { pə'tas·ē·əm li'nō·lēˌāt }

potassium manganate   [INORG CHEM] K_2MnO_4 Water-soluble dark-green crystals, decomposing at 190°C; used as an analytical reagent, bleach, oxidizing agent, disinfectant, mordant for dyeing wool and in photography, printing, and water purification. { pə'tas·ē·əm 'maŋ·gə,nāt }

potassium metabisulfite [INORG CHEM] $K_2S_2O_5$ White granules or powder, decomposing at 150–190°C; used as an antiseptic, for winemaking, food preservation, and process engraving, and as a source for sulfurous acid. Also known as potassium pyrosulfite. { pə'tas·ē·əm ,med·ə·bī'səl,fīt }

potassium metarsenite See potassium arsenite. { pə'tas·ē·əm ¦med·ə'ärs·ən,īt }

potassium monophosphate See potassium phosphate. { pə'tas·ē·əm ,män·ō'fäs,fāt }

potassium muriate See potassium chloride. { pə'tas·ē·əm 'myùr·ē,āt }

potassium nitrate [INORG CHEM] KNO_3 Flammable, water-soluble, white crystals with saline taste; melts at 337°C; used in pyrotechnics, explosives, and matches, as a fertilizer, and as an analytical reagent. Also known as niter. { pə'tas·ē·əm 'nī,trāt }

potassium nitrite [INORG CHEM] KNO_2 White, deliquescent prisms, melting at 297–450°C; soluble in water, insoluble in alcohol; strong oxidizer, exploding at over 550°C; used as an analytical reagent, in medicine, organic synthesis, pyrotechnics, and explosives. { pə'tas·ē·əm 'nī,trīt }

potassium oxalate [ORG CHEM] $K_2C_2O_4 \cdot H_2O$ Odorless, efflorescent, water-soluble, colorless crystals; decomposes when heated; used in analytical chemistry and photography and as a bleach and oxalic acid source. { pə'tas·ē·əm 'äk·sə,lāt }

potassium oxide [INORG CHEM] K_2O Gray, water-soluble crystals; melts at red heat; forms potassium hydroxide in water. { pə'tas·ē·əm 'äk,sīd }

potassium percarbonate [INORG CHEM] $K_2C_2O_6 \cdot H_2O$ White, granular, water-soluble mass with a melting point of 200–300°C; used in microscopy, photography, and textile printing. { pə'tas·ē·əm pər'kär·bə,nāt }

potassium perchlorate [INORG CHEM] $KClO_4$ Explosive, oxidative, colorless crystals; soluble in water, insoluble in alcohol; decomposes at 400°C; used in explosives, medicine, pyrotechnics, analysis, and as a reagent and oxidizing agent. Also known as potassium hyperchlorate. { pə'tas·ē·əm pər'klòr,āt }

potassium permanganate [INORG CHEM] $KMnO_4$ Highly oxidative, water-soluble, purple crystals with sweet taste; decomposes at 240°C; and explodes in contact with oxidizable materials; used as a disinfectant and analytical reagent, in dyes, bleaches, and medicines, and as a chemical intermediate. Also known as purple salt. { pə'tas·ē·əm pər'man·gə,nāt }

potassium peroxide [INORG CHEM] K_2O_2 Yellow mass with a melting point of 490°C; decomposes with oxygen evolution in water; used as an oxidizing and bleaching agent. { pə'tas·ē·əm pə'räk,sīd }

potassium peroxydisulfate See potassium persulfate. { pə'tas·ē·əm pə¦räk·sē·dī'səl,fāt }

potassium persulfate [INORG CHEM] $K_2S_2O_8$ White, water-soluble crystals, decomposing below 100°C; used for bleaching and textile desizing, as an oxidizing agent and antiseptic, and in the manufacture of soap and pharmaceuticals. Also known as potassium peroxydisulfate. { pə'tas·ē·əm pər'səl,fāt }

potassium phosphate [INORG CHEM] Any one of three orthophosphates of potassium. The monobasic form, KH_2PO_4, consists of colorless, water-soluble crystals melting at 253°C; used in sonar transducers, optical modulation, medicine, baking powders, and nutrient solutions; also known as potassium acid phosphate, potassium dihydrogen phosphate (KDP), potassium diphosphate, potassium orthophosphate. The dibasic form, K_2HOP_4, consists of white, water-soluble crystals; used in medicine, fermentation, and nutrient solutions; also known as potassium hydrogen phosphate, potassium monophosphate. The tribasic form, K_3PO_4, is a water-soluble, hygroscopic white powder, melting at 1340°C; used to purify gasoline, to soften water, and to make liquid soaps and fertilizers; also known as neutral potassium phosphate, tripotassium orthophosphate. { pə'tas·ē·əm 'fäs,fāt }

potassium platinichloride See potassium chloroplatinate. { pə'tas·ē·əm ¦plat·ən·ə'klòr,īd }

potassium polymannuronate

potassium polymannuronate *See* potassium alginate. { pə'tas·ē·əm ¦päl·ē·man'yür· ə,nāt }

potassium polymetaphosphate [INORG CHEM] $(KPO_3)_n$ White powder with a molecular weight up to 500,000; used in foods as a fat emulsifier and moisture-retaining agent. { pə'tas·ē·əm päl·e,med·ə'fäs,fāt }

potassium pyrophosphate [INORG CHEM] $K_4P_2O_7 \cdot 3H_2O$ Water-soluble, colorless crystals; dehydrates below 300°C, melts at 1090°C; used in tin plating, china-clay purification, dyeing, oil-drilling muds, and synthetic rubber production. Also known as normal potassium pyrophosphate; tetrapotassium pyrophosphate. { pə'tas·ē·əm ,pī·rō'fäs,fāt }

potassium pyrosulfite *See* potassium metabisulfite. { pə'tas·ē·əm ,pī·rō'səl,fīt }

potassium silicate [INORG CHEM] $SiO_2 = K_2O$ A compound existing in two forms, solution and solid (glass); as a solution, it is colorless to turgid in water, and is used in paints and coatings, as an arc-electrode binder and catalyst and in detergents; as a solid, it is colorless and water-soluble solid, and is used in glass manufacture and for dyeing and bleaching. { pə'tas·ē·əm 'sil·ə,kāt }

potassium silicofluoride *See* potassium fluosilicate. { pə'tas·ē·əm ¦sil·ə·kō'flür,īd }

potassium sodium ferricyanide [INORG CHEM] $K_2NaFe(CN)_6$ Red, water-soluble crystals; used for blueprint paper and in photography. { pə'tas·ē·əm 'sōd·ē·əm ,fer· ə'sī·ə,nīd }

potassium sodium tartrate [INORG CHEM] $KNaC_4H_4O_6 \cdot 4H_2O$ Colorless, water-soluble, efflorescent crystals or white powder with a melting point of 70–80°C; used in medicine and as a buffer and sequestrant in foods. Also known as Rochelle salt; Seignette salt. { pə'tas·ē·əm 'sōd·ē·əm 'tär,trāt }

potassium sorbate [ORG CHEM] $C_6H_7KO_2$ A crystalline compound, more soluble in water than in alcohol; decomposes above 270°C; used to inhibit mold and yeast growth in food. { pə'tas·ē·əm 'sor,bāt }

potassium stannate [INORG CHEM] $K_2SnO_3 \cdot 3H_2O$ White crystals; soluble in water, insoluble in alcohol; used in textile printing and dyeing, and in tin-plating baths. { pə'tas· ē·əm 'stan,āt }

potassium stibnate *See* potassium antimonate. { pə'tas·ē·əm 'stib,nīt }

potassium sulfate [INORG CHEM] K_2SO_4 Colorless crystals with bitter taste; soluble in water, insoluble in alcohol; melts at 1072°C; used as an analytical reagent, medicine, and fertilizer, and in aluminum and glass manufacture. Also known as salt of Lemery. { pə'tas·ē·əm 'səl,fāt }

potassium sulfide [INORG CHEM] K_2S Moderately flammable, water-soluble, deliquescent red crystals; melts at 840°C; used in analytical chemistry, medicine, and depilatories. Also known as fused potassium sulfide; hepar sulfuris; potassium sulfuret. { pə'tas·ē·əm 'səl,fīd }

potassium sulfite [INORG CHEM] $K_2SO_3 \cdot 2H_2O$ Water-soluble, white crystals; used in medicine and photography. { pə'tas·ē·əm 'səl,fīt }

potassium sulfuret *See* potassium sulfide. { pə'tas·ē·əm 'səl·fə,ret }

potassium tetraiodocadmate [INORG CHEM] $K_2(CdI_4) \cdot 2H_2O$ A crystalline compound; used in analytical chemistry for alkaloids, amines, and other compounds. Also known as cadmium potassium iodide; potassium cadmium iodide. { pə'tas·ē·əm ¦te·trə¦ī·ə·dō'kad,māt }

potassium thiocyanate [INORG CHEM] KCNS Water- and alcohol-soluble, colorless, odorless hygroscopic crystals with saline taste; decomposes at 500°C; used as an analytical reagent and in freezing mixtures, chemicals manufacture, textile printing and dyeing, and photographic chemicals. { pə'tas·ē·əm ¦thī·ō'sī·ə,nāt }

potassium undecylenate [ORG CHEM] $CH_2:CH(CH_2)_8COOK$ A white, water-soluble powder, decomposing at about 250°C; used in pharmaceuticals and cosmetics as a fungistat and bacteriostat. { pə'tas·ē·əm ,ən,des·ə'le,nāt }

potassium xanthate [ORG CHEM] KC_2H_5OCSS Water- and alcohol-soluble, yellow crystals; used as an analytical reagent and soil-treatment fungicide. { pə'tas·ē·əm 'zan,thāt }

potential electrolyte [PHYS CHEM] A solid material composed of uncharged molecules

primary alcohol

that can react chemically with a solvent to yield some ions in solution. { pə'ten·
chəl i'lek·trə,līt }

potentiometric cell [ANALY CHEM] Container for the two electrodes and the electrolytic
solution being titrated potentiometrically. { pə¦ten·chē·ə¦me·trik 'sel }

potentiometric titration [ANALY CHEM] Solution titration in which the end point is
read from the electrode-potential variations with the concentrations of potential-
determining ions, following the Nernst concept. Also known as constant-current
titration. { pə¦ten·chē·ə¦me·trik tī'trā·shən }

Pr *See* praseodymium.

praseodymium [CHEM] A chemical element, symbol Pr, atomic number 59, atomic
weight 140.9077; a metallic element of the rare-earth group. { ‚prā·zē·ō'dim·ē·əm }

prebiotic molecule [ORG CHEM] A molecule that is believed to be involved in the
processes leading to the origin of life. { ‚pre·bī‚äd·ik 'mäl·ə‚kyül }

precipitant [CHEM] A chemical or chemicals that cause a precipitate to form when
added to a solution. { prə'sip·ət·əns }

precipitate [CHEM] **1.** A substance separating, in solid particles, from a liquid as the
result of a chemical or physical change; **2.** To form a precipitate. { prə'sip·ə‚tāt }

precipitation [CHEM] The process of producing a separable solid phase within a liquid
medium; represents the formation of a new condensed phase, such as a vapor or
gas condensing to liquid droplets; a new solid phase gradually precipitates within
a solid alloy as a result of slow, inner chemical reaction; in analytical chemistry,
precipitation is used to separate a solid phase in an aqueous solution. { prə‚sip·
ə'tā·shən }

precipitation indicator [ANALY CHEM] In a titration, a substance that precipitates from
solution in a clearly visible form at the end point. { prə‚sip·ə'tā·shən ‚in·də‚kād·ər }

precipitation number [ANALY CHEM] The number of milliliters of asphaltic precipitate
formed when 10 milliliters of petroleum-lubricating oil is mixed with 90 milliliters
of a special-quality petroleum naphtha, then centrifuged according to American
Society for Testing and Materials test conditions; used to determine the quantity of
asphalt in petroleum-lubricating oil. { prə‚sip·ə'tā·shən ‚nəm·bər }

precipitation titration [ANALY CHEM] Amperometric titration in which the potential of
a suitable indicator electrode is measured during the titration. { prə‚sip·ə'tā·shən
tī‚trā·shən }

predissociation [PHYS CHEM] The dissociation of a molecule that has absorbed energy
before it can lose energy by radiation. { ¦prē·di‚sō·sē'ā·shən }

Pregl procedure [ANALY CHEM] Microanalysis technique in which the sample is decomposed thermally, with subsequent oxidation of decomposition products. { 'prā·gəl
prə‚sē·jər }

preparing salt *See* sodium stannate. { prə'per·iŋ ‚sȯlt }

prepolymer [ORG CHEM] A reactive low-molecular-weight macromolecule or an oligomer, capable of further polymerization. { prē'päl·i·mər }

pressed-disk technique *See* potassium bromide-disk technique. { 'prest ¦disk tek‚nēk }

pressure broadening [SPECT] A spreading of spectral lines when pressure is increased,
due to an increase in collision broadening. { 'presh·ər ‚brȯd·ən·iŋ }

pressure effect [SPECT] The effect of changes in pressure on spectral lines in the
radiation emitted or absorbed by a substance; namely, pressure broadening and
pressure shift. { 'presh·ər i‚fekt }

pressure shift [SPECT] An increase in the wavelength at which a spectral line has
maximum intensity, which takes place when pressure is increased. { 'presh·ər ‚shift }

primary [CHEM] A term used to distinguish basic compounds from similar or isomeric
forms; in organic compounds, for example, RCH_2OH is a primary alcohol, R_1R_2CHOH
is a secondary alcohol, and $R_1R_2R_3COH$ is a tertiary alcohol; in inorganic compounds,
for example, NaH_2PO_4 is primary sodium phosphate, Na_2HPO_4 is the secondary form,
and Na_3PO_4 is the tertiary form. { 'prī ‚mer·ē }

primary alcohol [ORG CHEM] An alcohol whose molecular structure may be written as
RCH_2OH, rather than as R_1R_2CHOH (secondary) or $R_1R_2R_3COH$ (tertiary). { 'prī‚mer·
ē 'al·kə‚hȯl }

309

primary amine [ORG CHEM] An amine whose molecular structure may be written as RNH_2, instead of R_1R_2NH (secondary) or $R_1R_2R_3N$ (tertiary).   { 'prī,mer·ē 'am,ēn }

primary carbon atom [ORG CHEM] A carbon atom in a molecule that is singly bonded to only one other carbon atom.   { 'prī,mer·ē 'kär·bən ,ad·əm }

primary hydrogen atom [ORG CHEM] A hydrogen atom that is bonded to a primary carbon atom.   { 'prī,mer·ē 'hī·drə·jən ,ad·əm }

primary structure [ORG CHEM] The chemical structure of a polymer chain.   { 'prī,mer· ē 'strək·chər }

principal line [SPECT] That spectral line which is most easily excited or observed.   { 'prin·sə·pəl 'līn }

principal moments [PHYS CHEM] The three moments of inertia of a rigid molecule calculated with respect to the principal axes.   { 'prin·sə·pəl 'mō·məns }

principal series [SPECT] A series occurring in the line spectra of many atoms and ions with one, two, or three electrons in the outer shell, in which the total orbital angular momentum quantum number changes from 1 to 0.   { 'prin·sə·pəl 'sir·ēz }

priscol See tolazoline hydrochloride.   { 'pris,kòl }

prism spectrograph [SPECT] Analysis device in which a prism is used to give two different but simultaneous light wavelengths derived from a common light source; used for the analysis of materials by flame photometry.   { 'priz·əm 'spek·trə,graf }

pristane [ORG CHEM] $C_{19}H_{40}$ A liquid soluble in such organic solvents as ether, petroleum ether, benzene, chloroform, and carbon tetrachloride; used as a lubricant, as an oil in transformers, and as an anticorrosion agent.   Also known as norphytane.   { 'pri,stān }

procaine See procaine base.   { 'prō,kān }

procaine base [ORG CHEM] $C_6H_4NH_2COOCH_2CH_2N(C_2H_5)_2$ Water-insoluble, light-sensitive, odorless, white powder, melting at 60°C; soluble in alcohol, ether, chloroform, and benzene; used in medicine as a local anesthetic.   Also known as planocaine base; procaine.   { 'prō,kān ,bās }

procaine penicillin G [ORG CHEM] $C_{29}H_{38}N_4O_6S·H_2O$ White crystals or powder, fairly soluble in chloroform; used as an antibiotic in animal feed.   { 'prō,kān ,pen·ə'sil· ən 'jē }

process analytical chemistry [ANALY CHEM] A branch of analytical chemistry concerned with quantitative and qualitative information about a chemical process.   { 'prä,səs ,an·əl'it·i·kəl 'kem·ə·strē }

prochirality [ORG CHEM] The property displayed by a molecule or atom which contains (or is bonded to) two constitutionally identical ligands; Also known as prostereoisomerism.   { ¦prō·kī'ral·əd·ē }

product [CHEM] A substance formed as a result of a chemical reaction.   { 'präd·əkt }

proflavine sulfate [ORG CHEM] $C_{13}H_{11}N_3·H_2SO_4$ A reddish-brown, crystalline powder, soluble in alcohol and water; used in medicine.   { prō'flā,vēn 'səl,fāt }

promazine hydrochloride [ORG CHEM] $C_{17}H_{20}N_2S·HCl$ A white to slightly yellow, crystalline powder, melting at 172–182°C; used in medicine and as a food additive.   { 'präm· ə,zēn ,hi·drə'klòr,īd }

promethium [CHEM] A chemical element, symbol Pm, atomic number 61, produced artificially in nuclear reactors; atomic weight of the most abundant separated isotope is 147; a member of the rare-earth group of metals.   { prə'mē·thē·əm }

promoter [CHEM] A chemical which itself is a feeble catalyst, but greatly increases the activity of a given catalyst.   { prə'mōd·ər }

propadiene See allene.   { ,präp·ə'dī,ēn }

propagation rate [CHEM] The speed at which a flame front progresses through the body of a flammable fuel-oxidizer mixture, such as gas and air.   { ,präp·ə'gā·shən ,rāt }

propagation step [CHEM] In a chain reaction, one of the fundamental steps that take place repeatedly until the reaction is complete.   { ,präp·ə'gā·shən ,step }

propane [ORG CHEM] $CH_3CH_2CH_3$ A heavy, colorless, gaseous petroleum hydrocarbon gas of the paraffin series; boils at −44.5°C; used as a solvent, refrigerant, and chemical intermediate.   { 'prō,pān }

1-propanethiol See n-propyl mercaptan.   { ¦wən¦prō,pān 'thī,òl }

310

propanoic acid *See* propionic acid. { ¦prō·pə¦nō·ik 'as·əd }

propanol *See* propyl alcohol. { 'prō·ə,nȯl }

2-propanone *See* acetone. { ¦tü 'prō·pə,nōn }

propargyl alcohol [ORG CHEM] HCCCH$_2$OH Colorless, water- and alcohol-soluble liquid, boiling at 114°C; used as a chemical intermediate, stabilizer, and corrosion inhibitor. Also known as 2-propyn-1-ol. { prō'pär·jəl 'al·kə,hȯl }

propargyl bromide [ORG CHEM] C$_3$H$_3$Cl A flammable liquid with a boiling point range of 56.0–57.1°C; used as a soil fumigant. { prō'pär·jəl 'brō,mīd }

propargyl chloride [ORG CHEM] C$_3$H$_3$Cl A liquid miscible with benzene, carbon tetrachloride, ethanol, and ethylene glycol; used as an intermediate in organic synthesis. { prō'pär·jəl 'klȯr,īd }

propellant 23 *See* fluoroform. { prə'pel·ənt ¦twen·tē'thrē }

propenyl guaethol [ORG CHEM] C$_{11}$H$_{14}$O$_2$ A white powder with a vanilla flavor and a melting point of 85–86°C; soluble in fats, essential oils, and edible solvents; used for artificial vanilla flavoring. { 'prō·pə,nil 'gwē,thȯl }

propham [ORG CHEM] C$_{10}$H$_{13}$NO$_2$ A light brown solid with a melting point of 87–88°C; slightly soluble in water; used as a pre- and postemergence herbicide for vegetable crops. Abbreviated IPC (isopropyl-N-phenylcarbamate). { 'prō,fam }

β-propiolactone [ORG CHEM] C$_3$H$_4$O$_2$ Water-soluble liquid that decomposes rapidly at boiling point (155°C); miscible with ethanol, acetone, chloroform, and ether; reacts with alcohol; used as a chemical intermediate. { ¦bād·ə ,prō·pē·ə'lak,tōn }

propionaldehyde [ORG CHEM] C$_2$H$_5$CHO Flammable, water-soluble, water-white liquid, with suffocating aroma; boils at 48.8°C; used to manufacture acetals, plastics, and rubber chemicals, and as a disinfectant and preservative. { ¦prō·pē,än'al·də,hīd }

propionate [ORG CHEM] A salt of propionic acid, CH$_3$CH$_2$COOH; an example is sodium propionate, CH$_3$CH$_2$COONa. { 'prō·pē·ə,nāt }

propionic acid [ORG CHEM] CH$_3$CH$_2$COOH Water- and alcohol-soluble, clear, colorless liquid with pungent aroma; boils at 140.7°C; used to manufacture various propionates, in nickel-electroplating solutions, for perfume esters and artificial flavors, for pharmaceuticals, and as a cellulosics solvent. Also known as methylacetic acid; propanoic acid. { ¦prō·pē¦än·ik 'as·əd }

propionic anhydride [ORG CHEM] (CH$_3$CH$_2$CO)$_2$O A colorless liquid with a boiling point of 167–169°C; soluble in ether, alcohol, and chloroform; used as an esterifying agent and for dyestuffs and pharmaceuticals. { ¦prō·pē¦än·ik an'hī,drīd }

propionic ether *See* ethyl propionate. { ¦prō·pē¦än·ik 'ē·thər }

propionitrile *See* ethyl cyanide. { ,prō·pē'än·ə,tril }

propyl- [ORG CHEM] The CH$_3$CH$_2$CH$_2$– radical, derived from propane; found, for example, in 1-propanol. { 'prō·pəl }

n-propyl acetate [ORG CHEM] C$_3$H$_7$OOCCH$_3$ Colorless liquid with pleasant aroma; miscible with alcohols, ketones, esters, and hydrocarbons; boils at 96–102°C; used for flavors and perfumes, in organic synthesis, and as a solvent. { ¦en 'prō·pəl 'as·ə,tāt }

propylacetone *See* methyl butyl ketone. { ¦prō·pəl'as·ə,tōn }

propyl alcohol [ORG CHEM] CH$_3$CH$_2$CH$_2$OH A colorless liquid made by oxidation of aliphatic hydrocarbons; boils at 97°C; used as a solvent and chemical intermediate. Also known as ethyl carbinol; propanol. { 'prō·pəl 'al·kə,hȯl }

n-propylamine [ORG CHEM] C$_3$H$_7$NH$_2$ Colorless, flammable liquid, boiling at 46–51°C; used as a sedative. { ¦en ,prō'pil·ə,mēn }

propyl benzene [ORG CHEM] C$_6$H$_5$C$_3$H$_7$ Water-insoluble, colorless liquid, boiling at 158°C. Also known as phenylpropane. { 'prō·pəl 'ben,zēn }

propylene [ORG CHEM] CH$_3$CH=CH$_2$ Colorless unsaturated hydrocarbon gas, with boiling point of −47°C; used to manufacture plastics and as a chemical intermediate. Also known as methyl ethylene; propene. { 'prō·pə,lēn }

propylene aldehyde *See* crotonaldehyde. { 'prō·pə,lēn 'al·də,hīd }

propylene carbonate [ORG CHEM] C$_3$H$_6$CO$_3$ Odorless, colorless liquid, boiling at 242°C; miscible with acetone, benzene, and ether; used as a solvent, extractant, plasticizer, and chemical intermediate. { 'prō·pə,lēn 'kär·bə,nāt }

propylene dichloride

propylene dichloride [ORG CHEM] $CH_3CHClCH_2Cl$ Water-insoluble, colorless, moderately flammable liquid, with chloroform aroma; boils at 96.3°C; miscible with most common solvents; used as a solvent, dry-cleaning fluid, metal degreaser, and fumigant. { 'prō·pə,lēn dī'klȯr,īd }

propylene glycol [ORG CHEM] $CH_3CHOHCH_2OH$ A viscous, colorless liquid, miscible with water, alcohol, and many solvents; boils at 188°C; used as a chemical intermediate, antifreeze, solvent, lubricant, plasticizer, and bactericide. { 'prō·pə,lēn 'glī,kȯl }

propylene glycol alginate [ORG CHEM] $C_9H_{14}O_7$ A white, water-soluble powder; used as a stabilizer, thickener, and emulsifier. { 'prō·pə,lēn 'glī,kȯl 'al·jə,nāt }

propylene glycol monomethyl ether [ORG CHEM] $C_4H_{10}O_2$ A colorless liquid with a boiling point of 120.1°C; soluble in water, methanol, and ether; used as a solvent for cellulose, dyes, and inks. { 'prō·pə,lēn 'glī,kȯl ¦män·ō'meth·əl 'ē·thər }

propylene glycol monoricinoleate [ORG CHEM] $C_{21}H_{30}O_4$ A pale yellow, moderately viscous oily liquid, soluble in organic solvents; used as a plasticizer and lubricant and in dye solvents and cosmetics. { 'prō·pə,lēn 'glī,kȯl ¦män·ō,ris·ən'ō·lē,āt }

propyleneimine [ORG CHEM] C_3H_7N A clear, colorless liquid with a boiling point of 66–67°C; soluble in water and organic solvents; used as an intermediate in organic synthesis. { ,prō·pə'lēn·ə,mēn }

propylene oxide [ORG CHEM] C_3H_6O Colorless, flammable liquid, with etherlike aroma; soluble in water, alcohol, and ether; boils at 33.9°C; used as a solvent and fumigant, in lacquers, coatings, and plastics, and as a petrochemical intermediate. { 'prō·pə,lēn 'äk,sīd }

propylene tetramer *See* dodecane. { 'prō·pə,lēn 'te·trə·mər }

propyl formate [ORG CHEM] $C_4H_8O_2$ A flammable liquid with a boiling point of 81.3°C; used for flavoring. { 'prō·pəl 'fȯr,māt }

n-propyl furoate [ORG CHEM] $C_8H_{10}O_3$ A colorless, fragrant liquid with a boiling point of 210.9°C; soluble in alcohol and ether; used for flavoring. { ¦en 'prō·pəl 'fyür·ə,wāt }

propyl gallate [ORG CHEM] $C_3H_7OOCC_6H_2(OH)_3$ Colorless crystals with a melting point of 150°C; used to prevent or retard rancidity in edible fats and oils. { 'prō·pəl 'ga,lāt }

propyliodone [ORG CHEM] $C_{10}H_{11}O_3NI_2$ A white, crystalline powder with a melting point of 187–190°C; soluble in alcohol, acetone, and ether; used in medicine as a radiopaque medium. { ,prō·pəl'ī·ə,dōn }

n-propyl mercaptan [ORG CHEM] C_3H_7SH A liquid with an offensive odor and a boiling range of 67–73°C; used as a herbicide. Also known as 1-propanethiol. { ¦en 'prō·pəl mər'kap,tan }

N-propyl nitrate [ORG CHEM] $C_3H_7NO_3$ A white to straw-colored liquid with a boiling range of 104–127°C; used as a monopropellant rocket fuel. { ¦en 'prō·pəl 'nī,trāt }

propylparaben [ORG CHEM] $C_{10}H_{12}O_3$ Colorless crystals or white powder with a melting point of 95–98°C; soluble in acetone, ether, and alcohol; used in medicine and as a food preservative and fungicide. { ,prō·pəl'par·ə·bən }

1-propylphosphonic acid [ORG CHEM] $C_3H_9O_3P$ A white solid with a melting point of 68–69°C; soluble in water; used as a growth regulator for herbaceous and woody species. { ¦wən ¦prō·pəl·fä'sfän·ik 'as·əd }

propylthiopyrophosphate [ORG CHEM] $C_{12}H_{28}P_2S_2O$ A straw-colored to dark amber liquid with a boiling point of 148°C; used as an insecticide for chinch bugs in lawns and turf. { ¦prō·pəl¦thī·ō,pī·rə'fä,sfāt }

2-propyn-1-ol *See* propargyl alcohol. { ¦tü 'prō·pən ,wən ,ȯl }

prostereoisomerism *See* prochirality. { prō¦ster·ē·ō·ī'säm·ə,riz·əm }

protactinium [CHEM] A chemical element, symbol Pa, atomic number 91; the third member of the actinide group of elements; all the isotopes are radioactive; the longest-lived isotope is protactinium-231. { ¦prōd,ak'tin·ē·əm }

protective colloid [PHYS CHEM] A colloidal substance that protects other colloids from the coagulative effect of electrolytes and other agents. { prə¦tek·tiv 'kä,lȯid }

proteinometer *See* hand sugar refractometer. { ,prōt·ən'äm·əd·ər }

protogenic [CHEM] Strongly acidic. { ¦prōd·ə¦jen·ik }

proton acid *See* Brönsted acid. { 'prō,tän 'as·əd }

protonate [CHEM] To add protons to a base by a proton source. { 'prōt·ən,āt }

312

protonic acid *See* Brönsted acid.　{ prō'tän·ik 'as·əd }

proton-induced x-ray emission [ANALY CHEM] A method of elemental analysis in which the energy of the characteristic x-rays emitted when a sample is bombarded with a beam of energetic protons is used to identify the elements present in the sample. Abbreviated PIXE.　{ 'prō,tän in,düst 'eks,rā i,mish·ən }

proton resonance [SPECT] A phenomenon in which protons absorb energy from an alternating magnetic field at certain characteristic frequencies when they are also subjected to a static magnetic field; this phenomenon is used in nuclear magnetic resonance quantitative analysis technique.　{ 'prō,tän 'rez·ən·əns }

proton stability constant [PHYS CHEM] The reciprocal of the dissociation constant of a weak base in solution.　{ 'prō,tän stə'bil·əd·ē ,kän·stənt }

protophilic [CHEM] Strongly basic.　{ ¦prōd·ō¦fil·ik }

prototropy [ORG CHEM] A reversible interconversion of structural isomers that involves the transfer of a proton.　{ prō'tä·trə·pē }

protropic [CHEM] Pertaining to chemical reactions that are influenced by protons.　{ prō'träp·ik }

Prout's hypothesis [PHYS CHEM] The hypothesis that all atoms are built up from hydrogen atoms.　{ 'praùts hī,päth·ə·səs }

Prussian blue [INORG CHEM] $Fe_4[Fe(CN)_6]_3$ Ferric ferrocyanide, used as a blue pigment and in the removal of hydrogen sulfide from gases.　{ 'prəsh·ən 'blü }

prussic acid *See* hydrocyanic acid.　{ 'prəs·ik 'as·əd }

pyrrolidine [ORG CHEM] C_4H_9N A colorless to pale yellow liquid with a boiling point of 87°C; soluble in water and alcohol; used in the manufacture of pharmaceuticals, insecticides, and fungicides.　{ pə'räl·ə,dēn }

pseudocritical properties [CHEM] Effective (empirical) values for the critical properties (such as temperature, pressure, and volume) of a multicomponent chemical system.　{ ¦sü·dō'krid·ə·kəl 'präp·ərd·ēz }

pseudocumene [ORG CHEM] C_9H_{12} Water-insoluble, hydrocarbon liquid, boiling at 168°C; soluble in alcohol, benzene, and ether; used to manufacture perfumes and dyes, and as a catgut sterilant. Also known as pseudocumol; uns-trimethylbenzene.　{ ¦sü·dō'kyü·mēn }

pseudocumol *See* pseudocumene.　{ ¦sü·dō'kyü,mòl }

pseudohalogen [CHEM] Any one of a group of molecules that exhibit significant similarity to the halogens, for example, cyanogen (NCCN).　{ ,süd·ō'hal·ə·jən }

pseudoionone [ORG CHEM] $C_{13}H_{20}O$ A pale yellow liquid with a boiling point of 143–145°C; soluble in alcohol and ether; used for perfumes and cosmetics.　{ ¦sü·dō'ī·ə,nōn }

pseudoreduced compressibility [CHEM] The compressibility factor for a multicomponent gaseous system, calculated at reduced conditions using the pseudoreduced properties of the mixture.　{ ,sü·dō·ri'düst kəm,pres·ə'bil·əd·ē }

pseudoreduced properties [CHEM] Reduced-state relationships (such as reduced pressure, reduced temperature, and reduced volume) calculated for multicomponent chemical systems by using pseudocritical properties.　{ ,sü·dō·ri'düst 'präp·ərd·ēz }

pseudorotation [ORG CHEM] Twist conformation in aliphatic rings or ring structures containing five or more atoms.　{ ,süd·ō·rō'tā·shən }

pseudorotaxane [CHEM] A supramolecular species consisting of a linear molecular component (without bulky end groups) encircled by a macrocyclic component.　{ ,süd·ō·rō'tak,sān }

Pt *See* platinum.

PTA *See* phosphotungstic acid.

Pu *See* plutonium.

pulsating combustion [CHEM] Combustion that is accompanied by spontaneous pressure oscillations, which occur if the Rayleigh criterion is satisfied.　{ ¦pəl,sād·iŋ kəm'bus·chən }

pulse radiolysis [PHYS CHEM] A method of studying fast chemical reactions in which a sample is subjected to a pulse of ionizing radiation, and the products formed by the resulting reactions are studied spectroscopically.　{ 'pəls ,rād·ē·'äl·ə·səs }

pure substance

pure substance [CHEM] A sample of matter, either an element or a compound, that consists of only one component with definite physical and chemical properties and a definite composition. { 'pyür 'səb·stəns }

purity [CHEM] The degree to which the content of impurity can be detected by an analytical procedure in a sample of matter that is classified as a pure substance; the grade of purity is in inverse proportion to the amount of impurity present. Also known as chemical purity. { 'pyür·əd·ē }

purple of Cassius See gold tin purple. { 'pər·pəl əv 'kash·əs }

purple salt See potassium permanganate. { 'pər·pəl 'sȯlt }

purpurin [ORG CHEM] $C_{14}H_8O_5$ A compound crystallizing as long orange needles from dilute alcohol solutions; used in the manufacture of dyes, and as a reagent for the detection of boron. Also known as natural red. { 'pər·pyə·rən }

purpurogallin [ORG CHEM] $C_{11}H_8O_5$ A red, crystalline compound, the aglycon of several glycosides from nutgalls; decomposes at 274–275°C; soluble in boiling alcohol, methanol, and acetone; used as an antioxidant or to retard metal contamination in hydrocarbon fuels or lubricants. { ˌpər·pyə·rō'gal·ən }

PVA See polyvinyl acetate; polyvinyl alcohol.

PVAc See polyvinyl acetate.

PVC See polyvinyl chloride.

PVDC See polyvinyl dichloride.

PVI See polyvinyl isobutyl ether.

PVM See polyvinyl methyl ether.

PVP See polyvinyl pyrrolidone.

pyracetic acid See pyroligneous acid. { ¦pī·rə¦sēd·ik 'as·əd }

pyramidal molecule [CHEM] A molecular structure in the shape of a pyramid in which the central atom at the peak possesses either three or four valence bonds that are directed to the other atoms, which form the base of the pyramid. { ¦pir·ə‚mid·əl 'mäl·ə‚kyül }

pyrazolone dye [ORG CHEM] An acid dye containing both $-N=N-$ and $=C=C=$ chromophore groups, such as tartrazine; used for silk and wool. { pə'raz·ə‚lōn ‚dī }

pyridine [ORG CHEM] C_5H_5N Organic base; flammable, toxic yellowish liquid, with penetrating aroma and burning taste; soluble in water, alcohol, ether, benzene, and fatty oils; boils at 116°C; used as an alcohol denaturant, solvent, in paints, medicine, and textile dyeing. { 'pir·ə‚dēn }

pyro- [CHEM] A chemical prefix for compounds formed by heat, such as pyrophosphoric acid, an inorganic acid formed by the loss of one water molecule from two molecules of an ortho acid. { 'pī·rō, 'pī·rə }

pyrocatechuic acid See catechol. { ¦pī·rō¦kad·ə¦chü·ik 'as·əd }

pyrocellulose [ORG CHEM] Highly nitrated cellulose; used to make explosives; originally called guncotton in the United States, cordite in England. { ¦pī·rō'sel·yə‚lōs }

pyrogallic acid [ORG CHEM] $C_6H_3(OH)_3$ Lustrous, light-sensitive white crystals, melting at 133°C; soluble in alcohol, ether, and water; used for photography, dyes, drugs, medicines, and process engravings, and as an analytical reagent and protective colloid. Also known as pyrogallol. { ¦pī·rō'gal·ik 'as·əd }

pyrogallol See pyrogallic acid. { ‚pī·rō'ga‚lȯl }

pyrogallolphthalein See gallein. { ¦pī·rō¦gal·ō'thal·ē·ən }

pyroligneous acid [ORG CHEM] An impure acetic acid derived from destructive distillation of wood or pine tar. Also known as pyracetic acid; wood vinegar. { ¦pī·rō'lig·nē·əs 'as·əd }

pyrolithic acid See cyanuric acid. { ¦pī·rō¦lith·ik 'as·əd }

pyrolysate [CHEM] Any product of pyrolysis. { pī'räl·ə‚zāt }

pyrolysis [CHEM] The breaking apart of complex molecules into simpler units by the use of heat, as in the pyrolysis of heavy oil to make gasoline. Also known as thermolysis. { pə'räl·ə·səs }

pyromellitic acid [ORG CHEM] $C_6H_2(COOH)_4$ A white powder with a melting point of 257–265°C; used as an intermediate for polyesters and polyamides. Abbreviated PMA. { ‚pī·rō·mə'lid·ik 'as·əd }

pyromellitic dianhydride [ORG CHEM] $C_6H_2(C_2O_3)_2$ A white powder with a melting point of 286°C; soluble in some organic solvents; used for curing epoxy resins.   Abbreviated PMDA.   { ˌpī·rō·mə'lid·ik ˈdī·an'hī͵drīd }

pyromucic acid *See* furoic acid.   { ¦pī·rō'myü·sik 'as·əd }

pyrophoric material [CHEM] A material that spontaneously ignites in air below 113°C (45°C), such as fine metal powder, alkali metal, partially or fully alkylated metal or nonmetal hydride, and metal carbonyl.   { ˌpī·rə͵fȯr·ik mə'tir·ē·əl }

pyrophosphoric acid [INORG CHEM] $H_4P_2O_7$ Water-soluble, syrupy liquid melting at 61°C; used as a catalyst and to make organic phosphate esters.   { ¦pī·rō·fä'sfȯr·ik 'as·əd }

pyrosin *See* tetraiodofluorescein.   { ˌpī·rə·sən }

pyroxylin [ORG CHEM] $[C_{12}H_{16}O_6(NO_3)_4]_x$ Any member of the group of commercially available nitrocelluloses that are used for properties other than their combustibility; the term is commonly used to identify products that are principally made from nitrocellulose, such as pyroxylin plastic or pyroxylin lacquer.   Also known as collodion cotton; soluble guncotton; soluble nitrocellulose.   { pə'räk·sə·lən }

pyrrole [ORG CHEM] C_4H_5N Water-insoluble, yellowish oil, with pungent taste; soluble in alcohol, ether, and dilute acids; boils at 130°C; polymerizes in light; used to make drugs.   { 'pi͵rōl }

pyrrole ring [ORG CHEM] A five-member heterocycle containing one nitrogen atom and four carbon atoms in the ring; frequently found in structures of natural products occurring as joined rings or attached to straight chains.   { 'pī͵rōl ͵riŋ }

2-pyrrolidone [ORG CHEM] C_4H_7ON Combustible, light-yellow liquid, boiling at 245°C; soluble in ethyl alcohol, water, chloroform, and carbon disulfide; used as a plasticizer and polymer solvent, in insecticides and specialty inks, and as a nylon-4 precursor.   { ¦tü pə'räl·ə͵dēn }

pyrrone [ORG CHEM] A polyimidazopyrrolone synthesized from dianhydrides and tetramines; soluble in sulfuric acid; resists temperatures to 600°C.   { 'pi͵rōn }

Q

Q branch |SPECT| A series of lines in molecular spectra that correspond to changes in the vibrational quantum number with no change in the rotational quantum number. { 'kyü ‚branch }

quadridentate ligand |CHEM| A group which forms a chelate and has four points of attachment. { ¦kwä·drə¦den‚tāt 'līg·ənd }

quadruple point |PHYS CHEM| Temperature at which four phases are in equilibrium, such as a saturated solution containing an excess of solute. { kwə'drüp·əl 'pȯint }

quadrupole spectrometer |ANALY CHEM| A type of mass spectroscope in which ions pass along a line of symmetry between four parallel cylindrical rods; an alternating potential superimposed on a steady potential between pairs of rods filters out all ions except those of a predetermined mass. Also known as Massenfilter. { 'kwä·drə‚pōl spek'träm·əd·ər }

qualitative analysis |ANALY CHEM| The analysis of a gas, liquid, or solid sample or mixture to identify the elements, radicals, or compounds composing the sample. { 'kwäl·ə‚tād·iv ə‚nal·ə·səs }

quantitative analysis |ANALY CHEM| The analysis of a gas, liquid, or solid sample or mixture to determine the precise percentage composition of the sample in terms of elements, radicals, or compounds. { 'kwän·ə·tād·iv ə'nal·ə·səs }

quantum chemistry |PHYS CHEM| A branch of physical chemistry concerned with the explanation of chemical phenomena by means of the laws of quantum mechanics. { 'kwän·təm 'kem·ə·strē }

quantum limit |SPECT| The shortest wavelength present in a continuous x-ray spectrum. Also known as boundary wavelength; end radiation. { 'kwän·təm 'lim·ət }

quantum theory of valence |PHYS CHEM| The theory of valence based on quantum mechanics; it accounts for many experimental facts, explains the stability of a chemical bond, and allows the correlation and prediction of many different properties of molecules not possible in earlier theories. { 'kwän·təm ‚thē·ə·rē əv 'vā·ləns }

quantum yield |PHYS CHEM| For a photochemical reaction, the number of moles of a stated reactant disappearing, or the number of moles of a stated product produced, per einstein of light of the stated wavelength absorbed. { 'kwän·təm 'yēld }

quarterpolymer |CHEM| A polymer in which the repeating groups comprise four species of monomer. { ¦kwȯrd·ər¦päl·i·mər }

quaternary ammonium base |ORG CHEM| Ammonium hydroxide (NH_4OH) with the ammonium hydrogens replaced by organic radicals, such as $(CH_3)_4NOH$. { 'kwät·ən‚er·ē ə¦mō·nē·əm 'bās }

quaternary ammonium salt |ORG CHEM| A nitrogen compound in which a central nitrogen atom is joined to four organic radicals and one acid radical, for example, hexamethonium chloride; used as an emulsifying agent, corrosion inhibitor, and antiseptic. { 'kwät·ən‚er·ē ə¦mō·nē·əm 'sȯlt }

quaternary carbon atom |ORG CHEM| A carbon atom bonded to four other carbon atoms with single bonds. { 'kwät·ən‚er·ē ¦kär·bən 'ad·əm }

quaternary phase equilibria |PHYS CHEM| The solubility relationships in any liquid system with four nonreactive components with varying degrees of mutual solubility. { 'kwät·ən‚er·ē ¦fāz ‚ē·kwə'lib·rē·ə }

quaternary system [PHYS CHEM] An equilibrium relationship between a mixture of four (four phases, four components, and so on). { 'kwät·ən‚er·ē 'sis·təm }

quercimelin See quercitrin. { ‚kwer'sim·ə·lən }

quercitrin [ORG CHEM] $C_{21}H_{20}O_{11}$ The 3-rhamnoside of quercitin, forming yellow crystals from dilute ethanol or methanol solution, melting at 176–179°C, soluble in alcohol; used as a textile dye. Also known as quercimelin; quercitroside. { 'kwer·sə·trən }

quercitroside See quercitrin. { kwer'si·trə‚sīd }

Quevenne scale [CHEM] Arbitrary scale used with hydrometers or lactometers in the determination of the specific gravity of milk; degrees Quevenne = 1000 (specific gravity − 1). { kə'ven ‚skāl }

quicklime See calcium oxide. { 'kwik‚līm }

quicksilver See mercury. { 'kwik‚sil·vər }

quicksilver vermilion See mercuric sulfide. { 'kwik‚sil·vər vər'mil·yən }

quinaldine [ORG CHEM] $C_9H_6NCH_3$ A colorless, oily liquid with a boiling point of 246–247°C; soluble in alcohol, chloroform, and ether; used in medicine as an antimalarial. Also known as chinaldine. { 'kwin·əl‚dēn }

quinalizarin [ORG CHEM] $C_{14}H_8O_6$ A red, crystalline compound, soluble in water solutions of alkalies, and in acetic and sulfuric acid; used to dye cottons. { ‚kwin·ə'liz·ə·rən }

quinhydrone [ORG CHEM] $C_6H_4O_2·C_6H_4(OH)_2$ Green, water-soluble powder, subliming at 171°C; a compound of quinone and hydroquinone dissociating in solution. { kwin'hī‚drōn }

quinhydrone electrode [ANALY CHEM] A platinum wire in a saturated solution of quinhydrone; used as a reversible electrode standard in pH determinations. { kwin'hī‚drōn i'lek‚trōd }

quinic acid [ORG CHEM] $C_6H_7(OH)_4COOH·H_2O$ Ether-insoluble, white crystals with acid taste; melts at 162°C; soluble in alcohol, water, and glacial acetic acid; used in medicine. Also known as chinic acid; kinic acid. { 'kwin·ik 'as·əd }

quinidine [ORG CHEM] $C_{20}H_{24}N_2O_2$ A crystalline alkaloid that melts at 174–175°C (345–347°F) and that may be derived from the bark of cinchona; used as the salt in medicine. Also known as β-quinine. { 'kwin·ə‚dēn }

quinine [ORG CHEM] $C_{20}H_{24}N_2O_2·3H_2O$ White powder or crystals, soluble in alcohol, ether, carbon disulfide, chloroform, and glycerol; an alkaloid derived from cinchona bark; used as an antimalarial drug and in beverages. { 'kwī‚nīn }

β-quinine See quinidine. { ¦bād·ə 'kwī‚nīn }

quinoidine [ORG CHEM] A brownish-black mass consisting of a mixture of alkaloids which remain in solution after extracting crystallized alkaloids from cinchona bark; soluble in dilute acids, alcohol, and chloroform; used in medicine. Also known as chinoidine. { kwi'nō·ə‚dēn }

quinol See hydroquinone. { 'kwi‚nȯl }

quinoline [ORG CHEM] C_9H_7N Water-soluble, aromatic nitrogen compound; colorless, hygroscopic liquid; also soluble in alcohol, ether, and carbon disulfide; boils at 238°C; used in medicine and as a chemical intermediate. Also known as chinoline; leucoline; leukol. { 'kwin·ə‚lēn }

quinoline blue See cyanine dye. { 'kwin·ə‚lēn 'blü }

8-quinolinol See 8-hydroxyquinoline; oxine. { ¦āt kwi'näl·ə‚nȯl }

quinone [ORG CHEM] $CO(CHCH)_2CO$ Yellow crystalline compound with irritating aroma; melts at 116°C; soluble in alcohol, alkalies, and ether; used to make dyes and hydroquinone. Also known as benzoquinone; chinone. { 'kwi‚nōn }

quinoxaline [ORG CHEM] $C_8H_6N_2$ Bicyclic organic base; colorless powder, soluble in water and organic solvents; melts at 30°C; used in organic synthesis. { kwi'näk·sə‚lēn }

N′-2-quinoxalysulfanilimide [ORG CHEM] $C_{14}H_{12}N_4SO_2$ Crystals with a melting point of 247°C; almost insoluble in water; used as a rodenticide. Also known as sulfaquinoxaline. { ¦en‚prīm ¦tü kwi¦näk·sə·lē‚səl·fə'nil·ə‚mīd }

quintozene See pentachloronitrobenzene. { 'kwin·tə‚zēn }

R

Ra *See* radium.

racemate [ORG CHEM] An equimolar mixture of the two enantiomers (+ and −, or R and S) of a substance; it is optically inactive. { 'ras·ə‚māt }

racemic acid [ORG CHEM] $C_2H_4O_2(COOH)_2 \cdot H_2O$ Colorless crystals, melting at 205°C; soluble in water, slightly soluble in alcohol; used as a chemical intermediate. Also known as inactive tartaric acid. { rə'sēm·ik 'as·əd }

racemic compound [ORG CHEM] Crystals containing an equimolar, random and ordered mixture of enantiomers (heterochiral crystals). { rə‚sēm·ik 'käm‚paúnd }

racemic conglomerate [ORG CHEM] Spontaneous resolution of a racemate, through crystallization, into a mixture of pure enantiomers (homochiral crystals). { rə‚sēm·ik kən'gläm·ə·rət }

racemic mixture [ORG CHEM] According to the IUPAC, this usage is strongly discouraged, racemate is preferred. { rə'sēm·ik 'miks·chər }

racemic modification *See* racemic mixture. { rā¦sēm·ik ‚mäd·ə·fə'kā·shən }

racemization [ORG CHEM] A process by which an optically active form of a substance is converted into a racemic mixture. { ‚rā·sə·mə'zā·shən }

radial chromatography [ANALY CHEM] A circular disk of absorbent paper which has a strip (wick) cut from edge to center to dip into a solvent; the solvent climbs the wick, touches the sample, and resolves it into concentric rings (the chromatogram). Also known as circular chromatography; radial paper chromatography. { 'rād·ē·əl ‚krō·mə'täg·rə·fē }

radial distribution function [PHYS CHEM] A function $\rho(r)$ equal to the average over all directions of the number density of molecules at distance r from a given molecule in a liquid. { 'rād·ē·əl ‚dis·trə'byü·shən ‚fəŋk·shən }

radial paper chromatography *See* radial chromatography. { 'rād·ē·əl 'pā·pər ‚krō·mə'täg·rə·fē }

radiation catalysis [CHEM] The use of radiation (such as gamma, neutron, proton, electron, or x-ray) to activate or speed up a chemical or physical change; for example, radiation alone can initiate polymerization without heat, pressure, or chemical catalysts. { ‚rād·ē'ā·shən kə'tal·ə·səs }

radical *See* free radical. { 'rad·ə·kəl }

radical ion [CHEM] A charged compound that has an unpaired electron; it may be either a radical cation (positively charged) or radical anion (negatively charged). { ‚rad·ə·kəl 'ī·ən }

radical scavenger [CHEM] One of a group of molecules that combines with free radicals in a chemical or biochemical system to render them less active chemically. { 'rad·ə·kəl ‚skav·ən·jər }

radicofunctional name [ORG CHEM] A name for an organic compound that uses two key words; the first word corresponds to the group or groups involved and the second word indicates the functional group—for example, alkyl halide. { ¦rad·ə·kō'fəŋk·shən·əl 'nām }

radioassay [ANALY CHEM] An assay procedure involving the measurement of the radiation intensity of a radioactive sample. { ¦rād·ē·ō'a‚sā }

radiochemical laboratory

radiochemical laboratory [CHEM] A specially equipped and shielded chemical laboratory designed for conducting radiochemical studies without danger to the laboratory personnel. { ¦rad·ē·ō'kem·ə·kəl 'lab·rə,tȯr·ē }

radiochemistry [CHEM] That area of chemistry concerned with the study of radioactive substances. { ¦rad·ē·ō'kem·ə·strē }

radiochromatography [ANALY CHEM] An analytic process for quantitative or qualitative determination of radioactive substances in a mixture by measuring the radioactivity of various zones in the chromatogram. { ¦rād·ē·ō,krō·mə'täg·rə·fē }

radiocolloid [CHEM] A colloid having a component that consists of radioactive atoms. { ,rād·ē·ō'kä,lȯid }

radio-frequency spectrometer [SPECT] An instrument which measures the intensity of radiation emitted or absorbed by atoms or molecules as a function of frequency at frequencies from 10^5 to 10^9 hertz; examples include the atomic-beam apparatus, and instruments for detecting magnetic resonance. { 'rād·ē·ō ¦frē·kwən·sē spek 'träm·əd·ər }

radio-frequency spectroscopy [SPECT] The branch of spectroscopy concerned with the measurement of the intervals between atomic or molecular energy levels that are separated by frequencies from about 10^5 to 10^9 hertz, as compared to the frequencies that separate optical energy levels of about 6×10^{14} hertz. { 'rād·ē·ō ¦frē·kwən· sē spek'träs·kə·pē }

radioisotope assay [ANALY CHEM] An analytical technique including procedures for separating and reproducibly measuring a radioactive tracer. { ¦rād·ē·ō'ī·sə,tōp 'a,sā }

radiolysis [PHYS CHEM] The dissociation of molecules by radiation; for example, a small amount of water in a reactor core dissociates into hydrogen and oxygen during operation. { ,rād·ē'äl·ə·səs }

radiometric analysis [ANALY CHEM] Quantitative chemical analysis that is based on measurement of the absolute disintegration rate of a radioactive component having a known specific activity. { ¦rād·ē·ō¦me·trik ə'nal·ə·səs }

radiometric titration [ANALY CHEM] Use of radioactive indicator to track the transfer of material between two liquid phases in equilibrium, such as titration of $^{110}AgNO_3$ (silver nitrate, with the silver atom having mass number 110) against potassium chloride. { ¦rād·ē·ō¦me·trik tī'trā·shən }

radiomimetic substances [CHEM] Chemical substances which cause biological effects similar to those caused by ionizing radiation. { ¦rād·ē·ō·mi'med·ik 'səb·stəns·əz }

radio recombination line [SPECT] A radio-frequency spectral line that results from an electron transition between energy levels in an atom or ion having a large principal quantum number n, greater than 50. { 'rād·ē·ō rē,käm·bə'nā·shən ,līn }

radium [CHEM] **1.** A radioactive member of group II, symbol Ra, atomic number 88; the most abundant naturally occurring isotope has mass number 226 and a half-life of 1620 years. **2.** A highly toxic solid that forms water-soluble compounds; decays by emission of α, β, and γ-radiation; melts at 700°C, boils at 1140°C; turns black in air; used in medicine, in industrial radiography, and as a source of neutrons and radon. { 'rād·ē·əm }

radium bromide [INORG CHEM] $RaBr_2$ Water-soluble, poisonous, radioactive white powder, corrosive to skin or flesh; melts at 728°C; used in medicine, physical research, and luminous paint. { 'rād·ē·əm 'brō,mīd }

radium carbonate [INORG CHEM] $RaCO_3$ Water-insoluble, poisonous, radioactive, white powder; used in medicine. { 'rād·ē·əm 'kär·bə,nāt }

radium chloride [INORG CHEM] $RaCl_2$ Water- and alcohol-soluble, poisonous, radioactive, yellow-white crystals; corrosive effect on skin and flesh; melts at 1000°C; used in medicine, physical research, and luminous paint. { 'rād·ē·əm 'klȯr,īd }

radium sulfate [INORG CHEM] $RaSO_4$ Water-insoluble, radioactive, poisonous, white crystals; used in medicine. { 'rād·ē·əm 'səl,fāt }

radius ratio [PHYS CHEM] The ratio of the radius of a cation to the radius of an ion; relative ionic radii are pertinent to crystal lattice structure, particularly the determination of coordination number. { 'rād·ē·əs ,rā·shō }

320

radon |CHEM| A chemical element, symbol Rn, atomic number 86; all isotopes are radioactive, the longest half-life being 3.82 days for mass number 222; it is the heaviest element of the noble-gas group, produced as a gaseous emanation from the radioactive decay of radium. { 'rā‚dän }

Raman spectroscopy |SPECT| Analysis of the intensity of Raman scattering of monochromatic light as a function of frequency of the scattered light; the information obtained is useful for determining molecular structure. { 'räm·ən spek'träs·kə·pē }

Raman spectrum |SPECT| A display, record, or graph of the intensity of Raman scattering of monochromatic light as a function of frequency of the scattered light. { 'räm· ən ‚spek·trəm }

random coil |PHYS CHEM| Any of various irregularly coiled polymers that can occur in solution. Also known as cyclic coil. { 'ran·dəm 'kȯil }

random copolymer |ORG CHEM| Resin copolymer in which the molecules of each monomer are randomly arranged in the polymer backbone. { 'ran·dəm kō'päl·i·mər }

Raoult's law |PHYS CHEM| The law that the vapor pressure of a solution equals the product of the vapor pressure of the pure solvent and the mole fraction of solvent. { rä'ülz ‚lȯ }

rare-earth element |CHEM| The name given to any of the group of chemical elements with atomic numbers 58 to 71; the name is a misnomer since they are neither rare nor earths; examples are cerium, erbium, and gadolinium. { 'rer ‚ərth 'el·ə·mənt }

rare-earth salts |INORG CHEM| Salts derived from monazite, and with rare earths in similar proportions as in monazite; contains La, Ce, Pr, Nd, Sm, Gd, and Y as acetates, carbonates, chlorides, fluorides, nitrates, sulfates, and so on. { 'rer ‚ərth 'sȯls }

rare gas See noble gas. { 'rer 'gas }

Rast method |ANALY CHEM| The melting-point depression method often used for the determination of the molecular weight of organic compounds. { 'rast ‚meth·əd }

rate constant |PHYS CHEM| Numerical constant in a rate-of-reaction equation; for example, $r_A = kC_ACbBCcC$, where C_A,C_B, and C_C are reactant concentrations, k is the rate constant (specific reaction rate constant), and a, b, and c are empirical constants. { 'rāt ‚kän·stənt }

rate-determining step |CHEM| In a multistep chemical reaction, the step with the lowest velocity, which determines the rate of the overall reaction. { 'rāt di｜tər·mən· iŋ ‚step }

rate of reaction |CHEM| A measurement based on the mass of reactant consumed in a chemical reaction during a given period of time. { 'rāt əv rē'ak·shən }

rational synthesis |CHEM| The production of a compound using a sequence of chemical reaction steps strategically chosen. { 'rash·ən·əl 'sin·thə·səs }

ratio of specific heats |PHYS CHEM| The ratio of specific heat at constant pressure to specific heat at constant volume, $\gamma = C_p/C_v$. { 'rā·shō əv spə'sif·ik 'hēts }

Rayleigh criterion |CHEM| The criterion for spontaneous pressure oscillations to accompany combustion, namely, that combustion progresses more rapidly or efficiently during the compression phase of the pressure oscillation than during the rarefaction phase. { 'rā·lē krī‚tir·ē·ən }

Rayleigh line |SPECT| Spectrum line in scattered radiation which has the same frequency as the corresponding incident radiation. { 'rā·lē ‚līn }

Rb See rubidium.

R-branch |SPECT| A series of lines in molecular spectra that correspond, in the case of absorption, to a unit increase in the rotational quantum number J. { 'är ‚branch }

RBS See Rutherford backscattering spectrometry.

RDGE See resorcinol diglycidyl ether.

RDX See cyclonite.

Re See rhenium.

reactant |CHEM| A substance that reacts with another one to produce a new set of substances (products). { rē'ak·tənt }

reaction boundary See reaction line. { rē'ak·shən ‚baún·drē }

reaction curve See reaction line. { rē'ak·shən ‚kərv }

reaction enthalpy number |PHYS CHEM| A dimensionless number used in the study of

reaction kinetics

interphase transfer in chemical reactions, equal to the enthalpy of reaction per unit mass of a specified compound produced in a reaction, times the mass fraction of that compound, divided by the product of the specific heat at constant pressure and the temperature change during the reaction. { rē'ak·shən 'en‚thal·pē ‚nəm·bər }

reaction kinetics See chemical kinetics. { rē'ak·shən ki'ned·iks }

reaction line [PHYS CHEM] In a ternary system, a special case of the boundary line along which one of the two crystalline phases present reacts with the liquid, as the temperature is decreased, to form the other crystalline phase. Also known as reaction boundary; reaction curve. { rē'ak·shən ‚līn }

reaction mechanism [CHEM] The sequence of steps during which a chemical reaction occurs, including the transition state during which the reactants are converted into products. { rē'ak·shən ‚mek·ə‚niz·əm }

reaction path See mechanism. { rē'ak·shən ‚path }

reactive bond [CHEM] A bond between atoms that is easily invaded (reacted to) by another atom or radical; for example, the double bond in $CH_2=CH_2$ (ethylene) is highly reactive to other ethylene molecules in the reaction known as polymerization to form polyethylene. { rē'ak·tiv 'bänd }

reactive intermediate [CHEM] An unstable compound formed as an intermediate during a chemical reaction. { rē'ak·tiv ‚in·tər'mē·dē·ət }

reactivity [CHEM] The relative capacity of an atom, molecule, or radical to combine chemically with another atom, molecule, or radical. { ‚rē·ak'tiv·əd·ē }

reagent [ANALY CHEM] A substance, chemical, or solution used in the laboratory to detect, measure, or otherwise examine other substances, chemicals, or solutions; grades include ACS (American Chemical Society standards), reagent (for analytical reagents), CP (chemically pure), USP (U.S. Pharmacopeia standards), NF (National Formulary standards), and purified, technical (for industrial use). [CHEM] The compound that supplies the molecule, ion, or free radical which is arbitrarily considered as the attacking species in a chemical reaction. { rē'ā·jənt }

reagent chemicals [ANALY CHEM] High-purity chemicals used for analytical reactions, for testing of new reactions where the effect of impurities are unknown, and, in general, for chemical work where impurities must either be absent or at a known concentration. { rē'ā·jənt ‚kem·ə·kəlz }

rearrangement reaction [ORG CHEM] A chemical reaction involving a change in the bonding sequence within a molecule. Also known as molecular rearrangement. { ‚rē·ə'rānj·mənt rē‚ak·shən }

reconstructive processing [INORG CHEM] The spinning of an inorganic compound of an organic support or binder subsequently removed by oxidation or volatilization to form an inorganic polymer. { ‚rē·kən'strək·tiv 'prä‚ses·iŋ }

recording balance [ANALY CHEM] An analytical balance equipped to record weight results by electromagnetic or servomotor-driven accessories. { ri'kòrd·iŋ ‚bal·əns }

recrystallization [CHEM] Repeated crystallization of a material from fresh solvent to obtain an increasingly pure product. { rē‚krist·əl·ə'zā·shən }

red lead See lead tetroxide. { 'red 'led }

red mercury sulfide See mercuric sulfide. { 'red 'mər·kyə·rē 'səl‚fīd }

red ocher See ferric oxide. { 'red 'ō·kər }

red oil See oleic acid. { 'red 'òil }

redox polymer [ORG CHEM] A polymer whose structure contains functional groups that can be reversibly reduced or oxidized. Also known as electron exchanger. { 'rē‚däks ‚päl·ə·mər }

redox potential [PHYS CHEM] Voltage difference at an inert electrode immersed in a reversible oxidation-reduction system; measurement of the state of oxidation of the system. Also known as oxidation-reduction potential. { 'rē‚däks pə‚ten·chəl }

redox potentiometry [ANALY CHEM] Use of neutral electrode probes to measure the solution potential developed as the result of an oxidation or reduction reaction. { 'rē‚däks pə‚ten·chē'äm·ə·trē }

redox system [CHEM] A chemical system in which reduction and oxidation (redox) reactions occur. { 'rē‚däks ‚sis·təm }

322

redox titration [ANALY CHEM] A titration characterized by the transfer of electrons from one substance to another (from the reductant to the oxidant) with the end point determined colorimetrically or potentiometrically. { 'rē,däks tī'trā·shən }

red phosphorus [CHEM] An allotropic form of the element phosphorus; violet-red, amorphous powder subliming at 416°C, igniting at 260°; insoluble in all solvents,; nonpoisonous. { 'red 'fä·sfə·rəs }

red potassium chromate See potassium dichromate. { 'red pə'tas·ē·əm 'krō,māt }

red potassium prussiate See potassium ferricyanide. { 'red pə'tas·ē·əm 'prəs·ē,ät }

red precipitate See mercuric oxide. { 'red prə'sip·ə,tāt }

red prussiate of potash See potassium ferricyanide. { 'red 'prəs·ē,ät əv 'päd,ash }

red prussiate of soda See sodium ferricyanide. { 'red 'prəs·ē,ät əv 'sōd·ə }

red tetrazolium See triphenyltetrazolium chloride. { 'red ,te·trə'zäl·ē·əm }

reducer See reducing agent. { ri'dü·sər }

reducing agent [CHEM] Also known as reducer. **1.** A material that adds hydrogen to an element or compound. **2.** A material that adds an electron to an element or compound, that is, decreases the positiveness of its valence. { ri'düs·iŋ ,ā·jənt }

reducing atmosphere [CHEM] An atmosphere of hydrogen (or other substance that readily provides electrons) surrounding a chemical reaction or physical device; the effect is the opposite to that of an oxidizing atmosphere. { ri'düs·iŋ 'at·mə,sfir }

reducing flame [CHEM] A flame having excess fuel and being capable of chemical reduction, such as extracting oxygen from a metallic oxide. { ri'düs·iŋ ,flām }

reducing sugar [ORG CHEM] Any of the sugars that because of their free or potentially free aldehyde or ketone groups, possess the property of readily reducing alkaline solutions of many metallic salts such as copper, silver, or bismuth; examples are the monosaccharides and most of the disaccharides, including maltose and lactose. { ri'düs·iŋ ,shúg·ər }

reduction [ANALY CHEM] Preparation of one or more subsamples from a sample of material that is to be analyzed chemically. [CHEM] **1.** Reaction of hydrogen with another substance. **2.** Chemical reaction in which an element gains an electron (has a decrease in positive valence). { ri'dək·shən }

reduction cell [CHEM] A vessel in which aqueous solutions of salts or fused salts are reduced electrolytically. { ri'dək·shən ,sel }

reduction potential [PHYS CHEM] The potential drop involved in the reduction of a positively charged ion to a neutral form or to a less highly charged ion, or of a neutral atom to a negatively charged ion. { ri'dək·shən pə,ten·chəl }

reference electrode [PHYS CHEM] A nonpolarizable electrode that generates highly reproducible potentials; used for pH measurements and polarographic analyses; examples are the calomel electrode, silver-silver chloride electrode, and mercury pool. { 'ref·rəns i'lek,trōd }

reference material [ANALY CHEM] A material or substance whose properties are sufficiently well established to be used in calibrating an apparatus, assessing a measurement method, or assigning values to other materials. { 'ref·rəns mə,tir·ē·əl }

reflectance spectrophotometry [SPECT] Measurement of the ratio of spectral radiant flux reflected from a light-diffusing specimen to that reflected from a light-diffusing standard substituted for the specimen. { ri'flek·təns ,spek·trə·fə'täm·ə·trē }

Reformatsky reaction [ORG CHEM] A condensation-type reaction between ketones and α-bromoaliphatic acids in the presence of zinc or magnesium, such as R_2CO + $BrCH_2 \cdot COOR$ + $Zn \rightarrow (ZnO \cdot HBr)$ + $R_2C(OH)CH_2COOR$. { ,rif·ər'mat·skē rē,ak·shən }

refractory hard metals [CHEM] True chemical compounds composed of two or more metals in the crystalline form, and having a very high melting point and high hardness. { ri'frak·trē 'härd 'med·əlz }

refrigerant 23 See fluoroform. { ri'frij·ə·rənt ‚twen·tē'thrē }

regenerant [CHEM] A solution whose purpose is to restore the activity of an ion-exchange bed. { rē'jen·ə·rənt }

regeneration [CHEM] Restoration of the activity of a deactivated catalyst. { rē,jen·ə'rā·shən }

regioselective [ORG CHEM] Pertaining to a chemical reaction which favors a single positional or structural isomer, leading to its yield being greater than that of the other products in the reaction. Also known as regiospecific. { ¦rē·jē·ō·si'lek·tiv }

regiospecific *See* regioselective. { ¦rē·jē·ō·spə'sif·ik }

regular polymer [CHEM] A polymer whose molecules possess only one kind of constitutional unit in a single sequential structure. { 'reg·yə·lər 'päl·ə·mər }

Reichert-Meissl number [ANALY CHEM] An indicator of the measure of volatile soluble fatty acids. { 'rī·kərt 'mīs·əl ‚nəm·bər }

Reimer-Tiemann reaction [ORG CHEM] Formation of phenolic aldehydes by reaction of phenol with chloroform in the presence of an alkali. { 'rīm·ər ¦tē·mən rē‚ak·shən }

Reinecke's salt [ANALY CHEM] |(NH₃)₂Cr(SCN)₄|NH₄·H₂O A reagent to detect mercury (gives a red color or a precipitate), and to isolate organic bases (such as proline or histidine). { 'rīn·ə·kēz ‚sȯlt }

Reinsch test [ANALY CHEM] A test for detecting small amounts of arsenic, silver, bismuth, and mercury. { 'rīnsh ‚test }

relative atomic mass *See* atomic weight. { 'rel·əd·iv ə'täm·ik 'mas }

relative fugacity [PHYS CHEM] The ratio of the fugacity in a given state to the fugacity in a defined standard state. { 'rel·əd·iv fyü'gas·əd·ē }

relative molecular mass *See* molecular weight. { 'rel·əd·iv mə'lek·yə·lər 'mas }

relative stability test [ANALY CHEM] A color test using methylene blue that indicates when the oxygen present in a sewage plant's effluent or polluted water is exhausted. { 'rel·əd·iv stə'bil·əd·ē ‚test }

relative volatility [CHEM] The volatility of a standard material whose relative volatility is by definition equal to unity. { 'rel·əd·iv ‚väl·ə'til·əd·ē }

relaxation kinetics [PHYS CHEM] A branch of kinetics that studies chemical systems by disturbing their states of equilibrium and making observations as they return to equilibrium. { ‚rē‚lak'sā·shən ki‚ned·iks }

Renner-Teller effect [PHYS CHEM] The splitting, into two, of the potential function along the bending coordinate in degenerate electronic states of linear triatomic or polyatomic molecules. { 'ren·ər 'tel·ər i‚fekt }

repeating unit [ORG CHEM] The group of atoms that is derived from a monomer and repeats throughout a polymer. Also known as monomeric unit. { ri¦pēd·iŋ ‚yü·nət }

repellency [CHEM] Ability to repel water, or being hydrophobic; opposite to water wettability. { ri'pel·ən·sē }

replication [ANALY CHEM] The formation of a faithful mold or replica of a solid that is thin enough for penetration by an electron microscope beam; can use plastic (such as collodion) or vacuum deposition (such as of carbon or metals) to make the mold. { ‚rep·lə'kā·shən }

resbenzophenone *See* benzoresorcinol. { rez‚ben'zäf·ə‚nōn }

residual intensity [SPECT] The intensity of radiation at some wavelength in a spectral line divided by the intensity in the adjacent continuum. { rə'zij·yə·wəl in'ten·səd·ē }

resin [ORG CHEM] Any of a class of solid or semisolid organic products of natural or synthetic origin with no definite melting point, generally of high molecular weight; most resins are polymers. { 'rez·ən }

resin matrix [PHYS CHEM] The molecular network of an ion exchange material that carries the ionogenic groups. { 'rez·ən ‚mā‚triks }

resin of copper *See* cuprous chloride. { 'rez·ən əv 'käp·ər }

resinography [CHEM] Science of resins, polymers, plastics, and their products; includes study of morphology, structure, and other characteristics relatable to composition or treatment. { ‚rez·ən'äg·rə·fē }

resinoid [ORG CHEM] A thermosetting synthetic resin either in its initial (temporarily fusible) or in its final (infusible) state. { 'rez·ən‚ȯid }

resite *See* C stage. { 're‚zīt }

resolution [ORG CHEM] The process of separating a racemic mixture into the two component optical isomers. { ‚rez·ə'lü·shən }

resolving power [SPECT] A measure of the ability of a spectroscope or interferometer to separate spectral lines of nearly equal wavelength, equal to the average wavelength of two equally strong spectral lines whose images can barely be separated, divided by the difference in wavelengths; for spectroscopes, the lines must be resolved according to the Rayleigh criterion; for interferometers, the wavelengths at which the lines have half of maximum intensity must be equal. { ri'zälv·iŋ ˌpaü·ər }

resonance [PHYS CHEM] A feature of the valence-bond method that accounts for the anomalies in certain molecules by representing their structures with approximate resonance hybrid formulas; no single electronic formula conforms both to the observed properties and to the octet rule. Also known as mesomerism. { 'rez·ən·əns }

resonance hybrid [CHEM] A molecule that may be considered an intermediate between two or more valence bond structures. { 'rez·ən·əns ˌhī·brəd }

resonance ionization spectroscopy [SPECT] A technique capable of detecting single atoms or molecules of a given element or compound in a gas, in which an atom or molecule in its ground state is excited to a bound state when a photon is absorbed from a laser beam at a very well-controlled wavelength that is resonant with the excitation energy; a second photon removes the excited electron from the atom or molecule, and this electron is then accelerated by an electric field and collides with the gas molecules, creating additional ionization which is detected by a proportional counter. Abbreviated RIS. { 'rez·ən·əns ˌī·ə·nə'zā·shən spek'träs·kə·pē }

resonance line [SPECT] The line of longest wavelength associated with a transition between the ground state and an excited state. { 'rez·ən·əns ˌlīn }

resonance spectrum [SPECT] An emission spectrum resulting from illumination of a substance (usually a molecular gas) by radiation of a definite frequency or definite frequencies. { 'rez·ən·əns ˌspek·trəm }

resonance structure [ORG CHEM] Any of two or more possible structures of the same compound that have identical geometry but different arrangements of their paired electrons; none of the structures has physical reality or adequately accounts for the properties of the compound, which exists as an intermediate form. { 'rez·ən·əns ˌstrək·chər }

resonant ionization mass spectrometry [SPECT] An instrumental technique for quantitative identification of trace impurities (at or below the part-per-billion level), it begins with laser-induced or ion-induced desorption, followed by resonant laser ionization (usually from two or three lasers), and then analysis by time-of-flight mass spectrometry. Abbreviated RIMS. { ¦rez·ən·ənt ˌī·ə·nəˌzā·shən ¦mas spek'träm·ə·trē }

resorcin See resorcinol. { rə'zòrs·ən }

resorcinol [ORG CHEM] $C_6H_4(OH)_2$ Sweet-tasting, white, toxic crystals; soluble in water, alcohol, ether, benzene, and glycerol; melts at 111°C; used for resins, dyes, pharmaceuticals, and adhesives, and as a chemical intermediate. Also known as resorcin. { rə'zòrs·ənˌòl }

resorcinol acetate [ORG CHEM] $HOC_6H_4OCOCH_3$ A viscous, combustible, yellow to amber liquid with burning taste; soluble in alcohol; boils at 283°C; used in cosmetics and medicine. Also known as resorcinol monoacetate. { rə'zòrs·ənˌòl 'as·ə‚tāt }

resorcinol diglycidyl ether [ORG CHEM] $C_{12}H_{14}O_2$ A straw yellow liquid with a boiling point of 172°C (at 0.8 mmHg or 100 pascals); used for epoxy resins. Abbreviated RDGE. { rə'zòrs·ənˌòl di'glis·ə‚dil ˌē·thər }

resorcinol-formaldehyde resin [ORG CHEM] A phenol-formaldehyde resin, soluble in water, ketones, and alchol; used to make fast-curing adhesives for wood gluing. { rə'zòrs·ənˌòl fòr'mal·dəˌhīd 'rez·ən }

resorcinol monoacetate See resorcinol acetate. { rə'zòrs·ən‚òl ¦män·ō'as·ə‚tāt }

β-resorcylic acid [ORG CHEM] $(OH)_2C_6H_3COOH$ Combustible, white needles; soluble in alcohol and ether, very slightly soluble in water; decomposes at 220°C; used as a dyestuff and a pharmaceutical intermediate, and in the manufacture of fine chemicals. { ¦bād·ə ¦rē·zòr¦sil·ik 'as·əd }

restricted internal rotation [PHYS CHEM] Restrictions on the rotational motion of molecules or parts of molecules in some substances, such as solid methane, at certain temperatures. { ri'strik·təd in¦tərn·əl rō'tā·shən }

325

ret |CHEM| The reduction or digestion of fibers (usually linen) by enzymes. { ret }

retene |ORG CHEM| $C_{18}H_{18}$ A cyclic hydrocarbon, melting at 100.5–101°C, soluble in benzene and hot ethanol; used in organic syntheses. { 'rē,tēn }

retention index |ANALY CHEM| In gas chromatography, the relationship of retention volume with arbitrarily assigned numbers to the compound being analyzed; used to indicate the volume retention behavior during analysis. { ri'ten·chən ,in,deks }

retention time |ANALY CHEM| In gas chromatography, the time at which the center, or maximum, of a symmetrical peak occurs on a gas chromatogram. { ri'ten·chən ,tīm }

retention volume |ANALY CHEM| In gas chromatography, the product of retention time and flow rate. { ri'ten·chən ,väl·yəm }

rethrolone |ORG CHEM| A generic name for the five-member ring portion of a pyrethrin. { 'reth·rə,lōn }

retrogradation |CHEM| **1.** Generally, a process of deterioration; a reversal or retrogression to a simpler physical form. **2.** A chemical reaction involving vegetable adhesives, which revert to a simpler molecular structure. { ¦re·trō·grā'dā·shən }

retrograde condensation |ORG CHEM| Phenomenon associated with the behavior of a hydrocarbon mixture in the critical region wherein, at constant temperature, the vapor phase in contact with the liquid may be condensed by a decrease in pressure; or at constant pressure, the vapor is condensed by an increase in temperature. { 're·trə,grād ,kän·dən'sā·shən }

retrograde evaporation |ORG CHEM| Phenomenon associated with the behavior of a hydrocarbon mixture in the critical region wherein, at constant temperature, the liquid phase in contact with the vapor may be vaporized by an increase in pressure; or at constant pressure, the liquid is evaporated by a decrease in temperature. { 're·trə,grād i,vap·ə'rā·shən }

retrosynthetic analysis |ORG CHEM| A method for planning an organic chemical synthesis in which the desired product molecule is considered first, and then steps are considered one at a time leading back to the appropriate starting materials. { ,re·trō·sin¦thed·ik ə'nal·ə·səs }

reversal spectrum |SPECT| A spectrum which may be observed in intense white light which has traversed luminous gas, in which there are dark lines where there were bright lines in the emission spectrum of the gas. { ri'vər·səl ,spek·trəm }

reversal temperature |SPECT| The temperature of a blackbody source such that, when light from this source is passed through a luminous gas and analyzed in a spectroscope, a given spectral line of the gas disappears, whereas it appears as a bright line at lower blackbody temperatures, and a dark line at higher temperatures. { ri'vər·səl ,tem·prə·chər }

reverse bonded-phase chromatography |ANALY CHEM| A technique of bonded-phase chromatography in which the stationary phase is nonpolar and the mobile phase is polar. { ri'vərs 'bän·dəd ¦fāz ,krō·mə'täg·rə·fē }

reverse deionization |CHEM| A process in which an ion-exchange unit and a cation-exchange unit are used in sequence to remove all ions from a solution. { ri'vərs dē,ī·ə·nə'zā·shən }

reversed-phase partition chromatography |ANALY CHEM| Paper chromatography in which the low-polarity phase (such as paraffin, paraffin jelly, or grease) is put onto the support (paper) and the high-polarity phase (such as water, acids, or organic solvents) is allowed to flow over it. { ri'vərst ¦fāz pär'tish·ən ,krō·mə'täg·rə·fē }

reversible chemical reaction |CHEM| A chemical reaction that can be made to proceed in either direction by suitable variations in the temperature, volume, pressure, or quantities of reactants or products. { ri'vər·sə·bəl 'kem·ə·kəl rē'ak·shən }

reversible electrode |PHYS CHEM| An electrode that owes its potential to unit charges of a reversible nature, in contrast to electrodes used in electroplating and destroyed during their use. { ri'vər·sə·bəl i'lek,trōd }

Reychler's acid See d-camphorsulfonic acid. { 'rī·klərz ,as·əd }

Rf See rutherfordium.

Rh See rhodium.

rhenium [CHEM] A metallic element, symbol Re, atomic number 75, atomic weight 186.207; a transition element. { 'rē·nē·əm }

rhenium halide [INORG CHEM] Halogen compound of rhenium; examples are ReCl₃, ReCl₄, ReF₄, and ReF₆. { 'rē·nē·əm 'ha,līd }

rheopexy [PHYS CHEM] A property of certain sols, having particles shaped like rods or plates, which set to gel form more quickly when mechanical means are used to hasten the orientation of the particles. { 'rē·ə,pek·sē }

rhizoctol *See* methylarsinic sulfide. { rī'zäk,tȯl }

rhodamine B [ORG CHEM] C₂₈H₃₁ClN₂O₃ Red, green, or reddish-violet powder, soluble in alcohol and water; forms bluish red, fluorescent solution in water; used as red dye for paper, wool, and silk, and as an analytical reagent and biological stain. { 'rōd·ə,mēn 'bē }

rhodanic acid *See* thiocyanic acid. { rō'dan·ik 'as·əd }

rhodanine [ORG CHEM] C₃H₃NOS₂ A pale-yellow crystalline compound that may decompose violently when heated, giving off toxic by-products; used in organic synthesis. { 'rōd·ən,īn }

rhodium [CHEM] A chemical element, symbol Rh, atomic number 45, atomic weight 102.9055. { 'rōd·ē·əm }

rhodium chloride [INORG CHEM] RhCl₃ Water-insoluble, brown-red powder, soluble in cyanides and alkalies; decomposes at 450–500°C. Also known as rhodium trichloride. { 'rōd·ē·əm 'klȯr,īd }

rhodium trichloride *See* rhodium chloride. { 'rōd·ē·əm trī'klȯr,īd }

rhombic sulfur [CHEM] Crystalline sulfur with three unequal axes, all at right angles. { 'räm·bik 'səl·fər }

Rice's bromine solution [ANALY CHEM] Analytical reagent for the quantitative analysis of urea; has 12.5% bromine and sodium bromide in aqueous solution. { 'rīs·əz 'brō,mēn sə,lü·shən }

rich mixture [CHEM] An air-fuel mixture that is high in its concentration of combustible component. { 'rich ¦miks·chər }

ricinoleic acid [ORG CHEM] C₁₈H₃₄O₃ Unsaturated fatty acid; a combustible, water-insoluble, viscous liquid; soluble in most organic solvents; boils at 226°C (10 mmHg); used as a chemical intermediate, in soaps and Turkey red oils, and for textile finishing. Also known as castor oil acid. { ¦ris·ən·ō¦lē·ik 'as·əd }

ricinoleyl alcohol [ORG CHEM] C₁₈H₃₆O₂ Fatty alcohol of ricinoleic acid; a combustible, colorless, nondrying liquid, boiling at 170–328°C; used as a chemical intermediate, in protective coatings, surface-active agents, pharmaceuticals, and plasticizers. { ¦ris·ən·ō'lē·əl 'al·kə,hȯl }

Riegler's test [ANALY CHEM] Analytical technique for nitrous acid; uses sodium naphthionate and β-naphthol. { 'rēg·lərz ,test }

RIMS *See* resonant ionization mass spectrometry. { rimz *or* ¦är¦ī¦em'es }

ring [ORG CHEM] A closed loop of bonded atoms in a chemical structure, for example, benzene or cyclohexane. { riŋ }

ring closure [ORG CHEM] A chemical reaction in which one part of an open chain of a molecule reacts with another part to form a ring. { 'riŋ ,klō·zhər }

Ringer's solution [CHEM] A solution of 0.86 gram sodium chloride, 0.03 gram potassium chloride, and 0.033 gram calcium chloride in boiled, purified water, used topically as a physiological salt solution. { 'riŋ·ərz sə,lü·shən }

ring isomerism [ORG CHEM] A type of geometrical isomerism in which bond lengths and bond angles prevent the existence of a trans structure if substituents are attached to alkenic carbons which are part of a cyclic system, the ring of which contains fewer than eight members; for example, 1,2-dichlorocyclohexene. { 'riŋ ī'säm·ə,riz·əm }

ring structure [ORG CHEM] A cyclic chemical structure consisting of a chain whose ends are connected by bonds. { 'riŋ ,strək·chər }

ring system [ORG CHEM] Arbitrary designation of certain compounds as closed, circular structures, as in the six-carbon benzene ring; common rings have four, five, and six members, either carbon or some combination of carbon, nitrogen, oxygen, sulfur, or other elements. { 'riŋ ‚sis·təm }

ring whizzer [INORG CHEM] A fluxional molecule frequently encountered in organometallic chemistry in which rapid rearrangements occur by migrations about unsaturated organic rings. { 'riŋ ‚wiz·ər }

RIS *See* resonance ionization spectroscopy.

Ritter reaction [ORG CHEM] A procedure for the preparation of amides by reacting alkenes or tertiary alcohols with nitriles in an acidic medium. { 'rid·ər rē‚ak·shən }

Ritz's combination principle [SPECT] The empirical rule that sums and differences of the frequencies of spectral lines often equal other observed frequencies. Also known as combination principle. { 'rit·səz ‚käm·bə'nā·shən ‚prin·sə·pəl }

Rn *See* radon.

Rochelle salt *See* potassium sodium tartrate. { rō'shel‚sȯlt }

roentgen spectrometry *See* x-ray spectrometry. { 'rent·gən spek'träm·ə·trē }

Roese-Gottlieb method [ANALY CHEM] A solvent extraction method used to obtain an accurate determination of the fat content of milk. { 'rez·ə 'gät‚lēb ‚meth·əd }

rosaniline *See* fuchsin. { rōz'an·ə·lən }

Rosenmund reaction [ORG CHEM] Catalytic hydrogenation of an acid chloride to form an aldehyde; the reaction is in the presence of sulfur to prevent the subsequent hydrogenation of the aldehyde. { 'rōz·ən‚múnd rē‚ak·shən }

rosin ester *See* ester gum. { 'räz·ən 'es·tər }

rotating platinum electrode [ANALY CHEM] Platinum wire sealed in a soft-glass tubing and rotated by a constant-speed motor; used as the electrode in amperometric titrations. Abbreviated RPE. { 'rō‚tād·iŋ 'plat·ən·əm i'lek‚trōd }

rotational constant [PHYS CHEM] That constant inversely proportional to the moment of inertia of a linear molecule; used in calculations of microwave spectroscopy quantums. { rō'tā·shən·əl 'kän·stənt }

rotational energy [PHYS CHEM] For a diatomic molecule, the difference between the energy of the actual molecule and that of an idealized molecule which is obtained by the hypothetical process of gradually stopping the relative rotation of the nuclei without placing any new constraint on their vibration, or on motions of electrons. { rō'tā·shən·əl 'en·ər·jē }

rotational level [PHYS CHEM] An energy level of a diatomic or polyatomic molecule characterized by a particular value of the rotational energy and of the angular momentum associated with the motion of the nuclei. { rō'tā·shən·əl ‚lev·əl }

rotational quantum number [PHYS CHEM] A quantum number J characterizing the angular momentum associated with the motion of the nuclei of a molecule; the angular momentum is $(h/2\pi)$ $(h/2\pi \sqrt{J(J + 1)})$ and the largest component is $(h/2\pi)J$, where h is Planck's constant. { rō'tā·shən·əl 'kwän·təm ‚nəm·bər }

rotational spectrum [SPECT] The molecular spectrum resulting from transitions between rotational levels of a molecule which behaves as the quantum-mechanical analog of a rotating rigid body. { rō'tā·shən·əl 'spek·trəm }

rotational sum rule [SPECT] The rule that, for a molecule which behaves as a symmetric top, the sum of the line strengths corresponding to transitions to or from a given rotational level is proportional to the statistical weight of that level, that is, to $2J+1$, where J is the total angular momentum quantum number of the level. { rō'tā·shən·əl 'səm ‚rül }

rotational transition [PHYS CHEM] A transition between two molecular energy levels which differ only in the energy associated with the molecule's rotation. { rō'tā·shən·əl tran'zish·ən }

rotation spectrum [PHYS CHEM] Absorption-spectrum (absorbed electromagnetic energy) wavelengths produced if only the rotational energy of a molecule is affected during excitation. { rō'tā·shən ‚spek·trəm }

rotation-vibration spectrum [PHYS CHEM] Absorption-spectrum (absorbed electromagnetic energy) wavelengths produced when both the energy of vibration and energy of rotation of a molecule are affected by excitation. { rō'tā·shən vī'brā·shən ‚spek·trəm }

328

rotatory power [PHYS CHEM] The product of the specific rotation of an element or compound and its atomic or molecular weight. { 'rōd·ə,tȯr·ē ,paü·ər }

rotaxane [ORG CHEM] A compound with two or more independent portions not bonded to each other but linked by a linear portion threaded through a ring and maintained in this position by bulky end groups. { rō'tak,sān }

rotenone [ORG CHEM] $C_{23}H_{22}O_6$ White crystals with a melting point of 163°C; soluble in ether and acetone; used as an insecticide and in flea powders and fly sprays. Also known as tubatoxin. { 'rōt·ən,ōn }

rowland [SPECT] A unit of length, formerly used in spectroscopy, equal to 999.81/999.94 angstrom, or approximately 0.99987 × 10⁻¹⁰ meter. { 'rō,lənd }

Rowland circle [SPECT] A circle drawn tangent to the face of a concave diffraction grating at its midpoint, having a diameter equal to the radius of curvature of a grating surface; the slit and camera for the grating should lie on this circle. { 'rō,lənd ,sər·kəl }

Rowland ghost [SPECT] A false spectral line produced by a diffraction grating, arising from periodic errors in groove position. { 'rō,lənd ,gōst }

Rowland grating See concave grating. { 'rō,lənd ,grād·iŋ }

Rowland mounting [SPECT] A mounting for a concave grating spectrograph in which camera and grating are connected by a bar forming a diameter of the Rowland circle, and the two run on perpendicular tracks with the slit placed at their junction. { 'rō,lənd ,maünt·iŋ }

roxarsone See 4-hydroxy-3-nitrobenzenearsonic acid. { ¦räks¦är,sōn }

RPE See rotating platinum electrode.

Ru See ruthenium.

rubber [ORG CHEM] A natural, synthetic, or modified high polymer with elastic properties and, after vulcanization, elastic recovery. { 'rəb·ər }

rubber accelerator [ORG CHEM] A substance that increases the speed of curing of rubber, such as thiocarbanilide. { 'rəb·ər ak'sel·ə,rād·ər }

rubber hydrochloride [ORG CHEM] White, thermoplastic hydrochloric acid derivative of rubber; water-insoluble powder or clear film, soluble in aromatic hydrocarbons; softens at 110–120°C; used for protective coverings, food packaging, shower curtains, and rainwear. { 'rəb·ər ,hī·drə'klȯr,īd }

rubidium [CHEM] A chemical element, symbol Rb, atomic number 37, atomic weight 85.4678; a reactive alkali metal; salts of the metal may be used in glass and ceramic manufacture. { rü'bid·ē·əm }.

rubidium bromide [INORG CHEM] RbBr Colorless, regular crystals, melting at 683°C; soluble in water; used as a nerve sedative. { rü'bid·ē·əm 'brō,mīd }

rubidium chloride [INORG CHEM] RbCl A water-soluble, white, lustrous powder melting at 715°C; used as a source for rubidium metal, and as a laboratory reagent. { rü'bid·ē·əm 'klȯr,īd }

rubidium halide [INORG CHEM] Any of the halogen compounds of rubidium; examples are RbBr, RbCl, RbF, RbIBrCl, RbBr₂Cl, and RbIBr₂. { rü'bid·ē·əm 'ha,līd }

rubidium halometallate [INORG CHEM] Halogen-metal-containing compounds of rubidium; examples are Rb_2GeF_6 (rubidium hexafluorogermanate), Rb_2PtCl_6 (rubidium chloroplatinate), and Rb_2PdCl_5 (rubidium palladium chloride). { rü'bid·ē·əm ,ha·lō'med·əl,āt }

rubidium sulfate [INORG CHEM] Rb_2SO_4 Colorless, water-soluble rhomboid crystals, melting at 1060°C; used as a cathartic. { rü'bid·ē·əm 'səl,fāt }

ruling engine [SPECT] A machine operated by a long micrometer screw which rules equally spaced lines on an optical diffraction grating. { 'rül·iŋ ,en·jən }

ruthenic chloride See ruthenium chloride. { rü'then·ik 'klȯr,īd }

ruthenium [CHEM] A chemical element, symbol Ru, atomic number 44, atomic weight 101.07. { rü'thē·nē·əm }

ruthenium chloride [INORG CHEM] RuCl₃ Black, deliquescent, water-insoluble solid that decomposes in hot water and above 500°C; used as a laboratory reagent. Also known as ruthenic chloride; ruthenium sesquichloride. { rü'thē·nē·əm 'klȯr,īd }

ruthenium halide

ruthenium halide [INORG CHEM] Halogen compound of ruthenium; examples are RuCl$_2$, RuCl$_3$, RuCl$_4$, RuBr$_3$, and RuF$_5$. { rü'thē·nē·əm 'ha,līd }

ruthenium red [INORG CHEM] Ru$_2$(OH)$_2$Cl$_4$·7NH$_3$·3H$_2$O A water-soluble, brownish-red powder; used as an analytical reagent and stain. { rü'thē·nē·əm 'red }

ruthenium sesquichloride *See* ruthenium chloride. { rü'thē·nē·əm ,ses·kwi'klȯr,īd }

ruthenium tetroxide [INORG CHEM] RuO$_4$ A yellow, toxic solid, melting at 25°C; used as an oxidizing agent. { rü'thē·nē·əm te'träk,sīd }

Rutherford backscattering spectrometry [SPECT] A method of determining the concentrations of various elements as a function of depth beneath the surface of a sample, by measuring the energy spectrum of ions which are backscattered out of a beam directed at the surface. { 'rəth·ər·fərd 'bak,skad·ə·riŋ spek'träm·ə·trē }

rutherfordium [CHEM] A chemical element, symbolized Rf, atomic number 104, a synthetic element; the first element beyond the actinide series, and the twelfth transuranium element. { ,rəth·ər'fȯr·dē·əm }

rutin [ORG CHEM] C$_{27}$H$_{32}$O$_{16}$ A hydroxyflavone glucorhamnoside derived from cowslip and other plants; yellow needles melting at 190°C; used to treat capillary disorders. { 'rüt·ən }

R$_F$ value [ANALY CHEM] In chromatography, a measurement based on the relative distance traveled by a sample of a substance in a specific procedure; under standard conditions it is a characteristic property of the substance. { ,är'ef *or* ,är səb'ef ,val·yü }

rydberg *See* kayser. { 'rid,bərg }

Rydberg series formula [SPECT] An empirical formula for the wave numbers of various lines of certain spectral series such as neutral hydrogen and alkali metals; it states that the wave number of the nth member of the series is $\lambda_\infty - R/(n + a)^2$, where λ_∞ is the series limit, R is the Rydberg constant of the atom, and a is an empirical constant. { 'rid,bərg 'sir·ēz ,fȯr·myə·lə }

Rydberg spectrum [SPECT] An ultraviolet absorption spectrum produced by transitions of atoms of a given element from the ground state to states in which a single electron occupies an orbital farther from the nucleus. { 'rīd,bərg ,spek·trəm }

S

S *See* sulfur.

saccharin [ORG CHEM] $C_6H_4COSO_2NH$ A sweet-tasting, white powder, soluble in acetates, benzene, and alcohol; slightly soluble in water and ether; melts at 228°C; used as a sugar substitute for syrups, and in medicines, foods, and beverages. Also known as benzosulfimide; gluside. { 'sak·ə·rən }

saccharolactic acid *See* mucic acid. { ¦sak·ə·rō¦lak·tik 'as·əd }

saccharose *See* sucrose. { 'sak·ə‚rōs }

sacrificial anode [PHYS CHEM] A protective coating applied to a metal surface to act as an anode and be consumed in an electrochemical reaction, thereby preventing electrolytic corrosion of the metal. { ¦sak·rə‚fish·əl 'an‚ōd }

sacrificial metal [PHYS CHEM] A metal that can be used for a sacrificial anode. { ‚sak·rə'fish·əl 'med·əl }

saddle-point azeotrope [PHYS CHEM] A rarely occurring azeotrope which is formed in ternary systems and for which the boiling point is intermediate between the highest and lowest boiling mixture in the system. { 'sad·əl ¦pȯint 'ā·zē·ə‚trōp }

safranine [ORG CHEM] Any of a group of phenazine-based dyes; some are used as biological stains. { 'saf·rə‚nēn }

safrole [ORG CHEM] $C_3H_5C_6H_3O_2CH_2$ A toxic, water-insoluble, colorless oil that boils at 233°C; found in sassafras and camphorwood oils; used in medicine, perfumes, insecticides, and soaps, and as a chemical intermediate. { 'sa‚frōl }

sal acetosella *See* potassium binoxalate. { 'sal ¦as·ə·dō¦sel·ə }

salazosulfadimidine [ORG CHEM] $C_{19}H_{17}N_5O_5S$ A brown crystalline compound that melts at 207°C; used in medicine in cases of ulcerative colitis. { ¦sal·ā·zō‚səl·fə'dī·mə‚dēn }

sal ethyl *See* ethyl salicylate. { 'sal 'eth·əl }

salicin [ORG CHEM] $C_{13}H_{18}O_7$ A glucoside; colorless crystals, soluble in water, alcohol, alkalies, and glacial acetic acid; melts at 199°C; used in medicine and as an analytical reagent. { 'sal·ə·sən }

salicyl alcohol [ORG CHEM] $C_7H_8O_2$ A crystalline alcohol that forms plates or powder, melting at 86–87°C; used in medicine as a local anesthetic. { 'sal·ə·səl 'al·kə‚hȯl }

salicylaldehyde [ORG CHEM] C_6H_4OHCHO Clear to dark-red oily liquid, with burning taste and almond aroma; soluble in alcohol, benzene, and ether, very slightly soluble in water; boils at 196°C; used in analytical chemistry, in perfumery, and for synthesis of chemicals. Also known as helicin. { sal·ə·səl'al·də‚hīd }

salicylamide [ORG CHEM] $C_6H_4(OH)CONH_2$ Pinkish or white crystals; soluble in alcohol, ether, chloroform, and hot water; melts at 193°C; used in medicine as an analgesic, antipyretic, and antirheumatic drug. { ‚sal·ə'sī·lə·məd }

salicylate [ORG CHEM] A salt of salicylic acid with the formula $C_6H_4(OH)COOM$, where M is a monovalent metal; for example, $NaC_7H_5O_3$, sodium salicylate. { sə'lis·ə‚lāt }

salicylated mercury *See* mercuric salicylate. { sə'lis·ə‚lād·əd 'mər·kyə·rē }

salicylic acid [ORG CHEM] $C_6H_4(OH)COOH$ White crystals with sweetish taste; soluble in alcohol, acetone, ether, benzene, and turpentine, slightly soluble in water; discolored by light; melts at 158°C; used as a chemical intermediate and in medicine, dyes, perfumes, and preservatives. { ¦sal·ə¦sil·ik 'as·əd }

salicylic acid ethyl ether *See* ethyl salicylate. { ¦sal·ə¦sil·ik 'as·əd 'eth·əl 'ē·thər }

salicylic ether

salicylic ether See ethyl salicylate. { ¦sal·ə¦sil·ik 'ē·thər }

saligenol See salicyl alcohol. { sə'lij·ə¸nȯl }

salol [ORG CHEM] C₆H₄OHCOOC₆H₅ White powder with aromatic taste and aroma; soluble in alcohol, ether, chloroform, and benzene; slightly soluble in water; melts at 42°C; used in medicinals and as a preservative. Also known as phenyl salicylate. { 'sa¸lȯl }

sal soda [INORG CHEM] Na₂CO₃·10H₂O White, water-soluble crystals, insoluble in alcohol; melts and loses water at about 33°C; mild irritant to mucous membrane; used in cleansers and for washing textiles and bleaching linen and cotton. Also known as sodium carbonate decahydrate; washing soda. { 'sal ¸sō·də }

salt [CHEM] The reaction product when a metal displaces the hydrogen of an acid; for example, H₂SO₄ + 2NaOH → Na₂SO₄ (a salt) + 2H₂O. { sȯlt }

sal tartar See sodium tartrate. { 'sal 'tär·tər }

salt bridge [PHYS CHEM] A bridge of a salt solution, usually potassium chloride, placed between the two half-cells of a galvanic cell, either to reduce to a minimum the potential of the liquid junction between the solutions of the two half-cells or to isolate a solution under study from a reference half-cell and prevent chemical precipitations. { 'sȯlt ¸brij }

salt cake [INORG CHEM] Impure sodium sulfate; used in soaps, paper pulping, detergents, glass, ceramic glaze, and dyes. { 'sȯlt ¸kāk }

salt error [ANALY CHEM] An error introduced in an analytical determination of a saline liquid such as sea water; caused by the effect of the neutral ions in the solution on the color of the pH indicator, and hence upon the apparent pH. { 'sȯlt ¸er·ər }

salt fingers [CHEM] A close-packed array of rising and sinking columns of fluid that form in a liquid when a slow-diffusing solute is separated by an interface from another, lower solute that diffuses more rapidly. { 'sȯlt ¸fiŋ·gərz }

salt of Lemery See potassium sulfate. { 'sȯlt əv lem'rē }

salt of sorrel See potassium binoxalate. { 'sȯlt əv 'sär·əl }

salt of tartar See potassium carbonate. { 'sȯlt əv 'tär·tər }

salt pan [CHEM] A pool used for obtaining salt by the natural evaporation of sea water. { 'sȯlt ¸pan }

saltpeter See potassium nitrate. { sȯlt'pēd·ər }

samarium [CHEM] A rare-earth metal, atomic number 62, symbol Sm; melts at 1350°C, tarnishes in air, ignites at 200–400°C. { sə'mar·ē·əm }

samarium oxide [INORG CHEM] Sm₂O₃ A cream-colored powder with a melting point of 2300°C; soluble in acids; used for infrared-absorbing glass and as a neutron absorber. { sə'mar·ē·əm 'äk¸sīd }

SAN See styrene-acrylonitrile resin.

Sandmeyer's reaction [CHEM] Conversion of diazo compounds (in the presence of cuprous halogen salts) into halogen compounds. { 'san¸mī·ərz rē¸ak·shən }

Sanger's reagent See 1-fluoro-2,4-dinitrobenzene. { 'saŋ·ərz rē¸ā·jənt }

santalol [ORG CHEM] C₁₅H₂₄O A colorless liquid with a boiling point of 300°C; derived from sandalwood oil and used for perfumes. { 'san·tə¸lȯl }

santonin [ORG CHEM] C₁₅H₁₈O₃ A white powder with a melting point of 170–173°C; soluble in chloroform and alcohol; used in medicine. { 'sant·ən·ən }

saponification [CHEM] The process of converting chemicals into soap; involves the alkaline hydrolysis of a fat or oil, or the neutralization of a fatty acid. { sə¸pän·ə·fə'kā·shən }

saponification equivalent [CHEM] The quantity of fat in grams that can be saponified by 1 liter of normal alkalies. { sə¸pän·ə·fə'kā·shən i¸kwiv·ə·lənt }

saponification number [ANALY CHEM] Milligrams of potassium hydroxide required to saponify the fat, oil, or wax in a 1-gram sample of a given material, using a specific American Society for Testing and Materials test method. { sə¸pän·ə·fə'kā·shən ¸nəm·bər }

saponin [ORG CHEM] Any of numerous plant glycosides characterized by foaming in water and by producing hemolysis when water solutions are injected into the bloodstream; used as beverage foam producer, textile detergent and sizing, soap substitute, and emulsifier. { 'sap·ə·nən }

sarcosine [ORG CHEM] CH_3NHCH_2COOH Sweet-tasting, deliquescent crystals; soluble in water, slightly soluble in alcohol; decomposes at 210–215°C; used in toothpaste manufacture. Also known as methyl glycocol. { 'sär·kə,sēn }

SAS See aluminum sodium sulfate.

satellite infrared spectrometer [SPECT] A spectrometer carried aboard satellites in the Nimbus series which measures the radiation from carbon dioxide in the atmosphere at several different wavelengths in the infrared region, giving the vertical temperature structure of the atmosphere over a large part of the earth. Abbreviated SIRS. { 'sad·əl,īt ¦in·frə'red spek'träm·əd·ər }

saturated ammonia [CHEM] **1.** Liquid ammonia in a state in which adding heat at constant pressure causes the liquid to vaporize at constant temperature, and in which removing heat at constant pressure causes the temperature of the liquid to drop immediately. **2.** Ammonia vapor in a state in which adding heat at constant pressure causes an immediate temperature rise (superheating) and in which removing heat at constant pressure starts immediate condensation at constant temperature. { 'sach·ə,rād·əd ə'mō·nyə }

saturated calomel electrode [PHYS CHEM] A reference electrode of mercury topped by a layer of mercury (I) chloride paste with potassium chloride solution placed above; easier to assemble than the normal and the one-tenth normal (referring to the concentration of KCl) calomel electrodes. { 'sach·ə,rād·əd 'kal·ə,mel i'lek,trōd }

saturated compound [ORG CHEM] An organic compound with all carbon bonds satisfied; it does not contain double or triple bonds and thus cannot add elements or compounds. { 'sach·ə,rād·əd 'käm,paúnd }

saturated hydrocarbon [ORG CHEM] A saturated carbon-hydrogen compound with all carbon bonds filled; that is, there are no double or triple bonds as in olefins and acetylenics. { 'sach·ə,rād·əd ¦hī·drə'kär·bən }

saturated interference spectroscopy [SPECT] A version of saturation spectroscopy in which the gas sample is placed inside an interferometer that splits a probe laser beam into parallel components in such a way that they cancel on recombination; intensity changes in the recombined probe beam resulting from changes in absorption or refractive index induced by a laser saturating beam are then measured. { 'sach·ə,rād·əd ,in·tər,fir·əns spek'träs·kə·pē }

saturated liquid [CHEM] A solution that contains enough of a dissolved solid, liquid, or gas so that no more will dissolve into the solution at a given temperature and pressure. { 'sach·ə,rād·əd 'lik·wəd }

saturation [PHYS CHEM] The condition in which the partial pressure of any fluid constituent is equal to its maximum possible partial pressure under the existing environmental conditions, such that any increase in the amount of that constituent will initiate within it a change to a more condensed state. { ,sach·ə'rā·shən }

saturation spectroscopy [SPECT] A branch of spectroscopy in which the intense, monochromatic beam produced by a laser is used to alter the energy-level populations of a resonant medium over a narrow range of particle velocities, giving rise to extremely narrow spectral lines that are free from Doppler broadening; used to study atomic, molecular, and nuclear structure, and to establish accurate values for fundamental physical constants. { ,sach·ə'rā·shən spek'träs·kə·pē }

Sb See antimony.

s-block element [CHEM] A chemical element whose valence shell contains s-electrons only; found in groups 1 and 2 of the periodic table. { 'es ,bläk ,el·ə·mənt }

Sc See scandium.

scale See boiler scale. { skāl }

scandia See scandium oxide. { 'skan·dē·ə }

scandium [CHEM] A transition element, symbol Sc, atomic number 21; melts at 1200°C; found associated with rare-earth elements. { 'skan·dē·əm }

scandium halide

scandium halide [INORG CHEM] A compound of scandium and a halogen; for example, scandium chloride, $ScCl_3$. { 'skan·dē·əm 'ha,līd }

scandium oxide [INORG CHEM] Sc_2O_3 White powder, soluble in hot acids; used to prepare scandium. Also known as scandia. { 'skan·dē·əm 'äk,sīd }

scandium sulfate [INORG CHEM] $Sc_2(SO_4)_3$ Water-soluble, colorless crystals. { 'skan·dē·əm 'səl,fāt }

scandium sulfide [INORG CHEM] Sc_3S_3 Yellowish powder; decomposes in dilute acids and boiling water to give off hydrogen sulfide. { 'skan·dē·əm 'səl,fīd }

scarlet See scarlet red. { 'skär·lət }

scarlet red [ORG CHEM] $CH_3C_6H_4H:NC_6H_3CH_3N:NC_{10}H_{15}OH$ A brown, water-insoluble powder, used as a dye in ointments. Also known as Biebrich red; scarlet. { 'skär·lət 'red }

scattering plane [PHYS CHEM] In a quasielastic light-scattering experiment performed with the use of polarizers, the plane containing the incident and scattered beams. { 'skad·ə·riŋ ,plān }

scavenger [CHEM] A substance added to a mixture or other system to remove or inactivate impurities. Also known as getter. { 'skav·ən·jər }

Schaeffer's salt [ORG CHEM] $HOC_{10}SO_3Na$ A light-yellow to pink, water-soluble powder; the sodium salt formed from 2-naphthol-6-sulfonic acid; used as an intermediate in synthesis of organic compounds. { 'shā·fərz ,sólt }

Scheele's green See copper arsenite. { 'shā·ləz 'grēn }

Schiff base [ORG CHEM] $RR'C=NR''$ Any of a class of derivatives of the condensation of aldehydes or ketones with primary amines; colorless crystals, weakly basic; hydrolyzed by water and strong acids to form carbonyl compounds and amines; used as chemical intermediates and perfume bases, in dyes and rubber accelerators, and in liquid crystals for electronics. { 'shif ,bās }

Schiff's reagent [ANALY CHEM] An aqueous solution of rosaniline and sulfurous acid; used in the Schiff test. { 'shifs rē,ā·jənt }

Schiff test [ANALY CHEM] A test for aldehydes by using an aqueous solution of rosaniline and sulfurous acid. { 'shif ,test }

schiller layer [PHYS CHEM] One of a series of layers formed by sedimenting particles that exhibit bright colors in reflected light, because the layers are separated by approximately equal distances, with the distances being of the same order of magnitude as the wavelength of visible light. Also known as iridescent layer. { 'shil·ər ,lā·ər }

Schmidt number 2 See Semenov number 1. { 'shmit ,nəm·bər 'tü }

Schmidt number 3 [PHYS CHEM] A dimensionless number used in electrochemistry, equal to the product of the dielectric susceptibility and the dynamic viscosity of a fluid divided by the product of the fluid density, electrical conductivity, and the square of a characteristic length. Symbolized Sc_3. { 'shmit ,nəm·bər 'thrē }

Schoelkopf's acid [ORG CHEM] A dye of the following types: 1-naphthol-4,8-disulfonic acid, 1-naphthylamine-4,8-disulfonic acid, and 1-naphthylamine-8-sulfonic acid; may be toxic. { 'shəl,kòpfs ,as·əd }

Schotten-Baumann reaction [ORG CHEM] An acylation reaction that uses an acid chloride in the presence of dilute alkali to acylate the hydroxyl and amino group of organic compounds. { 'shät·ən 'baů,män rē,ak·tən }

Schuermann series [CHEM] A list of metals so arranged that the sulfide of one is precipitated at the expense of the sulfide of any lower metal in the series. { 'shòi·ər,män ,sir·ēz }

Schultz-Hardy rule [CHEM] The sensitivity of lyophobic colloids to coagulating electrolytes is governed by the charge of the ion opposite that of the colloid, and the sensitivity increases more rapidly than the charge of the ion. { 'shúlts 'här·dē ,rül }

Schulze's reagent [ANALY CHEM] An oxidizing mixture consisting of a saturated aqueous solution of $KClO_3$ and varying amounts of concentrated HNO_3; commonly used in palynologic macerations. { 'shúlt·səz rē,ā·jənt }

Schuster method [SPECT] A method for focusing a prism spectroscope without using a distant object or a Gauss eyepiece. { 'shús·tər ,meth·əd }

Schweitzer's reagent [CHEM] An ammoniacal solution of cupric hydroxide; used to dissolve cellulose, silk, and linen, and to test for wool. { 'shvīt·sərz rē,ā·jənt }

scopoline [ORG CHEM] $C_8H_{13}O_2H$ A white crystalline alkaloid that melts at 108–109°C, soluble in water and ethanol; used in medicine. Also known as oscine. { 'skō·pə,lēn }

screening agent [ANALY CHEM] A nonchelating dye used to improve the colorimetric end point of a complexometric titration; a dye addition forms a complementary pair of colors with the metalized and unmetalized forms of the end-point indicator. { 'skrēn·iŋ ,ā·jənt }

SDDC See sodium dimethyldithiocarbamate.

Se See selenium.

seaborgium [CHEM] A chemical element, symbolized Sg, atomic number 106, a synthetic element; the fourteenth transuranium element. { sē'bȯrg·ē·əm }

sebacic acid [ORG CHEM] COOH(CH₂)₈COOH Combustible, white crystals; slightly soluble in water, soluble in alcohol and ether; melts at 133°C; used in perfumes, paints, and hydraulic fluids and to stabilize synthetic resins. { si'bas·ik 'as·əd }

sebacylic acid See sebacic acid. { ¦seb·ə¦sil·ik 'as·əd }

secondary alcohol [ORG CHEM] An organic alcohol with molecular structure R_1R_2CHOH, where R_1 and R_2 designate either identical or different groups. { 'sek·ən,der·ē 'al·kə,hȯl }

secondary amine [ORG CHEM] An organic compound that may be written R_1R_2NH, where R_1 and R_2 designate either identical or different groups. { 'sek·ən,der·ē 'am,ēn }

secondary carbon atom [ORG CHEM] A carbon atom that is singly bonded to two other carbon atoms. { 'sek·ən,der·ē 'kär·bən 'ad·əm }

secondary hydrogen atom [ORG CHEM] A hydrogen atom that is bonded to a secondary carbon atom. { 'sek·ən,der·ē 'hī·drə·jən ,ad·əm }

second boiling point [PHYS CHEM] In certain mixtures, the temperature at which a gas phase develops from a liquid phase upon cooling. { 'sek·ənd 'bȯil·iŋ ,pȯint }

second-order reaction [PHYS CHEM] A reaction whose rate of reaction is determined by the concentration of two chemical species. { 'sek·ənd ¦ȯr·dər rē'ak·shən }

sedimentation [CHEM] The settling of suspended particles within a liquid under the action of gravity or a centrifuge. { ,sed·ə·mən'tā·shən }

sedimentation balance [ANALY CHEM] A device to measure and record the weight of sediment (solid particles settled out of a liquid) versus time; used to determine particle sizes of fine solids. { ,sed·ə·mən'tā·shən ,bal·əns }

sedimentation coefficient [PHYS CHEM] In the sedimentation of molecules in an accelerating field, such as that of a centrifuge, the velocity of the boundary between the solution containing the molecules and the solvent divided by the accelerating field. (In the case of a centrifuge, the accelerating field equals the distance of the boundary from the axis of rotation multiplied by the square of the angular velocity in radians per second.) { ,sed·ə·mən'tā·shən ,kō·i'fish·ənt }

sedimentation constant [PHYS CHEM] A quantity used in studying the behavior of colloidal particles subject to forces, especially centrifugal forces; it is equal to $2r^2(\rho - \rho')/9\eta$, where r is the particle's radius, ρ and ρ' are reciprocals of partial specific volumes of particle and medium respectively, and η is the medium's viscosity. { ,sed·ə·mən'tā·shən ,kän·stənt }

sedimentation equilibrium [ANALY CHEM] The equilibrium between the forward movement of a sample's liquid-sediment boundary and reverse diffusion during centrifugation; used in molecular-weight determinations. { ,sed·ə·mən'tā·shən ,ē·kwə,lib·rē·əm }

sedimentation velocity [ANALY CHEM] The rate of movement of the liquid-sediment boundary in the sample holder during centrifugation; used in molecular-weight determinations. { ,sed·ə·mən'tā·shən və,läs·əd·ē }

seed [CHEM] A small, single crystal of a desired substance added to a solution to induce crystallization. { sēd }

seed charge [CHEM] A small amount of material added to a supersaturated solution to initiate precipitation. { 'sēd ‚chärj }

seeding [CHEM] The adding of a seed charge to a supersaturated solution, or a single crystal of a desired substance to a solution of the substance to induce crystallization. { 'sēd·iŋ }

segment [ANALY CHEM] A specific, demarcated portion of a lot of a substance that is to be chemically analyzed. { 'seg·mənt }

Seignette salt See potassium sodium tartrate. { sen'yet ‚sȯlt }

selective inhibition See selective poisoning. { si'lek·tiv ‚in·ə'bish·ən }

selective poisoning [CHEM] Retardation of the rate of one catalyzed reaction more than that of another by the use of a catalyst poison. Also known as selective inhibition. { si¦lek·tiv 'pȯiz·ən·iŋ }

selectivity [ANALY CHEM] The ability of a type of method or instrumentation to respond to a specified substance or constituent and not to others. { sə‚lek'tiv·əd·ē }

selectivity coefficient [ANALY CHEM] Ion equilibria relationship formula for ion-exchange-resin systems. { sə‚lek'tiv·əd·ē ‚kō·i‚fish·ənt }

selenic acid [INORG CHEM] H_2SeO_4 A highly toxic, water-soluble, white solid, melting point 58°C, decomposing at 260°C. { sə'len·ik 'as·əd }

selenide [INORG CHEM] M_2Se A binary compound of divalent selenium, such as Ag_2Se, silver selenide. [ORG CHEM] An organic compound containing divalent selenium, such as $(C_2H_5)_2Se$, ethyl selenide. { 'sel·ə‚nīd }

selenious acid See selenous acid. { sə'lē·nē·əs 'as·əd }

selenium [CHEM] A highly toxic, nonmetallic element in group 16, symbol Se, atomic number 34; steel-gray color; soluble in carbon disulfide, insoluble in water and alcohol; melts at 217°C; and boils at 690°C; used in analytical chemistry, metallurgy, and photoelectric cells, and as a lube-oil stabilizer and chemicals intermediate. { sə'lē·nē·əm }

selenium bromide [INORG CHEM] Any of three compounds of selenium and bromine: Se_2Br_2, a red liquid that melts at −46°C, also known as selenium monobromide; $SeBr_2$, a brown liquid, also known as selenium dibromide; and $SeBr_4$, orange, carbon-disulfide-soluble crystals, also known as selenium tetrabromide. { sə'lē·nē·əm 'brō‚mīd }

selenium dibromide See selenium bromide. { sə'lē·nē·əm dī'brō‚mīd }

selenium dioxide [INORG CHEM] SeO_2 Water- and alcohol-soluble, white to reddish, lustrous crystals; melts at 340°C; used in medicine, and as an oxidizing agent and catalyst. Also known as selenous acid anhydride; selenous anhydride; selenium oxide. { sə'lē·nē·əm dī'äk‚sīd }

selenium halide [INORG CHEM] A compound of selenium and a halogen, for example, Se_2Br_2, $SeBr_2$, $SeBr_4$; Se_2Cl_2, $SeCl_2$, $SeCl_4$; Se_2I_2, SeI_4. { sə'lē·nē·əm 'ha‚līd }

selenium monobromide See selenium bromide. { sə'lē·nē·əm ‚män·ō'brō‚mīd }

selenium nitride [INORG CHEM] Se_2N_2 A water-insoluble, yellow solid that explodes at 200°C. { sə'lē·nē·əm 'nī‚trīd }

selenium oxide See selenium dioxide. { sə'lē·nē·əm 'äk‚sīd }

selenium tetrabromide See selenium bromide. { sə'lē·nē·əm ‚te·trə'brō‚mīd }

selenone [ORG CHEM] A group of organic selenium compounds with the general formula R_2SeO_2. { 'sel·ə‚nōn }

selenonic acid [ORG CHEM] Any organic acid containing the radical −SeO_3H; analogous to a sulfonic acid. { ‚sel·ə'nän·ik 'as·əd }

selenous acid [INORG CHEM] H_2SeO_3 Colorless, transparent crystals; soluble in water and alcohol, insoluble in ammonia; decomposes when heated; used as an analytical reagent. Also spelled selenious acid. { sə'lē·nəs 'as·əd }

selenous acid anhydride See selenium dioxide. { sə'lē·nəs 'as·əd an'hī‚drīd }

selenous anhydride See selenium dioxide. { sə'lē·nəs an'hī‚drīd }

selenoxide [ORG CHEM] A group of organic selenium compounds with the general formula R_2SeO. { ‚sel·ə'näk‚sīd }

self-absorption [SPECT] Reduction of the intensity of the center of an emission line caused by selective absorption by the cooler portions of the source of radiation. Also known as self-reduction; self-reversal. { ¦self əb¦sȯrp·shən }

self-organization [CHEM] The capability of a system to spontaneously generate a well-defined supramolecular entity by self-assembling from components in a given set of conditions. { ˌself ˌȯr·gə·nə'zā·shən }

self-poisoning [CHEM] Inhibition of a chemical reaction by a product of the reaction. Also known as autopoisoning. { ˌself 'pȯiz·ən·iŋ }

self-reversal *See* self-absorption. { ¦self ri¦vər·səl }

Seliwanoff's test [ANALY CHEM] A color test helpful in the identification of ketoses, which develop a red color with resorcinol in hydrochloric acid. { sə'liv·ə‚nȯfs ‚test }

sellite [INORG CHEM] A solution of sodium sulfite (Na_2SO_3) used in the purification of 2,4,6-trinitrotoluene to remove unsymmetrical isomers. { 'se‚līt }

Semenov number 1 [PHYS CHEM] A dimensionless number used in reaction kinetics, equal to a mass transfer constant divided by a reaction rate constant. Symbolized S_m. Formerly known as Schmidt number 2. { 'se·mə‚nȯf 'nəm·bər 'wən }

semicarbazide [ORG CHEM] $H_2N-NHCONH_2$ A reagent used to produce semicarbazones by reaction with aldehydes or ketones. { ¦sem·i'kär·bə‚zīd }

semicarbazide hydrochloride [ORG CHEM] $CH_5ON_3·HCl$ Colorless prisms, soluble in water, decomposing at 175°C; used as an analytical reagent for aldehydes and ketones, and to recover constituents of essential oils. { ¦sem·i 'kär·bə‚zīd ‚hī·drə'klȯr‚īd }

semicarbazone [ORG CHEM] $R_2C:N_2HCONH_2$ A condensation product of an aldehyde or ketone with semicarbazide. { ˌsem·i'kär·bə‚zōn }

semiempirical computation [PHYS CHEM] Computation of the geometry of a molecule by using parameters that have been experimentally determined for similar molecules. { ˌsem·ē·em¦pir·ə·kəl ‚käm·pyə'tā·shən }

semiforbidden line [SPECT] A spectral line associated with a semiforbidden transition. { ¦sem·i·fər'bid·ən 'līn }

semimetal *See* metalloid. { ¦sem·ē'med·əl }

semimicroanalysis [ANALY CHEM] A chemical analysis procedure in which the weight of the sample is between 10 and 100 milligrams. { ¦sem·i‚mī·krō·ə'nal·ə·səs }

semioxamazide [ORG CHEM] $H_2NCOCONHNH_2$ A crystalline compound that decomposes at 220°C; soluble in hot water, acids, and alkalies; used as a reagent for aldehydes and ketones. { ¦sem·ē‚äk¦sam·ə·zəd }

semiquinone [ORG CHEM] A radical ion intermediate formed in the oxidation of a hydroquinone to a quinone. { ˌsem·ē·kwə'nōn }

sensing zone technique [ANALY CHEM] Particle-size measurement in a dilute solution, with fine particles passed through a small zone (opening) so that individual particles may be observed and measured by electrolytic, photic, or sonic methods. { 'sens· iŋ ¦zōn tek‚nēk }

separatory funnel [CHEM] A funnel-shaped device used for the careful and accurate separation of two immiscible liquids; a stopcock on the funnel stem controls the rate and amount of outflow of the lower liquid. { 'sep·rə‚tȯr·ē 'fən·əl }

sequestering agent [CHEM] A substance that removes a metal ion from a solution system by forming a complex ion that does not have the chemical reactions of the ion that is removed; can be a chelating or a complexing agent. { si'kwes·tə·riŋ ‚ā·jənt }

series [ANALY CHEM] A group of results of repeated analyses completed by using a single analytical method on samples of a homogeneous substance. [SPECT] A collection of spectral lines of an atom or ion for a set of transitions, with the same selection rules, to a single final state; often the frequencies have the general formula $[R/(a + c_1)^2] - [R/(n + c_2)^2]$, where R is the Rydberg constant for the atom, a and c_1 and c_2 are constants, and n takes on the values of the integers greater than a for the various lines in the series. { 'sir·ēz }

sesquioxide [CHEM] A compound composed of a metal and oxygen in the ratio 2:3; for example, Al_2O_3. { ¦ses·kwē'äk‚sīd }

sesquiterpene [ORG CHEM] Any terpene with the formula $C_{15}H_{24}$; that is, $1\frac{1}{2}$ times the terpene formula. { ¦ses·kwē'tər‚pēn }

set |CHEM| The hardening or solidifying of a plastic or liquid substance. { set }

SFS See sodium formaldehyde sulfoxylate.

Sg See seaborgium.

sharp series |SPECT| A series occurring in the line spectra of many atoms and ions with one, two, or three electrons in the outer shell, in which the total orbital angular momentum quantum number changes from 0 to 1. { 'shärp 'sir·ēz }

shift |SPECT| A small change in the position of a spectral line that is due to a corresponding change in frequency which, in turn, results from one or more of several causes, such as the Doppler effect. { shift }

Shpol'skii effect |SPECT| The occurrence of very narrow fluorescent lines in the spectra of certain compounds from molecules frozen at low temperatures. { 'shpōl·skē i‚fekt }

Si See silicon.

side chain |ORG CHEM| A grouping of similar atoms (two or more, generally carbons, as in the ethyl radical, C_2H_5-) that branches off from a straight-chain or cyclic (for example, benzene) molecule. Also known as branch; branched chain. { 'sīd ‚chān }

side reaction |CHEM| A secondary or subsidiary reaction that takes place simultaneously with the reaction of primary interest. { 'sīd rē‚ak·shən }

siderophile element |CHEM| An element with a weak affinity for oxygen and sulfur and that is readily soluble in molten iron; includes iron, nickel, cobalt, platinum, gold, tin, and tantalum. { 'sid·ə·rə‚fīl 'el·ə·mənt }

siegbahn |SPECT| A unit of length, formerly used to express wavelengths of x-rays, equal to 1/3029.45 of the spacing of the (200) planes of calcite at 18°C, or to (1.00202 ± 0.00003) × 10^{-13} meter. Also known as x-ray unit; X-unit. Symbolized X; XU. { 'sēg‚bän }

sigma bond |PHYS CHEM| The chemical bond resulting from the formation of a molecular orbital by the end-on overlap of atomic orbitals. { 'sig·mə ‚bänd }

sigmatropic shift |ORG CHEM| A rearrangement reaction that consists of the migration of a sigma bond (that is, the sigma electrons) and the group of atoms that are attached to it from one position in a chain or ring into a new position. { 'sig·mə‚träp·ik 'shift }

silane |INORG CHEM| Si_nH_{2n+2} A class of silicon-based compounds analogous to alkanes, that is, straight-chain, saturated paraffin hydrocarbons; they can be gaseous or liquid. Also known as silicon hydride. { 'si‚lān }

silanol |CHEM| A member of the family of compounds whose structure contains a silicon atom that is bound directly to one or more hydroxyl groups. { 'sī·lə‚nȯl }

silica gel |INORG CHEM| A colloidal, highly absorbent silica used as a dehumidifying and dehydrating agent, as a catalyst carrier, and sometimes as a catalyst. { 'sil·ə·kə ‚jel }

silicate |INORG CHEM| The generic term for a compound that contains silicon, oxygen, and one or more metals, and may contain hydrogen. { 'sil·ə·kət }

silicate of soda See sodium silicate. { 'sil·ə·kət əv 'sōd·ə }

silicic acid |INORG CHEM| $SiO_2 \cdot nH_2O$ A white, amorphous precipitate; used to bleach fats, waxes, and oils. Also known as hydrated silica. { sə'lis·ik 'as·əd }

silicide |CHEM| A binary compound in which silicon is bonded with a more electropositive element. { 'sil·ə‚sīd }

silicon |CHEM| A group 14 nonmetallic element, symbol Si, with atomic number 14, atomic weight 28.086; dark-brown crystals that burn in air when ignited; soluble in hydrofluoric acid and alkalies; melts at 1410°C; used to make silicon-containing alloys, as an intermediate for silicon-containing compounds, and in rectifiers and transistors. { 'sil·ə·kən }

silicon bromide See silicon tetrabromide. { 'sil·ə·kən 'brō‚mīd }

silicon carbide |INORG CHEM| SiC Water-insoluble, bluish-black crystals, very hard and iridescent; soluble in fused alkalies; sublimes at 2210°C; used as an abrasive and a heat refractory, and in light-emitting diodes to produce green or yellow light. { 'sil·ə·kən 'kär‚bīd }

silicon chloride See silicon tetrachloride. { 'sil·ə·kən 'klȯr‚īd }

silicon dioxide [INORG CHEM] SiO_2 Colorless, transparent crystals, soluble in molten alkalies and hydrofluoric acid; melts at 1710°C; used to make glass, ceramic products, abrasives, foundry molds, and concrete. { 'sil·ə·kən dī'äk,sīd }

silicon fluoride See silicon tetrafluoride. { 'sil·ə·kən 'flùr,īd }

silicon halide [INORG CHEM] A compound of silicon and a halogen; for example, $SiBr_4$, Si_2Br_6, $SiCl_4$, Si_2Cl_6, Si_3Cl_8, SiF_4, Si_2F_6, SiI_4, and Si_2F_6. { 'sil·ə·kən 'ha,līd }

silicon hydride See silane. { 'sil·ə·kən 'hī,drīd }

silicon monoxide [INORG CHEM] SiO A hard, abrasive, amorphous solid used as thin surface films to protect optical parts, mirrors, and aluminum coatings. { 'sil·ə·kən mə'näk,sīd }

silicon nitride [INORG CHEM] Si_3N_4 A white, water-insoluble powder, resistant to thermal shock and to chemical reagents; used as a catalyst support and for stator blades of high-temperature gas turbines. { 'sil·ə·kən 'nī,trīd }

silicon tetrabromide [INORG CHEM] $SiBr_4$ A fuming, colorless liquid that yellows in air; disagreeable aroma; boils at 153°C. Also known as silicon bromide. { 'sil·ə·kən ¦te·trə'brō,mīd }

silicon tetrachloride [INORG CHEM] $SiCl_4$ A clear, corrosive, fuming liquid with suffocating aroma; decomposes in water and alcohol; boils at 57.6°C; used in warfare smoke screens, to make ethyl silicate and silicones, and as a source of pure silicon and silica. Also known as silicon chloride; tetrachlorosilane. { 'sil·ə·kən ¦te·trə'klȯr,īd }

silicon tetrafluoride [INORG CHEM] SiF_4 A colorless, suffocating gas absorbed readily by water, in which it decomposes; boiling point, −86°C; used in chemical analysis and to make fluosilicic acid. Also known as silicon fluoride. { 'sil·ə·kən ¦te·trə'flùr,īd }

siloxane [ORG CHEM] R_2SiO Any of a family of silica-based polymers in which R is an alkyl group, usually methyl; these polymers exist as oily liquids, greases, rubbers, resins, or plastics. Also known as oxosilane. { si'läk,sān }

silver [CHEM] A white metallic transition element, symbol Ag, with atomic number 47; soluble in acids and alkalies, insoluble in water; melts at 961°C, boils at 2212°C; used in photographic chemicals, alloys, conductors, and plating. { 'sil·vər }

silver acetate [ORG CHEM] CH_3COOAg A white powder, moderately soluble in water and nitric acid; used in medicine. { 'sil·vər 'as·ə,tāt }

silver acetylide [INORG CHEM] Ag_2C_2 A white explosive powder used in detonators. { 'sil·vər ə'sed·əl,īd }

silver arsenite [INORG CHEM] Ag_3AsO_3 A poisonous, light-sensitive, yellow powder; soluble in acids and alkalies, insoluble in water and alcohol; decomposes at 150°C; used in medicine. { 'sil·vər 'ärs·ən,īt }

silver bromate [INORG CHEM] $AgBrO_3$ A poisonous, light- and heat-sensitive, white powder; soluble in ammonium hydroxide, slightly soluble in hot water; decomposed by heat. { 'sil·vər 'brō,māt }

silver bromide [INORG CHEM] AgBr Yellowish, light-sensitive crystals; soluble in potassium bromide and potassium cyanide, slightly soluble in water; melts at 432°C; used in photographic films and plates. { 'sil·vər 'brō,mīd }

silver carbonate [INORG CHEM] Ag_2CO_3 Yellowish, light-sensitive crystals; insoluble in water and alcohol, soluble in alkalies and acids; decomposes at 220°C; used as a reagent. { 'sil·vər 'kär·bə·nət }

silver chloride [INORG CHEM] AgCl A white, poisonous, light-sensitive powder; slightly soluble in water, soluble in alkalies and acids; melts at 445°C; used in photography, photometry, silver plating, and medicine. { 'sil·vər 'klȯr,īd }

silver chromate [INORG CHEM] Ag_2CrO_4 Dark-colored crystals insoluble in water, soluble in acids and in solutions of alkali chromates; used as an analytical reagent. { 'sil·vər 'krō,māt }

silver cyanide [INORG CHEM] AgCN A poisonous, white, light-sensitive powder; insoluble in water, soluble in alkalies and acids; decomposes at 320°C; used in medicine and in silver plating. { 'sil·vər 'sī·ə,nīd }

silver fluoride [INORG CHEM] $AgF·H_2O$ A light-sensitive, yellow or brownish solid, soluble in water; dehydrated form melts at 435°C; used in medicine. Also known as tachiol. { 'sil·vər 'flùr,īd }

silver halide [INORG CHEM] A compound of silver and a halogen; for example, silver bromide (AgBr), silver chloride (AgCl), silver fluoride (AgF), and silver iodide (AgI). { 'sil·vər 'ha,līd }

silver iodate [INORG CHEM] AgIO₃ A white powder, soluble in ammonium hydroxide and nitric acid, slightly soluble in water; melts above 200°C; used in medicine. { 'sil·vər 'ī·ə,dāt }

silver iodide [INORG CHEM] AgI A pale-yellow powder, insoluble in water, soluble in potassium iodide-sodium chloride solutions and ammonium hydroxide; melts at 556°C; used in medicine, photography, and artificial rainmaking. { 'sil·vər 'ī·ə,dīd }

silver lactate [ORG CHEM] CH₃CHOHCOOAg·H₂O Gray-to-white, light-sensitive crystals; slightly soluble in water and in alcohol; used in medicine. Also known as actol. { 'sil·vər 'lak,tāt }

silver nitrate [INORG CHEM] AgNO₃ Poisonous, corrosive, colorless crystals; soluble in glycerol, water, and hot alcohol; melts at 212°C; used in external medicine, photography, hair dyeing, silver plating, ink manufacture, and mirror silvering, and as a chemical reagent. { 'sil·vər 'nī,trāt }

silver nitrite [INORG CHEM] AgNO₂ Yellow or grayish-yellow needles which decompose at 140°C; soluble in hot water; used in organic synthesis and in testing for alcohols. { 'sil·vər 'nī,trīt }

silver orthophosphate See silver phosphate. { 'sil·vər ¦ȯr·thō'fäs,fāt }

silver oxide [INORG CHEM] Ag₂O An odorless, dark-brown powder with a metallic taste; soluble in nitric acid and ammonium hydroxide, insoluble in alcohol; decomposes above 300°C; used in medicine and in glass polishing and coloring, as a catalyst, and to purify drinking water. { 'sil·vər 'äk,sīd }

silver permanganate [INORG CHEM] AgMnO₄ Water-soluble, violet crystals that decompose in alcohol; used in medicine and in gas masks. { 'sil·vər pər'maŋ·gə,nāt }

silver phosphate [INORG CHEM] Ag₃PO₄ A poisonous, yellow powder; darkens when heated or exposed to light; soluble in acids and in ammonium carbonate, very slightly soluble in water; melts at 849°C; used in photographic emulsions and in pharmaceuticals, and as a catalyst. Also known as silver orthophosphate. { 'sil·vər 'fä,sfāt }

silver picrate [ORG CHEM] C₆H₂O(NO₂)₃Ag·H₂O Yellow crystals, soluble in water, insoluble in ether and chloroform; used in medicine. { 'sil·vər 'pi,krāt }

silver potassium cyanide [ORG CHEM] KAg(CN)₂ Toxic, white crystals soluble in water and alcohol; used in silver plating and as a bactericide and antiseptic. Also known as potassium argentocyanide; potassium cyanoargentate. { 'sil·vər pə'tas·ē·əm 'sī·ə,nīd }

silver protein [ORG CHEM] A brown, hygroscopic powder containing 7.5–8.5% silver; made by reaction of a silver compound with gelatin in the presence of an alkali; used as an antibacterial. { 'sil·vər 'prō,tēn }

silver selenide [INORG CHEM] Ag₂Se A gray powder, insoluble in water, soluble in ammonium hydroxide; melts at 880°C. { 'sil·vər 'sel·ə,nīd }

silver suboxide [INORG CHEM] AgO A charcoal-gray powder that crystallizes in the cubic or orthorhombic system, and has diamagnetic properties; used in making silver oxide-zinc alkali batteries. Also known as argentic oxide. { 'sil·vər səb'äk,sīd }

silver sulfate [INORG CHEM] Ag₂SO₄ Light-sensitive, colorless, lustrous crystals; soluble in alkalies and acids, insoluble in alcohol; melts at 652°C; used as an analytical reagent. Also known as normal silver sulfate. { 'sil·vər 'səl,fāt }

silver sulfide [INORG CHEM] Ag₂S A dark, heavy powder, insoluble in water, soluble in concentrated sulfuric and nitric acids; melts at 825°C; used in ceramics and in inlay metalwork. { 'sil·vər 'səl,fīd }

silylene [CHEM] A divalent silicon species (R₂Si, with two nonbonding electrons, where R = alkyl, aryl, or hydrogen); analogous to a carbene in carbon chemistry. { sə'lī,lēn }

simple salt [CHEM] One of four classes of salts in a classification system that depends on the character of completeness of the ionization; examples are NaCl, NaHCO₃, and Pb(OH)Cl. { 'sim·pəl 'sȯlt }

single-replacement reaction |CHEM| A chemical reaction in which an element replaces one element in a compound. { ¦siŋ·gəl ri'plās·mənt rē,ak·shən }

singlet |SPECT| A spectral line that cannot be resolved into components at even the highest resolution. { 'siŋ·glət }

SIRS See satellite infrared spectrometer. { sərz }

skatole |ORG CHEM| C_9H_9N A white, crystalline compound that melts at 93–95°C, dissolves in hot water, and has an unpleasant feceslike odor. Also known as 3-methylindole. { 'ska,tōl }

Skraup synthesis |ORG CHEM| A method for the preparation of commercial synthetic quinoline by heating aniline and glycerol in the presence of sulfuric acid and an oxidizing agent to form pyridine unsubstituted quinolines. { skraúp 'sin·thə·səs }

slaked lime See calcium hydroxide. { 'slākt 'līm }

slitlet mask |SPECT| A metal plate used in astronomical spectroscopy, with several small slits at locations corresponding to astronomical objects of interest. { 'slit· lət ,mask }

Sm See samarium.

smectic-A |PHYS CHEM| A subclass of smectic liquid crystals in which molecules are free to move within layers and are oriented perpendicular to the layers. { 'smek· dik 'ā }

smectic-B |PHYS CHEM| A subclass of smectic liquid crystals in which molecules in each layer are arranged in a close-packed lattice and are oriented perpendicular to the layers. { 'smek·dik 'bē }

smectic-C |PHYS CHEM| A subclass of smectic liquid crystals in which molecules are free to move within layers and are oriented with their axes tilted with respect to the normal to the layers. { 'smek·dik 'sē }

smectic phase |PHYS CHEM| A form of the liquid crystal (mesomorphic) state in which molecules are arranged in layers that are free to glide over each other with relatively small viscosity. { 'smek·dik ,fāz }

smectogenic solid |PHYS CHEM| A solid which will form a smectic liquid crystal when heated. { ¦smek·tə¦jen·ik 'säl·əd }

smoldering |CHEM| Combustion of a solid without a flame, often with emission of smoke. { 'smōl·driŋ }

Sn See tin.

snow point |PHYS CHEM| Referring to a gas mixture, the temperature at which the vapor pressure of the sublimable component is equal to the actual partial pressure of that component in the gas mixture; analogous to dew point. { 'snō ,pȯint }

soda See sodium carbonate. { 'sōd·ə }

soda alum See aluminum sodium sulfate. { 'sōd·ə 'al·əm }

soda ash |INORG CHEM| Na_2CO_3 The commercial grade of sodium carbonate; a powder soluble in water, insoluble in alcohol; used in glass manufacture and petroleum refining, and for soaps and detergents. Also known as anhydrous sodium carbonate; calcined soda. { 'sōd·ə 'ash }

soda crystals See metahydrate sodium carbonate. { 'sōd·ə ,krist·əlz }

sodamide See sodium amide. { 'sōd·ə,mīd }

sodide |INORG CHEM| A member of a class of alkalides in which the metal anion is sodium (Na^-). { 'sä,dīd }

sodium |CHEM| A metallic element of group 1, symbol Na, with atomic number 11, atomic weight 22.98977; silver-white, soft, and malleable; oxidizes in air; melts at 97.6°C; used as a chemical intermediate and in pharmaceuticals, petroleum refining, and metallurgy; the source of the symbol Na is natrium. { 'sōd·ē·əm }

sodium acetate |ORG CHEM| $NaC_2H_3O_2$ Colorless, efflorescent crystals, soluble in water and ether; melts at 324°C; used as a chemical intermediate and for pharmaceuticals, dyes, and dry colors. { 'sōd·ē·əm 'as·ə,tāt }

sodium acid carbonate See sodium bicarbonate. { 'sōd·ē·əm 'as·əd 'kär·bə·nət }

sodium acid chromate See sodium dichromate. { 'sōd·ē·əm 'as·əd 'krō,māt }

sodium acid fluoride See sodium bifluoride. { 'sōd·ē·əm 'as·əd 'flür,īd }

sodium acid sulfate See sodium bisulfate. { 'sōd·ē·əm 'as·əd 'səl,fāt }

sodium acid sulfite

sodium acid sulfite *See* sodium bisulfite.   { 'sōd·ē·əm 'as·əd 'səl,fīt }

sodium alginate   [ORG CHEM] $C_6H_7O_6Na$ Colorless or light yellow filaments, granules, or powder which forms a viscous colloid in water; used in food thickeners and stabilizers, in medicine and textile printing, and for paper coating and water-base paint.   Also known as algin; alginic acid sodium salt; sodium polymannuronate.   { 'sōd·ē·əm 'al·jə,nāt }

sodium aluminate   [INORG CHEM] $Na_2Al_2O_4$ A white powder soluble in water, insoluble in alcohol; melts at 1800°C; used as a zeolite-type of material and a mordant, and in water purification, milkglass manufacture, and cleaning compounds.   { 'sōd·ē·əm ə'lü·mə,nāt }

sodium aluminosilicate   [INORG CHEM]   White, amorphous powder or beads of variable stoichiometry, partially soluble in strong acids and alkali hydroxide solutions between 80 and 100°C; used in food as an anticaking agent.   Also known as sodium silicoaluminate.   { 'sōd·ē·əm ə¦lü·mə·nō'sil·ə,kāt }

sodium aluminum phosphate   [INORG CHEM] $NaAl_3H_{14}(PO_4)_8·4H_2O$ or $Na_3Al_2H_{15}(PO_4)_8$ White powder, soluble in hydrochloric acid; used as a food additive for baked products.   { 'sōd·ē·əm ə'lü·mə·nəm 'fä,sfāt }

sodium aluminum silicofluoride   [INORG CHEM] $Na_5Al(SiF_6)_4$ A toxic, white powder, used for mothproofing and in insecticides.   { 'sōd·ē·əm ə'lü·mə·nəm ,sil·ə·kō'flür,īd }

sodium aluminum sulfate *See* aluminum sodium sulfate.   { 'sōd·ē·əm ə'lü·mə·nəm 'səl,fāt }

sodium amalgam   [INORG CHEM] Na_xHg_x A fire-hazardous, silver-white crystal mass that decomposes in water; used to make hydrogen and as an analytical reagent.   { 'sōd·ē·əm ə'mal·gəm }

sodium amide   [INORG CHEM] $NaNH_2$ White crystals that decompose in water; melts at 210°C; a fire hazard; used to make sodium cyanide.   Also known as sodamide.   { 'sōd·ē·əm 'am,īd }

sodium antimonate   [INORG CHEM] $NaSbO_3$ A white, granular powder, used as an enamel opacifier and high-temperature oxidizing agent.   Also known as antimony sodiate.   { 'sōd·ē·əm an'tim·ə,nāt }

sodium arsanilate   [ORG CHEM] $C_6H_4NH_2(AsO·OH·ONa)$ A white, water-soluble, poisonous powder with a faint saline taste, used in medicine and as a chemical intermediate.   { 'sōd·ē·əm är'san·əl,āt }

sodium arsenate   [INORG CHEM] $Na_3AsO_4·12H_2O$ Water-soluble, poisonous, clear, colorless crystals with a mild alkaline taste; melts at 86°C; used in medicine, insecticides, dry colors, and textiles, and as a germicide and a chemical intermediate.   { 'sōd·ē·əm 'ärs·ən,āt }

sodium arsenite   [INORG CHEM] $NaAsO_2$ A poisonous, water-soluble, grayish powder; used in antiseptics, dyeing, insecticides, and soaps for taxidermy.   { 'sōd·ē·əm 'ärs·ən,īt }

sodium ascorbate   [ORG CHEM] $CH_2OH(CHOH)_2COHCOHCOONa$ White, odorless crystals; soluble in water, insoluble in alcohol; decomposes at 218°C; used in therapy for vitamin C deficiency.   { 'sōd·ē·əm ə'skȯr,bāt }

sodium azide   [INORG CHEM] NaN_3 Poisonous, colorless crystals; soluble in water and liquid ammonia; decomposes at 300°C; used in medicine and to make lead azide explosives.   { 'sōd·ē·əm 'ā,zīd }

sodium barbiturate   [ORG CHEM] $C_4H_3N_2O_3Na$ White to slightly yellow powder, soluble in water and dilute mineral acid; used in wood-impregnating solutions.   { 'sōd·ē·əm bär'bich·ə·rət }

sodium benzoate   [ORG CHEM] $NaC_7H_5O_2$ Water- and alcohol-soluble, white, amorphous crystals with a sweetish taste; used as a food preservative and an antiseptic and in tobacco, pharmaceuticals, and medicine.   { 'sōd·ē·əm 'ben·zə,wāt }

sodium benzoylacetone dihydrate   [ORG CHEM]   A metal chelate with low melting point (115°C) and slight solubility in acetone.   { 'sōd·ē·əm ¦ben·zə,wil'as·ə,tōn dī'hī,drāt }

sodium bicarbonate   [INORG CHEM] $NaHCO_3$ White, water-soluble crystals with an alkaline taste; loses carbon dioxide at 270°C; used as a medicine and a butter preservative, in food preparation, in effervescent salts and beverages, in ceramics, and to prevent timber mold. Also known as baking soda; bicarbonate of soda; sodium acid carbonate. { 'sōd·ē·əm bī'kär·bə,nət }

sodium bichromate See sodium dichromate. { 'sōd·ē·əm bī'krō,māt }

sodium bifluoride [INORG CHEM] NaHF₂ Poisonous, water-soluble, white crystals; decomposes when heated; used as a laundry-rinse neutralizer, preservative, and antiseptic, and in glass etching and tinplating. Also known as sodium acid fluoride. { 'sōd·ē·əm bi'flúr·īd }

sodium bismuthate [INORG CHEM] NaBiO₃ A yellow to brown amorphous powder; used as an analytical reagent and in pharmaceuticals. { 'sōd·ē·əm 'biz·mə,thāt }

sodium bisulfate [INORG CHEM] NaHSO₄ Colorless crystals, soluble in water; the aqueous solution is strongly acidic; decomposes at 315°C; used for flux to decompose minerals, as a disinfectant, and in dyeing and manufacture of magnesia, cements, perfumes, brick, and glue. Also known as niter cake; sodium acid sulfate. { 'sōd·ē·əm bī'səl,fāt }

sodium bisulfide See sodium hydrosulfide. { 'sōd·ē·əm bī'səl,fīd }

sodium bisulfite [INORG CHEM] NaHSO₃ A colorless, water-soluble solid; decomposes when heated. Also known as sodium acid sulfite. { 'sōd·ē·əm bī'səl,fīt }

sodium bisulfite test [ANALY CHEM] A test for aldehydes in which aldehydes form a crystalline salt upon addition of a 40% aqueous solution of sodium bisulfite. { 'sōd·ē·əm bī'səl,fīt ,test }

sodium bitartrate [ORG CHEM] NaHC₄H₅O₆·H₂O A white, combustible, water-soluble powder that loses water at 100°C, decomposes at 219°C; used in effervescing mixtures and as an analytical reagent. Also known as acid sodium tartrate. { 'sōd·ē·əm bī'tär,trāt }

sodium borate [INORG CHEM] Na₂B₄O₇·10H₂O A water-soluble, odorless, white powder; melts between 75 and 200°C; used in glass, ceramics, starch and adhesives, detergents, agricultural chemicals, pharmaceuticals, and photography; the impure form is known as borax. Also known as sodium pyroborate; sodium tetraborate. { 'sōd·ē·əm 'bȯ,rāt }

sodium boroformate [ORG CHEM] NaH₂BO₃·2HCOOH·2H₂O Water-soluble, white crystals, melting at 110°C; used in textile treating and in tanning, and as a buffering agent. { 'sōd·ē·əm ,bȯr·ō'fȯr,māt }

sodium borohydride [INORG CHEM] NaBH₄ A flammable, hygroscopic, white to gray powder; soluble in water, insoluble in ether and hydrocarbons; decomposes in damp air; used as a hydrogen source, a chemical reagent, and a rubber foaming agent. { 'sōd·ē·əm ,bȯr·ō'hī,drīd }

sodium bromate [INORG CHEM] NaBrO₃ Odorless, white crystals; soluble in water, insoluble in alcohol; decomposes at 381°C; a fire hazard, used as an analytical reagent. { 'sōd·ē·əm 'brō,māt }

sodium bromide [INORG CHEM] NaBr White, water-soluble, crystals with a bitter, saline taste; absorbs moisture from air; melts at 758°C; used in photography and medicine, as a chemical intermediate, and to make bromides. { 'sōd·ē·əm 'brō,mīd }

sodium cacodylate [ORG CHEM] C₂H₆AsNaO₂ A herbicide used as a harvest aid. Also known as bollseye (trade name). { ,sōd·ē·əm ka'käd·əl,āt }

sodium carbolate See sodium phenate. { 'sōd·ē·əm 'kär·bə,lāt }

sodium carbonate [INORG CHEM] Na₂CO₃ A white, water-soluble powder that decomposes when heated to about 852°C; used as a reagent; forms a monohydrate compound, Na₂CO₃·H₂O, and a decahydrate compound, Na₂CO₃·10H₂O. Also known as soda. { 'sōd·ē·əm 'kär·bə·nət }

sodium carbonate decahydrate See sal soda. { 'sōd·ē·əm 'kär·bə·nət ,dek·ə'hī,drāt }

sodium carbonate peroxide [INORG CHEM] 2Na₂CO₃·3H₂O A white, crystalline powder; used in household detergents, in dental cleansers, and for bleaching and dyeing. { 'sōd·ē·əm 'kär·bə·nət pə'räk,sīd }

sodium caseinate [ORG CHEM] A tasteless, odorless, water-soluble, white powder; used in medicine, foods, emulsification, and stabilization; formed by dissolving casein in sodium hydroxide and then evaporating. Also known as casein sodium; nutrose. { 'sōd·ē·əm 'kā·sē·ə,nāt }

sodium chlorate

sodium chlorate [INORG CHEM] NaClO₃ Water- and alcohol-soluble, colorless crystals with a saline taste; melts at 255°C; used as a medicine, weed killer, defoliant, and oxidizing agent, and in matches, explosives, and bleaching.   { 'sōd·ē·əm 'klȯr‚āt }

sodium chloride [INORG CHEM] NaCl Colorless or white crystals; soluble in water and glycerol, slightly soluble in alcohol; melts at 804°C; used in foods and as a chemical intermediate and an analytical reagent.   Also known as common salt; table salt.   { 'sōd·ē·əm 'klȯr‚īd }

sodium chlorite [INORG CHEM] NaClO₂ An explosive, white, mildly hygroscopic, water-soluble powder; decomposes at 175°C; used as an analytical reagent and oxidizing agent.   { 'sōd·ē·əm 'klȯr‚īt }

sodium chloroacetate [ORG CHEM] ClCH₂COONa A white, water-soluble powder; used as a defoliant and in the manufacture of weed killers, dyes, and pharmaceuticals.   { 'sōd·ē·əm ‚klȯr·ō'as·ə‚tāt }

sodium chloroplatinate [INORG CHEM] Na₂PtCl₆·4H₂O A yellow powder, soluble in alcohol and water; used for zinc etching, indelible ink, plating, and mirrors, and in photography and medicine.   Also known as platinic sodium chloride; platinum sodium chloride; sodium platinichloride.   { 'sōd·ē·əm ‚klȯr·ō'plat·ən‚āt }

sodium chromate [INORG CHEM] Na₂CrO₄·10H₂O Water-soluble, translucent, yellow, efflorescent crystals that melt at 20°C; used as a rust preventive and in inks, dyeing, and leather tanning.   { 'sōd·ē·əm 'krō‚māt }

sodium citrate [ORG CHEM] C₆H₅Na₃O₇·2H₂O A white powder with the taste of salt; soluble in water, slightly soluble in alcohol; has an acid taste; loses water at 150°C; decomposes at red heat; used in medicine as an anticoagulant, in soft drinks, cheesemaking, and electroplating.   Also known as trisodium citrate.   { 'sōd·ē·əm 'sī‚trāt }

sodium cobaltinitrite [INORG CHEM] Na₃Co(NO₂)₆·¹/₂H₂O Purple, water-soluble, hygroscopic crystals; used as a reagent for analysis of potassium.   { 'sōd·ē·əm ‚kō‚bȯl·tə'nī‚trīt }

sodium cyanate [INORG CHEM] NaOCN A poisonous, white powder; soluble in water, insoluble in alcohol and ether; used as a chemical intermediate and for the manufacture of medicine and the heat-treating of steels.   { 'sōd·ē·əm 'sī·ə‚nāt }

sodium cyanide [INORG CHEM] NaCN A poisonous, water-soluble, white powder melting at 563°C; decomposes rapidly when standing; used to manufacture pigments, in heat treatment of metals, and as a silver- and gold-ore extractant.   { 'sōd·ē·əm 'sī·ə‚nīd }

sodium cyclamate [ORG CHEM] C₆H₁₁NHSO₃Na White, water-soluble crystals; sweetness 30 times that of sucrose; formerly used as an artificial sweetener for foods, but now prohibited.   { 'sōd·ē·əm 'sī·klə‚māt }

sodium dehydroacetate [ORG CHEM] C₈H₇NaO₄·H₂O A tasteless, white powder, soluble in water and propylene glycol; used as a fungicide and plasticizer, in toothpaste, and for pharmaceuticals.   { 'sōd·ē·əm dē‚hī·drō'as·ə‚tāt }

sodium diacetate [ORG CHEM] CH₃COONa·x(CH₃COOH) Combustible, white, water-soluble crystals with an acetic acid aroma; decomposes above 150°C; used to inhibit mold, and as a buffer, varnish hardener, sequestrant, and food preservative, and in mordants.   { 'sōd·ē·əm dī'as·ə‚tāt }

sodium diatrizoate [ORG CHEM] C₁₁H₈NO₄I₃Na White, water-soluble crystals which give a radiopaque solution; used in medicine as a radiopaque medium.   { 'sōd·ē·əm ‚dī·ə'trī·zə‚wāt }

sodium dichloroisocyanate [ORG CHEM] HC₃N₃O₃NaCl A white, crystalline compound, soluble in water; used as a bactericide and algicide in swimming pools.   { 'sōd·ē·əm dī‚klȯr·ō‚ī·sō'sī·ə‚nāt }

sodium dichloroisocyanurate [ORG CHEM] C₃N₃O₃Cl₂Na A white, crystalline powder; used in dry bleaches, detergents, and cleaning compounds, and for water and sewage treatment.   { 'sōd·ē·əm dī‚klȯr·ō‚ī·sō‚sī·ə'nȯr‚āt }

sodium dichromate [INORG CHEM] Na₂Cr₂O₇·2H₂O Poisonous, red to orange deliquescent crystals; soluble in water, insoluble in alcohol; melts at 320°C; loses water of hydration upon prolonged heating at 105°C; used as a chemical intermediate and

344

corrosion inhibitor and in the manufacture of pigments, leather tanning, and electroplating. Also known as bichromate of soda; sodium acid chromate; sodium bichromate. { 'sōd·ē·əm dī'krō,māt }

sodium diethyldithiocarbamate [ORG CHEM] $(C_2H_5)_2NCS_2Na$ A solid that is soluble in water and in alcohol; the trihydrate is used to determine small amounts of copper and to separate copper from other metals. { 'sōd·ē·əm dī¦eth·əl·dī,thī·ō'kär·bə,māt }

sodium dimethyldithiocarbamate [ORG CHEM] $(CH_3)_2NCS_2Na$ Amber to light green liquid; used as a fungicide, corrosion inhibitor, and rubber accelerator. Abbreviated SDDC. { 'sōd·ē·əm dī¦meth·əl·dī,thī·ō'kär·bə,māt }

sodium dinitro-*ortho*-cresylate [ORG CHEM] $CH_3C_6H_2(NO_2)_2ONa$ A toxic, orange-yellow dye, used as a herbicide and fungicide. { 'sōd·ē·əm dī¦nī,trō ¦ȯr·thō 'kres·ə,lat }

sodium dithionite *See* sodium hydrosulfite. { 'sōd·ē·əm dī'thī·ə,nīt }

sodium diuranate [ORG CHEM] $Na_2U_2O_7·6H_2O$ A yellow-orange solid, soluble in dilute acids; used for colored glazes on ceramics and in the manufacture of fluorescent uranium glass. Also known as uranium yellow. { 'sōd·ē·əm dī'yur·ə,nāt }

sodium dodecylbenzenesulfanate [ORG CHEM] $C_{18}H_{29}SO_3Na$ Biodegradable, white to yellow flakes, granules, or powder, used as a synthetic detergent. { 'sōd·ē·əm ¦dō·də,sil¦ben,zēn'səl·fə,nat }

sodium ethoxide *See* sodium ethylate. { 'sōd·ē·əm e'thäk,sīd }

sodium ethylate [ORG CHEM] C_2H_5ONa A white powder formed from ethanol by replacement of the hydroxyl groups' hydrogen by monovalent sodium; used in organic synthesis. Also known as caustic alcohol; sodium ethoxide. { 'sōd·ē·əm 'eth·ə,lāt }

sodium 2-ethylhexyl sulfoacetate [ORG CHEM] $C_{10}H_{19}O_2SO_3Na$ Cream-colored, water-soluble flakes, used as a stabilizing agent in soapless shampoos. { 'sōd·ē·əm ¦tü ¦eth·əl'hek·səl ,səl·fō'as·ə,tāt }

sodium ethylxanthate [ORG CHEM] $C_2H_5OC(S)SNa$ A yellowish powder, soluble in water and alcohol; used as an ore flotation agent. Also known as sodium xanthate; sodium xanthogenate. { 'sōd·ē·əm ,eth·əl'zan,thāt }

sodium ferricyanide [INORG CHEM] $Na_3Fe(CN)_6·H_2O$ A poisonous, deliquescent, red powder; soluble in water, insoluble in alcohol; used in printing and for the manufacture of pigments. Also known as red prussiate of soda. { 'sōd·ē·əm ,fer·ə'sī·ə,nīd }

sodium ferrocyanide [INORG CHEM] $Na_4Fe(CN)_6·10H_2O$ Semitransparent crystals, soluble in water; insoluble in alcohol; used in photography, dyes, tanning, and blueprint paper. Also known as yellow prussiate of soda. { 'sōd·ē·əm ,fer·ə'sī·ə,nīd }

sodium fluoborate [INORG CHEM] $NaBF_4$ A white powder with a bitter taste; soluble in water, slightly soluble in alcohol; decomposes when heated, fuses below 500°C; used in electrochemical processes, as flux for nonferrous metals refining, and as an oxidation inhibitor. { 'sōd·ē·əm ,flü·ə'bȯr,āt }

sodium fluorescein *See* uranine. { 'sōd·ē·əm flü'res·ē·ən }

sodium fluoride [INORG CHEM] NaF A poisonous, water-soluble, white powder, melting at 988°C; used as an insecticide and a wood and adhesive preservative, and in fungicides, vitreous enamels, and dentistry. { 'sōd·ē·əm 'flur,īd }

sodium fluoroacetate [ORG CHEM] $C_2H_2FO_2Na$ A white powder, hygroscopic and nonvolatile; decomposes at 200°C; very soluble in water; used as a repellent for rodents and predatory animals. { 'sōd·ē·əm ,flur·ō'as·ə,tāt }

sodium fluosilicate [INORG CHEM] Na_2SiF_6 A poisonous, white, amorphous powder; slightly soluble in water; decomposes at red heat; used to fluoridate drinking water and to kill rodents and insects. Also known as sodium silicofluoride. { 'sōd·ē·əm ,flü·ə'sil·ə·kət }

sodium folate [ORG CHEM] $C_{19}H_{18}N_7NaO_6$ A yellow to yellow-orange liquid; used in medicine for folic acid deficiency. Also known as folic acid sodium salt. { 'sōd·ē·əm 'fō,lāt }

sodium formaldehyde sulfoxylate [ORG CHEM] $HCHO·HSO_2Na·2H_2O$ A white solid with a melting point of 64°C; soluble in water and alcohol; used as a textile stripping agent and a bleaching agent for soap and molasses. Abbreviated SFS. { 'sōd·ē·əm fȯr'mal·də,hīd səl'fäk·sə,lāt }

sodium formate [ORG CHEM] HCOONa A mildly hygroscopic, white powder, soluble in

sodium glucoheptonate

water; has a formic acid aroma; melts at 245°C; used in medicine and as a chemical intermediate and reducing agent. { 'sōd·ē·əm 'fȯr,māt }

sodium glucoheptonate [ORG CHEM] $C_7H_{13}O_8Na$ A light tan, crystalline powder; used for cleaning metal, mercerizing, paint stripping, and aluminum etching. { 'sōd·ē·əm ¦glü·kō'hep·tə,nāt }

sodium gluconate [ORG CHEM] $C_6H_{11}NaO_7$ A water-soluble, yellow to white, crystalline powder, produced by fermentation; used in food and pharmaceutical industries, and as a metal cleaner. Also known as gluconic acid sodium salt. { 'sōd·ē·əm 'glü·kə,nāt }

sodium glutamate [ORG CHEM] $COOH(CH_2)_2CH(NH_2)COONa$ A salt of an amino acid; a white powder, soluble in water and alcohol; used as a taste enhancer. Also known as monosodium glutamate (MSG). { 'sōd·ē·əm 'glüd·ə,māt }

sodium gold chloride [INORG CHEM] $NaAuCl_4·2H_2O$ Yellow crystals, soluble in water and alcohol; used in photography, fine glass staining, porcelain decorating, and medicine. Also known as gold salt; gold sodium chloride. { 'sōd·ē·əm 'gōld 'klȯr,īd }

sodium gold cyanide [ORG CHEM] $NaAu(CN)_2$ A yellow, water-soluble powder; used for gold plating radar and electric parts, jewelry, and tableware. Also known as gold sodium cyanide. { 'sōd·ē·əm 'gōld 'sī·ə,nīd }

sodium halide [INORG CHEM] A compound of sodium with a halogen; for example, sodium bromide (NaBr), sodium chloride (NaCl), sodium iodide (NaI), and sodium fluoride (NaF). { 'sōd·ē·əm 'ha,līd }

sodium halometallate [INORG CHEM] A compound of sodium with halogen and a metal; for example, sodium platinichloride, $Na_2PtCl_6·6H_2O$. { 'sōd·ē·əm ,hal·ō'med·əl,āt }

sodium hexylene glycol monoborate [ORG CHEM] $C_6H_{12}O_3BNa$ An amorphous, white solid with a melting point of 426°C; used as a corrosion inhibitor, flame retardant, and lubricating-oil additive. { 'sōd·ē·əm 'hek·sə,lēn 'glī,kȯl ¦män·ə'bȯr,āt }

sodium hydrate See sodium hydroxide. { 'sōd·ē·əm 'hī,drāt }

sodium hydride [INORG CHEM] NaH A white powder, decomposed by water, and igniting in moist air; used to make sodium borohydride and as a drying agent and a reagent. { 'sōd·ē·əm 'hī,drīd }

sodium hydrogen phosphate [INORG CHEM] $NaH_2PO_4·H_2O$ Hygroscopic, transparent, water-soluble crystals; used as a purgative, reagent, and buffer. { 'sōd·ē·əm 'hī·drə·jən 'fä,sfāt }

sodium hydrogen sulfide See sodium hydrosulfide. { 'sōd·ē·əm 'hī·drə·jən 'səl,fīd }

sodium hydrosulfide [INORG CHEM] $NaSH·2H_2O$ Toxic, colorless, water-soluble needles, melting at 55°C; used in pulping of paper, processing dyestuffs, hide dehairing, and bleaching. Also known as sodium bisulfide; sodium hydrogen sulfide; sodium sulfhydrate. { 'sōd·ē·əm ,hī·drə'səl,fīd }

sodium hydrosulfite [INORG CHEM] $Na_2S_2O_4$ A fire-hazardous, lemon to whitish-gray powder; soluble in water, insoluble in alcohol; melts at 55°C; used as a chemical intermediate and catalyst and in ore flotation. Also known as sodium dithionite. { 'sōd·ē·əm ,hī·drə'səl,fīt }

sodium hydroxide [INORG CHEM] NaOH White, deliquescent crystals; absorbs carbon dioxide and water from air; soluble in water, alcohol, and glycerol; melts at 318°C; used as an analytical reagent and chemical intermediate, in rubber reclaiming and petroleum refining, and in detergents. Also known as sodium hydrate. { 'sōd·ē·əm hī'dräk,sīd }

sodium hypochlorite [INORG CHEM] NaOCl Air-unstable, pale-green crystals with sweet aroma; soluble in cold water, decomposes in hot water; used as a bleaching agent for paper pulp and textiles, as a chemical intermediate, and in medicine. { 'sōd·ē·əm ¦hī·pō'klȯr,īt }

sodium hypophosphite [INORG CHEM] $NaH_2PO_2·H_2O$ Colorless, pearly, water-soluble crystalline plates or a white, granular powder; used in medicine and electroless nickel plating of plastic and metal. { 'sōd·ē·əm ¦hī·pō'fä,sfīt }

sodium hyposulfite See sodium thiosulfate. { 'sōd·ē·əm ¦hī·pō'səl,fīt }

346

sodium iodate [INORG CHEM] NaIO₃ A white, water- and acetone-soluble powder; used as a disinfectant and in medicine. { 'sōd·ē·əm 'ī·ə,dāt }

sodium iodide [INORG CHEM] NaI A white, air-sensitive powder, deliquescent, with bitter taste; soluble in water, alcohol, and glycerin; melts at 653°C; used in photography and in medicine and as an analytical reagent. { 'sōd·ē·əm 'ī·ə,dīd }

sodium isopropylxanthate [ORG CHEM] C₅H₇ONaS₂ Light yellow, crystalline compound that decomposes at 150°C; soluble in water; used as a postemergence herbicide and as a flotation agent for ores. { 'sōd·ē·əm ¦ī·sə¦prō·pəl'zan,thāt }

sodium lactate [ORG CHEM] CH₃CHOHCOONa A water-soluble, hygroscopic, yellow to colorless, syrupy liquid; solidifies at 17°C; used in medicine, as a corrosion inhibitor in antifreeze, and a hygroscopic agent. { 'sōd·ē·əm 'lak,tāt }

sodium lauryl sulfate [ORG CHEM] CH₃(CH₂)₁₀CH₂OSO₃Na A water-soluble salt, produced as a white or cream powder, crystals, or flakes; used in the textile industry as a wetting agent and detergent. Also known as dodecyl sodium sulfate. { 'sōd·ē·əm 'lȯr·əl 'səl,fāt }

sodium lead hyposulfate See lead sodium thiosulfate. { 'sōd·ē·əm 'led ¦hī·pō'səl,fāt }

sodium lead thiosulfate See lead sodium thiosulfate. { 'sōd·ē·əm 'led ¦thī·ə'səl,fāt }

sodium metaborate [INORG CHEM] NaBO₂ Water-soluble, white crystals, melting at 966°C; the aqueous solution is alkaline; made by fusing sodium carbonate with borax; used as an herbicide. { 'sōd·ē·əm ¦med·ə'bȯr,āt }

sodium metaphosphate [INORG CHEM] (NaPO₃)ₓ Sodium phosphate groupings; cyclic forms range from x = 3 for the trimetaphosphate, to x = 10 for the decametaphosphate; sodium hexametaphosphate with x = 10 to 20 is probably a polymer; used for dental polishing, building detergents, and water softening, and as a sequestrant, emulsifier, and food additive. { 'sōd·ē·əm ¦med·ə'fä,sfāt }

sodium metasilicate See sodium silicate. { 'sōd·ē·əm ¦med·ə'sil·ə,kāt }

sodium metavanadate [INORG CHEM] NaVO₃ Colorless crystals or a pale green, crystalline powder with a melting point of 630°C; soluble in water; used in inks, fur dyeing, and photography, and as a corrosion inhibitor in gas scrubbers. { 'sōd·ē·əm ¦med·ə'van·ə,dāt }

sodium methiodal [ORG CHEM] ICH₂SO₃Na A white, crystalline powder, soluble in water and methanol; used in medicine as a radiopaque medium. { 'sōd·ē·əm me'thī·ə,dal }

sodium methoxide [ORG CHEM] CH₃ONa A salt produced as a free-flowing powder, soluble in methanol and ethanol; used as an intermediate in organic synthesis. Also known as sodium methylate. { 'sōd·ē·əm me'thäk,sīd }

sodium methylate See sodium methoxide. { 'sōd·ē·əm 'meth·ə,lāt }

sodium N-methyldithiocarbamate dihydrate [ORG CHEM] CH₃NHC(S)SNa·2H₂O A white, water-soluble, crystalline solid; used as a fungicide, insecticide, nematicide, and weed killer. { 'sōd·ē·əm ¦en ¦meth·əl·dī¦thī·ə'kär·bə,māt dī'hī,drāt }

sodium molybdate [INORG CHEM] Na₂MoO₄ Water-soluble crystals, melting at 687°C; used as an analytical reagent, corrosion inhibitor, catalyst, and zinc-plating brightening agent, and in medicine. { 'sōd·ē·əm mə'lib,dāt }

sodium 12-molybdophosphate [INORG CHEM] Na₃PMo₁₂O₄₀ Yellow, water-soluble crystals; used in neuromicroscopy and photography, and as a water-resisting agent in plastic adhesives and cements. { 'sōd·ē·əm ¦twelv mə,lib·dō'fä,sfāt }

sodium monoxide [INORG CHEM] Na₂O A strong basic white powder soluble in molten caustic soda; forms sodium hydroxide in water; used as a dehydrating and polymerization agent. Also known as sodium oxide. { 'sōd·ē·əm mə'näk,sīd }

sodium naphthalenesulfonate [ORG CHEM] C₁₀H₇SO₃Na Yellow, water-soluble crystalline plates or white scales; used as a liquefying agent in animal glue. { 'sōd·ē·əm ¦naf·thə,lēn'səl·fə,nāt }

sodium naphthionate [ORG CHEM] NaC₁₀H₆(NH₂)SO₃·4H₂O White, light-sensitive crystals, soluble in water and insoluble in ether; used in analysis (Riegler's reagent) for nitrous acid. { 'sōd·ē·əm naf'thī·ə,nāt }

sodium nitrate [INORG CHEM] NaNO₃ Fire-hazardous, transparent, colorless crystals with bitter taste; soluble in glycerol and water; melts at 308°C; decomposes when

sodium nitrite

heated; used in manufacture of glass and pottery enamel and as a fertilizer and food preservative. { 'sōd·ē·əm 'nī,trāt }

sodium nitrite [INORG CHEM] $NaNO_2$ A fire-hazardous, air-sensitive, yellowish powder, soluble in water; decomposes above 320°C; used as an intermediate for dyestuffs and for pickling meat, textiles dyeing, and rust-proofing, and in medicine. { 'sōd· ē·əm 'nī,trīt }

sodium nitroferricyanide [INORG CHEM] $Na_2Fe(CN)_5NO\cdot2H_2O$ Water-soluble, transparent, reddish crystals; slowly decomposes in water; used as an analytical reagent. { 'sōd·ē·əm ¦nī·trō,fer·ə'sī·ə,nīd }

sodium oleate [ORG CHEM] $C_{17}H_{33}COONa$ A white powder with a tallow aroma; soluble in alcohol and water, with partial decomposition; used in medicine and textile waterproofing. { 'sōd·ē·əm 'ō·lē,āt }

sodium oxalate [ORG CHEM] $Na_2C_2O_4$ A poisonous, white powder; soluble in water, insoluble in alcohol; used for leather tanning and as an analytical reagent. { 'sōd· ē·əm 'äk·sə,lāt }

sodium oxide See sodium monoxide. { 'sōd·ē·əm 'äk,sīd }

sodium paraperiodate [INORG CHEM] $Na_3H_2IO_6$ White, crystalline solid, soluble in concentrated sodium hydroxide solutions; used to wet-strengthen paper and to aid in tobacco combustion. { 'sōd·ē·əm ¦par·ə·pər'ī·ə,dāt }

sodium pentaborate [INORG CHEM] $Na_2B_{10}O_{16}\cdot10H_2O$ A white, water-soluble powder; used in glassmaking, weed killers, and fireproofing compositions. { 'sōd·ē·əm ˌpen·tə'bòr,āt }

sodium pentachlorophenate [ORG CHEM] C_6Cl_5ONa A white or tan powder, soluble in water, ethanol, and acetone; used as a fungicide and herbicide. { 'sōd·ē·əm ¦pen· tə,klòr·ə'fe,nāt }

sodium perborate [INORG CHEM] $NaBO_2\cdot H_2O_2\cdot3H_2O$ A white powder with a saline taste; slightly soluble in water, decomposes in moist air; used in deodorants, in dental compositions, and as a germicide. Also known as peroxydol. { 'sōd·ē·əm pər'bòr,āt }

sodium perchlorate [INORG CHEM] $NaClO_4$ Fire-hazardous, white, deliquescent crystals; soluble in water and alcohol; melts at 482°C; explosive when in contact with concentrated sulfuric acid; used in jet fuel, as an analytical reagent, and for explosives. { 'sōd·ē·əm pər'klòr,āt }

sodium permanganate [INORG CHEM] $NaMnO_4\cdot3H_2O$ A fire-hazardous, water-soluble, purple powder; decomposes when heated; used to make saccharin, as a disinfectant, and as an oxidizing agent. { 'sōd·ē·əm pər'maŋ·gə,nāt }

sodium peroxide [INORG CHEM] Na_2O_2 A fire-hazardous, white powder that yellows with heating; decomposes when heated; causes ignition when in contact with water; used as an oxidizing agent and a bleach, and in medicinal soap. { 'sōd·ē·əm pə'räk,sīd }

sodium persulfate [INORG CHEM] $Na_2S_2O_8$ A white, water-soluble, crystalline powder; used as a bleaching agent and in medicine. { 'sōd·ē·əm pər'səl,fāt }

sodium phenate [ORG CHEM] C_6H_5ONa White, deliquescent crystals, soluble in water and alcohol; decomposed by carbon dioxide in air; used as a chemical intermediate, antiseptic, and military gas absorbent. Also known as sodium carbolate; sodium phenolate. { 'sōd·ē·əm 'fe,nāt }

sodium phenolate See sodium phenate. { 'sōd·ē·əm 'fen·əl,āt }

sodium phenylacetate [ORG CHEM] $C_6H_5CH_2\cdot COONa$ Pale yellow, 50% aqueous solution which crystallizes at 15°C; used in the manufacture of penicillin G. { 'sōd·ē· əm ˌfen·əl'as·ə,tāt }

sodium phenylphosphinate [ORG CHEM] $C_6H_5PH(O)(ONa)$ Crystals with a melting point of 355°C; used as an antioxidant and as a heat and light stabilizer. { 'sōd·ē·əm ˌfen·əl'fä·sfə,nāt }

sodium phosphate [INORG CHEM] A general term encompassing the following compounds: sodium hexametaphosphate, sodium metaphosphate, dibasic sodium phosphate, hemibasic sodium phosphate, monobasic sodium phosphate, tribasic sodium phosphate, sodium pyrophosphate, and acid sodium pyrophosphate. { 'sōd·ē·əm 'fä,sfāt }

sodium phosphite [INORG CHEM] $Na_2HPO_3 \cdot 5H_2O$ White, hygroscopic crystals, melting at 53°C; soluble in water, insoluble in alcohol; used in medicine. { 'sōd·ē·əm 'fä,sfīt }

sodium phosphotungstate *See* sodium tungstophosphate. { 'sōd·ē·əm ,fä·sfō'təŋ,stāt }

sodium phytate [ORG CHEM] $C_6H_9O_{24}P_6Na_9$ A hygroscopic, water-soluble powder; used as a chelating agent for trace metals and in medicine. { 'sōd·ē·əm 'fī,tāt }

sodium picramate [ORG CHEM] $NaOC_6H_2(NO_2)_2NH_2$ A yellow salt, soluble in water; used in the manufacture of dye intermediates. { 'sōd·ē·əm 'pik·rə,māt }

sodium platinichloride *See* sodium chloroplatinate. { 'sōd·ē·əm ¦plat·ən·ə'klȯr,īd }

sodium plumbite [INORG CHEM] $Na_2PbO_2 \cdot 3H_2O$ A toxic, corrosive solution of lead oxide (litharge) in sodium hydroxide; used (as doctor solution) to sweeten gasoline. { 'sōd·ē·əm 'pləm,bīt }

sodium polymannuronate *See* sodium alginate. { 'sōd·ē·əm ¦päl·i·mə'nyür·ə,nāt }

sodium polysulfide [INORG CHEM] Na_2S_x Yellow-brown granules, used to make dyes and colors, and insecticides, as a petroleum additive, and in electroplating. { 'sōd·ē·əm ¦päl·i'səl,fīd }

sodium propionate [ORG CHEM] CH_3CH_2COONa Deliquescent, transparent crystals; soluble in water, slightly soluble in alcohol; used as a fungicide, and mold preventive. { 'sōd·ē·əm 'prō·pē·ə,nāt }

sodium pyroborate *See* sodium borate. { 'sōd·ē·əm ¦pī·rō'bȯr,āt }

sodium pyrophosphate [INORG CHEM] $Na_4P_2O_7$ A white powder; soluble in water, insoluble in alcohol and ammonia; melts at 880°C; used as a water softener and newsprint deinker, and to control drilling-mud viscosity. Also known as normal sodium pyrophosphate; tetrasodium pyrophosphate (TSPP). { 'sōd·ē·əm ¦pī·rō'fä,sfāt }

sodium saccharin [ORG CHEM] $C_7H_4NNaO_3S \cdot 2H_2O$ White crystals or a crystalline powder, soluble in water and slightly soluble in alcohol; used in medicine and as a nonnutritive food sweetener. { 'sōd·ē·əm 'sak·ə·rən }

sodium salicylate [ORG CHEM] HOC_6H_4COONa A shiny, white powder with sweetish taste and mild aromatic aroma; soluble in water, glycerol, and alcohol; used in medicine and as a preservative. { 'sōd·ē·əm sə'lis·ə,lāt }

sodium selenate [INORG CHEM] $Na_2SeO_4 \cdot 10H_2O$ White, poisonous, water-soluble crystals; used as an insecticide. { 'sōd·ē·əm 'sel·ə,nāt }

sodium selenite [INORG CHEM] $Na_2SeO_3 \cdot 5H_2O$ White, water-soluble crystals; used in glass manufacture, as a bacteriological reagent, and for decorating porcelain. { 'sōd·ē·əm 'sel·ə,nīt }

sodium sesquicarbonate [INORG CHEM] $Na_2CO_3 \cdot NaHCO_3 \cdot 2H_2O$ White, water-soluble, needle-shaped crystals; used as a detergent, an alkaline agent for water softening and leather tanning, and a food additive. { 'sōd·ē·əm ¦ses·kwē'kär·bə,nāt }

sodium sesquisilicate [INORG CHEM] $Na_6Si_2O_7$ A white, water-soluble powder; used for metals cleaning and textile processing. { 'sōd·ē·əm ¦ses·kwē'sil·ə,kāt }

sodium silicate [INORG CHEM] Na_2SiO_3 A gray-white powder; soluble in alkalies and water, insoluble in alcohol and acids; used to fireproof textiles, in petroleum refining and corrugated paperboard manufacture, and as an egg preservative. Also known as liquid glass; silicate of soda; sodium metasilicate; soluble glass; water glass. { 'sōd·ē·əm 'sil·ə,kāt }

sodium silicoaluminate *See* sodium aluminosilicate. { 'sōd·ē·əm ¦sil·ə·kō·ə'lü·mə,nāt }

sodium silicofluoride *See* sodium fluosilicate. { 'sōd·ē·əm ¦sil·ə·kō'flür,īd }

sodium stannate [INORG CHEM] $Na_2SnO_3 \cdot 3H_2O$ Water- and alcohol-insoluble, whitish crystals; used in ceramics, dyeing, and textile fireproofing, and as a mordant. Also known as preparing salt. { 'sōd·ē·əm 'sta,nāt }

sodium stearate [ORG CHEM] $NaC_{18}H_{35}O_2$ A white powder with a fatty aroma; soluble in hot water and alcohol; used in medicine and toothpaste and as a waterproofing agent. { 'sōd·ē·əm 'stir,āt }

sodium subsulfite *See* sodium thiosulfate. { 'sōd·ē·əm ¦səb'səl,fīt }

sodium succinate [ORG CHEM] $Na_2C_4H_4O_4 \cdot 6H_2O$ Water-soluble, white crystals; loses water at 120°C; used in medicine. { 'sōd·ē·əm 'sək·sə,nāt }

sodium sulfate

sodium sulfate [INORG CHEM] Na$_2$SO$_4$ Crystalline compound, melts at 888°C, soluble in water; used to make paperboard, kraft paper, glass, and freezing mixtures.   { 'sōd·ē·əm 'səl‚fāt }

sodium sulfhydrate See sodium hydrosulfide.   { 'sōd·ē·əm ‚səlf'hī‚drāt }

sodium sulfide [INORG CHEM] Na$_2$S An irritating, water-soluble, yellow to red, deliquescent powder; melts at 1180°C; used as a chemical intermediate, solvent, photographic reagent, and analytical reagent. Also known as sodium sulfuret.   { 'sōd·ē·əm 'səl‚fīd }

sodium sulfite [INORG CHEM] Na$_2$SO$_3$ White, water-soluble, crystals with a sulfurous, salty taste; decomposes when heated; used as a chemical intermediate and food preservative, in medicine and paper manufacturing, and for dyes and photographic developing.   { 'sōd·ē·əm 'səl‚fīt }

sodium sulfocyanate See sodium thiocyanate.   { 'sōd·ē·əm ‚səl·fō'sī·ə‚nāt }

sodium sulfuret See sodium sulfide.   { 'sōd·ē·əm 'səl·fyə‚ret }

sodium tartrate [ORG CHEM] Na$_2$C$_4$H$_4$O$_6$·2H$_2$O White, water-soluble crystals or granules; loses water at 150°C; used in medicine and as a food stabilizer and sequestrant. Also known as disodium tartrate; sal tartar.   { 'sōd·ē·əm 'tär‚trāt }

sodium TCA See sodium trichloroacetate.   { 'sōd·ē·əm ‚tē‚sē'ā }

sodium tetraborate See sodium borate.   { 'sōd·ē·əm ‚te·trə'bór‚āt }

sodium tetrafluorescein See easin.   { 'sōd·ē·əm ‚te·trə'flùr·ə‚sēn }

sodium tetraphenylborate [ORG CHEM] [(C$_6$H$_5$)$_4$B]Na A snow-white, crystalline compound, soluble in water and acetone; used as a reagent in the determination of the following ions: potassium, ammonium, rubidium, and cesium.   { 'sōd·ē·əm ‚tre·trə‚fen·əl'bór‚āt }

sodium tetrasulfide [INORG CHEM] Na$_2$S$_4$ Hygroscopic, yellow or dark-red crystals, melting at 275°C; used for insecticides and fungicides, ore flotation, and dye manufacture, and as a reducing agent.   { 'sōd·ē·əm ‚tet·rə'səl‚fīd }

sodium thiocyanate [INORG CHEM] NaSCN A poisonous, water- and alcohol-soluble, deliquescent, white powder; melts at 287°C; used as an analytical reagent, solvent, and chemical intermediate, and for rubber treatment and textile dyeing and printing. Also known as sodium sulfocyanate.   { 'sōd·ē·əm ‚thī·ə'sī·ə‚nāt }

sodium thioglycolate [ORG CHEM] C$_2$H$_3$NaO$_3$S A water-soluble compound produced as hygroscopic crystals; used as an ingredient in bacteriology media, and in hair-waving solutions.   { 'sōd·ē·əm ‚thī·ə'glī·kə‚lāt }

sodium thiosulfate [INORG CHEM] Na$_2$S$_2$O$_3$·5H$_2$O White, translucent crystals or powder with a melting point of 48°C; soluble in water and oil of turpentine; used as a fixing agent in photography, for extracting silver from ore, in medicine, and as a sequestrant in food. Also known as sodium hyposulfite; sodium subsulfite.   { 'sōd·ē·əm ‚thī·ə'səl‚fāt }

sodium trichloroacetate [ORG CHEM] CCl$_3$COONa A toxic material, used in herbicides and pesticides.   Abbreviated sodium TCA.   { 'sōd·ē·əm trī‚klòr·ō'as·ə‚tāt }

sodium 2,4,5-trichlorophenate [ORG CHEM] C$_6$H$_2$Cl$_3$ONa·1^1/$_2$H$_2$O Buff to light brown flakes, soluble in water, methanol, and acetone; used as a bactericide and fungicide.   { 'sōd·ē·əm ‚tü ‚fòr ‚fīv trī‚klòrō'fe‚nāt }

sodium tripolyphosphate [INORG CHEM] Na$_5$P$_3$O$_{10}$ A white powder with a melting point of 622°C; used for water softening and as a food additive and texturizer.   Abbreviated STPP.   { 'sōd·ē·əm trī‚päl·i'fä‚sfāt }

sodium tungstate [INORG CHEM] Na$_2$WO$_4$·2H$_2$O Water-soluble, colorless crystals; lose water at 100°C, melts at 692°C; used as a chemical intermediate analytical reagent, and for fireproofing. Also known as sodium wolframate.   { 'sōd·ē·əm 'təŋ‚stāt }

sodium tungstophosphate [INORG CHEM] Approximately 2Na$_2$O·P$_2$O$_5$·12WO$_3$·18H$_2$O A yellowish-white powder, soluble in water and alcohols; used to manufacture organic pigments, as an antistatic agent for textiles, in leather tanning, and as a water-resistant agent in plastic films, adhesives, and cements. Also known as sodium phosphotungstate.   { 'sōd·ē·əm ‚twelv ‚təŋ·stō'fä‚sfāt }

sodium undecylenate [ORG CHEM] C$_{11}$H$_{19}$O$_2$Na A white, water-soluble powder that

350

decomposes above 200°C; used in cosmetics and pharmaceuticals as a bacteriostat and fungistat.   { 'sōd·ē·əm ˌən‚de·sə'le‚nāt }

sodium wolframate See sodium tungstate.   { 'sōd·ē·əm 'wúl·frə‚mīt }

sodium xanthate See sodium ethylxanthate.   { 'sōd·ē·əm 'zan‚thāt }

sodium xanthogenate See sodium ethylxanthate.   { 'sōd·ē·əm zan'thä·jə‚nāt }

soft electrophile   [PHYS CHEM] A molecule that readily accepts electrons during a primary reaction step.   { 'sóft i'lek·trə‚fīl }

soft water   [CHEM] Water that is free of magnesium or calcium salts.   { 'sóft 'wód·ər }

soft-x-ray absorption spectroscopy   [SPECT] A spectroscopic technique which is used to get information about unoccupied states above the Fermi level in a metal or about empty conduction bands in an inoculator.   { 'sóft ¦eks‚ra əb'sōrp·shən spek'träs·kə·pē }

soft-x-ray appearance potential spectroscopy   [SPECT] A branch of electron spectroscopy in which a solid surface is bombarded with monochromatic electrons, and small but abrupt changes in the resulting total x-ray emission intensity are detected as the energy of the electrons is varied.   Abbreviated SXAPS.   { 'sóft ¦eks‚ra ə'pir·əns pə¦ten·chəl spek'träs·kə·pē }

sol   [CHEM] A colloidal solution consisting of a suitable dispersion medium, which may be gas, liquid, or solid, and the colloidal substance, the disperse phase, which is distributed throughout the dispersion medium.   { säl }

solation   [PHYS CHEM] The change of a substance from a gel to a sol.   { sə'lā·shən }

sol-gel glass   [PHYS CHEM] An optically transparent amorphous silica or silicate material produced by forming interconnections in a network of colloidal, submicrometer particles under increasing viscosity until the network becomes completely rigid, with about one-half the density of glass.   { säl 'jel 'glas }

solid-liquid equilibrium   [PHYS CHEM] **1.** The interrelation of a solid material and its melt at constant vapor pressure.   **2.** The concentration relationship of a solid with a solvent liquid other than its melt.   Also known as liquid-solid equilibrium.   { 'säl·əd 'lik·wəd ‚ē·kwə'lib·rē·əm }

solidus   [PHYS CHEM] In a constitution or equilibrium diagram, the locus of points representing the temperature below which the various compositions finish freezing on cooling, or begin to melt on heating.   { 'säl·əd·əs }

solidus curve   [PHYS CHEM] A curve on the phase diagram of a system with two components which represents the equilibrium between the liquid phase and the solid phase.   { 'säl·əd·əs ‚kərv }

soliquid   [PHYS CHEM] A system in which solid particles are dispersed in a liquid.   { ¦sä'lik·wəd }

solubility   [PHYS CHEM] The ability of a substance to form a solution with another substance.   { ‚säl·yə'bil·əd·ē }

solubility coefficient   [PHYS CHEM] The volume of a gas that can be dissolved by a unit volume of solvent at a specified pressure and temperature.   { ‚säl·yə'bil·əd·ē ‚kō·i‚fish·ənt }

solubility curve   [PHYS CHEM] A graph showing the concentration of a substance in its saturated solution in a solvent as a function of temperature.   { ‚säl·yə'bil·əd·ē ‚kərv }

solubility product constant   [PHYS CHEM] A type of simplified equilibrium constant, K_{sp}, defined for and useful for equilibria between solids and their respective ions in solution; for example, the equilibrium

$$AgCl(s) \rightleftarrows Ag^+ + Cl^-, |Ag^+||Cl^-| \cong K_{sp}$$

where $|Ag^+|$ and $|Cl^-|$ are molar concentrations of silver ions and chloride ions.   { ‚säl·yə'bil·əd·ē 'präd·əkt ‚kän·stənt }

solubility test   [ANALY CHEM] **1.** A test for the degree of solubility of asphalts and other bituminous materials in solvents, such as carbon tetrachloride, carbon disulfide, or petroleum ether.   **2.** Any test made to show the solubility of one material in another (such as liquid-liquid, solid-liquid, gas-liquid, or solid-solid).   { ‚säl·yə'bil·əd·ē ‚test }

soluble   [CHEM] Capable of being dissolved.   { 'säl·yə·bəl }

soluble glass

soluble glass See sodium silicate. { 'säl·yə·bəl 'glas }

soluble guncotton See pyroxylin. { 'säl·yə·bəl 'gən‚kat·ən }

soluble indigo blue See indigo carmine. { 'säl·yə·bəl 'in·də·gō 'blü }

soluble nitrocellulose See pyroxylin. { 'säl·yə·bəl ‚nī·trō'sel·yə‚lōs }

solute [CHEM] The substance dissolved in a solvent. { 'säl·yüt }

solution [CHEM] A single, homogeneous liquid, solid, or gas phase that is a mixture in which the components (liquid, gas, solid, or combinations thereof) are uniformly distributed throughout the mixture. { sə'lü·shən }

solution pressure [PHYS CHEM] **1.** A measure of the tendency of molecules or atoms to cross a bounding surface between phases and to enter into a solution. **2.** A measure of the tendency of hydrogen, metals, and certain nonmetals to pass into solution as ions. { sə'lü·shən ‚presh·ər }

solutrope [CHEM] A ternary mixture with two liquid phases and a third component distributed between the phases, or selectively dissolved in one or the other of the phases; analogous to an azeotrope. { 'säl·yə‚trōp }

solvation [CHEM] The process of swelling, gelling, or dissolving of a material by a solvent; for resins, the solvent can be a plasticizer. { säl'vā·shən }

solvent [CHEM] That part of a solution that is present in the largest amount, or the compound that is normally liquid in the pure state (as for solutions of solids or gases in liquids). { 'säl·vənt }

solvent front [ANALY CHEM] In paper chromatography, the wet moving edge of the solvent that progresses along the surface where the separation of the mixture is occurring. { 'säl·vənt ‚frənt }

solvolysis [CHEM] A reaction in which a solvent reacts with the solute to form a new substance. { säl'väl·ə·səs }

solvus [PHYS CHEM] In a phase or equilibrium diagram, the locus of points representing the solid-solubility temperatures of various compositions of the solid phase. { 'säl·vəs }

Sommelet process [ORG CHEM] The preparation of thiophene aldehydes by treatment of thiophene with hexamethylenetetramine. { ‚sȯ·məl'yā ‚prä·səs }

Sonnenschein's reagent [ANALY CHEM] A solution of phosphomolybdic acid that forms a yellow precipitate with alkaloid sulfates. { 'zȯn·ən‚shīnz rē‚ā·jənt }

sonocatalysis [CHEM] **1.** Initiation of a catalytic reaction by irradiation with sound or ultrasound. **2.** Use of sound to impart catalytic activity to a chemical compound. { ‚sän·ə·kə'tal·ə·səs }

sonochemistry [CHEM] Any chemical change, such as in reaction type or rate, that occurs in response to sound or ultrasound. { ‚sän·ə'kem·ə·strē }

sonolysis [PHYS CHEM] The breaking of chemical bonds or formation of radicals using ultrasound. { sō'näl·ə·səs }

sorbate [CHEM] A substance that has been either adsorbed or absorbed. [ORG CHEM] A salt or an ester of sorbic acid. { 'sȯr‚bāt }

sorbic acid [ORG CHEM] $CH_3CH=CHCH=CHCOOH$ A white, crystalline compound; soluble in most organic solvents, slightly soluble in water; melts at 135°C; used as a fungicide and food preservative, and in the manufacture of plasticizers and lubricants. { 'sȯr·bik 'as·əd }

sorbide [CHEM] The generic term for anhydrides derived from sorbitol. { 'sȯr‚bīd }

sorbitol [ORG CHEM] $C_6H_8(OH)_6$ Combustible, white, water-soluble, hygroscopic crystals with a sweet taste; melt at 93 to 97.5°C (depending on the form); used in cosmetic creams and lotions, toothpaste, and resins; as a food additive; and for ascorbic acid fermentation. { 'sȯr·bə‚tȯl }

Sörensen titration [ANALY CHEM] Titration with one of the Sörensen hydrogen-ion-concentration indicators. { 'sȯr·ən·sən tī‚trā·shən }

sorption [PHYS CHEM] A general term used to encompass the processes of adsorption, absorption, desorption, ion exchange, ion exclusion, ion retardation, chemisorption, and dialysis. { 'sȯrp·shən }

sosoloid [PHYS CHEM] A system consisting of particles of a solid dispersed in another solid. { 'säs·ə‚lȯid }

sour |CHEM| Containing large amounts of malodorous sulfur compounds (such as mercaptans or hydrogen sulfide), as in crude oils, naphthas, or gasoline. { saür }

source |SPECT| The arc or spark that supplies light for a spectroscope. { sòrs }

Soxhlet extractor |CHEM| A flask and condenser device for the continuous extraction of alcohol- or ether-soluble materials. { 'säks·lət ik,strak·tər }

Spanish white See bismuth subnitrate. { 'span·ish 'wīt }

spark excitation |SPECT| The use of an electric spark (10,000 to 30,000 volts) to excite spectral line emissions from otherwise hard-to-excite samples; used in emission spectroscopy. { 'spärk ‚ek‚sī'tā·shən }

spark explosion method |ANALY CHEM| A technique for the analysis of hydrogen; the sample is mixed with an oxidant and exploded by a spark or hot wire, and the combustion products are then analyzed. { 'spärk ik'splō·zhən ‚meth·əd }

spark spectrum |SPECT| The spectrum produced by a spark discharging through a gas or vapor; with metal electrodes, a spectrum of the metallic vapor is obtained. { 'spärk ‚spek·trəm }

sparteine |ORG CHEM| $C_{15}H_{26}N_2$ A poisonous, colorless, oily alkaloid; soluble in alcohol and ether, slightly soluble in water; boils at 173°C; used in medicine. Also known as lupinidine. { 'spärd·ē‚ēn }

special cause |ANALY CHEM| A cause of variance or bias in a measurement process that is external to the system. { ¦spesh·əl ¦kóz }

species |CHEM| A chemical entity or molecular particle, such as a radical, ion, molecule, or atom. Also known as chemical species. { 'spē·shēz }

specific catalysis |CHEM| The acceleration of a given chemical reaction by a unique catalyst rather than by a family of related substances. { spə¦sif·ik kə'tal·ə·səs }

specificity |CHEM| The selective reactivity that occurs between substances, such as between an antigen and its corresponding antibody. { ‚spes·ə'fis·əd·ē }

specific retention volume |ANALY CHEM| The relationship among retention volume, void volume, and adsorbent weight, used to standardize gas chromatography adsorbents by the elution of a standard solute by a standard eluent from the adsorbent under test. { spə'sif·ik ri'ten·chən ‚väl·yəm }

specific susceptibility See mass susceptibility. { spə'sif·ik sə‚sep·tə'bil·əd·ē }

spectator ion |CHEM| An ion that serves to balance the electrical charges in a reaction environment without participating in product formation. { 'spek‚tād·ər 'ī‚än }

spectral bandwidth |SPECT| The minimum radiant-energy bandwidth to which a spectrophotometer is accurate; that is, 1–5 nanometers for better models. { 'spek·trəl 'band‚width }

spectral directional reflectance factor |ANALY CHEM| In spectrophotometric colorimetry, the ratio of the energy diffused in any desired direction by the object under analysis to that energy diffused in the same direction by an ideal perfect (energy) diffuser. { 'spek·trəl di'rek·shən·əl ri'flek·təns ‚fak·tər }

spectral line |SPECT| A discrete value of a quantity, such as frequency, wavelength, energy, or mass, whose spectrum is being investigated; one may observe a finite spread of values resulting from such factors as level width, Doppler broadening, and instrument imperfections. Also known as spectrum line. { 'spek·trəl ‚līn }

spectral radiance factor |ANALY CHEM| A situation when the desired directions for analysis of energy diffused from (reflected from) an object under spectrophotometric colorimetric analysis are all substantially the same (a solid angle of nearly zero steradians). { 'spek·trəl 'rād·ē·əns ‚fak·tər }

spectral reflectance |ANALY CHEM| Situation when the desired directions for analysis of energy from (reflected from) an object under spectrophotometric colorimetric analysis is diffused in all directions (not directed as a single beam). { 'spek·trəl ri'flek·təns }

spectral regions |SPECT| Arbitrary ranges of wavelength, some of them overlapping, into which the electromagnetic spectrum is divided, according to the types of sources that are required to produce and detect the various wavelengths, such as x-ray, ultraviolet, visible, infrared, or radio-frequency. { 'spek·trəl ‚rē·jənz }

spectral series

spectral series |SPECT| Spectral lines or groups of lines that occur in sequence. { 'spek·trəl ,sir·ēz }

spectrobolometer |SPECT| An instrument that measures radiation from stars; measurement can be made in a narrow band of wavelengths in the electromagnetic spectrum; the instrument itself is a combination spectrometer and bolometer. { ¦spek·trō·bō'läm·əd·ər }

spectrofluorometer |SPECT| A device used in fluorescence spectroscopy to increase the selectivity of fluorometry by passing emitted fluorescent light through a monochromator to record the fluorescence emission spectrum. { ¦spek·trō·flü'räm·əd·ər }

spectrogram |SPECT| The record of a spectrum produced by a spectrograph. Also known as measured spectrum. { 'spek·trə,gram }

spectrograph |SPECT| A spectroscope provided with a photographic camera or other device for recording the spectrum. { 'spek·trə,graf }

spectrography |SPECT| The use of photography to record the electromagnetic spectrum displayed in a spectroscope. { spek'träg·rə·fē }

spectrometer |SPECT| **1.** A spectroscope that is provided with a calibrated scale either for measurement of wavelength or for measurement of refractive indices of transparent prism materials. **2.** A spectroscope equipped with a photoelectric photometer to measure radiant intensities at various wavelengths. { spek'träm·əd·ər }

spectrometry |SPECT| The use of spectrographic techniques for deriving the physical constants of materials. { spek'träm·ə·trē }

spectrophone |ANALY CHEM| A cell containing the sample in the optoacoustic detection method; equipped with windows through which the laser beam enters the cell and a microphone for detecting sound. { 'spek·trə,fōn }

spectrophotometer |SPECT| An instrument that consists of a radiant-energy source, monochromator, sample holder, and detector, used for measurement of radiant flux as a function of wavelength and for measurement of absorption spectra. { ¦spek·trō·fə'täm·əd·ər }

spectrophotometric titration |ANALY CHEM| An analytical method in which the radiant-energy absorption of a solution is measured spectrophotometrically after each increment of titrant is added. { ¦spek·trō,fōd·ə'me·trik tī'trā·shən }

spectrophotometry |ANALY CHEM| A method of chemical analysis based on the absorption or attenuation by matter of electromagnetic radiation of a specified wavelength or frequency. The radiation interacts with specific features of the molecular species being determined, such as the vibrational or rotational motions of the chemical bonds. The radiation can also interact with specific atoms or the whole molecule, for example, by causing the molecule to change its electronic energy state. { ¦spek·trō·fə'täm·ə·trē }

spectropolarimetry |SPECT| The measurement of the polarization of light that has been dispersed into a continuum or line spectrum as a function of wavelength. { ,spek·trə,pō·lə'rim·ə·trē }

spectropyrheliometer |SPECT| An astronomical instrument used to measure distribution of radiant energy from the sun in the ultraviolet and visible wavelengths. { ¦spek·trō¦pīr,hē·lē'äm·əd·ər }

spectroscope |SPECT| An optical instrument consisting of a slit, collimator lens, prism or grating, and a telescope or objective lens which produces a spectrum for visual observation. { 'spek·trə,skōp }

spectroscopic displacement law |SPECT| The spectrum of an un-ionized atom resembles that of a singly ionized atom of the element one place higher in the periodic table, and that of a doubly ionized atom two places higher in the table, and so forth. { ¦spek·trə¦skäp·ik di'splās·mənt ,lȯ }

spectrum line See spectral line. { 'spek·trəm ,līn }

spherand |ORG CHEM| A macrocyclic compound capable of completely enveloping a cation, having donor atoms (O, N, S) arranged such that they provide a solvation sphere to the encapsulated cation. { 'sfir·ənd }

sphere of attraction |PHYS CHEM| The distance within which the potential energy arising from mutual attraction of two molecules is not negligible with respect to the molecules' average thermal energy at room temperature. { 'sfir əv ə'trak·shən }

spin label [PHYS CHEM] A molecule which contains an atom or group of atoms exhibiting an unpaired electron spin that can be detected by electron spin resonance (ESR) spectroscopy and can be bonded to another molecule. { 'spin ‚lā·bəl }

spinning-band column [ANALY CHEM] An analytical distillation column inside of which is a series of driven, spinning bands; centrifugal action of the bands throws a layer of liquid onto the inner surface of the column; used as an aid in liquid-vapor contact. { 'spin·iŋ ¦band ‚käl·əm }

spin-polarized atomic hydrogen [PHYS CHEM] A system of hydrogen atoms cooled to a very low temperature in a very high magnetic field so that electron spins in almost all the atoms are antiparallel to the magnetic field, with the result that the atoms interact only through the weak triplet-state interaction so that no hydrogen molecules are formed. { 'spin ¦pō·lə‚rīzd ə'täm·ik 'hī·drə·jən }

spiral wire column [ANALY CHEM] An analytical rectification (distillation) column with a wire spiral the length of the inside of the column to serve as a liquid-vapor contact surface. { 'spī·rəl ¦wīr 'käl·əm }

spiran [ORG CHEM] A polycyclic compound containing a carbon atom which is a member of two rings. { 'spī‚ran }

spirit [ORG CHEM] A solution of alcohol and a volatile substance, such as an essential oil. { 'spir·ət }

spiro atom [ORG CHEM] A single atom that is the only common member of two ring structures. { 'spir·ō ‚ad·əm }

spiro ring system [ORG CHEM] A molecular structure with two ring structures having one atom in common; for example, spiropentane. { 'spī·rō 'riŋ ‚sis·təm }

spontaneous combustion [CHEM] Ignition that can occur when certain materials such as tung oil are stored in bulk, resulting from the generation of heat, which cannot be readily dissipated; often heat is generated by microbial action. Also known as spontaneous ignition. { spän'tā·nē·əs kəm'bəs·chən }

spontaneous heating [CHEM] The slow reaction of material with atmospheric oxygen at ambient temperatures; liberated heat, if undissipated, accumulates so that in the presence of combustible substances a fire will result. { spän'tā·nē·əs 'hēd·iŋ }

spontaneous ignition See spontaneous combustion. { spän'tā·nē·əs ig'nish·ən }

spot test [ANALY CHEM] The addition of a drop of reagent to a drop or two of sample solution to obtain distinctive colors or precipitates; used in qualitative analysis. { 'spät ‚test }

square planar molecule [CHEM] A molecule in which a central atom possesses four valence bonds directed to the corners of a square, with all atoms lying in the same plane. { 'skwer ¦plā·nər ‚mäl·ə‚kyül }

Sr See strontium.

SRMS See structure resonance modulation spectroscopy.

SSD See steady-state distribution.

stability [CHEM] The property of a chemical compound which is not readily decomposed and does not react with other compounds. { stə'bil·əd·ē }

stability constant [CHEM] Refers to the equilibrium reaction of a metal cation and a ligand to form a chelating mononuclear complex; the absolute-stability constant is expressed by the product of the concentration of products divided by the product of the concentrations of the reactants; the apparent-stability constant (also known as the conditional- or effective-stability constant) allows for the nonideality of the system because of the combination of the ligand with other complexing agents present in the solution. { stə'bil·əd·ē ‚kän·stənt }

standard calomel electrode [PHYS CHEM] A mercury-mercurous chloride electrode used as a reference (standard) measurement in polarographic determinations. { 'stan·dərd 'kal·ə·məl i'lek‚trōd }

standard electrode potential [PHYS CHEM] The reversible or equilibrium potential of an electrode in an environment where reactants and products are at unit activity. { 'stan·dərd i'lek‚trōd pə‚ten·chəl }

standardization [ANALY CHEM] A process in which the value of a potential standard is fixed by a measurement made with respect to a standard whose value is known. { ˌstan·dər·dəˈzā·shən }

standard potential [PHYS CHEM] The potential of an electrode composed of a substance in its standard state, in equilibrium with ions in their standard states compared to a hydrogen electrode. { ˈstan·dərd pəˈten·chəl }

standard reference material [ANALY CHEM] A reference material distributed and certified by the appropriate national institute for standardization. { ˈstan·dərd ˈref·rəns məˌtir·ē·əl }

standard solution See titrant. { ˈstan·dərd səˈlü·shən }

stannane See tin hydride. { ˈstaˌnān }

stannic acid See stannic oxide. { ˈstan·ik ˈas·əd }

stannic anhydride See stannic oxide. { ˈstan·ik anˈhīˌdrīd }

stannic bromide [INORG CHEM] SnBr₄ Water- and alcohol-soluble, white crystals that fume when exposed to air, and melt at 31°C; used in mineral separations. Also known as tin bromide; tin tetrabromide. { ˈstan·ik ˈbrōˌmīd }

stannic chloride [INORG CHEM] SnCl₄ A colorless, fuming liquid; soluble in cold water, alcohol, carbon disulfide, and oil of turpentine; decomposed by hot water; boils at 114°C; used as a conductive coating and a sugar bleach, and in drugs, ceramics, soaps, and blueprinting. Also known as tin chloride; tin tetrachloride. { ˈstan·ik ˈklȯrˌīd }

stannic chromate [INORG CHEM] Sn(CrO₄)₂ Toxic, brownish-yellow crystals, slightly soluble in water; used to decorate porcelain and china. Also known as tin chromate. { ˈstan·ik ˈkrōˌmāt }

stannic iodide [INORG CHEM] SnI₄ Yellow-reddish crystals; insoluble in water, soluble in alcohol, ether, chloroform, carbon disulfide, and benzene; decomposed by water, melt at 144°C, sublime at 180°C. Also known as tin iodide; tin tetraiodide. { ˈstan·ik ˈī·əˌdīd }

stannic oxide [INORG CHEM] SnO₂ A white powder; insoluble in water, soluble in concentrated sulfuric acid; melts at 1127°C; used in ceramic glazes and colors, special glasses, putty, and cosmetics, and as a catalyst. Also known as flowers of tin; stannic acid; stannic anhydride; tin dioxide; tin oxide; tin peroxide. { ˈstan·ik ˈäkˌsīd }

stannic sulfide [INORG CHEM] SnS₂ A yellow-brown powder; insoluble in water, soluble in alkaline sulfides; decomposes at red heat; used as a pigment and for imitation gilding. Also known as artificial gold; mosaic gold; tin bisulfide. { ˈstan·ik ˈsəlˌfīd }

stannous bromide [INORG CHEM] SnBr₂ A yellow powder; soluble in water, alcohol, acetone, ether, and dilute hydrochloric acid; browns in air; melts at 215°C. Also known as tin bromide. { ˈstan·əs ˈbrōˌmīd }

stannous chloride [INORG CHEM] SnCl₂ White crystals; soluble in water, alcohol, and alkalies; oxidized in air to the oxychloride; melt at 247°C; used as a chemical intermediate, reducing agent, and ink-stain remover, and for silvering mirrors. Also known as tin chloride; tin crystals; tin dichloride; tin salts. { ˈstan·əs ˈklȯrˌīd }

stannous chromate [INORG CHEM] SnCrO₄ A brown powder; very slightly soluble in water; used to decorate porcelain. Also known as tin chromate. { ˈstan·əs ˈkrōˌmāt }

stannous 2-ethylhexoate [ORG CHEM] Sn(C₈H₁₅O₂)₂ A light yellow liquid, soluble in benzene, toluene, and petroleum ether; used as a lubricant, a vulcanizing agent, and a stabilizer for transformer oil. { ˈstan·əs ˈtü ˌeth·əlˈhek·səˌwāt }

stannous fluoride [INORG CHEM] SnF₂ A white, lustrous powder; slightly soluble in water; used to fluoridate toothpaste and as a medicine. { ˈstan·əs ˈflu̇rˌīd }

stannous oxalate [ORG CHEM] SnC₂O₄ A white, crystalline powder that decomposes at about 280°C; soluble in acids; used in textile dyeing and printing. Also known as tin oxalate. { ˈstan·əs ˈäk·səˌlāt }

stannous oxide [INORG CHEM] SnO An air-unstable, brown to black powder; insoluble in water, soluble in acids and strong bases; decomposes when heated; used as a reducing agent and chemical intermediate, and for glass plating. Also known as tin oxide; tin protoxide. { ˈstan·əs ˈäkˌsīd }

stannous sulfate [INORG CHEM] SnSO₄ Heavy light-colored crystals; decomposes rapidly in water, loses SO₂ at 360°C; used for dyeing and tin plating.   Also known as tin sulfate.   { 'stan·əs 'səl,fāt }

stannous sulfide [INORG CHEM] SnS Dark crystals; insoluble in water, soluble (with decomposition) in concentrated hydrochloric acid; melts at 880°C; used as an analytical reagent and catalyst, and in bearing material.   Also known as tin monosulfide; tin protosulfide; tin sulfide.   { 'stan·əs 'səl,fīd }

stannum [CHEM] The Latin name for tin, thus the symbol Sn for the element.   { 'stan·əm }

starburst polymer See dendrimer.   { ,stär,bərst 'päl·ə·mər }

starch nitrate See nitrostarch.   { 'stärch 'nī,trāt }

Stark effect [SPECT] The effect on spectrum lines of an electric field which is either externally applied or is an internal field caused by the presence of neighboring ions or atoms in a gas, liquid, or solid.   Also known as electric field effect.   { 'stärk i,fekt }

Stark-Einstein law See Einstein photochemical equivalence law.   { 'stärk 'īn,stīn ,lȯ }

star polymer [ORG CHEM] A macromolecule having a small core of molecules (branch point) with branches radiating from the core.   { 'stär |päl·ə·mər }

stationary phase [ANALY CHEM] In chromatography, the nonmobile phase contained in the chromatographic bed.   { 'stā·shə,ner·ē 'fāz }

statistical control [ANALY CHEM] In an analytical procedure, a state that exists when the means of a large number of individual values in the output of a measurement process tend to approach a limiting value known as the limiting mean.   { stə'tis· tə·kəl kən'trōl }

steady-state distribution [ANALY CHEM] The equilibrium condition between phases in each step of a multistage, countercurrent liquid-liquid extraction.   Abbreviated SSD.   { 'sted·ē |stāt ,dis·trə'byü·shən }

stearamide [ORG CHEM] CH₃(CH₂)₁₆CONH₂ Colorless leaflets with a melting point of 109°C; used as a corrosion inhibitor in oil wells.   { 'stir·ə·məd }

stearate [ORG CHEM] C₁₇H₃₅COOM A salt or ester of stearic acid where M is a monovalent radical, for example, sodium stearate, C₁₇H₃₅COONa.   { 'stir,āt }

stearic acid [ORG CHEM] CH₃(CH₂)₁₆COOH Nature's most common fatty acid, derived from natural animal and vegetable fats; colorless, waxlike solid, insoluble in water, soluble in alcohol, ether, and chloroform; melts at 70°C; used as a lubricant and in pharmaceuticals, cosmetics, and food packaging.   { 'stir·ik 'as·əd }

stearin [ORG CHEM] C₃H₅(C₁₈H₃₅O₂)₃ A colorless combustible powder; insoluble in water, soluble in alcohol, chloroform, and carbon disulfide; melts at 72°C; used in metal polishes, pastes, candies, candles, and soap, and to waterproof paper.   Also known as glyceryl tristearate; tristearin.   { 'stir·ən }

stearyl alcohol [ORG CHEM] CH₃(CH₂)₁₆CH₂OH Oily white, combustible flakes; insoluble in water, soluble in alcohol, acetone, and ether; melt at 59°C; used in lubricants, resins, perfumes, and cosmetics, and as a surface-active agent.   { 'sti,ril 'al·kə,hȯl }

step See elementary reaction.   { step }

stepwise reaction [CHEM] A chemical reaction in which at least one reactive intermediate is produced and at least two elementary reactions are involved.   { |step,wīz rē|ak·shən }

sterane [ORG CHEM] A cycloalkane derived from a sterol.   { 'sti,rān }

stereocenter [ORG CHEM] A (chiral) carbon atom that has four different substituents bonded to it.   Also known as a stereogenic atom.   { 'ster·ē·ō,sen·tər }

stereochemistry [PHYS CHEM] The study of the spatial arrangement of atoms in molecules and the chemical and physical consequences of such arrangement.   { |ster·ē· ə'kem·ə·strē }

stereogenic atom See stereocenter.   { ,ster·ē·ə,jen·ik 'ad·əm }

stereogenic center See asymmetric carbon atom.   { ,ster·ē·ə|jen·ik 'sen·tər }

stereoisomers [ORG CHEM] Compounds whose molecules have the same number and kind of atoms and the same atomic arrangement, but differ in their spatial relationship.   { |ster·ē·ō'ī·sə·mərz }

stereoregular polymer See stereospecific polymer.   { |ster·ē·ə'reg·yə·lər 'päl·i·mər }

stereorubber

stereorubber [ORG CHEM] Synthetic rubber, *cis*-polyisoprene, a polymer with stereospecificity. { 'ster·ō·ə‚rəb·ər }

stereoselective reaction [ORG CHEM] A chemical reaction in which one stereoisomer is produced or decomposed more rapidly than another. Also known as enantioselective reaction. { ¦ster·ē·ə·si'lek·tiv rē'ak·shən }

stereospecificity [ORG CHEM] The condition of a polymer whose molecular structure has a fixed spatial (geometric) arrangement of its constituent atoms, thus having crystalline properties; for example, synthetic natural rubber, *cis*-polyisoprene. { ¦ster· ē·ō·spə·sə'fis·ə d·ē }

stereospecific polymer [ORG CHEM] A polymer with specific or definite order of arrangement of molecules in space, as in isotactic polypropylene; permits close packing of molecules and leads to a high degree of polymer crystallinity. Also known as stereoregular polymer. { ¦ster·ē·ō·spə'sif·ik 'päl·i·mər }

stereospecific synthesis [ORG CHEM] Catalytic polymerization of monomer molecules to produce stereospecific polymers, as with Ziegler or Natta catalysts (derived from a transition metal halide and a metal alkyl). { ¦ster·ē·ō·spə'sif·ik 'sin·thə·səs }

steric effect [PHYS CHEM] The influence of the spatial configuration of reacting substances upon the rate, nature, and extent of reaction. { 'ster·ik i‚fekt }

steric hindrance [ORG CHEM] The prevention or retardation of chemical reaction because of neighboring groups on the same molecule; for example, ortho-substituted aromatic acids are more difficult to esterify than are the meta and para substitutions. { 'ster·ik 'hin·drəns }

stern layer [PHYS CHEM] One of two electrically charged layers of electrolyte ions, the layer of ions immediately adjacent to the surface, in the neighborhood of a negatively charged surface. { 'stərn ‚lā·ər }

stibide *See* antimonide. { 'sti‚bīd }

stibium [CHEM] The Latin name for antimony, thus the symbol Sb for the element. { 'stib·ē·əm }

stibnate *See* potassium antimonate. { 'stib‚nāt }

sticking coefficient [PHYS CHEM] The fraction of all atoms incident on a surface that are adsorbed on the surface. { 'stik·iŋ ‚kō·i‚fish·ənt }

stilbene [ORG CHEM] $C_6H_5CH:CHC_6H_5$ Colorless crystals soluble in ether and benzene, insoluble in water; melts at 124°C; used to make dyes and bleaches and as phosphors. Also known as diphenylethylene; toluylene. { 'stil‚bēn }

Stobbe reaction [ORG CHEM] A type of aldol condensation reaction represented by the reaction of benzophenone with dimethyl succinate and sodium methoxide to form monoesters of an α-alkylidene (or arylidene) succinic acid. { 'shtȯb·ə rē‚ak·shən }

stoichiometry [PHYS CHEM] The numerical relationship of elements and compounds as reactants and products in chemical reactions. { ‚stȯi·kē'äm·ə·trē }

Stokes' law [SPECT] The wavelength of luminescence excited by radiation is always greater than that of the exciting radiation. { 'stōks ‚lȯ }

Stokes line [SPECT] A spectrum line in luminescent radiation whose wavelength is greater than that of the radiation which excited the luminescence, and thus obeys Stokes' law. { 'stōks ‚līn }

Stokes shift [SPECT] The displacement of spectral lines or bands of luminescent radiation toward longer wavelengths than those of the absorption lines or bands. { 'stōks ‚shift }

stopped-flow method [CHEM] A method for studying chemical reactions in which the reactants are rapidly mixed, then abruptly stopped after a very short time. { 'stäpt ¦flō ‚meth·əd }

STPP *See* sodium tripolyphosphate.

stratified film [PHYS CHEM] A film in which two thicknesses are present in a fixed configuration for a significant period of time. { 'strad·ə‚fīd 'film }

stripping analysis [ANALY CHEM] An analytic process of solutions or concentrations containing ions, in which the ions are electrodeposited onto an electrode, stripped (dissolved) from the material from the electrode, and weighed. { 'strip·iŋ ə‚nal· ə·səs }

strong acid [CHEM] An acid with a high degree of dissociation in solution, for example, mineral acids, such as hydrochloric acid, HCl, sulfuric acid, H_2SO_4, or nitric acid, HNO_3. { 'strȯŋ 'as·əd }

strong base [CHEM] A base with a high degree of dissociation in solution, for example, sodium hydroxide, NaOH, potassium hydroxide, KOH. { 'strȯŋ 'bās }

strontia *See* strontium oxide. { 'strän·chə }

strontium [CHEM] A metallic element in group II, symbol Sr, with atomic number 38, atomic weight 87.62; flammable, soft, pale-yellow solid; soluble in alcohol and acids, decomposes in water; melts at 770°C, boils at 1380°C; chemistry is similar to that of calcium; used as electron-tube getter. { 'strän·tē·əm }

strontium acetate [ORG CHEM] $Sr(C_2H_3O_2)_2 \cdot {}^1/_2H_2O$ White, water-soluble crystals, loses water at 150°C; used for catalysts, as a chemical intermediate, and in medicine. { 'strän·tē·əm 'as·ə,tāt }

strontium bromide [INORG CHEM] $SrBr_2 \cdot 6H_2O$ A white, hygroscopic powder soluble in water and alcohol; loses water at 180°C, melts at 643°C; used in medicine and as an analytical reagent. { 'strän·tē·əm 'brō,mīd }

strontium carbonate [INORG CHEM] $SrCO_3$ A white powder slightly soluble in water, decomposes at 1340°C; used to make TV-tube glass, strontium salts, and ceramic ferrites, and in pyrotechnics. { 'strän·tē·əm 'kär·bə,nāt }

strontium chlorate [INORG CHEM] $Sr(ClO_3)_2$ Shock-sensitive, highly combustible, white, water-soluble crystals that decompose at 120°C; used in pyrotechnics and tracer bullets. { 'strän·tē·əm 'klȯr,āt }

strontium chloride [INORG CHEM] $SrCl_2$ Water- and alcohol-soluble white crystals, melts at 872°C; used in medicine and pyrotechnics and to make strontium salts. { 'strän·tē·əm 'klȯr,īd }

strontium chromate [INORG CHEM] $SrCrO_4$ A light yellow, rust- and corrosion-resistant pigment used in metal coatings and for pyrotechnics. { 'strän·tē·əm 'krō,māt }

strontium dioxide *See* strontium peroxide. { 'strän·tē·əm dī'äk,sīd }

strontium fluoride [INORG CHEM] SrF_2 A white powder, soluble in hydrochloric acid and hydrofluoric acid; used in medicine and for single crystals for lasers. { 'strän·tē·əm 'flùr,īd }

strontium hydrate *See* strontium hydroxide. { 'strän·tē·əm 'hī,drāt }

strontium hydroxide [INORG CHEM] $Sr(OH)_2$ Colorless deliquescent crystals that absorb carbon dioxide from air, soluble in hot water and acids, melts at 375°C; used by the sugar industry, in lubricants and soaps, and as a plastic stabilizer. Also known as strontium hydrate. { 'strän·tē·əm hī'dräk,sīd }

strontium iodide [INORG CHEM] SrI_2 Air-yellowing, white crystals that decompose in moist air, melts at 515°C; used in medicine and as a chemicals intermediate. { 'strän·tē·əm 'ī·ə,dīd }

strontium monosulfide *See* strontium sulfide. { 'strän·tē·əm ¦män·ə'səl,fīd }

strontium nitrate [INORG CHEM] $Sr(NO_3)_2$ A white, water-soluble powder melting at 570°C; used in pyrotechnics, signals and flares, medicine, and matches, and as a chemicals intermediate. { 'strän·tē·əm 'nī,trāt }

strontium oxalate [INORG CHEM] $SrC_2O_4 \cdot H_2O$ A white powder that loses water at 150°C; used in pyrotechnics and tanning. { 'strän·tē·əm 'äk·sə,lāt }

strontium oxide [INORG CHEM] SrO A grayish powder, melts at 2430°C, becomes the hydroxide in water; used in medicine, pyrotechnics, pigments, greases, soaps, and as a chemicals intermediate. Also known as strontia. { 'strän·tē·əm 'äk,sīd }

strontium peroxide [INORG CHEM] SrO_2 A strongly oxidizing, fire-hazardous, white, alcohol-soluble powder that decomposes in hot water; used in medicine, bleaching, and fireworks. Also known as strontium dioxide. { 'strän·tē·əm pər'äk,sīd }

strontium salicylate [ORG CHEM] $Sr(C_7H_5O_3)_2 \cdot 2H_2O$ White crystals or powder with a sweet saline taste; soluble in water and alcohol; used in medicine and manufacture of pharmaceuticals. { 'strän·tē·əm sə'lis·ə,lāt }

strontium sulfate [INORG CHEM] $SrSO_4$ White crystals insoluble in alcohol, slightly soluble in water and concentrated acids, melts at 1605°C; used in paper manufacture, pyrotechnics, ceramics, and glass. { 'strän·tē·əm 'səl,fāt }

strontium sulfide

strontium sulfide [INORG CHEM] SrS A gray powder with a hydrogen sulfide aroma in moist air, slightly soluble in water, soluble (with decomposition) in acids, melts above 2000°C; used in depilatories and luminous paints and as a chemicals intermediate. Also known as strontium monosulfide. { 'strän·tē·əm 'səl,fīd }

strontium titanate [INORG CHEM] SrTiO₃ A solid material, insoluble in water and melting at 2060°C; used in electronics and electrical insulation. { 'strän·tē·əm 'tīt·ən,āt }

structural formula [CHEM] A system of notation used for organic compounds in which the exact structure, if it is known, is given in schematic representation. { 'strək·chə·rəl 'fȯr·myə·lə }

structural isomers See constitutional isomers. { 'strək·chə·rəl 'ī·sə·mərz }

structure resonance [SPECT] An extremely narrow resonance exhibited by a small aerosol particle at a natural electromagnetic frequency at which the dielectric sphere oscillates, observed in the particle's scattered light excitation spectrum. { 'strək·chər ,rez·ən·əns }

structure resonance modulation spectroscopy [SPECT] The infrared modulation of visible scattered light near a structure resonance to determine the absorption spectrum of an aerosol particle. Abbreviated SRMS. { 'strək·chər ,rez·ən·əns ,mäj·ə¦lā·shən spek'träs·kə·pē }

strychnine [ORG CHEM] C₂₁H₂₂O₂N₂ An alkaloid obtained primarily from the plant nux vomica, formerly used for therapeutic stimulation of the central nervous system. { 'strik,nīn }

styphnic acid [ORG CHEM] C₆H(OH)₂(HO₂)₃ An explosive, yellow, crystalline compound, melting at 179–180°C, slightly soluble in water; used in explosives as a priming agent. { 'stif·nik 'as·əd }

styrene [ORG CHEM] C₆H₅CH:CH₂ A colorless, toxic liquid with a strong aroma; insoluble in water, soluble in alcohol and ether; polymerizes rapidly, can become explosive; boils at 145°C; used to make polymers and copolymers, polystyrene plastics, and rubbers. Also known as phenylethylene; styrene monomer; vinylbenzene. { 'stī,rēn }

styrene-acrylonitrile resin [ORG CHEM] A thermoplastic copolymer of styrene and acrylonitrile with good stiffness and resistance to scratching, chemicals, and stress. Also known as SAN. { 'stī,rēn ,ak·rə'län·ə·trəl 'rez·ən }

styrene monomer See styrene. { 'stī,rēn 'män·ə·mər }

styrene oxide [ORG CHEM] C₈H₈O A moderately toxic, combustible, colorless or straw-colored liquid miscible in acetone, ether, and benzene, and melts at 195°C; used as a chemical intermediate. { 'stī,rēn 'äk,sīd }

styrene plastic [ORG CHEM] A plastic made by the polymerization of styrene or the copolymerization of styrene with other unsaturated compounds. { 'stī,rēn 'plas·tik }

subcompound [CHEM] A compound, generally in the vapor phase, in which an element exhibits a valency lower than that exhibited in its ordinary compounds. { ¦səb 'käm,paúnd }

suberic acid [ORG CHEM] HOOC(CH₂)₆COOH A colorless, crystalline compound that melts at 143°C, and dissolves slightly in cold water; used in organic synthesis. Also known as octanedioic acid. { sü'ber·ik 'as·əd }

sublimatography [ANALY CHEM] A procedure of fractional sublimation in which a solid mixture is separated into bands along a condensing tube with a temperature gradient. { ,səb·lə·mə'täg·rə·fē }

sublimator [CHEM] Device used for the heating of solids (usually under vacuum) to the temperature at which the solid sublimes. { 'səb·lə,mād·ər }

subsample [ANALY CHEM] A portion taken from a sample of material for which a chemical analysis has been specified. { ,səb'sam·pəl }

subsolvus [PHYS CHEM] A range of conditions in which two or more solid phases can form by exsolution from an original homogeneous phase. { ¦səb'säl·vəs }

substituent [ORG CHEM] An atom or functional group substituted for another in a chemical structure. { səb'stich·ə·wənt }

substitution reaction [CHEM] Replacement of an atom or radical by another one in a chemical compound. { ,səb·stə'tü·shən rē,ak·shən }

substitutive nomenclature [ORG CHEM] A system in which the name of a compound is derived by using the functional group (the substituent) as a prefix or suffix to the name of the parent compound to which it is attached; for example, in 2-chloropropane a chlorine atom has replaced a hydrogen atom on the central carbon of the propane chain. { 'səb·stə,tüd·iv 'nō·mən,klā·chər }

substrate [ORG CHEM] A compound with which a reagent reacts. { 'səb,strāt }

succinate [ORG CHEM] A salt or ester of succinic acid; for example, sodium succinate, $Na_2C_4H_4O_4·6H_2O$, the reaction product of succinic acid and sodium hydroxide. { 'sək·sə,nāt }

succinic acid [ORG CHEM] $CO_2H(CH_2)_2CO_2H$ Water-soluble, colorless crystals with an acid taste; melts at 185°C; used as a chemical intermediate, in medicine, and to make perfume esters. { sək'sin·ik 'as·əd }

succinic acid 2,2-dimethylhydrazide [ORG CHEM] $C_6H_{12}O_3N_2$ White crystals with a melting point of 154–156°C; soluble in water; used as a growth regulator for many crops and ornamentals. Also known as aminocide. { sək'sin·ik 'as·əd ¦tü ¦tü dī,meth·əl'hī·drə,zīd }

succinic anhydride [ORG CHEM] $C_4H_4O_3$ Colorless or pale needles soluble in alcohol and chloroform; converts to succinic acid in water; melts at 120°C; used as a chemical and pharmaceutical intermediate and a resin hardener. { sək'sin·ik an'hī,drīd }

succinimide [ORG CHEM] $C_4H_5O_2N·H_2O$ Colorless or tannish water-soluble crystals with a sweet taste; melts at 126°C; used to make plant growth stimulants and as a chemical intermediate. { sək'sin·ə,mīd }

succinonitrite See ethylene cyanide. { ¦sək·sə·nō'nī,trāt }

succinylcholine chloride [ORG CHEM] $[Cl(CH_3)_3N(CH_2)_2OOCH_2]_2·2H_2O$ Water-soluble white crystals with a bitter taste; melts at 162°C; used in medicine. { ¦sək·sən·əl'kō,lēn 'klòr,īd }

sucrochemical [ORG CHEM] A chemical made from a feedstock derived from sucrose extracted from sugarcane or sugarbeet. { 'sü·krō,kem·i·kəl }

sucrochemistry [ORG CHEM] A type of chemistry based on sucrose as a starting point. { ¦sü·krō'kem·i·strē }

sucrose [ORG CHEM] $C_{12}H_{22}O_{11}$ Combustible, white crystals soluble in water, decomposes at 160 to 186°C; derived from sugarcane or sugarbeet; used as a sweetener in drinks and foods and to make syrups, preserves, and jams. Also known as saccharose; table sugar. { 'sü,krōs }

sucrose octoacetate [ORG CHEM] $C_{28}H_{38}O_{19}$ A bitter crystalline compound that forms needles from alcohol solution, melts at 89°C, and breaks down at 286°C or above; used as an adhesive, to impregnate and insulate paper, and in lacquers and plastics. { 'sü,krōs ¦äk·tō'as·ə,tāt }

sugar alcohol [ORG CHEM] Any of the acyclic linear polyhydric alcohols; may be considered sugars in which the aldehydic group of the first carbon atom is reduced to a primary alcohol; classified according to the number of hydroxyl groups in the molecule; sorbitol (D-glucitol, sorbite) is one of the most widespread of all the naturally occurring sugar alcohols. { 'shùg·ər 'al·kə,hòl }

sugar of lead See lead acetate. { 'shùg·ər əv 'led }

sulfallate [ORG CHEM] $C_8H_{14}NS_2Cl$ An oily liquid, used as a preemergence herbicide for vegetable crops and ornamentals. Also known as 2-chloroallyl diethyldithiocarbamate (CDEC). { səl'fa,lāt }

sulfamate [CHEM] A salt of sulfamic acid; for example, calcium sulfamate, $Ca(SO_3NH_2)_2·4H_2O$. { 'səl·fə,māt }

sulfamic acid [INORG CHEM] HSO_3NH_2 White, nonvolatile crystals slightly soluble in water and organic solvents, decomposes at 205°C; used to clean metals and ceramics, and as a plasticizer, fire retardant, chemical intermediate, and textile and paper bleach. { ¦səl¦fam·ik 'as·əd }

sulfanilic acid [ORG CHEM] $C_6H_4NH_2·SO_3H·H_2O$ Combustible, grayish-white crystals slightly soluble in water, alcohol, and ether, soluble in fuming hydrochloric acid; chars at 280–300°C; used in medicine and dyestuffs and as a chemical intermediate. { ¦səl·fə¦nil·ik 'as·əd }

sulfaquinoxaline

sulfaquinoxaline *See* N'-2-quinoxalysulfanilimide. { ¦səl·fə·kwə'näk· sə,lēn }

sulfate [CHEM] **1.** A compound containing the −SO₄ group, as in sodium sulfate, Na₂SO₄. **2.** A salt of sulfuric acid. { 'səl,fāt }

sulfation [CHEM] The conversion of a compound into a sulfate by the oxidation of sulfur, as in sodium sulfide, Na₂S, oxidized to sodium sulfate, Na₂SO₄; or the addition of a sulfate group, as in the reaction of sodium and sulfuric acid to form Na₂SO₄. { səl'fā·shən }

sulfenic acid [ORG CHEM] An oxy acid of sulfur with the general formula RSOH, where R is an alkyl or aryl group such as CH₃; known as the esters and halides. { ¦səl¦fen·ik 'as·əd }

sulfenyl chloride [ORG CHEM] Any of a group of well-known organosulfur compounds with the general formula RSCl; although highly reactive compounds, they can generally be synthesized and isolated; examples are trichloromethanesulfenyl chloride and 2,4-dinitrobenzenesulfenyl chloride. { ,səl'fen·əl 'klȯr,īd }

sulfhydryl compound [CHEM] A compound with a −SH group. Also known as a mercapto compound. { ,səlf'hī·drəl 'käm,paúnd }

sulfidation [CHEM] The chemical insertion of a sulfur atom into a compound. { ,səl·fə'dā·shən }

sulfide [CHEM] Any compound with one or more sulfur atoms in which the sulfur is connected directly to a carbon, metal, or other nonoxygen atom; for example, sodium sulfide, Na₂S. { 'səl,fīd }

sulfide dye [ORG CHEM] A dye containing sulfur and soluble in a 0.25–0.50% sodium sulfide solution, and used to dye cotton; the dyes are manufactured from aromatic polyamines or hydroxy amines; the amine group is primary, secondary, or tertiary, or may be an equivalent nitro, nitroso, or imino group; an example is the dye sulfur blue. Also known as sulfur dye. { 'səl,fīd ,dī }

sulfinate [ORG CHEM] **1.** A compound containing the R₂SX₂ grouping, where X is a halide. **2.** A salt of sulfinic acid having the general formula R·OH·S:O. { ,səl·fə'nāt }

sulfinic acid [ORG CHEM] Any of the monobasic organic acids of sulfur with the general formula RS:O(OH); for example, ethanesulfinic acid, C₂H₅SO₂H. { ¦səl¦fin·ik 'as·əd }

sulfinyl bromide *See* thionyl bromide. { səl·fə,nil 'brō,mīd }

sulfite [INORG CHEM] M₂SO₃ A salt of sulfurous acid, for example, sodium sulfite, Na₂SO₃. { 'səl,fīt }

sulfo- [CHEM] Prefix for a compound with either a divalent sulfur atom, or the presence of −SO₃H, the sulfo group in a compound. Also spelled sulpho-. { 'səl·fō *or* 'səl·fə }

sulfocarbanilide *See* thiocarbanilide. { ¦səl·fō·kär'ban·ə·ləd }

sulfocarbimide *See* isothiocyanate. { ¦səl·fō¦kär·bə,mīd }

sulfocyanate *See* thiocyanate. { ¦səl·fō'sī·ə,nāt }

sulfocyanic acid *See* thiocyanic acid. { ¦səl·fō·sī¦an·ik 'as·əd }

sulfocyanide *See* thiocyanate. { ¦səl·fō'sī·ə,nīd }

sulfolane [ORG CHEM] C₄H₈SO₂ A liquid with a boiling point of 285°C and outstanding solvent properties; used for extraction of aromatic hydrocarbons, fractionation of fatty acids, and textile finishing, and as a solvent and plasticizer. { 'səl·fə,lān }

sulfonamide [ORG CHEM] One of a group of organosulfur compounds, RSO₂NH₂, prepared by the reaction of sulfonyl chloride and ammonia; used for sulfa drugs. { ,səl'fän·ə,mīd }

sulfonate [CHEM] A sulfuric acid derivative or a sulfonic acid ester containing a −SO₃− group. [ORG CHEM] Any of a group of petroleum hydrocarbons derived from sulfuric-acid treatment of oils, used as synthetic detergents, emulsifying and wetting agents, and chemical intermediates. { 'səl·fə,nāt }

sulfonation [CHEM] Substitution of −SO₃H groups (from sulfuric acid) for hydrogen atoms, for example, conversion of benzene, C₆H₆, into benzenesulfonic acid, C₆H₅SO₃H. { ,səl·fə'nā·shən }

sulfone [ORG CHEM] R₂SO₂ (or RSOOR) A compound formed by the oxidation of sulfides, for example, ethyl sulfone, C₄H₁₀SO₂, from ethyl sulfide, C₄H₁₀S; the use of sulfones, particularly 4,4'-sulfonyldianiline (dapsone) in the treatment of leprosy

362

leads to apparent improvement; relapses associated with sulfone-resistant strains have been encountered.   { 'səl͵fōn }

sulfonic acid   [ORG CHEM] A compound with the radical —SO_2OH, derived by the sulfuric acid replacement of a hydrogen atom; for example, conversion of benzene, C_6H_6, to the water-soluble benzenesulfonic acid, $C_6H_5SO_3H$, by treatment with sulfuric acid; used to make dyes and drugs.   { ¦səl¦fän·ik 'as·əd }

sulfonyl   [CHEM] Also known as sulfuryl. **1.** A compound containing the radical —SO_2—. **2.** A prefix denoting the presence of a sulfone group.   { 'səl·fə͵nil }

sulfonyl chloride See sulfuryl chloride.   { 'səl·fə͵nil 'klȯr͵īd }

sulfosalicylic acid   [ORG CHEM] $C_7H_6O_6S$ A trifunctional aromatic compound whose dihydrate is in the form of white crystals or crystalline powder; soluble in water and alcohol; melting point is 120°C; used as an indicator for albumin in urine and as a reagent for the determination of ferric ion; it also has industrial uses.   { ¦səl·fō¦sal·ə¦sil·ik 'as·əd }

sulfoxide   [ORG CHEM] R_2SO A compound with the group =SO; derived from oxidation of sulfides, the proportion of oxidant, such as hydrogen peroxide, and temperature being set to avoid excessive oxidation; an example is dimethyl sulfoxide, $(CH_3)_2SO$.   { ͵səl'fäk͵sīd }

sulfur   [CHEM] A nonmetallic element in group 16, symbol S, atomic number 16, atomic weight 32.06, existing in a crystalline or amorphous form and in four stable isotopes; used as a chemical intermediate and fungicide, and in rubber vulcanization.   { 'səl·fər }

sulfurated lime See calcium sulfide.   { ͵səl·fə'rād·əd 'līm }

sulfuration   [CHEM] The chemical act of combining an element or compound with sulfur.   { ͵səl·fə'rā·shən }

sulfur bichloride See sulfur dichloride.   { 'səl·fər bī'klȯr͵īd }

sulfur bromide   [INORG CHEM] S_2Br_2 A toxic, irritating, yellow liquid that reddens in air, soluble in carbon disulfide, decomposes in water, boils at 54°C. Also known as sulfur monobromide.   { 'səl·fər 'brō͵mīd }

sulfur chloride   [INORG CHEM] S_2Cl_2 A combustible, water-soluble, oily, fuming, amber to yellow-red liquid with an irritating effect on the eyes and lungs, boils at 138°C; used to make military gas and insecticides, in rubber substitutes and cements, to purify sugar juices, and as a chemical intermediate. Also known as sulfur subchloride.   { 'səl·fər 'klȯr͵īd }

sulfur dichloride   [INORG CHEM] SCl_2 A red-brown liquid boiling (when heated rapidly) at 60°C, decomposes in water; used to make insecticides, for rubber vulcanization, and as a chemical intermediate and a solvent. Also known as sulfur bichloride.   { 'səl·fər dī'klȯr͵īd }

sulfur dioxide   [INORG CHEM] SO_2 A toxic, irritating, colorless gas soluble in water, alcohol, and ether; boils at −10°C; used as a chemical intermediate, in artificial ice, paper pulping, and ore refining, and as a solvent. Also known as sulfurous acid anhydride.   { 'səl·fər dī'äk͵sīd }

sulfur dye See sulfide dye.   { 'səl·fər ͵dī }

sulfur hexafluoride   [INORG CHEM] SF_6 A colorless gas soluble in alcohol and ether, slightly soluble in water, sublimes at −64°C; used as a dielectric in electronics.   { 'səl·fər ¦hek·sə'flùr͵īd }

sulfuric acid   [INORG CHEM] H_2SO_4 A toxic, corrosive, strongly acid, colorless liquid that is miscible with water and dissolves most metals, and melts at 10°C; used in industry in the manufacture of chemicals, fertilizers, and explosives, and in petroleum refining. Also known as dipping acid; oil of vitriol, vitriolic acid.   { ¦səl¦fyùr·ik 'as·əd }

sulfuric chloride See sulfuryl chloride.   { ͵səl¦fyùr·ik 'klȯr͵īd }

sulfur iodide See sulfur iodine.   { ¦səl¦fyùr·ik 'ī·ə͵dīd }

sulfur iodine   [INORG CHEM] I_2S_2 A gray-black brittle mass with an iodine aroma and a metallic luster, insoluble in water, soluble in carbon disulfide; used in medicine. Also known as iodine bisulfide; iodine disulfide; sulfur iodide.   { ͵səl¦fyùr·ik 'ī·ə͵dīn }

sulfur monobromide See sulfur bromide.   { 'səl·fər ¦män·ə·ə'brō͵mīd }

sulfur monoxide [INORG CHEM] SO A gas at ordinary temperatures; produces an orange-red deposit when cooled to temperatures of liquid air; prepared by passing an electric discharge through a mixture of sulfur vapor and sulfur dioxide at low temperature. { 'səl·fər mə'näk,sīd }

sulfur number [ANALY CHEM] The number of milligrams of sulfur per 100 milliliters of sample, determined by electrometric titration; used in the petroleum industry for oils. { 'səl·fər ,nəm·bər }

sulfurous acid [INORG CHEM] H_2SO_3 An unstable, water-soluble, colorless liquid with a strong sulfur aroma; derived from absorption of sulfur dioxide in water; used in the synthesis of medicine and chemicals, manufacture of paper and wine, brewing, metallurgy, and ore flotation, as a bleach and analytic reagent, and to refine petroleum products. { 'səl·fə·rəs 'as·əd }

sulfurous acid anhydride See sulfur dioxide. { 'səl·fə·rəs 'as·əd an'hī,drīd }

sulfurous oxychloride See thionyl chloride. { 'səl·fə·rəs ¦äk·sē'klȯr,īd }

sulfur oxide [INORG CHEM] An oxide of sulfur, such as sulfur dioxide, SO_2, and sulfur trioxide, SO_3. { 'səl·fər 'äk,sīd }

sulfur oxychloride See thionyl chloride. { 'səl·fər ¦äk·sē'klȯr,īd }

sulfur subchloride See sulfur chloride. { 'səl·fər ¦səb'klȯr,īd }

sulfur test [ANALY CHEM] **1.** Method to determine the sulfur content of a petroleum material by combustion in a bomb. **2.** Analysis of sulfur in petroleum products by lamp combustion in which combustion of the sample is controlled by varying the flow of carbon dioxide and oxygen to the burner. { 'səl·fər ,test }

sulfur trioxide [INORG CHEM] SO_3 A toxic, irritating liquid in three forms, α, β, γ, with respective melting points of 62°C, 33°C, and 17°C; a strong oxidizing agent and fire hazard; used for sulfonation of organic chemicals. { 'səl·fər trī'äk,sīd }

sulfuryl See sulfonyl. { 'səl·fə,ril }

sulfuryl chloride [INORG CHEM] SO_2Cl_2 A colorless liquid with a pungent aroma, boils at 69°C, decomposed by hot water and alkalies; used as a chlorinating agent and solvent and for pharmaceuticals, dyestuffs, rayon, and poison gas. Also known as sulfonyl chloride; sulfuric chloride. { 'səl·fə,ril 'klȯr,īd }

sulfuryl fluoride [INORG CHEM] SO_2F_2 A colorless gas with a melting point of -136.7°C and a boiling point of 55.4°C; used as an insecticide and fumigant. { 'səl·fə,ril 'flúr,īd }

Sullivan reaction [ORG CHEM] The formation of a red-brown color when cysteine is reacted with 1,2-naphthoquinone-4-sodium sulfate in a highly alkaline reducing medium. { 'səl·ə·vən rē,ak·shən }

sulpho- See sulfo-. { 'səl·fō }

superacid [CHEM] **1.** An acidic medium that has a proton-donating ability equal to or greater than 100% sulfuric acid. **2.** A solution of acetic or phosphoric acid. { ¦sü·pər'as·əd }

supercritical-fluid chromatography [ANALY CHEM] Any chemical separation technique using chromatography in which a supercritical fluid is used as the mobile phase. { ¦sü·pər,krid·ə·kəl ¦flü·əd ,krō·mə'täg·rə·fē }

superheavy element [INORG CHEM] A chemical element with an atomic number of 110 or greater. { ¦sü·pər'hev·ē 'el·ə·mənt }

supermolecule [PHYS CHEM] A single quantum-mechanical entity presumably formed by two reacting molecules and in existence only during the collision process; a concept in the hard-sphere collision theory of chemical kinetics. { ¦sü·pər'mäl·ə,kyül }

superoxide dismutase [ORG CHEM] An enzymatic antioxidant that removes the potentially toxic superoxide ion (O^{2-}) by disproportionating it to O_2 and hydrogen peroxide (H_2O_2). { ¦sü·pər,äk,sīd 'diz·myü,tās }

superoxide ion [CHEM] O_2^- An ion formed by the combination of one molecule of dioxygen (O_2) and one electron (e^-). { ,¦sü·pər'äk,sīd ,ī·ən }

supersaturation [PHYS CHEM] The condition existing in a solution when it contains more solute than is needed to cause saturation. Also known as supersolubility. { ¦sü·pər,sach·ə'rā·shən }

supersolubility *See* supersaturation.

supertransuranics [INORG CHEM] A group of relatively stable elements, with atomic numbers around 114 and mass numbers around 298, that are predicted to exist beyond the present periodic table of known elements. { ¦sü·pər,tranz·yü'ran·iks }

support-coated capillary column [ANALY CHEM] A capillary column that utilizes a fine-granular solid support to disperse the stationary liquid. { sə'pȯrt ¦kōd·əd 'kap·ə,ler· ē ,käl·əm }

suppressor [SPECT] In an analytical procedure, a substance added to the analyte to reduce the extraneous emission, absorption, or light scattering caused by the presence of an impurity. { sə'pres·ər }

suprafacial [ORG CHEM] The stereochemistry when, simultaneously, two sigma bonds are formed or broken on the same face of the component pi systems, such as in a cycloaddition reaction. { ¦sü·prə'fā·shəl }

supramolecular chemistry [CHEM] A highly interdisciplinary field covering the chemical, physical, and biological features of complex chemical species held together and organized by means of intermolecular (noncovalent) bonding interactions such as hydrogen bonds, van der Waals forces, and hydrophobic interactions. { ,sü·prə· mə,lek·yə·lər 'kem·ə·strē }

supramolecule [PHYS CHEM] A stable system formed by two or more molecules held together and organized by intermolecular (noncovalent bonding) interactions. { ,sü· prə'mäl·ə,kyül }

surface chemistry [PHYS CHEM] The study and measurement of the forces and processes that act on the surfaces of fluids (gases and liquids) and solids, or at an interface separating two phases; for example, surface tension. { 'sər·fəs ,kem·ə·strē }

surface orientation [PHYS CHEM] Arrangement of molecules on the surface of a liquid with one part of the molecule turned toward the liquid. { 'sər·fəs ,ȯr·ē·ən'tā·shən }

surface reaction [CHEM] A chemical reaction carried out on a surface as on an adsorbent or solid catalyst. { 'sər·fəs rē,ak·shən }

suspended solids *See* suspension. { sə'spen·dəd 'säl·ədz }

suspension [CHEM] A mixture of fine, nonsettling particles of any solid within a liquid or gas, the particles being the dispersed phase, while the suspending medium is the continuous phase. Also known as suspended solids. { sə'spen·shən }

svedberg [PHYS CHEM] A unit of sedimentation coefficient, equal to 10^{-13} second. { 'sfed,bərg }

Swarts reaction [ORG CHEM] The reaction of chlorinated hydrocarbons with metallic fluorides to form chlorofluorohydrocarbons, such as CCl_2F_2, which is quite inert and nontoxic. { 'svärts rē,ak·shən }

sweat [CHEM] Exudation of nitroglycerin from dynamite due to separation of nitroglycerin from its adsorbent. { swet }

sweet spirits of niter *See* ethyl nitrite. { 'swēt 'spir·əts əv 'nī·tər }

swep [ORG CHEM] $C_8H_7Cl_2NO_2$ A white, crystalline compound with a melting point of 112–114°C; insoluble in water; used as a pre- and postemergence herbicide for rice, carrots, potatoes, and cotton. Also known as methyl-N-(3,4-dichlorophenyl)carbamate. { swep }

SXAPS *See* soft-x-ray appearance potential spectroscopy.

sym- [ORG CHEM] A chemical prefix; denotes structure of a compound in which substituents are symmetrical with respect to a functional group or to the carbon skeleton. { sim }

symbol [CHEM] Letter or combination of letters and numbers that represent various conditions or properties of an element, for example, a normal atom, O (oxygen); with its atomic weight, ^{16}O; its atomic number, $_8{}^{16}O$; as a molecule, O_2; as an ion, O^{2+}; in excited state, O^*; or as an isotope, ^{18}O. { 'sim·bəl }

symclosene *See* trichloroisocyanuric acid. { 'sim·klə,zēn }

symmetric top molecule [PHYS CHEM] A nonlinear molecule which has one and only one axis of threefold or higher symmetry. { sə'me·trik ¦täp 'mäl·ə,kyül }

symmetry number [PHYS CHEM] The number of indistinguishable orientations that a molecule can exhibit by being rotated around symmetry axes. { 'sim·ə,trē ,nəm·bər }

syn [ORG CHEM] In stereochemistry, on the same side of a reference plane; for example, the stereochemical outcome of an addition reaction where the new bonds are on the same side of the original pi bond is called syn addition. { sin }

syndiotactic polymer [ORG CHEM] A vinyl polymer in which the side chains alternate regularly above and below the plane of the backbone. { ¦sin·dē·ə¦tak·tik 'päl·i·mər }

syneresis [CHEM] Spontaneous separation of a liquid from a gel or colloidal suspension due to contraction of the gel. { sə'ner·ə·səs }

synthesis [CHEM] Any process or reaction for building up a complex compound by the union of simpler compounds or elements. { 'sin·thə·səs }

synthetic resin [ORG CHEM] Amorphous, organic, semisolid, or solid material derived from the polymerization of unsaturated monomers such as ethylene, butylene, propylene, and styrene. { sin'thed·ik 'rez·ən }

systematic nomenclature [CHEM] A system for naming chemical compounds according to a specific set of rules, usually those developed by the International Union of Pure and Applied Chemistry. { ¦sis·tə,mad·ik 'nō·mən,klā·chər }

T

2,4,5-T See 2,4,5-trichlorophenoxyacetic acid.

2,4,6-T See trichlorophenol.

T_c See critical temperature.

Ta See tantalum.

table salt See sodium chloride. { 'tā·bəl ,sȯlt }

table sugar See sucrose. { 'tā·bəl ,shùg·ər }

tabun [ORG CHEM] (CH₃)₂NP(O)(C₂H₅O)(CN) A toxic liquid with a boiling point of 240°C; soluble in organic solvents; used as a nerve gas. { 'tä,bùn }

tachiol See silver fluoride. { 'tak·ē,ȯl }

tactic polymer [ORG CHEM] A polymer with regularity or symmetry in the structural arrangement of its molecules, as in a stereospecific polymer such as some types of polypropylene. { 'tak·tə·kəl 'päl·i·mər }

Tag closed-cup tester [ANALY CHEM] A laboratory device used to determine the flash point of mobile petroleum liquids flashing below 175°F (79.4°C). Also known as Tagliabue closed tester. { 'tag ¦klōzd ¦kəp 'tes·tər }

tagged molecule [CHEM] A molecule having one or more atoms which are either radioactive or have a mass which differs from that of the atoms which normally make up the molecule. { 'tagd 'mäl·ə,kyül }

Tagliabue closed tester See Tag closed-cup tester. { ,täl·yə'bü·ē 'klōsd 'tes·tər }

tannic acid [ORG CHEM] **1.** C₁₄H₁₀O₉ A yellowish powder with an astringent taste; soluble in water and alcohol, insoluble in acetone and ether; derived from nutgalls; decomposes at 210°C; used as an alcohol denaturant and a chemical intermediate, and in tanning and textiles. Also known as digallic acid; gallotannic acid; gallotannin; tannin. **2.** C₇₆H₅₂O₄₆ Yellowish-white to light-brown amorphous powder or flakes; decomposes at 210–215°C; very soluble in alcohol and acetone; used as a mordant in dyeing, in photography, as a reagent, and in clarifying wine or beer. Also known as pentadigalloylglucose. { 'tan·ik 'as·əd }

tannin See tannic acid. { 'tan·ən }

tantalic acid anhydride See tantalum oxide. { tan'tal·ik 'as·əd an'hī,drīd }

tantalic chloride See tantalum chloride. { tan'tal·ik 'klȯr,īd }

tantalum [CHEM] A metallic transition element, symbol Ta, atomic number 73, atomic weight 180.9479; black powder or steel-blue solid soluble in fused alkalies, insoluble in acids (except hydrofluoric and fuming sulfuric); melts about 3000°C. { 'tant·əl·əm }

tantalum carbide [INORG CHEM] TaC Hard, chemical-resistant crystals melting at 3875°C; used in cutting tools and dies. { 'tant·əl·əm 'kär,bīd }

tantalum chloride [INORG CHEM] TaCl₅ A highly reactive, pale-yellow powder decomposing in moist air; soluble in alcohol and potassium hydroxide; melts at 221°C; used to produce tantalum and as a chemical intermediate. Also known as tantalic chloride; tantalum pentachloride. { 'tant·əl·əm 'klȯr,īd }

tantalum nitride [INORG CHEM] TaN A very hard, black, water-insoluble solid, melting at 3360°C. { 'tant·əl·əm 'nī,trīd }

tantalum oxide [INORG CHEM] Ta₂O₅ Prisms insoluble in water and acids (except for hydrofluoric); melts at 1800°C; used to make tantalum, in optical glass and electronic

tantalum pentachloride

equipment, and as a chemical intermediate. Also known as tantalic acid anhydride; tantalum pentoxide. { 'tant·əl·əm 'äk,sīd }

tantalum pentachloride See tantalum chloride. { 'tant·əl·əm ¦pen·tə'klȯr,īd }

tantalum pentoxide See tantalum oxide. { 'tant·əl·əm pen'täk,sīd }

tar base [CHEM] A basic nitrogen compound found in coal tar, for example, pyridine and quinoline. { 'tär ,bās }

tar camphor See naphthalene. { 'tär ,kam·fər }

target compound [ORG CHEM] In chemical synthesis, the molecule of interest. { 'tär·gət ,käm,paùnd }

tartar emetic [ORG CHEM] $K(SbO)C_4H_4O_6 \cdot {}^1/_2 H_2O$ A transparent crystalline compound, soluble in water; used to attract and kill moths, wasps, and yellow jackets. Also known as antimony potassium tartrate; potassium antimonyl tartrate. { 'tärd·ər i'med·ik }

tartaric acid [ORG CHEM] $HOOC(CHOH)_2COOH$ Water- and alcohol-soluble colorless crystals with an acid taste, melts at 170°C; used as a chemical intermediate and a sequestrant and in tanning, effervescent beverages, baking power, ceramics, photography, textile processing, mirror silvering, and metal coloring. { tär'tar·ik 'as·əd }

tartrate [ORG CHEM] A salt or ester of tartaric acid, for example, sodium tartrate, $Na_2C_4H_4O_6$. { 'tär,trāt }

tartrazine [ORG CHEM] $C_{16}H_9N_4O_9S_2$ A bright orange-yellow, water-soluble powder, used as a food, drug, and cosmetic dye. { 'tär·trə,zēn }

Tauber test [ANALY CHEM] A color test for identification of pentose sugars; the sugars produce a cherry-red color when heated with a solution of benzidine in glacial acetic acid. { 'taù·bər ,test }

taurine [ORG CHEM] $NH_2CH_2CH_2SO_3H$ A crystalline compound that decomposes at about 300°C; present in bile combined with cholic acid. { 'tȯ,rēn }

tautomerism [CHEM] The reversible interconversion of structural isomers of organic chemical compounds; such interconversions usually involve transfer of a proton. { tȯ'täm·ə,riz·əm }

Tb See terbium.

TBH See 1,2,3,4,5,6-hexachlorocyclohexane.

TBP See tributyl phosphate.

TBT See tetrabutyl titanate.

Tc See technetium.

TCA See trichloroacetic acid.

TCP See tricresyl phosphate.

TDE See 2,2-bis(para-chlorophenyl)-1,1-dichloroethane.

Te See tellurium.

TEA chloride See tetraethylammonium chloride. { ,tē,ē'ā 'klȯr,īd }

technetium [CHEM] A transition element, symbol Tc, atomic number 43; derived from uranium and plutonium fission products; chemically similar to rhenium and manganese; isotope ^{99}Tc has a half-life of 200,000 years; used to absorb slow neutrons in reactor technology. { tek'nē·shē·əm }

tectoquinone [ORG CHEM] $C_{15}H_{10}O_2$ A white compound with needlelike crystals; sublimes at 177°C; insoluble in water; used as an insecticide to treat wood. Also known as 2-methyl anthraquinone. { ¦tek·tō·kwə'nōn }

TEG See tetraethylene glycol; triethylene glycol.

TEL See tetraethyllead.

Teller-Redlich rule [PHYS CHEM] For two isotopic molecules, the product of the frequency ratio values of all vibrations of a given symmetry type depends only on the geometrical structure of the molecule and the masses of the atoms, and not on the potential constants. { 'tel·ər 'red·lik ,rül }

telluric acid [INORG CHEM] H_6TeO_6 Toxic white crystals, slightly soluble in cold water, soluble in hot water and alkalies; melts at 136°C; used as an analytical reagent. Also known as hydrogen tellurate. { tə'lùr·ik 'as·əd }

telluric line [SPECT] Any of the spectral bands and lines in the spectrum of the sun and stars produced by the absorption of their light in the atmosphere of the earth. { tə'lür·ik 'līn }

tellurinic acid [ORG CHEM] A compound of tellurium with the general formula R_2TeOOH; an example is methanetellurinic acid, C_6H_5TeOOH. { ¦tel·yə¦rin·ik 'as·əd }

tellurium [CHEM] A member of group 16, symbol Te, atomic number 52, atomic weight 127.60; dark-gray crystals, insoluble in water, soluble in nitric and sulfuric acids and potassium hydroxide; melts at 452°C, boils at 1390°C; used in alloys (with lead or steel), glass, and ceramics. { tə'lür·ē·əm }

tellurium dibromide [INORG CHEM] $TeBr_2$ Toxic, hygroscopic, green- or gray-black crystals with violet vapor, soluble in ether, decomposes in water, and melts at 210°C. { tə'lür·ē·əm dī'brō,mīd }

tellurium dichloride [INORG CHEM] $TeCl_2$ A toxic, amorphous, black or green-yellow powder decomposing in water, melting at 209°C. { tə'lür·ē·əm dī'klòr,īd }

tellurium dioxide [INORG CHEM] TeO_2 The most stable oxide of tellurium, formed when tellurium is burned in oxygen or air or by oxidation of tellurium with cold nitric acid; crystallizes as colorless, tetragonal, hexagonlike crystals that melt at 452°C. { tə'lür·ē·əm dī'äk,sīd }

tellurium disulfide [INORG CHEM] TeS_2 A toxic, red powder, insoluble in water and acids. Also known as tellurium sulfide. { tə'lür·ē·əm dī'səl,fīd }

tellurium hexafluoride [INORG CHEM] TeF_6 A colorless gas which is formed from the elements tellurium and fluorine; it is slowly hydrolyzed by water. { tə'lür·ē·əm ¦hek·sə'flùr,īd }

tellurium monoxide [INORG CHEM] TeO A black, amorphous powder, stable in cold dry air; formed by heating the mixed oxide $TeSO_3$. { tə'lür·ē·əm mə'näk,sīd }

tellurium sulfide See tellurium disulfide. { tə'lür·ē·əm 'səl,fīd }

telluroketone [ORG CHEM] One of a group of compounds with the general formula R_2CTe. { ¦tel·yə·rō'kē,tōn }

telluromercaptan [ORG CHEM] One of a group of compounds with the general formula RTeH. { ¦tel·yə·rō·mər'kap,tan }

tellurous acid [INORG CHEM] H_2TeO_3 Toxic, white crystals, soluble in alkalies and acids, slightly soluble in water and alcohol; decomposes at 40°C. { 'tel·yə·rəs 'as·əd }

telvar [ORG CHEM] The common name for the herbicide 3-(para-chlorophenyl)-1,1-dimethylurea; used as a soil sterilant. { 'tel,vär }

TEM See triethylenemelamine.

temporary hardness [CHEM] The portion of the total hardness of water that can be removed by boiling whereby the soluble calcium and magnesium bicarbonate are precipitated as insoluble carbonates. { 'tem·pə,rer·ē 'härd·nəs }

TEP See triethyl phosphate.

terbacil [ORG CHEM] $C_9H_{13}ClN_2O_2$ A colorless, crystalline compound with a melting point of 175–177°C; used as an herbicide to control weeds in sugarcane, apples, peaches, citrus, and mints. { 'tər·bə,sil }

terbia See terbium oxide. { tər·bē·ə }

terbium [CHEM] A rare-earth element, symbol Tb, in the yttrium subgroup of the transition elements, atomic number 65, atomic weight 158.9254. { 'tər·bē·əm }

terbium chloride [INORG CHEM] $TbCl_3\cdot6H_2O$ Water- and alcohol-soluble, hygroscopic, colorless, transparent prisms; anhydrous form melts at 588°C. { 'tər·bē·əm 'klòr,īd }

terbium nitrate [INORG CHEM] $Tb(NO_3)_3\cdot6H_2O$ A colorless, fire-hazardous (strong oxidant) powder, soluble in water; melts at 89°C. { 'tər·bē·əm 'nī,trāt }

terbium oxide [INORG CHEM] Tb_2O_3 A slightly hygroscopic, dark-brown powder soluble in dilute acids, absorbs carbon dioxide from air. Also known as terbia. { 'tər·bē·əm 'äk,sīd }

terbutol [ORG CHEM] The common name for the herbicide 2,6-di-tert-butyl-p-tolylmethylcarbamate; used as a selective preemergence crabgrass herbicide for turf. { 'tər·byə,tòl }

terbutryn [ORG CHEM] $C_{13}H_{19}N_5S$ A colorless powder with a melting point of 104–105°C; used for weed control for wheat, barley, and grain sorghum. { tər'byü·trən }

terbutylhylazine [ORG CHEM] $C_9H_{16}N_5Cl$ A white solid with a melting point of 177–179°C; used as a preemergence herbicide. { ‚tər‚byüd·əl'hī·lə‚zēn }

terephthalic acid [ORG CHEM] $C_6H_4(COOH)_2$ A combustible white powder, insoluble in water, soluble in alkalies, sublimes above 300°C; used to make polyester resins for fibers and films and as an analytical reagent and poultry-feed additive. Also known as *para*-phthalic acid; TPA. { ¦ter·əf¦thal·ik 'as·əd }

terephthaloyl chloride [ORG CHEM] $C_6H_4(COCl)_2$ Colorless needles with a melting point of 82–84°C; soluble in ether; used in the manufacture of dyes, synthetic fibers, resins, and pharmaceuticals. { ‚ter·əf'thal·ə‚wil 'klȯr‚īd }

term [SPECT] A set of $(2S+1)(2L+1)$ atomic states belonging to a definite configuration and to definite spin and orbital angular momentum quantum numbers S and L. { tərm }

termination [CHEM] The steps that end a chain reaction by destroying or rendering inactive the reactive intermediates. { ‚tər·mə'nā·shən }

termination step [CHEM] In a chain reaction, the mechanism that halts the reaction. { ‚tər·mə'nā·shən ‚steps }

ternary compound [CHEM] A molecule consisting of three different types of atoms; for example, sulfuric acid, H_2SO_4. { 'tər·nə·rē 'käm‚paůnd }

ternary system [CHEM] Any system with three nonreactive components; in liquid systems, the components may or may not be partially soluble. { 'tər·nə·rē 'sis·təm }

terpene [ORG CHEM] **1.** $C_{10}H_{16}$ A moderately toxic, flammable, unsaturated hydrocarbon liquid found in essential oils and plant oleoresins; used as an intermediate for camphor, menthol, and terpineol. **2.** A class of naturally occurring compounds whose carbon skeletons are composed exclusively of isopentyl (isoprene) C_5 units. Also known as isoprenoid. { 'tər‚pēn }

terpene alcohol [ORG CHEM] A generic name for an alcohol related to or derived from a terpene hydrocarbon, such as terpineol or borneol. { 'tər‚pēn 'al·kə‚hȯl }

terpene hydrochloride [ORG CHEM] $C_{10}H_{16}\cdot HCl$ A solid, water-insoluble material melting at 125°C; used as an antiseptic. Also known as artificial camphor; dipentene hydrochloride; pinene hydrochloride; turpentine camphor. { 'tər‚pēn ¦hī·drə'klȯr‚īd }

terpenoid [ORG CHEM] Any compound with an isoprenoid structure similar to that of the terpene hydrocarbons. { 'tər·pə‚nȯid }

para-terphenyl [ORG CHEM] $(C_6H_5)_2C_6H_4$ A combustible, toxic liquid boiling at 405°C; crystals are used for scintillation counters; polymerized with styrene to make plastic phosphor. { ¦par·ə ¦tər'fen·əl }

terpineol [ORG CHEM] $C_{10}H_{17}OH$ A combustible, colorless liquid with a lilac scent, derived from pine oil, soluble in alcohol, slightly soluble in water, boils at 214–224°C; used in medicine, perfumes, soaps, and disinfectants, and as an antioxidant, a flavoring agent, and a solvent; isomeric forms are alpha-, beta- and gamma-. { tər'pin·ē‚ȯl }

terpin hydrate [ORG CHEM] $CH_3(OH)C_6H_9C(CH_3)_2OH\cdot H_2O$ Combustible, efflorescent, lustrous white prisms soluble in alcohol and ether, slightly soluble in water; melts at 116°C; used for pharmaceuticals and to make terpineol. Also known as dipentene glycol. { 'tər·pən 'hī‚drāt }

terpinolene [ORG CHEM] $C_{10}H_{16}$ A flammable, water-white liquid insoluble in water, soluble in alcohol, ether, and glycols, boils at 184°C; used as a solvent and as a chemical intermediate for resins and essential oils. { tər'pin·ə‚lēn }

terpinyl acetate [ORG CHEM] $C_{10}H_{17}OOCCH_3$ A combustible, colorless, liquid slightly soluble in water and glycerol, soluble in water, boils at 220°C; used in perfumes and flavors. { 'tər·pən·əl 'as·ə‚tāt }

terpolymer [ORG CHEM] A polymer that contains three distinct monomers; for example, acrylonitrile-butadiene-styrene terpolymer, ABS. { ¦tər'päl·i·mər }

terrachlor *See* pentachloronitrobenzene. { 'ter·ə‚klȯr }

tert- [ORG CHEM] Abbreviation for tertiary; trisubstituted methyl radical with the central carbon attached to three other carbons $(R_1R_2R_3C-)$. { tərt }

tertiary alcohol [ORG CHEM] A trisubstituted alcohol in which the hydroxyl group is

attached to a carbon that is joined to three carbons; for example, *tert*-butyl alcohol.
{ 'tər·shē,er·ē 'al·kə,hȯl }

tertiary amine [ORG CHEM] R_3N A trisubstituted amine in which the hydroxyl group is attached to a carbon that is joined to three carbons; for example, trimethylamine, $(CH_3)_3N$. { 'tər·shē,er·ē 'am,ēn }

tertiary carbon atom [ORG CHEM] A carbon atom bonded to three other carbon atoms with single bonds. { 'tər·shē,er·ē 'kär·bən 'ad·əm }

tertiary hydrogen atom [ORG CHEM] A hydrogen atom that is bonded to a tertiary carbon atom. { 'tər·shē,er·ē 'hī·drə·jən 'ad·əm }

tertiary sodium phosphate *See* trisodium phosphate. { 'tər·shē,er·ē 'sōd·ē·əm 'fä,sfāt }

tetraamylbenzene [ORG CHEM] $(C_5H_{11})_4C_6H_2$ A colorless liquid with a boiling range of 320–350°C; used as a solvent. { ¦te·trə¦am·əl'ben,zēn }

tetrabromobisphenol A [ORG CHEM] $(C_6H_2Br_2OH)_2C(CH_3)_2$ An off-white powder with a melting point of 180–184°C; soluble in methanol and ether; used as a flame retardant for plastics, paper, and textiles. { ¦te·trə¦brō·mō,bis'fē,nȯl 'ā }

tetrabromophthalic anhydride [ORG CHEM] $C_6Br_4C_2O_3$ A pale yellow, crystalline solid with a melting point of 280°C; used as a flame retardant for paper, plastics, and textiles. { ¦te·trə,brōm·äf'thal·ik an'hī,drīd }

tetrabutylthiuram monosulfide [ORG CHEM] $[(C_4H_9)_2NCS]_2S$ A brown liquid, soluble in acetone, benzene, gasoline, and ethylene dichloride; used as a rubber accelerator. { ¦te·trə¦byüd·əl'thī·yə,ram ¦män·ə'səl,fīd }

tetrabutyltin [ORG CHEM] $(C_4H_9)_4Sn$ A colorless or slightly yellow, oily liquid with a boiling point of 145°C; soluble in most organic solvents; used as a stabilizing agent and rust inhibitor for silicones, and as a lubricant and fuel additive. { ,te·trə'byüd·əl·tən }

tetrabutyl titanate [ORG CHEM] $Ti(OC_4H_9)_4$ A combustible, colorless to yellowish liquid soluble in many solvents, boils at 312°C; decomposes in water; used in paints, surface coatings, and heat-resistant paints. Abbreviated TBT. { ¦te·trə¦byüd·əl 'tīt·ən,āt }

tetrabutyl urea [ORG CHEM] $(C_4H_9)_4N_2CO$ A liquid with a boiling point of 305°C; used as a plasticizer. { te·trə¦byüd·əl yü'rē·ə }

tetracaine hydrochloride [ORG CHEM] $C_{15}H_{24}O_2N_2·HCl$ Bitter-tasting, water-soluble crystals melting at 148°C; used as a local anesthetic. { 'te·trə,kān ,hī·drə'klȯr,īd }

tetracene *See* naphthacene. { 'te·trə,sēn }

tetrachlorobenzene [ORG CHEM] $C_6H_2Cl_4$ Water-insoluble, combustible white crystals that appear in two forms: 1,2,3,4-tetrachlorobenzene which melts at 47°C and is used in chemical synthesis and in dielectric fluids; and 1,2,4,5-tetrachlorobenzene which melts at 138°C and is used to make herbicides, defoliants, and electrical insulation. { ¦te·trə¦klȯr·ə'ben,zēn }

sym-tetrachlorodifluoroethane [ORG CHEM] CCl_2FCCl_2F A white, toxic liquid with a camphor aroma, soluble in alcohol, insoluble in water, boils at 93°C; used for metal degreasing. { ¦sim ¦te·trə¦klȯr·ə·dī¦flür·ō'eth,ān }

sym-tetrachloroethane [ORG CHEM] $CHCl_2CHCl_2$ A colorless, corrosive, toxic liquid with a chloroform scent, soluble in alcohol and ether, slightly soluble in water, boils at 147°C; used as a solvent, metal cleaner, paint remover, and weed killer. { ¦sim ¦te·trə¦klȯr·ō'eth,ān }

tetrachloroethylene *See* perchloroethylene. { ¦te·trə¦klȯr·ō'eth·ə,lēn }

tetrachlorophenol [ORG CHEM] C_6HCl_4OH Either of two toxic compounds: 2,3,4,6-tetrachlorophenol comprises brown flakes, soluble in common solvents, melting at 70°C, and is used as a fungicide; 2,4,5,6-tetrachlorophenol is a brown solid, insoluble in water, soluble in sodium hydroxide, has a phenol scent, melts at about 50°C, and is used as a fungicide and for wood preservatives. { ¦te·trə¦klȯr·ə'fē,nȯl }

tetrachlorophthalic acid [ORG CHEM] $C_6Cl_4(CO_2H)_2$ Colorless plates, soluble in hot water; used in making dyes. { ¦te·trə¦klȯr·ə¦thal·ik 'as·əd }

tetrachlorophthalic anhydride [ORG CHEM] $C_6Cl_4(CO)_2O$ A white powder with a melting point of 254–255°C; slightly soluble in water; used in the manufacture of dyes and pharmaceuticals and as a flame retardant for epoxy resins. { ¦te·trə¦klȯr·ə¦thal·ik an'hī,drīd }

371

tetrachlorosilane

tetrachlorosilane *See* silicon tetrachloride. { ,te·tra,klȯr·ə'sī,lān }

tetracosane [ORG CHEM] $C_{24}H_{50}$ Combustible crystals insoluble in water, soluble in alcohol, melts at 52°C; used as a chemical intermediate. { ¦te·trə'kō,sān }

tetracyanoethylene [ORG CHEM] $(CN)_2C:C(CN)_2$ A member of the cyanocarbon compounds; colorless crystals with a melting point of 198–200°C; used in dye manufacture. { ¦te·trə¦sī·ə·nō'eth·ə,lēn }

n-tetradecane [ORG CHEM] $C_{14}H_{30}$ A combustible, colorless, water-insoluble liquid boiling at 254°C; used as a solvent and distillation chaser and in organic synthesis. { ¦en ¦te·trə'de,kān }

1-tetradecene [ORG CHEM] $CH_2:CH(CH_2)_{11}CH_3$ A combustible, colorless, water-insoluble liquid boiling at 256°C; used as a solvent for perfumes and flavors and in medicine. { ¦wən ¦te·trə'de,sēn }

tetradecylamine [ORG CHEM] $C_{14}H_{29}NH_2$ A white solid with a melting point of 37°C; soluble in alcohol and ether; used in making germicides. { ¦te·trə·də'sil·ə,mēn }

tetradecyl mercaptan [ORG CHEM] $CH_3(CH_2)_{13}SH$ A combustible liquid with a boiling point of 176–180°C; used for processing synthetic rubber. Also known as myristyl mercaptan. { ¦te·trə'des·əl mər'kap,tan }

tetradentate ligand [INORG CHEM] A chelating agent which has four groups capable of attachment to a metal ion. Also known as quadridentate ligand. { ¦te·trə'den,tāt 'līg·ənd }

tetraethanolammonium hydroxide [ORG CHEM] $(HOCH_2CH_2)_4NOH$ A white, water-soluble, crystalline solid with a melting point of 123°C; used as a dye solvent and in metal-plating solutions. { ¦te·trə¦eth·ə,nȯl·ə'mō·nē·əm hī'dräk,sīd }

tetraethylammonium chloride [ORG CHEM] $(C_2H_5)_4NCl$ Colorless, hygroscopic crystals with a melting point of 37.5°C; soluble in water, alcohol, acetone, and chloroform; used in medicine. Abbreviated TEAC. Also known as TEA chloride. { ¦te·trə¦eth·əl·ə'mō·nē·əm 'klȯr,īd }

tetra-(2-ethylbutyl)silicate [ORG CHEM] $[(C_2H_5)C_4H_8O]_4Si$ A colorless liquid with a boiling point of 238°C at 50 mmHg (6660 pascals); used as a lubricant and hydraulic fluid. { ¦te·trə ¦tü ¦eth·əl¦byüd·əl 'sil·ə,kāt }

tetraethylene glycol [ORG CHEM] $HO(C_2H_4O)_3C_2H_4OH$ A combustible, hygroscopic, colorless, water-soluble liquid, boils at 327°C; used as a nitrocellulose solvent and plasticizer and in lacquers and coatings. Abbreviated TEG. { ¦te·trə'eth·ə,lēn 'glī,kȯl }

tetraethylene glycol dimethacrylate [ORG CHEM] A colorless to pale straw-colored liquid with a boiling point of 200°C at 1 mmHg (133.32 pascals); soluble in styrene and some esters and aromatics; used as a plasticizer. { ¦te·trə'eth·ə,lēn 'glī,kȯl ,dī·mə'thak·rə,lāt }

tetraethylenepentamine [ORG CHEM] $C_8H_{23}N_2$ A toxic, viscous liquid with a boiling point of 333°C and a freezing point of −30°C; soluble in water and organic solvents; used as a motor oil additive, in the manufacture of synthetic rubber, and as a solvent for dyes, acid gases, and sulfur. { ¦te·trə¦eth·ə,lēn'pent·ə,mēn }

tetraethyllead [ORG CHEM] $Pb(C_2H_5)_4$ A highly toxic lead compound that, when added in small proportions to gasoline, increases the fuel's antiknock quality. Abbreviated TEL. { ¦te·tre¦eth·əl'led }

tetraethylpyrophosphate [ORG CHEM] $C_8H_{20}O_7P_2$ A hygroscopic corrosive liquid miscible with although decomposed by water, and miscible with many organic solvents; inhibits the enzyme acetylcholinesterase; used as an insecticide in place of nicotine sulfate. { ¦te·trə¦eth·əl¦pī·rō'fä,sfät }

tetrafluoroethylene [ORG CHEM] $F_2C:CF_2$ A flammable, colorless, heavy gas, insoluble in water, boils at 78°C; used as a monomer to make polytetrafluoroethylene polymers, for example, Teflon. Abbreviated TFE. { ¦te·trə¦flür·ō'eth·ə,lēn }

tetrafluorohydrazine [INORG CHEM] F_2NNF_2 A colorless liquid or gas with a calculated boiling point of −73°C; used as an oxidizer in rocket fuels. { ¦te·trə¦flür·ō'hī·drə,zēn }

tetrafluoromethane *See* carbon tetrafluoride. { ¦te·trə¦flür·ō'meth,ān }

tetrafunctional molecule [ORG CHEM] A chemical structure that possesses four highly reactive sites. { ,te·trə¦fəŋk·shən·əl 'mäl·ə,kyül }

372

tetrahedral molecule [CHEM] A molecule whose structure forms a tetrahedron with a central atom possessing four valence bonds that are directed toward the four points of the tetrahedron. { ˌte·trə¦hē·drəl 'mäl·ə͵kyül }

tetrahydrocannabinol [ORG CHEM] $C_{21}H_{30}O_2$ Any member of a group of isomers that are active components of marijuana. Abbreviated THC. { ¦te·trə¦hī·drə·kə'nab·ə͵nȯl }

tetrahydrofuran [ORG CHEM] C_4H_8O A clear, colorless liquid with a boiling point of 66°C; soluble in water and organic solvents; used as a solvent for resins and in adhesives, printing inks, and polymerizations. Abbreviated THF. { ¦te·trə͵hī·drə'fyu̇͵rän }

tetrahydrofurfuryl acetate [ORG CHEM] $C_7H_{12}O_3$ A colorless liquid with a boiling point of 194–195°C; soluble in water, alcohol, ether, and chloroform; used for flavoring. { ¦te·trə¦hī·drə'fər·fə͵ril 'as·ə͵tāt }

tetrahydrofurfuryl alcohol [ORG CHEM] $C_4H_7OCH_2OH$ A hygroscopic, colorless liquid, miscible with water, boils at 178°C; used as a solvent for resins, in leather dyes, and in nylon. { ¦te·trə¦hī·drə'fər·fə͵ril 'al·kə͵hȯl }

tetrahydrofurfurylamine [ORG CHEM] $C_4H_7OCH_2NH_2$ A colorless to light yellow liquid with a distillation range of −150 to −156°C; used for fine-grain photographic development and to accelerate vulcanization. { ¦te·trə¦hī·drə͵fər·fə'ril·ə͵ mēn }

tetrahydrofurfuryl oleate [ORG CHEM] $C_{23}H_{42}O_3$ A colorless liquid with a boiling point of 240°C at 2 mmHg (266.64 pascals); used as a plasticizer. { ¦te·trə¦hī·drə͵fər·fə'ril 'ōl·ē͵āt }

tetrahydrofurfuryl phthalate [ORG CHEM] $C_6H_4(COOCH_2C_4H_7O)_2$ A colorless liquid with a melting point below 15°C; used as a plasticizer. { ¦te·trə¦hī·drə͵fər·fə'ril 'tha͵lāt }

tetrahydrolinalool [ORG CHEM] $C_{10}H_{21}OH$ A colorless liquid with a floral odor, used in perfumery and flavoring. { ¦te·trə¦hī·drə·lə'nal·ō͵ȯl }

tetrahydronaphthalene [ORG CHEM] $C_{10}H_{12}$ A colorless, oily liquid that boils at 206°C, and is miscible with organic solvents; used as an intermediate in chemical synthesis and as a solvent. { ¦te·trə¦hī·drə'naf·thə͵lēn }

tetrahydroxyadipic acid See mucic acid. { ¦te·trə·hī¦dräk·sē·ə¦dip·ik 'as·əd }

tetraiodoethylene [ORG CHEM] $I_2C:CI_2$ Light yellow crystals with a melting point of 187°C; soluble in organic solvents; used in surgical dusting powder and antiseptic ointments, and as a fungicide. Also known as iodoethylene. { ¦te·trə͵ī·ə͵dō'eth·ə͵lēn }

tetraiodofluorescein [ORG CHEM] $C_{20}H_8O_5I_4$ A yellow, water-insoluble, crystalline compound; used as a dye. Also known as pyrosin. { ¦te·trə¦ī·ə͵dō'flu̇r·ə͵sēn }

tetrakis(hydroxymethyl)phosphonium chloride [ORG CHEM] $(HOCH_2)_4PCl$ A crystalline compound made from phosphine, formaldehyde, and hydrochloric acid; used as a flame retardant for cotton fabrics. Abbreviated THPC. { ¦te·trə·kəs·hī¦dräk·sē ¦meth·əl·f aua'sfō·nē·əm 'klȯr͵īd }

tetralite See tetryl. { 'te·trə͵līt }

tetramer [ORG CHEM] A polymer that results from the union of four identical monomers; for example, the tetramer C_8H_8 forms from union of four molecules of C_2H_2. { 'te·trə·mər }

tetramethyldiaminobenzophenone [ORG CHEM] $[(CH_3)_2NC_6H_4]_2CO$ White to greenish, crystalline leaflets with a melting point of 172°C; soluble in alcohol, ether, water, and warm benzene; used in the manufacture of dyes. Also known as Michler's ketone. { ¦te·trə¦meth·əl·dī¦am·ə·nō·ben'zäf·ə͵nōn }

tetramethylene See cyclobutane. { ¦te·trə'meth·ə͵lēn }

tetramethylethylenediamine [ORG CHEM] $(CH_3)_4N_2(CH_2)_2$ A colorless liquid with a boiling point of 121–122°C; soluble in organic solvents and water; used in the formation of polyurethane, as a corrosion inhibitor, and for textile finishing agents. { ¦te·trə¦meth·əl¦eth·ə͵lēn'dī·ə͵mēn }

tetramethyllead [ORG CHEM] $Pb(CH_3)_4$ An organic compound of lead that, when added in small amounts to motor gasoline, increases the antiknock quality of the fuel; not widely used. { ¦te·trə¦meth·əl'led }

tetramethylsilane [ORG CHEM] $(CH_3)_4Si$ A colorless, volatile, toxic liquid with a boiling

tetramethylthiuram monosulfide

point of 26.5°C; soluble in organic solvents; used as an aviation fuel. { ¦te·trə¦meth·əl'sī₁län }

tetramethylthiuram monosulfide [ORG CHEM] [(CH₃)₂NCS]₂S A yellow powder with a melting point of 104–107°C; soluble in acetone, benzene, and ethylene dichloride; used as a rubber accelerator, fungicide, and insecticide. { ¦te·trə¦meth·əl'thī·yə,ram ¦män·ə'səl,fīd }

tetramethylurea [ORG CHEM] C₅H₁₂N₂O A liquid that boils at 176.5°C, and is miscible in water and organic solvents; used as a reagent and solvent. { ¦te·trə¦meth·əl·yú'rē·ə }

tetranitromethane [ORG CHEM] C(NO₂)₄ A powerful oxidant; toxic, colorless liquid with a pungent aroma, insoluble in water, soluble in alcohol and ether, boils at 126°C; used in rocket fuels and as an analytical reagent. { ¦te·trə¦nī·trō'meth,än }

tetraphenyltin [ORG CHEM] (C₆H₅)₄Sn A white powder with a melting point of 225–228°C; soluble in hot benzene, toluene, and xylene; used for mothproofing. { ,te·trə'fen·əl·tən }

tetraphosphorus trisulfide See phosphorus sesquisulfide. { ¦te·trə'fä·sfə·rəs trī'səl,fīd }

tetrapotassium pyrophosphate See potassium pyrophosphate. { ¦te·trə·pə'tas·ē·əm ¦pī·rō'fä,sfāt }

tetrapropylene See dodecane. { ¦te·trə'prō·pə,lēn }

tetrapyrrole [ORG CHEM] A chemical structure in which four pyrrole rings are joined in straight chains, as in a phycobilin, or as joined rings, as in a chlorophyll. { ,te·trə'pī,rōl }

tetrasodium pyrophosphate See sodium pyrophosphate. { ¦te·trə¦sōd·ē·əm ¦pī·rō'fä ,sfāt }

tetraterpene [ORG CHEM] A class of terpene compounds that contain isoprene units; best known are the carotenoid pigments from plants and animals, such as lycopene, the red coloring matter in tomatoes. { ¦te·trə'tər,pēn }

tetrazene [ORG CHEM] H₂NC(NH)₃N₂C(NH)₃NO An explosive, colorless to yellowish solid practically insoluble in water and alcohol; used as an explosive initiator and in detonators. { 'te·trə,zēn }

tetrol See furan. { 'te,tról }

tetryl [ORG CHEM] (NO₂)₃C₆H₂N(NO₂)CH₃ A yellow, water-insoluble, crystalline explosive material stable at 130°C; used in explosives and ammunition. Also known as tetralite. { 'te·trəl }

TFE See tetrafluoroethylene.

Th See thorium.

thalline [ORG CHEM] C₉H₆N(OCH₃)H₄ Colorless rhomboids soluble in water and melting at 40°C. { 'tha,lēn }

thallium [CHEM] A metallic element in group 13, symbol Tl, atomic number 81, atomic weight 204.383; insoluble in water, soluble in nitric and sulfuric acids, melts at 302°C, boils at 1457°C. { 'thal·ē·əm }

thallium acetate [ORG CHEM] TlOCOCH₃ Toxic, white, deliquescent crystals, soluble in water and alcohol, melts at 131°C; used as an ore-flotation solvent and in medicine. { 'thal·ē·əm 'as·ə,tāt }

thallium bromide [INORG CHEM] TlBr A toxic, yellowish powder soluble in alcohol, slightly soluble in water, melts at 460°C; used in infrared radiation transmitters and detectors. Also known as thallous bromide. { 'thal·ē·əm 'brō,mīd }

thallium carbonate [INORG CHEM] Tl₂CO₃ Toxic, shiny, colorless needles soluble in water, insoluble in alcohol, melts at 272°C; used as an analytical reagent and in artificial gems. Also known as thallous carbonate. { 'thal·ē·əm 'kär·bə,nāt }

thallium chloride [INORG CHEM] TlCl A white, toxic, light-sensitive powder, slightly soluble in water, insoluble in alcohol, melts at 430°C; used as a chlorination catalyst and in medicine and suntan lamps. Also known as thallous chloride. { 'thal·ē·əm 'klór,īd }

thallium hydroxide [INORG CHEM] TlOH·H₂O Toxic yellow, water- and alcohol-soluble needles, decomposes at 139°C; used as an analytical reagent. Also known as thallous hydroxide. { 'thal·ē·əm hī'dräk,sīd }

374

thallium iodide [INORG CHEM] TlI A toxic, yellow powder, insoluble in alcohol, slightly soluble in water, melts at 440°C; used in infrared radiation transmitters and in medicine. Also known as thallous iodide. { 'thal·ē·əm 'ī·ə,dīd }

thallium monoxide [INORG CHEM] Tl_2O A black, toxic, water- and alcohol-soluble powder, melts at 300°C; used as an analytical reagent and in artificial gems and optical glass. Also known as thallium oxide; thallous oxide. { 'thal·ē·əm mə'näk,sīd }

thallium nitrate [INORG CHEM] $TlNO_3$ Colorless, toxic, fire-hazardous crystals soluble in hot water, insoluble in alcohol, melts at 206°C, decomposes at 450°C; used as an analytical reagent and in pyrotechnics. Also known as thallous nitrate. { 'thal·ē·əm 'nī,trāt }

thallium oxide See thallium monoxide. { 'thal·ē·əm 'äk,sīd }

thallium sulfate [INORG CHEM] Tl_2SO_4 Toxic, water-soluble, colorless crystals melting at 632°C; used as an analytical reagent and in medicine, rodenticides, and pesticides. Also known as thallous sulfate. { 'thal·ē·əm 'səl,fāt }

thallium sulfide [INORG CHEM] Tl_2S Lustrous, toxic, blue-black crystals insoluble in water, alcohol, and ether, soluble in mineral acids, melts at 448°C; used in infrared-sensitive devices. Also known as thallous sulfide. { 'thal·ē·əm 'səl,fīd }

thallous bromide See thallium bromide. { 'thal·əs 'brō,mīd }

thallous carbonate See thallium carbonate. { 'thal·əs 'kär·bə,nāt }

thallous chloride See thallium chloride. { 'thal·əs 'klór,īd }

thallous hydroxide See thallium hydroxide. { 'thal·əs hī'dräk,sīd }

thallous iodide See thallium iodide. { 'thal·əs 'ī·ə,dīd }

thallous nitrate See thallium nitrate. { 'thal·əs 'nī,trāt }

thallous oxide See thallium monoxide. { 'thal·əs 'äk,sīd }

thallous sulfate See thallium sulfate. { 'thal·əs 'səl,fāt }

thallous sulfide See thallium sulfide. { 'thal·əs 'səl,fīd }

THAM See tromethamine.

THC See tetrahydrocannabinol.

thenyl [ORG CHEM] $C_4H_3SCH_2-$ An organic radical based on methylthiophene; thus thenyl alcohol is also known as thiophenemethanol. { 'then·əl }

theobromine [ORG CHEM] $C_7H_8N_4O_2$ A toxic alkaloid found in cocoa, chocolate products, tea, and cola nuts; closely related to caffeine. { ‚thē·ə'brō,mēn }

theophylline [ORG CHEM] $C_7H_8N_4O_2 \cdot H_2O$ Alkaloid from tea leaves; bitter-tasting white crystals slightly soluble in water and alcohol, melts at 272°C; used in medicine. { ‚thē·ə'fī,lēn }

thermal analysis [ANALY CHEM] Analytical techniques developed to continuously monitor physical or chemical changes of a sample which occur as the temperature of a sample is increased or decreased. Thermogravimetry, differential thermal analysis, and differential scanning calorimetry are the principal thermoanalytical methods. { 'thər·məl ə'nal·ə·səs }

thermal black [CHEM] A type of carbon black made by a thermal process using natural gas; used in the rubber industry. { 'thər·məl 'blak }

thermal degradation [CHEM] Molecular deterioration of materials (usually organics) because of overheat; can be avoided by low-temperature or vacuum processing, as for foods and pharmaceuticals. { 'thər·məl ‚deg·rə'dā·shən }

thermal diffusion [PHYS CHEM] A phenomenon in which a temperature gradient in a mixture of fluids gives rise to a flow of one constituent relative to the mixture as a whole. Also known as thermodiffusion. { 'thər·məl di'fyü·zhən }

thermal titration See thermometric titration. { 'thər·məl tī'trā·shən }

thermoanalysis See thermal analysis. { ¦thərmo·ə'nal·ə·səs }

thermobalance [ANALY CHEM] An analytical balance modified for thermogravimetric analysis, involving the measurement of weight changes associated with the transformations of matter when heated. { ¦thər·mō'bal·əns }

thermochemistry [PHYS CHEM] The measurement, interpretation, and analysis of heat changes accompanying chemical reactions and changes in state. { ¦thər·mō'kem·ə·strē }

thermodiffusion See thermal diffusion. { ¦thər·mō·di'fyü·zhən }

thermoelectric diffusion potential [PHYS CHEM] A potential difference across an electrolyte that results when a temperature gradient causes one constituent to attempt to flow relative to the other. { ˌthər·mō·iˌlek·trik də'fyü·zhən pəˌten·chəl }

thermogravimetric analysis [ANALY CHEM] Chemical analysis by the measurement of weight changes of a system or compound as a function of increasing temperature. { ˌthər·mōˌgrav·ə'me·trik ə'nal·ə·səs }

thermokinetic analysis [ANALY CHEM] A type of enthalpimetric analysis which uses kinetic titrimetry; involves rapid and continuous automatic delivery of a suitable titrant, under judiciously controlled experimental conditions with temperature measurement; the end points obtained are converted by mathematical procedures into valid stoichiometric equivalence points and used for determining reaction rate constants. { ˌthər·mō·kiˌned·ik ə'nal·ə·səs }

thermolysis See pyrolysis.

thermometric analysis [PHYS CHEM] A method for determination of the transformations a substance undergoes while being heated or cooled at an essentially constant rate, for example, freezing-point determinations. { ˌthər·məˌme·trik ə'nal·ə·səs }

thermometric titration [ANALY CHEM] A titration in an adiabatic system, yielding a plot of temperature versus volume of titrant; used for neutralization, precipitation, redox, organic condensation, and complex-formation reactions. Also known as calorimetric titration; enthalpy titration; thermal titration. { ˌthər·məˌme·trik tī'trā·shən }

thermoplastic elastomer [ORG CHEM] A polymer that can be processed as a thermoplastic material but also possesses the properties of a conventional thermoset rubber. Abbreviated TPE. { ˌthər·məˌpla·stik i'las·tə·mər }

thermotropic liquid crystal [PHYS CHEM] A liquid crystal prepared by heating the substance. { ˌthər·mōˌträp·ik 'lik·wəd 'krist·əl }

THF See tetrahydrofuran.

thiabendazole [ORG CHEM] $C_{10}H_7N_3S$ A white powder with a melting point of 304–305°C; controls fungi on citrus fruits, sugarbeets, turf, and ornamentals, and roundworms of cattle and other animals. Also known as 2-(4-thiazolyl)benzimidazole. { ˌthī·ə'ben·dəˌzōl }

thiacetic acid See thioacetic acid. { ˌthī·əˌsēd·ik 'as·əd }

thiamine hydrochloride [ORG CHEM] $C_{12}H_{17}·ON_4S·HCl$ White, hygroscopic crystals soluble in water, insoluble in ether, with a yeasty aroma and a salty, nutlike taste, decomposes at 247°C; the form in which thiamine is generally employed. { 'thī·ə·mən ˌhī·drə'klòrˌīd }

thianaphthene [ORG CHEM] C_8H_6S A crystalline compound with a melting point of 32°C; soluble in organic solvents; used in the production of pharmaceuticals. Also known as benzothiofuran. { ˌthī·ə'naf,thēn }

thiazole [ORG CHEM] C_3H_3NS A colorless to yellowish liquid with a pyridinelike aroma, slightly soluble in water, soluble in alcohol and ether; used as an intermediate for fungicides, dyes, and rubber accelerators. { 'thī·ə,zōl }

thiazole dye [ORG CHEM] One of a family of dyes in which the chromophore groups are =C=N–, –S–C=, and used mainly for cotton; an example is primuline. { 'thī·ə,zōl 'dī }

2-(4-thiazolyl)benzimidazole See thiabendazole. { ˌtü ˌfȯr thīˌaz·ə,wil ˌbenz,im·ə'da,zōl }

Thiele melting-point apparatus [ANALY CHEM] A stirred, specially shaped test-tube device used for the determination of the melting point of a crystalline chemical. { 'tēl·ə 'melt·iŋ ˌpȯint ˌap·ə,rad·əs }

thin-layer chromatography [ANALY CHEM] Chromatographing on thin layers of adsorbents rather than in columns; adsorbent can be alumina, silica gel, silicates, charcoals, or cellulose. { 'thin ˌlā·ər ˌkrō·mə'täg·rə·fē }

thio- [CHEM] A chemical prefix derived from the Greek theion, meaning sulfur; indicates the replacement of an oxygen in an acid radical by sulfur with a negative valence of 2. { 'thī·ō }

thioacetamide [ORG CHEM] C_2H_5NS A crystalline compound with a melting point of 113–114°C; soluble in water and ethanol; used in laboratories in place of hydrogen sulfide. { ˌthī·ō·ə'sed·ə·mīd }

thioacetic acid [ORG CHEM] CH₃COSH A toxic, clear-yellow liquid with an unpleasant aroma, soluble in water, alcohol, and ether, boils at 82°C; used as an analytical reagent and a lacrimator. Also known as thiacetic acid. { ¦thī·ō·ə¦sēd·ik 'as·əd }

thioaldehyde [ORG CHEM] An organic compound that contains the −CHS radical and has the suffix -thial; for example, ethanethial, CH₃CHS. { ¦thī·ō'al·də,hīd }

thiobarbituric acid [ORG CHEM] C₆H₄N₂O₂S Malonyl thiourea, the parent compound of the thiobarbiturates; represents barbituric acid in which the oxygen atom of the urea component has been replaced by sulfur. { ¦thī·ō,bär·bə'tùr·ik 'as·əd }

thiocarbamide See thiourea. { ¦thī·ō'kär·bə,mīd }

thiocarbanilide [ORG CHEM] CS(NHC₆H₅)₂ A gray powder with a melting point of 148°C; soluble in alcohol and ether; used for making dyes, and as a vulcanization accelerator and ore flotation agent. Also known as sulfocarbanilide. { ¦thī·ō,kär·bə'ni,līd }

thiocyanate [INORG CHEM] A salt of thiocyanic acid that contains the −SCN radical; for example, sodium thiocyanate, NaSCN. Also known as sulfocyanate; sulfocyanide; thiocyanide. { ¦thī·ō'sī·ə,nāt }

thiocyanic acid [INORG CHEM] HSC:N A colorless, water-soluble liquid decomposing at 200°C; used to inhibit paper deterioration due to the action of light, and (in the form of organic esters) as an insecticide. Also known as rhodanic acid; sulfocyanic acid. { ¦thī·ō·sī¦an·ik 'as·əd }

thiocyanide See thiocyanate. { ¦thī·ō'sī·ə,nīd }

thiocyanogen [INORG CHEM] NCSSCN White, light-unstable rhombic crystals melting at −2°C. { ¦thī·ō·sī·'an·ə·jən }

thiodiglycol [ORG CHEM] (CH₂CH₂OH)₂S A combustible, colorless, syrupy liquid soluble in water, alcohol, acetone, and chloroform, boils at 283°C; used as a chemical intermediate, textile-dyeing solvent, and antioxidant. { ¦thī·ō·dī'glī,kól }

thiodiglycolic acid [ORG CHEM] HOOCCH₂SCH₂COOH Combustible, colorless, water- and alcohol-soluble crystals melting at 128°C; used as an analytical reagent. { ¦thī·ō,dī·glī'käl·ik 'as·əd }

3,3′-thiodiproprionic acid [ORG CHEM] (CH₂CH₂COOH)₂S A crystalline compound with a melting point of 134°C; soluble in hot water, acetone, and alcohol; used as an antioxidant for soap products and polymers of ethylene. { ¦thrē ¦thre,prīm ¦thī·ō,prō·prē'än·ik 'as·əd }

thioether [ORG CHEM] RSR A general formula for colorless, volatile organic compounds obtained from alkyl halides and alkali sulfides; the R groups can be the same, or different as in methylthioethane (CH₃SC₂H₅). { ¦thī·ō'ē·thər }

thioethyl alcohol See ethyl mercaptan. { ¦thī·ō'eth·əl 'al·kə,hól }

thioflavine T [ORG CHEM] C₁₆H₁₇N₂Cl A yellow basic dye, used for textile dyeing and fluorescent sign paints. { ¦thī·ō'flā·vən 'tē }

thiofuran See thiophene. { ¦thī·ō'fyù,ran }

thioglycolic acid [ORG CHEM] HSCH₂COOH A liquid with a strong unpleasant odor; used as a reagent for metals such as iron, molybdenum, silver, and tin, and in bacteriology. Also known as mercaptoacetic acid. { ¦thī·ō·glī'käl·ik 'as·əd }

2-thiohydantoin [ORG CHEM] NHC(S)NHC(O)CH₂ Crystals or a tan powder with a melting point of 230°C; used in the manufacture of pharmaceuticals, rubber accelerators, and copper-plating brighteners. Also known as glycolythiourea. { ¦tü ¦thī·ō·hī'dänt·ə,win }

thiol See mercaptan. { 'thī,ól }

thiolactic acid [ORG CHEM] CH₃CH(SH)COOH An oil with a disagreeable odor; used in toiletry preparation. Also known as 2-mercaptopropionic acid; 2-thiolpropionic acid. { ¦thī·ō¦lak·tik 'as·əd }

thiomalic acid [ORG CHEM] C₄H₆O₄S White crystals or powder with a melting point of 149–150°C; soluble in water, alcohol, and acetone; used as a sealer for fuel cells and machine and electrical parts, for caulking compounds, and as a propellant binder. Also known as mercaptosuccinic acid. { ¦thī·ō¦mal·ik 'as·əd }

thionic acid [INORG CHEM] H₂SₓO₆, where x varies from 2 to 6. [ORG CHEM] An organic acid with the radical −CSOH. { thī'än·ik 'as·əd }

thionyl bromide

thionyl bromide [ORG CHEM] $SOBr_2$ A red liquid boiling at 68°C (40 mmHg). Also known as sulfinyl bromide. { 'thī·ən·əl 'brō,mīd }

thionyl chloride [INORG CHEM] $SOCl_2$ A toxic, yellowish to red liquid with a pungent aroma, soluble in benzene, decomposes in water and at 140°C; boils at 79°C; used as a chemical intermediate and catalyst. Also known as sulfur oxychloride; sulfurous oxychloride. { 'thī·ən·əl 'klȯr,īd }

thiopental sodium [ORG CHEM] $C_{11}H_{17}O_2N_2NaS$ Yellow, water-soluble crystals with a characteristic aroma; used in medicine as a short-acting anesthetic. Also known as thiopentone sodium. { ˈthī·ō'pen,tal 'sōd·ē·əm }

thiopentone sodium See thiopental sodium. { ˈthī·ō'pen,tēn 'sōd·ē·əm }

thiophanate [ORG CHEM] $C_{14}H_{18}N_4O_4S_2$ A tan to colorless solid that decomposes at 195°C; slightly soluble in water; used to control fungus diseases of turf. { thī·ō'fa,nāt }

thiophene [ORG CHEM] C_4H_4S A toxic, flammable, highly reactive, colorless liquid, insoluble in water, soluble in alcohol and ether, boils at 84°C; used as a chemical intermediate and to make condensation copolymers. Also known as thiofuran. { 'thī·ə,fēn }

thiophenol [ORG CHEM] C_6H_5SH A toxic, fire-hazardous, water-white liquid with a disagreeable aroma, insoluble in water, soluble in alcohol and ether, boils at 168°C; used to make pharmaceuticals. Also known as phenyl mercaptan. { ˈthī·ō'fē,nȯl }

thiosalicylic acid [ORG CHEM] $HOOCC_6H_4SH$ A yellow solid with a melting point of 164–165°C; soluble in alcohol, ether, and acetic acid; used for making dyes. Also known as 2-mercaptobenzoic acid. { ˈthī·ō,salˈəˈsil·ik 'as·əd }

thiosemicarbazide [ORG CHEM] $NH_2CSNHNH_2$ A white, water- and alcohol-soluble powder melting at 182°C; used as an analytical reagent in photography and rodenticides. { ˈthī·ō,sem·iˈkär·bə,zīd }

thiosulfate [INORG CHEM] $M_2S_2O_3$ A salt of thiosulfuric acid and a base; for example, reaction of sodium hydroxide and thiosulfuric acid to produce sodium thiosulfate. { ˈthī·ə'səl,fāt }

thiosulfonic acid [ORG CHEM] Name for a group of oxy acids of sulfur, with the general formula RS_2O_2H; they are known as esters and salts. { ˈthī·ə,səl'fän·ik 'as·əd }

thiosulfuric acid [INORG CHEM] $H_2S_2O_3$ An unstable acid that decomposes readily to form sulfur and sulfurous acid. { ˈthī·ə,səl'fyu̇r·ik 'as·əd }

thiourea [ORG CHEM] $(NH_2)_2CS$ Bitter-tasting, white crystals with a melting point of 180–182°C; soluble in cold water and alcohol; used in photography and photocopying, as a rubber accelerator, and as an antithyroid drug in treating hyperthyroidism. Also known as thiocarbamide. { ˈthī·ō·yu̇'rē·ə }

thiram [ORG CHEM] $(CH_3)_2NCSSSCSN(CH_3)_2$ (tetramethylthioperoxydicarbonic diamide) A fungicide, bacteriostat (in soap), antimicrobial agent (chemotherapeutic for plants), seed disinfectant, and vulcanizing agent. { 'thī,ram }

third-order reaction [PHYS CHEM] A chemical reaction in which the rate of reaction is determined by the concentration of three reactants. { 'thərd ˈȯr·dər rē'ak·shən }

thiuram [ORG CHEM] A chemical compound containing a R_2NCS radical; occurs mainly in disulfide compounds; the most common monosulfide compound is tetramethylthiuram monosulfide. { 'thī·yə,ram }

thixotropy [PHYS CHEM] Property of certain gels which liquefy when subjected to vibratory forces, such as ultrasonic waves or even simple shaking, and then solidify again when left standing. { thik'sä·trə·pē }

Thomson-Berthelot principle [PHYS CHEM] The assumption that the heat released in a chemical reaction is directly related to the chemical affinity, and that, in the absence of the application of external energy, that chemical reaction which releases the greatest heat is favored over others; the principle is in general incorrect, but applies in certain special cases. { 'täm·sən ber·tə'lō ,prin·sə·pəl }

thoria See thorium dioxide. { 'thȯr·ē·ə }

thorium [CHEM] An element of the actinium series, symbol Th, atomic number 90, atomic weight 232; soft, radioactive, insoluble in water and alkalies, soluble in acids, melts at 1750°C, boils at 4500°C. { 'thȯr·ē·əm }

thorium anhydride See thorium dioxide. { 'thȯr·ē·əm an'hī,drīd }

thorium carbide [INORG CHEM] ThC$_2$ A yellow solid melting at above 2630°C, decomposes in water; used in nuclear fuel. { 'thȯr·ē·əm 'kär,bīd }

thorium chloride [INORG CHEM] ThCl$_4$ Hygroscopic, toxic colorless crystal needles soluble in alcohol, melts at 820°C, decomposes at 928°C; used in incandescent lighting. Also known as thorium tetrachloride. { 'thȯr·ē·əm 'klȯr,īd }

thorium dioxide [INORG CHEM] ThO$_2$ A heavy, white powder soluble in sulfuric acid, insoluble in water, melts at 3300°C; used in medicine, ceramics, flame spraying, and electrodes. Also known as thoria; thorium anhydride; thorium oxide. { 'thȯr·ē·əm dī'äk,sīd }

thorium fluoride [INORG CHEM] ThF$_4$ A white, toxic powder, melts at 1111°C; used to make thorium metal and magnesium-thorium alloys and in high-temperature ceramics. { 'thȯr·ē·əm 'flu̇r,īd }

thorium nitrate [INORG CHEM] Th(NO$_3$)$_4$·4H$_2$O Explosive white crystals soluble in water and alcohol, strong oxidizer; the anhydrous form decomposes at 500°C; used in medicine and as an analytical reagent. { 'thȯr·ē·əm 'nī,trāt }

thorium oxalate [ORG CHEM] Th(C$_2$O$_4$)$_2$·2H$_2$O A white, toxic powder soluble in alkalies and ammonium oxalate, insoluble in water and most acids, decomposes to thorium dioxide, ThO$_2$, above 300–400°C; used in ceramics. { 'thȯr·ē·əm 'äk·sə,lāt }

thorium oxide *See* thorium dioxide. { 'thȯr·ē·əm 'äk,sīd }

thorium sulfate [INORG CHEM] Th(SO$_4$)$_2$·8H$_2$O A white powder soluble in ice water, loses water at 42° and 400°C. Also known as normal thorium sulfate. { 'thȯr·ē·əm 'səl,fāt }

thorium tetrachloride *See* thorium chloride. { 'thȯr·ē·əm ¦te·trə'klȯr,īd }

Thorpe reaction [ORG CHEM] The reaction by which, in presence of lithium amides, α,ω-dinitriles undergo base-catalyzed condensation to cyclic iminonitriles, which can be hydrolyzed and decarboxylated to cyclic ketones. { 'thȯrp rē,ak·shən }

THPC *See* tetrakis(hydroxymethyl)phosphonium chloride.

thulia *See* thulium oxide. { 'thü·lē·ə }

thulium [CHEM] A rare-earth element, symbol Tm, of the lanthanide group, atomic number 69, atomic weight 168.9342; reacts slowly with water, soluble in dilute acids, melts at 1550°C, boils at 1727°C; the dust is a fire hazard; used as x-ray source and to make ferrites. { 'thü·lē·əm }

thulium chloride [INORG CHEM] TmCl$_3$·7H$_2$O Green, deliquescent crystals soluble in water and alcohol; melts at 824°C. { 'thü·lē·əm 'klȯr,īd }

thulium oxalate [ORG CHEM] Tm$_2$(C$_2$O$_4$)$_3$·6H$_2$O A toxic, greenish-white solid, soluble in aqueous alkali oxalates, loses one water at 50°C; used for analytical separation of thulium from common metals. { 'thü·lē·əm 'äk·sə,lāt }

thulium oxide [INORG CHEM] Tm$_2$O$_3$ A white, slightly hygroscopic powder that absorbs water and carbon dioxide from the air, and is slowly soluble in strong acids; used to make thulium metal. Also known as thulia. { 'thü·lē·əm 'äk,sīd }

thyme camphor *See* thymol. { 'tīm ,kam·fər }

thymol [ORG CHEM] C$_{10}$H$_{14}$O A naturally occurring crystalline phenol obtained from thyme or thyme oil, melting at 515°C; used to kill parasites in herbaria, to preserve anatomical specimens, and in medicine as a topical antifungal agent. Also known as thyme camphor. { 'thī,mȯl }

thymol blue [ORG CHEM] C$_6$H$_4$SO$_2$OC[C$_6$H$_2$(CH$_3$)(OH)CH(CH$_3$)$_2$]$_2$ Brown-green crystals soluble in alcohol and dilute alkalies, insoluble in water, decomposes at 223°C; used as acid-base pH indicator. { 'thī,mȯl 'blü }

thymol iodide [ORG CHEM] [C$_6$H$_2$(CH$_3$)(OI)(C$_3$H$_7$)]$_2$ A red-brown, light-sensitive powder with an aromatic aroma, soluble in ether and chloroform, insoluble in water; used in medicine and as a feed additive. { 'thī,mȯl 'ī·ə,dīd }

thymolphthalein [ORG CHEM] C$_6$H$_4$COOC[C$_6$H$_2$(CH$_3$)(OH)CH(CH$_3$)$_2$]$_2$ A white powder insoluble in water, soluble in alcohol and acetone, melts at 245°C; used in medicine and as an acid-base titration indicator. { ¦thī,mȯl¦thā,lēn }

Ti *See* titanium.

tiba [ORG CHEM] C$_7$H$_3$I$_3$O$_2$ A colorless solid with a melting point of 226–228°C; insoluble in water; used as a growth regulator for fruit. { 'tī·bə }

tie line

tie line [PHYS CHEM] A line on a phase diagram joining the two points which represent the composition of systems in equilibrium. Also known as conode. { 'tī ,līn }

tight ion pair [ORG CHEM] An ion pair composed of individual ions which keep their stereochemical configuration; no solvent molecules separate the cation and anion. Also known as contact ion pair; intimate ion pair. { 'tīt 'ī,än ,per }

time-of-flight mass spectrometer [SPECT] A mass spectrometer in which all the positive ions of the material being analyzed are ejected into the drift region of the spectrometer tube with essentially the same energies, and spread out in accordance with their masses as they reach the cathode of a magnetic electron multiplier at the other end of the tube. { ¦tīm əv ¦flīt 'mas spek'träm·əd·ər }

time-resolved laser spectroscopy [SPECT] A method of studying transient phenomena in the interaction of light with matter through the exposure of samples to extremely short and intense pulses of laser light, down to subnanosecond or subpicosecond duration. { 'tīm ri¦zälvd 'lā·zər spek'träs·kə·pē }

tin [CHEM] Metallic element in group IV, symbol Sn, atomic number 50, atomic weight 118.69; insoluble in water, soluble in acids and hot potassium hydroxide solution; melts at 232°C, boils at 2260°C. { tin }

tin bisulfide See stannic sulfide. { 'tin bī'səl,fīd }

tin bromide See stannic bromide; stannous bromide.

tin chloride See stannic chloride; stannous chloride. { 'tin 'klȯr,īd }

tin chromate See stannic chromate; stannous chromate. { 'tin 'krō,māt }

tin crystals See stannous chloride. { 'tin ,krist·əlz }

tin dichloride See stannous chloride. { 'tin dī'klȯr,īd }

tin difluoride See stannous fluoride. { 'tin dī'flu̇r,īd }

tin dioxide See stannic oxide. { 'tin dī'äk,sīd }

tin fluoride See stannous fluoride. { 'tin 'flu̇r,īd }

tin hydride [INORG CHEM] SnH₄ A gas boiling at −52°C. Also known as stannane. { 'tin 'hī,drīd }

tin iodide See stannic iodide. { 'tin 'ī·ə,dīd }

tin monosulfide See stannous sulfide. { 'tin ¦män·ə'səl,fīd }

tin oxalate See stannous oxalate. { 'tin 'äk·sə,lāt }

tin oxide See stannic oxide; stannous oxide. { 'tin 'äk,sīd }

tin peroxide See stannic oxide. { 'tin pə'räk,sīd }

tin protosulfide See stannous sulfide. { 'tin ¦prōd·ō'səl,fīd }

tin protoxide See stannous oxide. { 'tin prə'täk,sīd }

tin salts See stannous chloride. { 'tin ,sȯlts }

tin sulfate See stannous sulfate. { 'tin 'səl,fāt }

tin sulfide See stannous sulfide. { 'tin 'səl,fīd }

tin tetrabromide See stannic bromide. { 'tin ¦te·trə'brō,mīd }

tin tetrachloride See stannic chloride. { 'tin ¦te·trə'klȯr,īd }

tin tetraiodide See stannic iodide. { 'tin ¦te·trə'ī·ə,dīd }

Tischenko reaction [ORG CHEM] The formation of an ester by the condensation of two molecules of aldehyde utilizing a catalyst of aluminum alkoxides in the presence of a halide. { ti'shəŋ·kō rē,ak·shən }

titanate [INORG CHEM] A salt of titanic acid; titanates of the M₂TiO₃ type are called metatitanates, those of the M₄TiO₄ type are called orthotitanates; an example is sodium titanate, (Na₂O)₂Ti₂O₅. { 'tīt·ən,āt }

titanellow See titanium trioxide. { ¦tīt·ən'el·ō }

titania See titanium dioxide. { tī'tā·nē·ə }

titanic acid [INORG CHEM] H₂TiO₃ A white, water-insoluble powder; used as a dyeing mordant. Also known as metatitanic acid; titanic hydroxide. { tī'tan·ik 'as·əd }

titanic anhydride See titanium dioxide. { tī'tan·ik an'hī,drīd }

titanic chloride See titanium tetrachloride. { tī'tan·ik 'klȯr,īd }

titanic hydroxide See titanic acid. { tī'tan·ik hī'dräk,sīd }

titanic sulfate See titanium sulfate. { tī'tan·ik 'səl,fāt }

titanium [CHEM] A metallic transition element, symbol Ti, atomic number 22, atomic weight 47.90; ninth most abundant element in the earth's crust; insoluble in water, melts at 1660°C, boils above 3000°C.   { tī′tā·nē·əm }

titanium boride   [INORG CHEM]   TiB₂ A hard solid that resists oxidation at elevated temperatures and melts at 2980°C; used as a refractory and in alloys, high-temperature electrical conductors, and cermets.   { tī′tā·nē·əm ′bȯr,īd }

titanium carbide   [INORG CHEM]   TiC Very hard gray crystals insoluble in water, soluble in nitric acid and aqua regia, melts at about 3140°C; used in cermets, arc-melting electrodes, and tungsten-carbide tools.   { tī′tā·nē·əm ′kär,bīd }

titanium chloride   See titanium dichloride.   { tī′tā·nē·əm ′klȯr,īd }

titanium dichloride   [INORG CHEM]   TiCl₂ A flammable, alcohol-soluble, black powder that decomposes in water, and in vacuum at 475°C, and burns in air.   Also known as titanium chloride.   { tī′tā·nē·əm dī′klȯr,īd }

titanium dioxide   [INORG CHEM]   TiO₂ A white, water-insoluble powder that melts at 1560°C, and which is produced commercially from the titanium dioxide minerals ilmenite and rutile; used in paints and cosmetics.   Also known as titania; titanic anhydride; titanium oxide; titanium white.   { tī′tā·nē·əm dī′äk,sīd }

titanium hydride   [INORG CHEM]   TiH₂ A black metallic powder whose dust is an explosion hazard and which dissociates above 288°C; used in powder metallurgy, hydrogen production, foamed metals, glass solder, and refractories, and as an electronic gas getter.   { tī′tā·nē·əm ′hī,drīd }

titanium nitride   [INORG CHEM]   TiN Golden-brown brittle crystals melting at 2927°C; used in refractories, alloys, cermets, and semiconductors.   { tī′tā·nē·əm ′nī,trīd }

titanium oxalate   [ORG CHEM]   Ti₂(C₂O₄)₃·10H₂O Toxic, yellow prisms soluble in water, insoluble in alcohol; used to make titanic acid and titanium metal.   Also known as titanous oxalate.   { tī′tā·nē·əm ′äk·sə,lāt }

titanium oxide   See titanium dioxide; titanium trioxide.   { tī′tā·nē·əm ′äk,sīd }

titanium peroxide   See titanium trioxide.   { tī′tā·nē·əm pə′räk,sīd }

titanium sesquisulfate   See titanous sulfate.   { tī′tā·nē·əm ¦ses·kwə′səl,fāt }

titanium sulfate   [INORG CHEM]   Ti(SO₄)₂·9H₂O Caked solid, soluble in water, toxic, highly acidic; used as a dye stripper, reducing agent, laundry chemical, and in treatment of chrome yellow colors.   Also known as titanic sulfate; titanyl sulfate.   { tī′tā·nē·əm ′səl,fāt }

titanium tetrachloride   [INORG CHEM]   TiCl₄ A colorless, toxic liquid soluble in water, fumes when exposed to moist air, boils at 136°C; used to make titanium and titanium salts, as a dye mordant and polymerization catalyst, and in smoke screens and pigments.   Also known as titanic chloride.   { tī′tā·nē·əm ¦te·trə′klȯr,īd }

titanium trichloride   [INORG CHEM]   TiCl₃ Toxic, dark-violet, deliquescent crystals soluble in alcohol and some amines, decomposes in water with heat evolution, decomposes above 440°C; used as a reducing agent, chemical intermediate, polymerization catalyst, and laundry stripping agent.   Also known as titanous chloride.   { tī′tā·nē·əm trī′klȯr,īd }

titanium trioxide   [INORG CHEM]   TiO₃ Yellow titanium oxide used to make ivory shades in ceramics.   Also known as titanellow; titanium oxide; titanium peroxide.   { tī′tā·nē·əm trī′äk,sīd }

titanium white   See titanium dioxide.   { tī′tā·nē·əm ′wīt }

titanous chloride   See titanium trichloride.   { tī′tan·əs ′klȯr,īd }

titanous oxalate   See titanium oxalate.   { tī′tan·əs ′äk·sə,lāt }

titanous sulfate   [INORG CHEM]   Ti₂(SO₄)₃ Green crystals soluble in dilute hydrochloric and sulfuric acids, insoluble in water and alcohol; used as a textile reducing agent. Also known as titanium sesquisulfate.   { tī′tan·əs ′səl,fāt }

titanyl sulfate   See titanium sulfate.   { ′tīt·ən·əl ′səl,fāt }

titer   [CHEM]   **1.** The concentration in a solution of a dissolved substance as shown by titration.   **2.** The least amount or volume needed to give a desired result in titration. **3.** The solidification point of hydrolyzed fatty acids.   { ′tī·tər }

titrand   [ANALY CHEM]   The substance that is analyzed in a titration procedure. { ′tī,trand }

titrant [ANALY CHEM] A solution of known concentration and composition used for analytical titrations. Also known as standard solution. { 'tī·trənt }

titration [ANALY CHEM] A method of analyzing the composition of a solution by adding known amounts of a standardized solution until a given reaction (color change, precipitation, or conductivity change) is produced. { ti'trā·shən }

titrimetric analysis See volumetric analysis. { ¦tī·trə‚me·trik ə'nal·ə·səs }

Tl See thallium.

Tm See thulium.

TMA See trimethylamine.

TNF See 2,4,7-trinitrofluorenone.

TNT See 2,4,6-trinitrotoluene.

tocopherol [ORG CHEM] Any of several substances having vitamin E activity that occur naturally in certain oils; α-tocopherol possesses the highest biological activity. { tə'käf·ə‚röl }

tolazoline hydrochloride [ORG CHEM] $C_{10}H_{12}N_2 \cdot HCl$ Water-soluble white crystals, melting at 173°C; used as a sympatholytic and vasodilator. Also known as priscol. { täl'az·ə‚lēn ¦hī·drə'klör‚īd }

tolerance interval [ANALY CHEM] That range of values within which it has been calculated that a specified percentage of individual values of measurements will lie with a stated confidence level. { 'täl·ə·rəns ‚in·tər·vəl }

ortho-tolidine [ORG CHEM] [C₆H₃(CH₃)NH₂] Light-sensitive, combustible white to reddish crystals soluble in alcohol and ether, slightly soluble in water, melts at 130°C; used as an anlytical reagent and a curing agent for urethane resins. { ¦ör·thō 'täl·ə‚dēn }

Tollen's aldehyde test [ANALY CHEM] A test that uses an ammoniacal solution of silver oxides to test for aldehydes and ketones. { 'täl·ənz 'al·də‚hīd ‚test }

tolnaftate [ORG CHEM] $C_{19}H_{17}NOS$ An agricultural fungicide; it is also used medically as an antifungal agent. { tōl'naf‚tāt }

toluene [ORG CHEM] $C_6H_5CH_3$ A colorless, aromatic liquid derived from coal tar or from the catalytic reforming of petroleum naphthas; insoluble in water, soluble in alcohol and ether, boils at 111°C; used as a chemical intermediate, for explosives, and in high-octane gasolines. Also known as methylbenzene; phenylmethane; toluol. { 'täl·yə‚wēn }

para-toluenesulfonic acid [ORG CHEM] $C_6H_4(SO_3H)(CH_3)$ Toxic, colorless, combustible crystals soluble in water, alcohol, and ether; melts at 107°C; used in dyes and as a chemical intermediate and organic catalyst. { ¦par·ə ¦täl·yə‚wēn¦səl¦fän·ik 'as·əd }

toluene 2,4-diisocyanate [ORG CHEM] $CH_3C_6H_3(NCO)_2$ A liquid (at room temperature) with a sharp, pungent odor; miscible with ether, acetone, and benzene; used to make polyurethane foams and other elastomers, and also as a protein cross-linking agent. { 'täl·yə‚wēn ¦tü ¦för dī¦ī·sō'sī·ə‚nāt }

α-toluic acid See phenylacetic acid. { ¦al·fə tə'lü·ik 'as·əd }

meta-toluic acid [ORG CHEM] $C_6H_4CH_3COOH$ White to yellow, combustible crystals soluble in alcohol and ether, slightly soluble in water, melts at 109°C; used as a chemical intermediate and base for insect repellants. Also known as meta-toluylic acid. { ¦med·ə tə'lü·ik 'as·əd }

ortho-toluic acid [ORG CHEM] $C_6H_4CH_3COOH$ White, combustible crystals soluble in alcohol and chloroform, slightly soluble in water, melts at 104°C; used as a bacteriostat. Also known as ortho-toluylic acid. { ¦ör·thō tə'lü·ik 'as·əd }

para-toluic acid [ORG CHEM] $C_6H_4CH_3COOH$ Transparent, combustible crystals soluble in alcohol and ether, slightly soluble in water, melts at 180°C; used in agricultural chemicals and as an animal feed supplement. Also known as para-toluylic acid. { ¦par·ə tə'lü·ik 'as·əd }

α-toluic aldehyde See phenylacetaldehyde. { ¦al·fə tə'lü·ik 'al·də‚hīd }

meta-toluidine [ORG CHEM] $CH_3C_6H_4NH_2$ A combustible, colorless, toxic liquid soluble in alcohol and ether, slightly soluble in water, boils at 203°C; used for dyes and as a chemical intermediate. { ¦med·ə tə'lü·ə‚dēn }

ortho-toluidine [ORG CHEM] $CH_3C_6H_4NH_2$ A light-green, light-sensitive, combustible, toxic liquid soluble in alcohol and ether, very slightly soluble in water, boils at 200°C; used for dyes and textile printing and as a chemical intermediate.   { ¦ȯr·thō tə'lü·ə‚dēn }

para-toluidine   [ORG CHEM] $CH_3C_6H_4NH_2$ Toxic, combustible, white leaflets soluble in alcohol and ether, very slightly soluble in water, boils at 200°C; used as an analytical reagent and in dyes.   { ¦par·ə tə'lü·ə‚dēn }

toluol *See* toluene.   { 'täl·yə‚wȯl }

toluylene *See* stilbene.   { 'täl·yə·wə‚lēn }

toluylene red *See* neutral red.   { 'täl·yə·wə‚lēn 'red }

***meta*-toluylic acid** *See meta*-toluic acid.   { ¦med·ə ¦täl·yə¦wil·ik 'as·əd }

***ortho*-toluylic acid** *See ortho*-toluic acid.   { ¦ȯr·thō ¦täl·yə¦wil·ik 'as·əd }

***para*-toluylic acid** *See para*-toluic acid.   { ¦par·ə ¦täl·yə¦wil·ik 'as·əd }

***para*-tolylsulfonylmethylnitrosamide**   [ORG CHEM] $C_8H_{10}N_2O_3S$ Yellow crystals with a melting point of 62°C; soluble in ether, petroleum ether, benzene, carbon tetrachloride, and chloroform; a precursor to diazomethane; a useful reagent for the preparation of a wide range of biologically active compounds for gas chromatography analysis.   { ¦par·ə ¦tä‚lil¦səl·fə‚nil¦meth·əl‚nī'trō·sə‚mīd }

tomatine   [ORG CHEM] $C_{50}H_{83}NO_{21}$ A glycosidal alkaloid obtained from the leaves and stems from the tomato plant; the crude extract is known as tomatin: white, toxic crystals; used as a plant fungicide and as a precipitating agent for cholesterol.   { 'täm·ə‚tēn }

topochemical control   [CHEM] In a chemical reaction, product formation that is determined by the orientation of molecules in the crystal.   { ¦täp·ə'kem·ə·kəl kən'trōl }

torsional angle   [PHYS CHEM] The angle between bonds on adjacent atoms.   { 'tȯr·shən·əl 'aŋ·gəl }

total heat of dilution *See* heat of dilution.   { 'tōd·əl 'hēt əv di'lü·shən }

total heat of solution *See* heat of solution.   { 'tōd·əl 'hēt əv sə'lü·shən }

total solids   [CHEM] The total content of suspended and dissolved solids in water.   { 'tōd·əl 'säl·ədz }

TPA *See* terephthalic acid.

TPE *See* thermoplastic elastomer.

trace analysis   [ANALY CHEM] Analysis of a very small quantity of material of a sample by such techniques as polarography or spectroscopy.   { 'trās ə‚nal·ə·səs }

trace element   [ANALY CHEM] An element in a sample that has an average concentration of less than 100 parts per million atoms or less than 100 micrograms per gram.   { 'trās ‚el·ə·mənt }

tracer   [CHEM] A foreign substance, usually radioactive, that is mixed with or attached to a given substance so the distribution or location of the latter can later be determined; used to trace chemical behavior of a natural element in an organism.   Also known as tracer element.   { 'trā·sər }

tracer element *See* tracer.   { 'trā·sər ‚el·ə·mənt }

transactinide elements   [CHEM] In the periodic table, elements with atomic numbers higher than 103.   { tranz'ak·tə‚nīd ‚el·ə·məns }

transamination   [CHEM] **1.** The transfer of one or more amino groups from one compound to another.   **2.** The transposition of an amino group within a single compound.   { tran‚sam·ə'nā·shən }

transesterification   [ORG CHEM] Conversion of an organic acid ester into another ester of that same acid.   { ¦trans·e¦ster·ə·fə'kā·shən }

transference number *See* transport number.   { tranz'fər·əns ‚nəm·bər }

transition boiling   [PHYS CHEM] A stage in the boiling process that follows fully developed nucleate boiling, precedes film boiling, and has features of both of those stages, in which a decrease in the heat flux accompanies an increase in wall superheat, making it highly unstable.   { tran‚zish·ən 'bȯil·iŋ }

transition element   [CHEM] One of a group of metallic elements in which the members have the filling of the outermost shell to 8 electrons interrupted to bring the penultimate shell from 8 to 18 or 32 electrons; includes elements 21 through 29 (scandium through copper), 39 through 47 (yttrium through silver), 57 through 79 (lanthanum through gold), and all known elements from 89 (actinium) on. Also known as transition metal. 〈 tran'zish·ən ˌel·ə·mənt 〉

transition interval [ANALY CHEM] In a titrimetric analysis, the range in concentration of the species being determined over which a variation in a chemical indicator can be observed visually. 〈 tran'zish·ən ˌin·tər·vəl 〉

transition metal See transition element. 〈 tran'zish·ən ˌmed·əl 〉

transition state See activated complex. 〈 tran'zish·ən ˌstāt 〉

transition temperature [CHEM] The temperature at which an enantiotropic polymorph is converted into a different form. 〈 tran'zish·ən ˌtem·prə·chər 〉

transition time [ANALY CHEM] The time interval needed for a working (nonreference) electrode to become polarized during chronopotentiometry (time-measurement electrolysis of a sample). 〈 tran'zish·ən ˌtīm 〉

translational energy [PHYS CHEM] The kinetic energy of gaseous or liquid molecules that is associated with their motion within their particular chemical systems. 〈 tranˈslā·shən·əl 'en·ər·jē 〉

transmission diffraction [ANALY CHEM] A type of electron diffraction analysis in which the electron beam is transmitted through a thin film or powder whose smallest dimension is no greater than a few tenths of a micrometer. 〈 tranz'mish·ən di,frak·shən 〉

transmittance [ANALY CHEM] During absorption spectroscopy, the amount of radiant energy transmitted by the solution under analysis. 〈 tranz'mid·əns 〉

transpassive region [PHYS CHEM] That portion of an anodic polarization curve in which metal dissolution increases as the potential becomes noble. 〈 tranz'pas·iv ˈrē·jən 〉

transplutonium element [INORG CHEM] An element having an atomic number greater than that of plutonium (94). 〈 ˈtranz·plə'tō·nē·əm 'el·ə·mənt 〉

transport number [PHYS CHEM] The fraction of the total current carried by a given ion in an electrolyte. Also known as transference number. 〈 'tranz,pȯrt ˌnəm·bər 〉

transuranic elements [CHEM] Elements that have atomic numbers greater than 92; all are radioactive and are products of artificial nuclear changes. Also known as transuranium elements. 〈 tranz·yü'ran·ik 'el·ə·məns 〉

transuranium elements See transuranic elements. 〈 ˈtranz·yü'rā·nē·əm 'el·ə·məns 〉

trapping [CHEM] A method for intercepting a reactive intermediate or molecule and removing it from the system or converting it to a more stable form for further study and identification. 〈 'trap·iŋ 〉

Traube's rule [PHYS CHEM] In dilute solutions, the concentration of a member of a homologous series at which a given lowering of surface tension is observed decreases threefold for each additional methylene group in a given series. 〈 'traů·bəz ˌrül 〉

tretamine See triethylenemelamine. 〈 'tred·ə,mēn 〉

triacetin [ORG CHEM] $C_3H_5(CO_2CH_3)_3$ A colorless, combustible oil with a bitter taste and a fatty aroma; found in cod liver and butter; soluble in alcohol and ether, slightly soluble in water; boils at 259°C; used in plasticizers, perfumery, cosmetics, and external medicine and as a solvent and food additive. 〈 trī'as·əd·ən 〉

triamcinolone [ORG CHEM] $C_{21}H_{27}FO_6$ White, toxic crystals; insoluble in water, soluble in dimethylformamide; melts at 266°C; used as an intermediate for ion-exchange resin, wetting and frothing agent, and photographic developer. 〈 ˌtrī·am'sin·əl,än 〉

triamylamine [ORG CHEM] $(C_5H_{11})_3N$ A combustible, colorless, toxic liquid; soluble in gasoline, insoluble in water; used to inhibit corrosion and in insecticides. 〈 ˌtrī· ə'mil·ə,mēn 〉

triamyl borate [ORG CHEM] $(C_5H_{11})_3BO_3$ A combustible, colorless liquid with an alcoholic aroma; soluble in alcohol and ether; boils at 220–280°C; used in varnishes. 〈 trī'am·əl 'bȯr,āt 〉

triatomic [CHEM] Consisting of three atoms. 〈 ˈtrī·əˈtäm·ik 〉

triazole [ORG CHEM] A five-membered chemical ring compound with three nitrogens in the ring; for example, $C_2H_3N_3$; proposed for use as a photoconductor and for copying systems. 〈 'trī·ə,zōl 〉

tribasic calcium phosphate See calcium phosphate. 〈 trī'bā·sik 'kal·sē·əm 'fä,sfāt 〉

tribasic zinc phosphate See zinc phosphate. 〈 trī'bā·sik 'ziŋk 'fä,sfāt 〉

tributoxyethyl phosphate [ORG CHEM] [CH$_3$(CH$_2$)$_3$O(CH$_2$)$_2$O]PO A light yellow, oily liquid with a boiling range of 215–228°C; soluble in organic solvents; used as a plasticizer and flame retardant, and in floor waxes. { ‚trī·byü¦täk·sē'eth·əl 'fä‚sfāt }

tributyl borate [ORG CHEM] (C$_4$H$_9$)$_3$BO$_3$ A combustible, water-white liquid miscible with common organic liquids; boils at 232°C; used in welding fluxes and as a chemical intermediate and textile flame-retardant. { trī'byüd·əl 'bȯr‚āt }

tributyl phosphate [ORG CHEM] (C$_4$H$_9$)$_3$PO$_4$ A combustible, toxic, stable liquid; soluble in most solvents, and very slightly soluble in water; boils at 292°C; used as a heat-exchange medium, pigment-grinding assistant, antifoam agent, and solvent. Abbreviated TBP. { trī'byüd·əl 'fä‚sfāt }

tributyltin acetate [ORG CHEM] (C$_4$H$_9$)$_3$SnOOCCH$_3$ An organic compound of tin, used as an antimicrobial agent in the paper, wood, plastics, leather, and textile industries. { trī'byüd·əl·tən 'as·ə‚tāt }

tributyltin chloride [ORG CHEM] (C$_4$H$_9$)$_3$SnCl A colorless liquid with a boiling point of 145–147°C; soluble in alcohol, benzene, and other organic solvents; used as a rodenticide. { trī'byüd·əl·tən 'klȯr‚īd }

tricaine [ORG CHEM] C$_{10}$H$_{15}$NO$_5$S Fine, needlelike crystals, soluble in water; used as an anesthetic for fish. { 'trī‚kān }

tricalcium phosphate See calcium phosphate. { trī'kal·sē·əm 'fä‚sfāt }

trichloroacetaldehyde See chloral. { trī‚klȯr·ō‚as·ə'tal·də‚hīd }

trichloroacetic acid [ORG CHEM] CCl$_3$COOH Toxic, deliquescent, colorless crystals with a pungent aroma; soluble in water, alcohol, and ether; boils at 198°C; used as a chemical intermediate and laboratory reagent, and in medicine, pharmacy, and herbicides. Abbreviated TCA. { trī¦klȯr·ō·ə¦sed·ik 'as·əd }

trichloroacetic aldehyde See chloral. { trī¦klȯr·ō·ə¦sed·ik 'al·də‚hīd }

trichlorobenzene [ORG CHEM] C$_6$H$_3$Cl$_3$ Either of two toxic compounds: 1,2,3-trichlorobenzene forms white crystals, soluble in ether, insoluble in water, boiling at 221°C, and is used as a chemical intermediate; 1,2,4-trichlorobenzene is a combustible, colorless liquid, soluble in most organic solvents and oils, insoluble in water, boiling at 213°C, and is used as a solvent and in dielectric fluids, synthetic transformer oils, lubricants, and insecticides. { trī¦klȯr·ō'ben‚zēn }

trichloroethanal See chloral. { trī¦klȯr·ō'eth·ə‚nal }

trichloroethane [ORG CHEM] C$_2$H$_3$Cl$_3$ Either of two nonflammable, irritating liquid isomeric compounds: 1,1,1-trichloroethane (CH$_3$CCl$_3$) is toxic, soluble in alcohol and ether, insoluble in water, and boils at 75°C; it is used as a solvent, aerosol propellant, and pesticide and for metal degreasing, and is also known as methyl chloroform; 1,1,2-trichloroethane (CHCl$_2$CH$_2$Cl) is clear and colorless, is soluble in alcohols, ethers, esters, and ketones, insoluble in water, has a sweet aroma, and boils at 114°C; it is used as a chemical intermediate and solvent, and is also known as vinyl trichloride. { trī¦klȯr·ō'eth‚ān }

trichloroethylene [ORG CHEM] CHCl:CCl$_2$ A heavy, stable, toxic liquid with a chloroform aroma; slightly soluble in water, soluble with greases and common organic solvents; boils at 87°C; used for metal degreasing, solvent extraction, and dry cleaning and as a fumigant and chemical intermediate. { trī¦klȯr·ō'eth·ə‚lēn }

trichlorofluoromethane [ORG CHEM] CCl$_3$F A toxic, noncombustible, colorless liquid boiling at 24°C; used as a chemical intermediate, solvent, refrigerant, aerosol prepellant, and blowing agent (plastic foams) and in fire extinguishers. Also known as fluorocarbon-11; fluorotrichloromethane. { trī¦klȯr·ō¦flür·ō'meth‚ān }

trichloroiminocyanuric acid See trichloroisocyanuric acid. { trī¦klȯr·ō¦im·ə·nō¦sī·ə¦nür·ik 'as·əd }

trichloroisocyanuric acid [ORG CHEM] C$_3$Cl$_3$N$_3$O$_3$ A crystalline substance that releases hypochlorous acid on contact with water; melting point is 246–247°C; soluble in chlorinated and highly polar solvents; used as a chlorinating agent, disinfectant, and industrial deodorant. Also known as symclosene; trichloroiminocyanuric acid. { trī¦klȯr·ō¦i·sō¦sī·ə¦nür·ik 'as·əd }

trichloromethane See chloroform. { trī¦klȯr·ō'meth‚ān }

trichloromethyl chloroformate [ORG CHEM] ClCOOCCl$_3$ A toxic, colorless liquid with a

trichloromethylsilane

boiling point of 127–128°C; soluble in alcohol, ether, and benzene; used in organic synthesis, and as a military poison gas during World War I. Also known as diphosgene. { trī'klȯr·ō'meth·əl ¦klȯr·ə'fȯr,mät }

trichloromethylsilane *See* methyltrichlorosilane. { trī,klȯr·ə,meth·əl'sī,lān }

trichloronate *See* ortho-ethyl(O-2,4,5-trichlorophenyl)ethylphosphonothioate. { trī'klȯr·ə,nāt }

trichloronitromethane *See* chloropicrin. { trī¦klȯr·ō¦nī·trō'meth,ān }

trichlorophenol [ORG CHEM] $C_6H_2Cl_3OH$ Either of two toxic nonflammable compounds with a phenol aroma: 2,4,5-trichlorophenol is a gray solid, is soluble in alcohol, acetone, and ether, melts at 69°C, and is used as a fungicide and bactericide; 2,4,6-trichlorophenol forms yellow flakes, is soluble in alcohol, acetone, and ether, boils at 248°C, and is used as a fungicide, defoliant, and herbicide; it is also known as 2,4,6-T. { trī¦klȯr·ō'fē,nȯl }

2,4,5-trichlorophenoxyacetic acid [ORG CHEM] $C_6H_2Cl_3OCH_2CO_2H$ A toxic, light-tan solid; soluble in alcohol, insoluble in water; melts at 152°C; used as a defoliant, plant hormone, and herbicide. Also known as 2,4,5-T. { ¦tü ¦fȯr ¦fīv trī¦klȯr·ō·fə¦näk·sē·ə¦sēd·ik 'as·əd }

1,2,3-trichloropropane [ORG CHEM] $CH_2ClCHClCH_2Cl$ A toxic, colorless liquid with a boiling point of 156.17°C; used as a paint and varnish remover and degreasing agent. { ¦wən ¦tü ¦thrē trī¦klȯr·ō'prō,pān }

1,1,2-trichloro-1,2,2-trifluoroethane [ORG CHEM] CCl_2FCClF_2 A colorless, volatile liquid with a boiling point of 47.6°C; used as a solvent for dry cleaning, as a refrigerant, and in fire extinguishers. Also known as trifluorotrichloroethane. { ¦wən ¦wən ¦tü trī¦klȯr·ō ¦wən ¦tü ¦tü trī¦flür·ō'eth,ān }

tricosane [ORG CHEM] $CH_3(CH_2)_{21}CH_3$ Combustible, glittering crystals; soluble in alcohol, insoluble in water; melts at 48°C; used as a chemical intermediate. { 'trī·kə,sān }

tricresyl phosphate [ORG CHEM] $(CH_3C_6H_4O)_3PO$ A combustible, colorless liquid; insoluble in water, soluble in common solvents and vegetable oils; boils at 420°C; used as a plasticizer, plastics fire retardant, air-filter medium, and gasoline and lubricant additive. Abbreviated TCP. { trī'kres·əl 'fä,sfāt }

tricyclic dibenzopyran *See* xanthene. { trī'sī·klik dī¦ben·zō'pī·rən }

n-tridecane [ORG CHEM] $CH_3(CH_2)_{11}CH_3$ A combustible liquid; soluble in alcohol, insoluble in water; boils at 226°C; used as a distillation chaser and chemical intermediate. { ¦en trī'de,kān }

tridecanol *See* tridecyl alcohol. { trī'dek·ə,nȯl }

tridecyl alcohol [ORG CHEM] $C_{12}H_{25}CH_2OH$ An isomer mixture; a white, combustible solid with a pleasant aroma; melts at 31°C; used in detergents and perfumery and to make synthetic lubricants. Also known as tridecanol. { trī'des·əl 'al·kə,hȯl }

tridentate ligand [INORG CHEM] A chelating agent having three groups capable of attachment to a metal ion. Also known as terdentate ligand. { trī'den,tāt 'līg·ənd }

triethanolamine [ORG CHEM] $(HOCH_2CH_2)_3N$ A viscous, hygroscopic liquid with an ammonia aroma, soluble in chloroform, water, and alcohol, and boiling at 335°C; used in dry-cleaning soaps, cosmetics, household detergents, and textile processing, for wool scouring, and as a corrosion inhibitor. { trī,eth·ə'näl·ə,mēn }

triethanolamine stearate *See* trihydroxyethylamine stearate. { trī,eth·ə'näl·ə,mēn 'stir,āt }

triethylamine [ORG CHEM] $(C_2H_5)_3N$ A colorless, toxic, flammable liquid with an ammonia aroma; soluble in water and alcohol; boils at 90°C; used as a solvent, rubber-accelerator activator, corrosion inhibitor, and propellant, and in penetrating and waterproofing agents. { trī'eth·ə·lə¦mēn }

triethylborane [ORG CHEM] $(C_2H_5)_3B$ A colorless liquid with a boiling point of 95°C; used as a jet fuel or igniter for jet engines and as a fuel additive. Also known as boron triethyl; triethylborine. { trī,eth·əl'bȯr,ān }

triethylborine *See* triethylborane. { trī,eth·əl'bȯr,ēn }

triethylene glycol [ORG CHEM] $HO(C_2H_4O)_3H$ A colorless, combustible, hygroscopic, water-soluble liquid; boils at 287°C; used as a chemical intermediate, solvent, bactericide, humectant, and fungicide. Abbreviated TEG. { trī'eth·ə,lēn 'glī,kȯl }

386

triethylenemelamine [ORG CHEM] NC[N(CH₂)₂]NC[N(CH₂)₂]NC[N(CH₂)₂] White crystals, soluble in water, alcohol, acetone, chloroform, and methanol; polymerizes at 160°C; used in medicine and insecticides and as a chemosterilant. Abbreviated TEM. Also known as tretamine. { trī¦eth·ə¦lēn'mel·ə,mēn }

triethylenetetramine [ORG CHEM] NH₂(C₂H₄NH)₂C₂H₄NH₂ A yellow, water-soluble liquid with a boiling point of 277.5°C; used in detergents and in the manufacture of dyes and pharmaceuticals. { trī¦eth·ə,lēn'te·trə,mēn }

triethylic borate See ethyl borate. { ‚trī·ə'thil·ik 'bȯr,āt }

triethyl phosphate [ORG CHEM] (C₂H₅)₃PO₄ A toxic, colorless liquid that acts as a cholinesterase inhibitor; boiling point is 216°C; soluble in organic solvents; used as a solvent and plasticizer and for pesticides manufacture. Abbreviated TEP. { trī'eth·əl 'fä,sfāt }

trifluorochlorethylene resin [ORG CHEM] A fluorocarbon used as a base for polychlorotrifluoroethylene resin. { trī¦flür·ō¦klȯr·ō'eth·ə,lēn ,rez·ən }

trifluoromethane See fluoroform. { trī¦flür·ō¦klȯr·ō'meth,ān }

trifluorotrichloroethane See 1,1,2-trichloro-1,2,2-trifluoroethane. { trī¦flür·ō·trī¦klȯr·ō 'eth,ān }

triformal See sym-trioxane. { trī'fȯr·məl }

triglyceride [ORG CHEM] CH₂(OOCR₁)CH(OOCR₂)CH₂(OOCR₃) A naturally occurring ester of normal, fatty acids and glycerol; used in the manufacture of edible oils, fats, and monoglycerides. { trī'glis·ə,rīd }

trigonal planar molecule [CHEM] A molecule having a central atom that is bonded to three other atoms, with all four atoms lying in the same plane. { trī¦gōn·əl ¦plā,när 'mäl·ə,kyül }

trihydroxyethylamine stearate [ORG CHEM] C₂₄H₄₉NO₅ A cream-colored, waxy solid with a melting point of 42–44°C; soluble in methanol, ethanol, mineral oil, and vegetable oil; used as an emulsifier in cosmetics and pharmaceuticals. Also known as triethanolamine stearate. { ‚trī·hī¦dräk·sē¦eth·ə·lə¦mēn 'stir,āt }

triiodomethane See iodoform. { trī,ī'ō·də'meth,ān }

triisobutylene [ORG CHEM] (C₄H₈)₃ A mixture of isomers; combustible liquid boiling at 348–354°C; used as a chemical and resin intermediate, lubricating oil additive, and motor-fuel alkylation feedstock. { trī,ī·sō'byüd·əl,ēn }

trimer [CHEM] An oligomer whose molecule is composed of three identical monomers. { 'trī·mər }

trimercuric orthophosphate See mercuric phosphate. { ‚trī·mər'kyür·ik ¦ȯr·thō'fä,sfāt }

trimercurous orthophosphate See mercurous phosphate. { ‚trə·mər'kyür·əs ¦ȯr·thō'fä,sfāt }

trimethylamine [ORG CHEM] (CH₃)₃N A colorless, liquefied gas with a fishy odor and a boiling point of −4°C; soluble in water, ether, and alcohol; used as a warning agent for natural gas, a flotation agent, and insect attractant. Abbreviated TMA. { trī¦meth·ə·lə¦mēn }

uns-trimethylbenzene See pseudocumene. { ¦əns trī¦meth·əl'ben,zēn }

trimethyl borate [ORG CHEM] B(OCH₃)₃ A water-white liquid, boiling at 67–68°C; used as a solvent for resins, waxes, and oils, and as a catalyst and a reagent in analysis of paint and varnish. Also known as methyl borate. { trī'meth·əl 'bȯr,āt }

trimethylchlorosilane [ORG CHEM] (CH₃)₃SiCl A colorless liquid with a boiling point of 57°C; soluble in ether and benzene; used as a water-repelling agent. { trī¦meth·əl¦klȯr·ō'si,lān }

trimethylethylene See methyl butene. { trī¦meth·əl'eth·ə,lēn }

trimethylol aminomethane See tromethamine. { trī'meth·ə,lȯl ə¦mē·nō'meth,ān }

trimethylolethane [ORG CHEM] CH₃C(CH₂OH)₃ Colorless crystals, soluble in alcohol and water; used in the manufacture of varnishes and drying oils. Also known as methyltrimethylolmethane; pentoglycerine. { trī¦meth·ə,lȯl'eth,ān }

trinitrobenzene [ORG CHEM] C₆H₃(NO₂)₃ A yellow crystalline compound, soluble in alcohol and ether; used as an explosive. { trī¦nī·trō'ben,zēn }

2,4,7-trinitrofluorenone [ORG CHEM] C₁₃H₅N₃O₇ A yellow, crystalline compound with a

trinitromethane

melting point of 175.2–176°C; forms crystalline complexes with indoles for identification by mass spectroscopy. Abbreviated TNF. { ¦tü ¦fȯr ¦sev·ən trī¦nī·trō'flúr·ə,nōn }

trinitromethane [ORG CHEM] $CH(NO_2)_3$ A crystalline compound, melting at 150°C, decomposing above 25°C; used to make explosives. Also known as nitroform. { trī¦nī·trō'meth,ān }

2,4,6-trinitrotoluene [ORG CHEM] $CH_3C_6H_2(NO_2)_3$ Toxic, flammable, explosive, yellow crystals; soluble in alcohol and ether, insoluble in water; melts at 81°C; used as an explosive and chemical intermediate and in photographic chemicals. Abbreviated TNT. { ¦tü ¦fȯr ¦siks trī¦nī·trō'täl·yə,wēn }

sym-trioxane [ORG CHEM] $(CH_2O)_3$ White, flammable, explosive crystals; soluble in water, alcohol, and ether; melts at 62°C; used as a chemical intermediate, disinfectant, and fuel. Also known as metaformaldehyde; triformol; trioxin. { ¦sim trī'äk,sān }

trioxygen See ozone. { trī'äk·sə·jən }

tripalmitin [ORG CHEM] $C_3H_5(OOCC_{15}H_{31})_3$ A white, water-insoluble powder that melts at 65.5°C; used in the preparation of leather dressings and soaps. { trī'päm·ə·tən }

triphenylmethane dye [ORG CHEM] A family of dyes with a molecular structure derived from $(C_6H_5)_3CH$, usually by NH_2, OH, or HSO_3 substitution for one of the C_6H_5 hydrogens; includes many coal tar dyes, for example, rosaniline and fuchsin. { trī¦fen·əl'meth,ān 'dī }

triphenylmethyl radical [ORG CHEM] A free radical in which three phenyl rings are bonded to a single carbon. Also known as trityl radical. { trī¦fen·əl¦meth·əl 'rad·ə·kəl }

triphenyl phosphate [ORG CHEM] $(C_6H_5O)_3PO$ A crystalline compound with a melting point of 49–50°C; soluble in benzene, chloroform, ether, and acetone; used as a substitute for camphor in celluloid, as a plasticizer in lacquers and varnishes, and to impregnate roofing paper. { trī'fen·əl 'fä,sfāt }

triphenylphosphine [ORG CHEM] $(C_6H_5)_3P$ A crystalline compound with a melting point of 80.5°C; soluble in ether, benzene, chloroform, and glacial acetic acid; used as an initiator of polymerization and in organic synthesis. { trī¦fen·əl'fä,sfēn }

triphenyltetrazolium chloride [ORG CHEM] $C_{19}H_{15}ClN_4$ A crystalline compound, soluble in water, alcohol, and acetone; used as a sensitive reagent for reducing sugars. Also known as red tetrazolium. { trī¦fen·əl¦te·trə'zō·lē·əm 'klȯr,īd }

triphenyltinacetate See fentinacetate. { trī¦fen·əl·tə'nas·ə,tāt }

triple phosphate [CHEM] A phosphate containing magnesium, calcium, and ammonium ions. { 'trip·əl 'fäs,fāt }

triple point [PHYS CHEM] A particular temperature and pressure at which three different phases of one substance can coexist in equilibrium. { 'trip·əl 'pȯint }

tripotassium orthophosphate See potassium phosphate. { ,trī·pə'tas·ē·əm ¦ȯr·thō'fä,sfāt }

triprene [ORG CHEM] $C_{18}H_{32}O_2S$ An amber liquid used as a growth regulator for crops. { 'trī,prēn }

triprotic acid [CHEM] An acid that has three ionizable hydrogen atoms in each molecule. { trī¦präd·ik 'as·əd }

triptane [ORG CHEM] C_7H_{16} A hydrocarbon compound made commercially in small quantities, but having one of the highest antiknock ratings known. { 'trip,tān }

TRIS See tromethamine. { tris }

trisamine See tromethamine. { 'tris·ə,mēn }

tris buffer See tromethamine. { 'tris ,bəf·ər }

tris[2-(2,4-dichlorophenoxy)ethyl]phosphite [ORG CHEM] $C_{24}H_{21}Cl_6O_6P$ A dark liquid that boils above 200°C; used as a preemergence herbicide for corn, peanuts, and strawberries. Abbreviated 2,4-DEP. { ¦tris ¦tü ¦tü ¦fȯr dī¦klȯr·ō·fə¦näk·sē ¦eth·əl 'fä,sfīt }

trisodium citrate See sodium citrate. { trī'sōd·ē·əm 'sī,trāt }

trisodium orthophosphate See trisodium phosphate. { trī'sōd·ē·əm ¦ȯr·thō'fä,sfāt }

trisodium phosphate [INORG CHEM] Na_3PO_4 A water-soluble crystalline compound; used as a cleaning compound and as a water softener. Abbreviated TSP. Also known as tertiary sodium phosphate; trisodium orthophosphate. { trī'sōd·ē·əm 'fä‚sfāt }

tristearin *See* stearin. { ‚trī'stir·ən }

trisulfide |CHEM| A binary chemical compound that contains three sulfur atoms in its molecule, for example, iron trisulfude, Fe_2S_3. { trī'səl‚fīd }

triterpene |ORG CHEM| One of a class of compounds having molecular skeletons containing 30 carbon atoms, and theoretically composed of six isoprene units; numerous and widely distributed in nature, occurring principally in plant resins and sap; an example is ambrein. { trī'tər‚pēn }

trithioacetaldehyde |ORG CHEM| $(C_4H_4S_2)_3$ A colorless, water-insoluble, crystalline compound; used as a hypnotic. { trī‚thī·ō‚as·ə'tal·də‚hīd }

tritiated |CHEM| Pertaining to matter in which tritium atoms have replaced one or more atoms of ordinary hydrogen. { 'trish·ē‚ād·əd }

trityl radical *See* triphenylmethyl radical. { 'trīd·əl ‚rad·ə·kəl }

triuranium octoxide |INORG CHEM| U_3O_8 Olive green to black crystals or granules, soluble in nitric acid and sulfuric acid; decomposes at 1300°C; used in nuclear technology and in the preparation of other uranium compounds. Also known as uranousuranic oxide; uranyl uranate. { ‚trī·yü'rā·nē·əm äk'täk‚sīd }

trivial name |ORG CHEM| A common name for a chemical compound derived from the names of the natural source of the compound at the time of its isolation and before anything is known about its molecular structure. { 'triv·ē·əl ‚nām }

tromethamine |ORG CHEM| $C_4H_{11}NO_3$ A crystalline compound with a melting point of 171–172°C; soluble in water, ethylene glycol, methanol, and ethanol; used to make surface-active agents, vulcanization accelerators, and pharmaceuticals, and as a titrimetric standard. Also known as THAM; trimethylol aminomethane; TRIS; trisamine; tris buffers. { trō'meth·ə‚mēn }

tropeoline 00 |ORG CHEM| $NaSO_3C_6H_4NNC_6H_4NHC_6H_5$ An acid-base indicator with a pH range of 1.4–3.0, color change (from acid to base) red to yellow; used as a biological stain. { trō'pē·ə·lən 'zir·ō 'zir·ō }

Trouton's rule |PHYS CHEM| An approximation rule for the derivation of molar heats of vaporization of normal liquids at their boiling points. { 'traút·ənz ‚rül }

true condensing point *See* critical condensation temperature. { 'trü kən'dens·iŋ ‚pòint }

true electrolyte |PHYS CHEM| A substance in the solid state that consists entirely of ions. { ‚trü i‚lek·trə‚līt }

true freezing point |PHYS CHEM| The temperature at which the liquid and solid forms of a substance exist in equilibrium at a given pressure (usually 1 standard atmosphere, or 101,325 pascals). { 'trü 'frēz·iŋ ‚pòint }

TSP *See* trisodium phosphate.

TSPP *See* sodium pyrophosphate.

tubatoxin *See* rotenone. { 'tü·bə‚täk·sən }

tungstate |INORG CHEM| M_2WO_4 A salt of tungstic acid; for example, sodium tungstate, Na_2WO_4. { 'təŋ‚stāt }

tungstate white *See* barium tungstate. { 'təŋ‚stāt 'wīt }

tungsten |CHEM| Also known as wolfram. A metallic transition element, symbol W, atomic number 74, atomic weight 183.85; soluble in mixed nitric and hydrofluoric acids; melts at 3400°C. { 'təŋ·stən }

tungsten boride |INORG CHEM| WB_2 A silvery solid; insoluble in water, soluble in aqua regia and concentrated acids; melts at 2900°C; used as a refractory for furnaces and chemical process equipment. { 'təŋ·stən 'bòr‚īd }

tungsten carbide |INORG CHEM| WC A hard, gray powder; insoluble in water; readily attacked by nitric-hydrofluoric acid mixture; melts at 2780°C; used in tools, dies, ceramics, cermets, and wear-resistant mechanical parts, and as an abrasive. { 'təŋ·stən 'kär‚bīd }

tungsten carbonyl *See* tungsten hexacarbonyl. { 'təŋ·stən 'kär·bə‚nil }

tungsten disulfide |INORG CHEM| WS_2 A grayish-black solid with a melting point above 1480°C; used as a lubricant and aerosol. { 'təŋ·stən dī'səl‚fīd }

tungsten hexacarbonyl

tungsten hexacarbonyl [INORG CHEM] $W(CO)_6$ A white, refractive, crystalline solid which decomposes at 150°C; used for tungsten coatings on base metals. Also known as tungsten carbonyl. { 'təŋ·stən ¦hek·sə'kär·bə,nil }

tungsten hexachloride [INORG CHEM] WCl_6 Dark blue or violet crystals with a melting point of 275°C; soluble in organic solvents; used for tungsten coatings on base metals and as a catalyst for olefin polymers. { 'təŋ·stən ¦hek·sə'klȯr,īd }

tungsten lake See phosphotungstic pigment. { 'təŋ·stən 'lāk }

tungsten oxychloride [INORG CHEM] $WOCl_4$ Dark red crystals with a melting point of approximately 211°C; soluble in carbon disulfide; used for incandescent lamps. { 'təŋ·stən ¦äk·sə'klȯr,īd }

tungstic acid [INORG CHEM] H_2WO_4 A yellow powder; insoluble in water, soluble in alkalies; used as a color-resist mordant for textiles, as an ingredient in plastics, and for the manufacture of tungsten metal products. Also known as orthotungstic acid; wolframic acid. { 'təŋ·stik 'as·əd }

tungstic acid anhydride See tungstic oxide. { 'təŋ·stik 'as·əd an'hī,drīd }

tungstic anhydride See tungstic oxide. { 'təŋ·stik an'hī,drīd }

tungstic oxide [INORG CHEM] WO_3 A heavy, canary-yellow powder; soluble in caustic, insoluble in water; melts at 1473°C; used in alloys, in fabric fireproofing, for ceramic pigments, and for the manufacture of tungsten metal. Also known as anhydrous wolframic acid; tungstic acid anhydride; tungstic anhydride; tungstic trioxide. { 'təŋ·stik 'äk,sīd }

tungstic trioxide See tungstic oxide. { 'təŋ·stik trī'äk,sīd }

turbidimetric analysis [ANALY CHEM] A scattered-light procedure for the determination of the weight concentration of particles in cloudy, dull, or muddy solutions; uses a device that measures the loss in intensity of a light beam as it passes through the solution. Also known as turbidimetry. { ¦tər·bə·də¦me·trik ə'nal·ə·səs }

turbidimetric titration [ANALY CHEM] Titration in which the end point is indicated by the developing turbidity of the titrated solution. { ¦tər·bə·də¦me·trik 'tī,trā·shən }

turbidimetry See turbidimetric analysis. { ,tər·bə'dim·ə·trē }

turbidity [ANALY CHEM] **1.** Measure of the clarity of an otherwise clear liquid by using colorimetric scales. **2.** Cloudy or hazy appearance in a naturally clear liquid caused by a suspension of colloidal liquid droplets or fine solids. { tər'bid·əd·ē }

Turnbull's blue [INORG CHEM] A blue pigment that precipitates from the reaction of potassium ferricyanide with a ferrous salt. { 'tərn,bùl 'blü }

turpentine camphor See terpene hydrochloride. { 'tər·pən,tīn 'kam·fər }

21-centimeter line [SPECT] A radio-frequency spectral line of neutral atomic hydrogen at a wavelength of approximately 21 centimeters and a frequency of approximately 1420 megahertz, that results from hyperfine transitions between states in which the spins of the electron and proton are parallel and antiparallel. { ¦twen·tē¦wən 'sen·tə,mēd·ər ,līn }

twinned double bonds See cumulative double bonds. { ¦twind ¦dəb·əl 'bänz }

Twitchell reagent [ORG CHEM] A catalyst for the acid hydrolysis of fats; a sulfonated addition product of naphthalene and oleic acid, that is, a naphthalenestearosulfonic acid. { 'twich·əl rē,ā·jənt }

two-dimensional chromatography [ANALY CHEM] A paper chromatography technique in which the sample is resolved by standard procedures (ascending, descending, or horizontal solvent movement) and then turned at right angles in a second solvent and re-resolved. { 'tü ¦di¦men·shən·əl ,krō·mə'täg·rə·fē }

two-fluid cell [PHYS CHEM] Cell having different electrolytes at the positive and negative electrodes. { 'tü ¦flü·əd 'sel }

two-photon absorption [PHYS CHEM] A relatively weak photon absorption and excitation process that occurs when a sufficiently intense light source, such as a laser beam, is used, making it possible for a molecule to absorb simultaneously two photons, each of approximately half the energy (twice the wavelength) normally required to reach an excited state. The probability of a molecule absorbing two photons simultaneously is proportional to the square of the intensity of the input beam. { ,tü ¦fō,tän əb'sȯrp·shən }

U

U *See* uranium.

UDMH *See uns*-dimethylhydrazine.

Ullmann reaction [ORG CHEM] A variation of the Fittig synthesis, using copper powder instead of sodium. { 'əl·mən rē͵ak·shən }

ultimate analysis [ANALY CHEM] The determination of the percentage of elements contained in a chemical substance. { 'əl·tə·mət ə'nal·ə·səs }

ultramarine blue [INORG CHEM] A blue pigment; a powder with heat resistance, used for enamels on toys and machinery, white baking enamels, printing inks, and cosmetics, and in textile printing. { ͵əl'trə·mə'rēn 'blü }

ultrasensitive mass spectrometry [ANALY CHEM] A form of mass spectrometry in which the ions to be detected are accelerated to megaelectronvolt energies in a particle accelerator and passed through a thin gas cell or foil, stripping away outer electrons, so that contaminating molecules dissociate into lower-mass fragments, and isobars can be distinguished by particle detectors that measure ionization rate and total energy. { ¦əl·trə'sen·səd·iv 'mas spek'träm·ə·trē }

ultraviolet absorption spectrophotometry [SPECT] The study of the spectra produced by the absorption of ultraviolet radiant energy during the transformation of an electron from the ground state to an excited state as a function of the wavelength causing the transformation. { ¦əl·trə'vī·lət əb'sȯrp·shən ͵spek·trō·fə'täm·ə·trē }

ultraviolet densitometry [SPECT] An ultraviolet-spectrophotometry technique for measurement of the colors on thin-layer chromatography absorbents following elution. { ¦əl·trə'vī·lət ͵den·sə'täm·ə·trē }

ultraviolet photoemission spectroscopy [SPECT] A spectroscopic technique in which photons in the energy range 10–200 electronvolts bombard a surface and the energy spectrum of the emitted electrons gives information about the states of electrons in atoms and chemical bonding. Abbreviated UPS. { ¦əl·trə'vī·lət ¦fōd·ō·i'mish·ən spek'träs·kə·pē }

ultraviolet spectrometer [SPECT] A device which produces a spectrum of ultraviolet light and is provided with a calibrated scale for measurement of wavelength. { ¦əl·trə'vī·lət spek'träm·əd·ər }

ultraviolet spectrophotometry [SPECT] A method of chemical analysis based on the absorption of electromagnetic radiation between 200 and 400 nanometers, used extensively for quantitative analysis of simple inorganic ions and their complexes, as well as organic molecules that have at least one double bond. { ¦əl·trə'vī·lət ¦spek·trō·fə'täm·ə·trē }

ultraviolet spectroscopy [SPECT] Absorption spectroscopy involving electromagnetic wavelengths in the range 4–400 nanometers. { ¦əl·trə'vī·lət spek'träs·kə·pē }

ultraviolet stabilizer *See* UV stabilizer. { ¦əl·trə'vī·lət 'stā·bə͵līz·ər }

uncharged species [CHEM] A chemical entity with no net electric charge. Also known as neutral species. { ¦ən'chärjd 'spē·shēz }

uncoupling phenomena [SPECT] Deviations of observed spectra from those predicted in a diatomic molecule as the magnitude of the angular momentum increases, caused by interactions which could be neglected at low angular momenta. { ¦ən'kəp·liŋ fə͵näm·ə·nä }

undecanal [ORG CHEM] $CH_3(CH_2)_9CHO$ A sweet-smelling, colorless liquid, soluble in oils and alcohol; used in perfumes and flavoring. Also known as hendecanal; *n*-undecyclic aldehyde. { ən'dek·ə,nal }

undecane [ORG CHEM] $CH_3(CH_2)_9CH_3$ A colorless, combustible liquid, boiling at 385°F (196°C), flash point at 149°F (65°C); used as a chemical intermediate and in petroleum research. Also known as hendecane. { ,ən'de,kān }

undecanoic acid [ORG CHEM] $CH_3(CH_2)_9COOH$ Colorless crystals, soluble in alcohol and ether, insoluble in water; melts at 29°C; used as a chemical intermediate. { ¦ən¦dek·ə¦nó·ik 'as·əd }

2-undecanone *See* methyl nonyl ketone. { ¦tü ,ən'dek·ə,nōn }

undecyl [ORG CHEM] $C_{11}H_{23}$ The radical of undecane. Also known as hendecyl. { ¦ən¦des·əl }

undecylenic acid [ORG CHEM] $CH_2 \cdot CH(CH_2)_8COOH$ A light-colored, combustible liquid with a fruity aroma; soluble in alcohol, ether, chloroform, and benzene, almost insoluble in water; used in medicine, perfumes, flavors, and plastics. { ¦ən¦des·ə¦lin·ik 'as·əd }

undecylenic alcohol [ORG CHEM] $C_{11}H_{22}O$ A colorless liquid with a citrus odor, soluble in 70% alcohol; used in perfumes. { ¦ən¦des·ə¦lin·ik 'al·kə,hol }

undecylenyl acetate [ORG CHEM] $C_{13}H_{24}O_2$ A colorless liquid with a floral-fruity odor, soluble in 80% alcohol; used in perfumes and for flavoring. { ¦ən¦des·əl¦en·əl 'as·ə,tāt }

***n*-undecylic aldehyde** *See* undecanal. { ¦en ¦ən·də¦sil·ik 'al·də,hīd }

undersaturated fluid [PHYS CHEM] Any fluid (liquid or gas) capable of holding additional vapor or liquid components in solution at specified conditions of pressure and temperature. { ¦ən·dər¦sach·ə,rād·əd 'flü·əd }

unidentate ligand [CHEM] A ligand that donates one pair of electrons in a complexation reaction to form coordinate bonds. { ¦yü·nē¦den,tāt 'līg·ənd }

unimolecular reaction [PHYS CHEM] A chemical reaction involving only one molecular species as a reactant; for example, $2H_2O \rightarrow 2H_2 + O_2$, as in the electrolytic dissociation of water. { ¦yü·nə·mə'lek·yə·lər rē'ak·shən }

uns-, unsym- [ORG CHEM] A chemical prefix denoting that the substituents of an organic compound are structurally unsymmetrical with respect to the carbon skeleton, or with respect to a function group (for example, double or triple bond). { əns, ¦ən·sim }

unsaturated compound [CHEM] Any chemical compound with more than one bond between adjacent atoms, usually carbon, and thus reactive toward the addition of other atoms at that point; for example, olefins, diolefins, and unsaturated fatty acids. { ¦ən'sach·ə,rād·əd 'käm,paúnd }

unsaturated hydrocarbon [ORG CHEM] One of a class of hydrocarbons that have at least one double or triple carbon-to-carbon bond that is not in an aromatic ring; examples are ethylene, propadiene, and acetylene. { ¦ən'sach·ə,rād·əd ¦hī·drə'kär·bən }

unsaturation [ORG CHEM] A state in which the atomic bonds of an organic compound's chain or ring are not completely satisfied (that is, not saturated); usually applies to carbon, but can include other ring or chain atoms; unsaturation usually results in a double bond (as for olefins) or a triple bond (as for the acetylenes). { ¦ən,sach·ə'rā·shən }

upflow [CHEM] In an ion-exchange unit, an operation in which solutions enter at the bottom of the unit and leave at the top. { 'əp,flō }

upper explosive limit *See* upper flammable limit. { ¦əp·ər ik'splō·siv ,lim·ət }

upper flammable limit [CHEM] The maximum percentage of flammable gas or vapor in the air above which ignition cannot take place because the ratio of the gas to oxygen is too high. Also known as upper explosive limit. { ¦əp·ər 'flam·ə·bəl ,lim·ət }

UPS *See* ultraviolet photoemission spectroscopy.

uranate [INORG CHEM] A salt of uranic acid; for example, sodium uranate, Na_2UO_4. { 'yùr·ə,nāt }

urania *See* uranium dioxide. { yə'rā·nē·ə }

uranic chloride *See* uranium tetrachloride. { yú'ran·ik 'klòr,īd }

uranic oxide *See* uranium dioxide.   { yu'ran·ik 'äk,sīd }
uranin *See* uranine.   { 'yur·ə·nən }
uranine [ORG CHEM] $Na_2C_{20}H_{10}O_5$ A brown or orange-red hygroscopic powder soluble in water; used as a yellow dye for silk and wool, a marker in the ocean to facilitate air and sea rescues, and as an analytical reagent.   Also known as sodium fluorescein; uranin; uranine yellow.   { 'yur·ə,nēn }
uranine yellow *See* uranine.   { 'yur·ə,nēn 'yel·ō }
uranium [CHEM] A metallic element in the actinide series, symbol U, atomic number 92, atomic weight 238.03; highly toxic and radioactive; ignites spontaneously in air and reacts with nearly all nonmetals; melts at 1132°C, boils at 3818°C; used in nuclear fuel and as the source of ^{235}U and plutonium.   { yə'rā·nē·əm }
uranium acetate *See* uranyl acetate.   { yə'rā·nē·əm 'as·ə,tāt }
uranium carbide [INORG CHEM] One of the carbides of uranium, such as uranium monocarbide; used chiefly as a nuclear fuel.   { yə'rā·nē·əm 'kär,bīd }
uranium dioxide [INORG CHEM] UO_2 Black, highly toxic, spontaneously flammable, radioactive crystals; insoluble in water, soluble in nitric and sulfuric acids; melts at approximately 3000°C; used to pack nuclear fuel rods and in ceramics, pigments, and photographic chemicals.   Also known as urania; uranic oxide; uranium oxide.   { yə'rā·nē·əm dī'äk,sīd }
uranium hexafluoride [INORG CHEM] UF_6 Highly toxic, radioactive, corrosive, colorless crystals; soluble in carbon tetrachloride, fluorocarbons, and liquid halogens; it reacts vigorously with alcohol, water, ether, and most metals, and it sublimes; used to separate uranium isotopes in the gaseous-diffusion process.   { yə'rā·nē·əm ¦hek·sə'flur,īd }
uranium hydride [INORG CHEM] UH_3 A highly toxic, gray to black powder that ignites spontaneously in air, and that conducts electricity; used for making powdered uranium metal, for hydrogen-isotope separation, and as a reducing agent.   { yə'rā·nē·əm 'hī,drīd }
uranium nitrate *See* uranyl nitrate.   { yə'rā·nē·əm 'nī,trāt }
uranium oxide *See* uranium dioxide. *See* uranium trioxide.   { yə'rā·nē·əm 'äk,sīd }
uranium sulfate *See* uranyl sulfate.   { yə'rā·nē·əm 'səl,fāt }
uranium tetrachloride [INORG CHEM] UCl_4 Poisonous, radioactive, hygroscopic, dark-green crystals; soluble in alcohol and water; melts at 590°C, boils at 792°C. Also known as uranic chloride.   { yə'rā·nē·əm ¦te·trə'klòr,īd }
uranium tetrafluoride [INORG CHEM] UF_4 Toxic, radioactive, corrosive green crystals; insoluble in water; melts at 1036°C; used in the manufacture of uranium metal. Also known as green salt.   { yə'rā·nē·əm ¦te·trə'flur,īd }
uranium trioxide [INORG CHEM] UO_3 A poisonous, radioactive, red to yellow powder; soluble in nitric acid, insoluble in water; decomposes when heated; used in ceramics and pigments and for uranium refining.   Also known as orange oxide; uranium oxide.   { yə'rā·nē·əm trī'äk,sīd }
uranium yellow *See* sodium diuranate.   { yə'rā·nē·əm yel·ō }
uranous-uranic oxide *See* triuranium octoxide.   { 'yur·ə·nəs yə'ran·ik 'äk,sīd }
uranyl acetate [INORG CHEM] $UO_2(C_2H_3O_2)_2 \cdot 2H_2O$ Poisonous, radioactive yellow crystals, decomposed by light; soluble in cold water, decomposes in hot water; loses water of crystallization at 110°C, decomposes at 275°C; used in medicine and as an analytical reagent and bacterial oxidant.   Also known as uranium acetate.   { 'yur·ə,nil 'as·ə,tāt }
uranyl nitrate [INORG CHEM] $UO_2(NO_3)_2 \cdot 6H_2O$ Toxic, explosive, unstable yellow crystals; soluble in water, alcohol, and ether; melts at 60°C and boils at 118°C; used in photography, in medicine, and for uranium extraction and uranium glaze.   Also known as uranium nitrate; yellow salt.   { 'yur·ə,nil 'nī,trāt }
uranyl salts [INORG CHEM] Salts of UO_3 that ionize to form UO_2^{2+} and that are yellow in solution; for example, uranyl chloride, UO_2Cl_2.   { 'yur·ə,nil ,sòls }
uranyl sulfate [INORG CHEM] $UO_2SO_4 \cdot 3^1/_2H_2O$ and $UO_2SO_4 \cdot 3H_2O$ Poisonous, radioactive yellow crystals; soluble in water and concentrated hydrochloric acid; used as an analytical reagent.   Also known as uranium sulfate.   { 'yur·ə,nil 'səl,fāt }

393

uranyl uranate *See* triuranium octoxide. { 'yùr·ə,nil 'yùr·ə,nāt }

urbacid [ORG CHEM] $C_7H_{15}AsN_2S_3$ A colorless, crystalline compound with a melting point of 144°C; insoluble in water; used to control apple scale and diseases of coffee trees. { 'ər·bə,sid }

urea [ORG CHEM] $CO(HN_2)_2$ A natural product of protein metabolism found in urine; synthesized as white crystals or powder with a melting point of 132.7°C; soluble in water, alcohol, and benzene; used as a fertilizer, in plastics, adhesives, and flameproofing agents, and in medicine. Also known as carbamide. { yù'rē·ə }

urea anhydride *See* cyanamide. { yù'rē·ə an'hī,drīd }

urea-formaldehyde resin [ORG CHEM] A synthetic thermoset resin derived by the reaction of urea (carbamide) with formaldehyde or its polymers. Also known as urea resin. { yù'rē·ə fòr'mal·də,hīd 'rez·ən }

urea nitrate [ORG CHEM] $CO(NH_2)_2 \cdot HNO_3$ Colorless, explosive, fire-hazardous crystals; soluble in alcohol, slightly soluble in water; decomposes at 152°C; used in explosives and to make urethane. { yù'rē·ə 'nī,trāt }

urea peroxide [ORG CHEM] $CO(NH_2)_2 \cdot H_2O_2$ An unstable, fire-hazardous white powder; soluble in water, alcohol, and ethylene glycol; decomposes at 75–85°C or by moisture; used as a source of water-free hydrogen peroxide, as a disinfectant, in cosmetics and pharmaceuticals, and for bleaching. { yù'rē·ə pə'räk,sīd }

urea resin *See* urea-formaldehyde resin. { yù'rē·ə 'rez·ən }

urethane [ORG CHEM] $CO(NH_2)OC_2H_5$ A combustible, toxic, colorless powder; soluble in water and alcohol; melts at 49°C; used as a solvent and chemical intermediate and in biochemical research and veterinary medicine. Also known as ethyl carbamate; ethyl urethane. { 'yùr·ə,thān }

uronic acid [ORG CHEM] One of the compounds that are similar to sugars, except that the terminal carbon has been oxidized from the alcohol to a carboxyl group; for example, galacturonic acid and glucuronic acid. { yə'rän·ik 'as·əd }

urotropin *See* cystamine. { yə'rä·trə·pən }

USP acid test [ANALY CHEM] A United States Pharmacopoeia test to determine the carbonizable substances present in petroleum white oils. { ¦yü¦es'pē 'as·əd ,test }

UV stabilizer [CHEM] Any chemical compound that, admixed with a thermoplastic resin, selectively absorbs ultraviolet rays; used to prevent ultraviolet degradation of polymers. Also known as ultraviolet stabilizer. { ¦yü¦vē 'stā·bə,līz·ər }

V

V *See* vanadium.

vacuum condensing point |CHEM| Temperature at which the sublimate (vaporized solid) condenses in a vacuum. Abbreviated vcp. { 'vak·yəm kən'dens·iŋ ˌpȯint }

vacuum thermobalance |ANALY CHEM| An instrument used in thermogravimetry consisting of a precision balance and furnace that have been adapted for continuously measuring or recording changes in weight of a substance as a function of temperature; used in many types of physicochemical reactions where rates of reaction and energies of activation for vaporization, sublimation, and chemical reaction can be obtained. { 'vak·yəm ˈthər·mō'bal·əns }

vacuum ultraviolet spectroscopy |SPECT| Absorption spectroscopy involving electromagnetic wavelengths shorter than 200 nanometers; so called because the interference of the high absorption of most gases necessitates work with evacuated equipment. { 'vak·yəm ˈəl·trə'vī·lət spek'träs·kə·pē }

valence |CHEM| A positive number that characterizes the combining power of an element for other elements, as measured by the number of bonds to other atoms which one atom of the given element forms upon chemical combination; hydrogen is assigned valence 1, and the valence is the number of hydrogen atoms, or their equivalent, with which an atom of the given element combines. { 'vā·ləns }

valence angle *See* bond angle. { 'vā·ləns ˌaŋ·gəl }

valence bond |PHYS CHEM| The bond formed between the electrons of two or more atoms. { 'vā·ləns ˌbänd }

valence-bond method |PHYS CHEM| A method of calculating binding energies and other parameters of molecules by taking linear combinations of electronic wave functions, some of which represent covalent structures, others ionic structures; the coefficients in the linear combination are calculated by the variational method. Also known as valence-bond resonance method. { 'vā·ləns ˈbänd ˌmeth·əd }

valence-bond resonance method *See* valence-bond method. { 'vā·ləns ˈbänd 'rez·ən·əns ˌmeth·əd }

valence-bond theory |CHEM| A theory of the structure of chemical compounds according to which the principal requirements for the formation of a covalent bond are a pair of electrons and suitably oriented electron orbitals on each of the atoms being bonded; the geometry of the atoms in the resulting coordination polyhedron is coordinated with the orientation of the orbitals on the central atom. { 'vā·ləns ˈbänd ˌthē·ə·rē }

valence number |CHEM| A number that is equal to the valence of an atom or ion multiplied by +1 or −1, depending on whether the ion is positive or negative, or equivalently on whether the atom in the molecule under consideration has lost or gained electrons from its free state. { 'vā·ləns ˌnəm·bər }

valence transition |PHYS CHEM| A change in the electronic occupation of the 4f or 5f orbitals of the rare-earth or actinide atoms in certain substances at a certain temperature, pressure, or composition. { 'vā·ləns tran,zish·ən }

valeral *See* n-valeraldehyde. { 'val·ə·rəl }

n-valeraldehyde |ORG CHEM| $CH_3(CH_2)_3CHO$ A flammable liquid, soluble in ether and alcohol, slightly soluble in water; boils at 102°C; used in flavors and as a rubber accelerator. { ˈen ˈval·ər'al·də,hīd }

valeramide

valeramide [ORG CHEM] $CH_3(CH_2)_3CONH_2$ Water-soluble, colorless crystals, melting at 127°C. Also known as pentanamide; valeric amide. { ¦val·ər'am·əd }

valerianic acid *See* valeric acid. { və¦lir·ē¦an·ik 'as·əd }

valeric acid [ORG CHEM] $CH_3(CH_2)_3COOH$ A combustible, toxic, colorless liquid with a penetrating aroma; soluble in water, alcohol, and ether; boils at 185°C; used to make flavors, perfumes, lubricants, plasticizers, and pharmaceuticals. Also known as valerianic acid. { və'lir·ik 'as·əd }

γ-valerolactone [ORG CHEM] $C_5H_8O_2$ A combustible, mostly immiscible, colorless liquid, boiling at 205°C; used as a dye-bath coupling agent, in brake fluids and cutting oils, and as a solvent for adhesives, lacquers, and insecticides. { ¦gam·ə və¦lir·ō'lak,tōn }

value of isotope mixture [CHEM] A measure of the effort required to prepare a quantity of an isotope mixture; it is proportional to the amount of the mixture, and also depends on the composition of the mixture to be prepared and the composition of the original mixture. { 'val·yü əv 'ī·sə,tōp ,miks·chər }

vamidothion [ORG CHEM] $C_7H_{16}NO_4PS_2$ A white wax with a melting point of 40°C; very soluble in water; used to control pests in orchards, vineyards, rice, cotton, and ornamentals. { ¦vam·əd·ō'thī,än }

vanadic acid [INORG CHEM] Any of various acids that do not exist in a pure state and are found in various alkali and other metal vanadates; forms are meta-(HVO_3), ortho-(H_3VO_4), and pyro-$(H_4V_2O_7)$. { və'nād·ik 'as·əd }

vanadic acid anhydride *See* vanadium pentoxide. { və'nād·ik 'as·əd an'hī,drīd }

vanadic sulfate *See* vanadyl sulfate. { və'nād·ik 'səl,fāt }

vanadic sulfide *See* vanadium sulfide. { və'nād·ik 'səl,fīd }

vanadium [CHEM] A metallic transition element, symbol V, atomic number 23; soluble in strong acids and alkalies; melts at 1900°C, boils about 3000°C; used as a catalyst. { və'nād·ē·əm }

vanadium carbide [INORG CHEM] VC Hard, black crystals, melting at 2800°C, boiling at 3900°C; insoluble in acids, except nitric acid; used in cutting-tool alloys and as a steel additive. { və'nād·ē·əm 'kär,bīd }

vanadium dichloride [INORG CHEM] VCl_2 Toxic, green crystals, soluble in alcohol and ether; decomposes in hot water; used as a reducing agent. Also known as vanadous chloride. { və'nād·ē·əm dī'klōr,īd }

vanadium oxide [INORG CHEM] A compound of vanadium with oxygen, for example, vanadium tetroxide (V_2O_4), vanadium trioxide or sesquioxide (V_2O_3), vanadium oxide (VO), and vanadium pentoxide (V_2O_5). { və'nād·ē·əm 'äk,sīd }

vanadium oxytrichloride [INORG CHEM] $VOCl_3$ A toxic, yellow liquid that dissolves or reacts with many organic substances; hydrolyzes in moisture; boils at 126°C; used as an olefin-polymerization catalyst and in organovanadium synthesis. { və'nād·ē·əm ¦äk·sē·trī'klōr,īd }

vanadium pentasulfide *See* vanadium sulfide. { və'nād·ē·əm ¦pen·tə'səl,fīd }

vanadium pentoxide [INORG CHEM] V_2O_5 A toxic, yellow to red powder, soluble in alkalies and acids, slightly soluble in water; melts at 690°C; used in medicine, as a catalyst, as a ceramics coloring, for ultraviolet-resistant glass, photographic developers, textiles dyeing, and nuclear reactors. Also known as vanadic acid anhydride. { və'nād·ē·əm ¦pen'täk,sīd }

vanadium sesquioxide *See* vanadium trioxide. { və'nād·ē·əm ¦ses·kwē'äk,sīd }

vanadium sulfate *See* vanadyl sulfate. { və'nād·ē·əm 'səl,fāt }

vanadium sulfide [INORG CHEM] V_2S_5 A toxic, black-green powder; insoluble in water, soluble in alkalies and acids; decomposes when heated; used to make vanadium compounds. Also known as vanadic sulfide; vanadium pentasulfide. { və'nād·ē·əm 'səl,fīd }

vanadium tetrachloride [INORG CHEM] VCl_4 A toxic, red liquid; soluble in ether and absolute alcohol; boils at 154°C; used in medicine and to manufacture vanadium and organovanadium compounds. { və'nād·ē·əm ¦te·trə'klōr,īd }

vanadium tetraoxide [INORG CHEM] V_2O_4 A toxic blue-black powder; insoluble in water, soluble in alkalies and acids; melts at 1967°C; used as a catalyst.   { vəˈnād·ē·əm ˈte·trəˈäkˌsīd }

vanadium trichloride   [INORG CHEM] VCl_3 Toxic, deliquescent, pink crystals; soluble in ether and absolute alcohol; decomposes in water and when heated; used to prepare vanadium and organovanadium compounds.   { vəˈnād·ē·əm ˈtrīˈklȯrˌīd }

vanadium trioxide   [INORG CHEM] V_2O_3 Toxic, black crystals; soluble in alkalies and hydrofluoric acid, slightly soluble in water; melts at 1970°C; used as a catalyst.   Also known as vanadium sesquioxide.   { vəˈnād·ē·əm trīˈäkˌsīd }

vanadous chloride See vanadium dichloride.   { vəˈnād·əs ˈklȯrˌīd }

vanadyl chloride   [INORG CHEM] $V_2O_2Cl_4·5H_2O$ Toxic, deliquescent, water- and alcohol-soluble green crystals; used to mordant textiles.   { vəˈnād·əl ˈklȯrˌīd }

vanadyl sulfate   [INORG CHEM] $VOSO_4·2H_2O$ Blue, toxic, water-soluble crystals; used as a reducing agent, catalyst, glass and ceramics colorant, and mordant.   Also known as vanadic sulfate; vanadium sulfate.   { vəˈnād·əl ˈsəlˌfāt }

Van Deemter rate theory   [ANALY CHEM] A theory that the sample phase in gas chromatography flows continuously, not stepwise.   { van ˈdāmˌtər ˈrāt ˌthē·ə·rē }

van der Waals adsorption   [PHYS CHEM] Adsorption in which the cohesion between gas and solid arises from van der Waals forces.   { ˈvan dər ˌwȯlz adˌsȯrp·shən }

van der Waals attraction See van der Waals force.   { ˈvan dər ˌwȯlz əˌtrak·shən }

van der Waals covolume   [PHYS CHEM] The constant b in the van der Waals equation, which is approximately four times the volume of an atom of the gas in question multiplied by Avogadro's number.   { ˈvan dər ˌwȯlz ˈkōˈvälˌyəm }

van der Waals equation   [PHYS CHEM] An empirical equation of state which takes into account the finite size of the molecules and the attractive forces between them: $p = [RT/(v - b)] - (a/v^2)$, where p is the pressure, v is the volume per mole, T is the absolute temperature, R is the gas constant, and a and b are constants.   { ˈvan dər ˌwȯlz iˌkwā·zhən }

van der Waals force   [PHYS CHEM] An attractive force between two atoms or nonpolar molecules, which arises because a fluctuating dipole moment in one molecule induces a dipole moment in the other, and the two dipole moments then interact.   Also known as dispersion force; London force; van der Waals attraction.   { ˈvan dər ˌwȯlz ˌfȯrs }

van der Waals molecule   [PHYS CHEM] A molecule that is held together by van der Waals forces.   { ˈvan·dər ˌwälz ˌmäl·ə·kyül }

van der Waals radius   [PHYS CHEM] One-half the distance between two atoms of an element that are as close to each other as possible without being formally bonded to each other except for van der Waals forces.   { ˈvan·dərˌwälz ˈrād·ē·əs }

vanillic aldehyde See vanillin.   { vəˈnil·ik ˈal·dəˌhīd }

vanillin   [ORG CHEM] $C_8H_8O_3$ A combustible solid, soluble in water, alcohol, ether, and chloroform; melts at 82°C; used in pharmaceuticals, perfumes, and flavors, and as an analytical reagent.   Also known as vanillic aldehyde.   { vəˈnil·ən }

van't Hoff equation   [PHYS CHEM] An equation for the variation with temperature T of the equilibrium constant K of a gaseous reaction in terms of the heat of reaction at constant pressure, ΔH: $d(\ln K)/dT = \Delta H/RT^2$, where R is the gas constant.   Also known as van't Hoff isochore.   { vanˈtȯf iˌkwā·zhən }

van't Hoff formula   [ORG CHEM] The expression that the number of stereoisomers of a sugar molecule is equal to 2^n, where n is the number of asymmetric carbon atoms.   { vanˈtȯf ˌfȯr·myə·lə }

van't Hoff isochore See van't Hoff equation.   { vanˈtȯf ˈīˈsəˌkȯr }

van't Hoff isotherm   [PHYS CHEM] An equation for the change in free energy during a chemical reaction in terms of the reaction, the temperature, and the concentration and number of molecules of the reactants.   { vanˈtȯf ˈīˈsəˌthərm }

vapor-liquid equilibrium See liquid-vapor equilibrium.   { ˈvā·pər ˈlik·wəd ˌe·kwəˈlib·rē·əm }

vapor-pressure osmometer   [ANALY CHEM] A device for the determination of molecular weights by the decrease of vapor pressure of a solvent upon addition of a soluble sample.   { ˈvā·pər ˈpresh·ər äzˈmäm·əd·ər }

V band [SPECT] Absorption bands that appear in the ultraviolet part of the spectrum due to color centers produced in potassium bromide by exposure of the crystal at temperature of liquid nitrogen (81 K) to intense penetrating x-rays.   { 'vē ,band }

vcp *See* vacuum condensing point.

Venetian red [INORG CHEM] A pigment with a true red hue; contains 15–40% ferric oxide and 60–80% calcium sulfate.   { və'nēsh·ən 'red }

verdigris *See* cupric acetate.   { 'vərd·ə,grēs }

vermilion *See* mercuric sulfide.   { vər'mil·yən }

vernolate [ORG CHEM] $C_{10}H_{21}NOS$ An amber liquid, used to control weeds in sweet potatoes, peanuts, soybeans, and tobacco.   { 'vərn·əl,āt }

vibration [PHYS CHEM] Oscillation of atoms about their equilibrium positions within a molecular system.   { vī'brā·shən }

vibrational energy [PHYS CHEM] For a diatomic molecule, the difference between the energy of the molecule idealized by setting the rotational energy equal to zero, and that of a further idealized molecule which is obtained by gradually stopping the vibration of the nuclei without placing any new constraint on the motions of electrons.   { vī'brā·shən·əl 'en·ər·jē }

vibrational level [PHYS CHEM] An energy level of a diatomic or polyatomic molecule characterized by a particular value of the vibrational energy.   { vī'brā·shən·əl 'lev·əl }

vibrational quantum number [PHYS CHEM] A quantum number v characterizing the vibrational motion of nuclei in a molecule; in the approximation that the molecule behaves as a quantum-mechanical harmonic oscillator, the vibrational energy is $h(v + 1/2)f$, where h is Planck's constant and f is the vibration frequency.   { vī'brā·shən·əl 'kwän·təm ,nəm·bər }

vibrational spectrum [SPECT] The molecular spectrum resulting from transitions between vibrational levels of a molecule which behaves like the quantum-mechanical harmonic oscillator.   { vī'brā·shən·əl 'spek·trəm }

vibrational sum rule [SPECT] **1.** The rule that the sums of the band strengths of all emission bands with the same upper state is proportional to the number of molecules in the upper state, where the band strength is the emission intensity divided by the fourth power of the frequency.   **2.** The sums of the band strengths of all absorption bands with the same lower state is proportional to the number of molecules in the lower state, where the band strength is the absorption intensity divided by the frequency.   { vī'brā·shən·əl 'səm ,rül }

vibrational transition [PHYS CHEM] A transition between two quantized levels of a molecule that have different vibrational energies.   Also known as vibronic transition.   { vī'brā·shən·əl tran'zish·ən }

vibronic transition *See* vibrational transition.   { vī'brän·ik tran'zish·ən }

vic- [ORG CHEM] A chemical prefix indicating vicinal (neighboring or adjoining) positions on a carbon structure (ring or chain); used to identify the location of substituting groups when naming derivatives.   { vik }

vicinal [ORG CHEM] Referring to neighboring or adjoining positions on a carbon structure (ring or chain).   { 'vis·ən·əl }

Victoria blue [ORG CHEM] $C_{33}H_{31}N_3 \cdot HCl$ Bronze crystals, soluble in hot water, alcohol, and ether; used as a dye for silk, wool, and cotton, as a biological stain, and to make pigment toners.   { vik'tòr·ē·ə 'blü }

Vigreaux column [ANALY CHEM] An obsolete apparatus used in laboratory fractional distillation; it is a long glass tube with indentation in its walls; a thermometer is placed at the top of the tube and a side arm is attached to a condenser.   { ve'grō ,käl·əm }

vinetine *See* oxyacanthine.   { 'vin·ə,tēn }

vinyl acetal resin [ORG CHEM] $[CH_2CH(OC_2H_5)]_x$ A colorless, odorless, light-stable thermoplastic that is unaffected by water, gasoline, or oils; soluble in lower alcohols, benzene, and chlorinated hydrocarbons; used in lacquers, coatings, and molded objects.   Also known as polyvinyl acetal resin.   { 'vīn·əl 'as·ə,təl 'rez·ən }

vinyl acetate [ORG CHEM] $CH_3COOCH:CH_2$ A colorless, water-insoluble, flammable liquid that boils at 73°C; used as a chemical intermediate and in the production of polymers and copolymers (for example, the polyvinyl resins). { 'vīn·əl 'as·ə,tāt }

vinyl acetate resin [ORG CHEM] $(CH_2:CHOOCCH_3)_x$ An odorless thermoplastic formed by the polymerization of vinyl acetate; resists attack by water, gasoline, and oils; soluble in lower alcohols, benzene, and chlorinated hydrocarbons; used in lacquers, coatings, and molded products. { 'vīn·əl 'as·ə,tāt 'rez·ən }

vinylacetylene [ORG CHEM] $H_2CCHCCH$ A combustible dimer of acetylene, boiling at 5°C; used for the manufacture of neoprene rubber and as a chemical intermediate. { ¦vīn·əl·ə'sed·əl,ēn }

vinyl alcohol [ORG CHEM] $CH_2:CHOH$ A flammable, unstable liquid found only in ester or polymer form. Also known as ethenol. { 'vīn·əl 'al·kə,hól }

vinylation [CHEM] Formation of a vinyl-derived product by reaction with acetylene; for example, vinylation of alcohols gives vinyl ethers, such as vinyl ethyl ether. { ,vīn·əl'ā·shən }

vinylbenzene See styrene. { ¦vīn·əl'ben,zēn }

vinyl chloride [ORG CHEM] $CH_2:CHCl$ A flammable, explosive gas with an ethereal aroma; soluble in alcohol and ether, slightly soluble in water; boils at −14°C; an important monomer for polyvinyl chloride and its copolymers; used in organic synthesis and in adhesives. Also known as chloroethene; chloroethylene. { 'vīn·əl 'klór,īd }

vinyl chloride resin [ORG CHEM] $(CH_2CHCl)_x$ A white-power polymer made by the polymerization of vinyl chloride; used to make chemical-resistant pipe (when unplasticized) or bottles and parts (when plasticized). { 'vīn·əl 'klór,īd ,rez·ən }

vinylcyanide See acrylonitrile. { ¦vīn·əl'sī·ə,nīd }

vinyl ether [ORG CHEM] $CH_2:CHOCH:CH_2$ A colorless, light-sensitive, flammable, explosive liquid; soluble in alcohol, acetone, ether, and chloroform, slightly soluble in water; boils at 39°C; used as an anesthetic and a comonomer in polyvinyl chloride polymers. Also known as divinyl ether; divinyl oxide. { 'vīn·əl 'ē·thər }

vinyl ether resin [ORG CHEM] Any of a group of vinyl ether polymers; for example, polyvinyl methyl ether, polyvinyl ethyl ether, and polyvinyl butyl ether. { 'vīn·əl 'ē·thər 'rez·ən }

vinyl group [ORG CHEM] $CH_2=CH-$ A group of atoms derived when one hydrogen atom is removed from ethylene. { 'vīn·əl ,grüp }

vinylidene chloride [ORG CHEM] $CH_2:CCl_2$ A colorless, flammable, explosive liquid, insoluble in water; boils at 37°C; used to make polymers copolymerized with vinyl chloride or acrylonitrile (Saran). { vī'nil·ə,dēn 'klór,īd }

vinylidene resin [ORG CHEM] A polymer made up of the $(-H_2CCX_2-)$ unit, with X usually a chloride, fluoride, or cyanide radical. Also known as polyvinylidene resin. { vī'nil·ə,dēn 'rez·ən }

vinylog [ORG CHEM] Any of the organic compounds that differ from each other by a vinylene linkage $(-CH =CH-)$; for example, ethyl crotonate is a vinylog of ethyl acetate and of the next higher vinylog, ethyl sorbate. { 'vīn·əl,äg }

vinyl plastic See polyvinyl resin. { 'vīn·əl 'plas·tik }

vinyl polymerization [ORG CHEM] Addition polymerization where the unsaturated monomer contains a $CH_2=C-$group. { 'vīn·əl pə,lim·ə·rə'zā·shən }

vinylpyridine [ORG CHEM] $C_5H_4NCH:CH_2$ A toxic, combustible liquid; soluble in water, alcohol, hydrocarbons, esters, ketones, and dilute acids; used to manufacture elastomers and pharmaceuticals. { ¦vīn·əl'pir·ə,dēn }

N-vinyl-2-pyrrolidone [ORG CHEM] C_6H_9ON A colorless, toxic, combustible liquid, boiling at 148°C (100 mmHg); used as a chemical intermediate and to make polyvinyl pyrrolidone. { ¦en ¦vīn·əl ¦tü pə'räl·ə,dōn }

vinylstyrene See divinylbenzene. { ¦vīn·əl'stī,rēn }

vinyltoluene [ORG CHEM] $CH_2:CHC_6H_4CH_3$ A colorless, flammable, moderately toxic liquid; soluble in ether and methanol, slightly soluble in water; boils at 170°C; used as a chemical intermediate and solvent. Also known as methyl styrene. { ¦vīn·əl'täl·yə,wēn }

vinyl trichloride

vinyl trichloride *See* trichloroethane. { 'vīn·əl trī'klòr,īd }

vinyl trichlorosilone [ORG CHEM] $CH_2CH_3SiCl_3$ A liquid that boils at 90.6°C and is soluble in organic solvents; used in silicones and adhesives. { 'vīn·əl trī¦klòr·ō'si,lōn }

viologen [CHEM] Any member of a group of chlorides of certain quaternary bases derived from γ,γ'-dipyridyl that are used as oxidation-reduction indicators; color is exhibited in the reduced form. { vī'äl·ə·jən }

virtual orbital [PHYS CHEM] An orbital that is either empty or unoccupied while in the ground state. { 'vər·chə·wəl 'òr·bəd·əl }

visible absorption spectrophotometry [SPECT] Study of the spectra produced by the absorption of visible-light energy during the transformation of an electron from the ground state to an excited state as a function of the wavelength causing the transformation. { 'viz·ə·bəl əb'sòrp·shən ¦spek·trō·fə'täm·ə·trē }

visible spectrophotometry [SPECT] In spectrophotometric analysis, the use of a spectrophotometer with a tungsten lamp that has an electromagnetic spectrum of 380–780 nanometers as a light source, glass or quartz prisms or gratings in the monochromator, and a photomultiplier cell as a detector. { 'viz·ə·bəl ¦spek·trō·fə'täm·ə·trē }

visible spectrum [SPECT] **1.** The range of wavelengths of visible radiation. **2.** A display or graph of the intensity of visible radiation emitted or absorbed by a material as a function of wavelength or some related parameter. { 'viz·ə·bəl 'spek·trəm }

visual colorimetry [ANALY CHEM] A procedure for the determination of the color of an unknown solution by visual comparison to color standards (solutions or color-tinted disks). { 'vizh·ə·wəl ˌkəl·ə'rim·ə·trē }

vitamin K₁ *See* phytonadione. { 'vīd·ə·mən ¦kā¦wən }

vitriolic acid *See* sulfuric acid. { 'vi·trē,äl·ik 'əs·ed }

volatile [CHEM] Readily passing off by evaporation. { 'väl·əd·əl }

volatile fluid [CHEM] A liquid with the tendency to become vapor at specified conditions of temperature and pressure. { 'väl·əd·əl 'flü·əd }

volatility product [CHEM] The product of the concentrations of two or more molecules or ions that react to form a volatile substance. { ˌväl·ə'til·əd·ē ˌpräd·əkt }

Volhard titration [ANALY CHEM] Determination of the halogen content of a solution by titration with a standard thiocyanate solution. { 'fōl,härt tī'trā·shən }

voltameter *See* coulometer. { väl'tam·əd·ər }

voltammetry [PHYS CHEM] Any electrochemical technique in which a faradaic current passing through the electrolysis solution is measured while an appropriate potential is applied to the polarizable or indicator electrode; for example, polarography. { väl 'täm·ə·trē }

Volta series *See* displacement series. { 'vōl·tə ˌsir·ēz }

volume susceptibility [PHYS CHEM] The magnetic susceptibility of a specified volume (for example, 1 cubic centimeter) of a magnetically susceptible material. { 'väl·yəm sə,sep·tə'bil·əd·ē }

volumetric analysis [ANALY CHEM] Quantitative analysis of solutions of known volume but unknown strength by adding reagents of known concentration until a reaction end point (color change or precipitation) is reached; the most common technique is by titration. Also known as titrimetric analysis. { ¦väl·yə¦me·trik ə'nal·ə·səs }

volumetric flask [ANALY CHEM] A laboratory flask primarily intended for the preparation of definite, fixed volumes of solutions, and therefore calibrated for a single volume only. { ¦väl·yə¦me·trik 'flask }

volumetric pipet [ANALY CHEM] A graduated glass tubing used to measure quantities of a solution; the tube is open at the top and bottom, and a slight vacuum (suction) at the top pulls liquid into the calibrated section; breaking the vacuum allows liquid to leave the tube. { ¦väl·yə¦me·trik pī'pet }

W

W *See* tungsten.

Wagner's reagent [ANALY CHEM] An aqueous solution of iodine and potassium iodide; used for microchemical analysis of alkaloids. Also known as Wagner's solution. { 'väg·nərz rē͵ā·jənt }

Wagner's solution *See* Wagner's reagent. { 'väg·nərz sə͵lü·shən }

Walden's rule [PHYS CHEM] A rule which states that the product of the viscosity and the equivalent ionic conductance at infinite dilution in electrolytic solutions is a constant, independent of the solvent; it is only approximately correct. { 'wȯl·dənz ͵rül }

Wallach transformation [ORG CHEM] By the use of concentrated sulfuric acid, an azoxybenzene is converted into a *para*-hydroxyazobenzene. { 'wäl·ək ͵tranz·fər͵mā·shən }

wall-coated capillary column [ANALY CHEM] A capillary column characterized by a layer of stationary liquid coated directly on the inner wall of a coiled capillary tube. { 'wȯl ͵kōd·əd 'kap·ə͵ler·ē ͵käl·əm }

washing [ANALY CHEM] **1.** In the purification of a laboratory sample, the cleaning of residual liquid impurities from precipitates by adding washing solution to the precipitates, mixing, then decanting, and repeating the operation as often as needed. **2.** The removal of soluble components from a mixture of solids by using the effect of differential solubility. { 'wäsh·iŋ }

washing soda *See* sal soda. { 'wäsh·iŋ ͵sōd·ə }

water [CHEM] H_2O Clear, odorless, tasteless liquid that is essential for most animal and plant life and is an excellent solvent for many substances; melting point $0°C$ ($32°F$), boiling point $100°C$ ($212°F$); the chemical compound may be termed hydrogen oxide. { 'wȯd·ər }

water absorption tube [ANALY CHEM] A glass tube filled with a solid absorbent (calcium chloride or silica gel) to remove water from gaseous streams during or after chemical analyses. { 'wȯd·ər əb'sȯrp·shən ͵tüb }

watercolor pigment [INORG CHEM] A permanent pigment used in watercolor painting, for example, titanium oxide (white). { 'wȯd·ər͵kəl·ər ͵pig·mənt }

water glass *See* sodium silicate. { 'wȯd·ər ͵glas }

water of crystallization *See* water of hydration. { 'wȯd·ər əv ͵krist·əl·ə'zā·shən }

water of hydration [CHEM] Water present in a definite amount and attached to a compound to form a hydrate; can be removed, as by heating, without altering the composition of the compound. Also known as water of crystallization. { 'wȯd·ər əv hī'drā·shən }

water saturation [CHEM] **1.** A solid adsorbent that holds the maximum possible amount of water under specified conditions. **2.** A liquid solution in which additional water will cause the appearance of a second liquid phase. **3.** A gas that is at or just under its dew point because of its water content. { 'wȯd·ər ͵sach·ə'rā·shən }

water softening [CHEM] Removal of scale-forming calcium and magnesium ions from hard water, or replacing them by the more soluble sodium ions; can be done by chemicals or ion exchange. { 'wȯd·ər ͵sȯf·ə·niŋ }

water-wettable [CHEM] Denoting the capability of a material to accept water, or of being hydrophilic or hydrophoric. { 'wȯd·ər 'wed·ə·bəl }

water white |CHEM| A grade of color for liquids that has the appearance of clear water; for petroleum products, a plus 21 in the scale of the Saybolt chromometer. { 'wȯd·ər 'wīt }

Watson equation |PHYS CHEM| Calculation method to extend heat of vaporization data for organic compounds to within 10 or 15°C of the critical temperature; uses known latent heats of vaporization and reduced temperature data. { 'wät·sən i,kwā·zhən }

wavelength standards |SPECT| Accurately measured lengths of waves emitted by specified light sources for the purpose of obtaining the wavelengths in other spectra by interpolating between the standards. { 'wāv,leŋkth ,stan·dərdz }

weak acid |CHEM| An acid that does not ionize greatly; for example, acetic acid or carbonic acid. { 'wēk 'as·əd }

wedge spectrograph |SPECT| A spectrograph in which the intensity of the radiation passing through the entrance slit is varied by moving an optical wedge. { 'wej 'spek·trə,graf }

weight titration |ANALY CHEM| A titration in which the amount of titrant required is determined in terms of the weight that must be added to reach the end point. { 'wāt tī'trā·shən }

Weisz ring oven |ANALY CHEM| A device for vaporization of solvent from filter paper, leaving the solute in a ring (circular) shape; used for qualitative analysis of very small samples. { 'vīs 'riŋ ,əv·ən }

Werner band |SPECT| A band in the ultraviolet spectrum of molecular hydrogen extending from 116 to 125 nanometers. { 'ver·nər ,band }

Werner complex *See* coordination compound. { 'ver·nər ,käm,pleks }

wet ashing |ORG CHEM| The conversion of an organic compound into ash (decomposition) by treating the compound with nitric or sulfuric acid. { 'wet 'ash·iŋ }

wettability |CHEM| The ability of any solid surface to be wetted when in contact with a liquid; that is, the surface tension of the liquid is reduced so that the liquid spreads over the surface. { ,wed·ə'bil·əd·ē }

wetted |CHEM| Pertaining to material that has accepted water or other liquid, either on its surface or within its pore structure. { 'wed·əd }

white copperas *See* zinc sulfate. { 'wīt 'käp·rəs }

white lead |INORG CHEM| Basic lead carbonate of variable composition, the oldest and most important lead paint pigment; also used in putty and ceramics. { 'wīt 'led }

white phosphorus |CHEM| The element phosphorus in its allotropic form, a soft, waxy, poisonous solid melting at 44.5°C; soluble in carbon disulfide, insoluble in water and alcohol; self-igniting in air. Also known as yellow phosphorus. { 'wīt 'fä·sfə·rəs }

white vitriol *See* zinc sulfate. { 'wīt 'vi·trē,ȯl }

Wiedemann's additivity law |PHYS CHEM| The law that the mass (or specific) magnetic susceptibility of a mixture or solution of components is the sum of the proportionate (by weight fraction) susceptibilities of each component in the mixture. { 'vēd·ə,mänz ,ad·ə'tiv·əd·ē ,lȯ }

Wien effect |PHYS CHEM| An increase in the conductance of an electrolyte at very high potential gradients. { 'vēn i,fekt }

Wijs' iodine monochloride solution |ANALY CHEM| A solution in glacial acetic acid of iodine monochloride; used to determine iodine numbers. Also known as Wijs' special solution. { vīs 'ī·ə,dīn ¦män·ə'klȯr,īd sə,lü·shən }

Wijs' special solution *See* Wijs' iodine monochloride solution. { vīs 'spesh·əl sə,lü·shən }

Williamson synthesis |ORG CHEM| The synthesis of ethers utilizing an alkyl iodide and sodium alcoholate. { 'wil·yəm·sən 'sin·thə·səs }

Winkler titration |ANALY CHEM| A chemical method for estimating the dissolved oxygen in seawater: manganous hydroxide is added to the sample and reacts with oxygen to produce a manganese compound which in the presence of acid potassium iodide liberates an equivalent quantity of iodine that can be titrated with standard sodium thiosulfate. { 'viŋ·klər tī,trā·shən }

wintergreen oil *See* methyl salicylate. { 'win·tər,grēn ,ȯil }

Wittig ether rearrangement [ORG CHEM] The rearrangement of benzyl and alkyl ethers when reacted with a methylating agent, producing secondary and tertiary alcohols. { 'vid·ik 'ē·thər ə,rānj·mənt }

Witt theory [CHEM] A theory of the mechanism of dyeing stating that all colored organic compounds (called chromogens) contain certain unsaturated chromophoric groups which are responsible for the color, and if these compounds also contain certain auxochromic groups, they possess dyeing properties. { 'wit ,thē·ə·rē }

Wolf-Kishner reduction [ORG CHEM] Conversion of aldehydes and ketones to corresponding hydrocarbons by heating their semicarbazones, phenylhydrazones, and hydrazones with sodium ethoxide or by heating the carbonyl compound with excess sodium ethoxide and hydrazine sulfate. { 'wùlf 'kish·nər ri'dək·shən }

wolfram See tungsten. { 'wùl·frəm }

wolframic acid See tungstic acid. { wùl'fram·ik 'as·əd }

wolfram white See barium tungstate. { 'wùl·frəm wīt }

wood alcohol See methyl alcohol. { 'wùd 'al·kə,hól }

wood ether See dimethyl ether. { 'wùd 'ē·thər }

wood vinegar See pyroligneous acid. { 'wùd 'vin·ə·gər }

Woodward-Hoffmann rule [ORG CHEM] A concept which can predict or explain the stereochemistry of certain types of reactions in organic chemistry; it is also described as the conservation of orbital symmetry. { 'wùd·wərd 'häf·mən ,rül }

Woodward's Reagent K See N-ethyl-5-phenylisoxazolium-3′-sulfonate. { 'wùd·wərdz rē¦ä·jənt 'kā }

working electrode [PHYS CHEM] The electrode used in corrosion testing by an electrochemical cell. { 'wərk·iŋ i'lek,trōd }

Wurtz-Fittig reaction [ORG CHEM] A modified Wurtz reaction in which an aromatic halide reacts with an aklyl halide in the presence of sodium and an anhydrous solvent to form alkylated aromatic hydrocarbons. { 'wərtz 'fid·ig rē,ak·shən }

Wurtz reaction [ORG CHEM] Synthesis of hydrocarbons by treating alkyl iodides in ethereal solution with sodium according to the reaction $2CH_3I + 2Na \rightarrow CH_3CH_3 + 2NaI$. { 'wərts rē,ak·shən }

X

X *See* siegbahn.

XAFS *See* x-ray absorption fine structure.

xanthan gum [ORG CHEM] A high-molecular-weight (5–10 million) water-soluble natural gum; a heteropolysaccharide made up of building blocks of D-glucose, D-mannose, and D-glucuronic acid residues; produced by pure culture fermentation of glucose with *Xanthomonas campestris*. { 'zan·thən ‚gəm }

xanthate [ORG CHEM] A water-soluble salt of xanthic acid, usually potassium or sodium; used as an ore-flotation collector. { 'zan‚thāt }

xanthene [ORG CHEM] CH₂(C₆H₄)₂O Yellowish crystals that are soluble in ether, slightly soluble in water and alcohol; melts at 100°C; used as a fungicide and chemical intermediate. Also known as tricyclic dibenzopyran. { 'zan‚thēn }

xanthene dye [ORG CHEM] Any of a family of dyes related to the xanthenes; the chromophore groups are (C₆H₄). { 'zan‚thēn 'dī }

9-xanthenone *See* genicide. { ¦nīn 'zan·thə‚nōn }

xanthine [ORG CHEM] C₅H₄N₄O₂ A toxic yellow-white purine base that is found in blood and urine, and occasionally in plants; it is a powder, insoluble in water and acids, soluble in caustic soda; sublimes when heated; used in medicine and as a chemical intermediate. Also known as dioxopurine. { 'zan‚thēn }

xanthone [ORG CHEM] CO(C₆H₄)₂O White needle crystals that are found in some plant pigments; insoluble in water, soluble in alcohol, chloroform, and benzene; melts at 173°C, sublimes at 350°C; used as a larvicide, as a dye intermediate, and in perfumes and pharmaceuticals. { 'zan‚thōn }

Xe *See* xenon.

xenon [CHEM] An element, symbol Xe, member of the noble gas family, group 0, atomic number 54, atomic weight 131.291; colorless, boiling point −108°C (1 atmosphere, or 101,325 pascals), noncombustible, nontoxic, and nonreactive; used in photographic flash lamps, luminescent tubes, and lasers, and as an anesthetic. { 'zē‚nän }

xenyl [ORG CHEM] The functional group C₆H₅C₆H₄−. { 'zen·əl }

xerogel [CHEM] **1.** A gel whose final form contains little or none of the dispersion medium used. **2.** An organic polymer capable of swelling in suitable solvents to yield particles possessing a three-dimensional network of polymer chains. { 'zer·ə‚jel }

XPS *See* x-ray photoelectron spectroscopy.

x-ray absorption fine structure [SPECT] The structure in the x-ray absorption spectrum of a substance at energies above the absorption edge, including both the x-ray absorption near-edge structure and the extended x-ray absorption fine structure. Abbreviated XAFS. { ‚eks ‚rā əb¦sorp·shən ¦fīn 'strək·chər }

x-ray crystal spectrometer [SPECT] An instrument designed to produce an x-ray spectrum and measure the wavelengths of its components, by diffracting x-rays from a crystal with known lattice spacing. { 'eks ‚rā 'krist·əl spek'träm·əd·ər }

x-ray fluorescence analysis [SPECT] A nondestructive physical method used for chemical elemental analysis in which a material is irradiated by photons or charged particles of sufficient energy to cause its elements to emit (fluoresce) characteristic x-ray line spectra. { 'eks ‚rā flù'res·əns ə‚nal·ə·səs }

x-ray fluorescent emission spectrometer [SPECT] An x-ray crystal spectrometer used to measure wavelengths of x-ray fluorescence; in order to concentrate beams of low

x-ray fluorimetry

intensity, it has bent reflecting or transmitting crystals arranged so that the theoretical curvature required can be varied with the diffraction angle of a spectrum line. { 'eks ,rā flü'res·ənt i¦mish·ən spek'träm·əd·ər }

x-ray fluorimetry See x-ray fluorescence analysis. { 'eks ,rā flü'räm·ə·trē }

x-ray image spectrography [SPECT] A modification of x-ray fluorescence analysis in which x-rays irradiate a cylindrically bent crystal, and Bragg diffraction of the resulting emissions produces a slightly enlarged image with a resolution of about 50 micrometers. { 'eks ,rā ¦im·ij spek'träg·rə·fē }

x-ray microprobe See microprobe. { 'eks ,rā 'mī·krə,prōb }

x-ray photoelectron spectroscopy [SPECT] A form of electron spectroscopy in which a sample is irradiated with a beam of monochromatic x-rays and the energies of the resulting photoelectrons are measured. Abbreviated XPS. Also known as electron spectroscopy for chemical analysis (ESCA). { 'eks ,rā ¦fōd·ō·i¦lek,trän spek'träs·kə·pē }

x-ray spectrograph [SPECT] An x-ray spectrometer equipped with photographic or other recording apparatus; one application is fluorescence analysis. { 'eks ,rā 'spek·trə,graf }

x-ray spectrometer [SPECT] An instrument for producing the x-ray spectrum of a material and measuring the wavelengths of the various components. { 'eks ,rā spek'träm·əd·ər }

x-ray spectrometry [SPECT] A technique for quantitative analysis of the elemental composition of specimens. Irradiation of a sample by high-energy electrons, protons, or photons ionizes some of the atoms, which then emit characteristic x-rays whose wavelength depends on the atomic number of the element and whose intensity is related to the concentration of that element. { 'eks ,rā spek'träm·ə·trē }

x-ray spectroscopy See x-ray spectrometry. { 'eks ,rā spek'träs·kə·pē }

x-ray spectrum [SPECT] A display or graph of the intensity of x-rays, produced when electrons strike a solid object, as a function of wavelengths or some related parameter; it consists of a continuous bremsstrahlung spectrum on which are superimposed groups of sharp lines characteristic of the elements in the target. { 'eks ,rā ,spek·trəm }

x-ray unit See siegbahn. { 'eks ,rā ,yü·nət }

XU See siegbahn.

X unit See siegbahn. { 'eks ,yü·nət }

xylene [ORG CHEM] $C_6H_4(CH_3)_2$ Any one of the family of isomeric, colorless aromatic hydrocarbon liquids; produced by the destructive distillation of coal or by the catalytic reforming of petroleum naphthenic fractions; used for high-octane and aviation gasolines, solvents, chemical intermediates, and the manufacture of polyester resins. Also known as dimethylbenzene; xylol. { 'zī,lēn }

***meta*-xylene** [ORG CHEM] $1,3\text{-}C_6H_4(CH_3)_2$ A flammable, toxic liquid; insoluble in water, soluble in alcohol and ether; boils at 139°C; used as an intermediate for dyes, a chemical intermediate, and a solvent, and in insecticides and aviation fuel. { ¦med·ə 'zī,lēn }

***ortho*-xylene** [ORG CHEM] $1,2\text{-}C_6H_4(CH_3)_2$ A flammable, moderately toxic liquid; insoluble in water, soluble in alcohol and ether; boils at 144°C; used to make phthalic anhydride, vitamins, pharmaceuticals, and dyes, and in insecticides and motor fuels. { ¦ór·thō 'zī,lēn }

***para*-xylene** [ORG CHEM] $1,4\text{-}C_6H_4(CH_3)_2$ A toxic, combustible liquid; insoluble in water, soluble in alcohol and ether; boils at 139°C; used as a chemical intermediate, and to synthesize terephthalic acid, vitamins, and pharmaceuticals, and in insecticides. { ¦par·ə 'zī,lēn }

xylenol [ORG CHEM] $(CH_3)_2C_6H_3OH$ Highly toxic, combustible crystals; slightly soluble in water, soluble in most organic solvents; melts at 20–76°C; used as a chemical intermediate, disinfectant, solvent, and fungicide, and for pharmaceuticals and dyestuffs. { 'zī·lə,nòl }

xylidine |ORG CHEM| $(CH_3)_2C_6H_3NH_2$ A toxic, combustible liquid; soluble in alcohol and ether, slightly soluble in water; boils about 220°C; used as a chemical intermediate and to make dyes and pharmaceuticals. { 'zī·lə,dēn }

xylite See xylitol. { 'zī,līt }

xylitol |ORG CHEM| $CH_2OH(CHOH)_3CH_2OH$ Pentahydric alcohols derived from xylose. Also known as xylite. { 'zī·lə,tól }

xylol See xylene. { 'zī,lól }

Y

Y *See* yttrium.

yacca gum *See* acaroid resin. { 'yak·ə ‚gəm }

Yb *See* ytterbium.

yellow phosphorus *See* white phosphorus. { 'yel·ō 'fä·sfə·rəs }

yellow precipitate *See* mercuric oxide. { 'yel·ō pri'sip·ə‚tāt }

yellow prussiate of potash *See* potassium ferrocyanide. { 'yel·ō 'prəs·ē‚āt əv 'päd‚ash }

yellow prussiate of soda *See* sodium ferrocyanide. { 'yel·ō 'prəs·ē‚āt əv 'sōd·ə }

yellow pyoktanin *See* auramine hydrochloride. { 'yel·ō pī'äk·tə·nən }

yellow salt *See* uranyl nitrate. { 'yel·ō 'sȯlt }

ylide [ORG CHEM] An organic compound which contains two adjacent atoms bearing formal positive and negative charges, and in which both atoms have full octets of electrons. { 'i·līd }

ylium ion *See* enium ion. { 'ī·lē·əm 'ī‚än }

ytterbia *See* ytterbium oxide. { i'tər·bē·ə }

ytterbium [CHEM] A rare-earth metal of the yttrium subgroup, symbol Yb, atomic number 70, atomic weight 173.04; lustrous, malleable, soluble in dilute acids and liquid ammonia, reacts slowly with water; melts at 824°C, boils at 1427°C; used in chemical research, lasers, garnet doping, and x-ray tubes. { i'tər·bē·əm }

ytterbium oxide [INORG CHEM] Yb_2O_3 A colorless compound, melts at 2346°C, dissolves in hot dilute acids; used to prepare alloys, ceramics, and special glasses. Also known as ytterbia. { i'tər·bē·əm 'äk‚sīd }

yttria *See* yttrium oxide. { 'i·trē·ə }

yttrium [CHEM] A rare-earth metal, symbol Y, atomic number 39, atomic weight 88.9059; dark-gray, flammable (as powder), soluble in dilute acids and potassium hydroxide solution, and decomposes in water; melts at 1500°C, boils at 2927°C; used in alloys and nuclear technology and as a metal deoxidizer. { 'i·trē·əm }

yttrium acetate [ORG CHEM] $Y(C_2H_3O_2)_3 \cdot 8H_2O$ Colorless, water-soluble crystals used as an analytical reagent. { 'i·trē·əm 'as·ə‚tāt }

yttrium chloride [INORG CHEM] $YCl_3 \cdot 6H_2O$ Reddish, transparent, water- and alcohol-soluble prisms; decomposes at 100°C; used as an analytical reagent. { 'i·trē·əm 'klȯr‚īd }

yttrium oxide [INORG CHEM] Y_2O_3 A yellowish powder, insoluble in water, soluble in dilute acids; used as television tube phosphor and microwave filters. Also known as yttria. { 'i·trē·əm 'äk‚sīd }

yttrium sulfate [INORG CHEM] $Y_2(SO_4)_3 \cdot 8H_2O$ Reddish crystals that are soluble in concentrated sulfuric acid, slightly soluble in water; decomposes at 700°C; used as an analytical reagent. { 'i·trē·əm 'səl‚fāt }

Z

ZAA spectrometry See Zeeman-effect atomic absorption spectrometry. { ¦zē¦ā¦ā spek 'träm·ə·trē }

Zeeman displacement [SPECT] The separation, in wave numbers, of adjacent spectral lines in the normal Zeeman effect in a unit magnetic field, equal (in centimeter-gram-second Gaussian units) to $e/4\pi mc^2$, where e and m are the charge and mass of the electron, or to approximately 4.67×10^{-5} (centimeter)$^{-1}$(gauss)$^{-1}$. { 'zā·mən di‚splās·mənt }

Zeeman effect [SPECT] A splitting of spectral lines in the radiation emitted by atoms or molecules in a static magnetic field. { 'zā·mən i‚fekt }

Zeeman-effect atomic absorption spectrometry [SPECT] A type of atomic absorption spectrometry in which either the light source or the sample is placed in a magnetic field, splitting the spectral lines under observation into polarized components, and a rotating polarizer is placed between the source and the sample, enabling the absorption caused by the element under analysis to be separated from background absorption. Abbreviated ZAA spectrometry. { ¦zē·mən i‚fekt ə¦täm·ik əp¦sȯrp·shən spek'träm·ə·trē }

zeolite catalyst [INORG CHEM] Hydrated aluminum and calcium (or sodium) silicates (for example, $CaO \cdot 2Al_2O_3 \cdot 5SiO_2$ or $Na_2O \cdot 2Al_2O_3 \cdot 5SiO_2$) made with controlled porosity; used as a catalytic cracking catalyst in petroleum refineries, or loaded with catalyst for other chemical reactions. { 'zē·ə‚līt 'kad·əl·əst }

zeotrope [PHYS CHEM] A nonazeotropic liquid mixture which may be separated by distillation, and in which the components are miscible in all proportions (homogeneous zeotrope or homozeotrope) or not miscible in all proportions (heterogeneous zeotrope or heterozeotrope). { 'zē·ə‚trōp }

Zerewitinoff reagent [ANALY CHEM] A light-colored methylmagnesium iodide-n-butyl ether solution that reacts rapidly with moisture and oxygen; used to determine water, alcohols, and amines in inert solvents. { ‚zir·ə'wit·ən‚ȯf rē‚ā·jənt }

zero branch [SPECT] A spectral band whose Fortrat parabola lies between two other Fortrat parabolas, with its vertex almost on the wave number axis. { 'zir·ō 'branch }

zerogel [CHEM] A gel which has dried until apparently solid; sometimes it will swell or redisperse to form a sol when treated with a suitable solvent. { 'zir·ō‚jel }

zero-order reaction [PHYS CHEM] A reaction for which reaction rate is independent of the concentrations of the reactants; for example, a photochemical reaction in which the rate is determined by the intensity of light. { 'zir·ō ¦ȯrd·ər rē'ak·shən }

zigzag nanotube [PHYS CHEM] A carbon nanotube formed from a graphite sheet that is rolled up so that it has a zigzag edge. { ‚zig‚zag 'nan·ō‚tüb }

Zimm plot [ANALY CHEM] A graphical determination of the root-square-mean end-to-end distances of coillike polymer molecules during scattered-light photometric analyses. { 'zim ‚plät }

zinc [CHEM] A metallic transition element, symbol Zn, atomic number 30, atomic weight 65.38; explosive as powder; soluble in acids and alkalies, insoluble in water; strongly electropositive; melts at 419°C; boils at 907°C. { ziŋk }

zinc acetate [ORG CHEM] $Zn(C_2H_3O_2)_2 \cdot 2H_2O$ Pearly-white crystals with an astringent taste; soluble in water and alcohol; decomposes at 200°C; used to preserve wood

zinc arsenite

in textile dyeing, and as an analytical reagent, a feed additive, and a polymer cross-linking agent.   { 'ziŋk 'as·ə‚tāt }

zinc arsenite [INORG CHEM] $Zn(AsO_2)_2$ A toxic white powder that is insoluble in water, soluble in alkalies; used as an insecticide and timber preservative. Also known as zinc meta-arsenite.   { 'ziŋk 'ärs·ən‚īt }

zincate [INORG CHEM] A reaction product of zinc with an alkali metal or with ammonia; for example, sodium zincate, Na_2ZnO_2.   { 'ziŋ‚kāt }

zinc borate [INORG CHEM] $3ZnO·2B_2O_3$ A white, amorphous powder that is soluble in dilute acids, slightly soluble in water; melts at 980°C; used in medicine, as a ceramics flux, as an inhibitor for mildew, and to fireproof textiles.   { 'ziŋk 'bȯr‚āt }

zinc bromide [INORG CHEM] $ZnBr_2$ Water- and alcohol-soluble, white crystals that melt at 294°C; used in medicine, manufacture of rayon, and photography, and in a radiation viewing screen.   { 'ziŋk 'brō‚mīd }

zinc carbonate [INORG CHEM] $ZnCO_3$ White crystals that are insoluble in water, soluble in alkalies and acids; used in ceramics and ointments, and as a fireproofing agent and feed additive.   { 'ziŋk 'kär·bə‚nāt }

zinc chloride [INORG CHEM] $ZnCl_2$ Water- and alcohol-soluble, white, fire-hazardous crystals that melt at 290°C, and are irritating to the skin; used as a catalyst and in electroplating, wood preservation, textile processing, petroleum refining, medicine, and feed additives.   { 'ziŋk 'klȯr‚īd }

zinc chromate [INORG CHEM] $ZnCrO_4$ A toxic, yellow powder that is insoluble in water, soluble in acids; used as a pigment in paints (artists', automotive, primer), varnishes, linoleum, and epoxy laminates.   { 'ziŋk 'krō‚māt }

zinc cyanide [INORG CHEM] $Zn(CN)_2$ A toxic, white powder that is insoluble in water and alcohol, soluble in alkalies and dilute acids; melts at 800°C; used as an analytical reagent and insecticide, and in medicine and metal plating.   { 'ziŋk 'sī·ə‚nīd }

zinc fluoride [INORG CHEM] ZnF_2 A toxic white powder that is slightly soluble in water and melts at 872°C; used in enamels, ceramic glazes, and galvanizing.   { 'ziŋk 'flȯr‚īd }

zinc formate [ORG CHEM] $Zn(CHO_2)_2·2H_2O$ Toxic, white crystals that are soluble in water, insoluble in alcohol; used as a catalyst, weatherproofing agent, and wood preservative.   { 'ziŋk 'fȯr‚māt }

zinc halide [INORG CHEM] A binary compound of zinc and a halogen; for example, $ZnBr_2$, $ZnCl_2$, ZnF_2, and ZnI_2.   { 'ziŋk 'ha‚līd }

zinc hydroxide [INORG CHEM] $Zn(OH)_2$ Colorless, water-soluble crystals that decompose at 125°C; used as a chemical intermediate and in rubber compounding and surgical dressings.   { 'ziŋk hī'dräk‚sīd }

zinc metaarsenite See zinc arsenite.   { 'ziŋk ‚med·ə'ärs·ən‚āt }

zinc naphthenate [ORG CHEM] $Zn(C_6H_9COO)_2$ A combustible, viscous, acetone-soluble solid; used in paints, varnishes, and resins, and as a drier and wetting agent, insecticide, fungicide, and mildewstat.   { 'ziŋk 'naf·thə‚nāt }

zinc orthoarsenate [INORG CHEM] $Zn_3(AsO_4)_2$ A toxic white powder that is insoluble in water, soluble in alkalies; used as an insecticide and wood preservative.   { 'ziŋk ‚ȯr·thō'ärs·ən‚āt }

zinc orthophosphate See zinc phosphate.   { 'ziŋk ‚ȯr·thō'fä‚sfāt }

zinc oxide [INORG CHEM] ZnO A bitter-tasting, white to gray powder that is insoluble in water, soluble in alkalies and acids; melts at 1978°C; used as a pigment, mold-growth inhibitor, and dietary supplement, and in cosmetics, electronics, and color photography. Also known as philosopher's wool.   { 'ziŋk 'äk‚sīd }

zinc phosphate [INORG CHEM] $Zn_3(PO_4)_2$ A white powder that is insoluble in water, soluble in acids and ammonium hydroxide; melts at 900°C; used in coatings for steel, aluminum, and other metals, and in dental cements and phosphors. Also known as tribasic zinc phosphate; zinc orthophosphate.   { 'ziŋk 'fä‚sfāt }

zinc phosphide [INORG CHEM] Zn_3P_2 A toxic, alcohol-insoluble, gray gritty powder that reacts violently with oxidizing agents; melts at over 420°C, decomposes in water; used as a rat poison and in medicine.   { 'ziŋk 'fä‚sfīd }

zinc selenide [INORG CHEM] ZnSe A water-insoluble, moderately toxic, yellow to reddish solid that is a fire hazard when in contact with water and acids; melts above 1100°C; used as infrared optical windows.   { 'ziŋk 'sel·ə,nīd }

zinc sulfate [INORG CHEM] ZnSO₄·7H₂O Efflorescent, water-soluble, colorless crystals with an astringent taste; used to preserve skins and wood and as a paper bleach, analytical reagent, feed additive, and fungicide.   Also known as white copperas; white vitriol; zinc vitriol.   { 'ziŋk 'səl,fāt }

zinc sulfide [INORG CHEM] ZnS A yellowish powder that is insoluble in water, soluble in acids; exists in two crystalline forms (alpha, or wurtzite, and beta, or sphalerite); beta becomes alpha at 1020°C, and sublimes at 1180°C; used as a pigment for paints and linoleum, in opaque glass, rubber, and plastics, for hydrosulfite dyeing process, as x-ray and television screen phosphor, and as a fungicide.   { 'ziŋk 'səl,fīd }

zinc telluride [INORG CHEM] ZnTe Moderately toxic, reddish crystals that melt at 1238°C and decompose in water.   { 'ziŋk 'tel·yə,rīd }

zinc vitriol See zinc sulfate.   { 'ziŋk 'vi·trē,òl }

zinc white See Chinese white.   { 'ziŋk 'wīt }

zirconia See zirconium oxide.   { ,zər'kō·nē·ə }

zirconic anhydride See zirconium oxide.   { ,zər'kän·ik an'hī,drīd }

zirconium [CHEM] A metallic transition element, symbol Zr, atomic number 40, atomic weight 91.22; occurs as crystals, flammable as powder; insoluble in water, soluble in hot, concentrated acids; melts at 1850°C, boils at 4377°C.   { ,zər'kō·nē·əm }

zirconium boride [INORG CHEM] ZrB₂ A hard, toxic, gray powder that melts at 3000°C; used as an aerospace refractory, in cutting tools, and to protect thermocouple tubes.   Also known as zirconium diboride.   { ,zər'kō·nē·əm 'bòr,īd }

zirconium carbide [INORG CHEM] ZrC Hard, gray crystals that are soluble in water, soluble in acids; as powder, it ignites spontaneously in air; melts at 3400°C, boils at 5100°C; used as an abrasive, refractory, and metal cladding, and in cermets, incandescent filaments, and cutting tools.   { ,zər'kō·nē·əm 'kär,bīd }

zirconium chloride See zirconium tetrachloride.   { ,zər'kō·nē·əm 'klòr,īd }

zirconium diboride See zirconium boride.   { ,zər'kō·nē·əm dī'bòr,īd }

zirconium dioxide See zirconium oxide.   { ,zər'kō·nē·əm dī'äk,sīd }

zirconium halide [INORG CHEM] A compound of zirconium with a halogen; for example, ZrBr₂, ZrCl₂, ZrCl₃, ZrCl₄, ZrBr₂, ZrBr₃, ZrF₄, and ZrI₄.   { ,zər'kō·nē·əm 'ha,līd }

zirconium hydride [INORG CHEM] ZrH₂ A flammable, gray-black powder; used in powder metallurgy and nuclear moderators, and as a reducing agent, vacuum-tube getter, and metal-foaming agent.   { ,zər'kō·nē·əm 'hī,drīd }

zirconium hydroxide [INORG CHEM] Zr(OH)₄ A toxic, amorphous white powder; insoluble in water, soluble in dilute mineral acids; decomposes at 550°C; used in pigments, glass, and dyes, and to make zirconium compounds.   { ,zər'kō·nē·əm hī'dräk,sīd }

zirconium nitride [INORG CHEM] ZrN A hard, brassy powder that is soluble in concentrated acids; melts at 2930°C; used in refractories, cermets, and laboratory crucibles.   { ,zər'kō·nē·əm 'nī,trīd }

zirconium orthophosphate See zirconium phosphate.   { ,zər'kō·nē·əm ¦òr·thō'fä,sfāt }

zirconium oxide [INORG CHEM] ZrO₂ A toxic, heavy white powder that is insoluble in water, soluble in mineral acids; melts at 2700°C; used in ceramic glazes, special glasses, and medicine, and to make piezoelectric crystals.   Also known as zirconia; zirconic anhydride; zirconium dioxide.   { ,zər'kō·nē·əm 'äk,sīd }

zirconium oxychloride [INORG CHEM] ZrOCl₂·8H₂O White crystals that are soluble in water, insoluble in organic solvents, and acidic in aqueous solution; used for textile dyeing and oil-field acidizing, in cosmetics and greases, and for antiperspirants and water repellents.   Also known as zirconyl chloride.   { ,zər'kō·nē·əm ¦ak·sē'klòr,īd }

zirconium phosphate [INORG CHEM] ZrO(H₂PO₄)₂·3H₂O A toxic, dense white powder that is insoluble in water, soluble in acids and organic solvents; decomposes on heating; used as an analytical reagent, coagulant, and radioactive-phosphor carrier.   Also known as zirconium orthophosphate.   { ,zər'kō·nē·əm 'fä,sfāt }

zirconium tetrachloride [INORG CHEM] ZrCl₄ Toxic, alcohol-soluble, white lustrous crystals; sublimes above 300°C and decomposes in water; used to make pure zirconium and for water-repellent textiles and as an analytical reagent. Also known as zirconium chloride. { ‚zər'kō·nē·əm ¦te·trə'klȯr‚īd }

zirconyl chloride See zirconium oxychloride. { 'zər·kən·əl 'klȯr‚īd }

Zn See zinc.

zone See band. { zōn }

Zr See zirconium.

Zsigmondy gold number [CHEM] The number of milligrams of protective colloid necessary to prevent 10 milliliters of gold sol from coagulating when 0.5 milliliter of 10% sodium chloride solution is added. { 'zig·mȯn·dē 'gōld ‚nəm·bər }

zwitterion See dipolar ion. { 'tsfid·ər‚ī‚än }

Appendix

Equivalents of commonly used units for the U.S. Customary System and the metric system

1 inch = 2.5 centimeters (25 millimeters)	1 centimeter = 0.4 inch	1 inch = 0.083 foot
1 foot = 0.3 meter (30 centimeters)	1 meter = 3.3 feet	1 foot = 0.33 yard (12 inches)
1 yard = 0.9 meter	1 meter = 1.1 yards	1 yard = 3 feet (36 inches)
1 mile = 1.6 kilometers	1 kilometer = 0.62 mile	1 mile = 5280 feet (1760 yards)
1 acre = 0.4 hectare	1 hectare = 2.47 acres	
1 acre = 4047 square meters	1 square meter = 0.00025 acre	
1 gallon = 3.8 liters	1 liter = 1.06 quarts = 0.26 gallon	1 quart = 0.25 gallon (32 ounces; 2 pints)
1 fluid ounce = 29.6 milliliters	1 milliliter = 0.034 fluid ounce	1 pint = 0.125 gallon (16 ounces)
32 fluid ounces = 946.4 milliliters		1 gallon = 4 quarts (8 pints)
1 quart = 0.95 liter	1 gram = 0.035 ounce	1 ounce = 0.0625 pound
1 ounce = 28.35 grams	1 kilogram = 2.2 pounds	1 pound = 16 ounces
1 pound = 0.45 kilogram	1 kilogram = 1.1×10^{-3} ton	1 ton = 2000 pounds
1 ton = 907.18 kilograms		

$$°F = (1.8 \times °C) + 32$$

$$°C = (°F - 32) \div 1.8$$

Appendix

Conversion factors for the U.S. Customary System, metric system, and International System

A. Units of length

Units	cm	m	in.	ft	yd	mi
1 cm =	1	0.01	0.3937008	0.03280840	0.01093613	6.213712×10^{-6}
1 m =	100.	1	39.37008	3.280840	1.093613	6.213712×10^{-4}
1 in. =	2.54	0.0254	1	0.08333333...	0.02777777...	1.578283×10^{-5}
1 ft =	30.48	0.3048	12.	1	0.3333333...	$1.893939... \times 10^{-4}$
1 yd =	91.44	0.9144	36.	3.	1	$5.681818... \times 10^{-4}$
1 mi =	1.609344×10^{5}	1.609344×10^{3}	6.336×10^{4}	5280.	1760.	1

B. Units of area

Units	cm^2	m^2	$in.^2$	ft^2	yd^2	mi^2
1 cm^2 =	1	10^{-4}	0.1550003	1.076391×10^{-3}	1.195990×10^{-4}	3.861022×10^{-11}
1 m^2 =	10^{4}	1	1550.003	10.76391	1.195990	3.861022×10^{-7}
1 $in.^2$ =	6.4516	6.4516×10^{-4}	1	$6.944444... \times 10^{-3}$	7.716049×10^{-4}	2.490977×10^{-10}
1 ft^2 =	929.0304	0.09290304	144.	1	0.1111111...	3.587007×10^{-8}
1 yd^2 =	8361.273	0.8361273	1296.	9.	1	3.228306×10^{-7}
1 mi^2 =	2.589988×10^{10}	2.589988×10^{6}	4.014490×10^{9}	2.78784×10^{7}	3.0976×10^{6}	1

C. Units of volume

Units	m^3	cm^3	liter	$in.^3$	ft^3	qt	gal
1 m^3	= 1	10^6	10^3	6.102374×10^4	35.31467	1.056688×10^3	264.1721
1 cm^3	= 10^{-6}	1	10^{-3}	0.06102374	3.531467×10^{-5}	1.056688×10^{-3}	2.641721×10^{-4}
1 liter	= 10^{-3}	1000.	1	61.02374	0.03531467	1.056688	0.2641721
1 $in.^3$	= 1.638706×10^{-5}	16.38706	0.01638706	1	5.787037×10^{-4}	0.01731602	4.329004×10^{-3}
1 ft^3	= 2.831685×10^{-2}	28316.85	28.31685	$1728.$	1	2.992208	7.480520
1 qt	= 9.46352×10^{-4}	946.359	0.946351	57.75	0.03342014	1	0.25
1 gal (U.S.)	= 3.785412×10^{-3}	3785.412	3.785412	$231.$	0.1336806	$4.$	1

D. Units of mass

Units	g	kg	oz	lb	$metric\ ton$	ton
1 g	= 1	10^{-3}	0.03527396	2.204623×10^{-3}	10^{-6}	1.102311×10^{-6}
1 kg	= 1000.	1	35.27396	2.204623	10^{-3}	1.102311×10^{-3}
1 oz (avdp)	= 28.34952	0.02834952	1	0.0625	2.834952×10^{-5}	3.125×10^{-5}
1 lb (avdp)	= 453.5924	0.4535924	$16.$	1	4.535924×10^{-4}	$5. \times 10^{-4}$
1 metric ton	= 10^6	1000.	35273.96	2204.623	1	1.102311
1 ton	= 907184.7	907.1847	$32000.$	$2000.$	0.9071847	1

Appendix

Conversion factors for the U.S. Customary System, metric system, and International System (cont.)

E. Units of density

Units	$g \cdot cm^{-3}$	$g \cdot L^{-1}, kg \cdot m^{-3}$	$oz \cdot in.^{-3}$	$lb \cdot in.^{-3}$	$lb \cdot ft^{-3}$	$lb \cdot gal^{-1}$
1 g·cm⁻³ = 1	1000.	0.5780365	0.03612728	62.42795	8.345403
1 g·L⁻¹, kg·m⁻³ = 10^{-3}	1	5.780365×10^{-4}	3.612728×10^{-5}	0.06242795	8.345403×10^{-3}
1 oz·in.⁻³ = 1.729994	1729.994	1	0.0625	108.	14.4375
1 lb·in.⁻³ = 27.67991	27679.91	16.	1	1728.	231.
1 lb·ft⁻³ = 0.01601847	16.01847	9.259259×10^{-3}	5.787037×10^{-4}	1	0.1136806
1 lb·gal⁻¹ = 0.1198264	119.8264	4.749536×10^{-3}	4.329004×10^{-3}	7.480519	1

F. Units of pressure

Units	$Pa, N \cdot m^{-2}$	$dyn \cdot cm^{-2}$	bar	atm	$kgf \cdot cm^{-2}$	$mmHg$ (torr)	$in.$ Hg	$lbf \cdot in.^{-2}$
1 Pa, 1 N·m⁻² = 1	10	10^{-5}	9.869233×10^{-6}	1.019716×10^{-5}	7.500617×10^{-3}	2.952999×10^{-4}	1.450377×10^{-4}
1 dyn·cm⁻² = 0.1	1	10^{-6}	9.869233×10^{-7}	1.019716×10^{-6}	7.500617×10^{-4}	2.952999×10^{-5}	1.450377×10^{-5}
1 bar = 10^5	10^6	1	0.9869233	1.019716	750.0617	29.52999	14.50377
1 atm = 101325	101325.0	1.01325	1	1.033227	760.	29.92126	14.69595
1 kgf·cm⁻² = 98066.5	980665	0.980665	0.9678411	1	735.5592	28.95903	14.22334
1 mmHg (torr) = 133.3224	1333.224	1.333224×10^{-3}	1.315789×10^{-3}	1.359510×10^{-3}	1	0.03937008	0.01933678
1 in. Hg = 3386.388	33863.88	0.03386388	0.03342105	0.03453155	25.4	1	0.4911541
1 lbf·in.⁻² = 6894.757	68947.57	0.06894757	0.06804596	0.07030696	51.71493	2.036021	1

G. Units of energy

Units	g mass (energy equiv)	l	eV	cal	cal$_{IT}$	Btu$_{IT}$	kWh	hp-h	ft-lbf	ft$^3 \cdot$ lbf $\cdot$ in.$^{-2}$	liter-atm
1 g mass (energy equiv)	= 1	8.987552 × 10^{13}	5.609589 × 10^{32}	2.148076 × 10^{13}	2.146640 × 10^{13}	8.518555 × 10^{10}	2.496542 × 10^7	3.347918 × 10^7	6.628878 × 10^{13}	4.603388 × 10^{11}	8.870024 × 10^{11}
1 l	= 1.112650 × 10^{-14}	1	6.241510 × 10^{18}	0.2390057	0.2388459	9.478172 × 10^{-4}	2.777777... × 10^{-7}	3.725062 × 10^{-7}	0.7375622	5.121960 × 10^{-3}	9.869233 × 10^{-3}
1 eV	= 1.782662 × 10^{-33}	1.602176 × 10^{-19}	1	3.829293 × 10^{-20}	3.826733 × 10^{-20}	1.518570 × 10^{-22}	4.450490 × 10^{-26}	5.968206 × 10^{-26}	1.181705 × 10^{-19}	8.206283 × 10^{-22}	1.581225 × 10^{-21}
1 cal	= 4.655328 × 10^{-14}	4.184	2.611448 × 10^{19}	1	0.9993312	3.965667 × 10^{-3}	1.1622222... × 10^{-6}	1.558562 × 10^{-6}	3.085960	2.143028 × 10^{-2}	0.04129287
1 cal$_{IT}$	= 4.658443 × 10^{-14}	4.1868	2.613195 × 10^{19}	1.000669	1	3.968321 × 10^{-3}	1.163 × 10^{-6}	1.559609 × 10^{-6}	3.088025	2.144462 × 10^{-2}	0.04132050
1 Btu$_{IT}$	= 1.173908 × 10^{-11}	1055.056	6.585141 × 10^{21}	252.1644	251.9958	1	2.930711 × 10^{-4}	3.930148 × 10^{-4}	778.1693	5.403953	10.41259
1 kWh	= 4.005540 × 10^{-8}	3600000.	2.246944 × 10^{25}	860420.7	859845.2	3412.142	1	1.341022	2655224.	18349.06	35529.24
1 hp-h	= 2.986931 × 10^{-8}	2384519.	1.675545 × 10^{25}	641615.6	641186.5	2544.33	0.7456998	1	1980000.	13750.	26494.15
1 ft-lbf	= 1.508551 × 10^{-14}	1.355818	8.462351 × 10^{18}	0.3240483	0.3238315	1.285067 × 10^{-3}	3.766161 × 10^{-7}	5.050505... × 10^{-7}	1	6.944444... × 10^{-3}	0.01338088
1 ft$^3 \cdot$ lbf $\cdot$ in.$^{-2}$	= 2.172313 × 10^{-12}	195.2378	1.218579 × 10^{21}	46.66295	46.63174	0.1850497	5.423272 × 10^{-5}	7.272727... × 10^{-5}	144.	1	1.926847
1 liter-atm	= 1.127393 × 10^{-12}	101.325	6.324210 × 10^{20}	24.21726	24.20106	0.09603757	2.814583 × 10^{-5}	3.774419 × 10^{-5}	74.73349	0.5189825	1

Appendix

Defining fixed points of the International Temperature Scale of 1990 (ITS-90)

Equilibrium state	Temperature		
	K	°C	°F
Vapor pressure equation of helium	3 to 5	−270.15 to −268.15	−454.27 to 450.67
Triple point of equilibrium hydrogen	13.80	−259.35	−434.81
Vapor pressure point of equilibrium hydrogen	≈17	≈−256.15	−447.09
(or constant-volume gas thermometer point of helium)	≈20.3	≈−252.85	−423.13
Triple point of neon	24.56	−248.59	−415.46
Triple point of oxygen	64.37	−218.79	361.82
Triple point of argon	83.81	−189.34	308.81
Triple point of mercury	234.32	−38.83	−37.89
Triple point of water	273.16	0.01	32.02
Melting point of gallium	302.91	29.78	85.60
Freezing point of indium	429.75	156.60	313.88
Freezing point of tin	505.08	232.93	449.47
Freezing point of zinc	692.68	419.53	787.15
Freezing point of aluminum	933.47	660.32	1220.58
Freezing point of silver	1234.93	961.78	1763.20
Freezing point of gold	1337.33	1064.18	1947.52
Freezing point of copper	1357.77	1084.62	1984.32

Primary thermometry methods

Method	Approximate useful range of T, K	Principal measured variables	Relation of measured variables to T	Remarks
Gas thermometry	1.3–950	Pressure P and volume V	Ideal gas law plus correction; $PV \propto k_B T$ plus corrections	Careful determination of corrections necessary, but capable of high accuracy
Acoustic interferometry	1.5–3000	Speed of sound W	$W^2 \propto k_B T$ plus corrections	
Magnetic thermometry 1. Electron paramagnetism	0.001–35	Magnetic susceptibility	Curie's law plus corrections: $\chi \propto 1/k_B T$ plus corrections	
2. Nuclear paramagnetism	0.000001–1			
Gamma-ray anisotropy or nuclear orientation thermometry	0.01–1	Spatial distribution of gamma-ray emission	Spatial distribution related to Boltzmann factor for nuclear spin states	Useful standard for T < 1 K
Thermal electric noise thermometry 1. Josephson junction point contact	0.001–1	Mean square voltage fluctuation $\overline{V}^2$	Nyquist's law: $\overline{V}^2 \propto k_B T$	Other sources of noise serious problem for T > 4 K
2. Conventional amplifier	4–1400			
Radiation thermometry (visual, photoelectric, or photodiode)	500–50,000	Spectral intensity J at wavelength λ	Planck's radiation law, related to Boltzmann factor for radiation quanta	Needs blackbody conditions or well-defined emittance
Infrared spectroscopy	100–1500	Intensity of rotational lines of light molecules	Boltzmann factor for rotational levels related to J	Also Doppler line broadening $(\propto \sqrt{k_B T})$ useful; principal applications to plasmas and astrophysical observations; proper sampling, lack of equilibrium, atmospheric absorption often problems
Ultraviolet and x-ray spectroscopy	5000–2,000,000	Emission spectra from ionized atoms—H, He, Fe, Ca, and so on	Boltzmann factor for electron states related to band structure and line density	

Appendix

Periodic table

(The atomic numbers are listed above the symbols identifying the elements. The heavy line separates metals from nonmentals.)

Group	1	2	3	4	5	6	7	8	9	10	11	12	13	14	15	16	17	18
	1 H Hydrogen																	2 He Helium
s	3 Li Lithium	4 Be Beryllium											5 B Boron	6 C Carbon	7 N Nitrogen	8 O Oxygen	9 F Fluorine	10 Ne Neon
	11 Na Sodium	12 Mg Magnesium											13 Al Aluminum	14 Si Silicon	15 P Phosphorus	16 S Sulfur	17 Cl Chlorine	18 Ar Argon
	19 K Potassium	20 Ca Calcium	21 Sc Scandium	22 Ti Titanium	23 V Vanadium	24 Cr Chromium	25 Mn Manganese	26 Fe Iron	27 Co Cobalt	28 Ni Nickel	29 Cu Copper	30 Zn Zinc	31 Ga Gallium	32 Ge Germanium	33 As Arsenic	34 Se Selenium	35 Br Bromine	36 Kr Krypton
	37 Rb Rubidium	38 Sr Strontium	39 Y Yttrium	40 Zr Zirconium	41 Nb Niobium	42 Mo Molybdenum	43 Tc Technetium	44 Ru Ruthenium	45 Rh Rhodium	46 Pd Palladium	47 Ag Silver	48 Cd Cadmium	49 In Indium	50 Sn Tin	51 Sb Antimony	52 Te Tellurium	53 I Iodine	54 Xe Xenon
	55 Cs Cesium	56 Ba Barium	71 Lu Lutetium	72 Hf Hafnium	73 Ta Tantalum	74 W Tungsten	75 Re Rhenium	76 Os Osmium	77 Ir Iridium	78 Pt Platinum	79 Au Gold	80 Hg Mercury	81 Tl Thallium	82 Pb Lead	83 Bi Bismuth	84 Po Polonium	85 At Astatine	86 Rn Radon
	87 Fr Francium	88 Ra Radium	103 Lr Lawrencium	104 Rf Rutherfordium	105 Db Dubnium	106 Sg Seaborgium	107 Bh Bohrium	108 Hs Hassium	109 Mt Meitnerium	110	111	112	113	114	115	116	117	118

f block:

57 La Lanthanum	58 Ce Cerium	59 Pr Praseodymium	60 Nd Neodymium	61 Pm Promethium	62 Sm Samarium	63 Eu Europium	64 Gd Gadolinium	65 Tb Terbium	66 Dy Dysprosium	67 Ho Holmium	68 Er Erbium	69 Tm Thulium	70 Yb Ytterbium
89 Ac Actinium	90 Th Thorium	91 Pa Protactinium	92 U Uranium	93 Np Neptunium	94 Pu Plutonium	95 Am Americium	96 Cm Curium	97 Bk Berkelium	98 Cf Californium	99 Es Einsteinium	100 Fm Fermium	101 Md Mendelevium	102 No Nobelium

Electrochemical series of the elements*

Element	Symbol	Element	Symbol
Lithium	Li	Zinc	Zn
Potassium	K	Chromium	Cr
Rubidium	Rb	Gallium	Ga
Cesium	Cs	Iron	Fe
Radium	Ra	Cadmium	Cd
Barium	Ba	Indium	In
Strontium	Sr	Thallium	Tl
Calcium	Ca	Cobalt	Co
Sodium	Na	Nickel	Ni
Lanthanum	La	Molybdenum	Mo
Cerium	Ce	Tin	Sn
Magnesium	Mg	Lead	Pb
Scandium	Sc	Germanium	Ge
Plutonium	Pu	Tungsten	W
Thorium	Be	Hydrogen	H
Beryllium	Th	Copper	Cu
Uranium	U	Mercury	Hg
Hafnium	Hf	Silver	Ag
Aluminum	Al	Gold	Au
Titanium	Ti	Rhodium	Rh
Zirconium	Zr	Platinum	Pt
Manganese	Mn	Palladium	Pd
Vanadium	V	Bromine	Br
Niobium	Nb	Chlorine	Cl
Boron	B	Oxygen	O
Silicon	Si	Fluorine	F
Tantalum	Ta		

*According to standard oxidation potentials E^0 at 25°C or 77°F.

Appendix

Average electronegativities from thermochemical data

Element	Value	Element	Value
H	2.20	Al	1.61
Li	0.98	Ga	1.81
Na	0.93	In	1.78
K	0.82	Tl	2.04
Rb	0.82	C	2.55
Cs	0.79	Si	1.90
Be	1.57	Ge	2.01
Mg	1.31	Sn	1.96
Ca	1.00	Pb	2.33
Sr	0.95	N	3.04
Ba	0.89	P	2.19
Sc	1.36	As	2.18
Ti	1.54	Sb	2.05
V	1.63	Bi	2.02
Cr	1.66	O	3.44
Mn	1.55	S	2.58
Fe	1.83	Se	2.55
Co	1.88	F	3.98
Ni	1.91	Cl	3.16
Cu	1.90	Br	2.96
Zn	1.65	I	2.66
B	2.04		

Standard atomic weights

Atomic number	Symbol	Name	Atomic weight*
89	Ac	Actinium	[227]
13	Al	Aluminum	26.981538(2)
95	Am	Americium	[243]
51	Sb	Antimony	121.760(1)
18	Ar	Argon	39.948(1)
33	As	Arsenic	74.92160(2)
85	At	Astatine	[210]
56	Ba	Barium	137.327(7)
97	Bk	Berkelium	[247]
4	Be	Beryllium	9.012182(3)
83	Bi	Bismuth	208.98038(2)
107	Bh	Bohrium	[264]
5	B	Boron	10.811(7)
35	Br	Bromine	79.904(1)
48	Cd	Cadmium	112.411(8)
20	Ca	Calcium	40.078(4)
98	Cf	Californium	[251]
6	C	Carbon	12.0107(8)
58	Ce	Cerium	140.116(1)
55	Cs	Cesium	132.90545(2)
17	Cl	Chlorine	35.453(2)
24	Cr	Chromium	51.9961(6)
27	Co	Cobalt	58.933200(9)
29	Cu	Copper	63.546(3)
96	Cm	Curium	[247]
105	Db	Dubnium	[262]
66	Dy	Dysprosium	162.500(1)
99	Es	Einsteinium	[252]
68	Er	Erbium	167.259(3)
63	Eu	Europium	151.964(1)
100	Fm	Fermium	[257]
9	F	Fluorine	18.9984032(5)
87	Fr	Francium	[223]
64	Gd	Gadolinium	157.25(3)
31	Ga	Gallium	69.723(1)
32	Ge	Germanium	72.64(1)
79	Au	Gold	196.96655(2)
72	Hf	Hafnium	178.49(2)
108	Hs	Hassium	[277]
2	He	Helium	4.002602(2)
67	Ho	Holmium	164.93032(2)
1	H	Hydrogen	1.00794(7)
49	In	Indium	114.818(3)
53	I	Iodine	126.90447(3)
77	Ir	Iridium	192.217(3)
26	Fe	Iron	55.845(2)
36	Kr	Krypton	83.798(2)
57	La	Lanthanum	138.9055(2)
103	Lr	Lawrencium	[262]
82	Pb	Lead	207.2(1)
3	Li	Lithium	[6.941(2)]
71	Lu	Lutetium	174.967(1)
12	Mg	Magnesium	24.3050(6)
25	Mn	Manganese	54.938049(9)
109	Mt	Meitnerium	[268]
101	Md	Mendelevium	[258]
80	Hg	Mercury	200.59(2)
42	Mo	Molybdenum	95.94(2)

Appendix

Standard atomic weights (cont.)

Atomic number	Symbol	Name	Atomic weight*
60	Nd	Neodymium	144.24(3)
10	Ne	Neon	20.1797(6)
93	Np	Neptunium	[237]
28	Ni	Nickel	58.6934(2)
41	Nb	Niobium	92.90638(2)
7	N	Nitrogen	14.0067(2)
102	No	Nobelium	[259]
76	Os	Osmium	190.23(3)
8	O	Oxygen	15.9994(3)
46	Pd	Palladium	106.42(1)
15	P	Phosphorus	30.973761(2)
78	Pt	Platinum	195.078(2)
94	Pu	Plutonium	[244]
84	Po	Polonium	[209]
19	K	Potassium	39.0983(1)
59	Pr	Praseodymium	140.90765(2)
61	Pm	Promethium	[145]
91	Pa	Protactinium	231.03588(2)
88	Ra	Radium	[226]
86	Rn	Radon	[222]
75	Re	Rhenium	186.207(1)
45	Rh	Rhodium	102.90550(2)
37	Rb	Rubidium	85.4678(3)
44	Ru	Ruthenium	101.07(2)
104	Rf	Rutherfordium	[261]
62	Sm	Samarium	150.36(3)
21	Sc	Scandium	44.955910(8)
106	Sg	Seaborgium	[266]
34	Se	Selenium	78.96(3)
14	Si	Silicon	28.0855(3)
47	Ag	Silver	107.8682(2)
11	Na	Sodium	22.989770(2)
38	Sr	Strontium	87.62(1)
16	S	Sulfur	32.065(5)
73	Ta	Tantalum	180.9479(1)
43	Tc	Technetium	[98]
52	Te	Tellurium	127.60(3)
65	Tb	Terbium	158.92534(2)
81	Tl	Thallium	204.3833(2)
90	Th	Thorium	232.0381(1)
69	Tm	Thulium	168.93421(2)
50	Sn	Tin	118.710(7)
22	Ti	Titanium	47.867(1)
74	W	Tungsten	183.84(1)
112	Uub	Ununbium	[285]
110	Uun	Ununnilium	[281]
111	Uuu	Unununium	[272]
92	U	Uranium	238.02891(3)
23	V	Vanadium	50.9415(1)
54	Xe	Xenon	131.293(6)
70	Yb	Ytterbium	173.04(3)
39	Y	Yttrium	88.90585(2)
30	Zn	Zinc	65.409(4)
40	Zr	Zirconium	91.224(2)

*Atomic weights are those of the most commonly available long-lived isotopes on the 1999 IUPAC Atomic Weights of the Elements. A value given in square brackets denotes the mass number at the longest-lived isotope.

Principal organic functional groups

Compound class	Group	Structure	Compound class	Group	Structure
Alkene	Double bond	$\overset{\diagdown}{}C = C\overset{\diagup}{}$	Ester	Alkoxy-carbonyl	$-\overset{\overset{\displaystyle O}{\|}}{C}OR$
Alkyne	Triple bond	$-C\equiv C-$			
Alcohol	Hydroxyl	$-OH$			
Amine	Amino	$-NH_2(-NR_2)^*$	Amide	Carbamoyl	$\overset{\overset{\displaystyle O}{\|}}{C}N\overset{\diagup}{}$
Aldehyde	Carbonyl	$-\overset{\overset{\displaystyle O}{\|}}{C}H$	Nitrite	Cyano	$-C\equiv N$
			Azide	Azido	$-N=N=N$
			Nitro		$-NO_2$
Ketone	Carbonyl	$-\overset{\overset{\displaystyle O}{\|}}{C}R$	Sulfide		$-S-$
Acid	Carboxyl	$-\overset{\overset{\displaystyle O}{\|}}{C}OH$	Sulfoxide		$-\overset{\overset{\displaystyle O}{\|}}{S}-$
			Sulfonic acid		$-SO_3H$

*R = any carbon group, for example, CH_3.

Appendix

Compounds containing functional groups

Group[*]	Suffix	Prefix	Structure	Name
$-\overset{\overset{\text{O}}{\|\|}}{\text{C}}\text{OH}$	-oic acid	carboxy-	$\text{CH}_3\text{CH}_2\overset{\overset{\text{O}}{\|\|}}{\text{C}}\text{OH}$	Propanoic acid
$-\overset{\overset{\text{O}'}{\|\|}}{\text{C}}\text{OR}$	alkyl -oate	alkoxycarbonyl-	$\text{CH}_3\text{CH}_2\overset{\overset{\text{O}}{\|\|}}{\text{C}}\text{OCH}_3$	Methyl propanoate
			$\text{CH}_3\text{O}\overset{\overset{\text{O}}{\|\|}}{\text{C}}\text{CH}_2\text{CH}_2\overset{\overset{\text{O}}{\|\|}}{\text{C}}\text{OH}$	3-Methoxycarbonyl propanoic acid
$-\text{C}\equiv\text{N}$	-nitrile	cyano-	$\text{CH}_3\text{CH}_2\text{C}\equiv\text{N}$	Propanenitrile
			$\text{CH}_3\text{CHCN}\overset{\overset{\text{O}}{\|\|}}{\text{C}}\text{OH}$	2-Cyanopropanoic acid
$-\overset{\overset{\text{O}}{\|\|}}{\text{C}}\text{H}$	-al	formyl-	$\text{CH}_3\text{CH}_2\overset{\overset{\text{O}}{\|\|}}{\text{C}}\text{H}$	Propanal
$-\overset{\overset{\text{O}}{\|\|}}{\text{C}}\text{R}$	-one	oxo-	$\text{CH}_3\text{CH}_2\overset{\overset{\text{O}}{\|\|}}{\text{C}}\text{CH}_3$	Butanone
			$\text{CH}_3\overset{\overset{\text{O}}{\|\|}}{\text{C}}\text{CH}_2\overset{\overset{\text{O}}{\|\|}}{\text{C}}\text{OH}$	3-Oxobutanoic acid
$-\text{OH}$	-ol	hydroxy-	$\text{CH}_3\text{CH}_2\text{CH}_2\text{OH}$	1-Propanol
			$\text{HOCH}_2\text{CH}_2\overset{\overset{\text{O}}{\|\|}}{\text{C}}\text{H}$	3-Hydroxypropanal
$-\text{NH}_2$	-amine	amino-	$\text{CH}_3\text{CH}_2\text{CH}_2\text{NH}_2$	1-Propanamine
$-\text{OR}$	—	alkoxy-	$\text{NH}_2\text{CH}_2\text{CH}_2\text{CH}_2\text{OCH}_3$	3-Methoxy-1-propanamine

[*]R = any alkyl group.

430

Physical properties of some organic solvents

Organic solvent	Boiling point, °C (°F)	Freezing point, °C (°F)	Viscosity, cgs, 25°C (77°F)	Dielectric constant, 25°C (77°F)
Benzene	80.100 (176.18)	5.533 (41.959)	0.6028	2.275
1,2-Dichloroethane	83.483 (182.269)	-35.66 (-32.19)	0.730, 30°C	10.36
Methanol	64.70 (148.46)	-97.68 (-143.82)	0.5445	32.70
1,2-Ethanediol	197.3 (387.14)	-13 (8.6)	13.55, 30°C	37.7
Acetic acid	117.90 (244.22)	16.66 (61.99)	1.040, 20°C	6.15, 20°C (68°F)
Phenol	181.839 (359.310)	40.90 (105.62)	4.076	9.78, 60°C (140°F)
Acetone	56.29 (133.32)	-94.7 (-138.46)	0.3040	20.70
2-Propanol	82.26 (180.07)	-88.0 (-126.4)	1.765, 30°C	19.92
Ethanol	78.29 (172.92)	-114.1 (-173.4)	1.078	24.55
1,2-Dimethoxybenzene	206.25 (403.25)	22.5 (72.5)	3.281	4.09
Fluorobenzene	84.734 (184.521)	-42.21 (-43.98)	0.517, 30°C	5.42
Pyridine	115.256 (239.47)	-41.55 (-42.79)	0.884	12.4
2-Ethoxyethanol	135.6 (276.1)	<-90 (-130)	1.85	29.6
N,N-Dimethylacetamide	166.1 (331.0)	-20 (-4)	0.838, 30°C	37.78
Dimethyl sulfoxide	189.0 (372.2)	18.54 (65.37)	1.996	46.68
2-Nitropropane	120.25 (248.45)	-91.32 (-132.38)	0.721	25.52